Engineering Heat Transfer

Engineering Heat Transfer

Edited by **Edgar Miller**

NY RESEARCH PRESS

New York

Published by NY Research Press,
23 West, 55th Street, Suite 816,
New York, NY 10019, USA
www.nyresearchpress.com

Engineering Heat Transfer
Edited by Edgar Miller

International Standard Book Number: 978-1-63238-520-8 (Hardback)

Printed in China.

Contents

	Preface	**IX**
Chapter 1	**The Study of Temperature Profile Inside Wax Deposition Layer of Waxy Crude Oil in Pipeline** Zhen Tian, Wenbo Jin, Lei Wang and Zhi Jin	**1**
Chapter 2	**A Drying and Thermoelastic Model for Fast Microwave Heating of Concrete** Benjamin Lepers, Aditya Putranto, Martin Umminger, Guido Link and John Jelonnek	**9**
Chapter 3	**A Comparative Study on Thermal Conductivity and Rheology Properties of Alumina and Multi-Walled Carbon Nanotube Nanofluids** Zan Wu, Zhaozan Feng, Bengt Sundén and Lars Wadsö	**20**
Chapter 4	**Sintering of Titanium and Nickel Nanopowders with a ND:YAG Nanosecond Laser** Jing Huang, Vitaly E. Gruzdev, Yuwen Zhang and J. K. Chen	**30**
Chapter 5	**Thermodynamic Analysis for the MHD Flow of Two Immiscible Micropolar Fluids Between Two Parallel Plates** J. Srinivas and J. V. Ramana Murthy	**35**
Chapter 6	**Convective Heat Transfer Analysis in an Arch Enclosure** Manoj Kr. Triveni, Dipak Sen and Raj Sekhar Panua	**46**
Chapter 7	**Ex Vivo Liver Tissue Radiofrequency Thermal Ablation: IR Measurements and Simulations** Edoardo Gino Macchi, Giovanni Braschi and Mario Gallati	**52**
Chapter 8	**3D Numerical Analysis on Flow Configurations and Heat Transfer Characteristics for Fin-and-Oval-Tube Heat Exchanger with V-Downstream Delta Winglet Vortex Generators** Amnart Boonloi and Withada Jedasadaratanachai	**60**
Chapter 9	**Effects of Geometric Parameters for Wavy Finned-Tube Heat Exchanger in Turbulent Flow: A CFD Modeling** Arafat A. Bhuiyan, M. Ruhul Amin, Jamal Naser and A. K. M. Sadrul Islam	**75**
Chapter 10	**Mixed Convection Boundary Layer Flow Over a Vertically Stretching Sheet with Convective Boundary Condition and Effect of Partial Slip** Mitiku Daba, P. Devaraj and S. V. Subhashini	**86**
Chapter 11	**Experimental Investigation on Heat Transfer and Friction Factor Characteristics of a Stationary Square Duct Roughened by V and Λ-shaped Ribs** Anand Shukla, Alok Chaube, Shailesh Gupta and Arvind Sirsath	**92**

Chapter 12 **Heat Transfer Enhancement of Heat Exchanger Inserted Twisted Plates for Carbon Dioxide Gas** 100
Makoto Shibahara

Chapter 13 **Recent Progress on Experimental Research of Cryogenic Transport Line Chilldown Process** 104
J. N. Chung and Kun Yuan

Chapter 14 **The Effect of Melting on Mixed Convection Heat and Mass Transfer in Non-Newtonian Nanofluid Saturated in Porous Medium** 121
R. R. Kairi and Ch. Ram Reddy

Chapter 15 **A Review of Hole Geometry and Coolant Density Effect on Film Cooling** 128
Srinath Ekkad and Je-Chin Han

Chapter 16 **Performance Assessment and Empirical Correlation in a Heat Exchanger Square Duct with Diagonal Inserted Generators** 142
Amnart Boonloi

Chapter 17 **Effect of Wall Thermal Conductivity on Hydrogen- Assisted Catalytic Ignition Characteristics of Propane- Air at Micro-Scales in Different Feeding Modes** 154
Junjie Chen, Xuhui Gao and Deguang Xu

Chapter 18 **Natural Convection on a Porous Vertical Plate in a Doubly Stratified Non-Darcy Porous Medium** 167
D. Srinivasacharya, A. J. Chamkha, O. Surender and A. M. Rashad

Chapter 19 **Numerical Simulation of Metal-Plastic Composite Heat Radiator with Hemispherical Microstructure Array** 174
Hui Jiang, Daming Wu, Jian Zhuang, Ying Liu and Changqing Huang

Chapter 20 **Heat Transfer Based Numerical Investigation of Aircraft Cabin Environment from Various Inlet Conditions** 180
Zhuohuan Hu, Lulu Wang, Hui Wang and Mo Yang

Chapter 21 **Analysis of MHD Transient Free Convection Flow of a Newtonian Fluid Past an Infinite Vertical Porous Plate** 188
M. Umamaheswar, M. C. Raju and S. V. K. Varma

Chapter 22 **Entropy Generation in Boundary Layer Flow of a Micro Polar Fluid Over a Stretching Sheet Embedded in a Highly Absorbing Medium** 195
M. Y. Abdollahzadeh Jamalabadi

Chapter 23 **Development of a 10 KW Microwave Applicator for Thermal Cracking of Lignite Briquettes** 208
Benjamin Lepers, Thomas Seitz, Guido Link, John Jelonnek and Mark Zink

Chapter 24 **Accelerating MHD Flow of a Generalized Oldroyd-B Fluid with Fractional Derivative** 214
Yaqing Liu and Jinyu Ma

Chapter 25 **Modelling and Experimental Validation of Combustion in Straight Inoculation Compression Ignition Engine Fuelled with Diesel and Jatropha Methyl Ester Blend** 219
Biswajit De and Rajsekhar Panua

Chapter 26 **Influences of the Punched Delta Winglet Vortex Generators in a Circular Tube Heat Exchanger on Thermo-Hydraulic Performance** 225
Withada Jedsadaratanachai and Amnart Boonloi

Chapter 27 **Thermal Diffusion and Radiation Effects on Unsteady Free Convection Flow in the Presence of Magnetic Field Fixed Relative to the Fluid or to the Plate** 244
B. Rushi Kumar, T. Sravan Kumar and A .G Vijaya Kumar

Chapter 28 **MHD Convective Boundary Layer Flow Towards a Vertical Surface in a Porous Medium with Radiation, Chemical Reaction and Internal Heat Generation** 253
Emmanuel Maurice Arthur, Timothy Ayando and Yakubu Ibrahim Seini

Chapter 29 **Transient Free Convection MHD Flow Past a Vertical Plate with Exponentially Decaying Wall Temperature and Radiation** 263
Rudra Kanta Deka, Ashish Paul and Nityajyoti Kalita

Permissions

List of Contributors

Preface

This book traces the progress of the field of heat transfer engineering and highlights some of its key concepts and applications. Heat transfer refers to the study and applications of engineering practices used to transfer and exchange thermal energy and heat from one physical system to other. It has various mechanisms like thermal radiation, thermal conduction, transfer of energy, thermal convection, etc. These studies are applied in different engineering subjects like automotive, materials processing, power station, climate engineering, etc. The topics introduced in this book are of utmost significance and are bound to provide in-depth knowledge about this topic to readers. It is a compilation of chapters that discuss most vital concepts and emerging trends in this field. Students, scientists, engineers, researchers and all those interested in heat transfer will find this book greatly beneficial.

Significant researches are present in this book. Intensive efforts have been employed by authors to make this book an outstanding discourse. This book contains the enlightening chapters which have been written on the basis of significant researches done by the experts.

Finally, I would also like to thank all the members involved in this book for being a team and meeting all the deadlines for the submission of their respective works. I would also like to thank my friends and family for being supportive in my efforts.

Editor

THE STUDY OF TEMPERATURE PROFILE INSIDE WAX DEPOSITION LAYER OF WAXY CRUDE OIL IN PIPELINE

Zhen Tian[a*], Wenbo Jin[a], Lei Wang[b], Zhi Jin[c]

[a] *Petroleum and Natural Gas Engineering College, Southwest Petroleum University, Chengdu, Sichuan, 610500, China*
[b] *The Second Oil Production Plant of Qinghai Oilfield, Haixi/Qinghai, 817500, China*
[c] *Research Institute of Shaanxi Yanchang Petroleum(Group) Co, LTD, Xi'an/Shaanxi, 710075, China*

ABSTRACT

Taking the axial heat conduction of wax deposition layer into account, a two-dimensional heat transfer model of calculating the temperature profile inside wax deposition layer was deduced and established based on the energy balance equation, the finite difference method was used to solve this model, and the influence of axial heat conduction on the distribution law of temperature profile inside the wax deposition layer under different boundary conditions and thickness were discussed. The results showed that: Temperature profile inside wax deposition layer in middle region of testing pipe section was mainly influenced by axial heat conduction under boundary conditions of constant wall temperature while the region near the inlet was mainly influenced under boundary conditions of variable wall temperature. With the increasing thickness of wax deposition layer, the influence of axial heat conduction became more conspicuous. All in all, in laboratory flow loop experiment of wax deposition, although axial heat conduction had an influence on the temperature profile inside wax deposition layer under different conditions, the influence was small, thus could be ignored.

Keywords: *wax deposition layer, two-dimensional heat transfer model, temperature profile, boundary conditions.*

1. INTRODUCTION

For waxy crude oil with high freezing point, high viscosity, and poor mobility, its transportation generally uses heated oil pipeline. During the transportation, because oil temperature at initial station is higher than the ambient temperature, the oil temperature continues to decrease, and when it decreases to the wax precipitation point of crude oil, wax crystal in crude oil starts to precipitate, grow, and deposit on the pipe wall finally. Sediments on pipeline will reduce the effective flow area of the pipeline, increase friction, and thus reduce transmission capacity of the pipeline. Therefore, accurate prediction of wax deposition on pipeline has an important significance for the formulation of pigging program and safe, economical transportation of waxy crude oil in pipeline (Hsu and Santamaria, 1994; Pedersen, 1995).

Wax deposition prediction has always been a hot and important research topic for scholars at home and abroad (Hunt, 1962; Matlach *et al.*,1983; Agrawal *et al.*,1990; Bern *et al.*, 1980; Burger and Perkins, 1981; Brown, Niesen and Erickson, 1993; Svendsen, 1993; Nazar *et al*, 2001; Mehrotra and Bhat, 2007; Zhang and Liu, 2010; Ramirez-Jaramillo *et al.*, 2004; Ribeiro *et al.*, 1997; Lee, 2008).Based on respective purposes of measurement, scholars developed many equipments for measuring the amount of sediments on pipeline, including cold finger, cold plate, rotating disk and flow loop which was used more often than others. Through these devices, scholars had acquired some achievements. Early studies suggested that, the mechanism of wax deposition could be summarized as the following four types: molecular diffusion, shear dispersion, Brownian diffusion and gravity settling, in which molecular diffusion was generally considered as the most important mechanism causing wax deposition. Through flow loop experiment Singh (2000) confirmed that shear dispersion, Brownian diffusion and gravity settling had little influence on wax deposition, namely molecular diffusion was dominant mechanism of wax deposition; Hamouda and Davidsen (1995) considered that shear dispersion had an influence on wax deposition, but the main cause was still molecular diffusion, and he ignored the influence of shear dispersion when establishing the prediction model of wax deposition. Establishing the prediction model of wax deposition rate to study wax deposition on pipeline was one of the effective methods. Based on the mechanism of molecular diffusion, Bern, Burger, and Brown put forward their own prediction model of wax deposition rate on pipeline respectively, and validated with field data, but each prediction model had their own application range and limitations. In addition, wax deposition was affected by many factors and, hence, wax deposition on actual pipeline couldn't be completely accurate predicted currently. Based on a large number of experimental and theoretical studies, scholars continued to study the mechanism of wax deposition.

Recently, scholars proposed the mechanism of "aging", which was widely recognized by the industry. Singh and Fogler (2000) considered that: the process of wax deposition could be described by the following steps: (1). Gelation of the waxy oil on the cold surface, (2). Diffusion of waxes towards the gel layer from the bulk, (3). Internal diffusion of these molecules through the trapped oil, (4). Precipitation of these molecules in the deposit, (5). Counterdiffusion of de-waxed oil out of the gel layer. This view was confirmed by measuring the carbon number distribution in the sample of wax deposition layer under different experimental time. Based on the mechanism of molecular diffusion, Fogler research group developed WMP (Michigan Wax Prediction) wax deposition prediction software. This software could predict the growth and wax content of the gel deposit on externally

* Corresponding Author. E-mail: believemetoo@sina.com.

cooled pipe walls. As we all know, radial concentration gradient of wax molecule is an important parameter in calculating the amount of wax deposition, and it is affected by the distribution law of temperature profile of crude oil in pipe. Therefore, the calculation of temperature profile from the center of oil flow in pipe to the surface of sediments and temperature profile inside wax deposition layer is particularly important. Currently, temperature profile from the center of oil flow in pipe to the surface of sediments was more studied by scholars (Svendsen, 1993; Zhang, 2009; Huang, 2000; Singh *et al.*, 2001; Hoffmann and Amundsen, 2009; Patankar, 1980; Flajolet *et al.*, 1994; Aiyejina *et al.*, 2011; Bidmus, 2003; Bidmus *et al.*, 2004; Banki *et al.*, 2008; Nazar *et al.*, 2005; Venkatesan and Fogler, 2004; Lee *et al.*, 1997), however, the special study of temperature profile inside wax deposition layer was rarely reported. Taking into account that the wax deposition layer was too thin, when calculating the temperature profile inside wax deposition layer, Singh assumed that the heat conduction inside wax deposition layer was one-dimensional based on past experience and practice (the axial heat conduction can be ignored when axial distance is much larger than radial distance). However, whether the axial heat conduction can be ignored or not when calculating temperature profile inside wax deposition layer is worthy of further verification. Therefore we believe that, abandoning previous experience and practice, studying this problem by using the related heat transfer theory of pipe is necessary.

In this article, aiming at this problem, on the basis of considering axial heat conduction inside wax deposition layer, a two-dimensional heat transfer model of calculating the temperature profile inside wax deposition layer in flow loop experiment at a time was deduced and established by using energy balance equations, and the influence of the axial heat conduction on the distribution law of temperature profile inside wax deposition layer under different boundary conditions and thickness of wax deposition layer was studied. Finally, a comparative analysis of the calculation results of one-dimensional heat transfer model and two-dimensional heat transfer model was made. The calculation results showed that: In laboratory flow loop experiment of wax deposition Although axial heat conduction had an influence on the distribution law of temperature profile inside wax deposition layer under different conditions, the influence was quite small, thus could be ignored.

2. CALCULATION MODEL OF TEMPERATURE PROFILE INSIDE WAX DEPOSITION LAYER

After a period of flow loop experiment, there would be sedimentary layer with certain thickness on the inner pipe wall. According to mechanism of "aging", the main component of sedimentary layer close to the inner pipe wall was paraffin wax while the main components of sedimentary layer near the center of the oil flow were asphaltenes, resins, pour point oil and other impurities (Singh and Fogler, 2000). As the temperature profile inside wax deposition layer influenced the growth rate of wax content inside wax deposition layer directly, thereby affecting the calculation of wax deposition rate and, hence, in this article, the distribution law of temperature profile inside wax deposition layer was the main research object.

2.1 One-dimensional heat transfer model

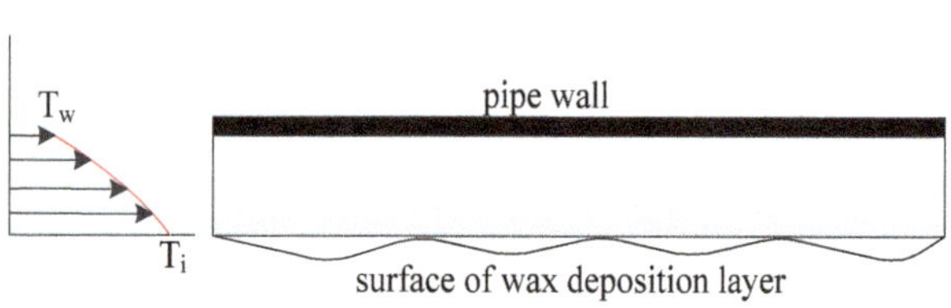

Fig. 1 Wax deposition layer on the pipe wall

Considering the wax deposition layer was too thin, when calculating the temperature profile from surface of wax deposition layer to the inner pipe wall, Singh made some assumptions as follows:

1) diffusion; particulate deposition mechanisms such as particle diffusion and gravity settling are neglected.
2) The extent of shear removal of the gel deposited in the flow loop is negligible.
3) All processes are assumed to be quasi-steady state.
4) Heat transfer from the oil-gel interface to the cold pipe-wall is a one-dimensional heat transfer.
5) There is no radial variation in the wax content of the thin gel deposit.
6) The thermal conductivity of the gel is a function of its wax content.

The schematic diagram of this model is shown in Fig. 1, and the energy equation is shown in equation (1).

$$\frac{d^2T}{dr^2}+\frac{1}{r}\frac{dT}{dr}=0 \tag{1}$$

2.2 Two-dimensional heat transfer model

In order to establish the two-dimensional heat transfer model, some assumptions are also needed to make as follows:

1) The dominant mechanism of wax deposition is molecular diffusion, ignoring the influence of other mechanisms.
2) The wax deposition layer on the pipe wall is uniform.
3) Wax deposition experiment is conducted under steady and laminar flow condition.

Physical model of calculating temperature profile inside wax deposition layer is shown in Fig. 2.

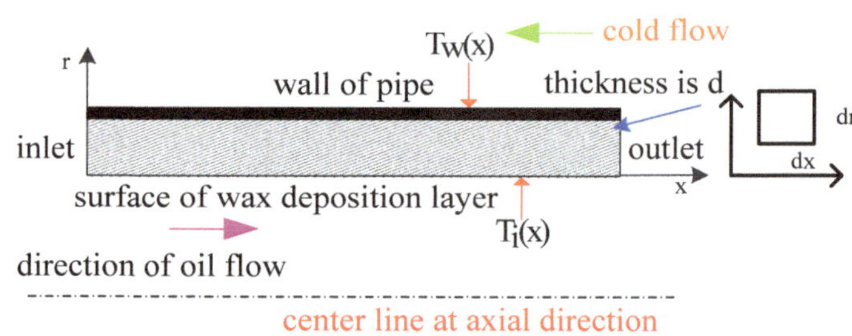

Fig. 2 Physical model of calculating the temperature profile

As for the micro unit in Fig. 2, the following equation is given based on the law of conservation of energy (Patankar, 1980):

Heat flow into micro unit + the power caused by external force on micro unit= Increasing amount of thermodynamic energy of the micro unit

By introducing Fourier heat conduction law, the energy equation expressed by specific enthalpy of fluid h and temperature T is:

$$\frac{\partial(\rho h)}{\partial t}+\frac{\partial(\rho uh)}{\partial x}+\frac{\partial(\rho vh)}{\partial y}+\frac{\partial(\rho wh)}{\partial z}=-p\,\mathrm{div}U+\mathrm{div}(\lambda\,\mathrm{grad}T)+\Phi+S \tag{2}$$

As for heat conduction inside wax deposition layer, the heat source S is zero; dissipated energy Φ is zero; the power ($p\,\mathrm{div}U$) done by surface forces on the micro unit can be ignored generally; h is taken as $h=C_pT$ and C_p as a constant. Therefore, formula (2) can be expressed as:

$$\mathrm{div}(UT)=\mathrm{div}\left(\frac{\lambda}{\rho C_p}\mathrm{grad}\,T\right)$$

In cylindrical coordinates, temperature variation along the circumferential direction is zero, therefore the formula can be written as:

$$u\frac{\partial T}{\partial x}=\frac{1}{r}\frac{\partial T}{\partial r}+\frac{\partial^2 T}{\partial r^2}+\frac{\partial^2 T}{\partial x^2}$$

As for the heat conduction inside wax deposition layer, u is zero, namely:

$$0=\frac{1}{r}\frac{\partial T}{\partial r}+\frac{\partial^2 T}{\partial r^2}+\frac{\partial^2 T}{\partial x^2} \tag{3}$$

Formula (3) is the equation of calculating temperature profile inside wax deposition layer that has been deduced in this article.

The difference between these two models is that the former has the item of axial heat conduction $\frac{\partial^2 T}{\partial x^2}$. Scholars generally consider that, as for the axial heat conduction $\frac{\partial^2 T}{\partial x^2}$, besides the fluid (e.g. liquid metal) whose Pr (Prandtl number) << 1, the axial heat conduction can be ignored (Zhang, 2009; Singh, 2001). But in this article, the study objective is wax deposition layer on pipe wall, which cannot be treated as fluid simply. Therefore, how it affected the distribution law of temperature profile inside wax deposition layer was explored when calculating temperature profile inside wax deposition layer in this article.

3. SOLUTION OF THE MODEL

3.1 The calculation of surface temperature of the wax deposition layer

When calculating the temperature profile inside wax deposition layer, the surface temperature of wax deposition layer should be obtained first. Experimental determination of the surface temperature of wax deposition layer is relatively difficult, so it is usually obtained by the calculation of heat balance equation. Steps of calculating the surface temperature of wax deposition layer are shown below.

It can be known from the heat balance that the heat flux of oil flow at the surface of wax deposition layer is equal to the heat flux of the wax deposition layer, namely:

$$\lambda_0\left.\frac{\partial T}{\partial r}\right|_{r=R-d}=\frac{T_w-T_i}{\frac{R-d}{\lambda}\ln\frac{R}{R-d}} \tag{4}$$

The calculation of oil temperature in pipe is given by Svendsen (1993):

$$T(r,x)=T_0+\frac{T_i(x)-T_0}{0.893}\int_\eta^\infty \exp(-\eta^3)\,d\eta$$

Therefore,

$$\frac{\partial T}{\partial r}=\frac{T_i(x)-T_0}{0.893}\frac{e^{-\eta^3}}{\sqrt[3]{9\beta x}} \tag{5}$$

where $\eta=\frac{R-r}{\sqrt[3]{9\beta x}}$, $\beta=\frac{\alpha R}{2u_{max}}$, $\alpha=\frac{\lambda}{\rho C_p}$.

Substituting formula (5) into (4) can obtain:

$$\frac{T_i(x)-T_0}{0.893}\frac{\lambda_0}{\sqrt[3]{9\beta x}}=\frac{T_w-T_i(x)}{\frac{R-d}{\lambda}\ln\frac{R}{R-d}}$$

Let $\frac{R-d}{\lambda}\ln\frac{R}{R-d}=A$, $\frac{\lambda_0}{0.893\sqrt[3]{9\beta x}}=B$, the above formula can be simplified as:

$$T_i(x)=\frac{T_w+A\cdot BT_0}{A\cdot B+1} \tag{6}$$

The surface temperature of wax deposition layer $T_i(x)$ can be obtained from equation (6).

3.2 Analytical solution of one-dimensional heat transfer model

As for the one-dimensional heat transfer model, the solution can be obtained by substituting boundary conditions into the model after integral. The general solution of equation (1) is:

$$T=C_1\ln r+C_2$$

Boundary conditions:

$$\begin{cases} r=r_1, T=T_i \\ r=r_2, T=T_w \end{cases}$$

The specific solution of equation (1) can be obtained by substituting the boundary conditions into equation (1), as shown in formula (7).

$$T=\frac{T_w-T_i}{\ln\frac{r_2}{r_1}}\ln r+\frac{T_i\ln r_2-T_w\ln r_1}{\ln\frac{r_2}{r_1}} \tag{7}$$

The solution of one-dimensional heat transfer model is relatively simple, the following focuses on the solution of two-dimensional heat transfer model established in this article.

3.3 The solution of two-dimensional heat transfer model

As for equation (3), it can be solved by separation of variables, obtaining its analytical solution, or solved numerically by finite difference method. Finite difference method is based on difference principle, through the discretization of the solved region and differential equations, linear equations of unknown function on network nodes in the solved region can be obtained, and then approximate solution of partial differential equations can be obtained by solving algebraic equations. Generally, finite difference method can solve related problems well when it is very difficult to obtain the analytical solution in solving a problem (Flajolet *et al*, 1994). Calculation steps of analytical solutions of the two-dimensional heat transfer model in this article were complicated and, hence, the finite difference method was used to solve the model here.

Steps of the solution of the two-dimensional heat transfer model through finite difference method are as follows:

1) The discretization of the calculation region. The calculation region is meshed by rectangle, as shown in Fig. 3.

2) The discretization of two-dimensional heat transfer equation

According to the characteristic of equation (3), the central difference is used to discrete the equation and the equation can be discretized into the following format according to the computational grid in Fig. 3:

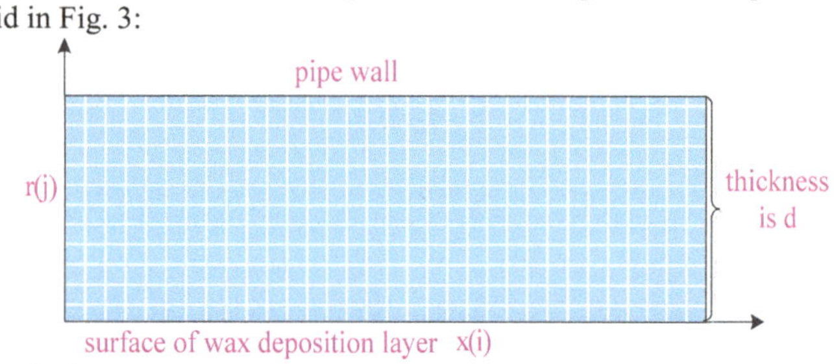

Fig. 3 The mesh generation of calculation region

$$\frac{1}{r_j}\frac{T_{i,j+1}-T_{i,j-1}}{2\Delta r}+\frac{T_{i,j+1}-2T_{i,j}+T_{i,j-1}}{\Delta r^2}+\frac{T_{i+1,j}-2T_{i,j}+T_{i-1,j}}{\Delta x^2}=0$$

Letbe $a=\frac{\Delta r^2}{\Delta x^2}$, $b=2+2a$, thus the above formula can be transformed into:

$$T_{i,j}=\frac{1+\frac{\Delta r}{2r_j}}{b}T_{i,j+1}+\frac{1-\frac{\Delta r}{2r_j}}{b}T_{i,j-1}+\frac{a}{b}T_{i-1,j}+\frac{a}{b}T_{i+1,j}$$

Letbe $C_1=\frac{1+\frac{\Delta r}{2r_j}}{b}$, $C_2=\frac{1-\frac{\Delta r}{2r_j}}{b}$, $C_3=\frac{a}{b}=C_4$, finally, the equation can be transformed into:

$$T_{i,j}=C_1T_{i,j+1}+C_2T_{i,j-1}+C_3T_{i-1,j}+C_4T_{i+1,j}$$

The boundary conditions can be discretized though the same method, that is, assign nodes at boundary according to the boundary conditions sequentially.

3) Solution of the equations

Formula (8) is the five-point difference format of Laplace equation. Formula (8) can be solved by substituting boundary conditions into the model.

In this article, the calculation region was evenly meshed by rectangle, and the axial grid number was 100, radial grid number was 80. First, assigned an initial value to each grid node, and then calculated equation (8) based on the differential equation, though multiple iterative calculations the temperature did not change on the whole (achieved the required accuracy) finally, the results at this point was the final calculation temperature.

4. RESULTS AND DISUSSION

In flow loop experiment of wax deposition, there are two types of boundary conditions which are often used generally, that is, the first type of boundary condition and the second type of boundary condition (Patankar, 1980).In the flow loop experiment of wax deposition in literature (Zhang,2009; Huang, 2000), the heat transfer coefficient of wax deposition layer and the temperature distribution of pipe wall could be measured by experiment, the radial temperature of wax deposition layer at the inlet and outlet could be measured though one-dimensional heat transfer model, that is, the temperature at each boundary of wax deposition layer were all known at a time, the boundary conditions at this situation could be regarded as the first type of boundary condition.

When calculating the temperature profile inside wax deposition layer, different boundary conditions have different influence on calculation results of temperature profile, and the continuously increasing thickness of wax deposition layer also has an influence on the distribution law of temperature profile inside wax deposition layer. The influence of axial heat conduction on the temperature profile inside wax deposition layer under different boundary conditions (constant wall temperature and variable wall temperature) and different thickness of wax deposition layer (0.2mm, 0.3mm, 0.5mm) were discussed in the followings. For the sake of discussion, the inlet and outlet of testing pipe section of the flow loop was defined as left boundary and right boundary respectively, surface of wax deposition layer was defined as down boundary, pipe wall was defined as up boundary.

4.1 The influence of axial heat conduction under boundary conditions of constant wall temperature

When Huang (2000) performed the flow loop experiment of wax deposition, temperature of the external pipe wall was maintained at a constant value, which meant the variation of wall temperature along the axial direction was small when the test pipe section was short, therefore the axial wall temperature could be considered as constant approximately, that was, the boundary conditions of constant wall temperature. The related experimental parameters are shown in Table 1

Table 1 Relative parameters of the flow loop experiment.

Parameters	T_0(°C)	T_{in}(°C)	T_{out}(°C)	*d*(mm)
Values	19	17	17	0.2, 0.3, 0.5

According to the above parameters and the assumptions made previously, the boundary conditions of these two models were determined first. As for one-dimensional heat transfer model, temperature at the left and right boundary could be determined though surface temperature of wax deposition layer calculated by formula (6) and experimentally measured temperature of pipe wall, and the calculation results were in line with logarithmic law. As for two-dimensional heat transfer model, since the influence of axial heat conduction was only considered inside wax deposition layer, temperature at each boundary was not affected, so the temperature at boundary of one-dimensional heat transfer model could still be used to calculate the two-dimensional heat transfer model, that was, temperature at boundary of these two models was consistent, so temperature at the left and the right boundary were still in line with logarithmic law.

Table 2 The up and down boundary when wall temperature is constant.

Position of boundary	*d*(mm)	*Temperature distribution*(°C)
Down boundary	0.2	$T_i(x)=0.1372x^5-0.5144x^4+0.679x^3-0.287x^2-0.2622x+17.31$
	0.3	$T_i(x)=0.1235x^3+0.0556x^2-0.4278x+17.44$, $0\le x\le 0.9$ $T_i(x)=0.2778x^2-0.8833x+17.76$, $0.9\le x\le 1.5$
	0.5	$T_i(x)=0.1461x^2-0.5331x+17.648$
Up boundary	0.2, 0.3, 0.5	$T_w(x)=17$

When thickness of wax deposition layer changes, surface temperature of wax deposition layer can be calculated by formula (6), namely the corresponding down boundary. Because of the boundary conditions of constant wall temperature, temperature of pipe wall (up boundary) is constant, as shown in Table 2.

According to the experimental data obtained by author and size parameters of flow loop, taking the thickness of 0.2mm of wax deposition layer as example, the distribution law of temperature profile inside wax deposition layer under boundary conditions of constant wall temperature was explored here.

The distribution law of temperature profile inside wax deposition layer was studied based on the coordinate shown in Fig. 2.When calculating the radial concentration gradient of wax molecule inside wax deposition layer, radial temperature gradient inside wax deposition layer often needed to be calculated, and this required to study the radial variation of temperature inside wax deposition layer at different axial positions. Because the nodes in calculation region were massive, in order to analyze the calculation results of one-dimensional heat transfer model and two-dimensional heat transfer model at different positions, temperature at three representative positions were taken along the axis direction of pipeline, which was at a distance of 0.135m (near the inlet), 0.735m (in the middle region) and 1.335m (near the outlet) from the inlet of test pipe section throughout the whole test pipe section (1.5m), then a comparative analysis of the distribution law of radial temperature at these three axial positions inside wax deposition layer was made, as shown in Fig. 4.

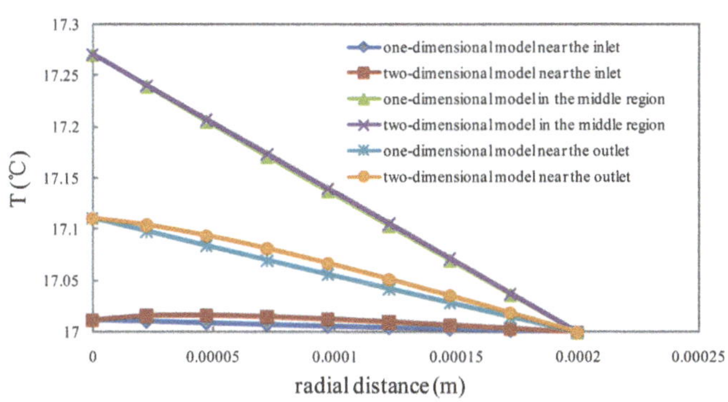

Fig. 4 Calculation results of radial temperature inside wax deposition layer at different axial positions.

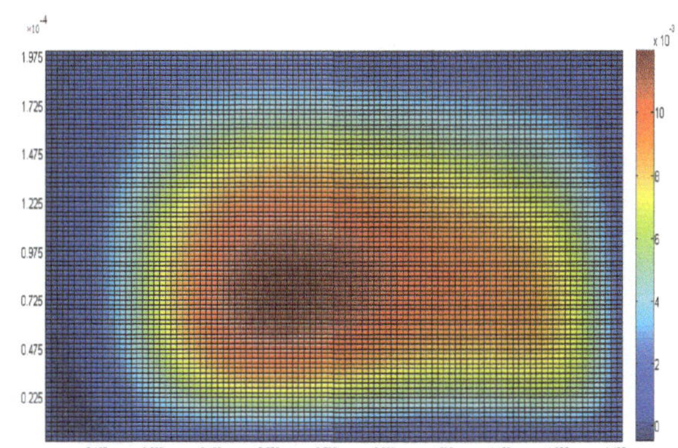

Fig. 5 Distribution of difference of the calculation temperature between the two models inside wax deposition layer.

It can be seen from Fig. 4 that radial temperature inside wax deposition layer at each axial position shows a decreasing trend. Near the inlet and outlet of the test pipe section, there is almost no difference between the results of the two models, while the difference is relatively larger in the middle region of wax deposition layer.

In the flow loop experiment, oil temperature was higher than temperature of cooling water, therefore axial temperature of pipe wall showed an increasing trend while the temperature of pipe wall was constant under boundary conditions of constant wall temperature, and calculation temperatures at different positions inside wax deposition layer were different compared with the boundary conditions of variable wall temperature in this situation. Due to the effect of cooling water, oil temperature decreased continually, there was a relatively bigger difference of the calculation temperature at somewhere inside wax deposition layer between the two models resulting from the combined effect of boundary conditions of constant wall temperature and axial heat conduction. It can be seen from the calculation results in this article that, it is in the middle region, but the difference is still small (the biggest difference is about 0.01°C).

It can be seen from Fig. 5 that the calculation results of the two models have little difference, that is, axial heat conduction has little effect on the distribution law of temperature profile inside wax deposition layer, thus can be ignored

4.2 The influence of axial heat conduction under boundary conditions of variable wall temperature

The experimental data obtained by Zhang (2009) is quoted in this article, and the related experimental parameters are shown in Table 3.

Table 3 Parameters of the flow loop experiment.

Parameters	T_0(°C)	T_{in}(°C)	T_{out}(°C)	*d*(mm)
Values	45	40.5	40	0.2

It was found that when fitting the temperature distribution at the up boundary by the experimentally measured temperature of pipe wall, high accuracy (R^2=1) could be achieved with linear fitting, but at the down boundary, in order to meet the fitting accuracy, segmental fitting was used here, with the fitting accuracy were R^2=0.9998 and R^2=1 respectively, as shown in Table 4.

Table 4 The up and down boundary when wall temperature is variable.

Position of boundary	*Temperature distribution*(°C)
Up boundary	$T_w=40.5-0.1053x$
Down boundary	$T_i=4.9978x^6-19.12x^5+29.792x^4-24.432x^3+11.455x^2-3.3241x+41.339, 0\le x\le 1$ $T_i=-0.0021x^3+0.0256x^2-0.2189x+40.899, 1\le x\le 4.75$

Like the analysis steps under boundary conditions of constant wall temperature, temperature at three representative positions were taken along the axis direction of pipeline, which was at a distance of 0.4275m (near the inlet), 2.3275m (in the middle region) and 4.2275m (near the outlet) from the inlet of test pipe section throughout the whole test pipe section (4.75m), then a comparative analysis of the distribution law of radial temperature at these three axial positions inside wax deposition layer was made, the results are shown in Fig. 6.

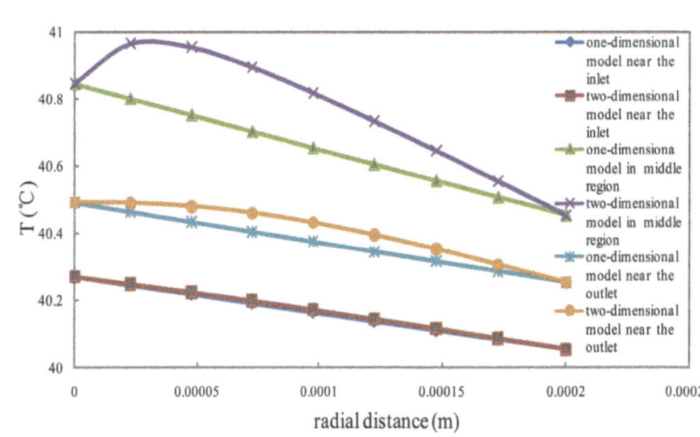

Fig. 6 Calculation results of radial temperature inside wax deposition layer at different axial positions

It can be seen from Fig. 6 that there are some differences near the inlet between the results of the two models while no difference at other regions. Therefore in this case, axial heat conduction mainly influences the distribution law of temperature profile near the inlet of wax deposition layer, which is the same with the influence of axial heat conduction on the heat transfer law when fluid flows with low Pr number.

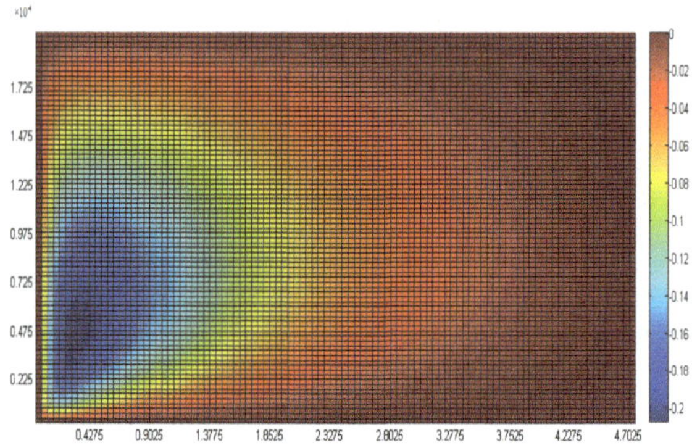

Fig. 7 Distribution of difference of the calculation temperature between the two models inside wax deposition layer

Near the inlet, the result of two-dimensional heat transfer model is larger than that of one-dimensional heat transfer model, the reasons for this difference are: in the flow loop experiment of wax deposition, because oil temperature near the inlet of testing pipe section was higher than the external cooling water temperature, hot oil in testing pipe section transferred heat to external cooling water continuously through the wax deposition layer and pipe wall under the effect of temperature difference, in the situation of considering axial heat conduction, besides radial heat conduction inside wax deposition layer, axial heat conduction was also ongoing, therefore the calculation temperature was higher than one-dimensional heat transfer model, and with the continuously increasing axial distance, the effect of axial heat conduction weakened gradually, calculation temperature of the two models at this point were of no difference, that is, the effect of axial heat conduction at other regions which were far away from the inlet was not conspicuous.

Difference of the calculation temperature between the two models inside wax deposition layer can be clearly seen from Fig. 7. The largest difference of the calculation temperature between the two models near the inlet is about 0.2°C, larger than other regions. Therefore, when calculating temperature profile inside wax deposition layer, special attention should be paid to the region near the inlet, where the influence of axial heat conduction is relatively larger. Overall, the difference of calculation temperature between the two models is small, but the two-dimensional heat transfer model reflects the actual heat transfer inside wax deposition layer better.

Under the boundary conditions of variable wall temperature, the influence of axial heat conduction on the distribution law of temperature profile inside wax deposition layer at the thickness of 0.3mm, 0.5mm of wax deposition layer is almost same with the thickness of 0.2mm, therefore no detailed discussion here.

4.3 The influence of axial heat conduction at different thickness of wax deposition layer

In order to discuss the influence of thickness of wax deposition layer on calculation results of temperature profile, temperature profile inside wax deposition layer at the thickness of 0.3mm and 0.5mm were also calculated. Under boundary conditions of constant wall temperature, like the analysis steps at the thickness of 0.2mm, temperature at three representative positions were taken along the axis direction of pipeline, which was at a distance of 0.135m (near the inlet), 0.735m (in the middle region) and 1.335m (near the outlet) from the inlet of test pipe section throughout the whole test pipe section (1.5m), then a comparative analysis of the calculation temperature at these three representative positions at different thickness of the two models was made, as shown in Fig. 8.

In can be seen from Fig. 8 that the distribution law of temperature profile inside wax deposition layer (the thickness of wax deposition layer is 0.3mm and 0.5mm) is consistent with that of the thickness of 0.2mm. At different axial positions (near the inlet, in the middle region, near the outlet), the influence of axial heat conduction on the distribution law of temperature profile inside wax deposition layer of different thickness is different, the differences are relatively larger in the middle region while little near the inlet and outlet. As for the two models, the radial temperature inside wax deposition layer at same axial position increases with the increasing thickness of wax deposition layer. This is because that when the thickness of wax deposition layer increases, the distance between surface of wax deposition layer and center of oil fluid becomes small, thus the temperature of wax deposition layer becomes higher. The larger the thickness of wax deposition layer, the more conspicuous of the effect of axial heat conduction, thereby the larger difference of calculation results of the two models.

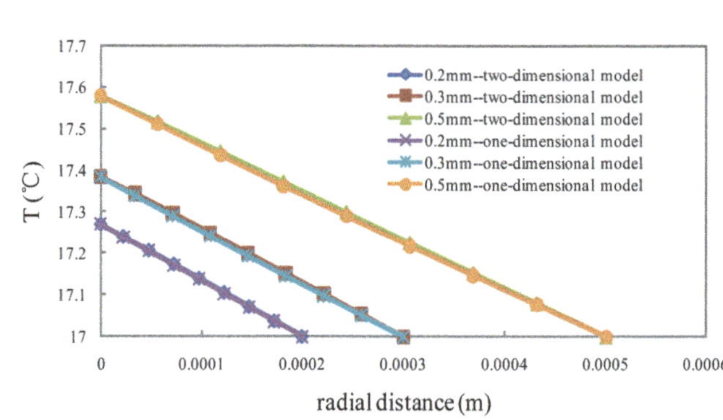

(a) near the inlet.

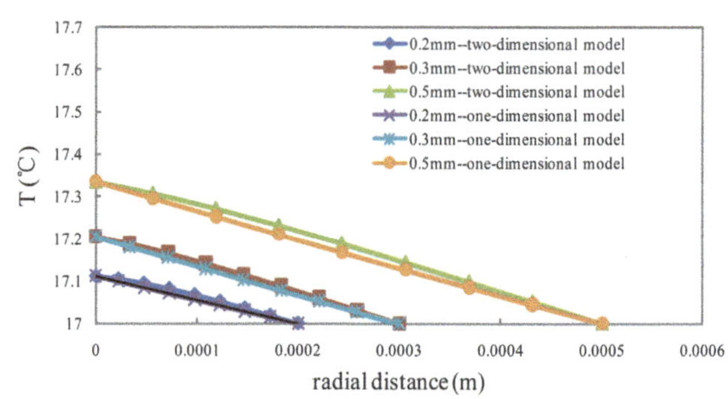

(b) in the middle region.

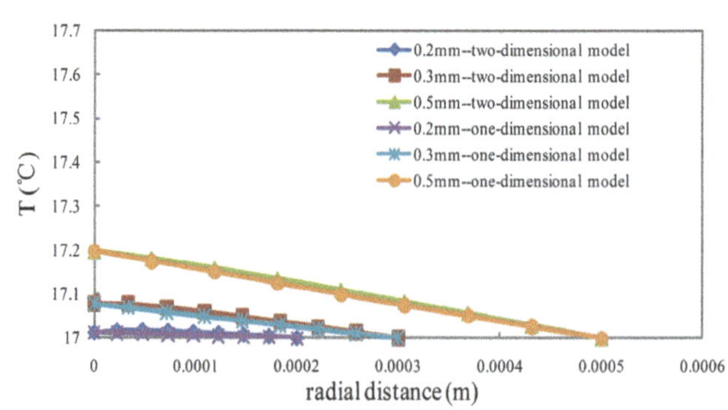

(c) near the inlet.

Fig. 8 The calculation temperature of the two models at different thickness and boundary conditions of constant wall temperature

In order to analyze the effect of axial heat conduction on the distribution law of temperature profile inside wax deposition layer at different thickness more clearly, the difference of calculation results of the two models are shown in Fig. 9. It can be seen from Fig. 9 that the differences of calculation results of the two models at each representative positions are relatively small, with the largest difference is less than 0.025°C (c in Fig. 9), therefore, the effect of axial heat conduction can still be ignored at different thickness of wax deposition layer under boundary conditions of constant wall temperature.

Under the boundary conditions of variable wall temperature, the influence of axial heat conduction on the distribution law of temperature profile inside wax deposition layer at different thickness of wax deposition layer is same with that under the boundary conditions of constant wall temperature, therefore no detailed discussion here.

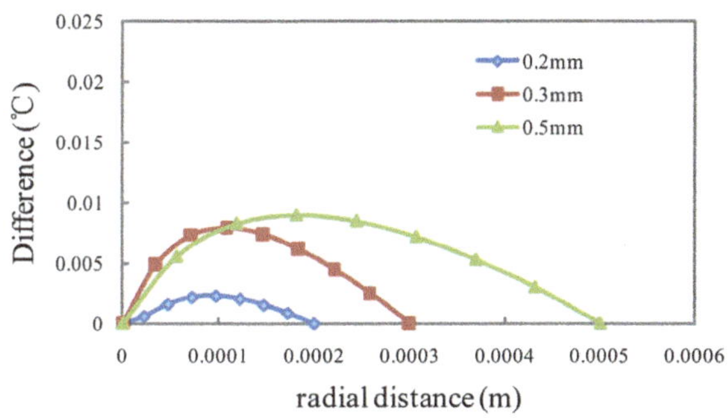

(a) near the inlet.

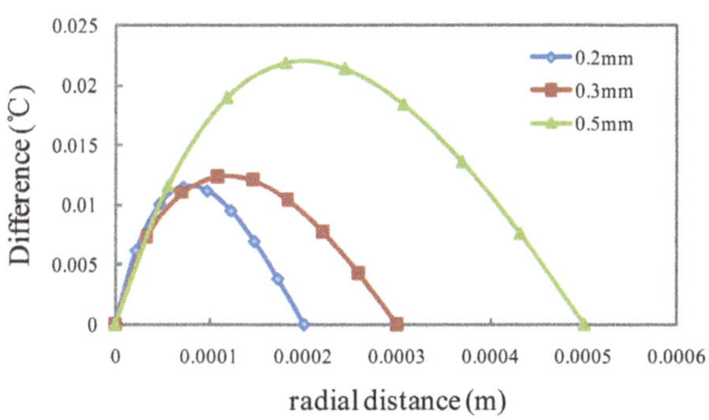

(b) in the middle region.

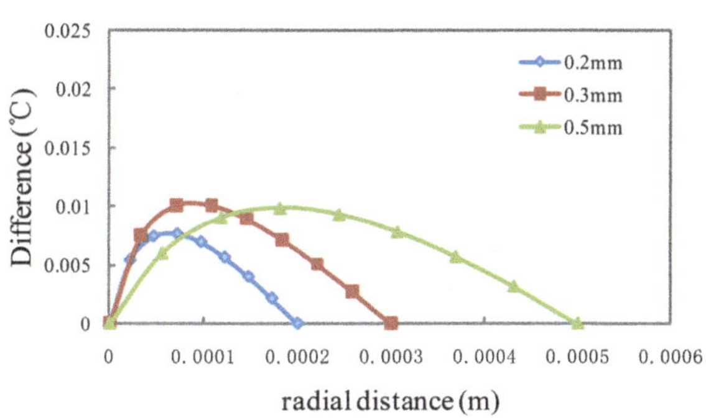

(c) near the outlet.

Fig. 9 The difference of calculation results of the two models at different thickness and axial positions

5. CONCLUSIONS

The distribution law of temperature profile inside wax deposition layer is the key to study wax deposition rate of crude oil in pipeline. As for this problem, the influence of axial heat conduction inside wax deposition layer was always ignored in previous studies, however, whether the influence on calculating temperature inside wax deposition layer was large or not was worthy of further verification when the axial heat conduction was ignored.

In this article, a two-dimensional heat transfer model of calculating temperature profile inside wax deposition layer in the flow loop experiment at a time was deduced and established by using energy balance equations, and the influence of the axial heat conduction on the distribution law of temperature profile inside wax deposition layer under different conditions was studied, then comparative analysis with the calculation results of one-dimensional heat transfer model was made. The calculation results showed that: the influence of axial heat conduction under boundary conditions of variable wall temperature was mainly embodied near the inlet of testing pipe section while the influence was relatively small in other regions; under boundary conditions of constant wall temperature, influenced by the combined effect of boundary conditions of constant wall temperature and axial heat conduction, the difference of calculation temperature of the two models was relatively larger in the middle region; at three types of thickness of the wax deposition layer given above, the influence of axial heat conduction became more conspicuous with the increasing thickness of wax deposition layer. Overall, in laboratory flow loop experiment, although axial heat conduction had an influence on the distribution law of temperature profile inside wax deposition layer, the influence was quite small. This method was also applicable to practical pipeline, and the concrete situations needed a further study based on field data and the two dimensional heat transfer model obtained in this work. From theoretical point of view, it was confirmed in this article that ignoring axial heat conduction in the calculation of temperature profile inside wax deposition layer was entirely feasible.

NOMENCLATURE

T_i	surface temperature of wax deposition layer, °C
T_w	temperature of pipe wall, °C
r_1	distance from the center of oil flow to surface of wax deposition layer, m
r_2	distance from surface of wax deposition layer to pipe wall, m
l	length of wax deposition layer (length of pipe), m
d	thickness of wax deposition layer at a time, mm
u	velocity of fluid (m/s)
C_p	specific heat (J/kg·K)
R	inner radius of pipe, m
T_0	oil temperature at the inlet, °C
T_{in}	wall temperature at the inlet, °C
T_{out}	wall temperature at the outlet, °C
C_p	specific heat (J/kg·K)

Greek Symbols

λ	heat conduction coefficient of wax deposition layer, W/(m·K)
λ_0	heat conduction coefficient of crude oil, W/ (m·K)
ρ	density (kg/m^3)

REFERENCES

Agrawal, K.M., Khan, H.U., Surianarayanan, M., Joshi, G.C., 1990, "Wax Deposition of Bombay High Crude Oil under Flowing Conditions," *Fuel*, **69**, 794-796. http://dx.doi.org/10.1016/0016-2361(90)90051-Q

Aiyejina, A., Dhurjati, P.C., Pilgrim, A., Sastry, M.K.S., 2011, "Wax Formation in Oil Pipelines: A Critical Review," *International Journal of Multiphase Flow*, **37**, 671-694. http://dx.doi.org/ 10.1016/j.ijmultiphaseflow.2011.02.007

Banki, R., Hoteit, H., Firoozabadi, A., 2008, "Mathematical Formulation and Numerical Modeling of Wax Deposition in Pipelines

from Enthalpy-porosity Approach and Irreversible Thermodynamics," *International Journal of Heat and Mass Transfer*, **51**, 3387-3398.
http://dx.doi.org/ 10.1016/j.ijheatmasstransfer.2007.11.012

Bern, P.A., Withers, V.R., Cairns, J.R., 1980, "Wax Deposition in Crude Oil Pipelines," 206-1980-MS, *European Offshore Petroleum Conference and Exhibition*, London.
http://dx.doi.org/10.2118/206-1980-MS

Bidmus, H.O., 2003, "A Thermal Study of Wax Deposition from Paraffinic Mixtures," *M.Sc. Thesis, University of Calgary, Calgary, Canada.*

Bidmus, H.O., Mehrotra, A.K., 2004, "Heat-Transfer Analogy for Wax Deposition from Paraffinic Mixtures," *Ind. Eng. Chem. Res*, **43**, 791-803.
http://dx.doi.org/ 10.1021/ie030573v

Brown, T.S., Niesen, V.G., Erickson, D.D., 1993, "Measurement and Prediction of Kinetics of Paraffin Deposition," 26548-MS, *SPE Annual Technical Conference and Exhibition*, Houston, Texas.
http://dx.doi.org/10.2118/26548-MS

Burger, E.D., Perkins, T.K., Striegler, J.H., 1981, "Studies of Wax Deposition in the Trans-Alaska Pipeline," *Journal of Petroleum Technology*, **33**, 1075-1086.
http://dx.doi.org/10.2118/8788-PA

Flajolet, P., Sedgewick, R., 1994, "Mellin Transforms and Asymptotics: Finite Differences and Rice's Integrals," *Theoretical Computer Science*, **144**, 101-124.
http://dx.doi.org/10.1016/0304-3975(94)00281-M

Hamouda, A.A., Davidsen, S., 1995, "An Approach for Simulation of Paraffin Deposition in Pipelines as a Function of Flow Characteristics with a Reference to Teesside Oil Pipeline," 28966-MS, *SPE International Symposium on Oilfield Chemistry*, San Antonio, Texas.
http://dx.doi.org/10.2118/28966-MS

Hoffmann, R., Amundsen, L., 2009, "Single-Phase Wax Deposition Experiments," *Energy and Fuels*, **24**, 1069-1080.
http://dx.doi.org/ 10.1021/ef900920x

Hsu, J.J.C., Santamaria, M.M., Brubaker, J.P., 1994, "Wax Deposition of Waxy Live Crudes under Turbulent Flow Condition," 28480-MS, *SPE Annual Technical Conference and Exhibition*, New Orleans, Louisiana.
http://dx.doi.org/10.2118/28480-MS

Hunt, E.B.Jr., 1962, "Laboratory Study of Paraffin Deposition," *Journal of Petroleum Technology*, **14**, 1605-1612.
http://dx.doi.org/ 10.2118/279-PA

Huang, Q.Y., 2000, "The Study of Dynamics Model of Wax Deposition of Waxy Crude Oil in Pipe," *Ph.D. Thesis, China University of Petroleum (Beijing), Beijing, China.*

Lee, H.S., 2008, "Computational and Rheological Study of Wax Deposition and Gelation in Subsea Pipelines," *Ph.D. Thesis, The University of Michigan, Michigan, United States.*

Lee, K.S., Kim, W.S., Lee, T.H., 1997, "A One-Dimensional Model for Frost Formation on a Cold Flat Surface," *International Journal of Heat and Mass Transfer*, **40**, 4359-4365.
http://dx.doi.org/ 10.1016/S0017-9310(97)00074-4

Matlach, W.J., Newberry, M.E., 1983, "Paraffin Deposition and Rheological Evaluation of High Wax Content Altamont Crude Oils," 11851-MS, *SPE Rocky Mountain Regional Meeting*, Salt Lake City, Utah.
http://dx.doi.org/ 10.2118/11851-MS

Mehrotra, A.K.., Bhat, N.V., 2007, "Modeling the Effect of Shear Stress on Deposition from "Waxy" Mixtures under Laminar Flow with Heat Transfer," *Energy and Fuels*, **21**, 1277-1286.
http://dx.doi.org/ 10.1021/ef060445u

Nazar, A.R.S, Dabir, B., Islam, M.R., 2005, "Experimental and Mathematical Modeling of Wax Deposition and Propagation in Pipes Transporting Crude Oil," *Energy Sources*, **27**, 185-207.
http://dx.doi.org/ 10.1080/00908310490448262

Nazar, A.R.S, Dabir, B., Islam, M.R., 2001, "Measurement and Modeling of Wax Deposition in Crude Oil Pipelines," 69425-MS, *SPE Latin American and Caribbean Petroleum Engineering Conference*, Buenos Aires, Argentina.
http://dx.doi.org/ 10.2118/69425-MS

Patankar, S.V., 1980, "Numerical Heat Transfer and Fluid Flow," Hemisphere Publishing Corporation, Washington, DC.

Pedersen, K.S., 1995, "Prediction of Cloud Point Temperatures and Amount of Wax Precipitation," *SPE Production and Facilities*, **10**, 46-49.
http://dx.doi.org/10.2118/27629-PA

Ramirez-Jaramillo, E., Lira-Galeana, C., Manero, O., 2004, "Modeling Wax Deposition in Pipelines," *Petroleum Science and Technology*, **22**, 821-861.
http://dx.doi.org/ 10.1081/LFT-120038726

Ribeiro, F.S., Mendes, P.R.S., Braga, S.L., 1997, "Obstruction of Pipelines due to Paraffin Deposition during the Flow of Crude Oils," *International Journal of Heat Mass Transfer*, **18**, 4319-4328.
http://dx.doi.org/ 10.1016/S0017-9310(97)00082-3

Singh, P., Venkatesan, R., Fogler, H.S., Nagarajan N., 2000, "Formation and Aging of Incipient Thin Film Wax-Oil Gels," *AIChE Journal*, **46**, 1059-1074.
http://dx.doi.org/ 10.1002/aic.690460517

Singh, P., Venkatesan, R., Fogler, H.S., Nagarajan N., 2001, "Morphological Evolution of Thick Wax Depositions during Aging," *AIChE Journal*, **47**, 6-18.
http://dx.doi.org/ 10.1002/aic.690470103

Svendsen, J.A., 1993, "Mathematical Modeling of Wax Deposition in Oil Pipeline Systems," *AIChE Journal*, **39**, 1377-1388.
http://dx.doi.org/ 10.1002/aic.690390815

Venkatesan, R., Fogler, H.S., 2004, "Comments on Analogies for Correlated Heat and Mass Transfer in Turbulent Flow," *AIChE Journal*, **50**, 1623-1626.
http://dx.doi.org/ 10.1002/aic.10146

Zhang, G.Z., Liu, G., 2010, "Study on the Wax Deposition of Waxy Crude in Pipelines and Its Application," *Journal of Petroleum Science and Engineering*, **70**, 1-9.
http://dx.doi.org/ 10.1016/j.petrol.2008.11.003

Zhang, Z.B., 2009, "Wax Deposition Property of Waxy Oil in Pipeline," *Ph.D. Thesis, China University of Petroleum (Huadong), Dongying, China.*

2

A DRYING AND THERMOELASTIC MODEL FOR FAST MICROWAVE HEATING OF CONCRETE

Benjamin Lepers[a,*], Aditya Putranto[b,c], Martin Umminger[d], Guido Link[a], John Jelonnek[a]

[a]*Karlsruhe Institute of Technology, Institute for Pulsed Power and Microwave Technology, Eggenstein Leopoldshafen, 76344, Germany*
[b]*Monash University, Dept. of Chemical Engineering, Clayton Campus, Melbourne, Victoria 3800, Australia*
[c]*Parahyangan Catholic University, Dept. of Chemical Engineering, Jolan Ciumbuleuit 94, Bandung, Indonesia*
[d]*Karlsruhe Institute of Technology, Institute of Concrete Structures and Building Materials, Karlsruhe, 76131, Germany*

ABSTRACT

The use of high power microwaves to perform explosive spalling of surface concrete is a promising technique with applications in the area of concrete facilities decommissioning. The mechanism that creates explosive spalling is an interactive process of the thermal stress from high temperature gradients and the pore pressure generated from the water vaporization. In order to better predict the total stress distribution, the temperature has to be calculated by including the effect of water vaporization and water transport through a porous medium. In this paper, a one dimensional model solving the heat and diffusion equation for liquid and vapor phase with COMSOL finite element software is presented. The modelling of the drying process is based on the Spatial Reaction Engineering Approach (SREA developed by X.D. Chen). This paper discusses the influence of the relative energy activation parameter and effective diffusion coefficients on the temperature, water content and pore pressure in the case of fast microwave heating of concrete. This model is then used for a 3-D geometry with a sealed insulated concrete block and a conical antenna to compute the thermal stress, pore pressure and total stress.
Keywords: *Concrete ablation, microwave heating, drying, SREA, thermal stress, pore pressure*

1. INTRODUCTION

In the case of rapid heating of concrete with high power microwaves, evaporation/condensation of water occurs inside the porous material and leads to high thermal stress and pore pressure. The explosive spalling mechanism is not modelled in itself, but the effect of temperature and liquid content on the thermal stress and pore pressure respectively is investigated with a coupled thermo mechanical model. To better predict the temperature field for the calculation of the thermal stress (Lepers *et al.*, 2013) and the pore pressure, a porous model with solid, liquid and gas phase is currently developed. The concrete pore pressure has been calculated by different authors using coupled heat transfer models with Lamberts's law for the microwave power distribution and an equation of state based on water content for the pore pressure (Bazant and Thonguthai, 1978; Akbarnezhad and Ong, 2011).

The pore pressure was measured by Phan (2008) with explosive spalling occuring around 2.1 MPa at $240\,^\circ C$. In reactive powder concrete (RPC), the measured peak pore pressure was about 3.3 MPa at a temperature of $350\,^\circ C$ (Ju *et al.*, 2013). Bazant and Goangseup (2003b) have calculated the pore pressure with a hydro thermo mechanical model for different microwave frequencies. At 2.45 GHz and temperature in the range of $140-180\,^\circ C$, the pore pressure was in the range $0.3-0.8$ MPa. At 10.6 GHz and temperatures $200-350\,^\circ C$, the pore pressure was in the range $1-3$ MPa. They have suggested that the pore pressure could possibly trigger the mechanism of explosive spalling. In the case of concrete exposed to high temperature (such as fire for example) for time scales up to 500 min, Mindeguia *et al.* (2010) have measured the pore pressure for concrete of different compactness and showed that the low permeability concrete leads to higher pore pressure. Gawin *et al.* (2011a,b) have developed a complex multiphysics model with many parameters taking into account the moisture transport, heat transfer, thermo chemical degradation and mechanical damage. For the case of concrete heated by fire, at a time scale up to 7 h, the maximum calculated pore pressure was in the order of 0.5 MPa.

The motivation of this work is to compare the magnitude of the pore pressure with the thermal stress. With a coupled electro thermo elastic model, it is possible to see the effect of the diffusivity on the temperature distribution, moisture content, pore pressure and total stress. The Spatial Reaction Engineering Approach (SREA) developed by Chen (2008) is used to model drying processes of materials. The method was successful in predicting liquid content and spatial temperature distributions for many different porous materials such as vegetables or fruits for conventional drying (Chen and Putranto, 2013b) or wood material for example (Putranto and Chen, 2013b). The local evaporation/condensation term is proportional to the difference between the internal saturated (or equilibrium) vapour concentration and the actual water vapour concentration. The internal vapour concentration is modelled with an Arrhenius type law with a relative activation energy function to be determined experimentally by

* *Corresponding author. Email: benjamin.lepers@kit.edu*

measuring the water loss rate profile versus time. To simplify our model, the local liquid and vapor concentration are calculated with the assumption of a closed pore volume system and an ideal gas law for the behaviour of the vapour and air phase. For the 3-D coupled electromagnetic model, the complex permittivity of concrete and the thermal properties are expressed as a function of the solid, liquid and gas phase. The computation of the temperature, liquid and water vapour concentration is necessary for the calculation of the total pressure. The pore pressure is combined with a thermoelastic model to compute the total stress. In the next section, the electromagnetic model of a conical antenna based on a standard WR 340 rectangular wave guide is briefly described. Then, the heat and mass transfer model together with the mechanical model used to compute the thermal stress and pore pressure of concrete are presented. The novelty of this work is the use the SREA approach (Chen and Putranto, 2013a) in order to compute the liquid and vapor content distribution in concrete during microwave heating.

2. MATHEMATICAL MODELLING

2.1. Microwave model

The microwave power density is obtained by solving the wave equation in the frequency domain at 2.45 GHz of a 3-D conical antenna including an air gap of 10 mm between the guide aperture and the concrete surface. The power density distribution is shown in Fig. 1. The conical antenna is based on a standard 86.6×43.3 mm rectangular wave guide. The length of the conical section is 68 mm and is fully filled with Teflon PTFE. The width, heights of the entry and exit of the conical section are respectively $\{54, 43.3, 21.6\}$ mm.

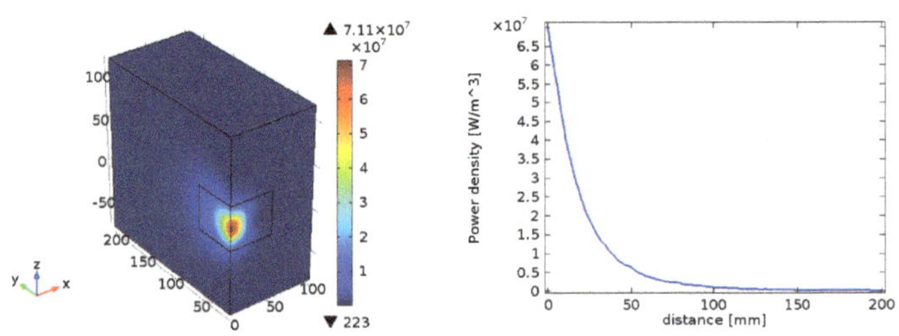

(a) (b)

Fig. 1 Power density distribution [$\mathrm{W/m^3}$] for the conical antenna with input power $P_{in} = 10$ kW and an air gap of 1 cm between the antenna aperture and the concrete bloc with a constant relative permittivity $\epsilon = 6 - 0.68j$ (a). Power density in the center along the central y axis (b).

A 1-D model allows the calculation of the temperature, water and vapour content along the thickness of the concrete wall by the solving the coupled heat and diffusion equations with much shorter computation time than a 3-D model. Following Lambert's law, an exponential decay of the E field and power density through the thickness of the concrete sample is used. The attenuation constant is obtained from the power density calculated with a 3-D conical antenna electromagnetic model as shown in Fig. 1(b) . For, $\epsilon_r = 6 - 0.68j$ at the frequency $f = 2.45$ GHz, the skin depth for the electric field is about $\delta = 4.18$ cm ($e^{-1} \simeq 0.36$ attenuation of the maximum electric field). The attenuation constant is $\beta = \frac{1}{\delta} = 23.9$ 1/m. The power density distribution used in the 1d model is:

$$q(y) = q_0 e^{-2\beta y} \tag{1}$$

with $q_0 = 70\ \mathrm{MW/m^3}$ obtained with the 3-D conical antenna model with 10 kW of input power.

2.2. Heat and mass transfer model

2.2.1. Definitions For a solid material such as concrete, the volume variation of the concrete sample (shrinkage or expansion) during heating is negligible and the system is assumed to be a sealed domain of constant volume V and constant porosity.

$$V = V_s + V_l + V_v + V_a \tag{2}$$

The subscript s, l, v, a designating the solid, liquid, vapour and air phase. The porosity ϵ_0 and the solid fraction are:

$$\epsilon_0 = \frac{V_l + V_v + V_a}{V} \tag{3}$$
$$1 - \epsilon_0 = \frac{V_s}{V}$$

The solid C_s and local liquid C_l, vapor concentration C_v and air concentration C_a in $\mathrm{kg/m^3}$ are:

$$\begin{aligned} C_s &= \frac{m_s}{V} = \frac{m_s}{V_s}\frac{V_s}{V} = \rho_s(1-\epsilon_0) \\ C_l &= \frac{m_l}{V} = \frac{m_l}{m_s}\frac{m_s}{V} = X_l C_s = \rho_l f_l \\ C_v &= X_v C_s = \rho_v f_v \\ C_a &= X_a C_s = \rho_a f_a \end{aligned} \tag{4}$$

X_l, X_v and X_a are the local mass of liquid, vapour and air per kg of dry solid and $f_i = \frac{V_i}{V}$, $i = \{l, v, a\}$ are respectively the volume fraction of liquid, vapour and air. For a closed volume, the air concentration is constant since the mass of air is constant.

2.2.2. Assumptions The assumptions used for the heat and mass transfer processes inside the porous model of concrete are:

- The effect of solid shrinkage or expansion on heat and mass transfer process is not taken into account, a sealed and constant volume is assumed.
- cracks formation and sudden increase of permeability are not modelled.
- The porous medium is homogeneous and isotropic (i.e. k, D_l, D_v are scalar quantities).
- The diffusion coefficients are assumed to be constant. In reality, they are usually function of temperature and water content.
- The transfer of liquid and vapour water occurs by diffusion, the "effective diffusion " coefficient includes all the contribution to the fluid movement such as temperature and pressure gradient (Chen, 2007).
- The evaporation/condensation rate is proportional to the difference between the internal saturated vapour concentration $C_{int,sat}$ and the current concentration inside the solid C_v.
- Because of fast microwave heating and for simplification of the analysis, the volume is sealed and insulated, no heat flux and mass flow (water or gas) cross its boundaries. In total, all the boundaries condition on the surface of the bloc are: $\mathbf{n}.\nabla T = 0$, $\mathbf{n}.\nabla C_l = 0$, $\mathbf{n}.\nabla C_v = 0$ with $\mathbf{n}$ the unit vector normal to the surface.
- The initial temperature and moisture content are uniformly distributed.
- The water vapour and air phase follow the ideal gas law.

2.2.3. Heat and mass transfer equations The temperature field is calculated with the heat equation for a porous medium

$$\rho c_p \frac{\partial T}{\partial t} = \nabla \cdot k \nabla T + q_{em} - I \Delta H \qquad (5)$$

with the thermal properties averaged with the solid, water and gas content (see appendix A). q_{em} is the power density source term from microwave heating and I the rate of evaporation/condensation. $\Delta H = 2257\ \mathrm{kJ/kg}$ is the latent heat of vaporization of liquid water and assumed to be constant in the range $20 - 300\,^{\circ}C$ independent of the temperature (in reality, $\Delta H = \Delta H(P, T)$). The initial temperature of the concrete block in the model is $T_i = 20\,^{\circ}C$. The spatial distribution of the liquid and vapour concentration at different times are calculated with the diffusion equations:

$$\begin{aligned} \frac{\partial C_l}{\partial t} &= \nabla \cdot D_l \nabla C_l - I \\ \frac{\partial C_v}{\partial t} &= \nabla \cdot D_v \nabla C_v + I \\ I &= h_{m,in} A_{in} (C_{int,sat} - C_v) \end{aligned} \qquad (6)$$

D_l and D_v are the effective diffusivity of the liquid and vapour phase and depends of the micro structure of the porous material, the temperature and pressure gradient (Chen, 2007). The effective diffusivity can be function of the temperature and water content (Zhang and Datta, 2004). It is assumed to be constant in this paper to facilitate the interpretation of the solution of the coupled heat and mass transfer equations. The heat Eq. (5) for temperature and the diffusion Eqs. (6) for liquid and vapour concentrations are coupled with the source/depletion term I modelling the evaporation/condensation process. This term is explained in more details in the subsection 2.2.4 with the description of the drying model. The concept of using the reaction engineering approach to model this evaporation/condensation term is based on the fact that this process can be seen as an activation process. $h_{m,in}\ [\mathrm{m/s}]$ is the internal mass transfer coefficient associated with the pore surfaces (Chen and Putranto, 2013a). $A_{in}\ [\mathrm{m^2/m^3}]$ is the internal surface area per cubic meter. It depends on the micro structure of the porous material (Putranto and Chen, 2013a). The product $h_{m,in} A_{in}$ depends on the micro structure of the porous media.

2.2.4. Drying model The drying model is based on the evaporation (or condensation) term proportional to the difference between the equilibrium vapour pressure (or concentration) and the actual vapour pressure. $I \propto (P_{v,sat} - P_v) \propto (C_{v,sat} - C_v)$. The equilibrium vapour pressure and concentration are commonly called the saturated vapour pressure $P_{v,sat}$ and concentration respectively $C_{v,sat}$. To understand this relation a constant volume with a mixture of water and vapour with three different states is considered:

Equilibrium, state 1

The closed system is in equilibrium with the environment, in this case the partial pressure of vapour is the equilibrium vapour pressure, $P_{v1} = P_{v1,sat}$ equivalent to $C_{v1} = C_{v1,sat}$. The number of gas molecules entering to the liquid equals the number of molecules leaving the liquid and entering into the gas phase . Macroscopically, there is no evaporation or condensation $I = 0$.

Heating, state 2

The system is suddenly heated, the temperature and total pressure increase. In this case the saturated vapour pressure reaches a new state $P_{v2,sat} > P_{v1,sat}$. The vapour pressure P_v of the mixture will increase until $P_{v2,sat}$ to reach this new equilibrium. During this process, $I > 0$ and evaporation occurs.

Cooling, state 3

The system is initially in state 1 and then suddenly cooled. In this case, the saturated vapour pressure drops to reach a new state $P_{v3,sat} < P_{v1,sat}$. The vapour pressure P_v of the mixture will decrease until $P_{v3,sat}$. During this transformation, $I < 0$ and condensation occurs. In this description of heating and cooling transformations of the liquid vapour mixture, the saturated (or equilibrium) vapour pressure and partial vapour pressure can be interchanged by the saturated vapour concentration and vapour concentration.

The SREA approach to model the drying of a porous medium considers that additional energy is required to evaporate the liquid inside the solid (compared to free space for example). The evaporation/condensation term is modified and proportional to the difference between the internal saturated vapour concentration $C_{int,sat}$ and the actual vapour concentration C_v. In summary, when the vapour concentration of the actual state is lower (higher) than the internal saturated vapour concentration, vapour is produced (consumed) with $I\Delta H$ being the power per unit volume absorbed during evaporation and released during condensation. For $C_{v,s} = C_v$, $I = 0$, the internal vapour concentration equals the vapour concentration, there is no evaporation or condensation and the system is in equilibrium. The main idea to model the drying process with the SREA approach is to consider evaporation as an activation process that can be described by an Arrhenius equation (Chen and Putranto, 2013a):

$$C_{int,sat} = \exp\left(-\frac{\Delta E_v}{RT}\right) C_{v,sat} \qquad (7)$$

This Eq. (7) shows that if the activation energy ΔE_v is 0, then the internal saturated concentration is the saturated concentration in free space. If the ΔE_v takes large values, then $C_{int,sat} << C_{v,sat}$ and evaporation will be small. The saturated vapour concentration $C_{v,sat}$ is not the same as the saturated partial vapour density of water $\rho_{v,sat}$ but can be linked to it by taking into account the porosity and the actual liquid concentration C_l:

$$\begin{aligned} C_{v,sat} &= \rho_{v,sat} \frac{V_g}{V} \\ &= \rho_{v,sat} (\epsilon_0 - \frac{V_l}{V}) \\ &= \rho_{v,sat} (\epsilon_0 - \frac{C_l}{\rho_l}) \end{aligned} \qquad (8)$$

With $V_g = V_v + V_a$ the volume of the gas phase. $\rho_{v,sat}$ is the partial density of the saturated vapour defined in respect to the gas volume and expressed with the Clapeyron relation (22) from appendix A. The partial density of air and vapour are function of the temperature and pressure described with the ideal gas law. ΔE_v is the apparent activation energy and can be expressed with a relative energy function:

$$\begin{aligned} \Delta E_v &= f(X_l, X_{l0}, X_\infty) \Delta E_\infty \\ \Delta E_\infty &= -RT_\infty ln(RH_\infty) \end{aligned} \qquad (9)$$

ΔE_∞ represents the equilibrium activation energy or the activation energy of the dry state. $X_l = \frac{C_l}{C_s}$, is the actual water content per unit mass of dry solid. X_{l0}, X_∞ are given and are the liquid content of the initial and final states respectively. RH_∞ and T_∞ are the relative humidity and temperature of the dry state. To model a drying behaviour of a particular porous sample the relative activation energy could be of the form for example:

$$\begin{aligned} f(X_l) &= \left(\frac{X_{l0} - X_l}{X_{l0} - X_{l\infty}} \right)^n \\ f(X_{l0}) &= 0 \\ f(X_{l\infty}) &= 1 \end{aligned} \qquad (10)$$

In principle, it should be obtained from a drying experiment but as explained in this section below, it was not possible to experimentally obtain this function. The activation energy expressed in Eq. (9) will be used in the heat and diffusion Eqs. (5) and (6). Using Eq. (7), the internal saturated vapour concentration can be expressed as:

$$C_{int,sat}\left(X_l, T\right) = RH_\infty^{\frac{T_\infty}{T}\left(\frac{X_{l0}-X_l}{X_{l0}-X_{l\infty}}\right)^n} C_{v,sat}\left(T\right) \tag{11}$$

in particular:

$$\begin{aligned} C_{int,sat}\left(X_{l\infty}, T\right) &= \left(RH_\infty\right)^{\frac{T_\infty}{T}} C_{v,sat}\left(T\right) \\ C_{int,sat}\left(X_{l\infty}, T_\infty\right) &= \left(RH_\infty\right) C_{v,sat}\left(T_\infty\right) \\ C_{int,sat}\left(X_{l0}, T\right) &= RH_\infty^0 C_{v,sat}(T) = C_{v,sat}(T) \end{aligned} \tag{12}$$

The Eq. (12) gives the definition of the relative humidity of the drying medium, i.e. the ratio between the vapour concentration and saturated vapour concentration of the dry state. The last equation shows that for the particular initial wet state with water concentration X_{l0}, the internal vapour concentration equals the saturated water vapour concentration in free space. For an accurate drying model of the system, the relative activation energy should be obtained from measurements. The time duration to achieve explosive spalling under high power microwave is in the order of 10 s. This short time scale gives some experimental difficulties for the measurement of the weight loss of water content during microwave heating. In this sense, it is not a classical drying problem. In operation, the antenna and concrete block are placed in a shield metallic box to avoid any microwave leakage and the high electric field in the vicinity of the concrete sample does not allow to place sensitive electrical equipment nearby. According to the spatial engineering approach, the measurement of the water loss rate versus time is the central point to extract the relative activation energy function. This process will capture the behaviour of the porous material under drying conditions. In the case of rapid heating with high power microwaves, the rate of water content loss versus time was not available and the relative activation energy could not be built. For this reason, it was decided to perform the simulations for the two extreme cases of the relative activation energy function $f = 0$ and $f = 1$ over the whole range, from wet to dry state $X_{l\infty} < X_l < X_{l0}$.

Maximum evaporation ($f = 0$)

In this case, the activation energy is $f = 0$, $C_{int,sat} = C_{v,sat}$ and $I = hA\left(C_{v,sat} - C_v\right)$.

Maximum restricted evaporation ($f = 1$)

For the case of maximum restricted evaporation, $f = 1$, $C_{int,sat} = RH_\infty^{\frac{T_\infty}{T}} C_{v,sat}$ and $I = h_{min}A_{in}(RH_\infty^{\frac{T_\infty}{T}} C_{v,sat} - C_v) \simeq 0$.

2.3. Mechanical model

With a 3-D mechanical model, it is possible to calculate the thermal stress from the temperature field. The total stress is calculated by adding the internal stress from the pore pressure on the diagonal components of the stress tensor.

2.3.1. Solid equilibrium

$$\rho\ddot{\mathbf{u}}(\mathbf{x}, t) - \nabla \cdot \boldsymbol{\sigma}(\mathbf{x}, t) - \boldsymbol{\kappa}(\mathbf{x}, t) = 0 \tag{13}$$

The mechanical model is based on force equilibrium (or Newton law extended to solids) with ρ the density of the solid, $\mathbf{u}$ the displacement vector, $\boldsymbol{\sigma}$ is the material stress tensor for small deformation. $\boldsymbol{\kappa}$ is the external volume forces such as gravity or magnetic forces. In our case, gravity can be neglected in comparison to internal thermal stress. In addition, the inertial term $\rho\ddot{\mathbf{u}}$ was neglected in comparison to the thermal stress because the time of the duration of the "heat pulse" (microwave heating about 10 s) is much larger than the characteristic time of propagation of stress waves inside the solid (speed of sound in the material over a characteristic length of the solid). The Eq. (13) without the inertial term is also called the quasi static approximation equilibrium of solids.

2.3.2. Linear elastic constitutive law

The constitutive law used to model the concrete material behaviour is a linear elastic model or generalised 3d Hooke's law as shown in Eq. (14). This behaviour is realistic in the elastic regime of the material. Three parameters are used to define the relationship between the strain $\boldsymbol{\epsilon}$ and the stress $\boldsymbol{\sigma}$ in the case of thermal expansion with a temperature difference ΔT. The young module E, which represents the stiffness of the material, the poisson ratio ν which is the ratio between longitudinal and lateral displacement, and the linear thermal expansion coefficient α. The pressure p is obtained from the heat diffusion model calculating the temperature, water and vapour content distribution inside the solid.

$$\begin{aligned} \boldsymbol{\sigma} &= \boldsymbol{\sigma}_0 + \frac{E}{1+\nu}\boldsymbol{\epsilon} + \frac{E\nu}{(1+\nu)(1-2\nu)} tr(\boldsymbol{\epsilon})\mathbf{I} - \alpha\frac{E}{1-2\nu}\Delta T\mathbf{I} \\ \boldsymbol{\sigma}_0 &= -p\boldsymbol{I} \end{aligned} \tag{14}$$

The effect of the pore pressure p is implemented as an initial stress $\boldsymbol{\sigma}_0$ operating only the diagonal components of the stress tensor $\boldsymbol{\sigma}$. The Von Mises stress $\sigma_m = \frac{1}{\sqrt{2}}\sqrt{(\sigma_1 - \sigma_2)^2 + (\sigma_2 - \sigma_2)^2 + (\sigma_3 - \sigma_1)^2}$, is used to evaluate the stress state inside the solid and to compare with the maximal material strength. σ_i are the main stress components of the stress tensor. For concrete material, the maximum tensile or compressive strength depends on the loading scenario and more realistic failure criterion as the Von Mises stress have been developed. The choice of the failure criterion for concrete material is a field of research and is not discussed in the present paper.

2.3.3. Boundary conditions

On all faces of the cube and the symmetry plane only tangential displacements are allowed, i.e $\mathbf{n} \cdot \mathbf{u} = 0$ with $\mathbf{n}$ the vector normal to the corresponding face. The face in front the aperture of the wave guide is not constrained, the displacements are allowed in all directions.

3. IMPLEMENTATION

3.1. 1-D model

The coupling term I between the heat and diffusion equations is modelled with the reaction engineering approach. Using Eqs. (7) and (8), it is written:

$$I = h_{m,in}A_{in}\left[\exp\left(\frac{-\Delta E_v}{RT}\right)\rho_{sat}\left(\epsilon_0 - \frac{C_l}{\rho_l}\right) - C_v\right] \tag{15}$$

with ρ_{sat} given in from appendix A. The heat transfer, and partial differential equations node are chosen to solve the heat and diffusion Eqs. (6) with COMSOL 4.4 multiphysics software. The resolution of these partial differential equations is performed with the finite element method and adequate numerical methods incorporated in the software. The calculated fields are the temperature T, the liquid concentration C_l and the vapor content C_v. The liquid and vapor content X_l and X_v are calculated from Eq. (4). With the assumption of a constant sealed pore volume $V_{gas} = (\epsilon_0 - \frac{C_l}{\rho_l})V$, the total pore pressure is calculated with the partial pressure of vapour and air:

$$\begin{aligned} p_v &= \frac{RT}{M_v}\frac{m_v}{V_{gas}} = \frac{RT}{M_v}\frac{C_v}{(\epsilon_0 - \frac{C_l}{\rho_l})} \\ p_a &= \frac{RT}{M_a}\frac{m_a}{V_{gas}} = \frac{RT}{M_a}\frac{C_a}{(\epsilon_0 - \frac{C_l}{\rho_l})} \\ p &= p_v + p_a \end{aligned} \tag{16}$$

with $R = 8.3$ J/(mol.K), the universal gas constant and $\{M_v, M_a\} = \{18, 29\}$ g/mol the molar mass of the vapour and air. ϵ_0 is the porosity previously defined in subsection 2.2.1 and used to define the volume occupied by the liquid and gas phase. The simulation was focused on the influence of the activation energy and the diffusion coefficient. Because the relative activation energy function of concrete is not known, the two extreme values 0 (evaporation) and 1 (no evaporation) are used. The diffusion coefficients for vapour chosen for the simulation are: $D_v = \{10^{-2}, 10^{-4}, 10^{-9}\}$ m^2/s. The initial water content used for the model is $X_{l0} = 0.05$ kg/kg dry solid.

3.2. 3-D model

The 3-D model solves the wave equation in the frequency domain, the heat and diffusion equations and the thermo elastic equations are solved every 2 s until 10 s. It is a non linear problem because the dielectric permittivity is function of the water content and needs to be updated at each time step. After solving this first problem, the temperature field and pore pressure are available and a second study is used to solve the mechanical model.

4. SIMULATION RESULTS AND DISCUSSION

4.1. Analytic estimation

4.1.1. Temperature increase For the maximum power density of $q = 70$ MW/m^3 with $\Delta t = 10$ s of heating, with constant thermal properties and an adiabatic system, the temperature increase can be estimated from energy conservation during this amount of time:

$$\begin{aligned}\Delta T &= \frac{q\Delta t}{\rho c_p} \\ &\simeq 364\,^\circ C\end{aligned} \quad (17)$$

with $\rho = 2400$ kg/m^3 and $c_p = 800$ J/(kg.K).

4.1.2. Thermal stress In the case of a 1d geometry and assuming a linear elastic behaviour with Hooke's law, the thermal stress is related to the thermal strain:

$$\sigma = E\epsilon = E\alpha\Delta T \quad (18)$$

with $E = 30\,GPa$ and $\Delta T = 300\,^\circ C$, if we assume that for fast heating the concrete solid is blocked, the order of magnitude for the stress is $\sigma \simeq 90\,MPa$ which exceeds the tensile strength of most concrete. This estimation gives an upper limit of the possible stress because it is based on the maximal temperature difference possible and because the solid is completely blocked.

4.1.3. Pore pressure Water can be in liquid or gas state depending on the pressure and temperature conditions. In a pressure-specific volume (P-v) or pressure-temperature (P-T) diagram of water, the curves delimiting the different states of water are called the saturated curves. For a pure substance in a closed volume, the saturated vapour pressure, is the pressure of the gas phase when in equilibrium with the liquid phase. The pressure evolution of a water vapour mixture is calculated for a constant volume (isochoric transformation) from an initial state of $T = 100°C$ and $P = 101.3$ kPa to the final state with temperature $T = \{200, 300\}°C$ in order to estimate the maximal possible pore pressure. The initial specific volume for this initial state of the saturated liquid water and vapour are $v_{vs1} = 1.67\,m^3/kg$, $v_{ls1} = 1.044\,dm^3/kg$. For the final states at $T = \{200, 300\}\,^\circ$C the specific volume of liquid and vapour are respectively $\{v_{ls200}, v_{ls300}\} = \{1.156, 1.404\}\,dm^3/kg$ and $\{v_{vs200}, v_{vs300}\} = \{127.36, 21.67\}\,dm^3/kg$. If we have an initial state made of a mixture of water and vapour, the mean specific volume of the mixture liquid and vapour is defined as:

$$\bar{v} = v_l + (v_v - v_l) \times x \quad (19)$$

with v_l, v_v the specific volume of the liquid and vapour phase. $x = \frac{m_v}{m}$ is called the quality factor of the liquid - vapor mixture, m_v and m are the mass of the vapour and the total mass of the fluid respectively. The final pressure is calculated for the liquid vapour mixture of different initial quality factor $x = \{0.01, 0.1, 0.4, 1\}$. In the table 1, the pressure of the water and vapour mixture is calculated for different quality factor

x	0.01	0.1	0.4	1
$\bar{v}$ $[m^3/kg]$	0.0178	0.1682	0.6698	1.6729
P_{200} [MPa]	1.553	1.208	0.325	0.130
P_{300} [MPa]	8.581	1.519	0.397	0.157

Table 1 Pressure of the liquid-vapour mixture for an isochoric transformation from initial state $T = 100\,^\circ$C and atmospheric pressure to the final temperatures $T = \{200, 300\}\,^\circ$C

and for the two final temperatures $T = \{200, 300\}\,^\circ$C for the isochoric transformation (specific volume is constant from state 1 to state 2 , $\bar{v}_2 = \bar{v}_1$).

From table 1, the specific volume of the mixture with $x = 0.01$ is $\bar{v} = 17.8$ dm^3/kg and bounded by saturated liquid and vapour specific volume $\{v_{ls200}, v_{ls300}\} < \bar{v} < \{v_{vs200}, v_{vs300}\}$ for the final state at $\{200, 300\}\,^\circ$C. In this case, the corresponding state is a mixture of liquid and vapour and the corresponding pressure is the saturated vapour pressure at these corresponding temperature, $T = \{200, 300\}\,^\circ$C. For the mixtures with higher vapour quality factor $x = \{0.1, 0.4, 1\}$, their specific volume are higher than the specific volume of the saturated vapour $\bar{v} = \{v_{x0.1}, v_{x0.4}, x_{x1}\} > \{v_{vs200}, v_{vs300}\}$, these mixtures are in the super heated vapour state. Looking into the corresponding table for super heated vapour and for the specific volume, the pressure can be obtained. For the same temperature increase, a high vapour mixture tends to give a smaller pressure than a high liquid mixture. The mixture with high vapour content behaves more closely to a perfect gas. The pressure ranges from 1.5 MPa for a mixture with $x = 0.01$ content to 0.13 MPa for pure vapour ($x = 1$) at $T = 200\,^\circ$C. The pressure ranges from 8.6 MPa for a mixture with $x = 0.01$ to 0.16 MPa for pure vapour at $T = 300\,^\circ$C.

The saturated (or equilibrium) vapour pressure of pure water can also be calculated with the Clapeyron equation:

$$\ln\frac{p_{sat}(T)}{p_0} = \frac{H}{R}\left(\frac{1}{T_0} - \frac{1}{T}\right) \quad (20)$$

if we assume that the latent heat of vaporisation of water is constant $H = 40$ kJ/mol. $R = 8.3$ J/(mol.K) is the universal gas constant. $T_0 = 373$ K with the corresponding saturated vapour pressure $p_0 = 101.3$ kPa can be chosen for example. For the case of very fast heating, an isochoric transformation is a logic assumption, since water does not have time to move through the porous medium. In reality the time scale of the mass transfer is larger or the same order than the time scale of heat transfer through the porous medium and so the transformation of the water vapour mixture during microwave heating is not isochoric. The volume of the mixture water and vapour may increase during heating, (for example if steam is moving through a open pore networks) causing a reduction of the pressure. Hence, the isochoric transformation assumption gives an upper limit of the pore pressure during fast heating. If the liquid-vapour mixture is mixed with air, the total pressure will be the sum of the partial pressure of the vapour and the air.

4.1.4. Time scale analysis With a characteristic length L of the micro structure of the medium and the thermal properties of the solid phase of the concrete taken from appendix A, $\kappa = \frac{k}{\rho c_p} \simeq 8.8 \times 10^{-7}$ m^2/s the thermal diffusivity of dry concrete, and D the mass diffusion coefficient, the time scales for the heat and mass diffusion are:

$$\begin{aligned} t_{heat} &= \frac{L^2}{\kappa} \\ t_{diff} &= \frac{L^2}{D} \\ \frac{t_{heat}}{t_{diff}} &= \frac{D}{\kappa} \\ D &= \frac{t_{heat}}{t_{diff}}\kappa \end{aligned} \quad (21)$$

If the porous medium has large open porosity or cracks, the time constant for the mass diffusion mechanism can be smaller than the thermal time constant. In this case, the diffusion coefficient can be much larger than the thermal diffusivity. On the contrary, for a sealed domain with closed porosity, there is no mass diffusion possible, $t_{diff} = \infty$ and the diffusion coefficient D is 0. In our first simulation, it was assumed that $D_l \simeq 0 << D_v$. The vapour diffusion coefficient is assumed to be constant (not the case in reality) and simulations were made for a large and a small vapour diffusion $D_v = \{10^{-2}, 10^{-9}\}$ m^2/s coefficient.

4.2. Numerical results

4.2.1. 1-D model In this section, the power density, temperature, liquid content, vapour content and pore pressure distribution through the first 10 cm thickness of the 40 cm sample are given for the extreme values of the relative activation energy corresponding to maximum evaporation ($f = 0$) and no evaporation ($f = 1$). The simulation are made for three different values of vapour diffusion coefficients $D_v = 10^{-2}, 10^{-4}, 10^{-9}$ m^2/s. All these calculated variables are represented along the depth of the concrete and after 10 s of microwave heating with 10 kW input power. A large vapour diffusion coefficient was chosen to evaluate the effect of diffusion on the time scale of 10 s. This could hypothetically correspond for example to an open cracks where a large passage through the porous medium is available for vapour transport.

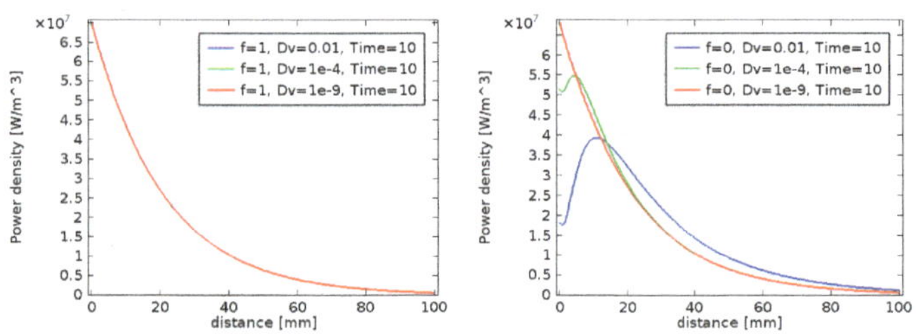

(a) $f = 1$ (b) $f = 0$

Fig. 2 Power density [W/m^3] for $D_v = \{10^{-2}, 10^{-4}, 10^{-9}\}$ m^2/s with no evaporation (a) and with evaporation (b)

Power density In case of no evaporation ($f = 1$), the Fig. 2(a) shows that the power density distribution is independent of the vapour diffusion coefficient. Because there is no evaporation process inside the material, no vapour is created and the local liquid and vapour content remains constant. For this reason, the power density profile is not modified by different diffusion coefficients. The Fig. 2(b) with maximum evaporation $f = 0$ shows that the power density is modified from the endothermic evaporation process. With larger vapour diffusion coefficient, vapour is transported through the material and more evaporation occurs. When the temperature of the water vapour mixture decreases in contact of the colder domains of the solid skeleton of the porous medium, the temperature can be lower than the saturation temperature of the mixture. In this case, condensation occurs and the vapour releases latent heat during the condensation process. The Fig. 2(b) shows that for depth above 1 cm the power density magnitude for $D_v = 0.01$ m^2/s case (blue curve) is higher than the case with limited vapour transport ($D_v = 10^{-9}$ m^2/s, red curve). In this case, heat from condensation has been released to the concrete in the colder part of the domain. It is also interesting to see that the maximum of the power density is located at $y \simeq \{4, 10\}$ mm for $D_v = \{10^{-4}, 10^{-2}\}$ m^2/s respectively.

Temperature The Fig. 3(a) shows that there is no influence of the vapor diffusivity because there is no evaporation. The power density is unchanged and so the thermal properties. The Fig. 3(b) shows the effect

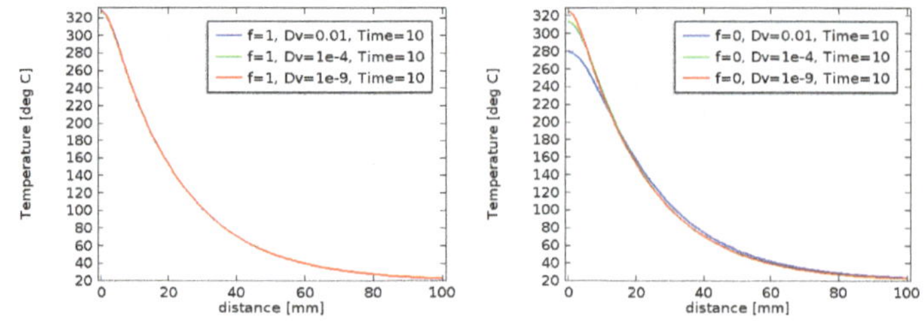

(a) $f = 1$ (b) $f = 0$

Fig. 3 Temperature [°C] for $D_v = \{10^{-2}, 10^{-4}, 10^{-9}\}$ m^2/s with no evaporation (a) and with evaporation (b)

of evaporation and vapour transport on the surface with a temperature difference of about 40 °C between limited (red curve) and large vapour transport (blue curve). A maximum surface temperature around 300 °C is realistic and was observed experimentally in our laboratory (See thermo camera picture in appendix B).

Liquid and vapor content Figures 4(a) and 4(b) show the liquid and vapor content for the case of maximum evaporation and for the 3 diffusion coefficients. The case of no evaporation is not of interest because the liquid and vapour content remain unchanged. More liquid is removed in the first 1 cm concrete depth when the vapour diffusivity increases. For the $D_v = 10^{-9}$ m^2/s the vapour increases locally near the surface. For $D_v = 10^{-2}$ m^2/s, the vapour diffuses through the thickness. It is also interesting to observe that locally the water content X_l can exceed the uniform initial water content X_{l0}. This is due to the transport and condensation of steam. These different vapour distribution through the thickness explains the different vapour and pore pressure distribution inside the sample.

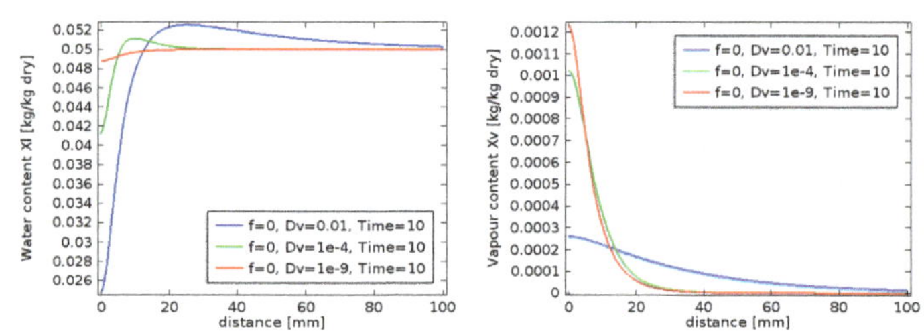

(a) X_l (b) X_v

Fig. 4 Liquid content X_l [−] (a) and vapor content X_v [−] (b) for $D_v = \{10^{-2}, 10^{-4}, 10^{-9}\}$ m^2/s

Total pressure The Fig. 5(a) shows that the peak pressure near the surface is about 0.2 MPa. For a sealed volume, with no vapour creation from the liquid phase, the effect of pressure build up with temperature is mostly due to the thermal expansion of air and vapour, according to the ideal gas model, the pressure is proportional to the temperature. The Fig. 5(b) shows that the peak pressures are $P \simeq \{2, 9, 14\}$ MPa for $D_v = \{10^{-2}, 10^{-4}, 10^{-9}\}$ m^2/s respectively. Near the surface, vapour is rapidly created and moves through the solid with a velocity function of the porosity of the material (or the diffusion coefficient). The partial pressure of vapour is proportional to the vapour concentration through the ideal gas law for a given temperature. At the maximum temperature of 320 °C, the saturated vapour pressure of liquid vapour water is 11.3 MPa. For the smallest diffusion coefficient, the calculated pressure of 14 MPa slightly exceeds the saturated vapour pressure. This means that the specific volume of the liquid vapour mixture exceeds the saturated specific volume, in this case the state of vapour is called superheated steam.

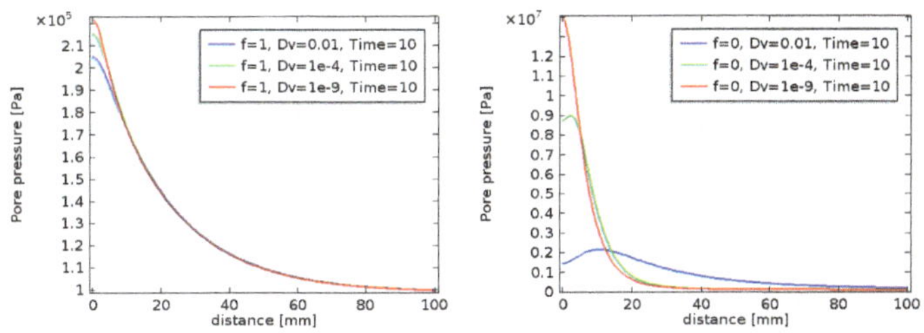

(a) $f = 1$ (b) $f = 0$

Fig. 5 Total pore pressure P [Pa] for $D_v = \{10^{-2}, 10^{-4}, 10^{-9}\}$ $\mathrm{m^2/s}$ with no evaporation (a) and with evaporation (b)

Materials properties The complex permittivity and thermal properties (thermal diffusivity and specific heat) expressed as a function of the solid, liquid and gas phase properties and weighted with the corresponding volume fraction (see appendix A), are plotted in Figs. 6 and 7. The value of the real part is between 6.7 and 8, the imaginary part is between 0.4 for the dry state and 0.7 for the wet state. The dielectric properties follow closely the liquid content distribution because it is the dominant factor that influence the dielectric properties for a constant temperature (At room temperature, free water has a real part of the relative permittivity of about 80, but in the case of concrete, the water is also bounded and for this reason the real part was taken to be 30). The dielectric properties are also strongly affected by temperature, but this effect is not investigated in this study.

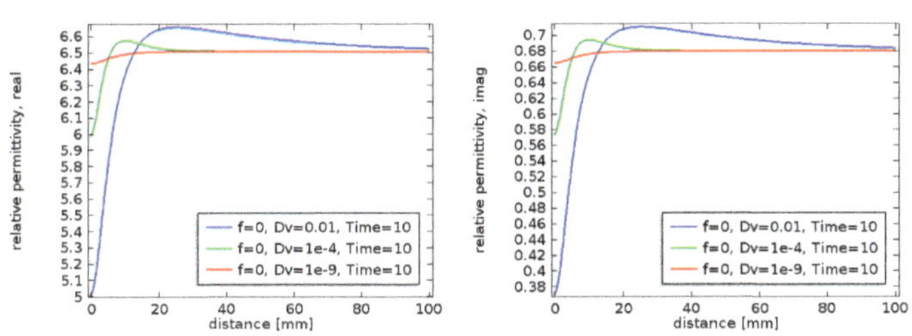

(a) ϵ^{real} (b) ϵ^{im}

Fig. 6 Relative permittivity, real part (a) and imaginary part (b) with evaporation for $D_v = \{10^{-2}, 10^{-4}, 10^{-9}\}$ $\mathrm{m^2/s}$

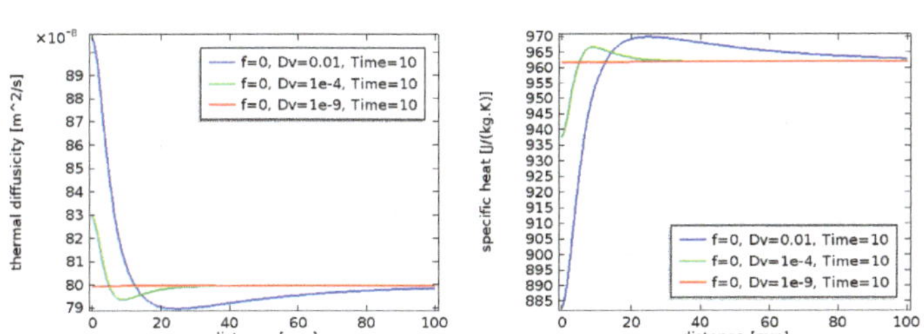

(a) κ (b) c

Fig. 7 Effective thermal diffusivity (a) and effective specific heat [J/(kg.K)] (b) with evaporation for $D_v = \{10^{-2}, 10^{-4}, 10^{-9}\}$ $\mathrm{m^2/s}$

The Fig. 7(a) shows the effective thermal diffusivity for the 3 different diffusion coefficients with values between $\{79, 91\} \times 10^{-8}$ $\mathrm{m^2/s}$. The Fig. 7(b) shows the effective thermal specific heat with values between $\{885, 970\}$ J/(kg.K). The effective specific heat decreases during microwave heating because of the decrease of the water content inside the first centimetre of the concrete.

4.2.2. 3-D model The 3-D model solves the wave equation in the frequency domain at 2.45 GHz. A time dependant solver is used to couple the electromagnetic model to the heat transfer and mechanical model. The complex dielectric permittivity of the concrete is expressed as a function of the solid, liquid and gas phase. The thermal properties are also depending on the liquid and gas content (see appendix A). Material properties are updated at each time steps and the electromagnetic model computed with the updated dielectric permittivity of the porous material. The mechanical model solves the elasticity equations to obtain the thermal stress from the temperature of the heat transfer model. The total stress is computed by adding the pore pressure as an initial stress contribution. The pore pressure is calculated with the distribution of the water and vapour content obtained from the coupled heat and mass transfer model. The results are presented for the case of no diffusion $D_v = 0$ $\mathrm{m^2/s}$ and with large diffusion $D_v = 10^{-2}$ $\mathrm{m^2/s}$ with $f = \{0, 1\}$.

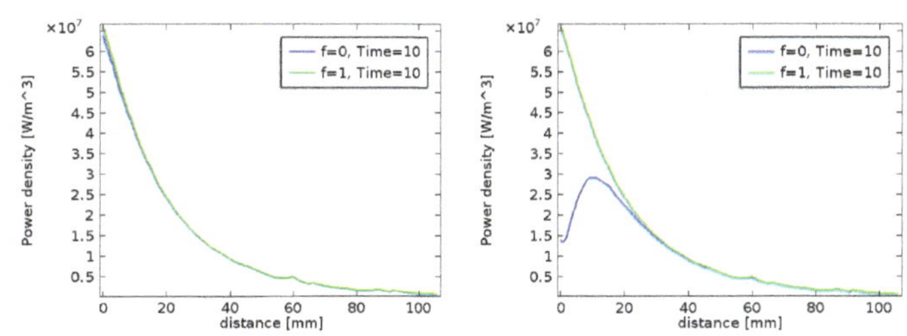

(a) (b)

Fig. 8 Power density [$\mathrm{W/m^3}$] for $D_v = 0\,\mathrm{m^2/s}$ (a) and $D_v = 10^{-2}\,\mathrm{m^2/s}$ (b) and with evaporation (blue) and no evaporation (green)

Power density The Fig. 8(a) shows that power density for $f = 0$ and $f = 1$ is nearly identical. The Fig. 8(b) shows that in the case of evaporation, the maximum of the power density is about 30 $\mathrm{MW/m^3}$ and located about 1 cm inside the material. The power profile is similar with the power profile from the 1-D model (see figure 2(b)). The discrepancies are due to some difference in the given initial power density and the variation of the dielectric permittivity with the 3 phases in the case of the 3-D model (See blue curves of Figs. 2(b) and 8(b)).

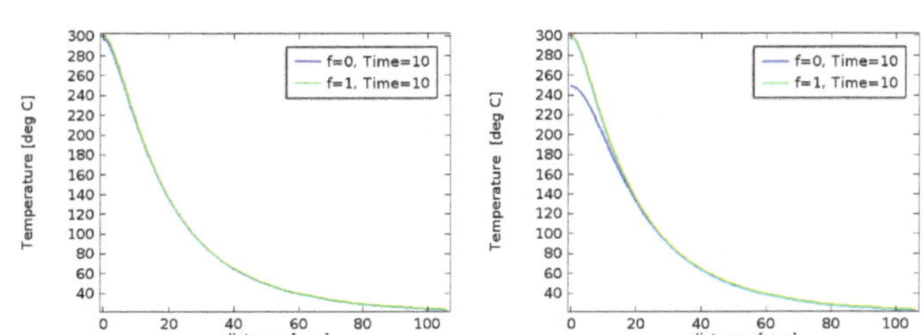

(a) (b)

Fig. 9 Temperature [°C] for $D_v = 0\,\mathrm{m^2/s}$ (a) and $D_v = 10^{-2}\,\mathrm{m^2/s}$ (b) and with evaporation (blue) and no evaporation (green)

Temperature The maximal temperature is about 300 °C (see figure 9(a)) near the surface for the case with no diffusion (320 °C for the 1-D case). The Fig. 9(b) shows that with high diffusion and evaporation, the maximum temperature is about 250 °C (280 °C for the 1-D case). The temperature calculated with the 1-D model are slightly higher because the microwave power density does not vary with the dielectric permittivity since the power density is approximated with Lambert's law. The comparison of Figs. 2(b) and 8(b) shows that the maximum power density with evaporation is about 40 $\mathrm{MW/m^3}$ and 30 $\mathrm{MW/m^3}$ for the 1-D and 3-D model respectively.

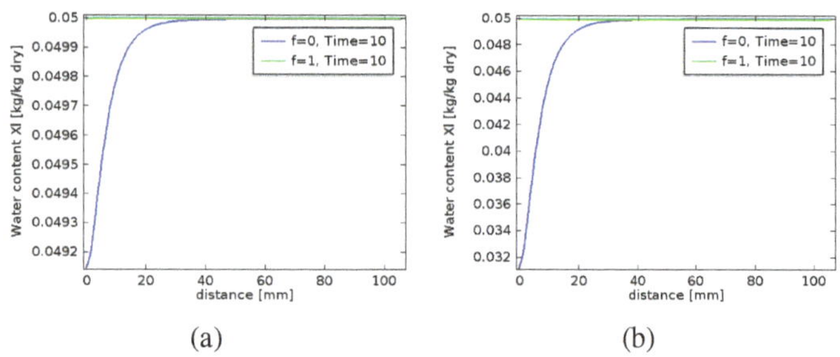

Fig. 10 Liquid content X_l [−] for $D_v = 0\,\mathrm{m^2/s}$ (a) and $D_v = 10^{-2}\,\mathrm{m^2/s}$ (b) and with evaporation (blue) and no evaporation (green)

Liquid content The liquid content shown in Fig. 10 decreases near the surface of the concrete. For the case of no diffusion $X_l \simeq 0.0492$ kg/kg at the surface an about 0.032 kg/kg for the case with large diffusion and is comparable to the 1-D case with $X_l \simeq 0.026$ kg/kg from Fig. 4(a).

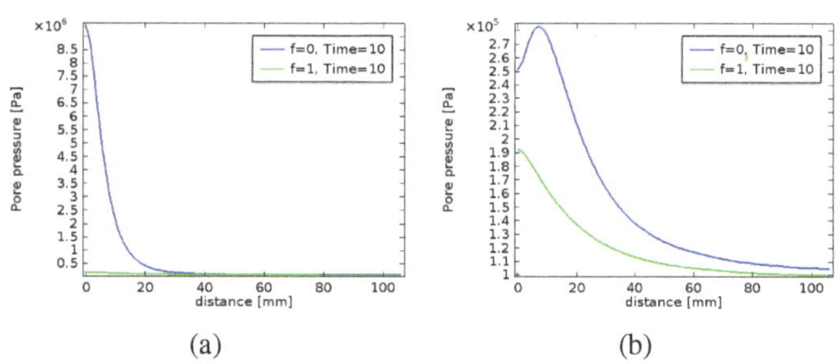

Fig. 11 Pore pressure P [Pa] for $D_v = 0\,\mathrm{m^2/s}$ (a) and $D_v = 10^{-2}\,\mathrm{m^2/s}$ (b) and with evaporation (blue) and no evaporation (green)

Total pressure Figure 11(a) shows that the maximum pressure near the surface is about 9.5 MPa when evaporation occurs. At 300 °C the saturated vapour pressure of water is 8.5 MPa. With large vapour diffusivity, the Fig. 11(b) shows lower pressure with a maximum of 0.3 MPa located 7 mm in depth. The pressure profile is similar to the pore pressure profile calculated by Bazant and Goangseup (2003a,b) with a maximum of 0.2 MPa at 2.45 GHz.

Von Mises stress In the case of high diffusion, the effect of the pressure on the Von Mises stress is negligible ($\frac{\sigma}{P} \simeq 100$) and therefore not visible in Fig. 12(b). In the case of no diffusion which could correspond to a high performance concrete), the difference in the Von Mises stress when the effect of the total pressure is included or not is about 8 MPa (see Fig. 12(a)). The total stress is mostly driven by the thermal stress function of the temperature gradients.

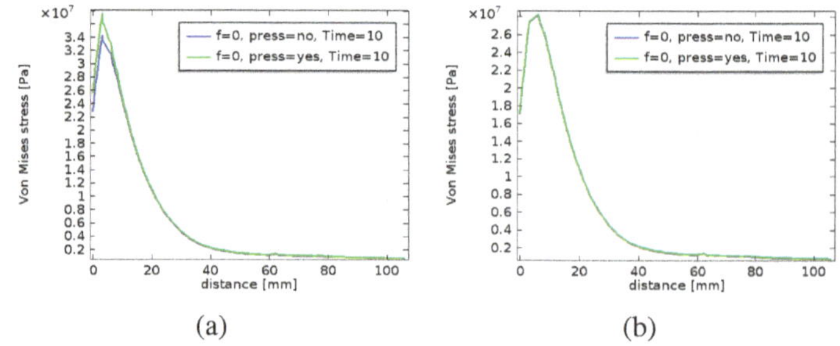

Fig. 12 Von Mises [Pa] for $D_v = 0\,\mathrm{m^2/s}$ (a) and $D_v = 10^{-2}\,\mathrm{m^2/s}$ (b) and with pressure (green) and no pressure included (blue)

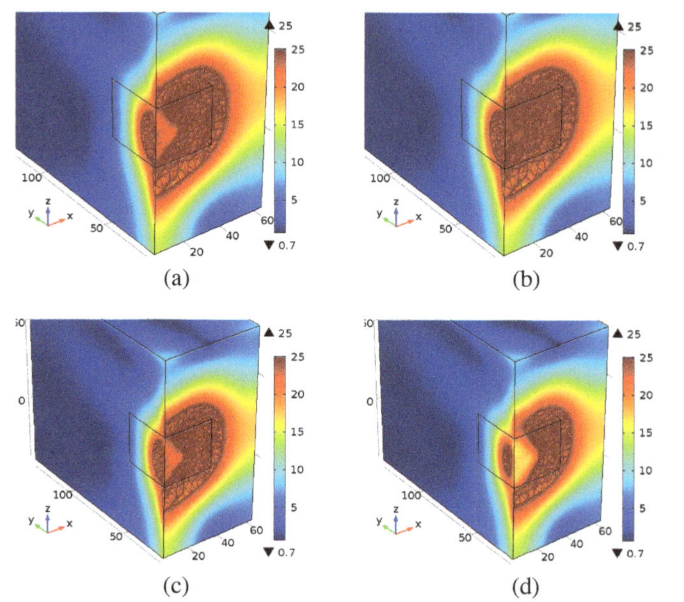

Fig. 13 Von Mises [MPa] for $D_v = 0\,\mathrm{m^2/s}$ for $f = 1$ (a) and for $f = 0$ (b). Von Mises stress for $D_v = 0.01\,\mathrm{m^2/s}$ for $f = 1$ (c) and for $f = 0$ (d). All the domains with a Von Mises stress above 25 MPa are not shown. $f = 1$ means no evaporation, $f = 0$ means maximum evaporation

The Figs. 13(a) and 13(b) show the Von Mises stress inside the material without and with evaporation respectively. With evaporation, higher stress occurs in the center with values above 25 MPa, the material with stress above this limit is not represented. For the case of high diffusion (see Figs. 13(c) and 13(d), with $D_v = 0.01\ \mathrm{m^2}s$, the effect of evaporation leads to a distribution of the vapour content through the domain, lowering the total pressure. Hence, the total stress is lower in case of evaporation, the transport of heat from vapour decrease the thermal gradients and so the thermal stress. The comparison between Fig. 13(a) and 13(c) in the case of no evaporation or Figs. 13(b) and 13(d) in the case of evaporation shows that a larger diffusion coefficient lowers the total stress. For comparison, typical strength of normal Portland cement concrete are 2 to 5 MPa for the tensile strength and 20 to 40 MPa for the compressive strength.

Materials properties The real and imaginary part of the relative permittivity of the 3-D model are plotted in Fig. 14(a) and 14(b) respectively. The dry state with lower dielectric losses will absorb less microwave energy. The real part is in the range 5.4 to 6.5 and the imaginary part takes values from 0.46 to 0.68. These values are consistent with the measured permittivity of concrete samples (Soldatov *et al.*, 2013).

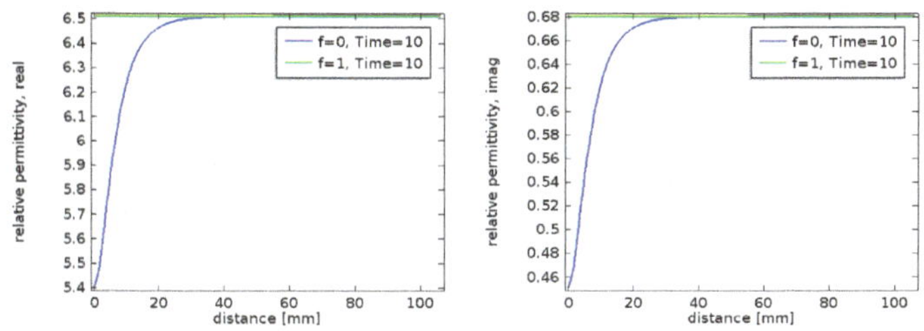

(a) ϵ^{real} (b) ϵ^{im}

Fig. 14 Relative permittivity for $D_v = 10^{-2}\,\mathrm{m}^2/\mathrm{s}$, real part (a) and imaginary part (b) with evaporation (green) and with no evaporation(blue)

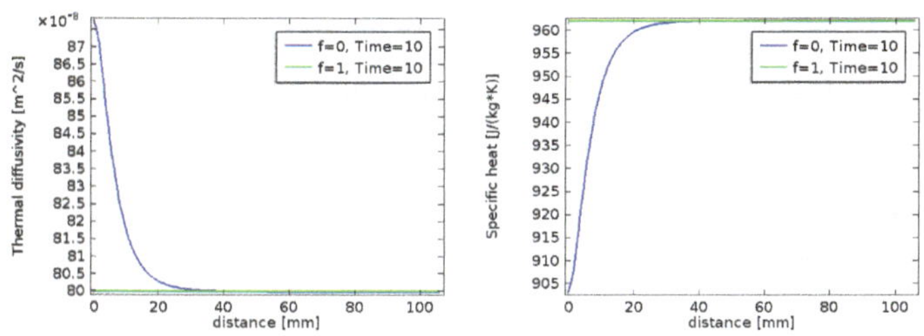

(a) κ (b) c

Fig. 15 Effective thermal diffusion [m^2/s] for $D_v = 10^{-2}\ \mathrm{m}^2/\mathrm{s}$ (a) and effective specific heat [$\mathrm{J}/(\mathrm{kg.K})$] (b) with evaporation (green) and with no evaporation (blue)

The Figs. 15(a) and 15(b) show the the variation of the thermal diffusivity and effective specific heat within depth for the 3-D model. As the surface is drying, less water content is present and the effective specific heat falls from 960 $J/(kg.K)$ (initial state) to 905 $J/(kg.K)$ (dry state).

5. CONCLUSION

A coupled multiphysics model based on heat an mass diffusion equations was developed in order to compute the total stress and to compare the thermal stress with the pore pressure. The deliberate use of constant and extremes values for the vapour diffusion coefficients is to facilitate the interpretation of the results and to help the understanding of the physical mechanisms involved during microwave heating of the porous concrete material. It was not possible to directly compare with experimental work, but the pressure values are consistent with literature and with the saturated vapour pressure. With a conical antenna operating at 2.45 GHz and an input power of 10 kW, the maximal temperature calculated near the surface are about 300 °C corresponding to temperatures measured experimentally. For this temperature, the calculated pore pressure is 9.5 MPa for the isochoric transformation (closed volume domain, no diffusion) and about 0.2 MPa with a large vapour diffusion coefficient $D_v = 0.01\ \mathrm{m}^2/\mathrm{s}$. The maximum Von Mises stress is in the range $28 - 37$ MPa and located about $7 - 8$ mm inside the concrete. From these first results, it seems clearly that the thermal stress is the dominant effect in case of explosive spalling of concrete with microwave heating. The pore pressure could play a role in triggering the spalling as suggested in the literature. It was observed experimentally that larger ablations occur in wet concrete than in dry concrete. One explanation could be that the microwave power density is higher in the case of wet concrete because of higher loss factor. The Spatial Reaction Engineering Approach used to calculate the water and vapour content distribution has the advantage of using a limited number of new parameters. Some large uncertainty exist for some parameters describing the porous medium such as the internal mass transfer coefficient $h_{m,in}$ or internal surface area A_{in}. In principle these parameters could be determined experimentally or estimated from the micro structure of the material. The relative activation energy was not available from experiment and therefore only two extreme cases were simulated, maximum and no evaporation. For further works, it would be useful to obtain this relative activation energy function. Secondly, a better estimation of the effective diffusion coefficient for liquid and vapour would be useful to improve the model. This is not a simple task because concrete under fast microwave heating is experiencing probably internal micro or macro cracks. The modification of the micro structure changes drastically the values of the diffusion coefficients. Finally, the calculated total stress exceeds the concrete strength and so these values are consistent with the apparition of the explosive spalling which occurs with our current antenna in the range from 3 to 10 s depending on the concrete initial water content and porosity.

NOMENCLATURE

q_0	peak power density ($\mathrm{W/m^3}$)
β	attenuation constant for a plane wave ($1/\mathrm{m}$)
V_i	Volume of the phase i ($i = s, l, v, a$) ($\mathrm{m^3}$)
ϵ_0	Porosity (-)
$f_i = V_i/V$	volume fraction of phase $i = s, l, v, a$ (-)
C_i	local concentration of phase $i = \{s, l, g, v, a\}$ ($\mathrm{kg/m^3}$)
ρ_i	density of phase $i = \{s, l, g, v, a\}$ ($\mathrm{kg/m^3}$)
ρ	effective density ($\mathrm{kg/m^3}$)
X_i	mass content of phase $i = \{l, v, a\}$ (kg/kg dry)
D_l	liquid effective diffusivity ($\mathrm{m^2/s}$)
D_v	vapor effective diffusivity ($\mathrm{m^2/s}$)
I	Evaporating / condensation rate ($\mathrm{kg/(s.m^3)}$)
$h_{m,in}$	internal mass transfer coefficient ($\mathrm{m/s}$)
A_{in}	internal surface area ($\mathrm{m^2/m^3}$)
ΔE_v	activation energy(J/kg)
ΔE_∞	equilibrium activation energy of the dry state (J/kg)
RH_∞	relative humidity of the dry state (-)
T_∞	temperature of hot air, dry state (K)
c	effective specific heat (J/kg $\cdot$ K)
c_i	specific heat of phase $i = \{s, l, g, v, a\}$ (J/kg $\cdot$ K)
k	effective thermal conductivity(W/(m.K))
k_i	conductivity of phase $i = \{s, l, g, v, a\}$ the solid(W/(m.K))
$C_{int,sat}$	internal saturated (or equilibrium) vapour concentration($\mathrm{kg/m^3}$)
$C_{v,sat}$	saturated (or equilibrium) vapour concentration ($\mathrm{kg/m^3}$)
$\rho_{v,sat}$	saturated vapour (or equilibrium) density(($\mathrm{kg/m^3}$)
I	source/depletion coupling term ($\mathrm{kg/s.m^3}$)
ΔH	latent heat of vaporisation of water (J/kg)
H	latent heat of vaporisation of water (J/mol)
σ	stress tensor (Pa)
σ_0	initial stress tensor (Pa)
E	Young modulus (Pa)
ν	poisson ratio (-)
α	linear expansion coefficient ($1/\mathrm{K}$)
M_v	molar mass of water (kg/kmol)
M_a	molar mass of air (kg/kmol)
R	Universal gas constant (J/kg K)
ϵ^{real}	real part of the relative effective permittivity
ϵ_i^{real}	real part of the relative permittivity of phase $i = \{s, l, g, v, a\}$
ϵ^{im}	imaginary part of the relative effective permittivity
ϵ_i^{im}	imag part of the relative permittivity of phase $i = \{s, l, g, v, a\}$
t_{heat}	time constant for the heat transfer model (s)
t_{diff}	time constant for diffusion model (s)
L	characteristic length (m)
κ	thermal diffusivity ($\mathrm{m^2/s}$)

ACKNOWLEDGEMENTS

This research and development project is funded by the German Federal Ministry of Education and Research (BMBF) within the funding action "Decommissioning / dismantling of nuclear facilities " (funding number 02S8719) and managed by the Project Management Agency Karlsruhe (PTKA).

REFERENCES

Akbarnezhad, A., and Ong, K., 2011, "Thermal Stress and Pore Pressure Development in Microwave Heated Concrete," *Computers and Concrete*, **8**(4), 425–443.
http://dx.doi.org/10.12989/cac.2011.8.4.425.

Bazant, Z.P., and Goangseup, Z., 2003a, "Decontamination of Radionuclides from Concrete by Microwave Heating.1: Theory," *Journal of engineering mechanics*, **129**(7), 777–784.
http://dx.doi.org/10.1061/(ASCE)0733-9399(2003)129:7(777).

Bazant, Z.P., and Goangseup, Z., 2003b, "Decontamination of Radionuclides from Concrete by Microwave Heating.2: Computations," *Journal of engineering mechanics*, **129**(7), 785–792.
http://dx.doi.org/10.1061/(ASCE)0733-9399(2003)129:7(785).

Bazant, Z.P., and Thonguthai, W., 1978, "Pore Pressure and Drying of Concrete at High Temperature," *Journal of the engineering mechanics division*, **52**, 1059–1079.

Chen, X., 2007, "Moisture Diffusivity in Food and Biological Materials," *Drying technology*, **25**, 1203–1213.
http://dx.doi.org/10.1080/07373930701438592.

Chen, X., 2008, "The Basics of a Reaction Engineering Approach to Modeling Air-drying of Small Droplets or Thin Layer Materials," *Drying technology*, **26**, 627–639.
http://dx.doi.org/10.1080/07373930802045908.

Chen, X.D., and Putranto, A., 2013a, "Modelling Drying Processes, a Reaction Engineering Approach," *Cambridge University Press*.
http://dx.doi.org/10.1017/cbo9780511997846.

Chen, X.D., and Putranto, A., 2013b, "Spatial Reaction Engineering Approach as an Alternative for Nonequilibrium Multiphase Mass Transfer Model for Drying of Food and Biological Materials," *American Institute of Chemical Engineers*, **59**(1), 55–67.
http://dx.doi.org/10.1002/aic.13808.

Gawin, D., Pesavento, F., and Schrefler, B.A., 2011a, "What Physical Phenomena can be Neglected when Modelling Concrete at High Temperature? A Comparative Study. Part 1: Physical Phenomena and Mathematical Model," *International Journal of Solids and Structures*, **48**, 1927–1944.
http://dx.doi.org/10.1016/j.ijsolstr.2011.03.004.

Gawin, D., Pesavento, F., and Schrefler, B.A., 2011b, "What Physical Phenomena can be Neglected when Modelling Concrete at High Temperature? A Comparative Study. Part 2: Comparison between Model," *International Journal of Solids and Structures*, **48**, 1927–1944.
http://dx.doi.org/10.1016/j.ijsolstr.2011.03.003.

Incropera, F.P., and DeWitt, D.P., 1996, "Fundamental of Heat and Mass Transfer," 839–843. ISBN 0-471-30460-3.

Ju, Y., Liu, H., Tian, K., Liu, J., Wang, L., and GE, Z., 2013, "An Investigation on Micro Pore Structures and the Vapor Pressure Mechanism of Explosive Spalling of RPC Exposed to High Temperature," *Science China, technological science*, **56**(2), 478–470.
http://dx.doi.org/10.1007/s11431-012-5110-4.

Lepers, B., Soldatov, S., Link, G., and Jelonnek, J., 2013, "A Thermoelastic Model for Microwave Ablation of Concrete," *14th International Conference on Microwave and high frequency heating, Nottingham UK.*

Mindeguia, J., Pimienta, P., Noumowe, A., and Kanema, M., 2010, "Temperature, Pore Pressure and Mass Variation of Concrete Subjected to High Temperature- Experimental and Numerical Discussion on Spalling Risk," *Cement and Concrete Research*, **40**, 477–487.
http://dx.doi.org/10.1016/j.cemconres.2009.10.011.

Phan, L., 2008, "Pore Pressure and Explosive Spalling in Concrete," *Materials and Structures*, **41**, 1623–1632.
http://dx.doi.org/10.1617/s11527-008-9353-2.

Putranto, A., and Chen, X.D., 2013a, "Multiphase Modeling of Intermittent Drying Using the Spatial Reaction Engineering Approach (S-REA)," *Chemical engineering and processing*, **70**, 169–183.
http://dx.doi.org/10.1016/j.cep.2013.04.005.

Putranto, A., and Chen, X.D., 2013b, "Spatial Reaction Engineering Approach as a Multiphase Drying Approach to Model the Heat Treatment of Wood under a Constant Heating Rate," *Industrial and engineering Chemistry research*, **52**, 6242–6252.
http://dx.doi.org/10.1021/ie4000689.

Soldatov, S., Kayser, T., Link, G., and Jelonnek, J., 2013, "Temperature Dependent Dielectric Measurements of Concrete with Cavity Perturbation Method," *14th International Conference on Microwave and High Frequency Heating, Nottingham UK.*

Zhang, J., and Datta, A., 2004, "Some Considerations in Modeling of Moisture transport in Heating Hydroscopic Materials," *Drying Technology*, **22**(8), 1983–2008.
http://dx.doi.org/10.1081/DRT-200032740.

APPENDIX A

The saturated vapour concentration of water can be expressed from the Clapeyron Eq. (20) and the ideal gas law for vapour with the following expression:

$$\rho_{v,sat}(T) = \frac{M_v}{RT} p_{sat}(T)$$

Material properties

Electric permittivity For a 3-D model solving the wave equations and computing the power density distribution inside the sample, the relative permittivity depends on the water content inside the material. The relative real and imaginary part of the dielectric permittivity can be expressed as a function of the dry solid, the water, vapor and the air.

$$\epsilon^{real,im} = f_s \epsilon_s^{real,im} + f_l \epsilon_l^{real,im} + (f_v + f_a)\epsilon_g^{real,im}$$

with $f_i = \frac{C_i}{\rho_i}$ for $i = \{s, l, v, a\}$ the volume fraction of the solid, liquid vapour and air phase. The effective permittivity can be written as:

$$\begin{aligned} \epsilon^{real,im} &= \frac{V_s}{V}\epsilon_s^{real,im} + \frac{V_l}{V}\epsilon_l^{real,im} + \frac{V_g}{V}\epsilon_g^{real,im} \\ &= \frac{C_s}{\rho_s}\epsilon_s^{real,im} + \frac{C_l}{\rho_l}\epsilon_l^{real,im} + (1 - \frac{C_s}{\rho_s} - \frac{C_l}{\rho_l})\epsilon_g^{real,im} \end{aligned}$$

with $\epsilon_g^{real} = \epsilon_v^{real} = \epsilon_a^{real} = 1$ and $\epsilon_g^{im} = \epsilon_v^{im} = \epsilon_a^{im} = 0$. The measurement of different samples of dry concrete (no more free water) from Soldatov *et al.* (2013) have given the following range for the loss tangent and real part of the permittivity $0.01 < \tan\delta < 0.03$, $4 < \epsilon^{real} < 5$. This gives for the relative imaginary part $0.04 < \epsilon^{im} < 0.15$. For our simulations, the following permittivity was chosen for the dry concrete

$e_s = 4 - 0.08j$. The free water inside the concrete is in some degree bounded to the concrete, and for realistic concrete permittivity value range, the permittivity of the liquid phase has to be lower than the free water. It was set to $\epsilon_l = 30 - 6j$. The intrinsic temperature dependence of the dielectric properties is not taken into account in this present work.

Thermal properties

Density The effective density is the mass of all the constituents over the total volume:

$$\begin{aligned} \rho_{eff} &= \frac{m_s + m_l + m_v + m_a}{V} \\ &= C_s + C_l + C_v + C_a \end{aligned}$$

The concentration are related to the density of the substance with Eq. (4). For the dry concrete and water the density are $\{\rho_s, \rho_l\} = \{2400, 1000\}$ $\mathrm{kg/m^3}$. The density of vapour and air function of temperature and pressure was calculated with the ideal gas law.

Specific heat The specific heat is calculated from energy conservation below:

$$m c_{eff} = m_s c_s + m_l c_l + m_g c_g$$

$$\frac{m c_{eff}}{V} = \rho_{eff} c_{eff} = C_s c_s + C_l c_l + C_g c_g$$

$$c_{eff} = \frac{C_s c_s + C_l c_l + C_g c_g}{\rho_{eff}}$$

The specific heat c_g of the mixture vapour and air assuming an ideal gas, follows Dalton law:

$$\begin{aligned} c_g(T) &= \frac{n_a M_a}{n_a M_a + n_v M_v} c_a(T) + \frac{n_v M_v}{n_a M_a + n_v M_v} c_v(T) \\ &= \frac{m_a}{m_a + m_v} c_a(T) + \frac{m_v}{m_a + m_v} c_v(T) \\ &= \frac{C_a}{C_a + C_v} c_a(T) + \frac{C_v}{C_a + C_v} c_v(T) \end{aligned}$$

with m, the total mass, m_i the mass of the constituent i ($i = s, l, a, v$). C_i and c_i, $i = s, l, v, a$. n_i is the number of moles of the constituent i and M_i is the molar mass ($M_a = 29$ g/mol and $M_v = 18$ g/mol). are respectively the concentration and specific heat of the constituent i at constant volume. $C_g = C_v + C_a$ is the concentration of the gas. For an ideal gas, the specific heat is only function of temperature. $c_v(T)$ and $c_a(T)$ are linearly interpolated from thermodynamic table. $\{c_s, c_l\} = \{800, 4200\}$ J/(kg.K) for dry concrete and liquid water.

Thermal conductivity The effective thermal conductivity is defined in the same way as the specific heat and weighted with the mass fraction of its constituents.

$$\begin{aligned} k_{eff} &= \frac{C_s k_s + C_l k_l + C_g k_g}{\rho_{eff}} \\ k_g &= \frac{C_a}{C_a + C_v} k_a(T) + \frac{C_v}{C_a + C_v} k_v(T) \end{aligned}$$

with V_i, k_i and ρ_i the volume, thermal conductivity and density of the constituent i. $k_a(T)$ and $k_v(T)$ are defined from Incropera and DeWitt (1996). The thermal conductivity of the gas is defined according to the mass ratio of air and vapour. $\{k_s, k_l\} = \{1.7, 0.6\}$ W/(m.K).

Mechanical properties For the mechanical model, standard properties of normal Portland cement concrete were used. The following values for the young modulus, poisson ratio and linear thermal expansion coefficient were used. $\{E, \nu, \alpha\} = \{30 \text{ GPa}, 0.12, 10^{-5} \text{ 1/K}\}$

APPENDIX B

Ablation tests were performed on concrete blocs of $33 \times 40 \times 10$ cm with a conical antenna of input power 10 kW operating at 2.45 GHz. An air gap of 1 cm between the surface of the concrete and the aperture of the antenna avoids direct heat conduction from the concrete surface to the antenna. Explosive spalling depends of the water content and the type of the concrete. With our antenna explosive spalling occurs after 3 to 10 s of high power microwave heating depending on the input power and displacement rate of the concrete blocs(see Fig. 16(b)).

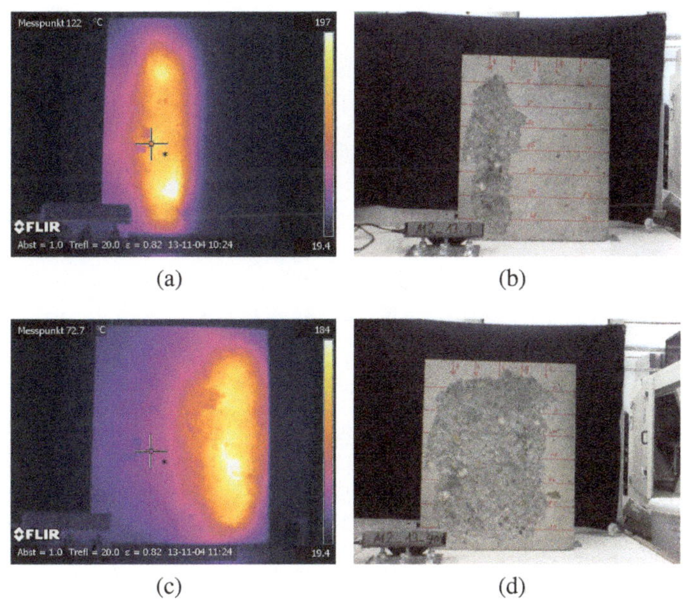

Fig. 16 Pictures of the surface ablation and temperature of the surface recorded with a thermo camera. The ablation time for one column takes about 30 s. Surface temperature was about 160 – 190 °C after some seconds of air natural convection cooling. Temperatures in the range of 200 – 300°C have been recorded with a pyrometer

3

A COMPARATIVE STUDY ON THERMAL CONDUCTIVITY AND RHEOLOGY PROPERTIES OF ALUMINA AND MULTI-WALLED CARBON NANOTUBE NANOFLUIDS

Zan Wu[a], Zhaozan Feng[b], Bengt Sundén[a,*], Lars Wadsö[c]

[a] *Department of Energy Sciences, Lund University, Lund, SE-22100, Sweden*
[b] *Department of Energy Engineering, Zhejiang University, Hangzhou, 310027, China*
[c] *Division of Building Materials, Lund University, Lund, SE-22100, Sweden*

ABSTRACT

Thermal conductivity and rheology behavior of two aqueous nanofluids, i.e., alumina and multi-walled carbon nanotube (MWCNT) nanofluids, were experimentally investigated and compared with previous analytical models. Information about the possible agglomeration size and interfacial thermal resistance in the nanofluids were obtained and partially validated. By incorporating the effects of interfacial thermal resistance, a revised model was found to accurately reproduce the experimental data based on the agglomeration size extracted from the rheology analysis. In addition, the thermal conductivity change of the alumina/water nanofluid with elapsed time was investigated. Thermal conductivity measurements were also conducted for alumina/water and MWCNT/water nanofluid mixtures.

Keywords: *Agglomeration; Interfacial thermal resistance; Nanofluid mixture; Property*

1. INTRODUCTION

Nanofluids are engineered colloidal suspensions of nanoparticles in a base fluid. They are more stable than microparticle colloids, with little particle setting, channel erosion and clogging. Nanofluids have distinctive features that offer potential for many applications in various fields including energy, bio and pharmaceutical processes, food industry, and chemical, electronic, environmental, material and thermal engineering etc. (Chung *et al.*, 2011; Wu and Sundén, 2014). For example, nanofluids can be considered as a new class of heat-transfer fluids as they generally provide higher thermal conductivity compared to their based conventional heat-transfer fluids (e.g., water and ethylene glycol). Various investigations (e.g., Feng and Li, 2013; Haghighi *et al.*, 2014; Wu *et al.*, 2014) on nanofluids flowing in tubes and heat exchangers indicate that conventional pressure drop and heat transfer correlations for the base fluid can accurately reproduce the flow and thermal behaviors for nanofluids by adopting the measured nanofluid properties in the analysis, respectively. That means, the flow structure and the convective heat transfer mechanisms were probably not modified by the addition of nanoparticles. The benefit of using nanofluids for heat transfer intensification mainly comes from the thermal conductivity enhancement. The generally accompanied higher nanofluid viscosity requires a higher pumping power which may counterbalance the benefit of the enhanced thermal conductivity. Therefore, thermal conductivity and viscosity of nanofluids need to be investigated and manipulated for possible heat transfer and other relevant applications.

Thermal conductivity and viscosity of nanofluids are strongly dependent on particle concentration, particle size and shape, the presence of agglomerations (i.e., nanoclusters, aggregates), the nature of the base fluid, temperature and nanofluid stability, etc. In nanofluids, nanoparticles tend to form agglomerations of different size due to the van der Waals attractive forces. Preparation methods such as addition of surfactants and ultrasonic vibration can reduce the size of the agglomerates substantially but are not able to break the agglomerates into primary particles. Existence of nanoparticle agglomerations has already been recognized by dynamic light scattering and SEM/TEM observations in the literature, e.g., Anoop *et al.* (2009), Hong *et al.* (2006), Karthikeyan *et al.* (2008), Keblinski *et al.* (2002) and Prasher *et al.* (2006). Agglomeration tends to enhance nanofluid viscosity due to the immobilized fluid trapped in the particle clusters and thus a higher effective volume fraction than the actual solid volume fraction. Anoop *et al.* (2009) considered the viscosity increase was primarily due to the agglomeration of particles in water-based and ethylene-based nanofluids.

There is still some controversy or discrepancy about the effects of nanoparticle agglomeration on thermal conductivity. On one hand, Hong *et al.* (2006) found that the thermal conductivity of Fe nanofluids by dispersing Fe particles of 10 ± 1 nm in ethylene glycol was closely related to nanoparticle agglomerations or nanoparticle clusters. The average agglomeration size in the prepared nanofluid was larger than 1 micrometer. Nanoparticles agglomerated rapidly right after preparation and formed larger clusters continuously. The thermal conductivity decreased with increasing agglomeration size. The cluster size is said to increase with increasing volume fraction because nanoparticle can agglomerate easily in concentrated nanofluids due to short inter-particle distance. Karthikeyan *et al.* (2008) investigated time-dependent thermal conductivity characteristics of aqueous CuO nanofluids (average particle diameter 8 nm) and showed that the thermal conductivity decreased with elapsed time due to the clustering of nanoparticles with time, as confirmed microscopically. On the other hand, Keblinski *et al.* (2002) stated that nanoparticle agglomeration might lead to local percolation behavior with paths for rapid heat transport and enhanced

* *Corresponding author. Email: bengt.sunden@energy.lth.se*

thermal conductivity. Similarly, Prasher *et al.* (2006) studied the effect of aggregation kinetics on the thermal conductivity of nanofluids and demonstrated that fractal agglomerations or aggregates lead to thermal conductivity enhancement due to the ability of the heat to move rapidly along the backbone of the cluster. In conclusion, information about agglomeration geometry, agglomeration size and relevant aggregate configuration in the nanofluid should be obtained to fully understand its effect on thermal conductivity. Proper particle morphology such as percolated chainlike agglomerations or fractal agglomerations of small size around several hundred nanometers can enhance thermal conductivity, while relatively large packed clusters may deteriorate thermal conductivity as it may induce sedimentation.

The lower and upper limits of the nanofluid thermal conductivity can be completely determined by the serial mode and the parallel mode (Nielsen, 1978) only using volume fractions and thermal conductivities of the two phases, respectively. Hashin and Shtrikman (1963) proposed strict bounds based on the classical effective medium theory, which is given below for the case of $k_p/k_f > 1$

$$k_f\left(1+\frac{3\varphi(k_p-k_f)}{3k_f+(1-\varphi)(k_p-k_f)}\right)\le k_{nf}\le k_p\left(1-\frac{3(1-\varphi)(k_p-k_f)}{3k_p-\varphi(k_p-k_f)}\right) \tag{1}$$

At the lower bound, nanoparticles are well-dispersed in the base fluid. At the upper bound, the nanoparticles form a continuous phase with linear or chainlike particle morphologies, and the base fluid becomes a dispersed phase. Besides the agglomeration morphology described above, liquid layering at the liquid-particle interface, Brownian motion and coupled transport can also influence the thermal conductivity located within the lower and upper bounds (Wang and Fan, 2011).

Mo *et al.* (2005) observed the presence of an ordered liquid layer near a nanoparticle surface by which the atomic structure of the liquid near the surface is significantly more ordered than that of bulk liquid. The thermal conductivity of ordered liquid layers tends to be larger than that of the bulk base fluid. Therefore, ordered liquid layers may enhance the effective thermal conductivity of the nanofluid by augmenting the particle effective volume fraction. Wang and Fan (2011) suggested the liquid layers offer insignificant conductivity enhancement for water-based nanofluids containing spherical particles as the liquid film thickness is only 0.28 nm for aqueous nanofluids. For nanoparticle size of the order of 10 nm, the increase in effective volume fraction induced by ordered liquid layers is only 0.1%, which contributes little. However, their presence may facilitate formation of interconnected particle agglomerations by relaxing the requirement of particle physical contact with each other. In addition, the liquid layering at the liquid-particle interface may present interfacial thermal resistance at the particle-fluid interface and among the particle-particle interface due to different nature of thermal conduction in nanoparticles and the base fluid, and thus decrease thermal conductivity enhancement.

Brownian motion can enable direct inter-particle transport of heat from one particle to another and induce surrounding fluid flow and thus so-called micro-convection. The ratio of the contribution to thermal conductivity by Brownian motion and micro-convection (k_{BM}) to the base fluid conductivity (k_f) was estimated based on the kinetic theory by Evans *et al.* (2006)

$$\frac{k_{BM}}{k_f}=\frac{\rho_p c_{p,f}\varphi k_B T}{3\pi\mu_f d_p k_f}+\frac{k_B T}{3\pi\mu_f d_p \alpha_f} \tag{2}$$

According to the above equation, the contribution of Brownian motion and micro-convection for nanofluids can be negligible in our case, as stated in Eapen *et al.* (2010) and Evans *et al.* (2006).

Eapen *et al.* (2010) believed that both the Soret effect (also known as thermodiffusion or thermophoresis) and the Dufour effect (an induced heat flow caused by the concentration gradient) do not directly influence the nanofluid thermal conductivity by analyzing the orders of magnitude, but the coupled or crossed transport between them or other transport processes might affect the thermal conductivity. As proposed in Wang and Fan (2011), the coupled transport could change the nature of heat conduction in nanofluids from a diffusion process to a wave process, thus giving a nanofluid thermal conductivity enhancement as high as 10%.

Based on the above statement, more experimental data are needed to better understand the nanoparticle behavior and identify possible underlying mechanisms for thermal conductivity enhancement in various nanofluids. The purpose of this study is to measure the thermal conductivity and viscosity of aqueous alumina and aqueous multi-walled carbon nanotube (MWCNT) nanofluids of different volume fractions, and to examine various analytical thermal conductivity and viscosity models to explain the possible related mechanisms in our tested nanofluids. For example, effects of particle agglomeration morphology and possible interfacial thermal resistance will be recognized and analyzed. In addition, thermal conductivity values of nanofluid mixtures formed by alumina/water nanofluid and MWCNT/water nanofluid will also be measured for the first time. Before providing the experimental data analysis, several selected analytical thermal conductivity and viscosity models are detailed below.

2. THERMAL CONDUCTIVITY AND VISCOSITY MODELS

2.1 Thermal conductivity models

The lower H-S bound equals to the Maxwell model (Maxwell, 1881), applicable for dilute suspensions of well-dispersed non-interacting spherical particles. Based on the effective medium theory, the Bruggeman model (BGM) (Bruggeman, 1935) is applicable up to relatively high volume fractions. A solution to the implicit form of the Bruggeman model is given below (Hui *et al.*, 1999)

$$\frac{k_{nf}}{k_f}=\frac{1}{4}\left[(-2+3\varphi)\left(\frac{k_p}{k_f}-1\right)+\frac{k_p}{k_f}\right]+\frac{1}{4}\sqrt{\left[(-2+3\varphi)\left(\frac{k_p}{k_f}-1\right)+\frac{k_p}{k_f}\right]^2+8\frac{k_p}{k_f}} \tag{3}$$

Davis (1986) presented a model to calculate the effective thermal conductivity of a composite material with spherical inclusions

$$\frac{k_{nf}}{k_f}=1+\frac{3(k_p/k_f-1)\varphi}{k_p/k_f+2-(k_p/k_f-1)\varphi}\left\{\varphi+f(k_p/k_f)\varphi^2+0(\varphi^3)\right\} \tag{4}$$

Values of the function $f(k_p/k_f)$ can be obtained from Davis (1986).

The Hamilton-Crosser (H-C) model (Hamilton and Crosser, 1962) can be applied to describe liquid-solid mixtures containing well-dispersed non-spherical particles. A shape factor n was introduced to consider the effects of particle shape.

$$\frac{k_{nf}}{k_f}=\frac{k_p+(n-1)k_f+(n-1)\varphi(k_p-k_f)}{k_p+(n-1)k_f-\varphi(k_p-k_f)} \tag{5}$$

The shape factor of the particle n can be calculated by

$$n = \frac{3}{\omega} \quad (6)$$

where ω is the particle sphericity defined as the ratio of the surface area of an equivalent sphere, to the real surface area of the particle. For sphere particles $n = 3$, Eq. (5) reduces to the Maxwell model. The H-C model reveals the importance of the particle shape, but it neglects the inter-particle interaction.

Recently, Deng *et al.* (2007) proposed an analytical model for carbon nanotube (CNT) based composites that considers the effects of volume fraction, anisotropic thermal conductivities, aspect ratio, non-straightness and interfacial thermal resistance of the CNT. The Deng *et al.* model can be expressed as:

$$\frac{k_{\mathrm{nf}}}{k_{\mathrm{f}}} = 1 + \frac{\eta\varphi}{\frac{3k_{\mathrm{f}}}{\eta k_{\mathrm{z}}} + \frac{3H\eta L_{\mathrm{p}}}{d_{\mathrm{p}}}} \quad (7)$$

The parameter H can take into account of the CNT aspect ratio:

$$H = \frac{1}{\left(\frac{L_{\mathrm{p}}}{d_{\mathrm{p}}}\right)^2 - 1}\left[\frac{\frac{L_{\mathrm{p}}}{d_{\mathrm{p}}}}{\sqrt{\left(\frac{L_{\mathrm{p}}}{d_{\mathrm{p}}}\right)^2 - 1}} \ln\left(\frac{L_{\mathrm{p}}}{d_{\mathrm{p}}} + \sqrt{\left(\frac{L_{\mathrm{p}}}{d_{\mathrm{p}}}\right)^2 - 1}\right) - 1\right] \quad (8)$$

As CNTs of large aspect ratio tend to bend very easily, the parameter η in Eq. (7) represents the effects of the non-straightness of the CNT, which is described as:

$$\eta = \frac{L_{\mathrm{actual}}}{L_{\mathrm{p}}} \quad (9)$$

where L_{actual} is the distance between the two ends of the non-straight CNT. The Deng *et al.* (2007) model suggests that larger aspect ratio and more straight CNTs can give better thermal conductivity enhancement. The Deng *et al.* (2007) model is applicable when the thermal conductivity along the longitudinal or axial direction k_z is much higher than that along the transverse direction k_x.

A methodology was proposed by Chen *et al.* (2009) to predict the effective thermal conductivity of nanofluids based on rheology by including particle agglomeration. The viscosity data was adopted to infer micro-structures of nanoparticles quantitatively, which was then incorporated into the conventional H-C model. Chen *et al.* (2009) modified the H-C model and the BGM as follows

$$\frac{k_{\mathrm{nf}}}{k_{\mathrm{f}}} = \frac{k_{\mathrm{a}} + (n-1)k_{\mathrm{f}} + (n-1)\varphi_{\mathrm{a}}(k_{\mathrm{a}} - k_{\mathrm{f}})}{k_{\mathrm{a}} + (n-1)k_{\mathrm{f}} - \varphi_{\mathrm{a}}(k_{\mathrm{a}} - k_{\mathrm{f}})} \quad (10)$$

$$\frac{k_{\mathrm{a}}}{k_{\mathrm{f}}} = \frac{1}{4}\left[(-2+3\varphi_{\mathrm{in}})\left(\frac{k_{\mathrm{p}}}{k_{\mathrm{f}}} - 1\right) + \frac{k_{\mathrm{p}}}{k_{\mathrm{f}}}\right] + \frac{1}{4}\sqrt{\left[(-2+3\varphi_{\mathrm{in}})\left(\frac{k_{\mathrm{p}}}{k_{\mathrm{f}}} - 1\right) + \frac{k_{\mathrm{p}}}{k_{\mathrm{f}}}\right]^2 + 8\frac{k_{\mathrm{p}}}{k_{\mathrm{f}}}} \quad (11)$$

Equation (11) is only applicable for spherical or near-spherical agglomeration morphology. For rod-like particles and nanotubes, the following model developed by Nan *et al.* (2004) can be used:

$$\frac{k_{\mathrm{a}}}{k_{\mathrm{f}}} = \frac{3 + \varphi_{\mathrm{in}}\left[2\beta_{\mathrm{x}}(1 - L_{\mathrm{x}}) + \beta_{\mathrm{z}}(1 - L_{\mathrm{z}})\right]}{3 - \varphi_{\mathrm{in}}\left[2\beta_{\mathrm{x}}L_{\mathrm{x}} + \beta_{\mathrm{z}}L_{\mathrm{z}}\right]} \quad (12)$$

The parameters β_x and β_z are calculated by

$$\beta_{\mathrm{x}} = \frac{k_{\mathrm{x}} - k_{\mathrm{f}}}{k_{\mathrm{f}} + L_{\mathrm{x}}(k_{\mathrm{c}} - k_{\mathrm{f}})} \quad (13)$$

$$\beta_{\mathrm{z}} = \frac{k_{\mathrm{z}} - k_{\mathrm{f}}}{k_{\mathrm{f}} + L_{\mathrm{z}}(k_{\mathrm{c}} - k_{\mathrm{f}})} \quad (14)$$

where k_x, k_z and k_c are the thermal conductivities of carbon nanotubes along transverse and longitudinal directions and isotropic thermal conductivity of the nanotube, respectively. L_x and L_z are geometrical factors dependent on the particle shape and given by

$$L_{\mathrm{x}} = \frac{p^2}{2(p^2 - 1)} - \frac{p}{2(p^2 - 1)^{3/2}}\cosh^{-1} p \quad \text{for } p > 1 \quad (15)$$

where p is the aspect ratio

$$L_{\mathrm{z}} = 1 - 2L_{\mathrm{x}} \quad (16)$$

2.2 Viscosity models

Batchelor (1977) considered the effects due to particle Brownian motion on the bulk stress of an approximately isotropic suspension of rigid and spherical particles and proposed the following model:

$$\frac{\mu_{\mathrm{nf}}}{\mu_{\mathrm{f}}} = 1 + 2.5\varphi + 6.5\varphi^2 \quad (17)$$

Chen *et al.* (2007) proposed to calculate the effective dynamic viscosity by integrating the aggregation mechanism into the Krieger and Dougherty (1959) model:

$$\frac{\mu_{\mathrm{nf}}}{\mu_{\mathrm{f}}} = \left(1 - \frac{\varphi_{\mathrm{a}}}{\varphi_{\mathrm{m}}}\right)^{-[\mu]\varphi_{\mathrm{m}}} = \left(1 - \frac{\varphi/\varphi_{\mathrm{in}}}{\varphi_{\mathrm{m}}}\right)^{-[\mu]\varphi_{\mathrm{m}}} \quad (18)$$

where φ_m is the maximum particle packing fraction taken as 0.605 for spherical particles at high shear rates, φ_{in} the volume fraction of the particles in the aggregates, φ_a the volume fraction of the aggregates in the entire nanofluid calculated by

$$\varphi_{\mathrm{a}} = \varphi\left(d_{\mathrm{a}}/d_{\mathrm{p}}\right)^{3-D} \quad (19)$$

The parameter $[\mu]$ is the intrinsic viscosity with a typical value of 2.5 for spherical nanoparticles. The fractal index D is typically 1.8 for spherical particles.

For non-spherical particles such as CNTs with high aspect ratios, Halelfadl *et al.* (2013) modified the Maron-Pierce model (Maron and Pierce, 1956)

$$\frac{\mu_{nf}}{\mu_f} = \left(1 - \frac{\varphi_a}{\varphi_m}\right)^{-2} \tag{20}$$

For fibers or CNTs, the maximum volume fraction depends also on the aspect ratio (Mueller *et al.*, 2010)

$$\varphi_m = \frac{2}{0.321p + 3.02} \tag{21}$$

A value of 0.0372 is here obtained for carbon nanotubes with an aspect ratio of 158. A value of 2.1 was taken for the fractal index D in Eq. (19) for aggregated nanofluids with nanorods or nanotubes (Halelfadl *et al.*, 2013).

3. RESULTS AND DISCUSSION

3.1 Alumina nanofluid

Untreated concentrated γ-Al_2O_3/water nanofluid with spherical alumina nanoparticles of 40-nm mean diameter was purchased from a commercial company (Nanophase Technologies Corporation, US). No surfactants were added in the nanofluid. Different amounts of concentrated nanofluid were diluted in water to obtain nanofluids with different concentrations. The diluted nanofluid mixture was mechanically stirred for 0.5 h followed by ultrasonic vibration for 3 h. Nine nanofluids with particle volume fractions, 1.0%, 1.51%, 1.89%, 2.99%, 4.99%, 6.0%, 7.82%, 9.84% and 18.4% were carefully prepared. For aqueous nanofluids, pH control, which has an important role in stability control, places the iso-electric point, far from the point of zero charge in order to avoid coagulation and instability. The pH value of the prepared alumina nanofluid is about 3 ~ 3.5, which is far away from the iso-electric point of alumina nanofluid.

A thermal constants analyzer (TPS 2500S from Hot Disk AB, Sweden) using the transient plane source method (TPS) was employed to measure the thermal conductivity of aqueous alumina and MWCNT nanofluids. The diameter of the used Hot Disk sensor is 2.0 mm. For each test, the measurement time and the total output of power were limited to 2 seconds and 0.015 W, respectively. Therefore, natural convection can be ignored due to the low temperature rise of the sensor. Before nanofluid measurements, several pure fluids were tested to check the accuracy of the TPS method. The thermal conductivity values of these pure fluids are listed in Table 1. The thermal conductivity uncertainty was estimated, from the standard deviations of experimental data and departures from literature data, to be lower than 3.0 %.

A rotational rheometer HAAKE RS6000 (Thermo Fisher Scientific Inc., US) was used to measure rheology behavior of alumina/water and MWCNT/water nanofluids. The standard deviation of the measured dynamic viscosity data of water (the base fluid) and departures from the literature data is less than 3.0%.

Figure 1 shows the relative thermal conductivity k_{nf}/k_f of alumina/water nanofluids of different volume fractions. The thermal conductivity increases with volume fraction. Previous data of Kim *et al.* (2007), Timofeeva *et al.* (2007) and Williams *et al.* (2008) were also shown for comparison. All the data sets give similar trends, especially at low volume fractions. The present data is higher than that of Timofeeva *et al.* (2007) when volume fraction is larger than 0.03, while lower than that of Williams *et al.* (2008) when the volume fraction is between 0.01 and 0.05. Overall, the difference among the four data sets at the same volume fraction is less than 10%.

Table 1 Experimental thermal conductivity for pure fluids at 20 °C.

Material	*k* (W m^{-1} K^{-1})	*Material*	*k* (W m^{-1} K^{-1})
Water	0.583	Ethylene glycol	0.244
Ethanol	0.1695	Glycerol	0.272
1-Butanol	0.1515	2-Propanal	0.1395
Olive oil	0.1691		

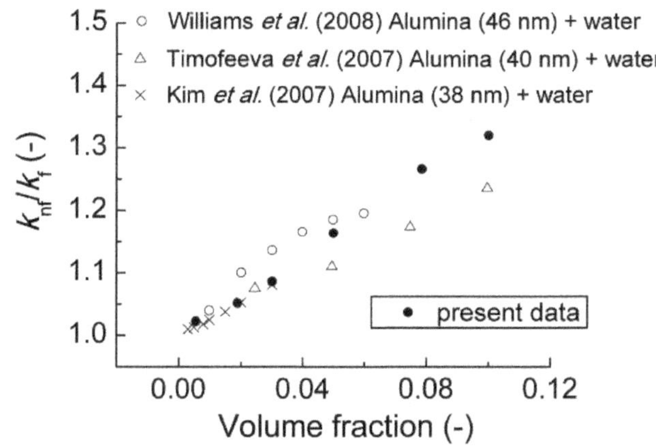

Fig. 1 Relative thermal conductivity of alumina/water nanofluids versus volume fraction at 20 °C.

Figure 2 presents the evaluation of the present data and several analytic models. The present data points are almost located within the lower and upper H-S bounds. The lower H-S bound or the Maxwell model can estimate the present data roughly, especially at volume fractions lower than 3.0%. The Bruggeman model (BGM) can predict the data well except those with volume fractions larger than 8%. The Davis model is located well below the lower H-S bound possibly because it is originally based on solid composites and the interfacial thermal resistance in solid composites is much higher than that in nanofluids.

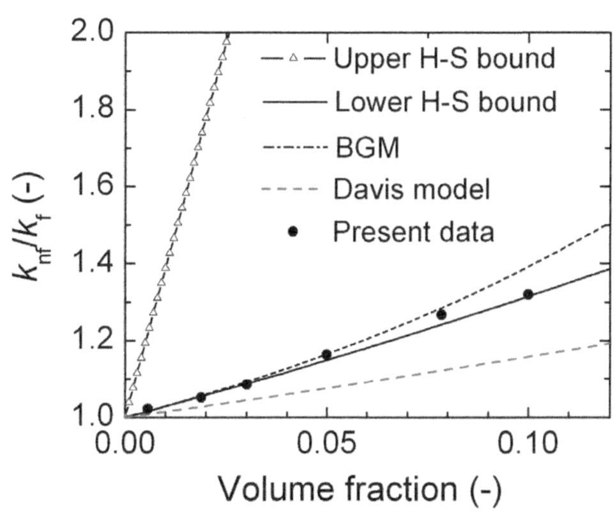

Fig. 2 Evaluation of the lower and upper H-S bounds (Hashin and Shtrikman, 1963), the Bruggeman model (BGM) (Bruggeman, 1935) and the Davis model (Davis, 1986) by the present data for alumina/water nanofluids.

Measurements of the change in thermal conductivity of an alumina/water nanofluid (7.82% volume fraction) with elapsed time were also performed at static conditions, as shown in Fig. 3. In general, thermal conductivity decreases with elapsed time. A 7.0% reduction in thermal conductivity averaged for each day was observed after 55 days. Possible reasons for thermal conductivity reduction are the formation of relatively large nanoparticle clusters and deposition. More experimental investigations are needed to better understand this phenomenon.

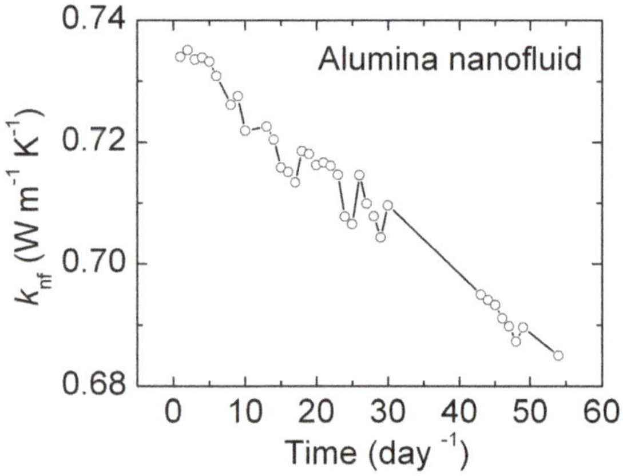

Fig. 3 Change in thermal conductivity averaged for each day with elapsed time for an alumina/water nanofluid (7.82% volume fraction) at 20 °C.

Rheology behavior of the alumina/water nanofluid is demonstrated in Fig. 4 for those data points with shear rates larger than 30 s^{-1}. The dynamic viscosity seems to be almost constant for the tested shear rates. Alumina nanofluids of high volume fractions may present non-Newtonian behavior, especially at very low shear rates. In our case, it is hard to see obvious non-Newtonian behavior for volume fractions up to 18.4% when the shear rate is larger than 50 s^{-1}. Therefore, the tested nanofluids can be regarded as Newtonian fluids when the shear rate is larger than 50 s^{-1}.

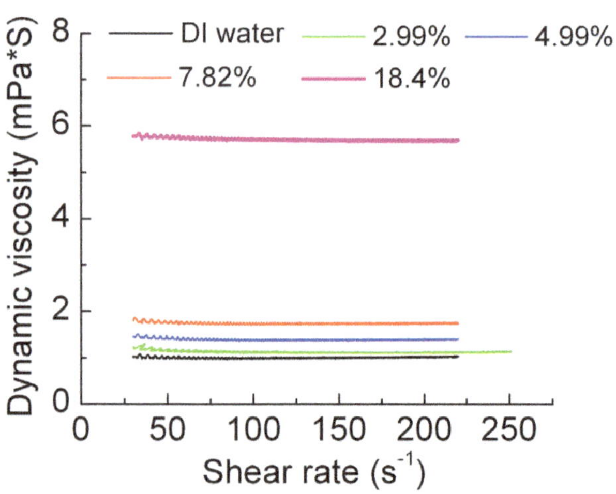

Fig. 4 Rheology behavior of alumina/water nanofluids at 20 °C.

Figure 5 shows the relative dynamic viscosity μ_{nf}/μ_f of alumina/water nanofluids versus volume fraction. For the present data, the measured dynamic viscosity values at a relatively high shear rate of 210 s^{-1} were used. The relative dynamic viscosity increases with volume fraction. The increase rate seems to become higher at larger volume fractions. Previous literature data of Anoop *et al.* (2009), Sahoo *et al.* (2009), Timofeeva *et al.* (2007) and Williams *et al.* (2008) are also shown for comparison. The five data sets present obvious discrepancy. The viscosity data of Williams *et al.* (2008) is much higher than that of the other four data sets. The present data and the data of Timofeeva *et al.* (2007) are the lowest compared to other data sets at the same volume fraction. Therefore, the present tested alumina nanofluid seems to give better convective heat transfer performance than that of Williams *et al.* (2008) and Timofeeva *et al.* (2007) as it has decent thermal conductivity enhancement and relatively low viscosity enhancement.

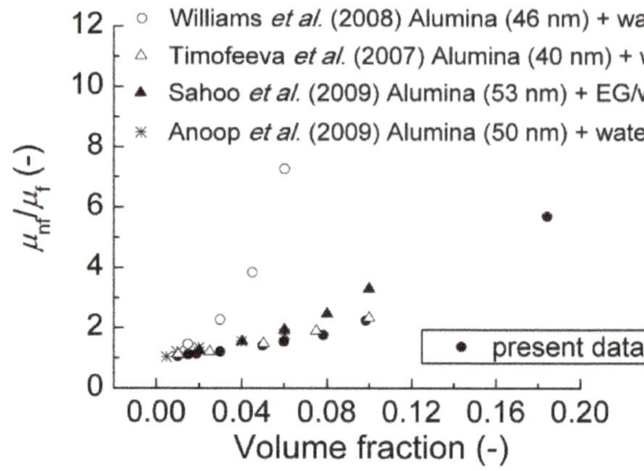

Fig. 5 Relative dynamic viscosity of alumina/water nanofluids versus volume fraction at 20 °C.

Comparison of the present data with that predicted by the Batchelor model (Batchelor, 1977) and the Chen *et al.* viscosity model (Chen *et al.*, 2007) is shown in Fig. 6. Different agglomeration sizes were used in the Chen *et al.* viscosity model (i.e., Eq. (18)) to guess the agglomeration size of the tested alumina/water nanofluid. The statement $d_a = d_p$ means no agglomeration. As shown in Fig. 6, both the Batchelor model and the Chen *et al.* viscosity model of no agglomeration under-predict the dynamic viscosity, especially at relatively high volume fractions. The Chen *et al.* viscosity model can accurately reproduce the present data for volume fractions less than about 15% when $d_a = 2.16\ d_p$. That means, nanoparticles might agglomerate in our tested alumina/water nanofluid and the agglomeration size is about 2.16 times of the alumina nanoparticle diameter. We further verified the agglomeration size through dynamic light scattering (DLS) measurements. Nanofluids of three different volume fractions, i.e., 1.0%, 1.89% and 4.99% were observed by DLS. Nanofluids were diluted carefully for DLS measurements. The three nanofluids give similar particle size distributions and have a peak at 80-95 nm, which confirmed the agglomeration size of 2.16 d_p. For an agglomeration size of 2.16 d_p, three nanoparticles might agglomerate and be closely packed together, with a small amount of liquid trapped in the agglomeration core.

According to the discussion by Chen *et al.* (2009), the effective thermal conductivity can be predicted accordingly based on the rheology analysis by including effects of particle agglomeration. As the approximate agglomeration size of the tested nanofluids has been extracted from Fig. 6, we can check if the measured thermal conductivity can be predicted by taking the effects of agglomeration into account. The blue dotted line in Fig. 7 without interfacial thermal resistance is the Chen *et al.* model (Chen *et al.*, 2009). The Chen *et al.* model over-estimates the present data largely. Therefore, interfacial thermal resistance may exist within the agglomerate and at the liquid/solid interfaces. We revised the Chen *et al.* model by incorporating the effects of interfacial thermal resistance in Eq. (10)

$$\frac{k_{nf}}{k_f} = \frac{k_a(1+2\cdot \mathrm{Bi})+(n-1)k_f+(n-1)\varphi_a\left[k_a(1-\mathrm{Bi})-k_f\right]}{k_a(1+2\cdot \mathrm{Bi})+(n-1)k_f-\varphi_a\left(k_a(1-\mathrm{Bi})-k_f\right)} \quad (22)$$

where Bi is the particle Biot number defined as $2R_k k_f/d_a$. As shown in Fig. 7, the revised model based on Chen *et al.* (2009) can reproduce our thermal conductivity data by using an interfacial thermal resistance R_k of 1.9 × 10^{-8} m^2 K W^{-1}. Thus, we can consider that the tested alumina nanofluids in this study contain agglomerations of an approximate size of 2.16 d_p, and an interfacial thermal resistance, which degrades thermal conductivity, exists in our case and cannot be neglected.

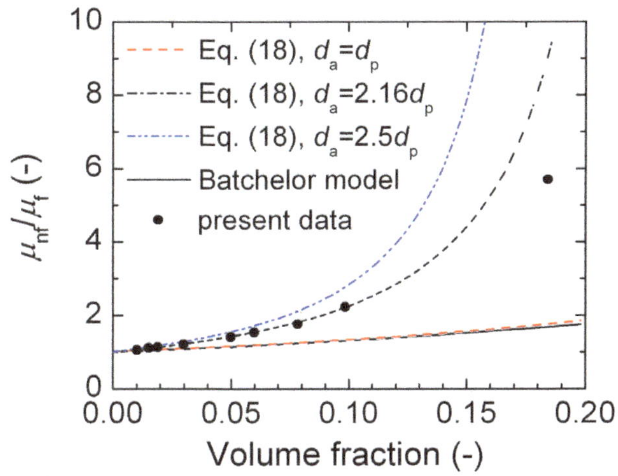

Fig. 6 Comparison of the present viscosity data for alumina/water nanofluids with that predicted by the Batchelor model (Batchelor, 1977) and the Chen *et al.* viscosity model (Chen *et al.*, 2007).

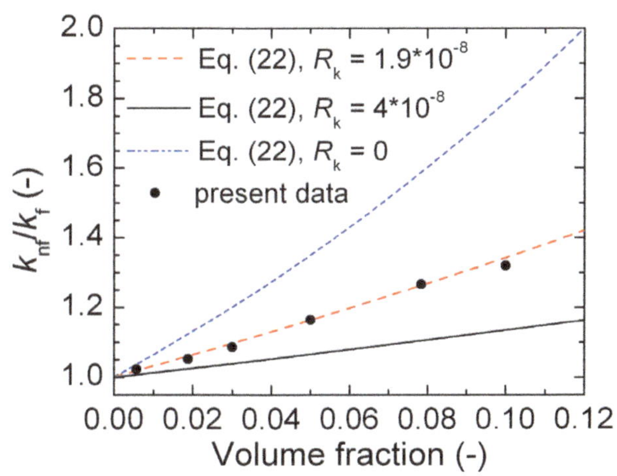

Fig. 7 Evaluation of the revised model based on Chen *et al.* (2009) at different values of interfacial thermal resistance for alumina/water nanofluids.

3.2 MWCNT/water nanofluid

An aqueous MWCNT suspension of 1.0% mass fraction was purchased from a commercial company (Nanocyl, Belgium). According to the vendor's specification, the suspension consists of thin MWCNTs dispersed in de-ionized (DI) water (97% mass fraction) and surfactant sodium dodecyl benzene sulfonate (SDBS, 2.0% mass fraction), and it is said to be stable for several months. The MWCNT, produced via the catalytic carbon vapor deposition (CCVD) process, has an average length of 1.5 μm and an average diameter of 9.5 nm, with an average aspect ratio of 158. The surface area of the MWCNT is 250-300 m^2/g. The carbon purity of the MWCNTs is 90%, while the remaining 10% is metal oxide. Similar to alumina nanofluid, different amounts of concentrated nanofluid were diluted in water to obtain MWCNT/water nanofluids with different concentrations. The diluted nanofluid mixture was mechanically stirred for 5 minutes followed by ultrasonic vibration for one hour. Four MWCNT/water nanofluids with carbon nanotube (CNT) volume fractions, 0.0278%, 0.0555%, 0.278% and 0.557% were carefully prepared. The pH values of the prepared MWCNT/water nanofluids are in the range 7.0 ~ 8.0.

Figure 8 shows the relative thermal conductivity k_{nf}/k_f of MWCNT/water nanofluids of different volume fractions. Basically, the thermal conductivity increases with volume fraction. Literature data of Cherkasova and Shan (2010), Ding *et al.* (2006), Garg *et al.* (2009), Meng *et al.* (2012) and Phuoc *et al.* (2011) are also shown in Fig. 8 for comparison. The MWCNTs used in Cherkasova and Shan (2010) are short, with nominal lengths of 0.5-2 μm and outer diameters of 30-50 nm. The mass ratio of the surfactant SDBS to MWCNT was five. The MWCNTs in Garg *et al.* (2009) had a specified average outer diameter of 10-20 nm, length of 0.5-40 μm and purity of 95%. The nanofluid was prepared by adding surfactant Gum Arabic (GA) of 0.25 wt%. No information about MWCNT dimension was provided in Ding *et al.* (2006), Meng *et al.* (2012) and Phuoc *et al.* (2011). MWCNTs in Meng *et al.* (2012) were treated with nitric acid before dispersion into glycol. GA of 0.25 wt% was added in the MWCNT/water nanofluid by Ding *et al.* (2006). The MWCNTs in Phuoc *et al.* (2011) have an outer diameter of 20-30 nm, inner diameter of 5-10 nm and length of 10-30 μm. Low molecular chitosan of 0.1 wt% was used for nanofluid preparation. The present relative thermal conductivity and the data of Garg *et al.* (2009), Meng *et al.* (2012) and Phuoc *et al.* (2011) give similar values and trends. The data sets of Cherkasova and Shan (2010) and Ding *et al.* (2006) are higher than the other four data sets and suggest that MWCNT/water nanofluids, if prepared properly, can provide decent thermal conductivity enhancements even at very low volume fractions.

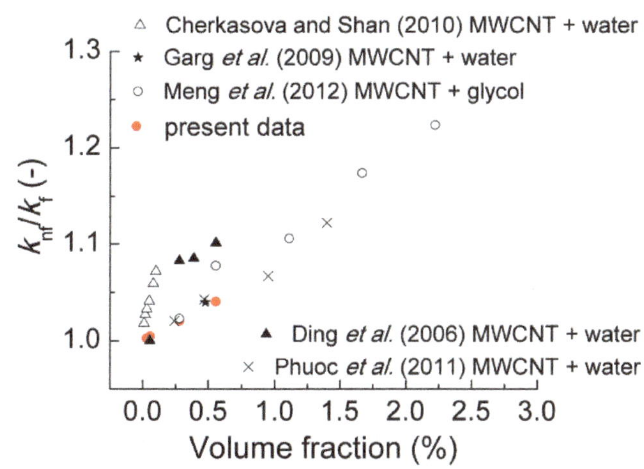

Fig. 8 Relative thermal conductivity of MWCNT/water nanofluids versus volume fraction at 20 °C.

The non-straightness of CNTs defined by Eq. (9) of the Deng *et al.* model (Deng *et al.*, 2007) may be used to represent the aggregated or entangled state of the CNTs. Figure 9 demonstrates the Deng *et al.* model at four different volume fractions. The relative thermal conductivity increases with volume fraction and non-straightness factor. The relative thermal conductivity increases with non-straightness factor very quickly when the non-straightness of CNTs is low, i.e., severe entanglement, and then gradually saturates. Therefore, methods should be figured out to avoid severe CNT entanglements and aggregations for thermal conductivity enhancement. As shown in Fig. 9, the non-straightness factor in our MWCNT/water nanofluids is in the range 0.11-0.13, indicating relatively severe CNT entanglement. This is one possible reason for the relatively low thermal conductivity of the tested MWCNT/water nanofluid.

Rheology behavior of MWCNT/water nanofluids with volume fractions of 0.0278% and 0.557% is demonstrated in Fig. 10. The horizontal solid line in Fig. 10 indicates the dynamic viscosity value for water at 20 °C. The 0.557% nanofluid behaves as a shear-thinning fluid (non-Newtonian fluid) as the dynamic viscosity decreases when the shear rate increases, especially at low shear rates. When the shear rate is larger than 150 s^{-1}, the viscosity is independent of the shear rate.

The 0.0278% nanofluid behaves like a Newtonian fluid, at least for shear rates larger than 50 s^{-1}.

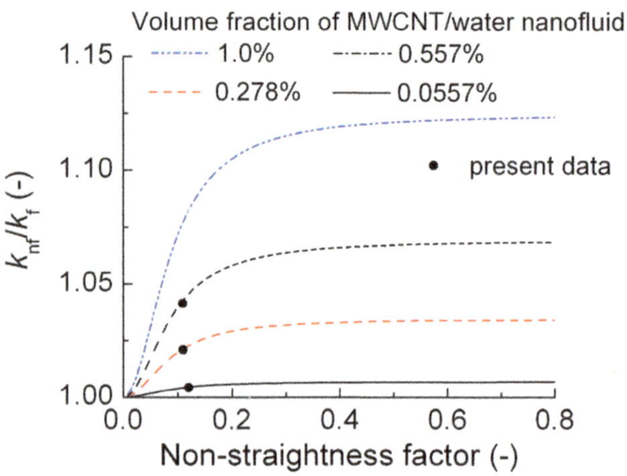

Fig. 9 Comparison of the Deng *et al.* model (Deng *et al.*, 2007) with the present data for MWCNT/water nanofluids.

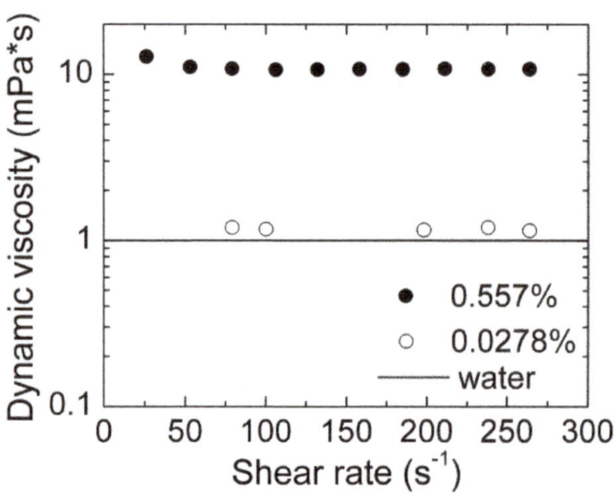

Fig. 10 Rheology behavior of MWCNT/water nanofluids at 20 °C.

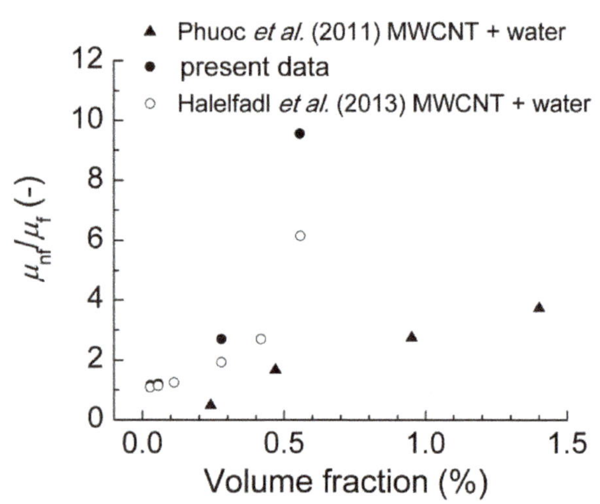

Fig. 11 The relative viscosity of MWCNT/water nanofluids versus volume fraction at 20 °C.

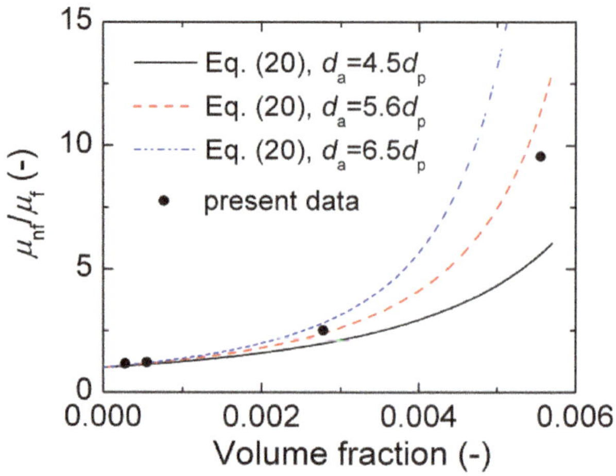

Fig. 12 Comparison of the present data with that predicted by the Halelfadl *et al.* (2013) model for MWCNT/water nanofluids.

Figure 11 shows the relative viscosity of the present MWCNT/water data and that of Halelfadl *et al.* (2013) and Phuoc *et al.* (2011). For the present data, measured viscosities at a high shear rate of 264 s^{-1} were used. The relative viscosity increases with volume fraction. The increase rate also increases with volume fraction for the present data and the data of Halelfadl *et al.* (2013). The data of Phuoc *et al.* (2011) is much lower than the present data and that of Halelfadl *et al.* (2013).

Comparison of the present data with that predicted by the Halelfadl *et al.* (2013) model (i.e., Eq. (20)) of different agglomeration sizes is shown in Fig. 12. The present data can be predicted well when $d_a = 5.6\, d_p$.

Based on the above analysis, the MWCNTs in the tested nanofluid are entangled and agglomerated in the base fluid. According to DLS observation, the particle size distributions of the tested nanofluids have peak values located within 190-230 nm. The non-straightness factor is in the range 0.11-0.13 as shown in Fig. 9, which corresponds to L_{actual} values from 165 to 195 nm. If we define L_{actual} as the distance between the two ends of an agglomeration or aggregation rather than the two ends of a non-straight CNT, the values from DLS observation and that obtained from the Deng *et al.* model (Deng *et al.*, 2007) are of the same order of magnitude. In this study, we assume that the MWCNT agglomeration is approximately in the form of rod-like particles with a diameter of about 5.6 d_p (53 nm) and a length of about 200 nm, thus with an aspect ratio of 3.8. Therefore we can check if the revised model based on Chen *et al.* (2009) can reproduce the thermal conductivity. During the calculation, thermal conductivities of carbon nanotubes along transverse and longitudinal directions and isotropic thermal conductivity of the nanotube are 5.6 W m^{-1} K^{-1}, 3000 W m^{-1} K^{-1} (Che *et al.*, 2000) and 2000 W m^{-1} K^{-1}, respectively. As shown in Fig. 13, the revised model based on Chen *et al.* (2009) can reproduce our thermal conductivity data by using an interfacial thermal resistance R_k of 1.2 × 10^{-8} m^2 K W^{-1}. Thus, the tested MWCNT nanofluid in this study contains agglomerations, and an interfacial thermal resistance, which degrades thermal conductivity, probably exists in our case and cannot be neglected.

For rod-like particle agglomerations, the thermal conductivity enhancement at flowing conditions may be lower than that at static conditions. As shown in Fig. 14, the randomly oriented agglomerations (might be percolated) at static conditions tend to be parallel with the flow direction at flowing conditions due to the liquid inertia. The long axis of the agglomeration is prone to be perpendicular to the temperature gradient under flowing conditions, and therefore it degrades the thermal conductivity enhancement.

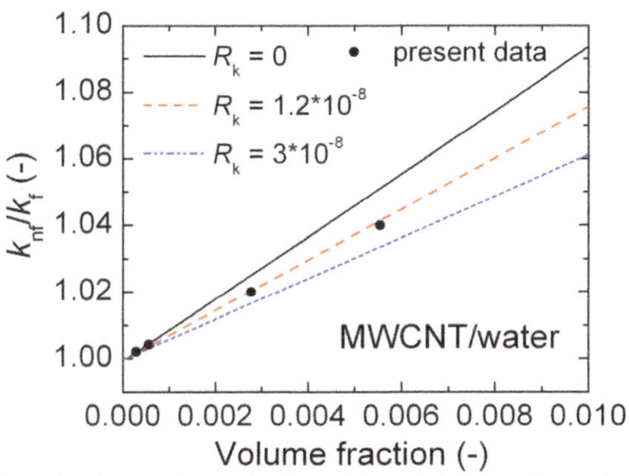

Fig. 13 Evaluation of the revised model at different values of interfacial thermal resistance for MWCNT/water nanofluids.

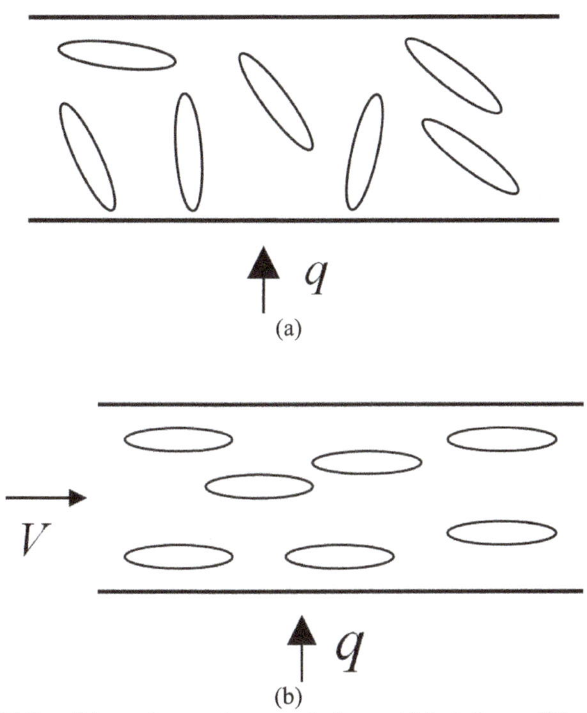

Fig. 14 Possible agglomeration morphology at (a) static conditions and (b) flowing conditions.

3.3 Nanofluid mixtures of alumina/water and MWCNT/water nanofluids

Two mixture samples listed in Table 2 were obtained and ultrasonically vibrated for 3 hours. The measured thermal conductivity values of the mixture are also listed in Table 2. The two samples have different alumina nanoparticle volume fractions. As shown in Table 2, the thermal conductivity enhancement for sample No. 1 is 7.2%, which is slightly larger than the addition of that of the two original nanofluids. One possible reason might be that the agglomerate morphology has been modified. For sample No. 2, the thermal conductivity enhancement of 7.9% is less than the addition of that of the two original nanofluids. The relative thermal conductivity of sample No. 2 is almost the same as sample No. 1. An obvious increase in the alumina volume fraction for No. 2 does not produce an obvious enhancement in the mixture thermal conductivity, which is probably due to large interfacial thermal resistances and cluster deposition.

Table 2 Thermal conductivity of two nanofluid mixtures.

Sample	*Mixture contents*	k_{nf} ($Wm^{-1}K^{-1}$)	k_{nf}/k_f (-)
No. 1	MWCNT/water 0.278% 10 mL Alumina/water 1.89% 25 mL	0.625	1.072
No. 2	MWCNT/water 0.278% 10 mL Alumina/water 5.0% 25 mL	0.629	1.079

4. CONCLUSIONS

In this work, thermal conductivity and rheology behavior of aqueous alumina and MWCNT nanofluids were measured and compared with several available analytical models. The thermal conductivity and viscosity increase with increasing volume fraction. The tested alumina nanofluid may provide a good convective heat transfer performance as it has a decent thermal conductivity enhancement but relatively low viscosity enhancement compared to literature data, while the tested MWCNT/water nanofluid is not efficient for convective heat transfer due to its large viscosity enhancement and relatively low thermal conductivity enhancement. The measured thermal conductivity values for both nanofluids are located near the lower H-S bound and far away from the upper H-S bound. Thus the thermal conductivity enhancement can be further increased. Further conductivity enhancement can be achieved by manipulating particle or agglomeration distribution and morphology in the nanofluid. The structure-property relationship was exemplified in this study. Information about possible agglomeration size and interfacial thermal resistance were obtained and partially validated. SEM/TEM observations are required for further validation. It is found that by incorporating the effects of interfacial thermal resistance, the revised model based on Chen *et al.* (2009) can reproduce the experimental data well based on the agglomeration size extracted from the rheology analysis.

The change in thermal conductivity of an alumina/water nanofluid of volume fraction 7.82% with elapsed time was also investigated. A reduction of 7.0% in average thermal conductivity was observed after 55 days. Possible reasons for the thermal conductivity reduction might be the formation of relatively large nanoparticle clusters and deposition. More experimental investigations are needed to better understand the change of thermal conductivity with time.

Besides, thermal conductivity measurements were conducted for nanofluid mixtures of alumina/water and MWCNT/water nanofluids. Proper proportions of the two nanofluids may give high thermal conductivity enhancement larger than the addition of that of the two original nanofluids by modifying agglomeration morphology. The obvious increase in the alumina volume fraction for sample No. 2 does not produce an obvious enhancement, which is probably due to large interfacial thermal resistances and cluster deposition.

ACKNOWLEDGEMENTS

Financial support from the Swedish Energy Agency and the Swedish Research Council (VR) are gratefully acknowledged. Special thanks to Prof. W. Li at Zhejiang University for his help on the viscosity measurement of nanofluids.

NOMENCLATURE

Bi	particle Biot number
$c_{p,f}$	specific heat at constant pressure (J kg^{-1} K^{-1})
d_a	agglomerate diameter (m)
d_p	particle diameter (m)
H	parameter defined in Eq. (8)
k	thermal conductivity (W m^{-1} K^{-1})
k_B	Boltzmann constant (m^2 kg s^{-2} K^{-1})

k_x, k_z, k_c thermal conductivities of carbon nanotubes along transverse and longitudinal directions and isotropic thermal conductivity of the nanotube (W m^{-1} K^{-1})
L_x, L_z geometrical factors defined in Eqs. (15) and (16)
L_p length of the paticle (m)
n shape factor
p aspect ratio
q heat flux (W m^{-2})
R_k interfacial thermal resistance (m^2 K W^{-1})
T temperature (K)
V velocity (m s^{-1})

Greek Symbols
β parameters defined in Eqs. (13) and (14)
η non-straightness factor
μ dynamic viscosity (Pa s)
ρ density (kg m^{-3})
φ volume fraction
ω particle sphericity

Subscripts
a agglomeration
f base fluid
m maximum
nf nanofluid
p nanoparticle

REFERENCES

Anoop, K.B., Kabelac, S., Sundararajan, T., and Das, S.K., 2009, "Rheological and Flow Characteristics of Nanofluids: Influence of Electroviscous Effects and Particle Agglomeration," *Journal of Applied Physics*, **106**, 034909.
http://dx.doi.org/10.1063/1.3182807

Batchelor, G.K., 1977, "The Effect of Brownian Motion on the Bulk Stress in a Suspension of Spherical Particles," *Journal of Fluid Mechanics*, **83**(1), 97-117.
http://dx.doi.org/10.1017/S0022112077001062

Bruggeman, D.A.G., 1935, "Dielectric Constant and Conductivity of Mixtures of Isotropic Materials," *Ann Phys (Leipzig)*, **24**, 636-679.

Che, J., Cagin, T., and Goddard III, W.A., 2000, "Thermal Conductivity of Carbon Nanotubes," *Nanotechnology*, **11**(2), 65.
http://dx.doi.org/10.1088/0957-4484/11/2/305

Chen, H., Ding, Y., and Tan, C., 2007, "Rheological Behaviour of Nanofluids," *New Journal of Physics*, **9**, 367.
http://dx.doi.org/10.1088/1367-2630/9/10/367

Chen, H., Witharana, S., Jin, Y., Kim, C., and Ding, Y., 2009, "Predicting Thermal Conductivity of Liquid Suspensions of Nanoparticles (Nanofluids) based on Rheology," *Particuology*, **7**(2) 151-157.
http://dx.doi.org/10.1016/j.partic.2009.01.005

Cherkasova, A.S., and Shan, J.W., 2010, "Particle Aspect-Ratio and Agglomeration-State Effects on the Effective Thermal Conductivity of Aqueous Suspensions of Multiwalled Carbon Nanotubes," *ASME Journal of Heat Transfer*, **132**(8), 082402.
http://dx.doi.org/10.1115/1.4001364

Chung, J.N., Chen, T., and Maroo, S.C., 2011, "A Review of Recent Progress on Nano/Micro Scale Nucleate Boiling Fundamentals," *Frontiers in Heat and Mass Transfer*, **2**, 023004.
http://dx.doi.org/10.5098/hmt.v2.2.3004

Davis, R.H., 1986, "The Effective Thermal Conductivity of a Composite Material with Spherical Inclusions," *International Journal of Thermophysics*, **7**(3), 609-620.
http://dx.doi.org/10.1007/BF00502394

Deng, F., Zheng, Q.S., Wang, L.F., and Nan, C.W., 2007, "Effects of Anisotropy, Aspect Ratio, and Nonstraightness of Carbon Nanotubes on Thermal Conductivity of Carbon Nanotube Composites," *Applied Physics Letters*, **90**(2), 021914.
http://dx.doi.org/10.1063/1.2430914

Ding, Y., Alias, H., Wen, D., and Williams, R.A., 2006, "Heat Transfer of Aqueous Suspensions of Carbon Nanotubes (CNT Nanofluids)," *International Journal of Heat and Mass Transfer*, **49**(1-2), 240-250.
http://dx.doi.org/10.1016/j.ijheatmasstransfer.2005.07.009

Eapen, J., Rusconi, R., Piazza, R., and Yip, S., 2010, "The Classical Nature of Thermal Conduction in Nanofluids," *ASME Journal of Heat Transfer*, **132**(10), 102402.
http://dx.doi.org/10.1115/1.4001304

Evans, W., Fish, J., and Keblinski, P., 2006, "Role of Brownian Motion Hydrodynamics on Nanofluid Thermal Conductivity," *Applied Physics Letters*, **88**, 093116.
http://dx.doi.org/10.1063/1.2179118

Feng, Z., and Li, W., 2013, "Laminar Mixed Convection of Large-Prandtl-Number in-Tube Nanofluid Flow, Part I: Experimental Study," *International Journal of Heat and Mass Transfer*, **65**, 919-927.
http://dx.doi.org/10.1016/j.ijheatmasstransfer.2013.07.005

Garg, P., Alvarado, J.L., Marsh. C., Carlson, T.A., Kessler, D.A., and Annamalai, K., 2009, "An Experimental Study on the Effect of Ultrasonication on Viscosity and Heat Transfer Performance of Multi-wall Carbon Nanotube-based Aqueous Nanofluids," *International Journal of Heat and Mass Transfer*, **52**(21-22), 5090-5101.
http://dx.doi.org/10.1016/j.ijheatmasstransfer.2009.04.029

Haghighi, E.B., Saleemi, M., Nikkam, N., Khodabandeh, R., Toprak, M.S., Muhammed, M., and Palm, B., 2014, "Accurate Basis of Comparison for Convective Heat Transfer in Nanofluids," *International Communications in Heat and Mass Transfer*, **52**, 1-7.
http://dx.doi.org/10.1016/j.ichcatmasstransfer.2014.01.002

Halelfadl, S., Estelle, P., Aladag, B., Doner, N., and Mare, T., 2013, "Viscosity of Carbon Nanotubes Water based Nanofluids: Influence of Concentration and Temperature," *International Journal of Thermal Sciences*, **71**, 111-117.
http://dx.doi.org/10.1016/j.ijthermalsci.2013.04.013

Hamilton, R.L., and Crosser, O.K., 1962, "Thermal Conductivity of Heterogeneous Two-Component Systems," *Industrial & Engineering Chemistry Fundamentals*, **1**(3), 187-191.
http://dx.doi.org/10.1021/i160003a005

Hashin, Z., and Shtrikman, S., 1963, "A Variational Approach to the Theory of the Elastic Behaviour of Multiphase Materials," *Journal of the Mechanics and Physics of Solids*, **11**(2), 127-140.
http://dx.doi.org/10.1016/0022-5096(63)90060-7

Hong, K.S., Hong, T.K., and Yang, H.S., 2006, "Thermal Conductivity of Fe Nanofluids Depending on the Cluster Size of Nanoparticles," *Applied Physics Letters*, **88**, 031901.
http://dx.doi.org/10.1063/1.2166199

Hui, P.M., Zhang, X., Markworth, A.J., and Stroud, D., 1999,

"Thermal Conductivity of Graded Composites: Numerical Simulations and an Effective Medium Approximation," *Journal of Materials Science*, **34**(22), 5497-5503.
http://dx.doi.org/10.1023/A:1004760427981

Karthikeyan, N.R., Philip, J., and Raj, B., 2008, "Effect of Clustering on the Thermal Conductivity of Nanofluids," *Materials Chemistry and Physics*, **109**(1), 50-55.
http://dx.doi.org/10.1016/j.matchemphys.2007.10.029

Keblinski, P., Phillpot, S.R., Choi, S.U.S., and Eastman, J.A., 2002, "Mechanisms of Heat Flow in Suspensions of Nano-Sized Particles (Nanofluids)," *International Journal of Heat and Mass Transfer*, **45**(4), 855-863.
http://dx.doi.org/10.1016/S0017-9310(01)00175-2

Kim, S.H., Choi, S.R., and Kim, D., 2007, "Thermal Conductivity of Metal-Oxide Nanofluids: Particle Size Dependence and Effect of Laser Irradiation," *ASME Journal of Heat Transfer*, **129**(3), 298-307.
http://dx.doi.org/10.1115/1.2427071

Krieger, I.M., and Dougherty, T.J., 1959, "Concentration Dependence of the Viscosity of Suspensions," *Transaction of the Society of Rheology*, **3**, 137-152.

Maron, S.H., and Pierce, P.E., 1956, "Application of Ree-Eyring Generalized Flow Theory to Suspensions of Spherical Particles," *Journal of Colloid Science*, **11**(1), 80-95.
http://dx.doi.org/10.1016/0095-8522(56)90023-X

Maxwell, J.C., 1881, *A Treatise on Electricity and Magnetism*, 2nd ed., Clarendon Press, Oxford.

Meng, Z., Wu, D., Wang, L., Zhu, H., and Li, Q., 2012, "Carbon Nanotube Glycol Nanofluids: Photo-Thermal Properties, Thermal Conductivities and Rheological Behavior," *Particuology*, **10**(5), 614-618.
http://dx.doi.org/10.1016/j.partic.2012.04.001

Mo, H., Evmenenko, G., and Dutta, P., 2005, "Ordering of Liquid Squalane Near a Solid Surface," *Chemical Physics Letters*, **415**(1-3), 106-109.
http://dx.doi.org/10.1016/j.cplett.2005.08.142

Mueller, S., Llewellin, E.W., and Mader, H.M., 2010, "The Rheology of Suspensions of Solid Particles," *Proceedings of the Royal Society A: Mathematical, Physical & Engineering Sciences*, **466**, 1201-1228.
http://dx.doi.org/10.1098/rspa.2009.0445

Nan, C.W., Liu, G., Lin, Y., and Li, M., 2004, "Interface Effect on Thermal Conductivity of Carbon Nanotube Composites," *Applied Physics Letters*, **85**, 3549-3551.
http://dx.doi.org/10.1063/1.1808874

Nielsen, L.E., 1978, *Predicting the Properties of Mixtures: Mixture Rules in Science and Engineering*, Marcel Dekker, New York.

Phuoc, T.X., Massoudi, M., and Chen, R.H., 2011, "Viscosity and Thermal Conductivity of Nanofluids Containing Multi-walled Carbon Nanotubes Stabilized by Chitosan," *International Journal of Thermal Sciences*, **50**(1), 12-18.
http://dx.doi.org/10.1016/j.ijthermalsci.2010.09.008

Prasher, R., Phelan, P.E., and Bhattacharya, P., 2006, "Effect of Aggregation Kinetics on the Thermal Conductivity of Nanoscale Colloidal Solutions (Nanofluid)," *Nano Letters*, **6**(7), 1529-1534.
http://dx.doi.org/10.1021/nl060992s

Sahoo, B.C., Vajjha, R.S., Ganguli, R., Chukwu, G.A., and Das, D.K., 2009, "Determination of Rheological Behavior of Aluminum Oxide Nanofluid and Development of New Viscosity Correlations," *Petroleum Science and Technology*, **27**(15), 1757-1770.
http://dx.doi.org/10.1080/10916460802640241

Timofeeva, E.V., Gavrilov, A.N., McCloskey, J.M., Tolmachev, Y.V., Sprunt, S., Lopatina, L.M., and Selinger, J.V., 2007, "Thermal Conductivity and Particle Agglomeration in Alumina Nanofluids: Experiment and Theory," *Physical Review E*, **76**, 061203.
http://dx.doi.org/10.1103/PhysRevE.76.061203

Wang, L., and Fan, J., 2011, "Toward Nanofluids of Ultra-High Thermal Conductivity," *Nanoscale Research Letters*, **6**, 153.
http://dx.doi.org/10.1186/1556-276X-6-153

Williams, W., Buongiorno, J., and Hu, L.W., 2008, "Experimental Investigation of Turbulent Convective Heat Transfer and Pressure Loss of Alumina/Water and Zirconia/Water Nanoparticle Colloids (Nanofluids) in Horizontal Tubes," *ASME Journal of Heat Transfer*, **130**(4), 042412.
http://dx.doi.org/10.1115/1.2818775

Wu, Z., and Sundén, B., 2014, "On Further Enhancement of Single-Phase and Flow Boiling Heat Transfer in Micro/Minichannels," *Renewable and Sustainable Energy Reviews*, **40**, 11-27.
http://dx.doi.org/10.1016/j.rser.2014.07.171

Wu, Z., Wang, L., and Sundén, B., 2013, "Pressure Drop and Convective Heat Transfer of Water and Nanofluids in a Double-Pipe Helical Heat Exchanger," *Applied Thermal Engineering*, **60**(1-2), 266-274.
http://dx.doi.org/10.1016/j.applthermaleng.2013.06.051

4

SINTERING OF TITANIUM AND NICKEL NANOPOWDERS WITH A ND:YAG NANOSECOND LASER

Jing Huang, Vitaly E. Gruzdev, Yuwen Zhang*, and J. K. Chen

Department of Mechanical and Aerospace Engineering, University of Missouri, Columbia MO 65211, USA

ABSTRACT

Experimental study on laser sintering of nanosized titanium and nickel powders using a nanosecond laser is reported in this paper. Pulse laser with a wavelength of 532 nm and pulse width (FWHM) of 23 nanoseconds is used. The diameters of the metal nanoparticles are between 35 to 50 nanometers. Pulse repetition rates range between 100 and 10,000 Hz and the average power is from 1 to 20 W. The powder bed is placed on a moving stage to control the scanning velocity. Sintered metal strips with 20 mm length are formed by focusing the laser beam on the surface of the moving power bed. Experiment results discover four different patterns of sintered metal strips. Nickel powder shows poor sintering ability due to severe shrinkage of the powder bed, while titanium powder forms solid sintered strips under specific parameters. However, scaly structure is observed in certain conditions which weakens the mechanical strength of the sintered strips.

Keywords: Laser Sintering; Nanopowder; Nanosecond laser.

1. INTRODUCTION

Since it was introduced in mid 1980s for the first time, theoretical and experimental studies on selective laser sintering (SLS) have been studied both theoretically and experimentally in a large number of reported works (Agarwala *et al.*, 1995; Kruth *et al.*, 2004, 2007; Peltola *et al.*, 2008; Wang *et al.*, 2006; Chen and Zhang, 2007; Xiao and Zhang, 2007; Shi and Zhang, 2008). The researchers developed various methods to improve the original idea of building working parts layer by layer through laser sintering of fine material particles. Various fabrication parameters were tested thoroughly, including laser wavelengths, intensity, laser beam size, scanning strategy, and so on. Different materials were used, single-component or mixture. The quality and efficiency of the technology has been improved significantly since then. Some commercial systems are also available now in the market.

One of the most attractive advantages of SLS is its ability to process a wide range of material: polymers, metals, ceramics, etc. (Kruth *et al.*, 2003). However, the mostly investigated materials in SLS are still commercially available metal powders, such as tool steel, titanium, etc. (Kruth *et al.*, 2007; Childs *et al.*, 2003; Fischer *at al.*, 2003). Fischer at al. (2003; 2004a,b; 2005a,b) carried out a large amount of works on the sintering of single phase metal powders. Their work showed that by optimizing the laser parameters, the degree of porosity, density and the microstructure of the finished work piece could be controlled. Some others used a combination of two different kinds of powders, or coated powder (such as nylon-coated or polymer coated metal particles) to achieve sintering of high melting point metal at a relatively low temperature (Zhang *et al.*, 2000; Kruth *et al.*, 2005).

Selective laser sintering was also applied in tissue engineering in which scaffolds fabricated in high precision plays an important role. Chua *et al.* reported to use SLS technology to process a biocomposite blend comprising of polyvinyl alcohol and hydroxyapatite and create scaffolds aiming at dealing with craniofacial and join defects (Chua *et al.*, 2004). Ciardelli *et al.* (2005) studied the influence of laser beam parameters, including beam scanning speed and power on the sintering of bioartificial blends of poly-ε-caprolactone and polysaccharide. The materials used in these works were mostly synthetic in considering the special requirement of biomedical issues such as degradability, biocompatibility, non-immunogenicity, etc. (Tang *et al.*, 2003).

In the aforementioned works, most were conducted with conventional metal powder, where the dimensions of size are in micrometers or larger. Recently researchers started to study the nano-liquid phase laser sintering of metal powder which aimed at developing the next generation large scale integrated circuit printing technology. Ko *et al.* (2007) reported their work on laser sintering of inkjet-printed metal nanoparticles. The advantage of high precision of SLS was fully utilized to achieve high resolution patterning which was crucial for the large-area, flexible electronics manufacturing. The heat affected zone and thermal damage to the substrate was well controlled to enhance the resolution.

As the basis of nanotechnology, nanopowders are consisted of solid particles that measured on the nanoscale, only the size of three to five molecules together. Because of the size, they showed many unique characteristics that are different from the materials in bulk size. An example is the self-heating of aluminium nanopowder with liquid water which can be utilized to produce hydrogen (Astankova *et al.*, 2008).

Another imporant factor in SLS technology is the laser. In most studies, continuous wave lasers were used because they are more mature and have the advantage of low cost. With the development of ultrafast laser, nanosecond, even femtosecond laser are more and more reliable and affordable. As pointed out by R. Glardon, *et. al*, nanosecond laser offers a wider range of control and better dimension accuracy than continuous wave lasers which are important for high precision SLS (Glardon *et al.*, 2001).

In this paper, we will report our experimental work on laser sintering of solid phase nanopowders. Titanium and Nickel

* Corresponding author. Email: zhangyu@missouri.edu

nanopowders and their mixture (1:1 volume ratio) were tested. A Nd-YAG nano-second laser was used to form sintered metal strips on the surface of the powder bed. Average laser power, pulse repetition rate and powder bed moving speed were changed to find their influence on the quality of the final pieces.

2. EXPERIMENTAL SETUP AND PROCEDURES

A nanosecond Nd:YAG laser from Clark-MXR is used in the experiments. The laser had the wavelength of 532 nm and pulse width (FWHM) of 23 nanoseconds. The repetition rate of laser pulses was adjustable between 10^2 and 10^4 Hz, and the average power was adjustable between 1 and 20W (depending on frequency.)

(a) Optical microscope

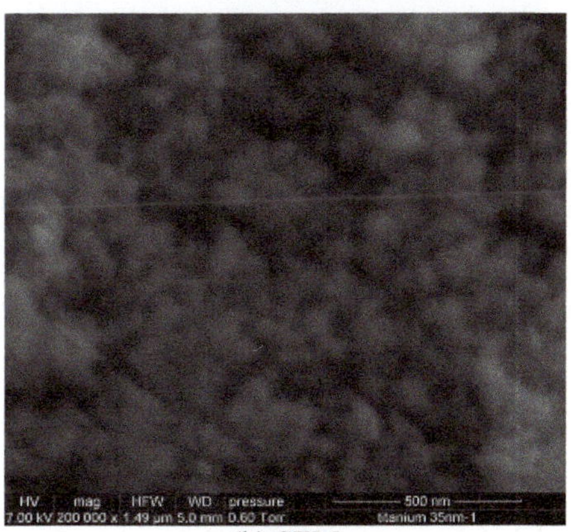

(b) SEM

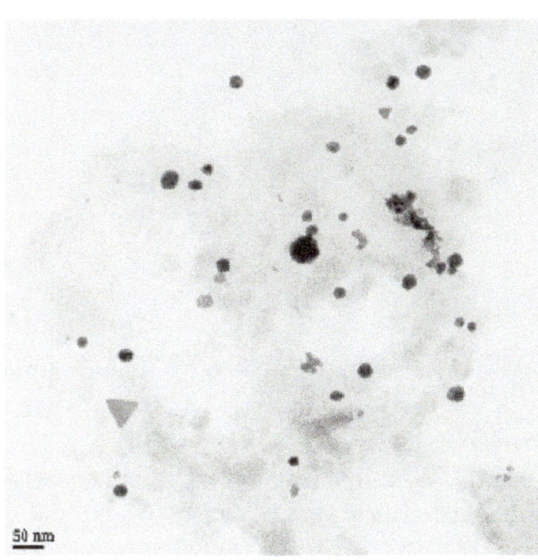

(c) TEM

Fig. 1 Titanium powder under different magnification factor

The nanopowder was provided by Nanostructured & Amorphous Materials. Two different materials were used: Ti (99%) and Ni (99.7%). The average size of the particles ranged from 30 to 50 nanometers. The titanium powders were characterized by optical microscope, transmission electron microscope (JEOL 1400) and scanning electron microscope (Hitachi S-4700). Figure 1 shows the pictures of the titanium powder under different amplification factors. Due to the high flammability of nickel powder in air, we did not observe it under microscopes. The average densities of powders were 1.114 g/mL and 0.872 g/mL for the titanium and nickel powder, in comparison to 4.506 g/cm^3 and 8.908 g/cm^3 of bulk values, respectively. This means the nickel power has a much higher porosity than the titanium powder.

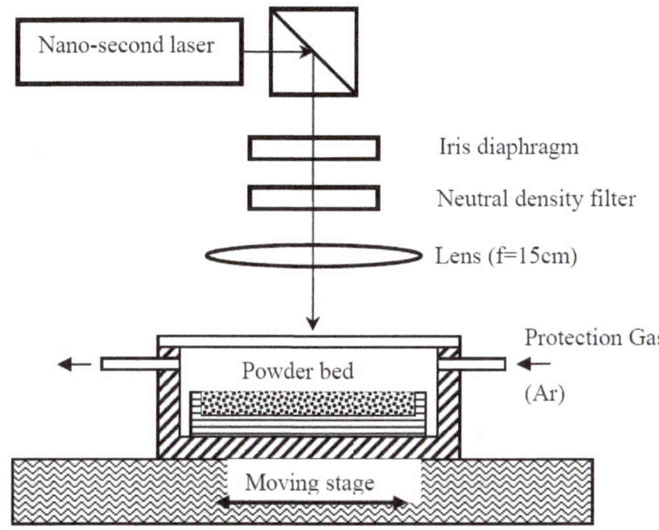

Fig. 2 Experiment platform and optical path

Figure 2 shows the schematic diagram of the test platform. Neutral density filters with different optical densities were used to change the average power without changing other parameters. Power density of the laser pulse train was measured with a pyroelectric energy sensor prior to each test. The distance between lens and powder bed surface was kept at a constant to maintain a laser beam diameter of 0.3 mm on the powder bed surface. The ceramic powder bed was placed in a cylindrical chamber with glass cover on the top. The chamber was infused with argon gas to prevent oxidation at high temperature. The whole chamber was placed on a moving stage that can be controlled by a computer to achieve different scanning speed. The parameters used in the experiment are listed in Table 1.

Table 1 Parameter range used in the experiment

Powder	Repetition rate (Hz)	Average power (W)	Moving speed (cm/s)
Nickel	700 - 5000	1.033 – 15.76	0.2 – 0.8
Titanium	1100 - 3000	3.51 – 13.63	0.2 – 1.0
Nickel + Titanium	3000	1.691 – 13.14	0.2 – 1.5

Before the experiments described below we conducted some tests with fixed single point of laser irradiation. The purposes of these tests were to find an appropriate range of laser parameters, protection gas flow rate, optical path adjustment steps and experiment procedures.

3. RESULTS AND DISCUSSIONS

3.1 Nickel powder

Fig. 3 Typical result by sintering nickel nanopowder

The nickel powder showed serious powder shrinkage during sintering under the above parameter ranges. Figure 3 shows a typical result, with the average power of 3.5 W, repetition rate 1500 Hz and scanning speed of 0.2 mm/s. The circles in the figure showed some nickel droplets caused by balling effects during the irradiation, with diameters around 1 mm. With such a problem, no usable working piece could be formed. The shrinkage of the powder bed was due to a relatively low powder density. As stated in section 2, the porosity of nickel powder is much smaller than titanium powder, which caused a more loose structure in the powder bed.

3.2 Titanium powder

The effect of laser power, repetition rate and scanning speed was tested by keeping all the other parameters constant while changing one of the parameter mentioned in the preceding subsection. Figure 4 shows the results with different average powers, while the repetition rate and scanning speed were kept at 3000 Hz and 0.2 mm/s, respectively. Since the aim of current work is to find the parameter range for further study, a simple strip pattern was chosen instead of a raster scanned area pattern. The strip length was kept constant at 20 mm. It can be seen that with higher laser power, the strip width increased because higher temperature led to further heat transfer.

(a) Average power 6.7 W

(b) Average power 8.5 W

(c) Average power 13.6 W

Fig. 4 Metal strip formed with titanium powder under different laser power (Scanning speed is 0.2 mm/s, and repetition rate is 3000 Hz)

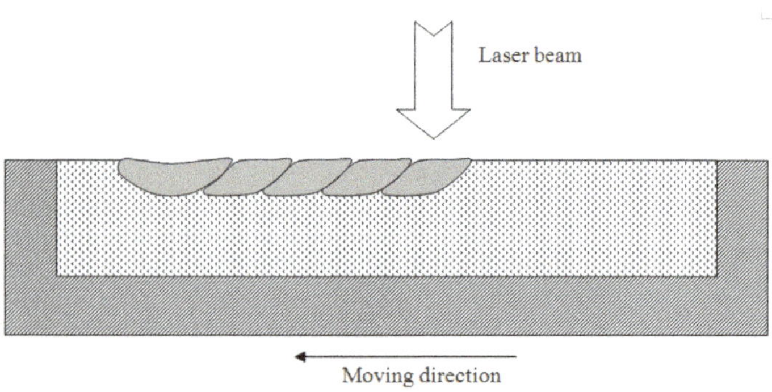

Fig. 5 Formation of a scale structure

The pictures showed that only power in a narrow range (Fig. 4b) would form ideal metal strip. When the laser power is too high or too low (Fig. 4a, 4c), a scaly interlayer structure would be formed. The mechanical strength of this structure was very poor as the strips were prone to breakage. Compared to the results reported in Fischer *et al.* (2003), which used titanium particles with the size of micrometer level, this optimum power density to achieve smooth flattened surface is about in the same range.

The process to cause the scale struture was analyzed based on the understanding of the sintering process. As shown in Fig. 5, when new melting pool took into being with the moving of the powder bed, melted metal fluid would flow under the old, already half-solidified melting pool. Then the continuous scanning of the laser beam would cause this scaly interlayer structure. Only in an appropriate combination of solidification time, which was related to laser power and moving speed, this effect would be alleviated to a low level.

Figure 6 shows the effect of moving speed of the powder bed, i.e., the scanning speed. It showed similar tendency as Fig. 5, only at a specific speed, the strip showed smooth appearance. In other cases, the scale structure would form and caused bad mechanical strength of the formed strips. At low scanning speed, each point on the power bed received more laser irradiation, thus the resolidification process became longer. As explained above, the heavier melted metal flew to the deeper part of the powder bed and formed scale structure and led to weak structure strength. When the scanning speed was too high, the metal particles received not enough energy to melt and sinter together appropriately. This is especially severe for the particles far away from the laser path because the energy received by these regions is not directly from the laser irradiation but from the heat transfer through the powder bed. As shown by the work of Xu *et.al.* (2007), the thermal conductivity of metal nanoparticle bed can be four orders of magnitude lower than that of bulk metal. This effect can be seen clearly in Fig. 6 (c) and (d). Different sintering regimes can be between these two limits and other parameters also influenced this balance, such as particle size and pulse frequency.

(a) Scanning speed 0.4 mm/s

(b) Scanning speed 0.6 mm/s

(c) Scanning speed 0.8 mm/s

(d) Scanning speed 1.5 mm/s

Fig. 6 The metal strip formed with titanium powder under different scanning speed (average power is 9.4 W, and the repetition rate 3000 Hz)

Different pulse frequencies (1000 Hz ~ 4000 Hz) were tested in the experiment to see its effect. However, the frequency range was not very wide because it was difficult to keep a constant average power while changing repetition rate. The experiment results showed little difference in the pulse frequency tested. This was because the intervals between two single pulses ranged from 0.25 to 1 ms, which were always several magnitudes larger than the duration of each single pulse

(23ns).

In addition to the above results, in the experiments we found that the quality of powder bed influenced sintering results significantly. The powder bed must be well vibrated to form a solid and smooth surface to ensure the repeatability of the experiment results. Although every effort has been taken to achieve best smoothness of the powder bed, the toughness of the bed surface cannot be compared to the size of individual particles.

3.3 Titanium/Nickel mixture

Finally, the titanium/nickel powder mixtures were tested. The powders were mixed with a volume ratio of 1:1. The results were similar to pure nickel powder, only some unconsolidated structures were formed. Figure 7 shows a typical result (Average power 5.4 W, scanning speed 0.1 mm/s, repetition rate 3000 Hz). More experiments are needed to find an appropriate mixing ratio of titanium and nickel nanopowders.

Fig. 7 Sintering result of titanium/nickel mixture (average power is 5.4 W, scanning speed is 0.1 mm/s, and repetition rate is 3000 Hz)

By comparing all the experiment results, we can summarize the formed pieces into 4 different sintered strip patterns: (A) Scale structure, such as shown in Fig. 4(b); (B) Well sintered, such as shown in Fig. 4(a); (C) Balling structure, such as shown in Fig. 3; (D) Porous structure as shown in Fig. (7). The corresponding parameters and resulting patterns are summarized in table 2. It is showed that in order to get a good sintered strip, the parameters should be sought under the following range: average power 5~8 W, repetition rate about 3000Hz, scanning speed 0.2~0.5 mm/s. However, it is also clear that there is still a lot blank area of parameters left unstudied in this table, which will be our work in the next step.

Table 2 Summary of experiment results in four patterns

Frequency (Hz)	Average Power (W)	Scanning Speed (mm/s)	Pattern
3000	0.2	13.63	A
3000	0.2	10.62	A
3000	0.2	9.64	A
3000	0.2	8.52	B
3000	0.2	6.78	B
3000	0.2	5.42	A
3000	0.2	3.51	D
3000	0.2	1.718	D
3000	0.1	9.61	A
3000	0.3	9.64	A
3000	0.4	9.63	A
3000	0.6	9.63	B
3000	0.8	9.61	B
3000	1.0	9.57	A
3000	1.5	9.6	A
3000	2.0	9.6	A
1300	0.2	5.44	D
1800	0.2	5.44	D
2300	0.2	5.42	D
4200	0.2	5.46	D
3500	0.2	5.41	D
3000	0.2	5.45	D
2000	0.2	5.44	D
1100	0.2	5.42	D

4. CONCLUSIONS

A nanosecond Nd-YAG pulsed laser was used to sinter titanium, nickel and their mixture nanopowders. Different laser powers, repetition rates, scanning speeds were used to study their influence on the metal strip formed by laser irradiation. By summarizing the experimental results, we can reach the following conclusions:

(1) For nickel powders, because of serious powder shrinkage after melting, there were no usable metal strips formed during the irradiation. It should be pointed out that the values of repetition rate and scanning speed between two tests are not very close in our experiments. We cannot completely rule out the possibility that better sintering results could be achieved in a very narrow range of parameters.

(2) For titanium powder, only a fine combination of laser power and scanning speed would lead to good quality of sintered strip. In other cases, scaly interlayer structure is obtained because of interaction between newly melted powder and resolidified pool. In the range we tested, pulse repetition rate showed little influence on the final results.

(3) With a volume ratio of 1:1, titanium/nickel mixture did not show a better property than pure titanium powder.

ACKNOWLEDGEMENTS

Support for this work by the U.S. National Science Foundation under Grant No. CBET-1066917 is gratefully acknowledged.

REFERENCES

Agarwala, M., Bourell, D., Beaman, J., Marcus, H., and Barlow, J., 1995, "Direct Selective Laser Sintering of Metals," *Rapid Prototyping Journal*, **1**(1), pp. 26-36. http://dx.doi.org/10.1108/13552549510078113

Astankova, A.P., Godymchuk, A.Y., Gromov, A.A., and Il'In, A.P., 2008, "The Kinetics of Self-heating in the Reaction Between Aluminum Nanopowder and Liquid Water," *Russian Journal of Physical Chemistry A*, **82**(11), pp. 1913-1920. http://dx.doi.org/10.1134/S0036024408110204

Chen, T., and Zhang, Y., 2007, "Thermal Modeling of Laser Sintering of Two-Component Metal Powder on Top of Sintered Layers via Multi-line Scanning," *Applied Physics A: Materials Science & Processing*, **86**(2), 213-220. http://dx.doi.org/10.1007/s00339-006-3739-1

Childs, T.H.G., Hauser, G., and Badrossamay, M., 2005, "Selective Laser Sintering (Melting) of Stainless and Tool Steel Powders: Experiments and Modelling," *Proceedings of the Institution of Mechanical Engineers, Part B: Journal of Engineering Manufacture*, **219**(4), pp. 339-357. http://dx.doi.org/10.1243/095440505X8109

Chua, C.K., Leong, K.F., Tan, K.H., Wiria, F.E., and Cheah, C.M., 2004, "Development of Tissue Scaffolds Using Selective Laser Sintering of Polyvinyl Alcohol/Hydroxyapatite Biocomposite for Craniofacial and Joint Defects," *Journal of Materials Science: Materials in Medicine*, **15**(10), pp. 1113-1121. http://dx.doi.org/10.1023/B:JMSM.0000046393.81449.a5

Ciardelli, G., Chiono, V., Vozzi, G., Pracella, M., Ahluwalia, A., Barbani, N., Cristallini, C., and Giusti, P., 2005, "Blends of Poly-(ε-caprolactone) and Polysaccharides in Tissue Engineering Applications," *Biomacromolecules*, **6**(4), pp. 1961-1976. http://dx.doi.org/10.1021/bm0500805

Fischer, P., Blatter, A., Romano, V., and Weber, H.P., 2005a, "Selective Laser Sintering of Amorphous Metal Powder," *Applied Physics A: Materials Science and Processing*, **80**(3), pp. 489-492. http://dx.doi.org/10.1007/s00339-004-3062-7

Fischer, P., Leber, H., Romano, V., Weber, H.P., Karapatis, N.P., André, C., and Glardon, R., 2004a, "Microstructure of Near-Infrared Pulsed Laser Sintered Titanium Samples," *Applied Physics A: Materials Science and Processing*, **78**(8), pp. 1219-1227. http://dx.doi.org/10.1007/s00339-003-2205-6

Fischer, P., Romano, V., Blatter, A., and Weber, H.P., 2005b, "Highly Precise Pulsed Selective Laser Sintering of Metallic Powders," *Laser Physics Letters*, **2**(1), pp. 48-55. http://dx.doi.org/10.1002/lapl.200410118

Fischer, P., Romano, V., Weber, H.P., Karapatis, N.P., Boillat, E., and Glardon, R., 2003, "Sintering of Commercially Pure Titanium Powder with a Nd:YAG Laser Source," *Acta Materialia*, **51**(6), pp. 1651-1662. http://dx.doi.org/10.1016/S1359-6454(02)00567-0

Fischer, P., Romano, V., Weber, H.P., and Kolossov, S., 2004b, "Pulsed Laser Sintering of Metallic Powders," *Thin Solid Films*, **453-454**, pp. 139-144. http://dx.doi.org/10.1016/j.tsf.2003.11.152

Glardon, R., Karapatis, N.P., and Romano, V., Levy, G.N., 2001, "Influence of Nd:YAG Parameters on the Selective Laser Sintering of Metallic Powders," *CIRP Annals - Manufacturing Technology*, **50**(1), pp. 133-136. http://dx.doi.org/10.1016/S0007-8506(07)62088-5

Ko, S.H., Pan, H., Grigoropoulos, C.P., Luscombe, C.K., Fréchet, J.M.J., and Poulikakos, D., 2007, "All-Inkjet-Printed Flexible Electronics Fabrication on A Polymer Substrate by Low-temperature High-Resolution Selective Laser Sintering of Metal Nanoparticles," *Nanotechnology*, **18**(34), 345202. http://dx.doi.org/10.1088/0957-4484/18/34/345202

Kruth, J.P., Froyen, L., Van Vaerenbergh, J., Mercelis, P., Rombouts, M., and Lauwers, B., 2004, "Selective Laser Melting of Iron-Based Powder," *Journal of Materials Processing Technology*, **149**(1-3), pp. 616-622. http://dx.doi.org/10.1016/j.jmatprotec.2003.11.051

Kruth, J.P., Levy, G., Klocke, F., and Childs, T. H. C., 2007, "Consolidation Phenomena in Laser and Powder-Bed Based Layered Manufacturing," *CIRP Annals - Manufacturing Technology*, **56**(2), pp. 730-759. http://dx.doi.org/10.1016/j.cirp.2007.10.004

Kruth, J.P., Mercelis, P., Van Vaerenbergh, J., Froyen, L., and Rombouts, M., 2005, "Binding Mechanisms in Selective Laser Sintering and Selective Laser Melting," *Rapid Prototyping Journal*, **11**(1), pp. 26-36. http://dx.doi.org/10.1108/13552540510573365

Kruth, J.P., Wang, X., Laoui, T., and Froyen, L., 2003, "Lasers and Materials in Selective Laser Sintering," *Assembly Automation*, **23**(4), pp. 357-371. http://dx.doi.org/10.1108/01445150310698652

Peltola, S.M., Grijpma, D.W. Melchels, F.P.W., and Kellomäki, M., 2008, "A Review of Rapid Prototyping Techniques for Tissue Engineering Purposes," *Annals of Medicine*, **40**(4), pp. 268-280. http://dx.doi.org/10.1080/07853890701881788

Shi, Y., and Zhang, Y., 2008, "Simulation of Random Packing of Spherical Particles with Different Size Distributions," *Applied Physics A: Materials Science & Processing,* **92**(3), 621-626. http://dx.doi.org/10.1007/s00339-008-4547-6

Tan, K.H., Chua, C.K., Leong, K.F., Cheah, C.M., Cheang, P., Abu Bakar, M.S., and Cha, S.W., 2003, "Scaffold Development Using Selective Laser Sintering of Polyetheretherketone-Hydroxyapatite Biocomposite Blends," *Biomaterials*, **24**(18), pp. 3115-3123. http://dx.doi.org/10.1016/S0142-9612(03)00131-5

Wang, Y., Bergstrom, J., and Burman, C., 2006, "Characterization of an Iron-Based Laser Sintered Material," *Journal of Materials Processing Technology*, **172**(1), pp. 77-87. http://dx.doi.org/10.1016/j.jmatprotec.2005.09.004

Xiao, B., and Zhang, Y., 2007, "Laser Sintering of Metal Powders on Top of Sintered Layers under Multiple-Line Laser Scanning," *Journal of Physics D: Applied Physics*, **40**(21), 6725-6734. http://dx.doi.org/10.1088/0022-3727/40/21/036

Xu, X., Prasher, R., and Lofgreen, K., 2007, Ultralow Thermal Conductivity of Nanoparticle Packed Bed," *Applied Physics Letter*, **91**(20), 203113. http://dx.doi.org/10.1063/1.2814959

Zhang, Y., Faghri, A., Buckley, C.W., and Bergman, T.L., 2000, "Three-Dimensional Sintering of Two-Component Metal Powders with Stationary and Moving Laser Beams," ASME *J. Heat Transfer*, **122**(1), pp. 150-158. http://dx.doi.org/10.1115/1.521445

5

THERMODYNAMIC ANALYSIS FOR THE MHD FLOW OF TWO IMMISCIBLE MICROPOLAR FLUIDS BETWEEN TWO PARALLEL PLATES

J. Srinivas* and J. V. Ramana Murthy

Department of Mathematics, NIT Warangal 506004, Telangana State, India.

ABSTRACT

The paper aims the heat transfer analysis for the flow of two immiscible micropolar fluids inside a horizontal channel, by the first and second laws of thermodynamics under the action of an imposed transverse magnetic field. The plates of the channel are maintained at constant temperatures higher than that of the fluid. The flow region consists of two zones, the flow of the heavier fluid taking place in the lower zone. The condition of hyper-stick is taken on the plates and continuity of velocity, micro-rotation, temperature, heat flux, shear stress and couple stress are imposed at the interface. The velocity, micro-rotation and temperature profiles are derived analytically and these are used to compute the dimensionless expressions for the entropy generation number and Bejan number. The results are presented graphically. It is observed that the imposed magnetic field reduces the entropy production rate near the walls.

Keywords: *MHD flow, Immiscible fluids, Micropolar fluids, Entropy generation number, Bejan number.*

1. INTRODUCTION

The optimal design of thermal systems can be achieved by minimizing the entropy generation rate in the systems. This issue has been the topic of great importance in many engineering fields such as heat exchangers, thermal insulation, petroleum and geothermal reservoirs, cooling of nuclear reactors, MHD power generators, cooling of electronic devices etc. Two decades earlier, measuring the entropy generation rate of such complex systems which have different geometries in the presence and absence of a magnetic field in a porous or non-porous channel was restricted to the first law of thermodynamics. In recent days, to estimate the entropy generation rate it is observed that use of the second law of thermodynamics, is more reliable than the calculations based on first law. In almost all thermal systems, based on the second law, the efficiency of the system can be defined in terms of the ratio of actual thermal efficiency to reversible thermal efficiency under the same conditions. In order to enhance the thermal performance of a system, i.e., to reduce the destruction of the available work, the entropy generation rate must be minimized (Bejan, 1980). The entropy generation rate is associated with thermodynamic irreversibility, which is common in all types of heat transfer processes. Different sources like heat transfer across finite temperature gradient, characteristic of convective heat transfer, fluid friction, viscosity effects within the fluid and at the solid boundaries, chemical reactions, body force effect, coupling effects between heat and mass transfer, viscous dissipation effect etc. are all responsible for entropy generation rate. Bejan (1982, 1979) discussed different reasons behind the entropy generation rate in many thermal systems. Generation of entropy destroys available work of a system. Therefore, it makes a good engineering sense to focus on the irreversibility of heat transfer and fluid flow processes and try to understand the mechanism of entropy generation rate. Also he performed an analytical study to show how the entropy generation rate can be calculated in fundamental convective heat transfer problems such as pipe flow, boundary layer flow over a flat plate, single cylinder in cross flow and flow in the entrance region of a flat rectangular duct. He demonstrated how the flow geometric parameters may be selected in order to minimize the irreversibility associated with a specific convective heat transfer.

There are many problems in the fields of hydrology and reservoir mechanics in which systems involving two or more immiscible fluids of different densities/viscosities flowing in same pipe or channel or through porous media are encountered. Typical fluid examples of these systems are: air-water, water-salt water, oil-water, gas-oil, and gas-oil-water systems. These are referred to as multi-phase flows in literature. Blood flow in arteries has been studied by many researchers considering the flow of blood as a two phase flow. Several investigations of multi-phase flows are reported by various researchers (e.g., Bird *et al.*, 1960; Kapur, 1964; Bhattacharya, 1968; Chaturani and Samy, 1985; Rao and Usha, 1979; Bakhtiyarov and Siginer, 1997). The flow and heat transfer in immiscible fluids are of special importance in the petroleum extraction and transport problem. Heat transfer in immiscible flows was discussed by many authors (e.g., Chamkha, 2000; Umavathi *et al.*, 2005; Malashetty *et al.*, 2006; Muthuraj and Srinivas, 2010; Prathap Kumar *et al.*, 2011).

In the last few decades, a lot of study has been developed in order to investigate the effect of a magnetic field on different geometries

† *Corresponding author. Email:j.srinivasnit@gmail.com*

(Chamkha, 1998; Chamkha and Quadri, 2001). Alpher (1961) discussed the heat transfer in magnetohydrodynamic flow between parallel plates. Nikodijevic *et al.* (2011) studied the MHD Couette two-fluid flow and heat transfer in the presence of a uniform inclined magnetic field. It is also to be noted that the analytical study of first and second laws of thermodynamics for the flow within the channel in the presence of a magnetic field is significant in many industrial applications, such as MHD marine propulsion, MHD generators, pumps, accelerators, flow meters, filtration, geothermal systems and has applications in nuclear reactors too. Several works have been reported on the effect of MHD flow on the entropy generation for various flows and geometries (e.g., Damseh *et al.*, 2008; Mahian *et al.*, 2012; Kiyasatfar *et al.*, 2012; Komurgoz *et al.*, 2012). In recent years, the fluid flow and entropy generation in two immiscible fluids in a channel have received considerable attention of researchers. Kamisli and Hakan (2008) considered the fluid flow and entropy generation in two immiscible fluids in a channel. These authors explained very nicely the thermodynamic interface conditions involved in a flow of immiscible fluids and made a significant observation that minimum temperature gradient in the transverse direction of the flow offers minimum entropy generation near the plates.

The theory of a micropolar fluid was initiated by Eringen (1966, 1960) in 1966. This theory accounts for the internal characteristics of the substructure particles with the assumption that they are allowed to undergo a rotation independent of their linear velocity. Micropolar fluids represent fluids consisting of rigid randomly oriented particles suspended in a viscous medium when the deformation of the particles is ignored. The equations of motion characterizing a micropolar fluid flow are nonlinear in nature (as in the case of Newtonian viscous fluids) and are constituted by a coupled system of vector differential equations in velocity $\bar{q}$ and micro-rotation $\bar{\nu}$ (which is independent of $\bar{q}$). To understand the departure from the viscous fluid flow model, several problems that were studied in viscous fluid flow theory, have also been studied in the realm of micropolar fluids. This theory can be used to explain the flow of colloidal fluids, liquid crystals, animal blood etc. An account of the early developments in polar fluid theory can be found in the book by Stokes (1984). Ariman and Cakmak (1968) discussed some basic viscous flows in micropolar fluids. A detailed survey of microcontinuum fluid mechanics with several applications in physiological fluid flows has been presented by Ariman *et al.* (1973). Interesting aspects of the theory and applications of micropolar fluids are dealt in the books by Eringen (1960) and Lukaszewicz (1999).

Thermodynamic analysis in the case of viscous fluids was carried out by many researchers (e.g., Kiyasatfar *et al.*, 2012; Kamisli and Hakan, 2008). But very less attention has been paid in this direction for the micropolar fluid flows. Since micropolar fluids are known to have rotational effects, these fluids may help to increase the exergy. Hence, the present study aims to study a comparatively involved problem dealing with the entropy generation characteristics in a channel of two immiscible micropolar fluid flows under the influence of a transverse magnetic field.

2. MATHEMATICAL FORMULATION

Consider the flow of two immiscible micropolar fluids between two parallel plates extending in the axial direction and distant $2h$ apart. X and Y are the axial and transverse coordinates respectively with the origin at the center of the channel (Fig. 2). The length of the plates is much greater than the distance between them so that the flow at any point in the X-direction is same. Fluid flow is generated due to a constant pressure gradient which acts at the mouth of the channel. A constant transverse magnetic field is applied on the plates. The fluid in the lower zone (viscosity μ_1, micropolarity κ_1, density ρ_1 and thermal conductivity k_1) occupies the region (i.e. $-h \leq Y \leq 0$) comprising the lower half of the channel and this region is named zone I. The fluid in the upper zone (viscosity μ_2, micropolarity κ_2, density $\rho_2 (< \rho_1)$ and thermal conductivity k_2) is assumed to occupy the upper half of the channel (i.e. $0 \leq Y \leq h$), and this region is called zone II. The equations for the flow and energy in zone I and II (i.e. $-h \leq Y \leq h$) are assumed to be governed by micropolar fluid flow equations as given by Eringen (1966).

$$\frac{\partial \rho}{\partial t} + \nabla \cdot (\rho \vec{\mathbf{q}}) = 0 \tag{1}$$

$$\rho \frac{d\vec{\mathbf{q}}}{dt} = \rho \vec{\mathbf{f}} - \nabla P + \kappa \nabla \times \vec{\nu} - (\mu + \kappa) \nabla \times \nabla \times \vec{q} + (\lambda + 2\mu + \kappa) \nabla (\nabla \cdot \vec{q}) + \vec{J} \times \vec{H} \tag{2}$$

$$\rho j \frac{d\vec{\nu}}{dt} = \rho \vec{\ell} - 2\kappa \vec{\nu} + \kappa \nabla \times \vec{q} - \gamma \nabla \times \nabla \times \vec{\nu} + (\alpha + \beta + \gamma) \nabla (\nabla \cdot \vec{\nu}) \tag{3}$$

$$\rho \frac{dE}{dt} = -P(\nabla \cdot \vec{q}) + \rho \Phi - (\nabla \cdot \vec{\mathbf{h}}) + \vec{J}^2 \tag{4}$$

where

$$\rho \Phi = \lambda (\nabla \cdot \vec{q})^2 + 2\mu (D : D) + 4\kappa \left(\frac{1}{2} \nabla \times \vec{q} - \vec{\nu} \right)^2 + \alpha (\nabla \cdot \vec{\nu})^2 + \gamma (\nabla \vec{\nu} : \nabla \vec{\nu}) + \beta \left(\nabla \vec{\nu} : (\nabla \vec{\nu})^T \right)$$

The equations (1)-(4) represent conservation of mass, balance of linear momentum, micro-rotation and energy equation respectively. The scalar quantities ρ and j are respectively the density and gyration coefficient and are assumed to be constants. The vectors $\vec{q}$, $\vec{\nu}$, $\vec{f}$ and $\vec{\ell}$ are the velocity, micro-rotation, body force per unit mass and body couple per unit mass respectively. P is the fluid pressure at any point. $\vec{J}$ is the current density and $\vec{H}$ is the total magnetic field ($\vec{H} = \vec{H}_o + \vec{h}_o$) where $\vec{H}_o$ is the applied magnetic field and h_o is the induced magnetic field ($h_o << H_o$). The material constants (λ, μ, κ) are viscosity coefficients and (α, β, γ) are gyro-viscosity coefficients. In the energy equation, Φ is the dissipation function of mechanical energy per unit mass, D denotes the deformation tensor, E is the specific internal energy and $\vec{\mathbf{h}}$ is the heat flux, where k is the thermal conductivity.

The current density $\vec{J}$, magnetic field $\vec{H}$ and electric field $\vec{E}$ are related by Maxwell's equations $\nabla \times \vec{E} = 0$, $\nabla \cdot \vec{H} = 0$, $\nabla \times \vec{H} = \mu^1 \vec{J}$, $\nabla \cdot \vec{J} = 0$, $\vec{J} = \sigma(\vec{E} + \vec{q} \times \vec{H})$ where μ^1 is the magnetic permeability and σ is the electric conductivity.

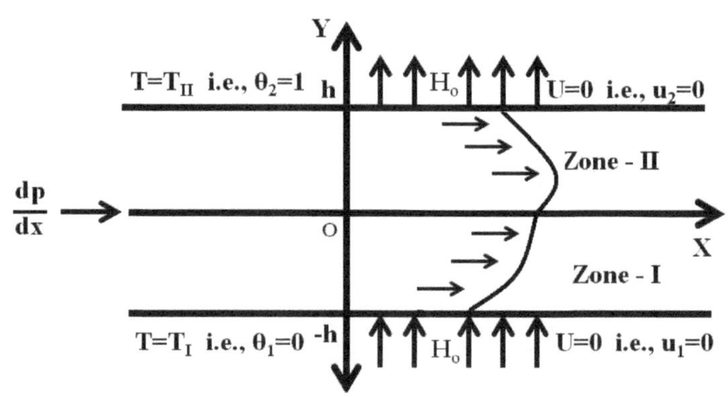

Fig. 1 Geometry of the problem

To develop the governing equations for the problem considered, the following assumptions are made:

1. The flow is assumed to be one-dimensional, steady, laminar, and incompressible.

2. The gravity effect is negligible and the Lorentz force ($\vec{J} \times \vec{H}$) is the only body force acting on the fluid.

3. The magnetic Reynolds number is assumed to be small, so that the induced magnetic field is neglected and the Hall-effect of magnetohydrodynamics is assumed to be negligible.

We assume that the velocity of the fluid is $\vec{q} = (U(Y), 0, 0)$ and microrotation vector as $\vec{\nu} = (0, 0, C(Y))$. The flow is subjected to a uniform applied magnetic field H_o in the positive y-direction.i.e., $\vec{H} = H_o\vec{j}$ which gives $\vec{J} \times \vec{H} = -\sigma H_o^2 \vec{q}$.

Introducing the non-dimensional variables: $x = \frac{X}{h}, y = \frac{Y}{h}, u = \frac{U}{U_o}$, $p = \frac{P}{\rho_1 U_o^2}$, $C = \frac{CU_o}{h}$ where U_o is the maximum velocity of the fluid in the channel, we get the following sets of non-dimensional form of governing equations (neglecting body couples) and boundary conditions corresponding to the flow in the two zones.

3. GOVERNING EQUATIONS

Zone I: $(-1 \leq y \leq 0)$
The governing equations in zone I are:

$$\frac{d^2u_1}{dy^2} + c_1\frac{dC_1}{dy} - Ha^2(1-c_1)u_1 = (1-c_1)Re\frac{dp}{dx} \tag{5}$$

$$\frac{d^2C_1}{dy^2} - s_1\frac{du_1}{dy} - 2s_1C_1 = 0 \tag{6}$$

Zone II: $(0 \leq y \leq 1)$
The governing equations in zone II are:

$$\frac{d^2u_2}{dy^2} + c_2\frac{dC_2}{dy} - \frac{n_\sigma}{n_\mu}Ha^2(1-c_2)u_2 = \frac{1}{n_\mu}(1-c_2)Re\frac{dp}{dx} \tag{7}$$

$$\frac{d^2C_2}{dy^2} - s_2\frac{du_2}{dy} - 2s_2C_2 = 0 \tag{8}$$

where $Re = \frac{\rho_1 U_o h}{\mu_1}$ is the Reynolds number, $Ha^2 = \frac{\sigma_1 H_o^2 h^2}{\mu_1}$ is the Hartmann number, $n_\mu = \frac{\mu_2}{\mu_1}$ is the viscosity ratio, $n_\sigma = \frac{\sigma_2}{\sigma_1}$ is the electric conductivity ratio, $c_i = \frac{\kappa_i}{\mu_i + \kappa_i}$ is the micropolarity parameter and $s_i = \frac{\kappa_i h^2}{\gamma_i}$ is the couple stress parameter (i=1,2).

4. BOUNDARY AND INTERFACE CONDITIONS

A characteristic feature of the two-layer flow problem is the coupling across liquid-liquid interfaces. The liquid layers are mechanically coupled via transfer of momentum across the interfaces. Transfer of momentum results from the continuity of tangential velocity and a stress balance across the interface.

To determine the velocity and micro-rotation components $u_1(y)$, $C_1(y)$, $u_2(y)$ and $C_2(y)$ in the zones I and II described above, we adopt the following boundary and interface conditions:

Zone I is constituted by the fixed lower plate given by $y = -1$ and a fluid-fluid interface defined by $y = 0$. Zone II is constituted by the fluid interface given by $y = 0$ and the fixed upper plate given by $y = 1$.

In view of the no-slip condition (Sandeep and Deshpande (2003)) on the static boundaries, we have to prescribe velocity as:

$$u_1(y) = 0 \text{ on } y = -1 \text{ and } u_2(y) = 0 \text{ on } y = 1 \tag{9}$$

which represent the no-slip condition.

The microrotation vector on the boundary = angular velocity of the fluid on the boundary. i.e., $C_{wall} = \frac{1}{2}(\nabla \times \vec{q}_{wall})$. A more general condition is taken as $C_{wall} = \frac{n}{2}(\nabla \times \vec{q}_{wall})$ where $0 \leq n \leq 1$ (Refer Lukaszewicz (1999), page 31). This value of n indicates the concentration of micropolarity or interaction of fluid particles with the boundary. The case $n = 0$ indicates $C = 0$ at the plates. It represents the flow of concentrated particles in which the microelements close to the wall surface are unable to rotate (Jena and Mathur, 1981). While there has been theoretical justification for no-slip condition for the velocity on the boundary, the boundary condition for the micro-rotation vector $\vec{\nu}$ needs to be established in a rigorous form. The hyper-stick condition on the boundary as given in Lukaszewicz (1999) is the one adopted here following the studies made by Rees and Bossom (1996); Bhattacharyya *et al.* (2012) and several workers in micropolar fluid flow and hence we take that the microrotation vanishes on the static boundaries.

$$C_1(y) = 0 \text{ on } y = -1 \text{ and } C_2(y) = 0 \text{ on } y = 1 \tag{10}$$

At the fluid-fluid interface $y = 0$, we assume that the velocity, microrotation, shear stress and couple stress components are continuous. This implies

$$u_{1(0-)} = u_{2(0+)},\ C_{1(0-)} = C_{2(0+)},\ \tau_{1xy}|_{(0-)} = \tau_{2xy}|_{(0+)} \text{ and } M_{1xy}|_{(0-)} = n_\beta\, M_{2xy}|_{(0+)} \tag{11}$$

The last two conditions of (11) give us

$$\tau_{1xy}|_{(0-)} = \tau_{2xy}|_{(0+)} \Rightarrow \left[\frac{\partial u_1}{\partial y}\Big|_{(0-)} + 2\,c_1\,C_1\Big|_{(0-)}\right] = n_\mu\left(\frac{1-c_1}{1-c_2}\right)\left[\frac{\partial u_2}{\partial y}\Big|_{(0+)} + 2\,c_2\,C_2\Big|_{(0+)}\right],$$

$$M_{1xy}|_{(0-)} = M_{2xy}|_{(0+)} \Rightarrow \frac{\partial C_1}{\partial y}\Big|_{(0-)} = n_\beta\frac{\partial C_2}{\partial y}\Big|_{(0-)}. \tag{12}$$

At the interface as there is no mass transfer, it is justified to assume that there is no velocity in normal direction and that the tangential velocity and microrotation are continuous. Also interface is assumed to be a smooth surface representing force balance and hence shear stress and couple stress are prescribed to be continuous.

5. SOLUTION OF THE PROBLEM

5.1. Velocity and Microrotation distributions

Solution of momentum equations subject to the conditions through (9)-(11) are obtained as:
Zone I: $(-1 \leq y \leq 0)$
Eliminating C_1 from (5), (6) we have

$$\frac{d^4u_1}{dy^4} - (Ha^2(1-c_1) + s_1(2-c_1))\frac{d^2u_1}{dy^2} + 2s_1(1-c_1)u_1 = -2s_1(1-c_1)ReB \tag{13}$$

substituting $\frac{dC_1}{dy}$ from (5) in (6) we have

$$C_1(y) = -\frac{1}{2c_1s_1}\left[(Ha^2(1-c_1) - c_1s_1)\frac{du_1}{dy} - \frac{d^3u_1}{dy^3}\right] \tag{14}$$

solving (13) we get,

$$u_1(y) = c_{11}\cosh\alpha_1 y + c_{12}\sinh\alpha_1 y + c_{13}\cosh\alpha_2 y + c_{14}\sinh\alpha_2 y + \frac{1}{Ha^2}ReB \tag{15}$$

and substituting $u_1(y)$ in (14) we have

$$C_1(y) = \frac{1}{2c_1s_1}[\alpha_1(Ha^2(1-c_1) - c_1s_1 - \alpha_1^2)(c_{11}\sinh\alpha_1 y + c_{12}\cosh\alpha_1 y) + \alpha_2(Ha^2(1-c_1) - c_1s_1 - \alpha_2^2)(c_{13}\sinh\alpha_2 y + c_{14}\cosh\alpha_2 y)] \tag{16}$$

Zone II: $(0 \le y \le 1)$
Eliminating $\mathcal{C}_2$ from (7), (8) we have

$$\frac{d^4u_2}{dy^4} - (Ha^2\frac{n_\sigma}{n_\mu}(1-c_2) + s_2(2-c_2))\frac{d^2u_2}{dy^2} + 2s_2(1-c_2)u_2 = -2s_2(1-c_2)\frac{1}{n_\mu}ReB \quad (17)$$

substituting $\frac{d\mathcal{C}_2}{dy}$ from (7) in (8) we have

$$\mathcal{C}_2(y) = -\frac{1}{2c_2s_2}\left[(Ha^2\frac{n_\sigma}{n_\mu}(1-c_2) - c_2s_2))\frac{du_2}{dy} - \frac{d^3u_2}{dy^3}\right] \quad (18)$$

solving (17) we get,

$$u_2(y) = c_{21}\cosh\alpha_3 y + c_{22}\sinh\alpha_3 y + c_{23}\cosh\alpha_4 y + c_{24}\sinh\alpha_4 y - \frac{1}{n_\mu}\frac{1}{Ha^2}ReB \quad (19)$$

and substituting $u_2(y)$ in (18), we have

$$\mathcal{C}_2(y) = \frac{1}{2c_2s_2}\Big[\alpha_3(Ha^2(1-c_2)\frac{n_\sigma}{n_\mu} - c_2s_2 - \alpha_3^2)(c_{21}\sinh\alpha_3 y + c_{22}\cosh\alpha_3 y) + \alpha_4(Ha^2(1-c_2)\frac{n_\sigma}{n_\mu} - c_2s_2 - \alpha_4^2)(c_{23}\sinh\alpha_4 y + c_{24}\cosh\alpha_4 y)\Big] \quad (20)$$

where α_1^2, α_2^2 are the roots of $x^2 - a_1x + a_2 = 0$ and α_3^2, α_4^2 are the roots of $x^2 - b_1x + b_2 = 0$ and $a_1 = \alpha_1^2 + \alpha_2^2 = Ha^2(1-c_1) + s_1(2-c_1)$, $a_2 = \alpha_1^2\alpha_2^2 = 4s_1(1-c_1)Ha^2$, $b_1 = \alpha_3^2 + \alpha_4^2 = \frac{n_\sigma}{n_\mu}Ha^2(1-c_2) + s_2(2-c_2)$, $b_2 = \alpha_3^2\alpha_4^2 = 4\frac{n_\sigma}{n_\mu}s_2(1-c_2)Ha^2$

The solutions $u_1(y), \mathcal{C}_1(y)$ and $u_2(y), \mathcal{C}_2(y)$ involve eight constants $c_{11}, c_{12}, c_{13}, c_{14}, c_{21}, c_{22}, c_{23}$ and c_{24}. These constants are found from the boundary conditions given in (9)-(11) and these are obtained using MATHEMATICA. As the expressions are cumbersome, they are not presented here.

5.2. Temperature distributions

In the present problem, it is assumed that the two walls are maintained at constant and different temperatures T_I and T_{II} $(T_I < T_{II})$.
The governing equation for the temperature T_1 of the conducting fluid in zone I is then given by

$$k_1\frac{d^2T_1}{dY^2} = -\left[\mu_1\left(\frac{dU_1}{dY}\right)^2 + \kappa_1\left(\frac{dU_1}{dY} + 2C_1\right)^2 + \beta_1\left(\frac{dC_1}{dY}\right)^2 + \sigma_1 H_o^2 U_1^2\right] \quad (21)$$

The governing equation for the temperature T_2 of the conducting fluid in zone II is then given by

$$k_2\frac{d^2T_2}{dY^2} = -\left[\mu_2\left(\frac{dU_2}{dY}\right)^2 + \kappa_2\left(\frac{dU_2}{dY} + 2C_2\right)^2 + \beta_2\left(\frac{dC_2}{dY}\right)^2 + \sigma_2 H_o^2 U_2^2\right] \quad (22)$$

In order to non-dimensionalize the above equations, the following transformation is used in addition to those already introduced in above: $\theta = \dfrac{T - T_I}{T_{II} - T_I}$.

The Equations (21) and (22) are then reduced to the following form:

$$\frac{d^2\theta_1}{dy^2} = -Br\left[\left(\frac{du_1}{dy}\right)^2 + \left(\frac{c_1}{1-c_1}\right)\left(\frac{du_1}{dy} + 2\mathcal{C}_1\right)^2 + \delta_1\left(\frac{d\mathcal{C}_1}{dy}\right)^2 + Ha^2u_1^2\right] \quad (23)$$

$$\frac{d^2\theta_2}{dy^2} = -\frac{Br}{n_k}\left[n_\mu\left(\left(\frac{du_2}{dy}\right)^2 + \left(\frac{c_2}{1-c_2}\right)\left(\frac{du_2}{dy} + 2\mathcal{C}_2\right)^2\right) + n_\beta\delta_1\left(\frac{d\mathcal{C}_2}{dy}\right)^2 + n_\sigma Ha^2u_2^2\right] \quad (24)$$

where $Br = Ek\ Pr$ is the Brinkman number, $Ek = \dfrac{U_o^2}{c_{p_1}(T_{II} - T_I)}$ is the Eckert number, $Pr = \dfrac{\mu_1 c_{p_1}}{k_1}$ is the Prandtl number, $n_k = \dfrac{k_2}{k_1}$ is the thermal conductivity ratio, $n_\beta = \dfrac{\beta_2}{\beta_1}$ is the couple stress co-efficient ratio and $\delta_1 = \dfrac{\beta_1}{\mu_1 h^2}$ is the couple stress parameter.
In the non-dimensional form, the boundary conditions for temperature are as below:
(i) at the lower and upper plate boundaries the temperatures are respectively,

$$\theta_1 = 0 \text{ at } y = -1 \text{ and } \theta_2 = 1 \text{ at } y = 1 \quad (25)$$

(ii) at the fluid interface temperature (θ) and heat flux $(\vec{\mathbf{h}})$ are continuous:

$$\theta_1 = \theta_2 \text{ and } \frac{d\theta_1}{dy} = n_k\frac{d\theta_2}{dy} \text{ at } y = 0 \quad (26)$$

The solutions of Eqns. (23) and (24) with boundary and interface conditions are solved analytically and they are lengthy and are not shown here. The solution involves four constants and they are found from the four boundary conditions (Eqns. (25) and (26)) and are obtained using MATHEMATICA.

5.3. Nusselt number

The heat transfer coefficient at the walls is given by Fourier's law of heat conduction $\vec{\mathbf{h}} = -k\nabla T$. In non-dimensional form this represents Nusselt number $Nu = -\frac{d\theta}{dy}\Big]_{y=\pm 1}$. The Nusselt number Nu is the ratio of convective to conductive heat transfer across (normal to) the wall.

6. ENTROPY GENERATION ANALYSIS

6.1. The volumetric entropy generation

If we take an infinitesimal fluid element in each zone and assume that the element as an open thermodynamic system subjected to mass fluxes, energy transfer and entropy transfer interactions through a fixed control surface, the volumetric rate of entropy generation for incompressible micropolar fluid is given as

$$(S_i)_G = \underbrace{\frac{k_i}{T_o^2}\left(\frac{\partial T_i}{\partial Y}\right)^2}_{\ge 0} + \underbrace{\frac{\mu_i}{T_o}\left(\frac{\partial U_i}{\partial Y}\right)^2}_{\ge 0} + \underbrace{\frac{\kappa_i}{T_o}\left(\frac{\partial U_i}{\partial Y} + 2C_i(Y)\right)^2}_{\ge 0} + \underbrace{\frac{\beta_i}{T_o}\left(\frac{\partial C_i}{\partial Y}\right)^2}_{\ge 0} + \underbrace{\frac{1}{T_o}\sigma_i H_o^2 U_i^2}_{\ge 0} \quad (27)$$

where the value of i can be either 1 or 2 that represents zone I or zone II, respectively. Equation (27) clearly shows contributions of three sources of entropy generation rate. The first term on the right-hand side of Equation (27) is the entropy production due to heat transfer across a finite temperature difference, the second, third, and fourth terms are due to viscous dissipation and the fifth term is due to the effect of the magnetic field.

6.2. The characteristic entropy generation rate

The characteristic entropy generation rate $S_{G,C}$ is defined as,

$$S_{G,C} = \left[\frac{(\vec{\mathbf{h}}_1)^2}{k_1 T_o^2}\right] = \left[\frac{k_1(\Delta T)^2}{h^2 T_o^2}\right] \quad (28)$$

In the above equation, $\vec{\mathbf{h}}_1$ is the heat flux, T_o is the average, characteristic, absolute reference temperature of the medium, $\Delta T = T_{II} - T_I$ and h is the half of transverse distance of the channel.

6.3. The entropy generation number

According to Bejan (1996), the dimensionless form of entropy generation is the entropy generation number Ns and which is, by definition, equal to the ratio of actual generation rate to a characteristic entropy transfer rate. The entropy generation number for each fluid is given by

$$Ns_1 = \left(\frac{d\theta_1}{dy}\right)^2 + \left(\frac{Br}{\Omega}\right)\left[\left(\frac{du_1}{dy}\right)^2 + \left(\frac{c_1}{1-c_1}\right)\left(\frac{du_1}{dy} + 2C_1\right)^2 + \delta_1\left(\frac{dC_1}{dy}\right)^2 + Ha^2u_1^2\right] \quad (29)$$

$$Ns_2 = n_k\left(\frac{d\theta_2}{dy}\right)^2 + \left(\frac{Br}{\Omega}\right)\left[n_\mu\left(\left(\frac{du_2}{dy}\right)^2 + \left(\frac{c_2}{1-c_2}\right)\left(\frac{du_2}{dy} + 2C_2\right)^2\right) + \delta_1 n_\beta\left(\frac{dC_2}{dy}\right)^2 + \frac{n_k n_\sigma}{n_\mu}Ha^2u_2^2\right] \quad (30)$$

where $Br = \left(\frac{\mu_1 U_o^2}{k_1 \Delta T}\right)$ is the Brinkman number, which determines importance of viscous dissipation because of the fluid frictions relative to the conduction heat flow resulting from the impressed temperature difference and $\Omega = \left(\frac{\Delta T}{T_o}\right)$ is the temperature difference number. This number is a dimensionless parameter always present in the second law treatment of heat transfer processes. In most applications of engineering interest, we find $\Omega \leq 1$ (Bejan, 1979).

6.4. The viscous dissipation parameter

The viscous dissipation parameter is an important dimensionless number for the irreversibility analysis. It determines the relative importance of the viscous effects for the entropy generation and it is equal to the ratio of Brinkman number to the dimensionless temperature difference i.e., (Br/Ω). It is desirable to consider the Ek and Pr in a group called the Brinkman number ($Br = Ek.Pr$) for evaluating the relative importance of the energy due to viscous dissipation to the energy due to heat conduction. It was reported that Br is much less than unity for many engineering processes (Bejan, 1979).

6.5. The Bejan number

The domination of the irreversibility mechanisms is physically important. The entropy generation originates from both heat transfer and viscous friction, these effects should be separately investigated to see the dominance of one over the other. The entropy generation number is unable to to give any idea about the relative importance of friction and heat transfer effects. An alternate parameter, called Bejan number (Be), is introduced for this purpose and it is achieving an increasing popularity among researchers studying the second law. This parameter was defined by Paoletti *et al.* (1989) as ratio of irreversibility due to heat transfer to the total irreversibility (sum of heat transfer irreversibility, fluid friction irreversibility and Joule dissipation irreversibility) and it is given by

$$Be = \frac{Ny_i}{Ns_i} = \frac{Ny_i}{Ny_i + Nf_i + Nm_i} \quad (31)$$

This is employed to understand the entropy generation mechanisms as proposed by Bejan (1996, 1979). The value of $Be \to 1$ indicates that the heat transfer irreversibility dominates. On the other hand, $Be \to 0$ indicates that the irreversibility due to fluid friction and magnetic field dominates over the irreversibility due to the heat transfer. It is obvious that $Be = 0.5$ is the case in which the irreversibility due to heat transfer is equal to sum of fluid friction and magnetic field contributions in the entropy generation.

7. RESULTS AND DISCUSSION

The closed form solutions for the flow of two immiscible micropolar fluids are obtained and reported in the previous section. These solutions are evaluated numerically and depicted graphically. The variations of velocity, micro-rotation, temperature, entropy generation number and Bejan number for different values of parameters are shown through figures.

7.1. Flow field

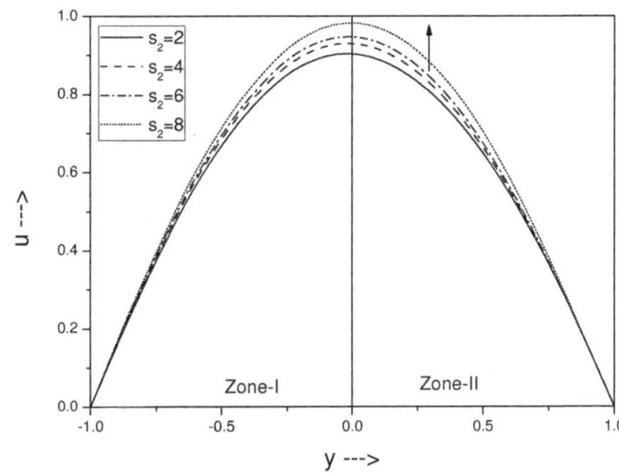

Fig. 2 Effect of s_2 on u for $B = -1.5$, $c_1 = 0.6$, $c_2 = 0.7$, $Ha = 1$, $n_\beta = 0.9$, $n_\mu = 0.9$, $n_\sigma = 0.9$, $Re = 2$, $s_1 = 5$.

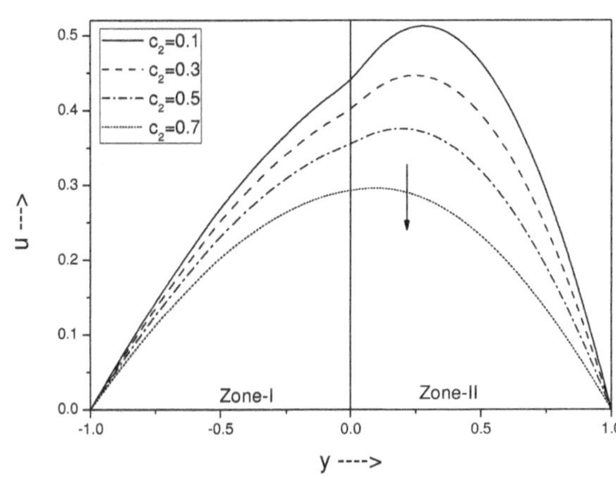

Fig. 3 Effect of c_2 on u for $B = -1.5$, $c_1 = 0.8$, $Ha = 0.5$, $n_\beta = 0.9$, $n_\mu = 0.9$, $n_\sigma = 0.5$, $Re = 1.5$, $s_1 = 2$, $s_2 = 3$.

The effect of the couple stress parameter s_2 on the velocity field is shown in Fig. 2. It is seen that as s_2 increases, the velocity increases in both the zones of the channel. As $s_2 \to \infty$, we get the case of Newtonian (viscous) fluids. The effect of the micropolarity parameter $c_2 (0 \leq c_2 < 1)$ on the velocity field is shown in Fig. 3. It is seen that as c_2 increases, the velocity decreases in both zones of the channel. The velocity in case of micropolar fluid is less than that of the viscous fluid ($c_2 \to 0$ corresponds to viscous fluid).

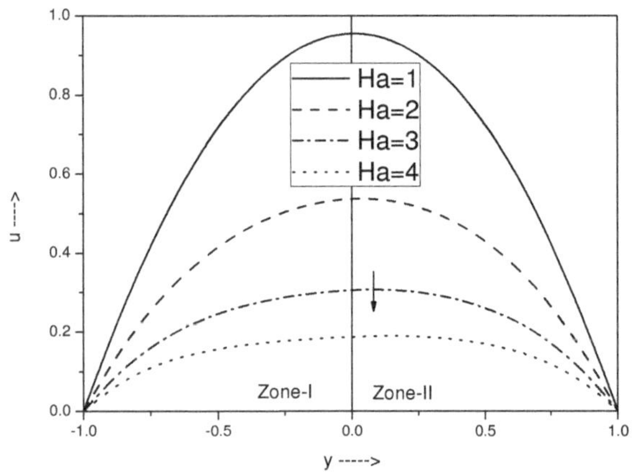

Fig. 4 Effect of Ha on u for $B = -3$, $c_1 = 0.9$, $c_2 = 0.9$, $Re = 4$, $n_\beta = 0.8$, $n_\mu = 0.9$, $n_\sigma = 0.8$, $s_1 = 8$, $s_2 = 8$.

Fig. 4 shows the effect of the Hartmann number Ha (magnetic parameter) on the velocity field. As we expected, the velocity profiles are symmetrical about the centerline of the channel. It is clear that, increasing the value of Ha have a tendency to slow down the fluid motion. This is because of the presence of the transverse magnetic field, which creates a resistive force similar to the drag force that acts in the opposite direction of the fluid motion, thus causing the velocity of the fluid to decrease. This shows that the imposed magnetic field has a retarding influence on the flow field.

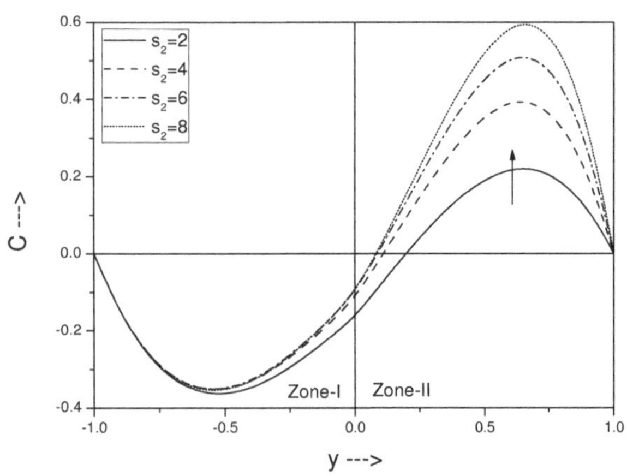

Fig. 5 Effect of s_2 on $\mathcal{C}$ for $B = -2$, $c_1 = 0.9$, $c_2 = 0.8$, $Ha = 0.5$, $n_\beta = 0.8$, $n_\mu = 0.8$, $n_\sigma = 0.8$, $Re = 6$, $s_1 = 5$.

The effect of the couple stress parameter s_2 is found to decrease the micro-rotational velocity drastically as seen in zone II and slightly in zone I as shown in Fig. 5. From Fig. 6, we observe that the micro-rotation changes sign from increasing nature to decreasing character in the zone II and in zone I the rotational velocity is increasing. The micro-rotation tends to zero as is expected that in the limit $\kappa \to 0$ i.e., $c_2 \to 0$, the eqns. (1) and (2) are uncoupled with (3) and they reduce to viscous fluid flow equations. This may be due to increase in rotation of particles in zone II.

Fig. 7 shows the effect of the Hartmann number Ha on the micro-rotation field. As the Hartmann number Ha increases, micro-rotation changes its increasing nature to decreasing nature in zone II and as Ha increases, micro-rotation decreases numerically in the channel.

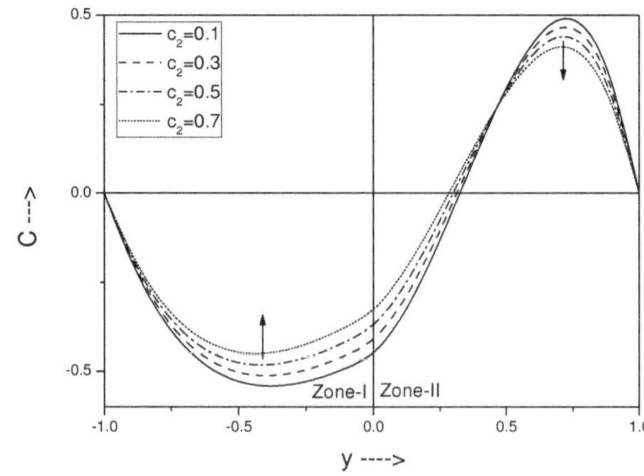

Fig. 6 Effect of c_2 on $\mathcal{C}$ for $B = -2$, $c_1 = 0.9$, $Ha = 1$, $n_\beta = 0.6$, $n_\mu = 0.6$, $n_\sigma = 0.8$, $Re = 6$, $s_1 = 4$, $s_2 = 4$.

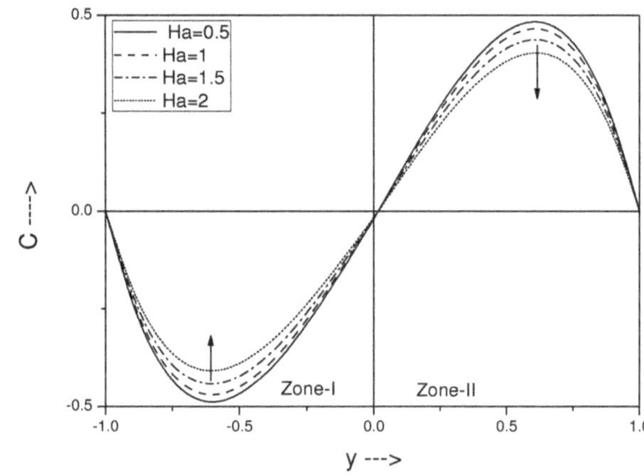

Fig. 7 Effect of Ha on $\mathcal{C}$ for $B = -4$, $c_1 = 0.9$, $c_2 = 0.9$, $n_\beta = 0.9$, $n_\mu = 0.9$, $n_\sigma = 0.9$, $Re = 6$, $s_1 = 5$, $s_2 = 5$.

7.2. Thermal field and Heat transfer

The effect of the couple stress parameter s_2 on temperature field is shown in Fig. 8. It is seen that as s_2 increases, the temperature increases in both zones of the channel. This may be due to the fact that velocity increases and hence dissipation increases (see fig 2). Fig. 9 depicts the effect of the micropolarity parameter c_2 on temperature field. It is observed that the temperature decreases with the increase of parameter c_2. Since velocity is decreasing, dissipation of energy due to velocity decreases and hence temperature decreases.

Fig. 10 displays the effect of the Hartmann number Ha on the temperature field. As the Hartmann number Ha increases, temperature decreases in both the regions of the channel. As Ha increases due to increasing magnetic field intensity, the fluid temperature decreases within the channel. This behavior is attributed to decrease the fluid velocity due to the magnetic field as shown in Fig. 4. Fig. 11 displays the effect of Brinkman number Br on temperature. It is observed that the temperature of both the fluids increases with increase in Brinkman number Br. This may be due to viscous dissipation.

Fig. 12 shows that as Reynolds number Re increases, Nusselt number Nu increases.

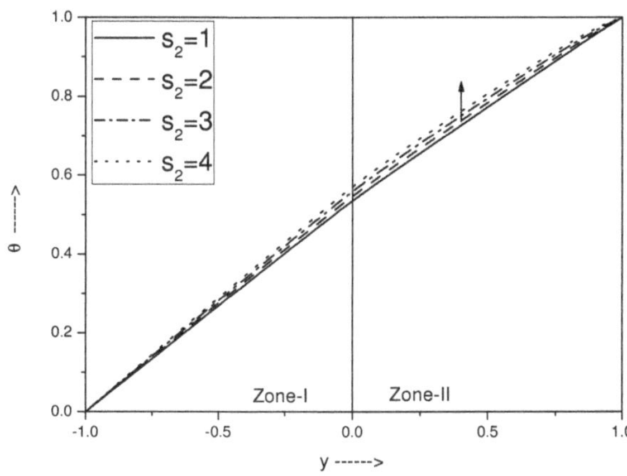

Fig. 8 Effect of s_2 on θ for $B = -1.5$, $Br = 0.5$, $c_1 = 0.7$, $c_2 = 0.7$, $\delta_1 = 0.7$, $Ha = 0.5$, $n_\beta = 0.9$, $n_k = 1.1$, $n_\mu = 0.6$, $n_\sigma = 0.8$, $Re = 2$, $s_1 = 6$.

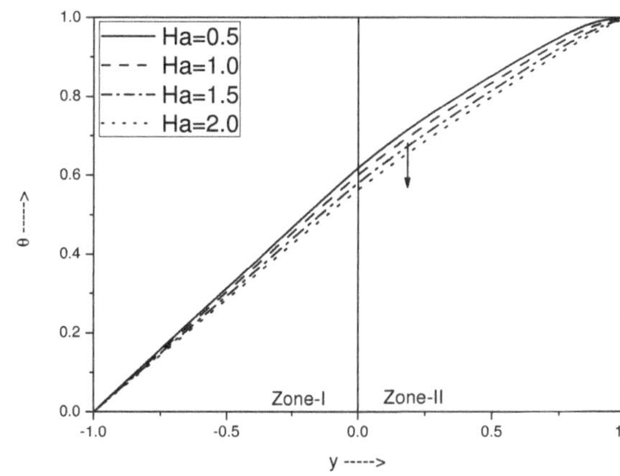

Fig. 10 Effect of Ha on θ for $B = -1.5$, $Br = 0.5$, $c_1 = 0.9$, $c_2 = 0.9$, $\delta_1 = 0.7$, $n_\beta = 0.9$, $n_k = 1.1$, $n_\mu = 0.6$, $n_\sigma = 0.8$, $Re = 2.5$, $s_1 = 8$, $s_2 = 8$.

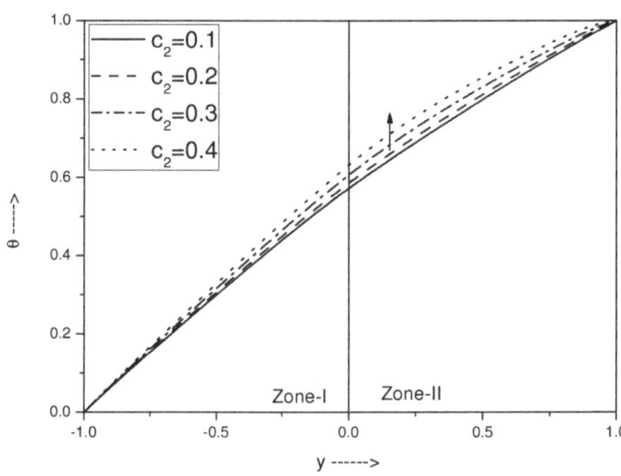

Fig. 9 Effect of c_2 on θ for $B = -1$, $Br = 0.5$, $c_1 = 0.9$, $\delta_1 = 0.7$, $Ha = 0.5$, $n_\beta = 0.9$, $n_k = 1.1$, $n_\mu = 0.5$, $n_\sigma = 0.8$, $Re = 0.5$, $s_1 = 6$, $s_2 = 6$.

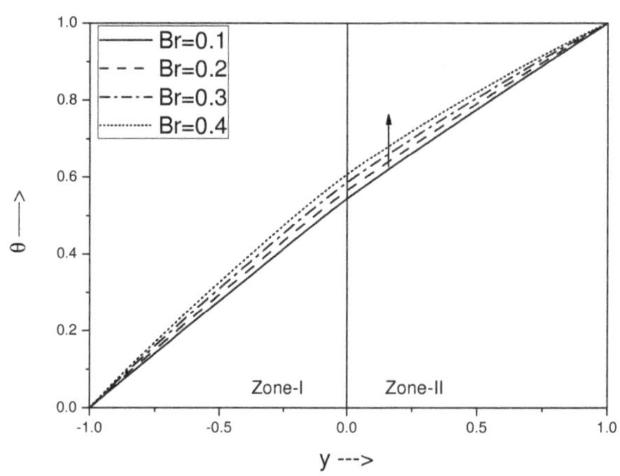

Fig. 11 Effect of Br on θ for $B = -1.5$, $c_1 = 0.9$, $c_2 = 0.9$, $\delta_1 = 0.7$, $Ha = 3$, $n_\beta = 0.9$, $n_k = 1.1$, $n_\mu = 0.6$, $n_\rho = 0.8$, $n_\sigma = 0.8$, $Re = 1.5$, $s_1 = 8$, $s_2 = 8$.

7.3. Entropy generation and Heat transfer irreversibility

The transverse variation in heat transfer irreversibility in terms of entropy generation number Ns is shown in Figs. 13 - 16 for different values of the parameters. Fig. 13 shows the effect of couple stress parameter s_2 on entropy generation number Ns. As s_2 increases entropy generation increases drastically near the plates in zone II. Fig. 14 shows the effect of micropolarity parameter c_2 on entropy generation number Ns. As c_2 increases entropy generation decreases in both the zones of the channel. The values of Ns when c_2 is near to zero are more than the values of Ns when c_2 is tending to 1. Since the case $c_2 \to 0$ represents viscous fluid case, the entropy generation rate in this special case is more than that of the micropolar fluid case.

Fig. 15 shows the variation of entropy generation number Ns as a function of y at different Hartmann numbers Ha. As Ha increases, Ns decreases drastically near the plates and has negligibly small effect at the interface. As Hartmann number increases, the curve for an entropy generation number becomes flat for a considerable distance in the middle of the channel.

The entropy generation number Ns is plotted in Fig. 16 as a function of transverse direction y for different values of the viscous dissipation parameter (Br/Ω). This parameter is an important dimensionless number in the entropy generation analysis. It determines the relative importance of viscous effects to temperature gradient entropy generation. The entropy generation near the plates increases more rapidly in the zone I than in the zone II. The more the viscosity of the fluid is, the more the entropy generation. It is observed that since Ns takes the least value at the interface, the fluid friction irreversibility dominates at the interface of the channel. Again the entropy generation near the plates increases with increasing the viscous dissipation parameter (Br/Ω). For $(Br/\Omega) = 0$, fluid friction irreversibility has no contribution to an entropy generator which falls to zero in the entire channel.

The transverse variation in heat transfer irreversibility in terms of Bejan number Be is shown in Figs. 17 - 20 for different values of the parameters. Fig. 17 shows the effect of couple stress parameter s_2 on Bejan number Be. As s_2 increases, Bejan number decreases. At the interface the variation in Be is rapid. Fig. 18 shows the effect of micropolarity parameter c_2 on Bejan number Be. As c_2 increases, Bejan number increases from 0.992 to 1 within a small range. Which implies c_2 almost

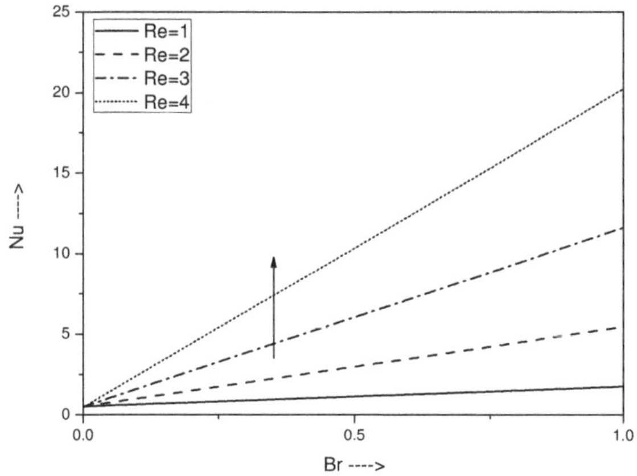

Fig. 12 Effect of Re on Nusselt number Nu as a function of Br for $B = -2$, $c_1 = 0.2$, $c_2 = 0.3$, $Ha = 0.5$, $n_\mu = 0.6$, $\delta_1 = 0.7$, $n_\beta = 0.9$, $n_k = 1.1$, $s_1 = 6$, $s_2 = 6$.

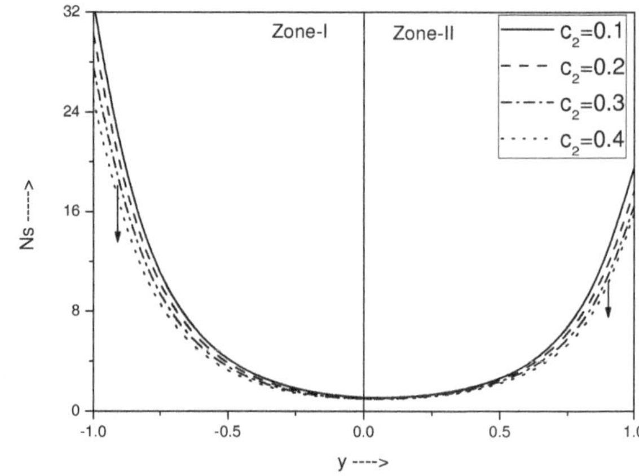

Fig. 14 Effect of c_2 on Ns for $B = -1$, $Br = 0.3$, $c_1 = 0.3$, $\delta_1 = 0.9$, $Ha = 0.3$, $Re = 2$, $n_\beta = 0.9$, $n_k = 1.1$, $n_\mu = 0.9$, $n_\sigma = 0.8$, $s_1 = 6$, $s_2 = 6$, $\Omega = 1$.

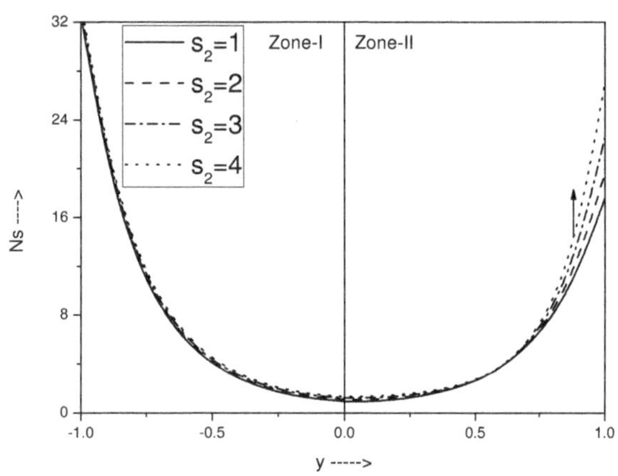

Fig. 13 Effect of s_2 on Ns for $B = -1$, $Br = 0.1$, $c_1 = 0.3$, $c_2 = 0.4$, $\delta_1 = 0.9$, $Ha = 0.1$, $Re = 1$, $n_\beta = 0.9$, $n_k = 1.1$, $n_\mu = 0.9$, $n_\sigma = 0.8$, $s_1 = 8$, $\Omega = 1$.

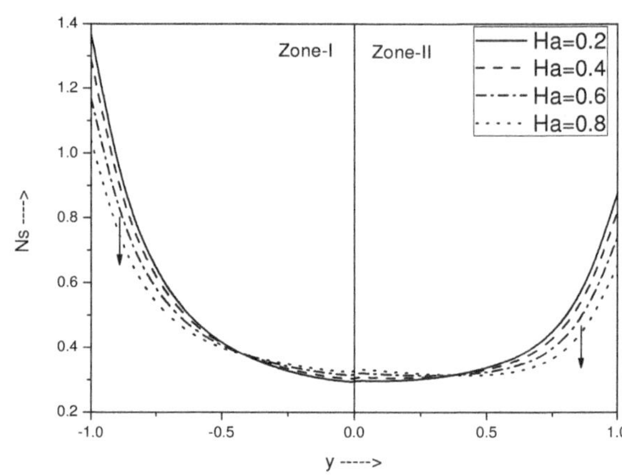

Fig. 15 Effect of Ha on Ns for $B = -0.1$, $Br = 0.1$, $c_1 = 0.2$, $c_2 = 0.2$, $\delta_1 = 0.9$, $Re = 5$, $n_\beta = 0.9$, $n_k = 1.1$, $n_\mu = 0.9$, $n_\sigma = 0.8$, $s_1 = 8$, $s_2 = 8$, $\Omega = 1$.

has no effect on Be.

Fig 19 displays the variation of Bejan number Be as a function of y at different Hartmann numbers Ha. As Ha increases, Be increases drastically near the plates and has negligibly small effect at the interface. As Hartmann number increases, the curve for Bejan number becomes flat for a considerable distance in the middle of the channel. This shows that magnetic effect decreases friction at the walls but increases the entropy generation rate at the interface. This is a significant observation that can be made from the present study in view of its practical applicability.

The Bejan number Be is shown in Fig. 20 as a function of transverse direction y for different values of viscous dissipation parameter (Br/Ω). At (Br/Ω)=0, the contribution of fluid friction irreversibility to overall entropy generation is absent, thus resulting in an invariant distribution of Bejan number. Bejan number shows its maximum value $Be = 1$ which means there is no entropy generation due to fluid friction. The Bejan number is highest at the interface of the channel and decreases as we move towards the channel walls on either direction. The Bejan number decreases as the viscous dissipation parameter (Br/Ω) increases.

8. CONCLUSION

The MHD flow of two immiscible incompressible micropolar fluids in a horizontal channel of width $2h$ with the plates at constant temperature, under the influence of a constant pressure gradient has been analyzed in terms of entropy generation rate. It is observed that

1. The higher values of the micropolarity parameter c_2 (i.e., the effect of micro-rotation becomes significant) result in lowering the velocity distribution and the presence of couple stresses in the fluid decreases the velocity.
2. The values of Ns near the plates are more than they are at the interface, indicating that friction due to surface on the fluids increases entropy generation rate.
3. The values of Ns in zone-I are more than they are in the zone-II near the plates. This indicates that the more is the viscosity of the fluid, the more is the entropy generation rate.
4. Increasing Hartmann number Ha (Magnetic parameter) tends to increase the values of Bejan number near the plates. The entropy generation number decreases near the plates with an increase in magnetic parameter and it attains almost flat shape in the middle of the channel. This implies

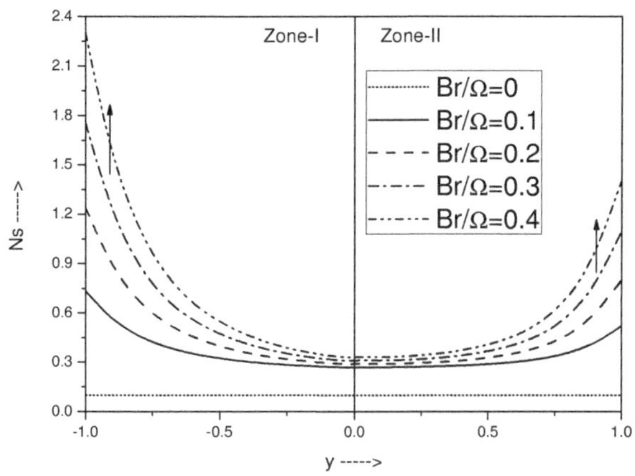

Fig. 16 Effect of (Br/Ω) on Ns for $B = -2$, $c_1 = 0.1$, $c_2 = 0.1$, $\delta_1 = 0.9$, $Ha = 0.2$, $Re = 2$, $n_\beta = 0.9$, $n_k = 1.1$, $n_\mu = 0.9$, $n_\sigma = 0.8$, $s_1 = 8$, $s_2 = 8$.

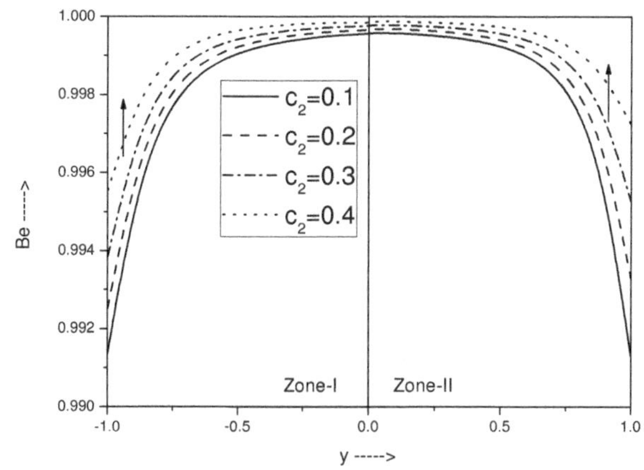

Fig. 18 Effect of c_2 on Be for $B = -0.3$, $Br = 0.2$, $c_1 = 0.2$, $\delta_1 = 0.9$, $Ha = 0.2$, $Re = 6$, $n_\beta = 0.9$, $n_k = 1.1$, $n_\mu = 0.9$, $n_\sigma = 1.1$, $s_1 = 6$, $s_2 = 6$, $\Omega = 1$.

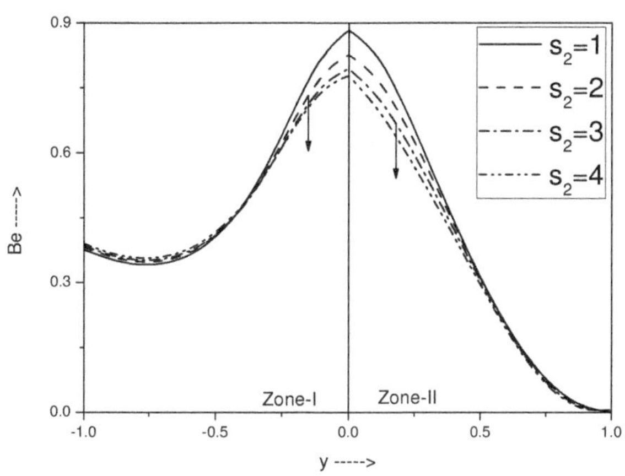

Fig. 17 Effect of s_2 on Be for $B = -0.1$, $Br = 0.1$, $c_1 = 0.2$, $c_2 = 0.2$, $\delta_1 = 0.9$, $Ha = 0.1$, $Re = 10$, $n_\beta = 0.9$, $n_k = 1.1$, $n_\mu = 0.9$, $n_\sigma = 1.1$, $s_1 = 8$, $\Omega = 1$.

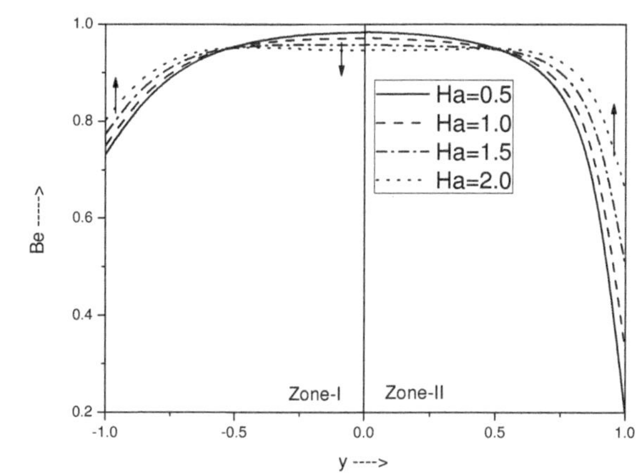

Fig. 19 Effect of Ha on Be for $B = -0.1$, $Br = 0.1$, $c_1 = 0.5$, $c_2 = 0.5$, $\delta_1 = 0.9$, $Re = 2$, $n_\beta = 0.9$, $n_k = 1$, $n_\mu = 0.9$, $n_\sigma = 1.1$, $s_1 = 8$, $s_2 = 8$, $\Omega = 1$.

that magnetic field reduces the fluid friction near the walls.

5. The walls act as strong sources of irreversibility since the viscous shear stress on the walls is maximum due to momentum transfer from the plate to the fluid. This shows an industrial application that micropolar fluids with high micropolarity and less couples stresses will act as good lubricants.

ACKNOWLEDGEMENTS

The authors thank the anonymous reviewers for their valuable comments which led to the improvement of the paper.

NOMENCLATURE

Be	Bejan number($= \frac{1}{1+\phi}$)
Br	Brinkman number($= Ek\ Pr$)
$\frac{Br}{\Omega}$	viscous dissipation parameter
c_1, c_2	micropolarity or coupling numbers
$\mathcal{C}$	non-dimensional micro-rotation in z-direction
d_{ij}	components of the strain
D	deformation tensor
E	specific internal energy
$\vec{\boldsymbol{E}}$	electric field
Ek	Eckert number
$\vec{\boldsymbol{f}}$	body forces per unit mass
h_o	induced magnetic field
$\vec{\boldsymbol{H}}$	total magnetic field
H_o	applied magnetic field
Ha	Hartmann number
2h	height of the free channel
$\vec{\mathbf{h}}$	heat flux
$\overline{J}$	current density
j	gyration coefficient
k_1, k_2	thermal conductivity of the fluid in zone-I,II
$\vec{\ell}$	body couple per unit mass
m_{ij}	couple stress tensor
Nf_i	entropy generation due to viscous dissipation
Ns_i	dimensionless total entropy generation number
Ny_i	entropy generation due to transverse conduction

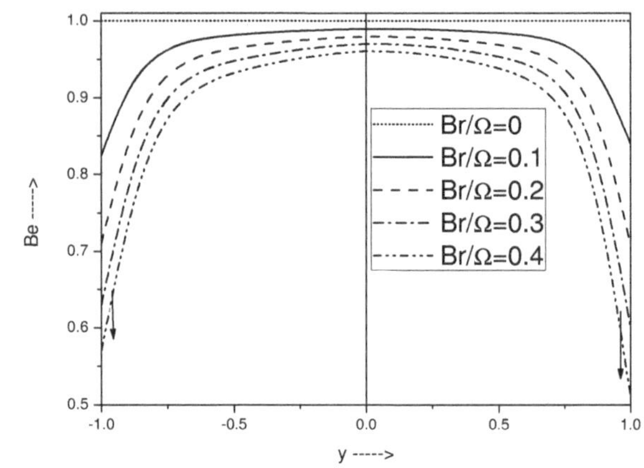

Fig. 20 Effect of (Br/Ω) on Be for $B = -0.5$, $c_1 = 0.5$, $c_2 = 0.5$, $\delta_1 = 0.9$, $Ha = 0.1$, $Re = 10$, $n_\beta = 0.9$, $n_k = 1.1$, $n_\mu = 0.9$, $n_\sigma = 1.1$, $s_1 = 10$, $s_2 = 10$.

Nu	Nusselt number
n_β	ratio of couple stress viscosity coefficients($= \frac{\beta_2}{\beta_1}$)
n_k	ratio of thermal conductivities($= \frac{k_2}{k_1}$)
n_μ	ratio of viscosities($= \frac{\mu_2}{\mu_1}$)
n_ρ	ratio of densities($= \frac{\rho_2}{\rho_1}$)
n_σ	ratio of electric conductivities($= \frac{\sigma_2}{\sigma_1}$)
p	pressure
Pr	Prandtl number
$\vec{q}$	velocity vector
q_1, q_2	non-dimensional volumetric flow rates in zones
Re	Reynolds number
s_1, s_2	couple stress parameters in zone I,II
$(S_i)_G$	entropy generation rate
$(S_i)_{G,C}$	characteristic entropy transfer rate
t_{ij}	stress tensor
T_1, T_2	non-dimensional temperatures of the plates
u	non-dimensional velocity in X-direction

GREEK SYMBOLS

α, β, γ	gyration viscosity coefficients
δ_{ij}	kronecker delta
δ_1	couplestress parameter ($= \frac{\beta_1}{\mu_1 h^2}$)
ϵ_{ijk}	Levi-Civita symbol or permutation symbol
$\vec{\nu}$	micro-rotation vector
Ω	dimensionless temperature difference($= \frac{\Delta T}{T_o}$)
Φ	dissipation function
ϕ	irreversibility ratio($= \frac{Nf}{Ny}$)
ρ	density
θ	non-dimensional temperature

REFERENCES

Alpher, R.A., 1961, "Heat transfer in magnetohydrodynamic flow between parallel plates," *Int J Heat Mass Transf*, **3**(2), 108–112.

Ariman, T., and Cakmak, A.S., 1968, "Some basic viscous flows in micropolar fluids," *Rheol Acta*, **7**(3), 236–242. http://dx.doi.org/10.1007/bf01985784.

Ariman, T., Turk, M.A., and Sylvester, N.D., 1973, "Application of microcontinuum fluid mechanics," *Int J Eng Sci*, **12**(4), 273–293.

Bakhtiyarov, S.I., and Siginer, D.A., 1997, "A note on the laminar core-annular flow of two immiscible fluids in a horizontal tube," *Symposium on liquid-liquid two phase flow and transport phenomena, Begell house, Inc. Santa Barbara*, 107–111.

Bejan, A., 1979, "A study of entropy generation in fundamental convective heat transfer," *ASME J Heat Transf*, **101**(4), 718–725. http://dx.doi.org/10.1115/1.3451063.

Bejan, A., 1980, "Second law analysis in heat transfer," *Energy*, **5**(8–9), 720–732.

Bejan, A., 1982, "Second law analysis in heat transfer and thermal design," *Adv Heat Transf*, **15**, 1–58.

Bejan, A., 1996, *Entropy Generation Minimization*, CRC Press, Boca Raton, New York.

Bhattacharya, R.N., 1968, "The flow of immiscible fluids between rigid plates with a time dependent pressure gradient," *Bull Calcutta Math Soc*, **1**, 129–137.

Bhattacharyya, K., Mukhopadhyay, S., Layek, G.C., and Pop, I., 2012, "Effects of thermal radiation on micropolar fluid flow and heat transfer over a porous shrinking sheet," *Int J Heat Mass Transf*, **55**(1), 2945–2952. http://dx.doi.org/10.1016/j.ijheatmasstransfer.2012.01.051.

Bird, R.B., Stewart, W.E., and Lightfoot, E.N., 1960, *Transport Phenomena*, John Wiley & Sons, New York.

Chamkha, A.J., 1998, "Mixed convection flow along a vertical permeable plate embedded in a porous medium in the presence of a transverse magnetic field," *Numer Heat Transf*, **34**(1), 93–103. http://dx.doi.org/10.1080/10407789808913979.

Chamkha, A.J., 2000, "Flow of two immiscible fluids in porous and nonporous channels," *J Fluids Eng*, **122**(1), 117–124. http://dx.doi.org/10.1115/1.483233.

Chamkha, A.J., and Quadri, M.A., 2001, "Heat and mass transfer from a permeable cylinder in a porous medium with magnetic field, and heat generation, absorption effect," *Numer Heat Transf*, **40**(4), 387–401. http://dx.doi.org/10.1080/104077801753238167.

Chaturani, P., and Samy, R.P., 1985, "A study of non-Newtonian aspects of blood flow through stenosed arteries and its applications in arterial diseases," *Biorheol*, **22**(6), 521–531.

Damseh, R.A., Al-Odat, M.Q., and Al-Nimr, M.A., 2008, "Entropy generation during fluid flow in a channel under the effect of transverse magnetic field," *Heat Mass Transf*, **44**(8), 897–904. http://dx.doi.org/10.1007/s00231-007-0342-8.

Eringen, A.C., 1960, *Microcontinuum field theories: II*, Fluent media, Springer, New York.

Eringen, A.C., 1966, "The theory of micropolar fluids," *J Math Mech*, **16**, 1–18.

Jena, S.K., and Mathur, M.N., 1981, "Similarity solutions for laminar free convection flow of a thermo-micropolar fluid past a non-isothermal vertical flat plate," *Int J Eng Sci*, **19**(11), 1431–1439.

Kamisli, F., and Hakan, F.O., 2008, "Second law analysis of the 2D laminar flow of two immiscible, incompressible viscous fluids in a channel," *Numer Heat Transf*, **44**(6), 751–761. http://dx.doi.org/10.1007/s00231-007-0308-x.

Kapur, J.N., 1964, "The flow of incompressible immiscible fluids between two parallel plates," *Appl Sci Res*, **13**, 55–60.

Kiyasatfar, M., Pourmahmoud, N., Golzan, M.M., and Mirzaee, I., 2012, "Thermal behavior and entropy generation rate analysis of a viscous flow in MHD micropumps," *J Mech Sci Technol*, **26**(6), 1949–1955. http://dx.doi.org/10.1007/s12206-012-0401-z.

Komurgoz, G., Arikoglu, A., and Ozkol, I., 2012, "Analysis of the magnetic effect on entropy generation in an inclined channel partially filled with a porous medium," *Numer Heat Transf*, **61**(10), 786–799. http://dx.doi.org/10.1080/10407782.2012.672890.

Lukaszewicz, G., 1999, *Micropolar fluids: Theory and applications*, Birkhauser, Springer, Boston.

Mahian, O., Mahmud, S., and Pop, I., 2012, "Analysis of first, second laws of thermodynamics between two isothermal cylinders with relative rotation in the presence of MHD flow," *Int J Heat Mass Transf*, **55**(17–18), 4808–4816. http://dx.doi.org/10.1016/j.ijheatmasstransfer.2012.04.048.

Malashetty, M.S., Umavathi, J.C., and Prathap Kumar, J., 2006, "Magnetoconvection of two immiscible fluids in vertical enclosure," *Heat Mass Transf*, **42**(11), 977–993. http://dx.doi.org/10.1007/s00231-005-0062-x.

Muthuraj, R., and Srinivas, S., 2010, "Fully developed MHD flow of a micropolar and viscous fluids in a vertical porous space using HAM," *Int J Appl Math Mech*, **6**(11), 55–78.

Nikodijevic, D., Milenkovic, D., and Stamenkovic, Z., 2011, "MHD Couette two fluid flow and heat transfer in presence of uniform inclined magnetic field," *Heat Mass Transf*, **47**(12), 1525–1535. http://dx.doi.org/10.1007/s00231-011-0815-7.

Paoletti, S., Rispoli, F., and Sciubba, E., 1989, "Calculation exergetic loses in compact heat exchanger passages," *ASME AES, Int J Heat Mass Transf*, **10**(2), 21–29.

Prathap Kumar, J., Umavathi, J.C., Chamkha, A.J., and Pop, I., 2011, "Fully developed free convective flow of micropolar and viscous fluids in a vertical channel," *Appl Math Model*, **34**(5), 1175–1186. http://dx.doi.org/10.1016/j.apm.2009.08.007.

Rao, A.R., and Usha, S., 1979, "Peristaltic transport of two immiscible viscous fluids in a circular tube," *J Fluid Mech*, **298**, 271–285. http://dx.doi.org/10.1017/S0022112095003302.

Rees, D.A.S., and Bossom, A.P., 1996, "The Blasius boundary layer flow of a micropolar fluid," *Int J Eng Sci*, **34**(1), 113–124.

Sandeep, P., and Deshpande, M.D., 2003, *A note on the no-slip boundary condition*, Technical Report, National Aerospace Laboratories, Bangalore, India.

Stokes, V.K., 1984, *Theories of Fluids with Microstructures*, Springer, New York.

Umavathi, J.C., Chamkha, A.J., Mateen, A., and Al-Mudhaf, A., 2005, "Unsteady two fluid flow and heat transfer in a horizontal channel," *Heat Mass Transf*, **42**(2), 81–90. http://dx.doi.org/10.1007/s00231-004-0565-x.

6

CONVECTIVE HEAT TRANSFER ANALYSIS IN AN ARCH ENCLOSURE

Manoj Kr. Triveni*, Dipak Sen, RajSekhar Panua

National Institute of Technology, Agartala, Tripura, 799055, India

ABSTRACT

A numerical investigation through laminar natural convection has been executed to illuminate the effect of curvature ratio in an arch enclosure filled with water. The left side wall of the cavity is maintained at a higher temperature than that of the right side wall while the other walls are kept insulated. The governing equations such as continuity, momentum and energy equation are solved by finite volume method. The effect of pertinent parameters such as curvature ratio ($1 \leq CR \leq 1.5$) and Rayleigh number ($1\times10^4 \leq Ra \leq 1\times10^6$) and) on heat transfer are calculated by commercial available computational fluid dynamics software, FLUENT 6.3. The fluid flow and heat transfer are shown for fixed Pandtl number 6.5 by streamlines, isotherms and velocity contour. From the investigation, it has been reported that the varying height of the arch cavity and Rayleigh number playing a significant role in heat transfer rate. The variation in heat transfer is calculated in terms of average Nusselt number.

Keywords: *Arch cavity, Curvature ratio, Natural convection.*

1. INTRODUCTION

A few decades past, researchers have been working on the forced convection heat transfer for high heat transfer aspect. But increasing the crisis of energy induces the researchers to look towards the natural convection process as wide applications in electronic equipment's, instruments, industries, solar collector, building insulation, nuclear reactor etc. However, among all these applications, heating or cooling from side walls are readily applicable in cooling of electronic goods, chips and transferring of heat in case of solar collector. Abu-Nada *et al.* (2010) worked for heat transfer enhancement in a differentially heated enclosure using Al_2O_3–water and CuO–water nanofluids. It was observed that enclosures having high aspect ratios, experience more deterioration in the average Nusselt number compared to enclosures having low aspect ratios. Asan *et al.* (2000) have done an experiment on a pitched roof of triangular cross section considering an adiabatic mid–plane wall numerically. From the analysis, it was observed that the heat transfer rate is decreased with increasing the value of aspect ratio and it increases with the increase of Rayleigh number. Braga and Lemos (2005) investigated a square cavity partially filled with a fixed amount of conducting solid material. It has been observed that the overall values of the average Nusselt number are slightly lower than those obtained with the square obstacles. Barletta *et al.* (2006) have done a numerical study in a 2-D rectangular enclosure. Results revealed that the elliptic boundaries enhance the mean Nusselt number and the dimensionless mean kinetic energy of the fluid as comparison than the rectangular cavity. Basak *et al.* (2006) have done a numerical analysis of steady laminar natural convection flow in a square cavity. It was examined that the non-uniform heating of the bottom wall produces greater heat transfer rates at the center of the bottom wall than the uniform heating case for all Rayleigh numbers. Basak *et al.* (2008) used a penalty finite element analysis with bi-quadratic elements which are carried out to investigate the effects of uniform and non-uniform heating conditions in an isosceles triangular enclosure. Results have been concluded that non-uniform heating produces high heat transfer rate at the center of the walls than the uniform heating. However, average Nusselt numbers show overall low heat transfer rate for the non-uniform heating case. Basak *et al.* (2012) studied the heat transfer analysis by heatlines in a square cavity. Results unfold that the larger enhancement of heat transfer rates were observed for all nanofluids at $Ra=10^5$ and Alumina–Water and Copper–Water exhibit larger enhancement of heat transfer rates. Kandaswamy *et al.* (2007) have done an analysis in a square cavity using a heating plate inside it. The effect of position and aspect ratio of heated plate on heat transfer and flow were addressed. Results signify that the increase of Gr increased the heat transfer rate in both vertical and horizontal positions of the plate. Khanafer *et al.* (2003) have investigated heat transfer enhancement in a two-dimensional enclosure utilizing nanofluids for various pertinent parameters. It was found that the suspended nanoparticles substantially increase the heat transfer rate at any given Grashof number. Nardini and Paroncini (2012) have done a work on square cavity with discrete sources. The temperature distribution in the air and the heat transfer coefficient are measured experimentally by holographic interferometry and compared the numerical results obtained with fluent. The size and position of the heat sources which influence the velocity and fluid flow is observed in this investigation. Nityadevi *et al.* (2007) performed an investigation in a rectangular cavity with partially active side walls. The result was found that the heat transfer increases with an increase in the aspect ratio. Oztop and Bilgen (2006) worked on a square cavity with internal heat generation. Results have been obtained for various geometrical parameters specifying the height, thickness and position of the partition and for Rayleigh numbers (10^3-10^6). It was concluded that the heat transfer is increasing for both internal and external Rayleigh number. Also, it decreases gradually with the increase in partition height and thickness. Paroncini *et al.* (2012) performed a numerical and experimental analysis in a square enclosure with partially active side walls. It was found that the heat transfer increases with the increase of number of strips. Pesso and Piva (2009) worked on laminar natural convection in a

* *Corresponding author. Email: triveni_mikky@yahoo.com*

square cavity. The investigation has done by both numerically and theoretically. Based on physical argument, heat transfer correlation is proposed and valid for all Prandtl and Gay-Lussac number. Roy and Basak (2005) have performed numerically using a Galerkin finite element method with bi-quadratic rectangular elements. It was noticed that in case of uniform heating, the heat transfer rate is very high at the right edge of the bottom wall and almost uniformly on the rest part of the bottom wall. In contrast, for the case of non-uniform heating the heat transfer rate reaches its maximum value at the center of both the heated walls. Sathiyamoorty *et al.* (2007) investigated heat transfer in a square cavity filled with porous medium. The results were found for the range of parameters Rayleigh number Ra, $10^3 \leq Ra \leq 10^6$, Darcy number Da, $10^{-5} \leq Da \leq 10^{-3}$, and Prandtl number Pr, $0.2 \leq Pr \leq 100$. the average Nusselt number in general increases with the increase of Da and Ra. Seikhzadeh *et al.* (2011) worked in square cavity with partially active side walls filled with Cu-water nanofluid. It was observed that the average Nusselt number increases with increasing both the Rayleigh number and the volume fraction of the nanoparticles. Moreover, the maximum average Nusselt number for the high and the low Rayleigh numbers occur for the bottom–middle and the middle–middle locations of the thermally active parts. Varol *et al.* (2011) have worked on natural convection heat transfer in an inclined fin attached square enclosure both experimentally and numerically. It was observed that the heat transfer can be controlled by attaching an inclined fin onto the wall.

The above brief survey illustrates that the many researchers are worked on triangular, rectangular, square, rhombus, quadrantal cavities. But no work has yet been done for heat transfer analysis in arch cavity. Hence, the main aim of this work is to present the effect of the shape of the cavity in fluid flow and temperature field and to calculate the heat transfer rate. It might be possible that the current work may be helpful to the other researchers or designer to short out the cooling problem from the complex shape of the electronic equipment's as well heat transfer from solar collectors.

2. PROBLEM DESCRIPTION AND MATHEMATICAL FORMULATION

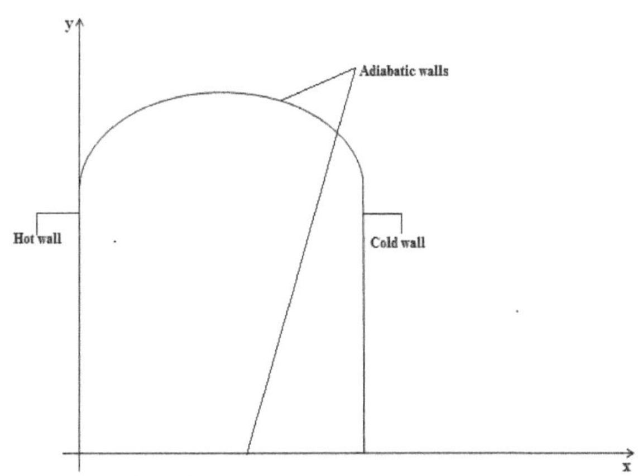

Fig. 1 Physical model of Arch cavity

A schematic diagram of the physical domain has been shown in Fig.1. The vertical walls are maintained at different isothermal temperature whereas other walls are kept insulated. Plexiglas is used as an insulating material for insulated walls and aluminum for thermally active walls. The curvature ratio of the cavity is varying from 1-1.5.

2.1 Governing equations

The problem has been solved on certain assumption that

I. There is no radiation heat transfer will takes place inside the enclosure.
II. The fluid filled in the enclosure is viscous, incompressible Newtonian fluid. Boussinesq approximation is used for fluid flow.
III. Gravitational force act over the entire volume in downward direction.
IV. All the fluid properties would be constant except density variation in the y-momentum equation.

The governing equation of 2-D, incompressible, steady state and laminar natural convection flow using conservation equation can be written in dimensional form:
as:

$$\frac{\partial u}{\partial x}+\frac{\partial v}{\partial y}=0 \quad (1)$$

$$u\frac{\partial u}{\partial x}+v\frac{\partial u}{\partial y}=-\frac{1}{\rho}\frac{\partial p}{\partial x}+\nu\left(\frac{\partial^2 u}{\partial x^2}+\frac{\partial^2 u}{\partial y^2}\right) \quad (2)$$

$$u\frac{\partial v}{\partial x}+v\frac{\partial v}{\partial y}=-\frac{1}{\rho}\frac{\partial p}{\partial y}+\nu\left(\frac{\partial^2 v}{\partial x^2}+\frac{\partial^2 v}{\partial y^2}\right)+g\beta(T-T_0) \quad (3)$$

$$u\frac{\partial T}{\partial x}+v\frac{\partial T}{\partial y}=\alpha\left(\frac{\partial^2 T}{\partial x^2}+\frac{\partial^2 T}{\partial y^2}\right) \quad (4)$$

By using dimensionless parameters such as

$$X=\frac{x}{H}, Y=\frac{y}{H}, U=\frac{uH}{\alpha}, V=\frac{vH}{\alpha}, \theta=\frac{T-T_c}{T_h-T_c},$$
$$P=\frac{pH^2}{\rho\alpha^2}, \Pr=\frac{\upsilon}{\alpha}, Ra=\frac{g\beta(T_h-T_c)H^3}{\alpha\upsilon} \quad (5)$$

The dimensionless governing equation can be written as

$$\frac{\partial U}{\partial X}+\frac{\partial V}{\partial Y}=0 \quad (6)$$

$$U\frac{\partial U}{\partial X}+V\frac{\partial U}{\partial Y}=-\frac{\partial P}{\partial X}+\Pr\left(\frac{\partial^2 U}{\partial X^2}+\frac{\partial^2 U}{\partial Y^2}\right) \quad (7)$$

$$U\frac{\partial V}{\partial X}+V\frac{\partial V}{\partial Y}=-\frac{\partial P}{\partial Y}+\Pr\left(\frac{\partial^2 V}{\partial X^2}+\frac{\partial^2 V}{\partial Y^2}\right)+Ra\Pr\theta \quad (8)$$

$$U\frac{\partial \theta}{\partial X}+V\frac{\partial \theta}{\partial Y}=\left(\frac{\partial^2 \theta}{\partial X^2}+\frac{\partial^2 \theta}{\partial Y^2}\right) \quad (9)$$

And dimensionless boundary conditions are:

$U(X, 0) = U(X, 1) = U(0, Y) = U(1, Y) = 0$

$V(X, 0) = V(X, 1) = V(0, Y) = V(1, Y) = 0$

$\theta(0, Y) = 1, \theta(1, Y) = 0$ (10)

$\frac{\partial\theta}{\partial Y}(X,0)=\frac{\partial\theta}{\partial Y}(X,1)=0 \; between \; 0<X<1$

3. NUMERICAL APPROACH

The problem has solved by commercial available software; FLUENT 6.3 which hire finite volume method. For fluid based problem, FVM is providing better flexibility and preciseness. SIMPLE algorithm coupled with pressure and velocity is used to solve the dimensionless governing equations (6-9) with the corresponding boundary conditions (10). Momentum and energy equations are discretized by second order upwind technique while the pressure interpolation is done by PRESTO scheme. The heat transfer coefficient in term of local Nusselt number can be expressed as:

$$Nu_x = \frac{hL}{k} = -\frac{\partial \theta}{\partial n} \quad (11)$$

Where h is heat transfer coefficient, θ is a dimensionless temperature and n denotes the normal direction to the plane.

The average nusselt number can be written as:

$$Nu = \frac{1}{L}\int_0^L Nu_x \, dy \quad (12)$$

6. GRID INDEPENDENCY TEST AND VALIDATION

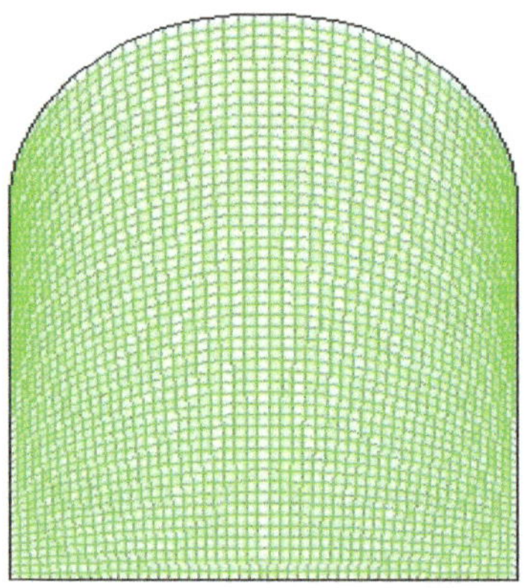

Fig. 2 Grid structure for arch cavity

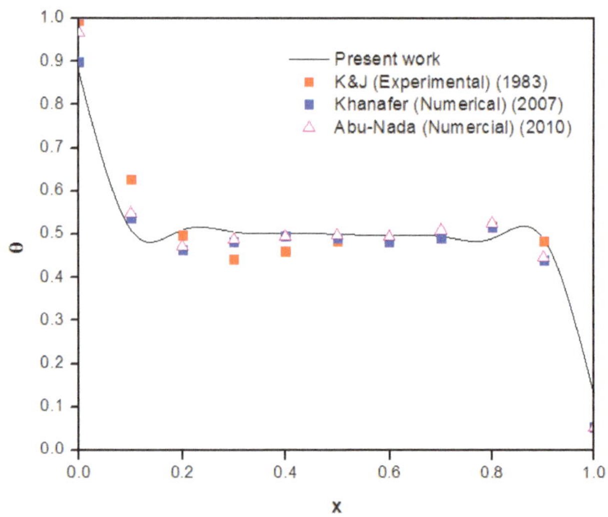

Fig. 3 Comparison between present work and other published data for temperature distribution along the width of the enclosure for nanofluid at Ra = 1×10^5 and Pr =0.7

In this study, five different grid sizes (41×41, 51×51, 61×61, 71×71 and 81×81) are adopted to check the mesh independence. A uniform grid structure in arch cavity is shown in fig 2. A detailed grid independency test has been performed for average Nusselt number at Ra = 1×10^5. The table 1 elucidate that the relative error value of the average Nusselt number is very less for grid size 51×51 and 61×61 compare to others. Since, at high grid size solution will take more time to converge the solution and the accuracy level is almost same. So 51×51 grid adopted for computational time limits.

Table 1: Relative error data at different grid size

Grid size	**41×41**	**51×51**	**61×61**	**71×71**	**81×81**
Nu	4.88	4.894	4.897	4.881	4.893
Relative Error (%)	0.31	0.061	0.33	0.25	

The present work is validated with experimental result of Krane and Jesse (1983) and numerical results of Khanefer (2007) and Abu-Nada (2010) at Ra = 1×10^5 and Pr = 0.7. The above dimensionless temperature and position graph shows the values of experimental and numerical results are fluctuating within ± 12 %.

5. RESULT AND DISCUSSIONS

A numerical simulation analysis has been conducted to examine the effects on flow regimes due to change of curvature ratio (1 to 1.5) and Rayleigh number (10^4 - 10^6).

Figure 4 demonstrates the fluid flow at different curvature ratio and Ra. The row of the figure 4 and 5 are showing the effect of shape of the cavity while column of the figure effect of Rayleigh number. Fig. 4 (a) present the case of Ra= 10^4 and CR =1 where the streamline is in parabolic shape at the center but as the CR increases it started to turn into the circular and at CR=1.5, the streamline turn into fully circular shape. The change in curvature ratio includes additional number of water molecules for heat transport. This increment in numbers of molecule increases the heat transfer rate. At Rayleigh number 10^5, the cell length increases and distinguished into the oval shape which is shown in fig 4b. Further, as the Rayleigh number increases to higher value, the cell length extended and turned into the peculiar shape inside the cavity which display in figure 4 (c-d). It is also noticeable that the streamlines are spreading throughout the cavity and concentrating towards the thermally active walls. The deformation and concentration of the streamlines cell signifies that the boundary layer diminishes by increasing the flow strength of the fluid with the increase in Rayleigh number.

Figure 5 illustrates the temperature distribution inside the cavity for various curvature ratio and Rayleigh number. The thickness of the thermal boundary layer is directly related to the viscous effect of the fluid. Higher the thickness of the thermal boundary layer induces high viscous effect and less the heat transfer. The thermal plumes are started from the left corner of the cavity, move upward, flow throughout the cavity and impinged to the right corner of the cavity. As the curvature ratio increases, shown in Figure 5a, the plumes are shifted towards the cold wall and at higher value of CR, it sticks to the cold wall. Also the motions of the isotherms are very slow at low Rayleigh number which indicates the weak flow strength of the fluid. Also the viscous effect is dominating over convection at low Ra. However, as the Rayleigh number increases the flow strength of the fluid increases and the temperature gradient between hot and cold wall become more severe. At high Rayleigh number, shown in figure 5d, the isotherms become parallel to each other at the center of the cavity. The viscous effect of the fluid decreases with the increment in Ra which leads to increase the flow strength of the fluid. The convective heat transfer started to dominate over viscosity of the fluid at high Rayleigh number because of increment in the flow strength of the fluid which leads to enhance the heat transfer rate. The velocity vector has been shown in figure 5e which signifies the flow direction and temperature distribution of the fluid between the hot and cold wall.

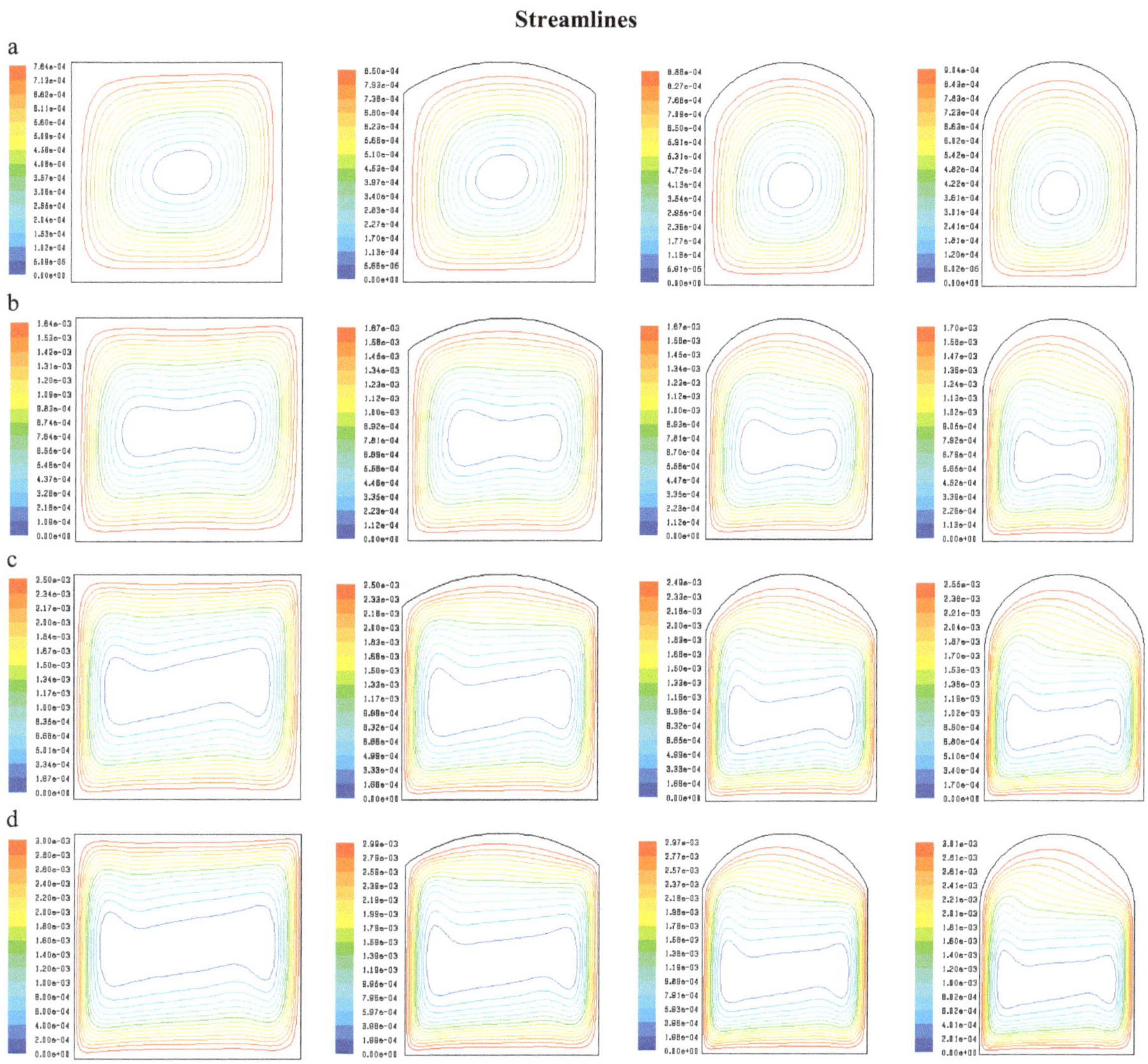

Fig. 4 Variation in Streamlines with curvature ratio 1- 1.5 (from left to right) at (a) Ra= 1×10^4 (b) Ra= 1×10^5 (c) Ra= 5×10^5 and (d) Ra= 1×10^6.

Table 2: Values of Nusselt number at different Ra and CR

	CR = 1	**CR = 1.17**	**CR = 1.33**	**CR = 1.5**
RA = 1×10^4	1.81	2.17	2.35	2.31
RA = 1×10^5	4.59	4.88	4.98	5.1
RA = 5×10^5	7.6	7.86	7.88	8.03
RA = 1×10^6	9.36	9.61	9.6	9.67

Table 2 illustrates that the variation of average Nusselt number with Rayleigh number. The thickness of the thermal boundary layer demonstrates the viscous effect of the fluid. More the thickness of the thermal boundary layer lesser will be the heat transfer. At low Rayleigh number the sluggish velocity and convection induced the increase of the thickness of boundary layer. Also, conduction heat transfer dominates over the convection heat transfer. As the Rayleigh number increases, the velocity of the fluid dominates over the viscous effect of fluid due to which convective heat transfer will be the prime source of heat transfer. The increment of buoyancy effect and reduction of the viscosity of the fluid attenuate the thickness of thermal boundary layer due to which the heat transfer increases. Hence, the numerical data shown in table 2 appears that the Nusselt number increases with the increase of Ra and CR.

6. CONCLUSIONS

A numerical study has been performed in an arch cavity. The results are validated with the published experimental and numerical results. The analysis is done for Ra = 10^4 - 10^6. The objective of the current investigation is to study the effect of Rayleigh number and curvature ratio on the fluid flow and heat transfer characteristics. Fluid flow and temperature distributions have shown by streamlines and isotherms as well as heat transfer is calculated in term of average Nusselt number. From the investigation it has been observed that the convective heat transfer is dominating over conduction at high Rayleigh number. The heat transfer rate is increasing with the increase in curvature ratio. As expected, increase in Rayleigh number leads to increase the average Nusselt number.

Fig. 5 Variation in isotherms with curvature ratio 1- 1.5 (from left to right) at (a) Ra= 1×10^4 (b) Ra= 1×10^5 (c) Ra= 5×10^5 and (d) Ra= 1×10^6.

NOMENCLATURE

g	acceleration due to gravity (m/s^2)
k	thermal conductivity of the fluid (W/m·K)
H	height of the enclosure (m)
L	width of the enclosure (m)
p	pressure (Pa)
P	non-dimensional pressure
u,v	velocity in x and y direction (m/s)
U, V	non-dimensional velocity in x and y direction
T	temperature (K)
Pr	prandtl number, (υ/α)
Ra	Rayleigh number
Nu_x	local Nusselt number
Nu	average Nusselt number
CR	curvature ratio
Greek Symbols	
α	thermal diffusivity (m^2/s)
β	thermal expansion coefficient (1/K)
θ	dimensionless temperature
ψ	stream function
Ψ	non-dimensional stream function
μ	dynamic viscosity (kg/m-s)
ν	kinematic viscosity (m^2/s)
ρ	density of the fluid (kg/m^3)
Subscripts	
h	hot wall
c	cold wal

ACKNOWLADGEMENT

The authors wish to express their very sincerely thanks to the reviewers for their valuable comments and suggestions.

REFERENCES

Abu-Nada, E., Masoud, Z., Oztop, H.F. and Campo, A., 2010, "Effect of Nanofluid Variable Properties on Natural Convection in Enclosure," *Int. J. of Thermal Science*, **49**, 479-491. http://dx.doi.org/10.1016/j.ijthermalsci.2009.09.002

Asan, H. and Namil, L., 2000, "Laminar Natural Convection in a Pitched Roof of Triangular Cross Section: Summer Day Boundary Condition," *Energy and Buildings*, **33**(1), 69-73. http://dx.doi.org/10.1016/S0378-7788(00)00066-9

Barga, E.J. and Lemos, 2005, "Laminar Natural Convection in Cavities Filled with Circular and Square Rods," *Int. Communication in Heat and Mass Transfer*, **32**, 1289-1297. http://dx.doi.org/10.1016/j.icheatmasstransfer.2005.07.014

Barletta, A., Nobile, E., Pinto, F., Schio, E.R. and Zanchini, E., 2006, "Natural Convection in A 2D- Cavity with Vertical Isothermal Walls: Cross Validation of Two Numerical Solutions," *Int. J. of Thermal Science*, **45**, 917-922. http://dx.doi.org/10.1016/j.ijthermalsci.2005.12.002

Basak, T., Roy, S. and Balakrishnan, A. R., 2006, "Effects of Thermal Boundary Conditions on Natural Convection Flows within a Square Cavity," *Int. J. Heat and Mass Transfer*, **49**, 4525-4535. http://dx.doi.org/10.1016/j.ijheatmasstransfer.2006.05.015

Basak, T., Roy, S., Babu, S.K. and Balakrishnan, A.R., 2008, "Finite Element Analysis of Natural Convection Flow in a Isosceles Triangular Enclosure due to Uniform and Non- Uniform Heating At The Side Walls," *Int. J. Heat and Mass Transfer*, **51**, 4496-4505. http://dx.doi.org/10.1016/j.ijheatmasstransfer.2007.12.018

Basak. T and Chamkha Ali J., 2012, "Heatline Analysis on Natural Convection For Nanofluids Confined Within Square Cavities With Various Thermal Boundary Conditions," *Int. J. Heat and Mass Transfer*, **55**, 5526-5543. http://dx.doi.org/10.1016/j.ijheatmasstransfer.2012.05.025

Kandaswamy, P., Lee, J. and Hakeen, A.K. Abdul, 2007, "Natural Convection in a Square Cavity in The Presence Of Heated Plate," *Nonlinear Analysis: Modelling and Control*, **12**(2), 203-212. http://www.lana.lt/journal/25/Kandaswamy.pdf

Khanafer, K., Vafai, K. and Lightstone, M., 2003, "Buoyancy-Driven Heat Transfer Enhancement in a Two Dimensional Enclosure Utilizing Nanofluids". *Int. J. Heat Mass Transfer*, **46**, 3639–3653. http://dx.doi.org/10.1016/S0017-9310(03)00156-X

Krane, R. J. and Jessee, J., 1983, "Some Detailed Field Measurements for a Natural Convection Flow In A Vertical Square Enclosure," *in: 1st ASME-JSME Thermal Engineering Joint Conference*, Vol. 1, 323–329.

Nardini, G. and Paroncini F., 2012, "Heat Transfer Experiment on Natural Convection in A Square Cavity with Discrete Sources," *Heat Mass Transfer*, **48**, 1855-1865. http://dx.doi.org/10.1007/s00231-012-1026-6

Nithyadevi, N., Kandaswamy, P. and Lee, J., 2007, "Natural Convection in a Rectangular Cavity with Partially Active Side Walls," *Int. J. Heat and Mass Transfer*, **50**, 4688-4697. http://dx.doi.org/10.1016/j.ijheatmasstransfer.2007.03.050

Paroncini, M., Carvo, F., Montucchiari, A. and Nardini, G., 2012, "A Numerical and Experimental Analysis on Natural Convection Heat Transfer in a square Enclosure with Partially Active Side Walls," *Experimental Thermal and Fluid Science*, **36**, 118-125. http://dx.doi.org/10.1016/j.expthermflusci.2011.09.004

Pesso, T. and Piva, S., 2009, "Laminar Natural Convection in a Square Cavity: Low Prandtl Numbers and Large Density Differences," *Int. J. Heat and Mass Transfer*, **52**, 1036-1043. http://dx.doi.org/10.1016/j.ijheatmasstransfer.2008.07.005

Roy, S. and Basak, T., 2005, "Finite Element Analysis of Natural Convection Flows in a Square Cavity with Non-Uniformly Heated Wall(S)," *Int. J. of Engineering Science*, **43**, 668-680. http://dx.doi.org/10.1016/j.ijengsci.2005.01.002

Sathiyamoorty, M., Basak, T., Roy, S. and Pop, I., 2007, "Steady Natural Convection Flow in a Square Cavity Filled with Porous Medium for Linearly Heated Side Wall(s)," *Int. J. Heat and Mass Transfer*, **50**, 1892.1901. http://dx.doi.org/10.1016/j.ijheatmasstransfer.2006.10.010

Sheikhzadeh, G. A., Arefmanesh, A., Kheirkhah, M. H. and Abdollahi, R. 2011, "Natural Convection of Cu-Water Nanofluid in a Cavity with Partially Active Side Walls," *European Journal of Mechanics B/Fluids*, **30**, 166-176. http://dx.doi.org/10.1016/j.euromechflu.2010.10.003

Varol, Y., Oztop, H.F., Ozgen, F. and Koca, A., 2011, "Experimental and Numerical Study on Laminar Natural Convection in a Cavity Heated from Bottom Due To an Inclined Fin," *Heat and Mass Transfer*, **48**, 61–70. http://dx.doi.org/10.1007/s00231-011-0843-3

7

EX VIVO LIVER TISSUE RADIOFREQUENCY THERMAL ABLATION: IR MEASUREMENTS AND SIMULATIONS

Edoardo Gino Macchi*, Giovanni Braschi, Mario Gallati

Università di Pavia, Dipartimento di Ingegneria Civile ed Architettura, via Ferrata 3, 27100 Pavia (PV), Italy

ABSTRACT

Radiofrequency thermal ablation (RFTA) is a medical procedure currently widely adopted for liver tumors treatment. Its outcome is strongly influenced by temperature distribution near the RF applicator therefore continuous measurements are required both to validate RFTA numerical models and to better control the outcome of the procedure. The space-time evolution of the thermal field during RFTA on *ex vivo* porcine liver tissue has been measured by infrared thermal imaging in different experimental setups. A three-dimensional simulation of the whole experiment reproduces all the features of the thermal field measurements and validates the proposed measurement methodology.

Keywords: *radiofrequency ablation, ex vivo liver tissue, IR temperature measurements, mathematical model, numerical simulations*

1. INTRODUCTION

In the last 10 years, many works in medical literature have highlighted the importance of radiofrequency thermal ablation (RFTA) in the treatment of liver tumors. The practice, pioneered by Rossi *et al.* (1990) and McGahan *et al.* (1992), consists in necrotizing the tumor tissue by means of a localized Joule heating produced by a RF electrical field applied between a small active electrode (needle), placed in the centroid of the tumor, and an external electrode with great surface area positioned on the skin. Above about 60°C proteins denaturation (and therefore cell death) takes place almost instantaneously (Goldberg *et al.*, 1995; van Sonnenberg *et al.*, 2005); when boiling temperature is reached the liquids start evaporating. Vapor bubble generation close to the active electrode breaks the electrical continuity between the needle and the tissue, and stops the procedure.

The literature has highlighted both the merits of the technique (e.g., ease of use, minimally invasive) and its limits (e.g., procedure duration limited by vaporization). The dimension of treatable tumors has progressively grown using cooling and extensible tips of various shapes and today lesions of 3-3.5 cm can be often achieved (van Sonnenberg *et al.*, 2005). Still several challenges must be faced. The main challenges deal with: (1) how to get larger lesions and control their development and outcome, (2) how to apply the method to tissues physically unfit for RFTA (poorly electrically conductive tissues), (3) how to input power efficiently and with even less invasive methods.

The numerical simulation of the phenomenon can help in such a context: rather than a tool to foresee the outcome of a particular treatment, the numerical simulation should be seen as a method useful to study the effects of particular biological situations and to preliminary test new ways to improve the RFTA procedure. RFTA modeling is a complex task since, in order to correctly represent the phenomena, liquid-vapor phase change must be modeled. Both the RF ablation procedure duration and the temperature distribution inside the tissue are fundamental data needed to validate any RFTA model.

This work is focused on the temperature distribution measurement during RFTA using thermal imaging. IR measurements are used both to qualitatively outline the heating dynamics around the active tip and to further validate a previously calibrated model. Numerical modeling is then employed to check both the physical consistency of the measurements and to validate the measurement methodology.

2. EXPERIMENTAL TESTING

2.1 Thermal field measurements strategies

The reference experiments must preserve the main features of the medical procedure such as space and time scales and the operational constraints: the RF generator, the power applied and the dimensions of the needle are similar to those employed in medical practice. The electromagnetic field frequency is 480 kHz, while the power switch off is automatic, when the electrical impedance Z exceeds 300 Ω. The porcine liver specimens came from a local slaughterhouse.

The ideal reference geometry to carry RFTA tests is usually 3D and axisymmetric with the active electrode placed at the center of mass of the cylindrical testing volume of liver tissue. Theoretically the use of traditional temperature sensors allows one to get the temperature time evolution at any specific point, however, from the practical point of view, it is nearly impossible to get a meaningful evaluation of the strong temperature gradients (up to 7°C/mm) with measurements in close points, due to the sensor size and placement uncertainty. Furthermore the intense EM field near the active electrode often produces a strong EM noise that may even make the measurements meaningless. For this reason in most experiments found in the literature the tissue temperature is measured only in a few points, usually far from the active electrode, to avoid both strong EM interferences and alterations to the RF field that produces the tissue heating (Yang *et al.*, 2007; Ai *et al.*, 2012; Trujillo *et al.*, 2012; Gallati and Braschi, 2013). While standard fiber optic sensors (fiber Bragg gratings) solve the

* *Corresponding author. Email: edoardogino.macchi01@ateneopv.it*

problems related to size and EM interferences, they provide a temperature measurement averaged over their sensing length (usually 0.5-1 cm) (Tosi *et al*., 2014), and therefore are not very suited for capturing the steep temperature gradients induced by most thermal ablation treatments. On the other hand, infrared (IR) imaging has the advantage to provide high resolution field measurements avoiding the thermal and electrical effects due to the sensors presence that affect the system under measurement and it is insensitive to electromagnetic noise. However the need for a clear line of sight to the emitting surface often fights against the energy balance preservation.

In the present investigation the 3D sample is sectioned along the axial plane; the sample surface can be left uncovered, to limit any energy filtering effect along the line of sight, or it can be protected with a thermally insulating lid. In case the "free surface" solution is being adopted significant thermal losses are expected, as heat and vapor can freely leave the sample, and, moreover, the surface emissivity might vary considerably as the water content dramatically changes during the procedure. On the other hand, with the insulating lid, thermal losses are substantially reduced and the surface emissivity is unchanged, however it is hard to find a material exhibiting at the same time the proper thermo-mechanical properties and the needed infrared total transparency. A compromise solution has been adopted: the temperature is measured only at some points, relying on an array of highly conductive thin copper nails (diameter 0.6 mm) piercing through an insulating balsa wood lid. It is expected that these nails produce a heat bypass from the tissue towards the outside only in a few points, therefore limiting any loss and preserving the overall thermal balance; the temperature of the nails is then measured by IR imaging.

The axial setup with free surface, allows one to simultaneously record the phenomena evolution during the RFTA in both visible and IR range: as stated above, due to the significant heat losses, the results can be used only from a qualitative point of view. On the other hand if the lid with nails is employed the losses are greatly reduced and the temperature measurements could be used from a quantitative point of view.

2.2 Experimental setup and measurements

The test box is shown in Fig. 1: it is a hollow half cylinder, having diameter *D* and length *sl*, carved out of a single block of insulating material (extruded polystyrene, XPS). Porcine liver tissue (*L*) fills the box cavity. The cavity wall is covered with a thin brass coating, acting as the passive electrode (*pe*). For the free surface experiment the active electrode (*ae*) is the tip of a thin brass pipe, having 4 mm outer diameter and 20 mm active length, while the rest of the pipe is electrically insulated (Fig. 1a).

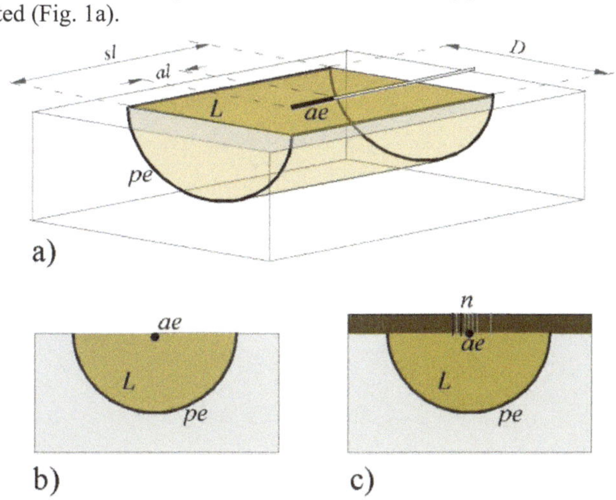

Fig. 1 RFTA test box: a) axonometric view, free-surface setup; b) vertical cross section, free-surface setup; c) vertical cross section, setup with nail-equipped lid. *ae*: active electrode, *pe*: passive electrode, *t*: thermistor, *L*: liver, *al*=20 mm, *D*=80 mm, *sl*= ~100 mm.

In the tests with nail-equipped lid the active electrode is directly fixed to the balsa wood lid (see Fig. 1c and Fig. 2); the cover also carries a NTC thermistor. A schematic of needle, nails numbering and NTC position is depicted in Fig. 2 together with a picture of the insulating lid near the active electrode (inner side). The outer surface of the lid (both nails and balsa wood) was painted black to get higher and more homogeneous surface emissivity. The emissivity was estimated using a small copper cube of 3 cm side, previously black-painted, and equipped with several micro-thermocouples (both on the surfaces and inside the cube). After heating the cube using a heat gun, temperature measurements were performed using both thermocouples (that give the actual temperature) and IR imaging. The surface emissivity was corrected until IR temperature measurements and thermocouple measurements (those recorded on the surface) matched. Using this method, black-painted copper emissivity, was estimated to be 0.86, with a computed uncertainty of about 1.5%.

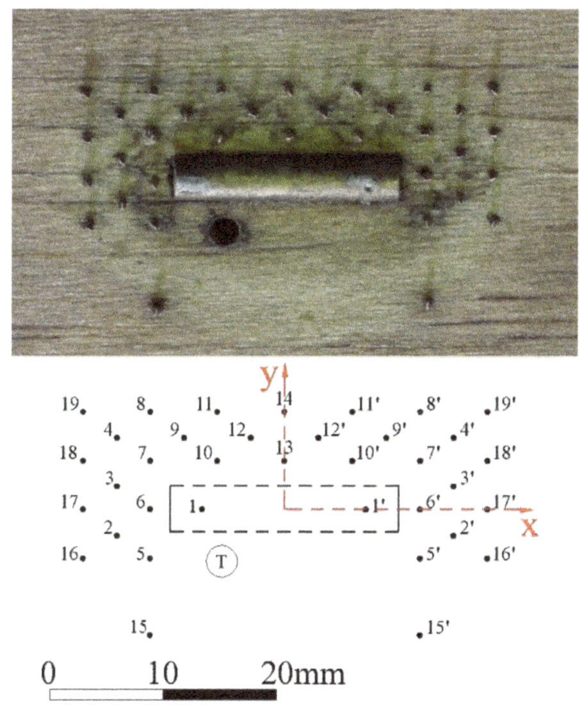

Fig. 2 Balsa wood lid with copper nails, detail near the active tip: picture (top) and schematic drawing (bottom) of nails pattern, active electrode and thermistor position (T).

With the above described box, laid down horizontally, videos were recorded from the top, by means of an IR camera. The main characteristics of the employed IR camera (Thermo Tracer TH7100 NEC, Tokyo, Japan) are: resolution 320x240 pixel, spectral band 7.5 to 13 μm, temperature range -20 to +120 °C, uncertainty ±2 °C or ±2% of reading, frame rate 10 Hz. The IR images were processed by means of custom developed Matlab® scripts. A temperature value was extracted from each pixel, the nails temperature was computed as the average of the pixels belonging to the area occupied by each nail. Since traditional temperature sensors (thermocouples, RTDs, *etc.*) can be more sensitive to EM noise than NTC thermistors, the latter (B57861 Epcos, Munich, Germany) were employed to check the IR measurements.

The temperature measurements obtained using the IR camera may be subject to many uncertainty sources which are part of the measurement process and not just related to the IR camera uncertainty. Being conscious of the potentially large bias between liver and measured temperatures a proper uncertainty analysis was carried out (JCGM 100, 2008). The approach followed consists in evaluating all the uncertainty contributions, laying down an uncertainty budget, and defining a combined uncertainty of the measurements system (IR camera plus nail-equipped balsa wood lid). The main uncertainty sources are related to: IR camera, emissivity of the surface, dynamic response of the measurement system (nails included) and temperature gradient along the nail. The resulting expanded uncertainty is about 5.5°C providing a level of confidence of approximately 95%. This depicts of course the worst scenario, but it provides at least a first uncertainty on the temperature estimation. Since the expanded uncertainty of the IR measurement system was quite high, a systematic check with NTC thermistors in spot points was carried out during each experiment.

The effects of two significant uncertainty sources (dynamic response of each nail and gradient along the nail) can be easily included in the numerical simulation, thus allowing to reduce the expanded uncertainty to about 3°C. The simulation presented in section 3 takes into account the heat transfer and the energy storage inside nails, balsa wood lid and active tip, other than the RF heating of the tissue. This also allows a direct comparison between IR surface measurements and simulated nails' temperatures.

The adopted RF generator (Invatec TAG 100W) works within the impedance range 20-300 Ω the procedure is automatically interrupted when the Z exceeds the maximum value. The generator nominal power is usually different from the real power supplied to the sample: the latter was directly measured with the aid of an oscilloscope, a voltage probe and a current probe. Checks performed with saline through calorimetric measurements allowed verifying the reliability of the power values given by the oscilloscope: a systematic 5% power reduction must be applied to obtain the real power supplied to the sample. During each test the power was set to 15 W, corresponding to a real power supplied to the sample of about 14.2 W and held constant throughout the test.

2.3 Experimental results

The results for the configuration without insulating lid are used to determine the qualitative evolution of the thermal field close to the active tip. On the other hand the results from the setup with balsa wood lid provided data useful for the validation of mathematical model of the RFTA procedure and for evaluating the reliability of the experimental measurements with nail-equipped lid. During the first test the aforementioned heat losses greatly affect the procedure duration that is doubled (289 s) with respect to the same test with insulating lid (145 s). In order to check replicability, the test with nail-equipped lid was repeated a few times obtaining similar results.

The frames of Fig. 3 show some IR images from the "free surface" test, selected at proper times, to qualitatively outline the thermal field development. The frames can be grouped into two sets: a), b) and c) describe the temperature evolution before the heating effects can be visually observed, while d), e) and f) depict situations also visually detectable from the tissue color change produced by protein denaturation.

Fig. 3(a) shows the situation at the heating onset (t=10 s): two separate areas (~30°C) appear, where the active tip ends are positioned. The temperature increase is slightly higher at the right-hand end, where the stem tip is located. This asymmetry is observed throughout the test. Fig. 3(b) reproduces the observed situation at t=42 s: the temperature distribution is getting more uniform along the active tip and the maxima at the two ends increase (~50°C). No tissue color change is yet noticeable. Fig. 3(c) gives the situation at t=64 s: the two temperature peaks (~60°C) expand to meet halfway, towards the middle of active electrode, with a temperature distribution assuming a ribbon shape. At this stage a slight tissue color change can be detected near the active tip ends through naked-eye observation. Fig. 3(d) shows the situation at t=201 s: the temperature peak reaches 70°C and its distribution has an external elliptical shape. The temperature maxima at the ends of the tip are still asymmetric. Fig. 3(e) gives the situation at t=265 s: vaporization has started near the active tip (~100°C). Fig. 3(f) gives the situation at t=289 s, when the electrical power is automatically switched off: two wider areas with temperature higher than 100°C are detected. They are centered on the ends of the active tip, where the vapor slightly inflates the tissue, as also noticed through naked-eye observation.

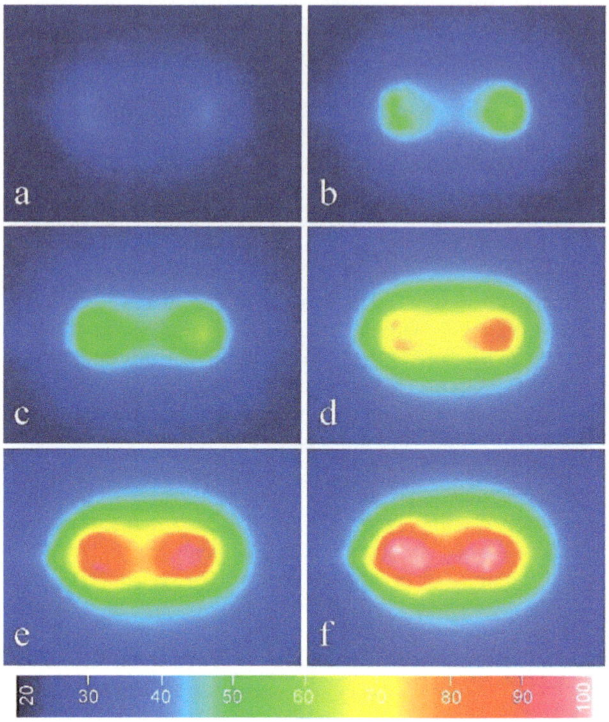

Fig. 3 Thermal field evolution ("free surface" test, IR images at different times): a) t=10 s; b) t=42 s; c) t=64 s; d) t=201 s; e) t=265 s; f) t=289 s (just before power off). Color bar represents the temperature in degree Celsius.

The results of the insulated lid test are presented in the following. The temperature histories of nails regarded as the most meaningful are reported in Fig. 4, together with the results of the numerical simulation that will be discussed later.

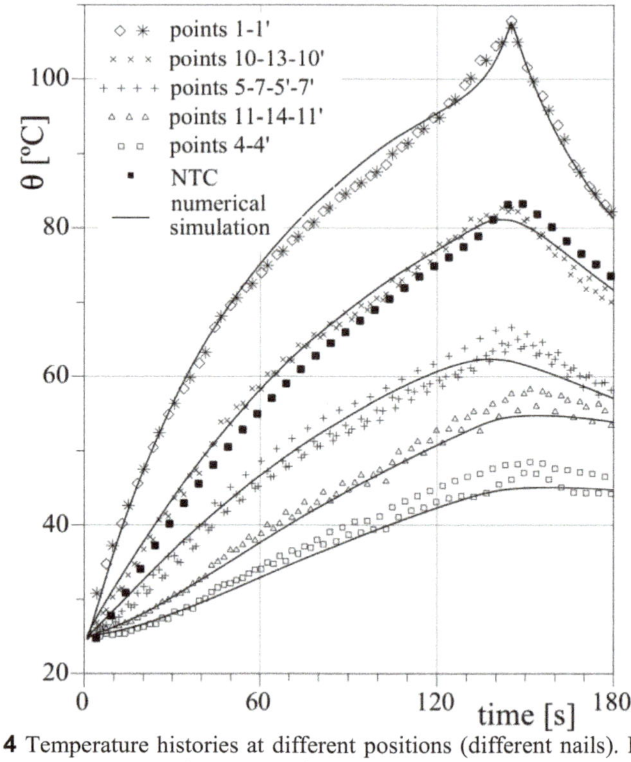

Fig. 4 Temperature histories at different positions (different nails). IR measurements (empty markers), NTC measurement (filled square markers) and numerical simulation results (lines).

In this test the active electrode was fixed to the balsa wood lid and directly connected to nails 1-1' as appears from Fig. 2. First of all a very good agreement of measurements from symmetrical points with respect to the nail pattern symmetry axis is observed (e.g., 1-1',10-10'). The overall symmetry is confirmed by comparing the NTC sensor measurements with the IR measurements of nails 10-10', whose position with respect to the symmetry axes of the active electrode is same (see Fig. 2). The symmetry of the thermal field distribution with respect to both symmetry axes of the active electrode highlights that effects of the heat losses due to the copper nails are negligible.

3. NUMERICAL SIMULATIONS

In this section, after briefly describing the mathematical model employed for the RFTA simulation, we will discuss the geometrical model used for the simulation of insulated lid test, the imposed boundary conditions and the model parameters, finally the simulations results will be presented and compared with experimental measurements.

3.1 Mathematical model

The temperature is computed solving the heat equation:

$$\rho C_p \frac{\partial \theta}{\partial t} = \nabla \cdot (k \nabla \theta) + q \tag{1}$$

where ρ, C_p and k are respectively density, specific heat at constant pressure and thermal conductivity. The source term q is the heat produced by Joule effect per unit volume and is computed as:

$$q = \frac{1}{\varepsilon_r} |\nabla V|^2 \tag{2}$$

where ε_r is the electrical resistivity.

Adopting the quasi-static approximation, the electric field is computed by solving a quasi-harmonic equation for the electric potential V, obtained combining the continuity principle for the electric current and the local Ohm's law:

$$\nabla \cdot \left(\frac{1}{\varepsilon_r} \nabla V \right) = 0 \tag{3}$$

In the temperature range typical of RFTA (37-120°C) the temperature dependence of liver tissue' physical parameters can be neglected (without losing accuracy) for the thermal parameters (ρ, C_p and k) while it must be taken into account for the electrical parameters, sensibly affecting the simulation results (Rossiet al., 2008; Trujillo and Berjano, 2013). Therefore, both Eq. (1) and Eq. (3) must be solved iteratively at each time-step until convergence is achieved. The solution allows one to compute the impedance Z, which is the variable used to stop the power supply as it occurs in the experimental situation. In order to obtain plausible results, the mathematical models used for the simulation of high-temperature hyperthermia treatments (e.g, RF, MW and laser ablation) must take into account the effects of vaporization. A review of the scientific literature suggests that Baldwin *et al.* (2001) were the first to investigate liquid-vapor phase change in thermal ablation treatments. The mathematical model proposed by Baldwin *et al.* (2001) was later reformulated and improved by Abraham and Sparrow (2007), using the enthalpy method to model the phase change, and applied to balloon thermal endometrial ablation, a method for the treatment of menorrhagia. Similar heuristic vaporization models were developed and applied to simulate high-temperature hyperthermia treatments (Yang *et al.*, 2007; Ai *et al.*, 2012; Gallati and Braschi, 2013). Since main purpose of the forthcoming numerical simulations is the validation of the measurement methodology, only the general features of the implemented phase change model will be discussed.

The liquid-vapor phase change of the tissue liquids is modeled using the Apparent Heat Capacity method (Bonacina *et al.*, 1973; Muhieddine *et al.*, 2011): the tissue specific heat is modified adding a symmetric Gaussian pulse to mimic the effects of the latent heat due to liquid vaporization. The pulse parameters μ and σ (mean and standard deviation) control respectively the temperature corresponding to the Gaussian' peak and the temperature range over which the specific heat is modified.

During the phase change the tissue electrical resistivity is defined with the following empirical law:

$$\varepsilon_r = \varepsilon_r(\theta_{vap}) + \chi\, \psi(\theta)^{\tau} \tag{4}$$

to model the effect of vapor bubbles that progressively insulate the active electrode leading to the procedure interruption. χ and τ are calibration parameters while $\psi(\theta)$ is the temperature integral of the Gaussian function (i.e., the relative amount of absorbed latent heat) that can be taken as the vapor volume fraction.

The thermal lesion boundary is approximated either using an Arrhenius damage model (Schutt and Haemmerich, 2008; Haemmerich and Schutt, 2011) or simply through a temperature isotherm; both 50°C (Mulier *et al.*, 2012; Viglianti *et al.*, 2014) and 60°C isotherms (Chang, 2010; Wood *et al.*, 2011) are tested for the purpose. In the Arrhenius model the thermal damage Ω and the concentration of living cells C are computed as follows (assuming $C(0) = 1$):

$$\Omega(t) = \int_0^t A e^{-\frac{E_a}{R^* \vartheta_k(t)}} d\xi \tag{5}$$

$$C(t) = \mathrm{e}^{-\Omega(t)} \tag{6}$$

where t is the time, θ_k is the absolute temperature (K), $A = 2.984 \cdot 10^{80}$ is the frequency factor (s^{-1}), $E_a = 5.064 \cdot 10^5$ is the activation energy ($Jmol^{-1}$) and R^* is the universal gas constant (8.314 $Jmol^{-1}K^{-1}$). A threshold value of 4.6 for $\Omega(t)$ (which corresponds to 99% cell death - $C(t) = 0.01$) is used to define the cell death boundary.

3.2 Numerical model

The equations of the mathematical model are discretized by means of the finite volume method using the C++ library OpenFOAM®. Equations (1) and (3) are solved with a fully implicit scheme to obtain the primary variables: θ and V. Second order accurate schemes are used for both time and space discretization.

Since one of main objectives was the validation of the experimental methodology we decided to build a complete three-dimensional model of the testing box, simulating not only the RFTA on liver tissue but also the heat transfer inside lid, nails and active electrode. Exploiting the symmetry of the experimental setup only half of the testing box has been modeled, this helped reducing the computational cost without loss of generality. An exploded view of the three-dimensional geometrical model is depicted in Fig. 5 while some details are depicted in Fig. 6.

The computational domain is discretized into about 1.2M tetrahedral cells using the Delaunay-based volume grid generation method implemented in Gmsh (Geuzaine and Remacle, 2009). The mesh is manually refined where strong gradients of the primary variables are expected (i.e., close to the active electrode) and near the interfaces between different materials. The time-step is set to 1 s, a sensitivity analysis of both mesh size and time-step has been carried out. The results later presented are obtained with a more refined mesh (about 2.8M cells) that gives smoother results close to the nail-lid interface. Equations (1) and (3) are solved in all the cells of the computational domain assigning appropriate electrical and thermal properties to the various materials.

The electric potential equation is solved assigning, as boundary conditions, a fixed electric potential to the passive electrode ($V = 0$) and to the cells belonging to the active electrode ($V = V^{ae}$), while a no-flux condition is imposed on the remaining boundaries. Since the impedance Z changes during the RFTA, an iterative procedure is needed to compute, at each time-step, the electric potential V^{ae} required to supply a constant power to the tissue; for details see (Rossi *et al.*, 2008). Furthermore a check is performed to control that all the electrical power is supplied to the liver tissue, avoiding erroneous thermal sources due to discretization and interpolation schemes.

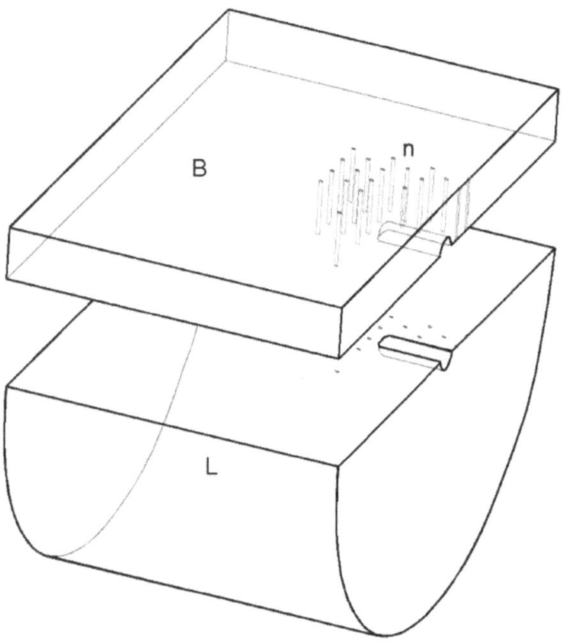

Fig. 5 3D geometrical model, exploded view. Liver tissue (L), balsa wood lid (B), nails (n).

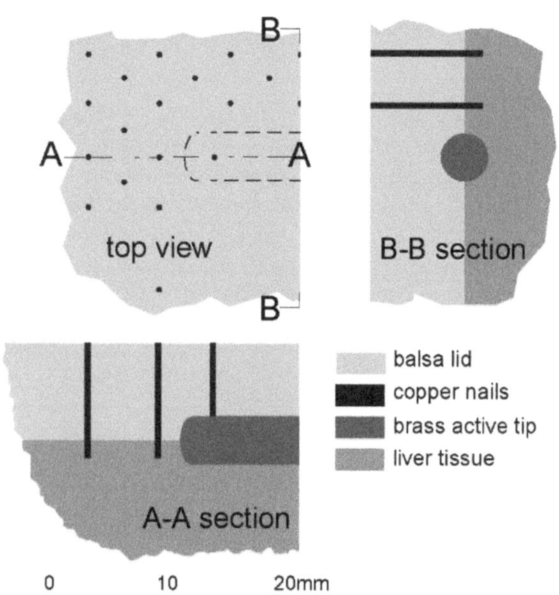

Fig. 6 Details of the computational model with material properties assignment.

Concerning the heat equation, no-flux boundary conditions are assigned to all boundaries except for the balsa lid and nails boundaries where the following convective boundary condition (accounting for the heat losses) is imposed:

$$-k\nabla\theta\cdot\boldsymbol{n} = \alpha\left(\theta - \theta^{amb}\right) \tag{7}$$

where θ^{amb} is the ambient temperature, α is a heat transfer coefficient and $\boldsymbol{n}$ is the outward unit normal vector.

In the simulation the initial conditions for Eq. (1) are set using the temperatures measured before the starting the RFTA experiment: 28°C for balsa wood lid and 23°C for liver tissue, nails and active electrode. Similarly the power set in the simulation is 14.2 W and the ambient temperature is 28°C.

3.3 Physical and model parameters

Regarding the physical parameters standard literature values have been used for all the modeled materials (liver tissue, balsa wood lid, copper nails and brass active electrode). Most physical parameters are kept constant during the simulation; only the temperature dependence of the tissue electrical resistivity is taken into account since it greatly affects the tissue heating. The resistivity is computed from the temperature using an experimental curve determined for *ex vivo* porcine liver tissue (Rossi *et al.*, 2008). It prescribes a significant variation (55% in the temperature range 30-85°C) following approximately a quadratic law (Zurbuchen *et al.*, 2010). The experimental curve, normalized with the liver tissue resistivity at 40°C, can be adequately interpolated by a 4th degree polynomial:

$$\varepsilon_r / \varepsilon_r^{40} = a_0 + a_1 x + a_2 x^2 + a_3 x^3 + a_4 x^4 \tag{8}$$

where: $x = (\theta - 40)$, $a_0 = 2.5$, $a_1 = -9.13\cdot10^{-2}$, $a_2 = 2.178\cdot10^{-3}$, $a_3 = -2.495\cdot10^{-5}$, $a_4 = 1.029\cdot10^{-7}$.

The RFTA model has been previously calibrated using the *ex vivo* porcine liver tissue RF ablations performed by Gallati and Braschi (2013). The physical and calibration parameters are reported respectively in Table 1 and Table 2. It is important to highlight that the values used for electrical resistivity of balsa wood, copper and brass are at least three orders of magnitude higher or lower than the liver tissue resistivity (Eq. 8 with $\varepsilon_r^{40} = 3.3\ \Omega$ m), their exact value does not affect the simulation results and therefore is not reported. The liquid-vapor phase change model employs the following physical parameters: water latent heat of vaporization (2300 kJ/kg) and water content of liver tissue (75%).

The heat transfer coefficient α is the only *ad hoc* parameter, nevertheless its value is low and therefore its effects are limited. It is introduced only in order to properly reproduce the heating and especially the cooling of the balsa wood lid. Because of their low value the losses do not sensibly affect the temperature distribution inside the tissue and the duration of the RFTA procedure.

During the phase change the resistivity must be significantly increased to model the electrical effect of vapor bubbles near the active electrode. The calibration parameters χ and τ control this increase; the sensitivity analysis proved that their variation affects even considerably the procedure duration and therefore also the temperature distribution. The calibration parameters σ and μ , that control the shape and position of the Gaussian function, affect both procedure duration (obviously also connected to the phase change) and the heating rate during vaporization; however only the latter has a significant effect. Several functions, other than the Gaussian adopted here, could have been used for the same purpose, with similar effects (provided that they subtend unit area). The sensitivity analysis proved that the main calibration parameters of the model are μ and τ. These parameters are related to the phase change and their variation affects even strongly the procedure duration, that in turn affects the temperature gradient. However, the model analysis also highlighted that relatively small differences in the calibration parameters values (even up to 20%) do not substantially affect the simulation outcome.

Table 1 Physical parameters: density, specific heat at constant density and thermal conductivity for each material considered in the simulation.

	ρ (kg/m^3)	C_p (J/kg·°C)	k (W/m·°C)
Liver tissue	1070	3500	0.65
Balsa wood lid	150	2900	0.09
Copper nails	8950	390	395
Brass active tip	8500	380	115

Table 2 Calibration parameters; the first four parameters are connected with the phase change and were previously calibrated using RFTA tests.

μ (°C)	σ (°C)	χ (Ω·m)	τ (-)	α (W/m^2·°C)
110	6	500	2	20

3.4 Simulation results

The numerical simulation results are compared with the experimental measurements in Fig. 4 that depicts the temperature history of selected nails. A good agreement can be observed for all the nails, in particular for those close to the active electrode, whose maximum temperature is higher. The simulation results also highlight that the temperature on the nails' external face (the one measured by IR imaging) is very close to the temperature inside the liver tissue as shown in Figs. 7-8 (maximum difference about 2°C). The model also reproduces quite well the temperature space-time evolution of balsa wood as can be seen in Fig. 9, where numerical simulation results and IR images are compared.

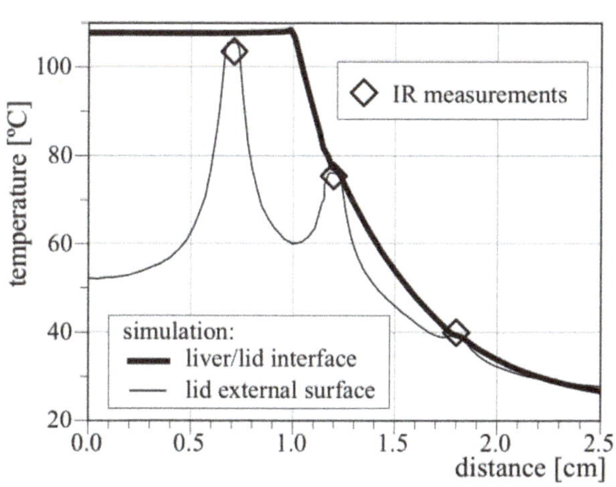

Fig. 7 Simulated axial temperature distribution (t=140 s, along x-axis of Fig. 2) in planes corresponding to the lid external surface (balsa wood lid with nails) and to the liver-lid interface. IR temperature measurements from nails 1', 6' and 17' are also plotted with empty markers.

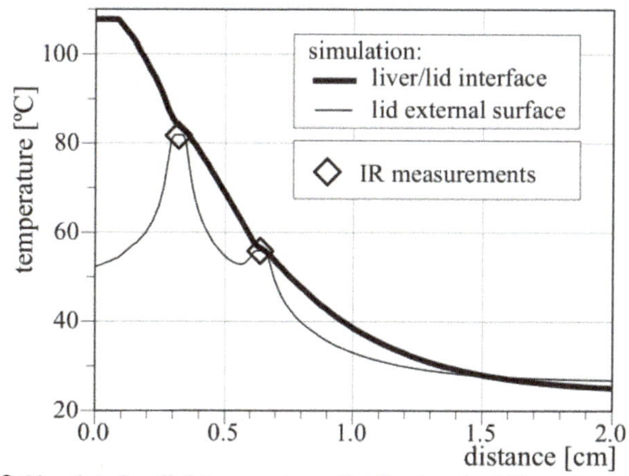

Fig. 8 Simulated radial temperature distribution (t=140 s, along y-axis of Fig. 2) in planes corresponding to the lid external surface (balsa wood lid with nails) and to the liver-lid interface. IR temperature measurements from the nails 13 and 14 are also plotted with empty markers.

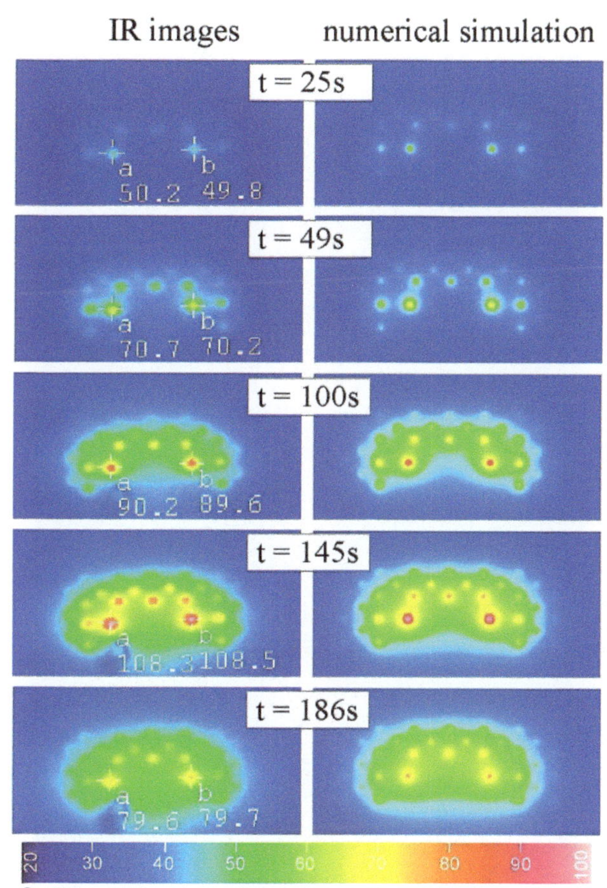

Fig. 9 Temperature space-time evolution (lid external surface): comparison between IR images (left-hand side) and numerical simulation results (right-hand side). The first four frames are during RFTA while the last frame (t=186 s) is during cooling.

In the IR images the nails appear to be larger and hotter due to the nails' heads (diameter 1.3 mm) that were not modeled because very thin. The difference may also be explained by the resolution of the IR measurements: the pixel dimensions are clearly higher than the cell size employed in the simulation for the nail discretization.

It is important to highlight that the asymmetry of the IR images (with respect to the active electrode longitudinal direction – x-axis in Fig. 2) is due to the enhancement of the thermal losses on the side of the lid with the nails. This is also reproduced by the model as appears from Fig. 9 and Fig. 10(a) and clearly due to the copper nails presence. This asymmetry however does not sensibly alter the heating of liver tissue whose temperature remains symmetrical as can be seen from Fig. 10(b). On the other hand, the temperature distribution is always symmetric with respect to the transverse direction of the active electrode (y-axis in Fig. 2), as appears from Fig. 9. This is also confirmed by comparing the numerical simulation results of Fig. 10(b) with the *ex vivo* liver tissue thermal lesion (approximated by the tissue color change) obtained after performing the RFTA (Fig. 10(c)). The simulated thermal lesion boundaries, defined using Arrhenius model (Eqs. 5-6), 50°C isotherm and 60°C isotherm are compared with the experimental lesion, identified though tissue color change, in Fig. 10(c). In this experiment the simulated lesion, obtained using the Arrhenius model (dashed white line in Fig. 10), appears to be a better indicator for the thermal lesion.

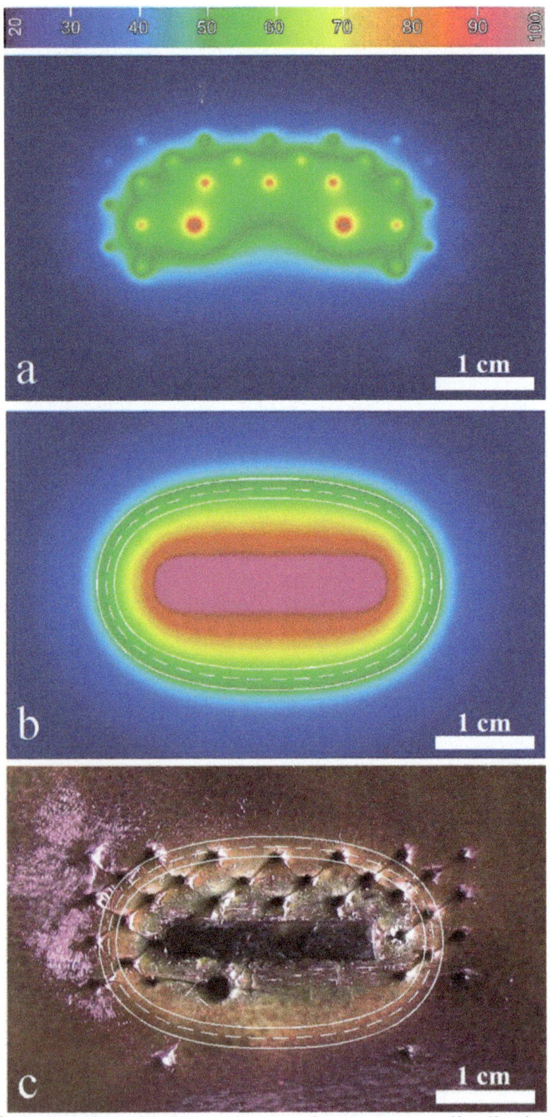

Fig. 10 Comparison between simulated temperature distribution (at the end of the procedure, t=140 s) and experimental thermal lesion: a) lid external surface, simulation; b) liver tissue-lid interface, simulation; c) experimental thermal lesion due to protein denaturation. The white continuous lines in frames b) and c) represent the 50°C and 60°C isotherms obtained from the numerical simulation. The white dashed line identifies the thermal lesion obtained using the Arrhenius model.

In conclusion the numerical simulations results are in good agreement with the IR experimental measurements both from a qualitative and quantitative point of view. These results allowed the validation of the proposed measurement methodology that exploits an array of highly conductive nails and an insulating lid to measure the temperature in multiple points without altering significantly the phenomena. In addition these results confirm the soundness of the mathematical model, further contributing to its validation.

4. CONCLUSIONS

The space-time evolution of the thermal field was measured during RFTA on *ex vivo* porcine liver using thermal imaging. The experimental setup exploits an insulating lid equipped with an array of highly conductive small copper nails to allow multiple temperature measurements without altering the phenomena. The three-dimensional numerical simulation of the experiment, including the nail-equipped lid, allowed validating the temperature measurement methodology and further verifying the calibration of the proposed RFTA model. In future studies the proposed experimental methodology will be used for a thorough *ex vivo* experimental campaign using a similar setup. The most recent generation of distributed sensors, which make use of an interferometric setup to detect the Rayleigh backscattering within a single-mode optical fiber, are also under testing for temperature distribution measurements during RFTA. Further studies are underway for the development of a more physically-based model of liquid-vapor phase change exploiting a multiphase porous media theoretical framework.

ACKNOWLEDGEMENTS

The financial support of Ordine degli Ingegneri della Provincia di Pavia and Fondazione Cura Mini-invasiva Tumori ONLUS is gratetully acknowledged. The authors wish to thank Prof. Alfredo Cigada and PhD Lorenzo Comolli (Politecnico di Milano) for their assistance during IR measurements.

NOMENCLATURE

A	frequency factor (s^{-1})
C_p	specific heat at constant pressure ($Jkg^{-1}°C^{-1}$)
E_a	activation energy ($Jmol^{-1}$)
k	thermal conductivity ($Wm^{-1}°C^{-1}$)
$\boldsymbol{n}$	outward unit normal vector (-)
q	Joule heat per unit volume (Wm^{-3})
t	time (s)
R^*	universal gas constant ($Jmol^{-1}K^{-1}$)
V	electric potential (V)
Z	electrical impedance (Ω)
Greek Symbols	
α	convective heat transfer coefficient ($Wm^{-2}°C^{-1}$)
ε_r	electrical resistivity (Ω m)
θ	temperature (°C)
θ_k	absolute temperature (K)
μ	Gaussian function mean (°C)
ρ	density ($kg\ m^{-3}$)
σ	Gaussian function standard deviation (°C)
τ	parameter for phase change electrical resistivity (-)
χ	parameter for phase change electrical resistivity (Ω m)
ψ	vapor volume fraction (-)
Superscripts	
amb	ambient environment
ae	active electrode
40	40°C value
Subscripts	
vap	vaporization

REFERENCES

Abraham, J.P. and Sparrow, E.M., 2007, "a Thermal Ablation Model Including Liquid-to-Vapor Phase Change, Necrosis-dependent Perfusion, and Moisture-dependent Properties," *International Journal Heat and Mass Transfer*, **50**(13), 2537-2544. http://dx.doi.org/10.1016/j.ijheatmasstransfer.2006.11.045

Ai, H., Wu, S., Gao, H., Zhao, L., Yang, C., Zeng, Y., 2012, "Temperature Distribution Analysis of Tissue Water Vaporization during Microwave Ablation: Experiments and Simulations," *International Journal of Hyperthermia*, **28**(7), 674-685. http://dx.doi.org/10.3109/02656736.2012.710769

Baldwin, S.A., Pelman, A., Bert, J.L., 2001, "a Heat Transfer Model of Thermal Balloon Endometrial Ablation," *Annals of Biomedical Engineering*, **29**(11), 1009–1018. http://dx.doi.org/10.1114/1.1415521

Bonacina, C., Comini, G., Fasano, A., Primicerio, M., 1973, "Numerical Solution of Phase-change Problems," *International Journal of Heat and Mass Transfer*, **16**(10), 1825-1832. http://dx.doi.org/10.1016/0017-9310(73)90202-0

Chang I.A., 2010, "Considerations for Thermal Injury Analysis for RF Ablation Devices," *The open biomedical engineering journal* **4**, 3-12. http://dx.doi.org/10.2174%2F1874120701004020003

Gallati, M., and Braschi, G., 2013, "On the Simulation of Radio Frequency Thermal Lesions in Porcine Liver," *Proceedings of the 32nd IASTED International Conference on Modelling, Identification and Control*, Innsbruck, Austria. http://dx.doi.org/10.2316/P.2013.794-032

Geuzaine, C., Remacle, J.-F., 2009, "Gmsh: a Three-dimensional Finite Element Mesh Generator with Built-in Pre- and Post-processing Facilities," *International Journal for Numerical Methods in Engineering*, **79**(11), 1309-1331. http://dx.doi.org/10.1002/nme.2579

Goldberg, S.N., Gazelle, G.S., Dawson, S.L., Rittman, W.J., Mueller, P.R., Rosenthal, D.I., 1995, "Tissue Ablation with Radiofrequency: Effect of Probe Size, Gauge, Duration, and Temperature on Lesion Volume," *Academic Radiology*, **2**(5), 399-404. http://dx.doi.org/10.1016/S1076-6332(05)80342-3

Haemmerich, D. and Schutt, D.J., 2011, "RF Ablation at Low Frequencies for Targeted Tumor Heating: In Vitro and Computational Modeling Results," *IEEE Transactions on Biomedical Engineering* **58**(2), 404-410. http://dx.doi.org/10.1109/TBME.2010.2085081

JCGM 100, 2008, *Evaluation of Measurement Data - Guide to the Expression of Uncertainty in Measurement*, BIPM, Sèvres, France. http://www.bipm.org/en/publications/guides/gum.html

McGahan, J.P., Brock, J.M., Tesluk, H., Gu, W.Z., Schneider, P., Browning, P.D., 1992, "Hepatic Ablation with Use of Radiofrequency Electrocautery in the Animal Model," *Journal of Vascular and Interventional Radiology*, **3**(2), 291-297. http://dx.doi.org/10.1016/S1051-0443(92)72028-4

Muhieddine, M., Canot, É., March, R.J., Delannay, R., 2011, "Coupling Heat Conduction and Water–steam Flow in a Saturated Porous Medium," *International Journal for Numerical Methods in Engineering*, **85**(11), 1390-1414. http://dx.doi.org/10.1002/nme.3022

Mulier, S., Jiang, Y., Wang, C., Jamart, J., Marchal, G., Michel, L., Ni, Y., 2012, "Bipolar Radiofrequency Ablation with Four Electrodes: Ex vivo Liver Experiments and Finite Element Method Analysis. Influence of Inter-electrode Distance on Coagulation Size and Geometry," *Internatonal Journal of Hyperthermia* **28**(7), 686-697. http://dx.doi.org/10.3109/02656736.2012.706729

Rossi, S., Fornari, F., Pathies, C., Buscarini, L., 1990, "Thermal Lesions Induced by 480 kHz Localized Current Field in Guinea Pig and Pig Liver," *Tumori*, **76**(1), 54-57.

Rossi, S., Gallati, M., Braschi, G., Raboni, M., 2008, "Modelling the Liver RF Thermal Lesions", *Proceedings of the 17th IASTED International Conference on Applied Simulation and Modelling*, Corfu, Greece.

Schutt, D.J. and Haemmerich, D., 2008, "Effects of Variation in Perfusion Rates and of Perfusion Models in Computational Models of Radio Frequency Tumor Ablation," *Medical Physics* **35**(8), 3462-3470. http://dx.doi.org/10.1118/1.2948388

Tosi, D., Macchi, E.G., Braschi, G., Gallati, M., Cigada, A., Poeggel, S., Leen, G., Lewis, E., 2014, "Monitoring of Radiofrequency Thermal Ablation in Liver Tissue Through Fibre Bragg Grating Sensors Array," *Electronics Letters,* **50**(14), 981–983. http://dx.doi.org/10.1049/el.2014.0620

Trujillo, M., and Berjano, E., 2013, "Review of the Mathematical Functions Used to Model the Temperature Dependence of Electrical and Thermal Conductivities of Biological Tissue in Radiofrequency Ablation," *International Journal of Hyperthermia*, **29**(6), 590-597. http://dx.doi.org/10.3109/02656736.2013.807438

Trujillo, M., Alba, J., Berjano, E., 2012, "Relationship Between Roll-off Occurrence and Spatial Distribution of Dehydrated Tissue during RF Ablation with Cooled Electrodes," *International Journal of Hyperthermia*, **28**(1), 62-68. http://dx.doi.org/10.3109/02656736.2011.631076

van Sonnenberg, E., McMullen, W., Solbiati, L., 2005, *Tumor Ablation: Principles and Practice*, Springer, New York, NY. http://dx.doi.org/10.1007/0-387-28674-8

Viglianti, B.L., Dewhirst, M.W., Gorman, J.M., Abraham, J.P., Sparrow, E.M., 2014, "Rationalization of Thermal Injury Quantification Methods: Application to Skin Burns," *Burns*, **40**(5), 896-902. http://dx.doi.org/10.1016/j.burns.2013.12.005

Wood, M., Goldberg, S., Lau, M., Goel, A., Alexander, D., Han, F., Feinstein, S., 2011, "Direct Measurement of the Lethal Isotherm for Radiofrequency Ablation of Myocardial Tissue," *Circulation: Arrhythmia and Electrophysiology* **4**(3), 373-378. http://dx.doi.org/10.1161/CIRCEP.110.961169

Yang, D., Converse, M.C., Mahvi, D.M., Webster, J.G., 2007, "Expanding the Bioheat Equation to Include Tissue Internal Water Evaporation during Heating," *IEEE Transactions on Biomedical Engineering*, **54**(8), 1382-1388. http://dx.doi.org/10.1109/TBME.2007.890740

Zurbuchen, U., Holmer, C., Lehmann, K.S., Stein, T., Roggan, A., Seifarth, C., Buhr, H.J., Ritz, J.P., 2010, "Determination of the Temperature-dependent Electric Conductivity of Liver Tissue Ex Vivo and In Vivo: Importance for Therapy Planning for the Radiofrequency Ablation of Liver Tumours," *International Journal of Hyperthermia*, **26**(1), 26-33. http://dx.doi.org/10.3109/02656730903436442

3D NUMERICAL ANALYSIS ON FLOW CONFIGURATIONS AND HEAT TRANSFER CHARACTERISTICS FOR FIN-AND-OVAL-TUBE HEAT EXCHANGER WITH V-DOWNSTREAM DELTA WINGLET VORTEX GENERATORS

Amnart Boonloi[a] and Withada Jedasadaratanachai[b,*]

[a]*Department of Mechanical Engineering Technology, College of Industrial Technology, King Mongkut's University of Technology North Bangkok, Bangkok 10800, Thailand*

[b]*Department of Mechanical Engineering, Faculty of Engineering, King Mongkut's Institute of Technology Ladkrabang, Bangkok 10520, Thailand*

ABSTRACT

3D numerical investigations for heat transfer characteristics and flow configurations in a fin–and-oval-tube heat exchanger with V-tip pointing downstream delta winglet pairs (*DDWP*) are examined. The *DDWP*s are placed on the fin surface with pointing downstream and the oval tube row number is set at three in a staggered arrangement. The flow attack angles (θ = 15°, 30°, 45° and 60°) and the distance from V-tip to the oval tube center in transverse axis (a = 3.77, 4.77 and 5.77 mm) are investigated for Reynolds number based on hydraulic diameter, Re = 500 – 2500. The numerical results are compared with the previous experimental results and the validations of the grid system on both heat transfer and friction factor are also reported. As the results, it is found that the use of the *DDWP* performs higher heat transfer rate and the friction factor than the smooth fin with no *DDWP* for all cases. The vortex flows and the impinging jet of the flows over the oval-tube walls lead to the increase in heat transfer rate, but also increase in pressure loss. The augmentations are maximum around 1.59 and 4.4 times higher than the smooth fin for heat transfer and friction factor, respectively. In addition, the optimum thermal enhancement factor, *TEF*, is around 1.09 at Re = 2500, a = 5.77 mm and θ = 15°.

Keywords: *Compact heat exchanger, Delta winglet, Fin-and-oval-tube, Laminar flow, Thermal enhancement factor*

1. INTRODUCTION

Compact heat exchangers as a fin-and-tube heat exchanger are widely used in many industrials and equipment; electronic equipment, air conditioning, automobiles, refrigeration, radiator, chemical industry, etc. Several investigators had been attempted to enhance thermal performance of the compact heat exchanger by using turbulators or vortex generators. Many types of the vortex generator; winglet, wing, baffle, rib, groove, etc., had been installed in the heat exchanger for augmenting heat transfer and their performance. The use of the vortex generators leads to increase in vortex strength or turbulence intensity, secondary flows and disrupting the boundary layer. The vortex generator can be divided into two types; longitudinal vortex generators (*LVG*) and transverse vortex generator (*TVG*) considered from the axes of vortices which created from the vortex generators.

Thermal performance enhancement in fin-and-tube heat exchangers with various types of vortex generators has been studied. The experimental and numerical investigations on the vortices in internal flow were reported by Fiebig, 1995. He found that the longitudinal vortices perform better heat transfer augmentation than transverse vortices with similar pressure loss values. He also summarized that the winglets give the nearly value of heat transfer rate for less pressure loss in comparison with the wings. Zhu *et al.*, 1995, numerical studied combined vortex generators; longitudinal vortex generators and rib turbulators, in a rectangular channel on heat transfer for turbulent flow regime. They reported that the combined vortex generators can increase the average Nusselt number around 450%. The literature reviews for heat transfer augmentation with longitudinal vortices were explained by Jacobi and Shah, 1995. Biswas *et al.*, 1996, investigated on both numerical and experimental for the influences of longitudinal vortices in a channel on heat transfer characteristics and flow configurations. They concluded that the flow configurations can be divided into three parts; main vortex flow, corner vortex flow and induced vortex flow. They also found that the appearance of the complex flow provided the increase in heat transfer and thermal performance in the test section. Lee *et al.*, 1999, numerical investigated a test channel with longitudinal vortices for turbulent regimes. They reported that the disturbance of the boundary layer nears the channel wall regime results in the heat transfer augmentation. The influences of the vortex generator configurations on heat transfer and fluid flow were studied by Liou *et al.* (2000). They found that the important factors for heat transfer enhancement are the direction and intensity of the secondary flow. Torii *et al.* (2002) experimentally studied on the effect of the delta winglet vortex generators in a fin-and-tube heat exchanger with in-line or staggered tube banks on heat transfer and pressure loss. They showed that the staggered tube banks case provides the heat transfer augmentation around 10 – 30% and the pressure loss is reduced by 34–55% for Re = 350 – 2100.

* *Corresponding author. Email: kjwithad@kmitl.ac.th*

Gentry and Jacobi (2002) studied the influences of the delta-wing-generated tip vortices in a flat-plate and developing channel flows. The enhancements were around 55% and 100% for heat transfer and friction factor, respectively, which described by Gentry and Jacobi, 2002. Leu *et al.* (2004) investigated a plate-fin and tube heat exchanger with inclined block shape vortex generators placed behind the tubes on heat transfer and flow configurations by using numerical and experimental methods. Sommers and Jacobi, 2005, experimental investigated the use of vortex generators on the air-side of the evaporator. They concluded that the use of the vortex generators can reduce the thermal resistance around 35 – 42% of Re = 500 – 1300. Pesteei *et al.*, 2005 found that the optimum heat transfer enhancement for winglet pairs is appeared when placed at the downstream regime. The numerical investigation on the winglet pair with zero thickness on heat transfer was described by Hiravennavar *et al.*, 2007. They found that the heat transfer enhancement is around 33% and 67% for single winglet and winglet pair, respectively.

The investigations on heat transfer augmentations in a finned oval tube with inline longitudinal vortex generators (Chen *et al.*, 1998) and with staggered longitudinal vortex generators (Chen *et al.*, 2000) were reported. Tiwari *et al.*, 2003, numerical investigated heat transfer and thermal performance in a channel with built-in oval tube and delta winglet vortex generators for laminar flow. They exposed that the use of the winglet pairs can significantly increase in heat transfer rate and thermal performance in the heat exchanger. O'Brien *et al.*, 2004, studied an elliptical tube with one or two delta-winglet pairs in a rectangular channel on thermal performance. They reported that the presence of the single winglet pair to the oval-tube geometry leads to the increase in heat transfer around 38% in comparison with no winglet case. However, the use of the delta-winglet pairs not only increase in heat transfer but also increase in friction factor which also concluded by O'Brien *et al.* (2004). Chu *et al.* (2009) studied a fin-and-oval-tube heat exchanger with longitudinal vortex generators (*LVGs*) on heat transfer configuration, thermal performance and flow behavior. They found that the enhancements are around 13.6–32.9% and 29.2–40.6% when compared with the baseline case for heat transfer and friction loss, respectively.

As previous investigations, it is found that the delta winglet vortex generators can help to improve the thermal performance and heat transfer rate in the heating system. Moreover, the flow attack angle and the installing position of the vortex generators are an important factor for enhancing the heat transfer rate. Therefore, in the present study, the three dimensional numerical investigations on the flow configurations and heat transfer characteristics in fin-and-oval-tube compact heat exchanger with V-tip pointing downstream delta-winglet pairs (*DDWP*) are presented. The influences of the flow attack angle (θ = 15°, 30°, 45° and 60°) and the distance between V-tip to the center of the oval tube in the transverse axis (a = 3.77, 4.77 and 5.77 mm) for the Reynolds number (Re = 500 – 2500) are studied.

2. FLOW DESCRIPTIONS

2.1 Physical models

According to Chu *et al.* (2009), Fig. 1 presents a fin-and-oval-tube compact heat exchanger which interesting in the current research work. The computational domain of the fin-and-oval-tube heat exchanger with *DDWPs* is shown in Figs. 2 – 4. The delta winglet pairs are placed in front of the oval tubes with pointing downstream (V-tip downstream). The winglet and the channel height are identical, H. The aspect ratio $\Lambda(4H/c)$ for delta winglet equals to 2. The influences of the flow attack angle, θ = 15°, 30°, 45° and 60°, and the distances between V-tip and the center of oval tube, a = 3.77, 4.77 and 5.77 mm are investigated for Re = 500 – 2500. The parameters of the computational domain are shown in table1.

To ensure a recirculation-free flow regime, the exit region is extended around $10H$. The entry region of the domain is developed by $10H$ to maintain the uniform velocity.

2.2 Boundary conditions and governing equations

The flow is assumed to be steady laminar flow due to the low inlet velocity flowing though the small fin pitches. The fin thickness is taken into account in heat conduction. The considered fluid is an incompressible with constant properties. The boundary conditions are shown as follows.

1. At the entrance extended region, uniform velocity and temperature boundary condition are used at the inlet of the computational domain. Pressure outlet condition is employed at the outlet of the exit extended region. Symmetry condition is used at side boundaries of both extended regions.
2. For the fin-and-oval-tube region, no slip condition and constant wall temperature are used for a fin and tube surface. Symmetry condition is used at the both side boundaries.

The temperatures at the inlet of the air and tube wall are set at 340K and 300K, respectively. The physical properties of the air are assumed to be constant at average bulk temperature. The inlet air velocities vary in the range 1.3 to 6.5 m/s depended on Reynolds number, Re = 500 – 2500.

The governing equations for continuity, momentum and energy can be expressed as follows

Continuity equation:

$$\frac{\partial}{\partial x_i}(\rho u_i) = 0 \quad (1)$$

Momentum equation:

$$\frac{\partial}{\partial x_i}(\rho u_i u_k) = -\frac{\partial P}{\partial x_k} + \frac{\partial}{\partial x_i}\left[\mu\left(\frac{\partial u_k}{\partial x_i}\right)\right] \quad (2)$$

Energy equation:

$$\frac{\partial}{\partial x_i}(\rho u_i T) = \frac{\partial}{\partial x_i}\left(\frac{k}{C_p}\frac{\partial T}{\partial x_i}\right) \quad (3)$$

The important parameters in the present study are the Reynolds number (Re), friction factor (f), Nusselt number (Nu), and thermal enhancement factor (TEF) which as follows

$$Re = \frac{\rho U_c D_h}{\mu} \quad (4)$$

$$f = \frac{(2\Delta P)}{\rho U_c^{\,2}}\left(\frac{D_h}{L}\right) \quad (5)$$

$$Nu = \frac{hD_h}{k} \quad (6)$$

$$TEF = \frac{\frac{Nu}{Nu_0}}{\left(\frac{f}{f_0}\right)^{\frac{1}{3}}} \quad (7)$$

where, Nu and Nu_0 are area-average Nusselt number and average Nusselt number of plain fin, respectively.

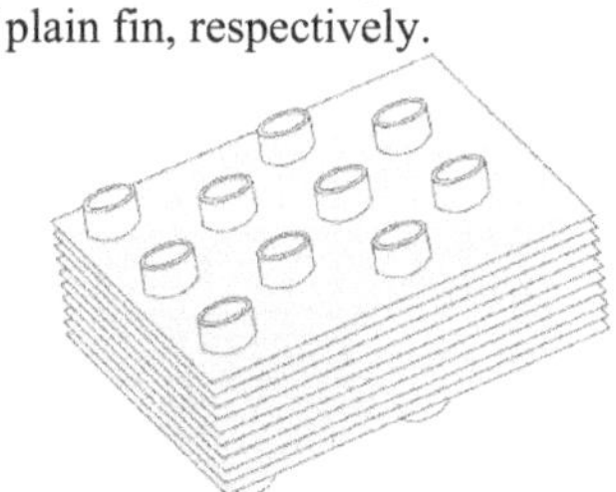

Fig. 1 Fin-and-oval-tube compact heat exchanger.

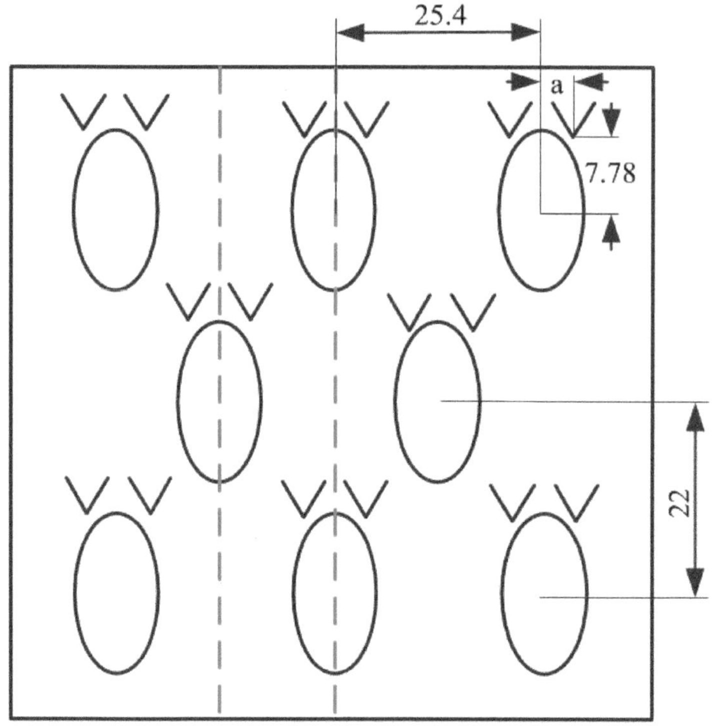

Fig. 2 Schematic diagram of computational domain.

Table 1 Parameters of the computational domains (Chu *et al.*, 2009).

Parameter	Symbol/unit	Value
Channel height	H/mm	3.2
Computational domain width	B/mm	12.7
Computational domain length	L/mm	64.4
Semi-major diameter of oval tube	R_a/mm	6.28
Semi- minor diameter of oval tube	R_b/mm	3.77
Winglet placed downstream from inlet	Y/mm	17.92
Longitudinal tube pitch	P_l/mm	22
Spanwise tube pitch	P_s/mm	25.4
Fin thickness	F_t/mm	0.33
Fin pitch	F_p/mm	3.2
Number of tube row	n	3
Hydraulic diameter	D_h/mm	3.63
Wall temperature	T_w/K	300
Inlet temperatures of air	T_{in}/K	340
Frontal velocity	u_{in}/m s-1	1.3 – 6.5

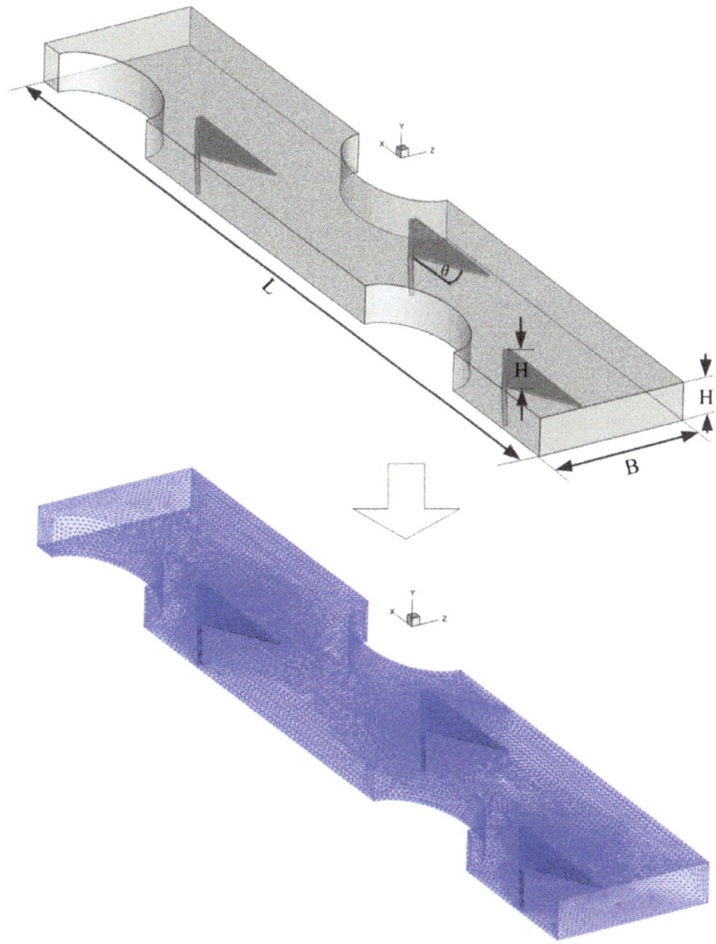

Fig. 3 Channel geometry and computational domain.

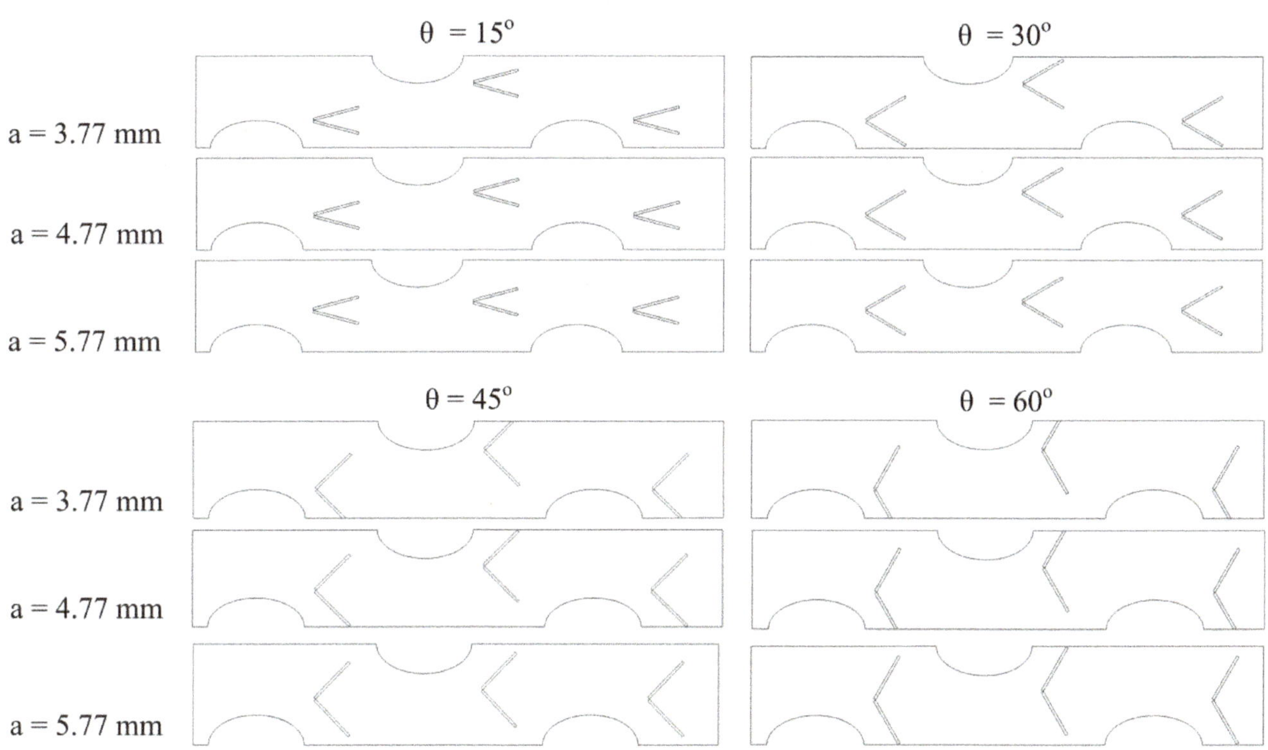

Fig. 4 Delta winglet configuration with various cases.

2.3 Method of research

The computational fluid dynamics code is used for solving previous governing equations and the boundary conditions. The convective terms in the governing equations for momentum and energy are discretized with Power law scheme and QUICK scheme, respectively. The SIMPLE algorithm has been applied to the coupling among pressure and velocity. The convergence criterions for the velocities and temperature are arranged by the residual less than 10^{-5} and 10^{-9}, respectively.

3. RESULTS AND DISCUSSIONS

3.1 Verification of the computational domain

The validations of the computational domain are separated into two parts; grid systems and compared with the experimental results. Grid independence is required to ensure the accuracy and validity of the numerical results. In order to validate the independency of solution on the grid, three difference grid systems which including 300000, 400000 and 500000 cells are investigated. The results on the three sets of grid found that the relative errors of the Nusselt number and friction factor are less than 1.1 % and 1.2 %, respectively. Thus, there is no advantage for increase grid cells; the computational domain is applied with 300000 cells of grid number.

Except from the investigations on the grid systems, the flow configurations and heat transfer characteristics are compared with the experimental results (Chu *et al*., 2009) in terms of Nusselt number and pressure loss, respectively, as depicted in Fig. 5*a* and *b*. As seen in the figures, the average discrepancy between the predicted and the experimental results is less than 9% and 4% for Nusselt number and the pressure loss, respectively. The good agreement with the validation results indicates that the computational domain is reliable to predict heat transfer characteristics and flow configurations in the fin-and-oval-tube heat exchangers.

3.2 Flow topology

The flow configurations for the *DDWP*s in the fin-and-oval-tube heat exchanger are presented in terms of streamlines in transverse planes and the three dimensional streamlines flow over the *DDWP*s as Figs. 6 – 8. The numerical results can be described into two parts; the influence of the flow attack angle and distance between oval tube center and V-tip. The effects of the flow attack angle, Figs. 6*a*, *b*, *c* and *d* illustrate the streamlines in transverse planes for flow attack angles, θ = 15°, 30°, 45° and 60°, at *Re* = 1000 and *a* = 3.77 mm, respectively. The streamlines in transverse planes for each case are located at the V - tip and the winglet leading edges (*WLE*). As figures, there are found that the *DDWP*s produce the vortex flows as clearly seen at the V-tip planes for all cases. The differences of the flow attack angle result in the variations of the vortex flow intensity and the vortex core locations. It is seen that the rise of the flow attack angle leads to the increase in vortex strength; the 60° *DDWP* performs the highest of the vortex intensity while the 15° *DDWP* provides the lowest.

Effects of the distance between V-tip and oval tube center, Figs. 7*a*, *b* and *c* show the streamlines in transverse planes for 30° *DDWP* at *Re* = 1500 and *a* = 3.77, 4.77 and 5.77, respectively. In general, the flow configurations are appearing similar, but, the cores of the vortex flows are changed depending on the positions of the *DDWP*s. Due to the streamlines in transverse planes cannot describe the influence of the distance between V-tip and oval tube center, therefore, the effect of the distance from V-tip to the center of the oval tube will be concluded by graph for the variance of Nusselt number and friction factor in the next part.

The streamlines flow over the *DDWP* in the test section is presented in Fig. 8 for the 30° of the flow attack angle at *Re* = 1000 and *a* = 3.77 mm. As seen from the figure, the fluid flows become swirling when the flow passing the *WLE* before impinge at the leading curve of the oval tube. This phenomenon is a key for enhancing the heat transfer rate and thermal performance in the compact heat exchanger.

In addition, the presences of the vortex flows which created by *DDWP*s can help with the mixing of the fluid flow that lead to the increase in heat transfer rate over the tube walls. However, the use of the *DDWP* on only increases in the heat transfer, but also increases in pressure loss in the fin-and-oval-tube heat exchanger.

3.3 Velocity fields

In this part, the velocity profiles in the fin-and-oval-tube heat exchanger with *DDWP*s are displayed in the form of the local distributions of the x-velocity in the middle cross section plane. The influences of the flow attack angles and the distance from V-tip to the center of the oval tube, *a*, values are also presented.

Figs. 9 and 10 illustrate the local distributions of the x-velocity on the middle cross section (parallel to the x-z plane) at various the flow attack angles and the *a* value, respectively. It can be seen in the figures that the velocity fields provide uniformly at the inlet regions of the test sections for all cases. When the fluid flows passing the *DDWP*s, the velocity performs higher, especially, at the middle regimes between the first pairs and the third pairs for the *DDWP*s and on both two sides regimes of the second pairs. The numerical results are according to the flow configurations of the previous part.

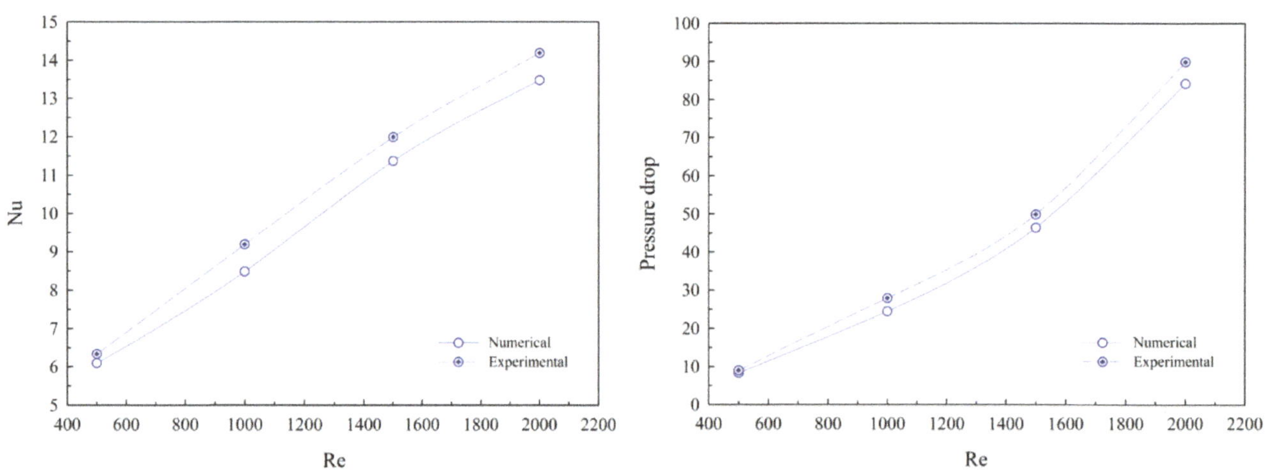

(a) (b)

Fig. 5 Experimental-numerical comparison for model validations (a) Nusselt number and (b) pressure drop

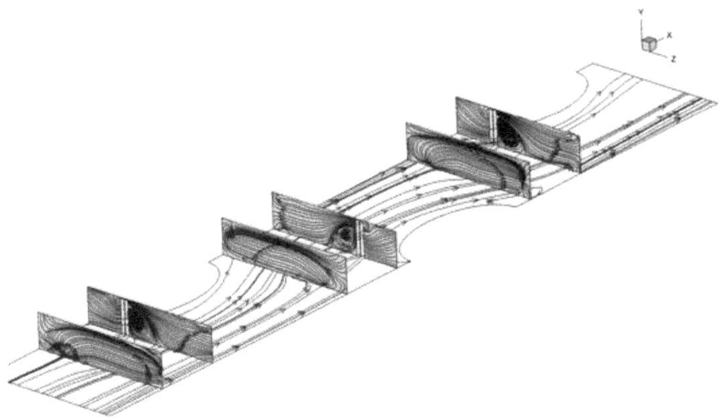

(a)

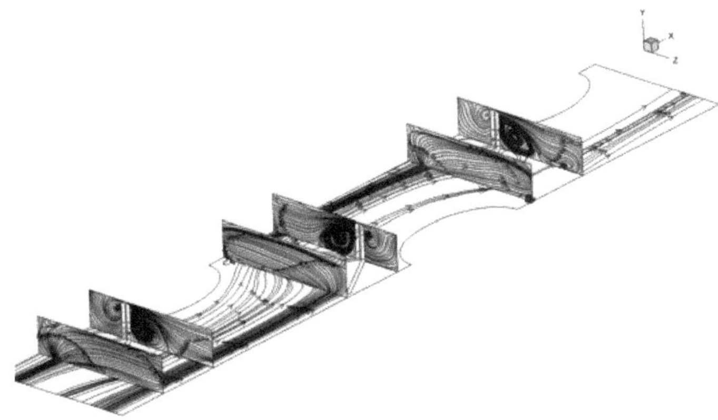

(b)

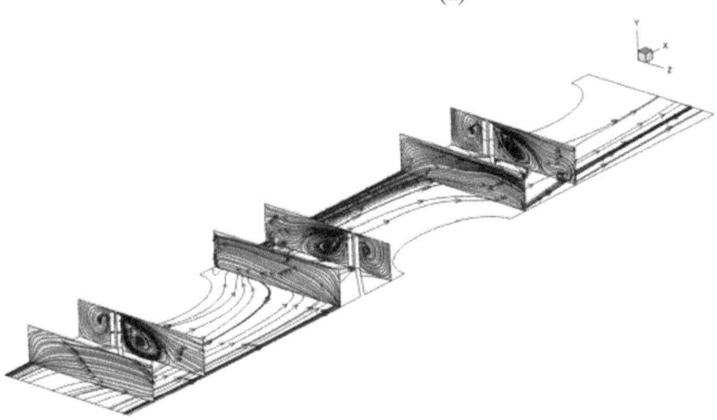

(c)

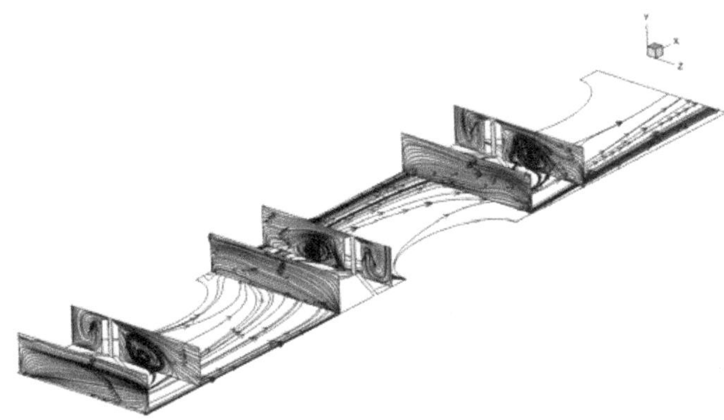

(d)

Fig. 6 Streamlines in transverse planes for (a) 15°, (b) 30°, (c) 45° and (d) 60° at *Re* = 1000 and *a* = 3.77 mm.

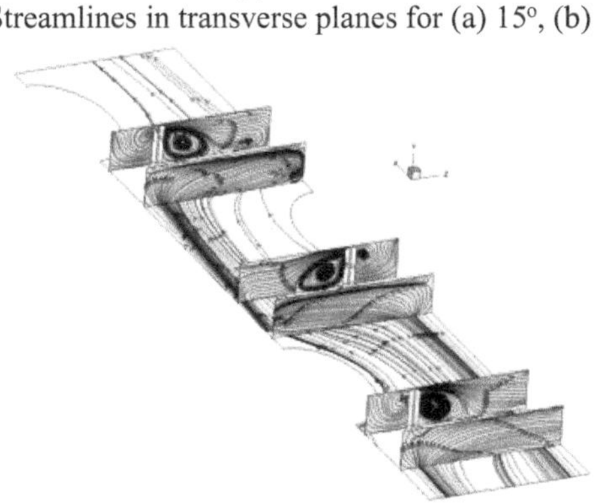

(a)

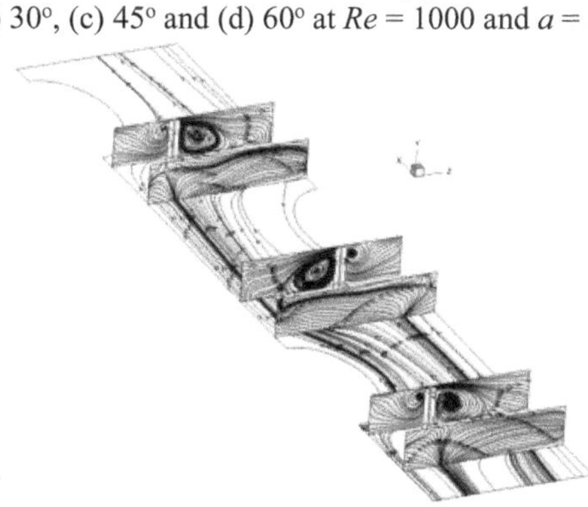

(b)

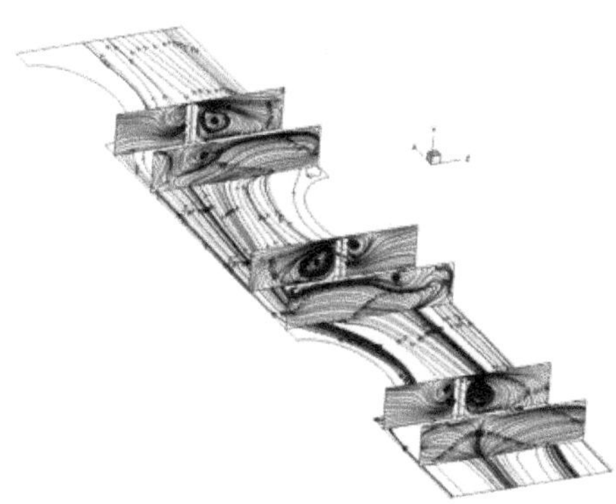

(c)

Fig. 7 Streamlines in transverse planes for (a) *a* = 3.77 mm, (b) *a* = 4.77 mm and (c) *a* = 5.77 mm at *Re* = 1500 and θ = 30°.

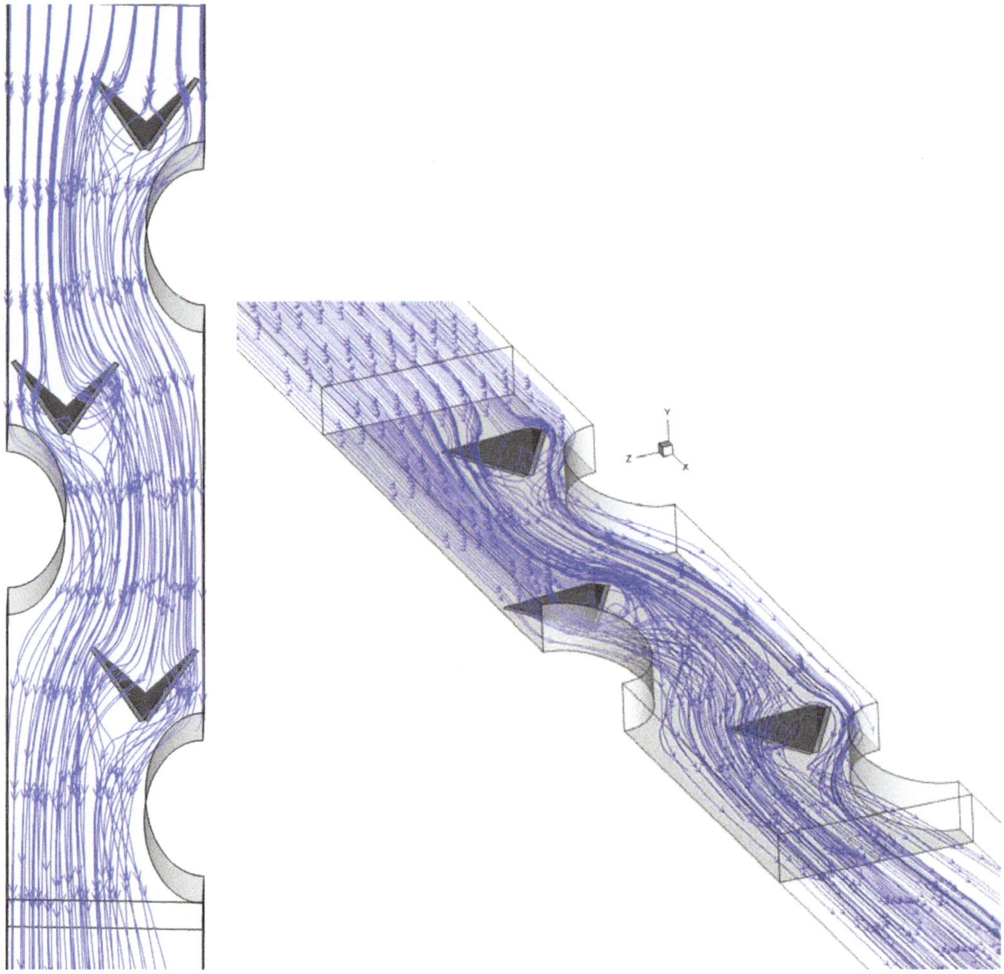

Fig. 8 Streamlines in three dimensions for $a = 3.77$ mm at $Re = 1500$ and $\theta = 30^o$.

The local distributions of the x-velocity on the middle cross section for the flow attack angle, $\theta = 15^o$, 30°, 45° and 60°, at $Re = 1000$ and $a = 3.77$ mm are presented in the Figs. 9*a*, *b*, *c* and *d*, respectively. It is found that the differences of the flow attack angle have an effect of the velocity fields. The higher of the flow attack angle lead to the increase in the velocity values. It is clearly seen that the $\theta = 60^o$ performs the most different of the velocity field, the large blue contours are found behind the *DDWP*s regimes when compared with other cases. The velocity distributions are found consistently in the case of $\theta = 15^o$. The rise of the flow attack angle leads to higher of the velocity values. The reason of this may be that the high flow attack angle effects for the increasing turbulence intensity or vortex strength. Although, the high flow attack angle, $\theta = 60^o$, leads to the increase in heat transfer rate, but the better velocity distributions are found at the low flow attack angle, $\theta = 15^o$.

Figs. 10*a*, *b* and *c* present the local distributions of the x-velocity at the middle cross section of the flow attack angle, $\theta = 30^o$ and $Re = 1500$ for $a = 3.77$, 4.77 and 5.77 mm, respectively. In general, the velocity distributions for all cases, perform nearly profile; the uniform velocity appears in the entry regions of the test sections and performs increase when passing the *DDWP*s. The variations of the velocity profiles depended on the position of the *DDWP* in the test section. It seems that the distance between V-tip and oval tube center, $a = 3.77$ mm gives the highest values of the velocity while the distance between V-tip and oval tube center, $a = 4.77$ and 5.77 mm show nearly values of the velocity. Additionally, the increase in the distance between V-tip and oval tube center results in the augmenting velocity values.

3.4 Heat transfer characteristics

The heat transfer characteristics in the fin-and-oval-tube heat exchanger with *DDWP*s are considered from the local distributions of the temperature and the local Nusselt number over the tube walls.

Figs. 11 and 12 show the temperature contours at the middle cross sections in the test section with *DDWP*s at various the flow attack angles and the distance between V-tip and oval tube center values, respectively. In general, the local distributions of the temperature for all cases are nearly profiles; the uniform contours of 340K are found in the inlet regimes of all test sections and the lower temperature values are found at behind the oval tube regimes.

Figs. 11*a*, *b*, *c* and *d* present the local distributions of the temperature in the test sections with *DDWP*s for the flow attack angle, $\theta = 15^o$, 30°, 45° and 60°, respectively, at $Re = 1000$ and $a = 3.77$mm. Considering at behind the oval tubes, the differences of the contour temperature are found, especially, for the $\theta = 15^o$. It is seen that the $\theta = 15^o$ performs the best mixing of the fluid flow at behind the oval tube while the $\theta = 30^o$, 45° and 60° give better mixing of the fluid flow when passing the third pairs of the *DDWP*s.

The temperature contours of the $\theta = 30^o$, $Re = 1500$ at the middle cross section for $a = 3.77$, 4.77 and 5.77 mm are illustrated in Figs. 12*a*, *b* and *c*, respectively. As seen, the case of $a = 3.77$ mm performs better mixing of the fluid flow at behind the third pairs of the *DDWP*s when comparing with the cases of $a = 4.77$ and 5.77 mm. The case of $a = 5.77$ mm, the better mixing of the fluid flow is found at behind the oval tube. Additionally, the case of $a = 3.77$ mm gives the best mixing of the fluid flow in comparison with other *a* values.

The local Nusselt number distributions over the tube walls for *DDWPs* in the test sections are shown in Figs. 13 and 14 at various the flow attack angles and the *a* value, respectively. It is found that the peaks regimes of the heat transfer rate are found at the leading curve of the oval tube walls for all cases due to the impinging of the jets in these areas as described in the Fig. 8 in the flow topology part. Additionally, the impinging jet of the fluid flow on the leading curve of the oval tube wall is a main factor for heat transfer augmentation.

Figs. 13*a*, *b*, *c* and *d* present the local Nusselt number contours of the *DDWPs* in the fin-and-oval-tube heat exchanger for the flow attack angle, $\theta = 15°$, $30°$, $45°$ and $60°$, respectively, at $Re = 1000$ and $a = 3.77$ mm. It is appeared that the $\theta = 60°$ performs the highest heat transfer rate while the $\theta = 15°$ gives the lowest values of the heat transfer rate. This means that the larger flow attack angle results in the higher vortex flow intensity that leads to the augmenting heat transfer rate in the compact heat exchanger.

Figs. 14*a*, *b* and *c* illustrate the contours of the Nusselt number at $\theta = 30°$, $Re = 1500$ for $a = 3.77$, 4.77 and 5.77 mm, respectively, for *DDWPs* in the compact heat exchanger. The similar profiles or very close configurations of the local Nusselt number contours for all cases are appearing, so, the effect of the distance values, *a*, on heat transfer characteristics cannot consider from the local Nusselt number contours. The influence of the distance values on heat transfer will be concluded in the form of the Nusselt number values in the part of the performance evaluation.

3.5 Performance evaluation

There are three parts in this section for considering; heat transfer, pressure drop and the thermal performance. The performance evaluations in the fin-and-oval-tube heat exchanger with *DDWPs* are presented in terms of the variations of the Nusselt number, the Nusselt number ratio (Nu/Nu_0), the friction factor, the friction factor ratio (f/f_0) and the thermal enhancement factor (*TEF*) with Reynolds number as depicted in Figs. 15, 16, 17, 18 and 19, respectively.

Figs. 15*a*, *b* and *c* show the variations of the Nusselt number with Reynolds number at various the flow attack angles for $a = 3.77$, 4.77 and 5.77 mm, respectively. The Nusselt number tends to increase with the increasing Reynolds number for all cases. The flow attack angle, $\theta = 60°$ case performs the highest values of the Nusselt number while the lowest values are found at $\theta = 15°$ for all a values. The Nusselt number values are around 6.6 – 24, 6.8 – 20 and 7 – 19 for $a = 3.77$, 4.77 and 5.77 mm, respectively, for *DDWP* in range $Re = 500 – 2500$ and $\theta = 15° – 60°$.

The variations of the Nu/Nu_0 are presented in the Figs. 16*a*, *b* and *c* for $a = 3.77$, 4.77 and 5.77 mm, respectively. In range $500 \leq Re \leq 2000$, the Nu/Nu_0 tends to increase with the rise of Reynolds number and then slightly decrease when $Re > 2000$ for $a = 3.77$ and 4.77 mm, and for $a = 5.77$ mm at $\theta = 15°$ and $30°$. At $a = 5.77$ mm for $\theta = 45°$ and $60°$, the Nu/Nu_0 increase in range $500 \leq Re \leq 1500$ after that the Nu/Nu_0 slightly decrease. The use of the *DDWP* can help to increase heat transfer in the fin-and-oval-tube heat exchanger around 1.1 – 1.59, 1.12 – 1.56 and 1.15 – 1.50 times over the plain fin with no *DDWP* for $a = 3.77$, 4.77 and 5.77 mm, respectively. The optimum points are found at $Re = 2000$ and $\theta = 60°$ for $a = 3.77$ and 4.77 mm while the maximum enhancing heat transfer is found at $Re = 1500$ and $\theta = 60°$ for $\theta = 5.77$mm. In addition, the increase in the flow attack angle and the decrease in *a* value perform higher heat transfer rate in the heat exchanger with *DDWPs*. This is because the higher turbulence intensity or vortex strength which occurring at a high flow attack angle and at the lower distance form V-tip to the center of the oval tube.

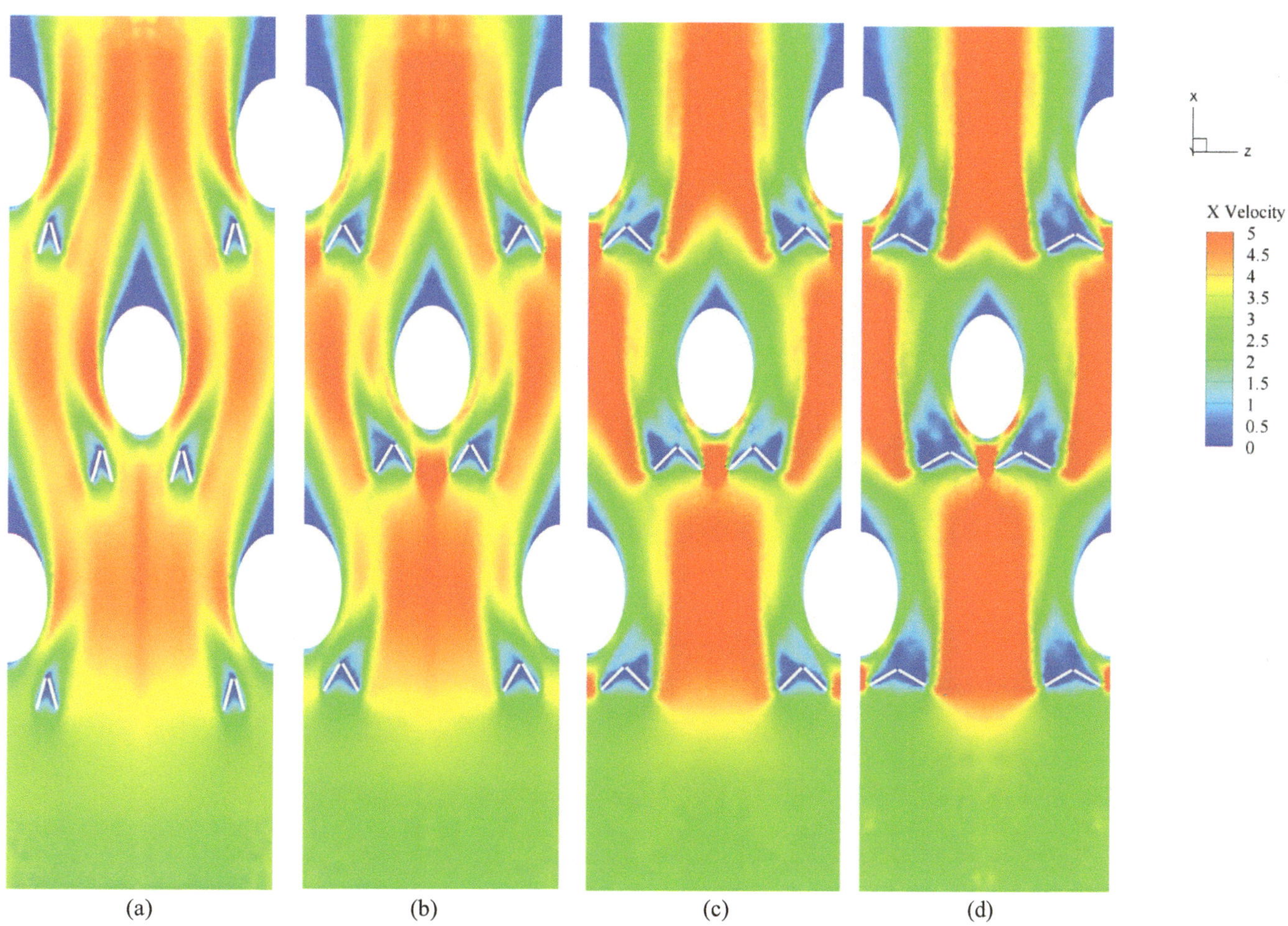

Fig. 9 Contours of x-velocity for (a) 15°, (b) 30°, (c) 45° and (d) 60° at $Re = 1000$ and $a = 3.77$ mm.

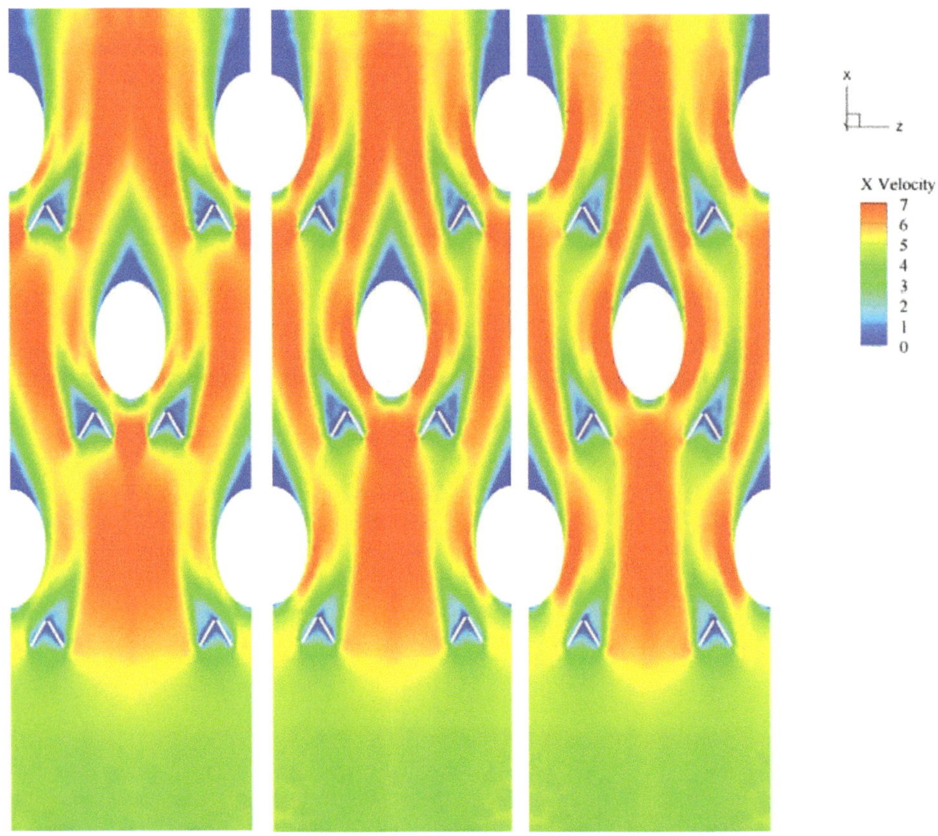

(a) (b) (c)

Fig. 10 Contours of x-velocity for (a) $a = 3.77$ mm, (b) $a = 4.77$ mm and (c) $a = 5.77$ mm at $Re = 1500$ and $\theta = 30°$.

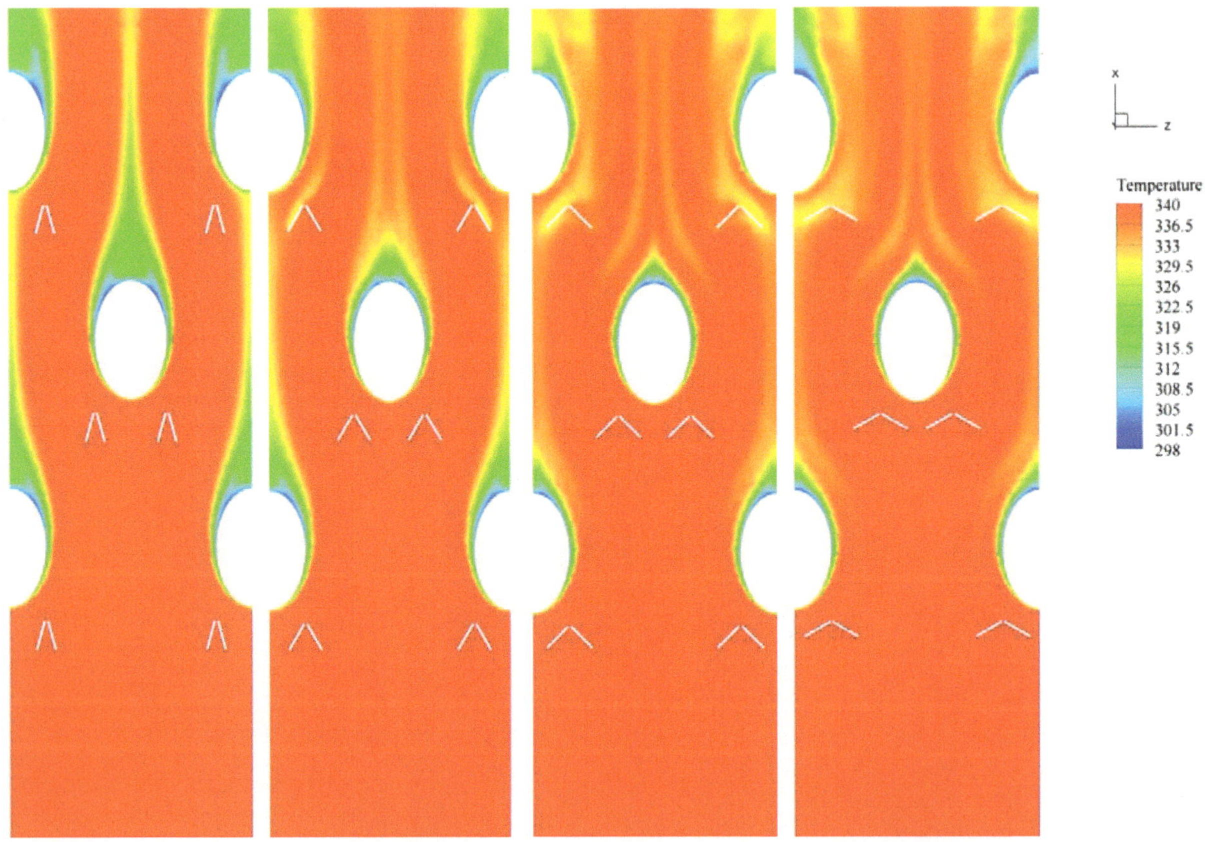

(a) (b) (c) (d)

Fig. 11 Contours of temperature for (a) 15°, (b) 30°, (c) 45° and (d) 60° at $Re = 1000$ and $a = 3.77$ mm.

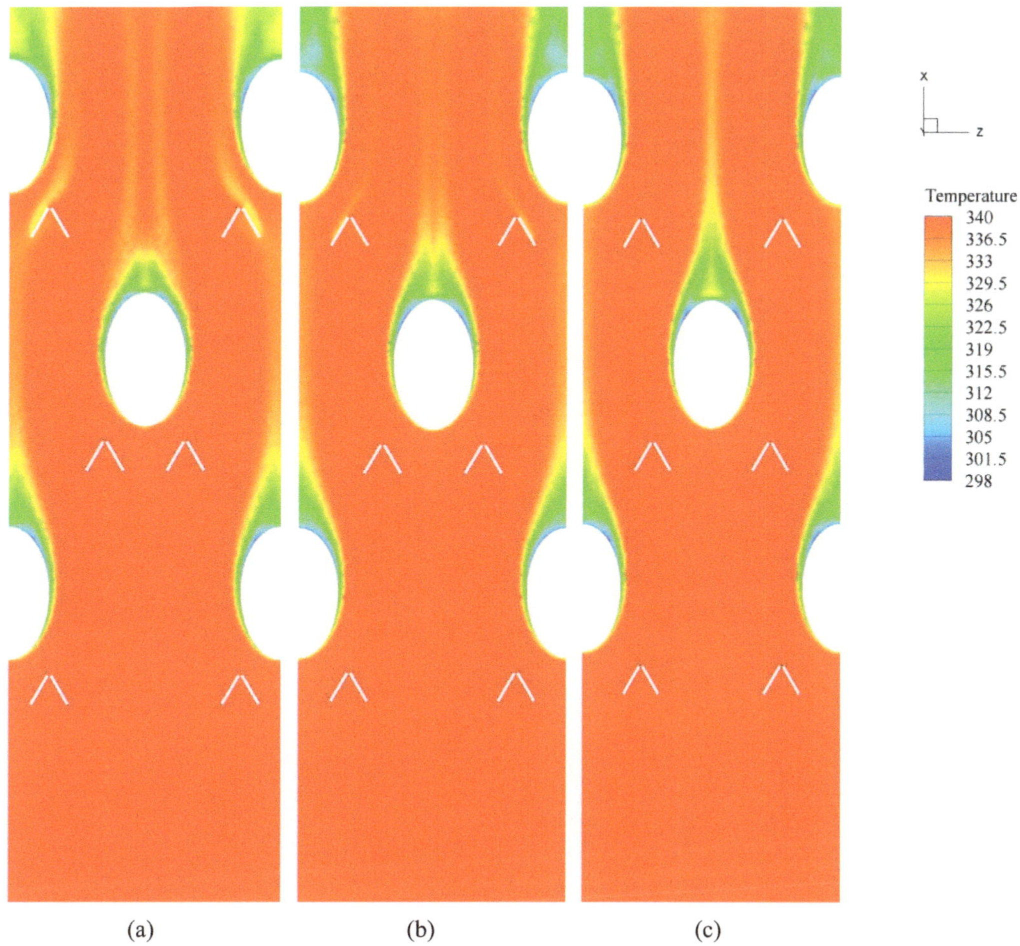

Fig. 12 Contours of temperature for (a) $a = 3.77$ mm, (b) $a = 4.77$ mm and (c) $a = 5.77$ mm at $Re = 1500$ and $\theta = 30°$.

The use of the *DDWP*s in the compact heat exchanger leads to the increasing in the friction factor over the smooth plain fin. Figs. 17*a*, *b* and *c* illustrate the variations of the friction factor with Reynolds number for fin-and-tube heat exchanger with *DDWP*s at $a = 3.77$, 4.77 and 5.77 mm, respectively. As seen in the figures, the friction factor tends to decrease with increasing the Reynolds number for all cases. In range studied, the friction factor is found to be about 0.06 – 0.24 depended on the flow attack angle, *a* value and the Reynolds number.

The f/f_0 values with the Reynolds number for $a = 3.77$, 4.77 and 5.77mm are presented in Figs. 18*a*, *b* and *c*, respectively, at various the flow attack angles. The use of the *DDWP*s performs higher friction factor than the smooth plain fin for all cases. For $\theta = 45°$ and 60°, the f/f_0 provides increased with the increasing Reynolds number for all *a* values. For $\theta = 15°$ and 30°, the f/f_0 tends to increase when the Reynolds number increasing from 500 to 1000, but performs slightly decrease when $Re > 1000$. The f/f_0 values are found to be around 1.6 – 4.4, 1.6 – 4.25 and 1.6 – 4.25 for $a = 3.77$, 4.77 and 5.77 mm, respectively. Additionally, the rise of the flow attack angle provides the increase in the friction factor while the difference between *a* value performs nearly values of the friction factor.

The thermal performance is considered in term of the thermal enhancement factor, *TEF*, which referred from Sriromreun *et al*., 2012. The *TEF* value is calculated from the increasing of the heat transfer rate and also the rising friction factor in the heating system when considering the current test section as rectangular channel. Figs. 19*a*, *b* and *c* show the variations of the *TEF* with the Reynolds number at various the flow attack angles for $a = 3.77$, 4.77 and 5.77 mm, respectively. The *TEF* varies in the range 0.91 – 1.09 depended on the flow attack angle, the Reynolds number and *a* value of the *DDWP* in the compact heat exchanger. The optimum *TEF* values are found to be about 1.075, 1.075 and 1.09 at $Re = 2500$, 2000 and 2500 and $\theta = 30°$, 30° and 15° for $a = 3.77$, 4.77 and 5.77mm, respectively. Although, the $\theta = 45°$ and 60° perform higher heat transfer rate than the $\theta = 15°$ and 30°, but also produces very enlarge pressure loss, therefore, almost the optimum point of the *TEF* is found at $\theta = 30°$ which provides the optimum value on both the heat transfer rate and the friction factor augmentation.

4. CONCLUSIONS

In this article, 3D numerical simulations are examined to study the flow configurations and heat transfer characteristics in the fin-and-oval-tube heat exchangers with *DDWP*s. The influences of the geometrical parameters with respect to fin-and-oval-tube heat exchangers with *DDWP*s; the flow attack angle ($\theta = 15°$, 30°, 45° and 60°) and the a value ($a = 3.77$, 4.77 and 5.77 mm) are studied for the Reynolds number based on the hydraulic diameter, $Re = 500 – 2500$. The major conclusions are drawn as follows:

- The use of the *DDWP*s in the fin-and-oval tube heat exchanger can help to create the vortex flow or swirl flow and also the impinging of the fluid flow at the leading curve of the oval tubes. The fluid flow phenomenon leads to the increase in the heat transfer rate in the compact heat exchanger but also enhance in the friction loss.
- The rise of the flow attack angle leads to the increase in the heat transfer rate and friction factor. The $\theta = 60°$ performs the highest on both heat transfer and the friction factor while the $\theta = 15°$ provides the lowest values. Due to the high of the flow blockage for $\theta = 60°$, when considering in cross sectional area of the test section, that leads to the difference of the pressure between before and after of the *DDWP*s. The higher different values of the pressure are the main reason to produce the strength vortex flow over the test section in comparison with other

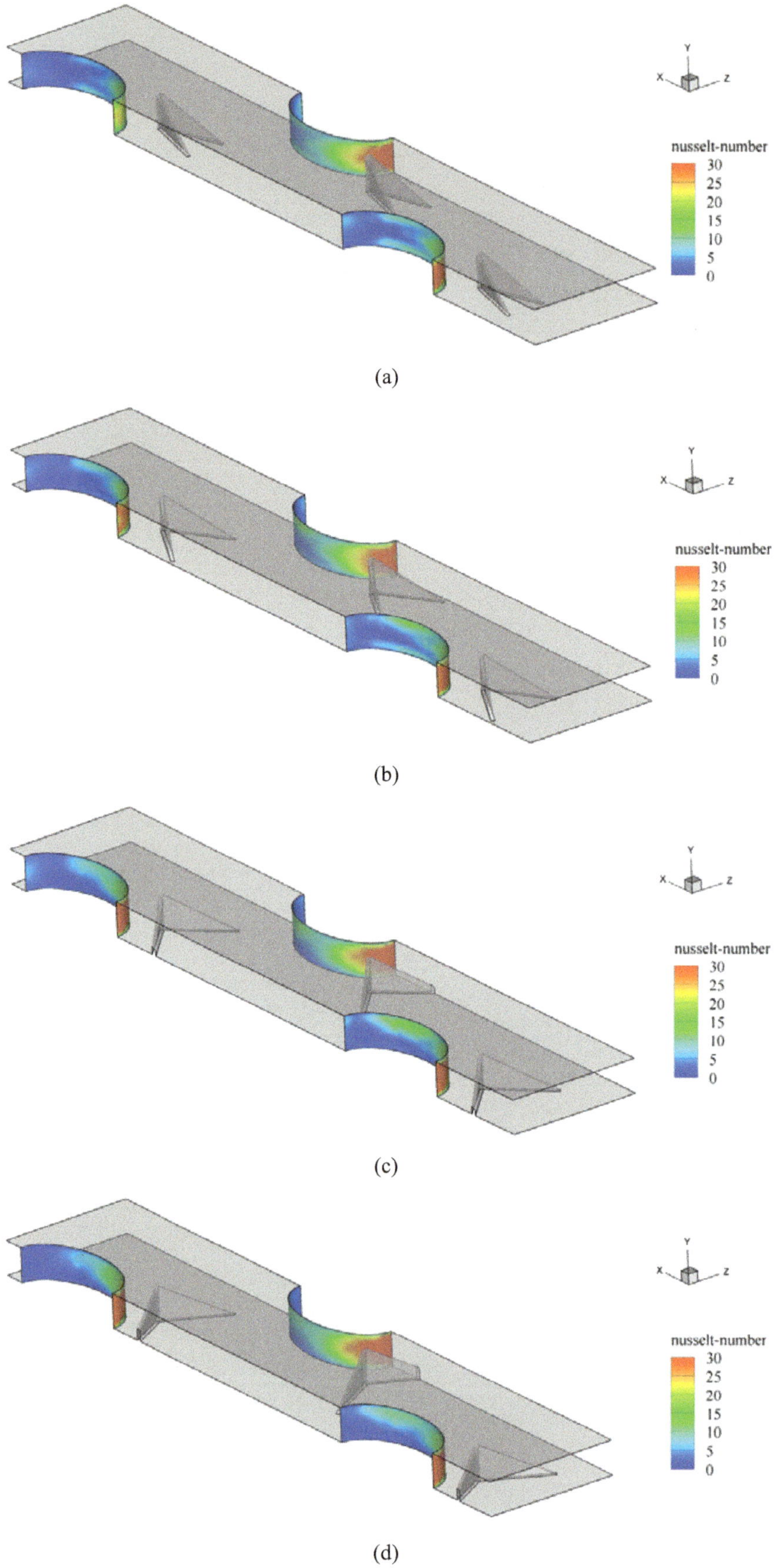

Fig. 13 Contours of local Nusselt number at oval tube walls for (a) 15°, (b) 30°, (c) 45° and (d) 60° at $Re = 1000$ and $a = 3.77$ mm.

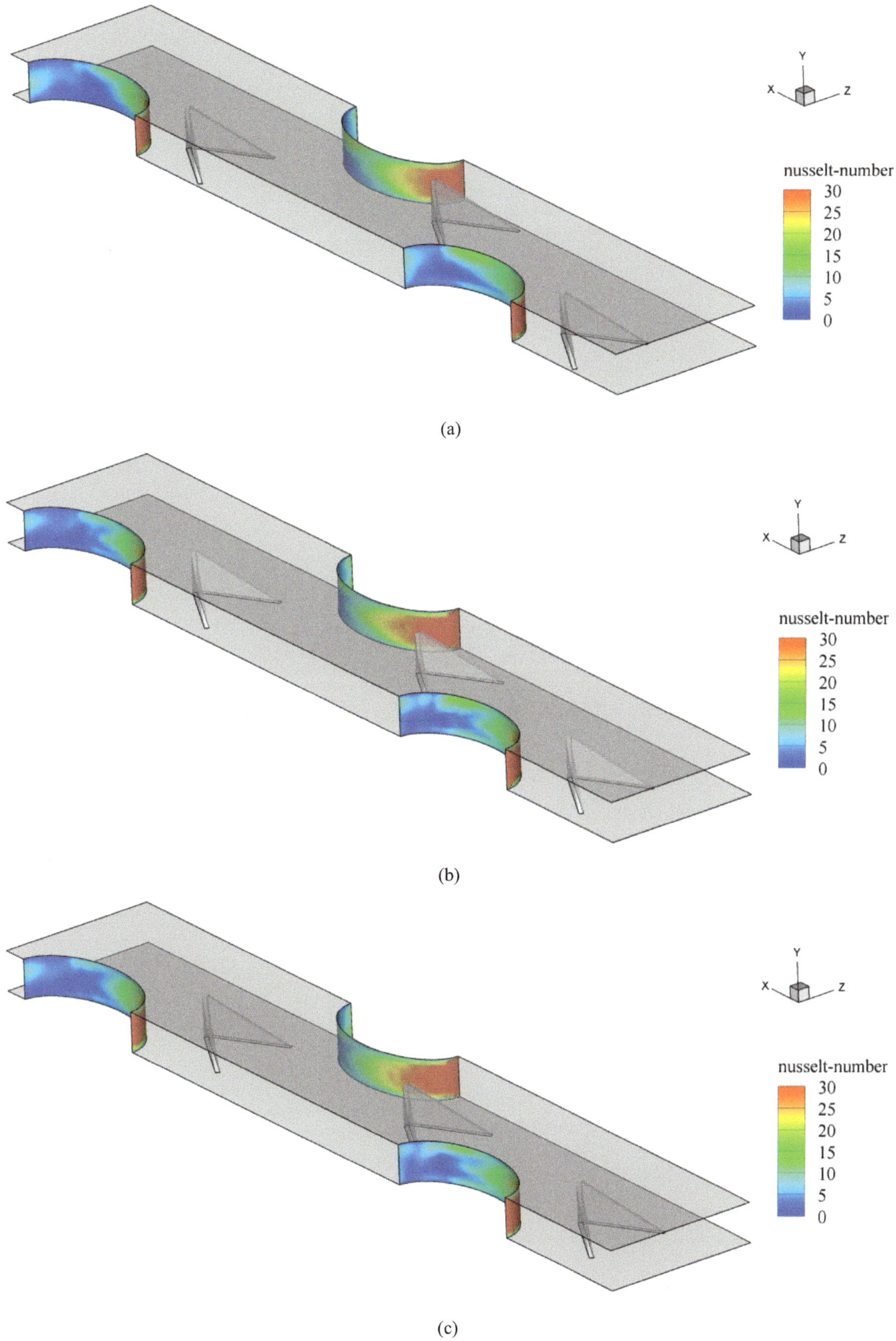

Fig. 14 Contours of local Nusselt number at oval tube walls for (a) $a = 3.77$ mm, (b) $a = 4.77$ mm and (c) $a = 5.77$ mm at $Re = 1500$ and $\theta = 30°$.

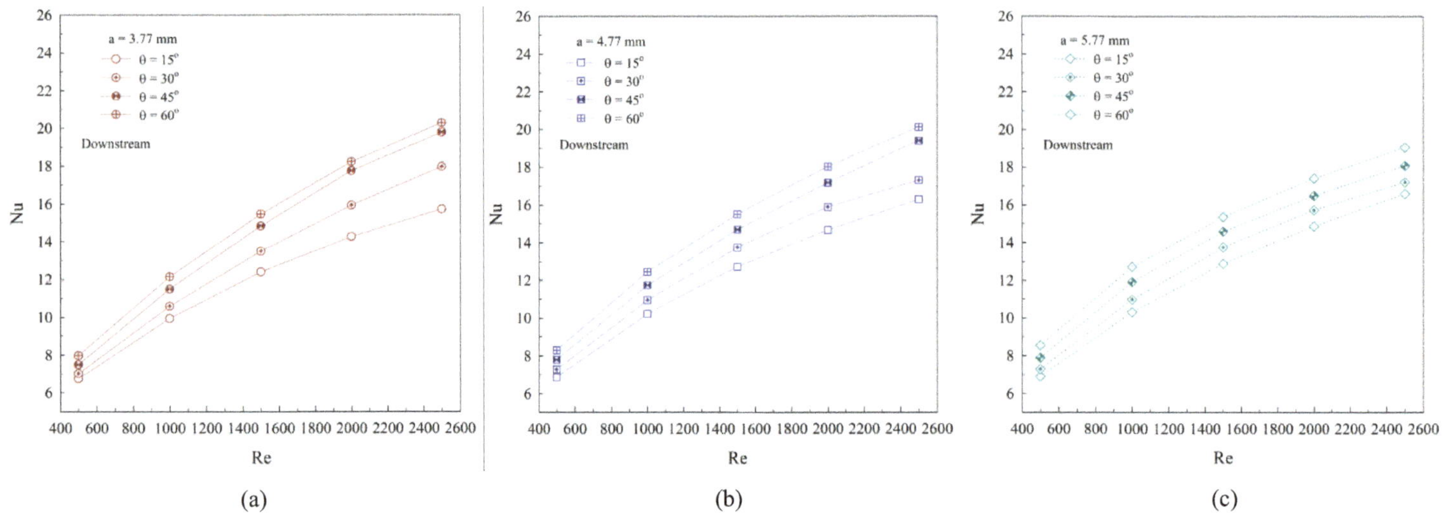

Fig. 15 The variations of Nusselt number with Reynolds number for (a) $a = 3.77$ mm, (b) $a = 4.77$ mm and (c) $a = 5.77$ mm.

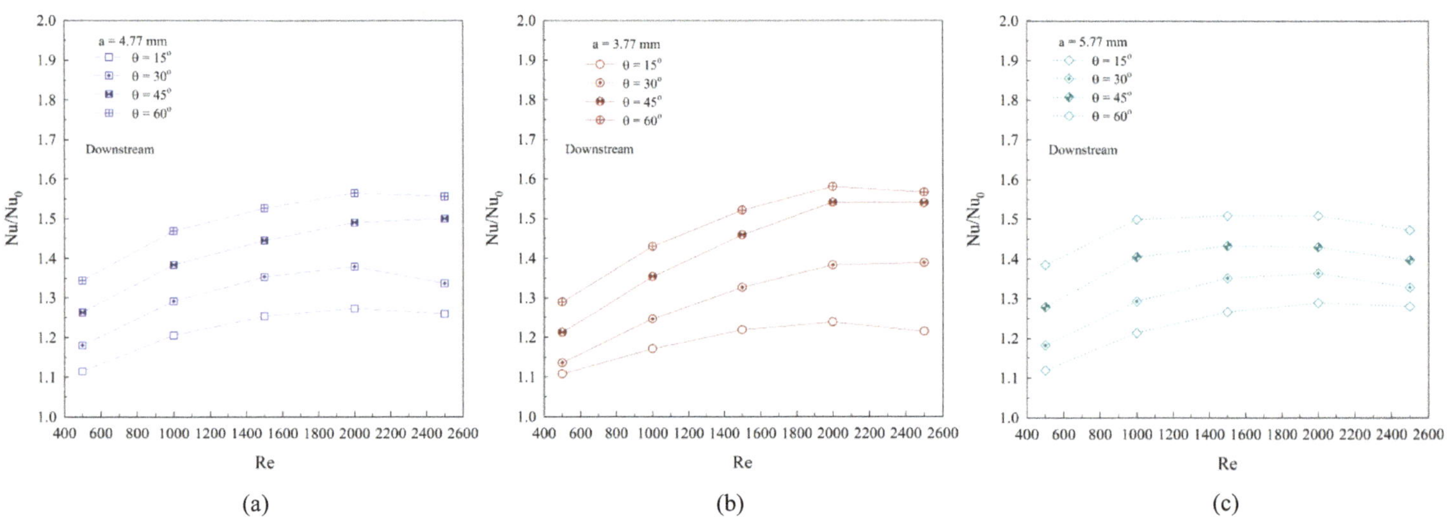

Fig. 16 The variations of Nusselt number ratio with Reynolds number for (a) $a = 3.77$ mm, (b) $a = 4.77$ mm and (c) $a = 5.77$ mm.

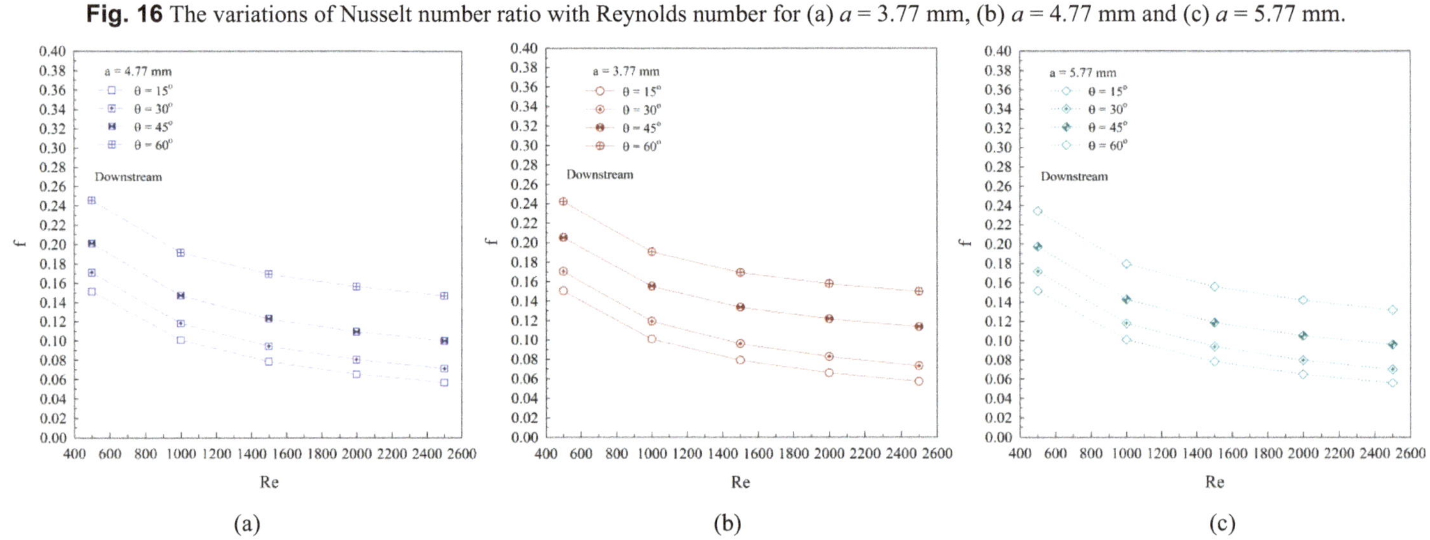

Fig. 17 The variations of friction factor with Reynolds number for (a) $a = 3.77$ mm, (b) $a = 4.77$ mm and (c) $a = 5.77$ mm.

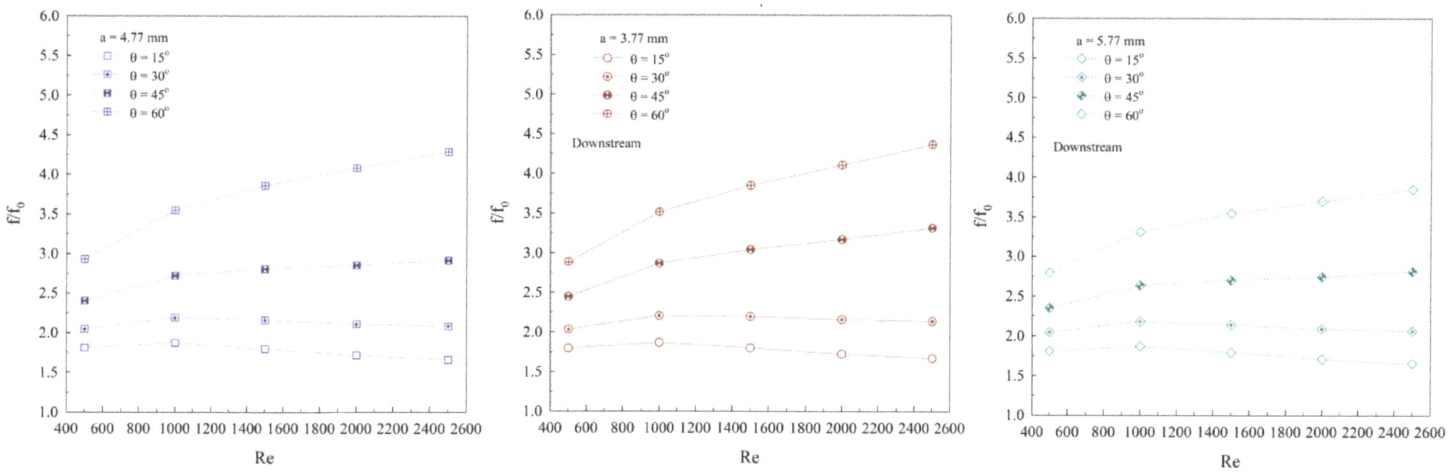

(a) (b) (c)

Fig. 18 The variations of friction factor ratio with Reynolds number for (a) $a = 3.77$ mm, (b) $a = 4.77$ mm and (c) $a = 5.77$ mm.

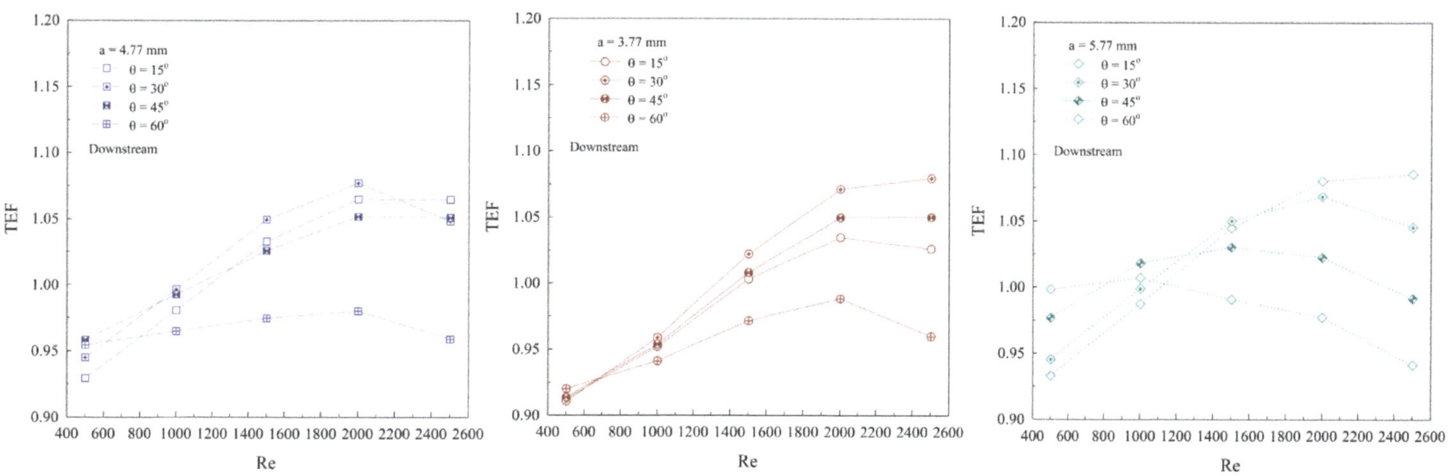

(a) (b) (c)

Fig. 19 The variations of thermal enhancement factor with Reynolds number for (a) $a = 3.77$ mm, (b) $a = 4.77$ mm and (c) $a = 5.77$ mm.

cases. The variations of a value result in the slightly different on both the heat transfer rate and friction factor.

- The presence of the *DDWP*s leads to the augment in the heat transfer rate and the friction factor over the smooth plain fin for all cases. The enhancement of the heat transfer rate is around 1.1 – 1.59 times over the smooth plain tube while the pressure loss is higher than the smooth fin around 1.6 – 4.4 times depended on the flow attack angle, the Reynolds number and the a value.

- Owing to the computational domain of the fin-and-oval-tube heat exchanger similar to the rectangular channel, therefore, the thermal performance is considered in term of thermal enhancement factor, *TEF*. The *TEF* varies in the range 0.91 – 1.09 depended on the Reynolds number, the flow attack angle and a value. The optimum *TEF* is found 1.09 at $Re = 2500$, $a = 5.77$ mm and $\theta = 15°$.

ACKNOWLEDGEMENTS

This research was funded by King Mongkut's University of Technology North Bangkok. Contract no. KMUTNB-GEN-58-52. The authors would like to thank Assoc. Prof. Dr.Pongjet Promvonge for the suggestions.

NOMENCLATURE

a	distance between V-tip to the center of the oval tube in transverse axis (mm)
B	channel width (m)
c	delta winglet chord length (m)
c_p	specific heat capacity (kJ/kg K)
D_h	hydraulic diameter
DDWP	downstream delta winglet pairs
f	friction factor
F_p	fin pitch (m)
F_t	fin thickness (m)
h	heat transfer coefficient (W/m^2 K)
H	channel height (m)
k	thermal conductivity (W/m K)
L	flow length (m)
LVG	longitudinal vortex generators
n	tube row number
Nu	Nusselt number, $(hD_h)/k$
P	pressure (Pa)
P_l	longitudinal tube pitch (m)
P_s	spanwise tube pitch (m)
ΔP	air-side pressure drop (Pa)
R_a	semi-major diameter
R_b	semi-minor diameter
Re	Reynolds number, $Re = U_c D_h \upsilon$
T	temperature (K)
TEF	thermal enhancement factor, $TEF = (Nu/Nu_0)(f/f_0)^{-(1/3)}$
TVG	transverse vortex generators

u, v, w velocity components (m/s)
u_{in} frontal velocity (m/s)
U_c velocity at the minimum cross sectional area, A_c (m/s)
WLE winglet leading edge
Y winglet placed downstream from inlet (mm)

Greek symbols
θ flow attack angle
μ dynamic viscosity (Pa s)
ρ density (kg/m^3)
Λ winglet aspect ratio

Subscripts
in inlet parameter
w wall
0 smooth fin

REFERENCES

Biswas, G., Torii, K., Fujii, D., Nishino, K., 1996, "Numerical and Experimental Determination of Flow Structure and Heat Transfer Effects of Longitudinal Vortices in a Channel Flow," *International Journal of Heat and Mass Transfer*, 39(16), 3441–3451.
http://dx.doi.org/10.1016/0017-9310(95)00398-3

Chen, Y., Fiebig, M., Mitra, N.K., 1998, "Heat Transfer Enhancement of a Finned Oval Tube with Punched Longitudinal Vortex Generators In-line," *International Journal of Heat and Mass Transfer*, 41, 4151–4166.
http://dx.doi.org/10.1016/S0017-9310(98)00130-6

Chen, Y., Fiebig, M., Mitra, N.K., 2000, "Heat Transfer Enhancement of Finned Oval Tube with Staggered Punched Longitudinal Vortex Generators," *International Journal of Heat and Mass Transfer*, 43, 417–435.
http://dx.doi.org/10.1016/S0017-9310(99)00157-X

Chu, P., He, Y.L., Lei, Y.G., Tian, L.T., Li, R., 2009, "Three-Dimensional Numerical Study on Fin-and-oval-tube Heat Exchanger with Longitudinal Vortex Generators," *Applied Thermal Engineering*, 29, 859–876.
http://dx.doi.org/10.1016/j.applthermaleng.2008.04.021

Fiebig, M., 1995, "Embedded Vortices in Internal Flow: Heat Transfer and Pressure Loss Enhancement," *International Journal of Heat and Fluid Flow*, 16, 376–388.
http://dx.doi.org/10.1016/0142-727X(95)00043-P

Gentry, M.C., Jacobi, A.M., 2002, "Heat Transfer Enhancement by Delta-Wing-Generated Tip Vortices in Flat-plate and Developing Channel Flows," *Journal of Heat Transfer*, 124, 1158–1168.
http://dx.doi.org/10.1115/1.1513578

Hiravennavar, S.R., Tulapurkara, E.G., Biswas, G., 2007, "A Note on the Flow and Heat Transfer Enhancement in a Channel with Built-in Winglet Pair," *International Journal of Heat and Fluid Flow*, 28, 299–305.
http://dx.doi.org/10.1016/j.ijheatfluidflow.2006.03.030

Jacobi, A.M., Shah, R.K., 1995, "Heat Transfer Surface Enhancement Through the Use of Longitudinal Vortices: a Review of Recent Progress," *Experimental Thermal and Fluid Science*, 11, 295–309.
http://dx.doi.org/10.1016/0894-1777(95)00066-U

Lee, S.H., Ryou, H.S., Choi, Y.K., 1999, "Heat Transfer in a Three-Dimensional Turbulent Boundary Layer with Longitudinal Vortices," *International Journal of Heat and Mass Transfer*, 42, 1521–1534.
http://dx.doi.org/10.1016/S0017-9310(98)00290-7

Leu, J.S., Wu, Y.H., Jang, J.Y., 2004, "Heat Transfer and Fluid Flow Analysis in Plate-fin and Tube Heat Exchangers with a Pair of Block Shape Vortex Generators," *International Journal of Heat and Mass Transfer*, 47, 4327–4338.
http://dx.doi.org/10.1016/j.ijheatmasstransfer.2004.04.031

Liou, T.M., Chen, C.C., Tsai, T.W., 2000, "Heat Transfer and Fluid Flow in a Square Duct with 12 Different Shaped Vortex Generators," *Journal of Heat Transfer*, 122, 327– 335.
http://dx.doi.org/10.115/1.521487

O'Brien, J.E., Sohal, M.S., Wallstedt, P.C., 2004, "Local Heat Transfer and Pressure Drop for Finned-tube Heat Exchangers using Oval Tubes and Vortex Generators," *Journal of Heat Transfer*, 126, 826–835.
http://dx.doi.org/10.1115/1.1795239

Pesteei, S.M., Subbarao, P.M.V., Agarwal, R.S., 2005, "Experimental Study of the Effect of Winglet Location on Heat Transfer Enhancement and Pressure Drop in Fin-tube Heat Exchangers," *Applied Thermal Engineering*, 25, 1684–1696.
http://dx.doi.org/10.1016/j.applthermaleng.2004.10.013

Sommers, A.D., Jacobi, A.M., 2005, "Air-side Heat Transfer Enhancement of a Refrigerator Evaporator using Vortex Generation," *International Journal of Refrigeration*, 28, 1006–1017.
http://dx.doi.org/10.1016/j.ijrefrig.2005.04.003

Sriromreun, P., Thianpong, C., Promvonge, P., 2012, "Experimental and Numerical Study on Heat Transfer Enhancement in a Channel with Z-shaped Baffles," *International Communications in Heat and Mass Transfer*, 39, 945–952.
http://dx.doi.org/10.1016/j.icheatmasstransfer.2012.05.016

Tiwari, S., Maurya, D., Biswas, G., Eswaran, V., 2003, "Heat Transfer Enhancement in Cross-flow Heat Exchangers using Oval Tubes and Multiple Delta Winglets," *International Journal of Heat and Mass Transfer*, 46, 2841–2856.
http://dx.doi.org/10.1016/S0017-9310(03)00047-4

Torii, K., Kwak, K.M., Nishino, K., 2002, "Heat Transfer Enhancement Accompanying Pressure-loss Reduction with Winglet-type Vortex Generators for Fin-tube Heat Exchangers," *International Journal of Heat and Mass Transfer*, 45, 3795–3801.
http://dx.doi.org/10.1016/S0017-9310(02)00080-7

Zhu, J.X., Fiebig, M., Mitra, N.K., 1995, "Numerical Investigation of Turbulent Flows and Heat Transfer in a Rib-roughened Channel with Longitudinal Vortex Generators, " *International Journal of Heat and Mass Transfer*, 38 (3), 495–501.
http://dx.doi.org/10.1016/0017-9310(94)00177-W

9

EFFECTS OF GEOMETRIC PARAMETERS FOR WAVY FINNED-TUBE HEAT EXCHANGER IN TURBULENT FLOW: A CFD MODELING

Arafat A. Bhuiyan[a,c], M. Ruhul Amin[b], Jamal Naser[c], A. K. M. Sadrul Islam[a]

[a]*Department of Mechanical Engineering, Islamic University of Technology (IUT), Board Bazar, Gazipur-1704, Bangladesh*
[b]*Department of Mechanical and Industrial Engineering, Montana State University, Bozeman, Montana, MT 59717, USA*
[c]*Faculty of Science Engineering and Technology (FSET), Swinburne University of Technology, Victoria-3122, Australia*

ABSTRACT

In this study, the effects of thermal and hydraulic characteristics of wavy fin and tube heat exchanger are investigated. Simulation has been carried out by a commercial computational fluid dynamics code, ANSYS CFX12.0. The main objective of this study is to investigate the flow characteristics in turbulent flow. Results are predicted for the turbulent flow regime ($2100 \leqslant Re \leqslant 7000$) and compared with author's previous work for laminar ($400 \leqslant Re \leqslant 1200$) and transitional ($1300 \leqslant Re \leqslant 2000$) flow regime. Regarding turbulence, the k-ω model was used to predict the turbulent flow characteristics with 5% turbulence intensity. Predicted results were compared with the experimental data for the purpose of validation and the discrepancy is found within 10% in error range. Parametric study was conducted for different pitches and wavy angles. Flow characteristics obtained for the turbulent range is in line with the pattern observed in laminar and transitional ranges. This study demonstrates a clear understanding and relationship of among different flow ranges and the effects of different geometric parameters on the performance of heat exchanger.

Keywords: *Wavy fin tube heat exchanger, CFD modeling, Friction factor, Colburn factor, Efficiency index, Reynolds number, Turbulent flow.*

1. INTRODUCTION

Finned-tube heat exchangers have been extensively used in several automotive applications for various purposes. The performance of heat exchanger is normally dominated by the thermal resistance. The fin geometry has become as increasingly important factor in the design of a plate-and-fin heat exchanger. Typical extended surfaces used for the plate-and-fin heat exchangers are: plain, wavy, offset strip, louvered, perforated and multi-louvered fin have been developed to minimize the thermal resistance. The wavy fin extended surfaces are recognized by the wavy corrugations which boost the heat transfer performance in contrast to the plain plate surfaces. In last decades, several efforts have been devoted to investigate the fundamentals and performance related issues in different experimental works (Kim, Youn et al. 1996, Wang, Fu et al. 1997, Wang 1999, Yan and Sheen 2000, Wang, Hwang et al. 2002, Pirompugd, Wongwises et al. 2006, Junqi, Jiangping et al. 2007, Kim, Ham et al. 2008, Wang, Liaw et al. 2011, Dong, Chen et al. 2012, Dong, Su et al. 2013, Du, Feng et al. 2013, Wu, Wu et al. 2014). Also a number of numerical attempts have been accomplished in 2D/3D analysis (Jang and Chen 1997, Tao, He et al. 2007, Tao, He et al. 2007, Bhuiyan 2010, Bhuiyan, Zaman et al. 2010, Bhuiyan, Islam et al. 2011, Bhuiyan, Islam et al. 2012, Bhuiyan, Amin et al. 2012, Glazar, Trp et al. 2012, Gong, Min et al. 2013, Bhuiyan 2014, Bhuiyan, Amin et al. 2014, Lin, Wang et al. 2014). But most of the earlier works are limited to laminar to transitional ranges (Chu, He et al. 2009, Bhuiyan, Zaman et al. 2010, Bhuiyan, Islam et al. 2011, Bhuiyan, Islam et al. 2012, Bhuiyan, Amin et al. 2012, He, Chu et al. 2013, Bhuiyan, Amin et al. 2014). In this study, simulation results for wavy fin considering different tube arrangements for turbulent flow range will be investigated. This numerical study is a continuation of the author's previous work for laminar flow (Bhuiyan, Islam et al. 2011) and transitional flow (Bhuiyan, Islam et al. 2012) to show the thermal and hydraulic performance of different flow ranges.

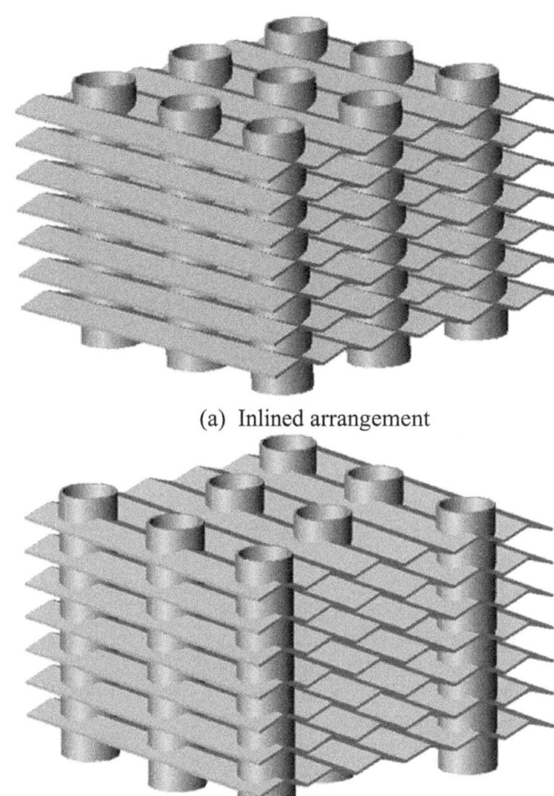

(a) Inlined arrangement

(b) Staggered arrangement

Fig. 1 Overview of typical wavy fin-and -tube heat exchanger model

Corresponding author. Email: arafat@iut-dhaka.edu

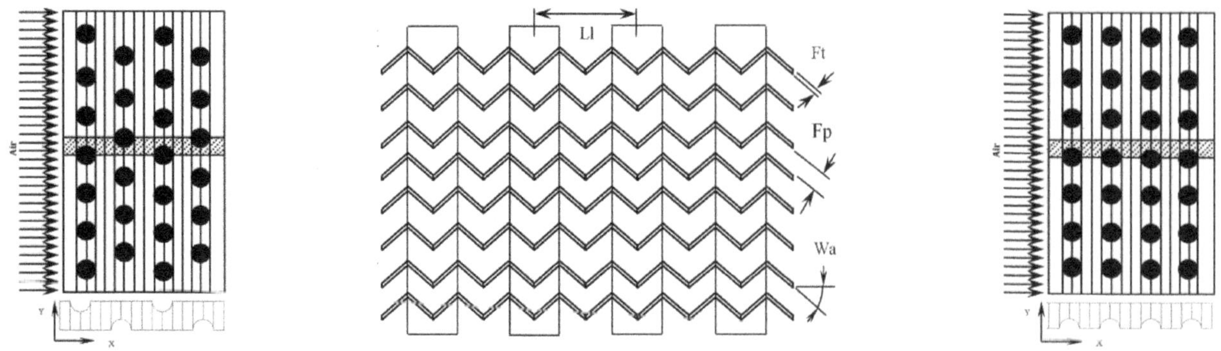

Fig. 2 Computational domain and co-ordinate system: staggered arrangement (left), in-lined arrangement (right), nomenclatures (middle)

Table 1. Geometric dimensions of heat exchanger model

Parameters	Symbols	Units	Dimensions
Tube diameter	D	(mm)	9.525
Longitudinal tube pitch	Ll	(mm)	19.05
Transverse tube pitch	Lt	(mm)	25.40
Fin Pitch	Fp	(mm)	3.530

Table 2. Cases considered for parametric study in turbulent flow range.

Parameters	Units	Case-I	Case-II	Case-III
Ll	(mm)	19.05	28.57	38.10
Lt	(mm)	25.40	30.40	35.40
Fp	(mm)	3.53	2.53	1.53
Wa	(°)	8	17	35

Simulation results for wavy fin considering staggered and inlined arrangements for turbulent (2100 ⩽ Re ⩽ 7000) flow range will be investigated to determine the effect on the heat transfer and pressure drop performance based on the available experimental data of Wang l(Wang, Fu et al. 1997)and as a continuation of the numerical work of the author's for laminar flow (Bhuiyan, Islam et al. 2011) and transitional flow (Bhuiyan, Islam et al. 2012). Also a detailed analysis of the effect of those geometrical parameters mentioned above will be presented. Overall, a comprehensive parametric study considering longitudinal (Ll) (Lu, Huang et al. 2011), transverse (Lt) (Bhuiyan, Islam et al. 2011, Lu, Huang et al. 2011, Bhuiyan, Islam et al. 2012, Bhuiyan, Amin et al. 2012, Bhuiyan 2014), fin pitch (Fp) (Bhuiyan, Islam et al. 2011, Bhuiyan, Islam et al. 2012, Bhuiyan, Amin et al. 2012, Bhuiyan 2014) and wavy angle (Wa) (Bhuiyan, Islam et al. 2012, Glazar, Trp et al. 2012) has been carried out for staggered tube arrangement in a typical wavy heat exchanger using a commercial computational fluid dynamics code ANSYS CFX 12.0.

2. MODEL DESCRIPTIONS

2.1 Physical model of the heat exchanger

The objective of this study is to understand the hydrodynamics of flow and the corresponding heat transfer and pressure drop as a function of the longitudinal pitch (Ll), transverse pitch(Lt), fin pitch(Fp) and Wavy angle (Wa) in turbulent flow range. In this study, two different wavy fin and tube heat exchanger were considered for the simulations. These are: (i). Wavy fin staggered arrangement and (ii). Wavy fin inlined arrangement. The typical wavy fin and tube arrangements are shown in figure 1. The detailed specifications of the selected heat exchanger are taken from the experimental work of Wang (Wang, Fu et al. 1997) as tabulated in Table 1. In brief, the computational domain for the present study is defined by $0 \leqslant X \leqslant 16.16D$, $0 \leqslant Y \leqslant 2.66D$ and $0 \leqslant Z \leqslant 0.37D$, where tube diameter, D=9.525mm. The computational domain and the coordinate system are shown in figure 2 for the two configurations such as wavy fin staggered configuration (left) and Wavy fin In-lined configuration (right). The dashed line in the figure designates the computational domain. Three dimensional computational domain with detailed boundary condition is given in author's previous work given in Ref (Bhuiyan, A., A. Islam and M. Amin (2012). Assuming symmetry condition on the mid plane between the two fins, the bottom and the top boundaries simulate the fin and the mid-plane respectively. The nomenclatures used for the present computation and the parametric study are shown in the middle of the fig. 2.

2.2 Boundary conditions

The boundary conditions of the present study are described in the author's previous work (Bhuiyan 2010, Bhuiyan, Zaman et al. 2010, Bhuiyan, Islam et al. 2011, Bhuiyan, Islam et al. 2012, Bhuiyan, Amin et al. 2012, Bhuiyan 2014). The boundaries of the domain consist of inlet and outlet boundaries, symmetry planes and solid walls. The volume representing the air which passes through the gap between the two fins is extended downstream from the outlet of the last row cylinder of the heat exchanger for more accurate applications of boundary conditions that is to ensure a representative flow in the computational domain of the actual heat exchanger and to reduce the numerical oscillations (Reddy and Gartling 2010). A typical staggered tube arrangement domain with different boundary conditions is given in fig. 3 which represents the computational domain considered for the present investigation taken from (Bhuiyan, A., A. Islam and M. Amin (2012).

Inlet Boundary condition: The inlet boundary condition is representative to the experimental study considered for the numerical modeling in turbulent flow range. Based on the Reynolds number (Re_H), at upstream boundary, uniform flow with constant velocity and constant temperature are assumed. This was considered to trigger the flow unsteadiness in the flow passage. In CFD, inlet flow field is applied normal to the inlet face of the computational domain. Other velocity components are assumed to be zero. A constant temperature of 25°C is set at the flow inlet to meet the room air conditions.

Wall boundary condition: No-slip boundary condition is used at the fins and the tube surfaces. These surfaces are assumed to be solid wall with no slip boundary condition and constant wall temperature T_{wall} set to 100°C. The fins and tubes are assumed to be made of aluminium. At symmetric wall, no-slip boundary condition is used with constant wall temperature. In this study, turbulent flow is considered having Reynolds number in the range of $2100 \leqslant Re_H \leqslant 7000$.

Outlet boundary condition: At the outlet, stream wise gradient (Neumann boundary conditions) for all the variables are set to zero for all the cases considered. As mentioned earlier, the computational domain is extended up to seven times of the tube diameter from the last tube column. Though, the full computational domain is not shown in the present paper. This is considered only to avoid or reduce numerical oscillation. It can be concluded that the computational domain is long enough in the flow direction for the flow and thermal field to be fully developed.

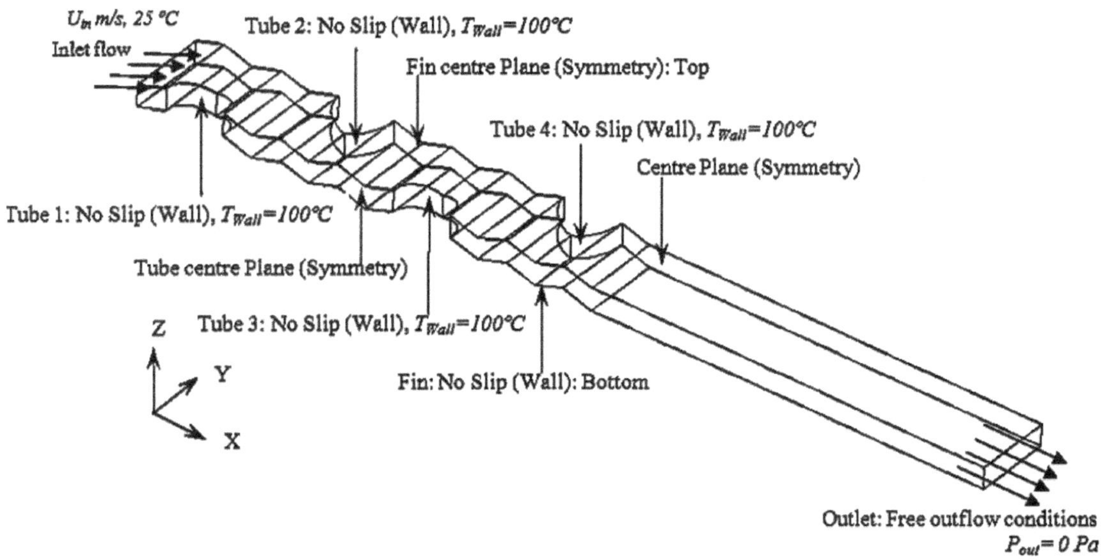

Fig. 3 A typical 3D computational domain with different boundary conditions (for Laminar flow) considered in the present study.

2.3 Investigated cases

As the main objective of this study is to investigate the effects of different geometric parameter on the overall performance of the heat exchanger, a detail parametric study has been conducted for turbulent flow range. For parametric study, three different cases are considered for a specific geometric parameter such as Lt, Lt, Fp and Wa as of previous (Bhuiyan, Islam et al. 2011, Bhuiyan, Islam et al. 2012, Bhuiyan, Amin et al. 2012, Bhuiyan 2014) and given in Table 2. Table 1 represents the baseline case of the study where wavy angle is assumed equal to 17 °. All other parameters were kept constant as baseline case when one parameter changing from case 1 to case 3.

3. COMPUTATIONAL METHODOLOGY

3.1 Mathematical modelling

The numerical simulations of the fin-tube heat exchanger were conducted for turbulent (2100≤ReH≤7000) case. 3D Navier-Stokes equations (Sohr 2012) were considered for the simulation of flow and heat transfer and energy equation for steady incompressible fluid with constant properties. The flow is described by the conservation laws for mass (continuity), momentum (Navier-Stokes) and by the energy equations are as follows (Hossain and Naser 2011, Bhuiyan and Naser 2014, Bhuiyan and Naser 2015):

$$\frac{\partial u_i}{\partial x_i} = 0 \tag{1}$$

$$\rho\left(\frac{\partial u_i}{\partial t} + u_j\frac{\partial u_i}{\partial x_j}\right) = -\frac{\partial p}{\partial x_i} + \frac{\partial}{\partial x_j}\left[(\mu + \mu_T)\left(\frac{\partial u_i}{\partial x_j} + \frac{\partial u_j}{\partial x_i}\right)\right] + \rho g_i \tag{2}$$

$$\rho C_p\left(\frac{\partial T}{\partial t} + u_j\frac{\partial T}{\partial x_j}\right) = \frac{\partial}{\partial x_j}\left[\left(\lambda + \frac{\mu_T C_p}{Pr_T}\right)\frac{\partial T}{\partial x_j}\right] \tag{3}$$

As suggested by Yuan (Yuan 2000), $Pr_T = 0.9$ was used in the current study. The value of μ_T is determined based on the specific turbulence model that is being used. In k- ω turbulent model the μ_T is linked to the turbulence kinetic energy (k) and turbulence frequency (ω). In this study, k-ω turbulence model is employed. Details of different turbulent models are documented in literature (Menter 1994, Panse 2005, Reddy and Gartling 2010). One of the advantages of the k-ω formulation is the near wall treatment for low Reynolds number computations. The model does not involve the complex non-linear wall treatment for low Reynolds number computations. The model does not involve the complex non-linear damping functions required for the k-ε model as is therefore more accurate and more robust. A low Reynolds k-ε model would typically require a near wall resolution of $y^+<2$, while a low Reynolds number k-ω model would require at least $y^+<2$, where y+ is the non-dimensional near wall spacing. In industrial flows, even $y^+<2$ cannot be guaranteed in most applications and for this reason; a new near wall treatment was developed for the k-ω models. It allows for smooth shift from a low Reynolds number form to a wall function formulation. Details of the used k-ω turbulent models are documented in literature (Menter 1994, Yang, Asako et al. 1997, Reddy and Gartling 2010, Amin and Ramachandran 2011). More details about the mathematical formulations considered in this study are documented in (Yuan 2000, Bhuiyan 2010, Bhuiyan, Zaman et al. 2010, Bhuiyan, Islam et al. 2011, Bhuiyan, Islam et al. 2012, Bhuiyan, Amin et al. 2012, Bhuiyan 2014).

3.2. Numerical descriptions

The commercial computational fluid dynamics code ANSYS CFX-12 was employed to carry out the numerical investigation. ICEM CFD, a meshing tool of ANSYS CFX was used to mesh the computational domain. FVM (Islam, Amin et al. 2009, Bhuiyan and Naser 2015, Bhuiyan and Naser 2015) was used in this CFD code. This code uses a coupled solver, which solves the hydrodynamic equations as a single system. For steady state problems, the time-step behaves like an acceleration parameter, which is responsible to guide the approximate solutions in a physically based manner to a steady state solution. In order to attain the desired accuracy it is recommended to define a target variable that will monitor the numerical error. The typical range used for the present convergence of the computation is about $1e^{-4}$ which is good enough for these types of simulations. The maximum number of iteration applied for the present investigation is 1000 and convergence is monitored and ensured by the following steps: (i) reduction of residuals by a pre-specified level and plotting of the residuals, (ii). Plotting the max residuals with iteration number, (iii). Plotting the target variables as function of iteration number or residual level.

Table 3: Different grid resolutions for the wavy fin configurations

Grid test	No of nodes	No of elements
Test 1	364722	890238
Test 2*	549669	1355049
Test 3	789954	2080026

*-used in present study.

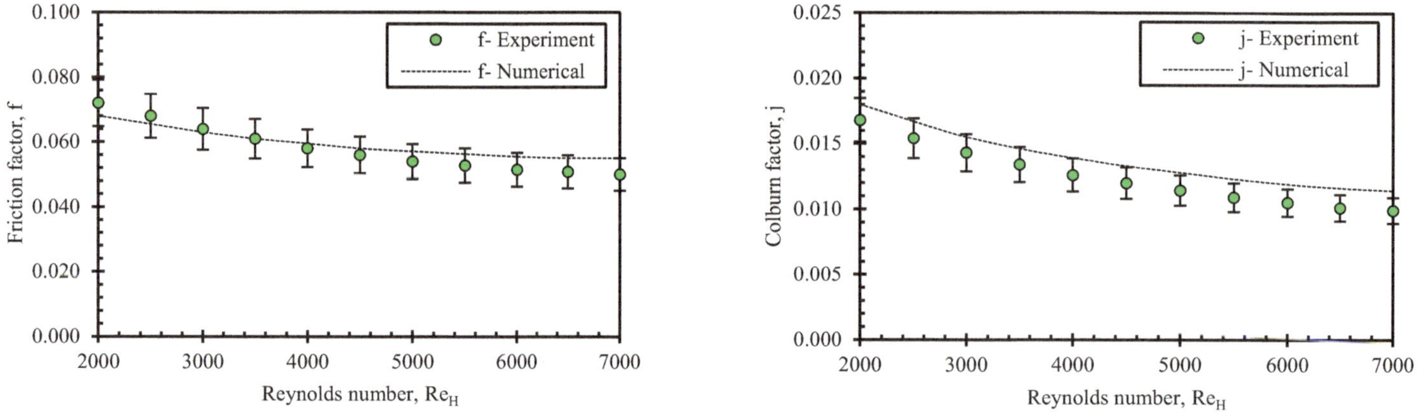

Fig. 4 Comparison of numerical prediction with the experimental data for (a). Friction factor (f) and (b). Colburn factor (j)

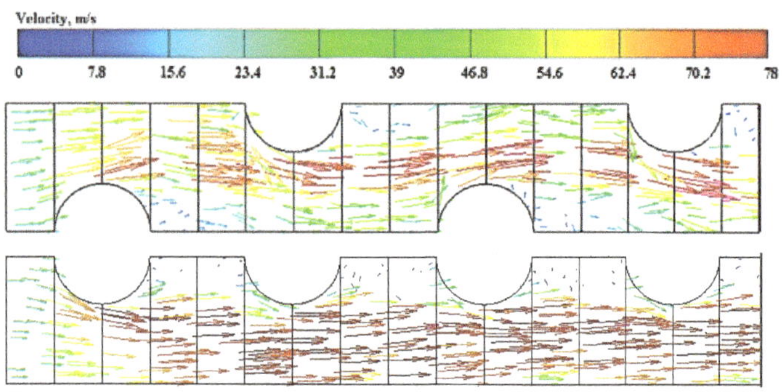

Fig. 5 Comparison of velocity (m/s) distribution for different tube arrangement at Re_H=7000

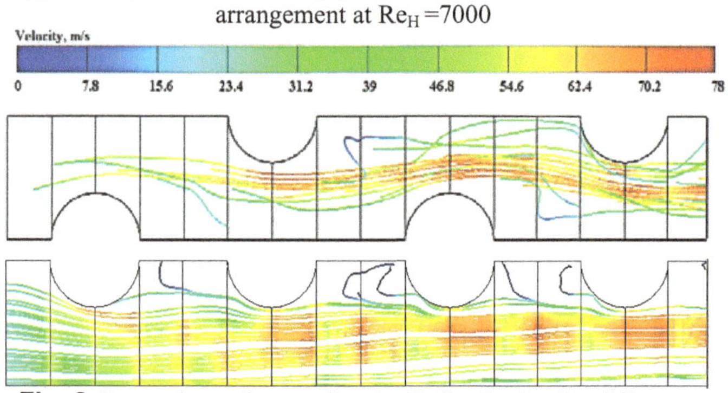

Fig. 6 Comparison of streamline (m/s) distribution for different tube arrangement at Re_H=7000

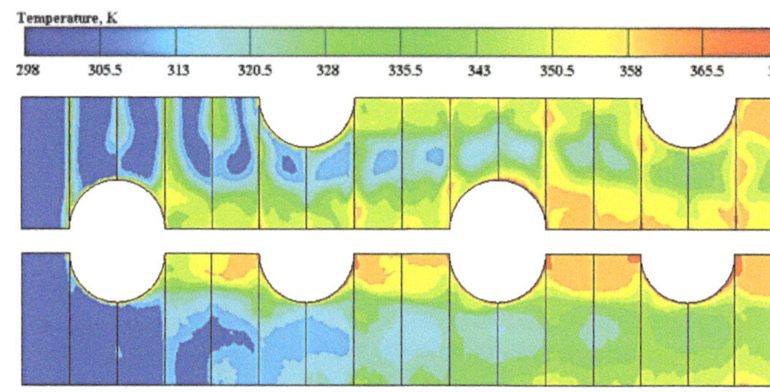

Fig. 7 Comparison of temperature (K) distribution for different tube arrangement at Re_H=7000

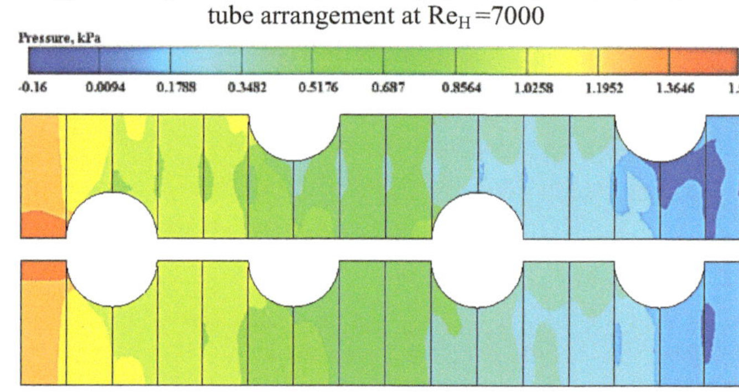

Fig. 8 Comparison of pressure (kPa) distribution for different tube arrangement at Re_H=7000

3.3. Grid systems

Grid sensitivity test is required to attain a desired level of accuracy with minimum computational time. Grid sensitivity tests with three different grid sizes were conducted to develop the acceptability of the computation. Grid was denser in the region of the tubes as the pressure, temperature and the velocity gradients increases with the flow. A comprehensive grid independency test is carried out presented in author's previous paper (Bhuiyan, Islam et al. 2012) and the grid having 183223 nodes with 451683 elements is chosen for laminar and transitional flow ranges. However, in this turbulent flow range, the maximum Reynolds number used is more than three times higher than the one used in authors' previous studies, so the grid must be different as well, and a new grid convergence study must be carried out: the higher the Reynolds number, the higher the gradients, and the finer the grid. In order to consider this effect, a new grid test is considered in this study with several times higher grid and convergence results were obtained. In this study, grid having 549669 nodes with 1355049 elements is chosen for transitional flow ranges. The statistics of the different grid used in the present study are given in Table 3.

3.4. Validation

In order to attain the confidence about the numerical result, a comprehensive validation study was conducted considering the similar conditions as in the experimental work considered in Ref. (Wang, Fu et al. 1997). To ensure the numerical results are reliable, calculations were first prepared to scrutinize the recital of fin geometry having 4 rows staggered circular tube configuration with the experimental data by Wang et al. (Wang, Fu et al. 1997). The detailed geometry of the examined heat exchanger is same as (Wang, Fu et al. 1997). The accuracy of the study was established by comparing the values for friction factor (f) and Colburn factor (j) by Wang et al. (Wang, Fu et al. 1997) with the present numerical modelling for turbulent flow range. Figure 4 shows the Comparison of numerical prediction with the experimental data for (a). Friction factor (f) and (b). Colburn factor (j) (Dotted line represents the numerical data & coloured filled point shows the experimental data extracted). To show the comparison between numerical and experimental data, a 10% error bar is set on the experimental values. From both the plots, the discrepancy is found within 10% in error range. The graphical presentation is shown in fig. 3

4. RESULT AND DISCUSSION

4.1. Flow variation in different tube arrangements

In wavy fin tube arrangements, flow interruption is generally caused by two constructive features such as the wavy corrugation and the tubes position. It is expected that the flow will maintain a re-orientation at each of the wavy corrugation starts. This phenomenon is responsible for less flow recirculation in the wake of the tubes. In order to differentiate among the inlined and staggered tube arrangement, flow velocity and streamline contours is presented in fig. 5 & 6 respectively. It is noticed that flow disruption proceeds on both sides of the computational structure for staggered configuration. As there are repeated disruptions of the flow direction because of tubes on both sides of the structure, comparatively less recirculation is detected. But for in-lined arrangements stream is congested only on one side of the heat exchanger as all the tubes are on one side. There are two flow regions can be observed such as free flow regions and the stagnant flow regions. This result contents the response of variation of recirculation zone. The temperature and pressure distribution for different tube arrangement are shown in figures 7 & 8 respectively. Similar behavior is observed as the streamline pattern and the velocity vector. It is seen from the temperature profile of in-lined arrangements, there is high temperature zones in the trailing edge of the tubes can be called warm zones because of the recirculation flow which stretches between two adjacent tubes. Figure 9 represents the flow characteristics for different tube arrangements in turbulent flow range compared with the laminar (Bhuiyan, Islam et al. 2011)and transitional (Bhuiyan, Islam et al. 2012) flow ranges. There is a definite variation of staggered and in-lined tube arrangement presented here. Comparatively higher heat transfer and pressured drop is observed in all flow ranges in staggered case. Though f and j both are lower in in-lined arrangement but higher efficiency is found in in-lined case. It is seen that the tube engagements play a vigorous role in the heat transfer and pressure drop features. In staggered arrangements, improved flow mixing is witnessed due to staggered tube designs and thus delivers higher heat transfer and pressure drop physiognomies than the in-lined arrangements. So wavy staggered configuration will be considered to investigate the effect of different geometrical parameter in the turbulent range.

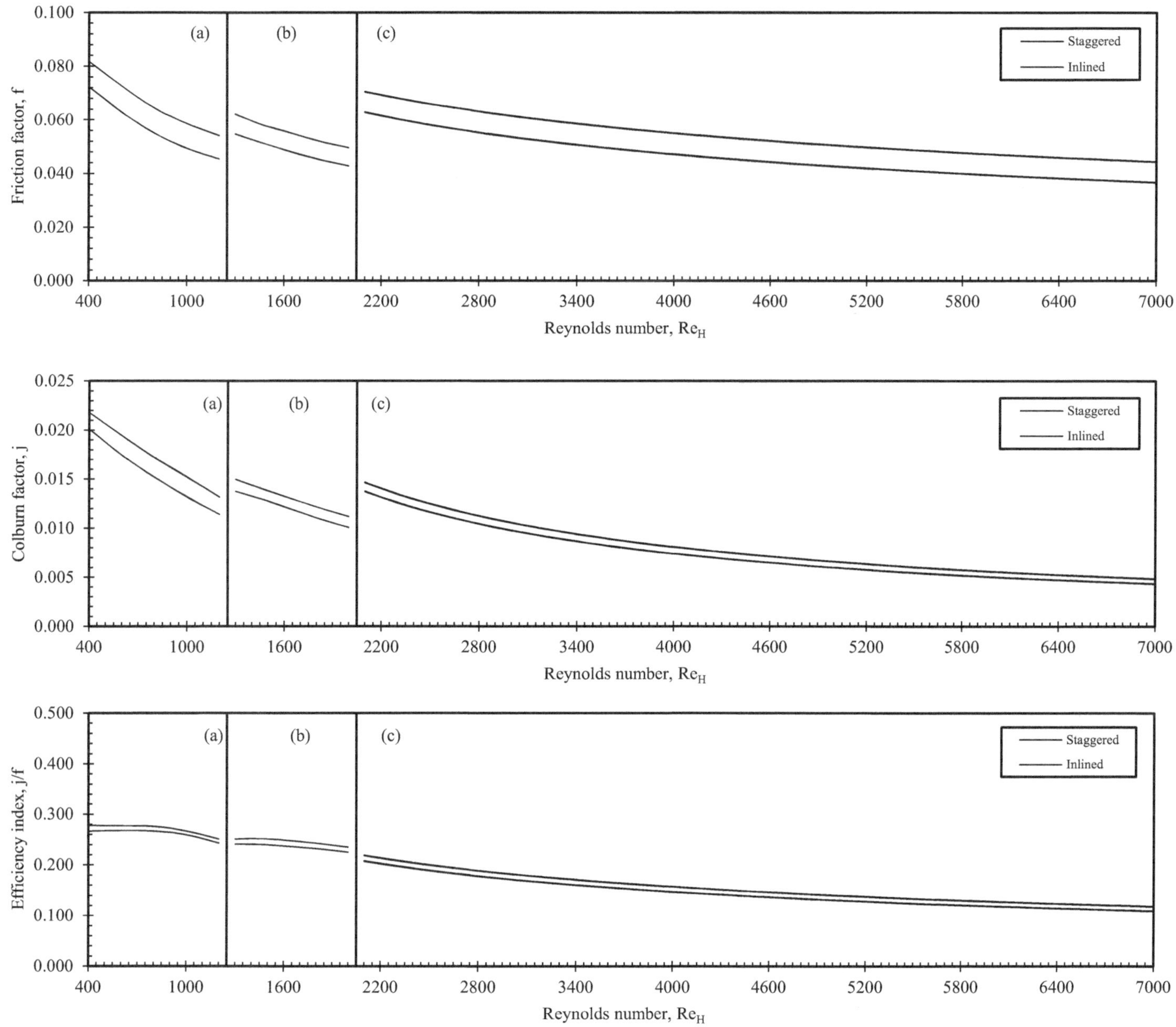

Fig. 9 Flow characteristics for different tube arrangement in (c). Turbulent flow range ($2100 \leqslant Re_H \leqslant 7000$) compared with the published (a). Laminar flow range ($400 \leqslant Re_H \leqslant 1200$) and (b). Transitional flow range ($1300 \leqslant Re_H \leqslant 2000$)

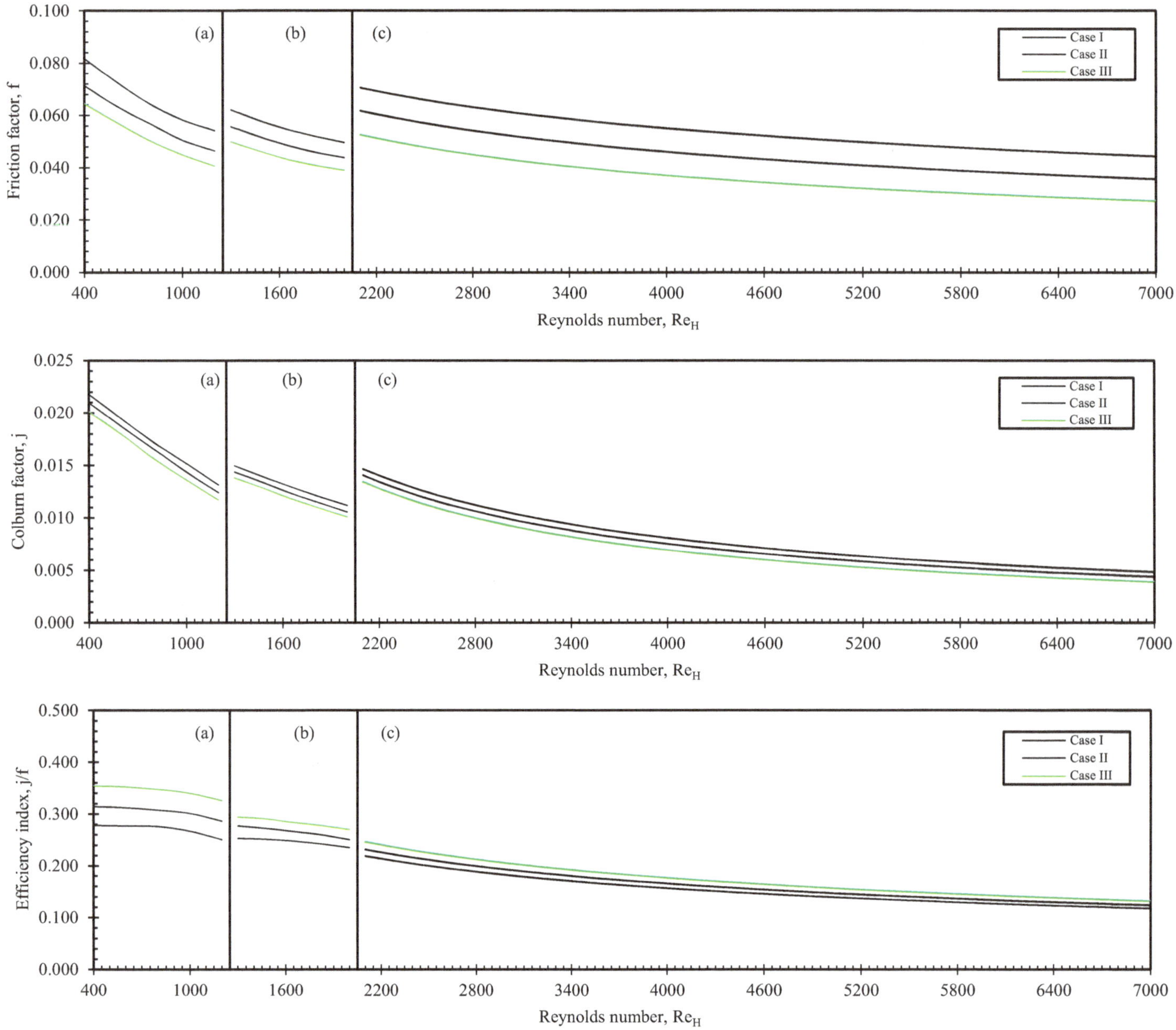

Fig. 10 Flow characteristics for the variation of different longitudinal pitches (Ll) for staggered tube arrangement in (c). Turbulent flow range ($2100 \leqslant Re_H \leqslant 7000$) compared with the published (a). Laminar flow range ($400 \leqslant Re_H \leqslant 1200$) (b). Transitional flow range ($1300 \leqslant Re_H \leqslant 2000$).

4.2. Effect of Longitudinal Pitches (Ll)

Flow characteristics for the variation of different longitudinal tube pitch (Ll) such as 19.05 mm, 28.575 mm and 38.10 mm in turbulent flow range are presented in figure 10. Also the comparisons with the laminar and transitional range are critically evaluated. It is found that with the increase of Ll, friction factor (f) and Colburn factor (j) decrease, but efficiency index increases. With the increase of Ll from 19.05 mm to 28.575 mm and 28.575 mm to 38.10 mm, f decreases 19.54% and 15.84% while j decreases 4.83% and 5.07% respectively in the turbulent range. This statement can be explained that higher the longitudinal pitch (Ll) the surface area is also increased. So the flow is freer and the airflow becomes distributed results in lower friction factor. The same behavior as f is found for heat transfer. In general, the increase in the heat transfer area would increase the heat transfer. But the current findings contradict this phenomenon. This can be explained that lower the longitudinal pitch, flow is more restricted and the airflow is dense due to close tube spacing and this enhance the heat transfer.

But the rate of increase in friction factor is higher than the Colburn factor and as a result the efficiency decrease with the decrease of Ll. But in laminar and transitional range, f increases 12.83% and 11.05% while j increases 4.20% and 4.55% respectively for the change of 19.05 mm to 28.575 mm. But for 28.575 mm to 38.10 mm pitch increase, f increase10.50% and 10.33%, while j changes 4.56% and 3.86% for laminar and transitional flow ranges respectively. Efficiency index increases 10% and 12% for the increase of Ll of 19.05mm to 28.575mm respectively. But it was seen previously that for laminar case this change was found as 9.74% and 6.67% while for transitional case, these change was observed as 7.3% and 7.2% respectively. While comparing the flow model, for laminar to transitional change, f increase 16.61% and 14.91% and 14.75% for the three longitudinal pitch cases respectively. But for the case of j, these changes are more significant. The change is around 25% in all cases. But for transitional to turbulent shift, f decrease 18.95%, 15.6% and 8.52 % for the selected case. But the change is nominal in these two flow changes.

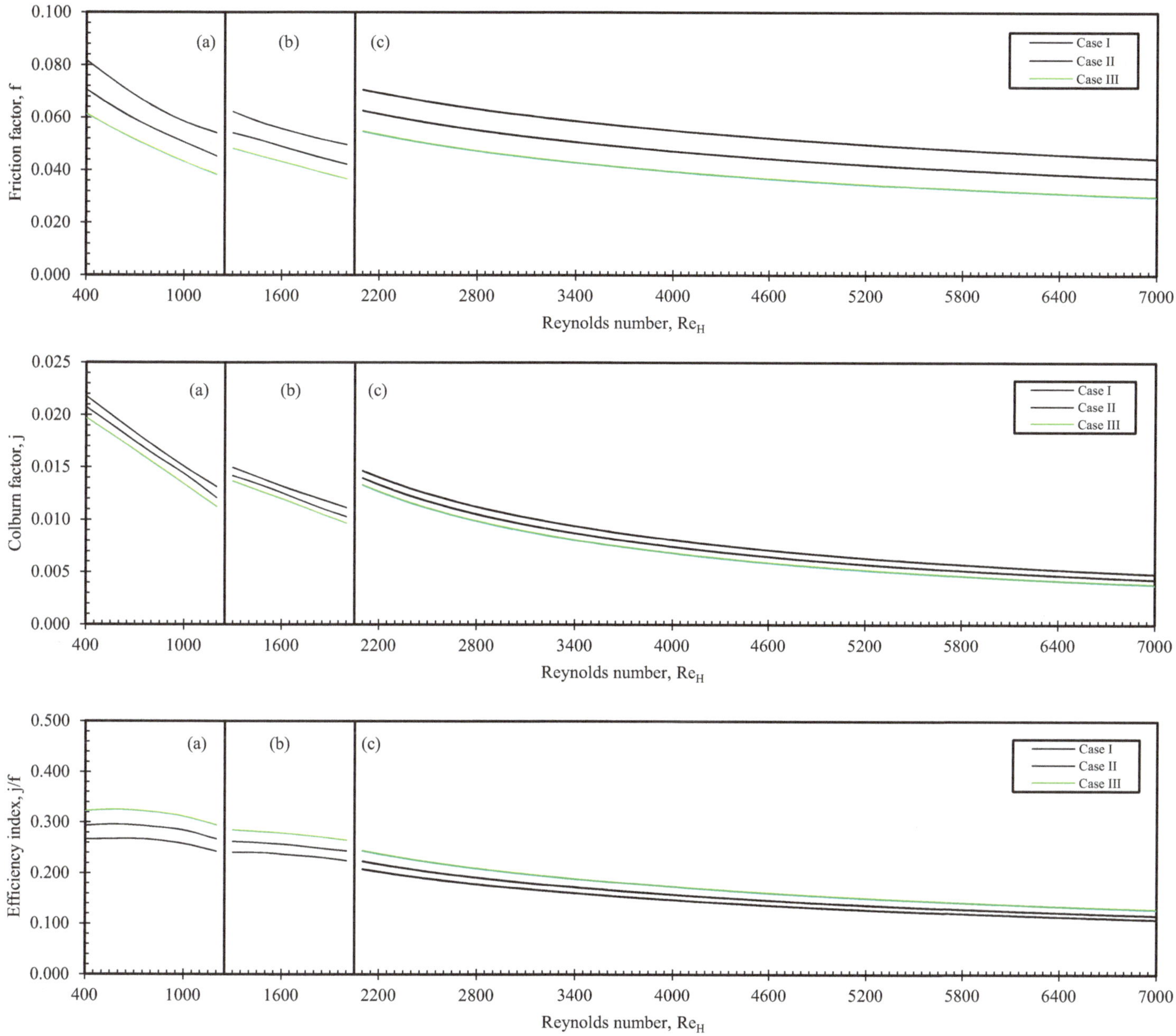

Fig. 11 Flow characteristics for the variation of different transverse pitches (Lt) for staggered tube arrangement in (c). Turbulent flow range ($2100 \leqslant Re_H \leqslant 7000$) compared with the published (a). Laminar flow range ($400 \leqslant Re_H \leqslant 1200$) (b). Transitional flow range ($1300 \leqslant Re_H \leqslant 2000$)

4.3. Effect of Transverse Pitches (Lt)

Flow characteristics for the variation of different transverse pitches (Lt) for staggered tube arrangement in turbulent flow range compared with the published laminar (Bhuiyan, Islam et al. 2011) and transitional flow range (Bhuiyan, Islam et al. 2012) is presented in this study. The effects of transverse pitch (Lt) on the pressure drop and heat transfer and efficiency for the staggered arrangements are shown in fig. 11. The effect is very much similar to that of longitudinal pitch. For the increase of Lt from 25.4 mm to 30.4 mm, f decrease 13.9%, 13.04% and 12.17% while j decrease 5.1%, 5.8% and 5.63% for laminar, transitional and turbulent flow ranges respectively. For the increase of Lt from 30.4 mm to 35.4 mm, f decrease 14.12%, 12.31% and13.86% while j decrease 5.82%, 5.05% and 5.97% for laminar, transitional and turbulent flow ranges respectively. Though f and j decrease with the increase of Lt, but j/f increase. The increase found as 10.12%, 8.28% and 7.39% for first set and 9.63%, 8.26% and 9.10% for the second set of Lt in the laminar, transitional and turbulent flow ranges respectively. While comparing the flow model, for laminar to transitional change, f increases 16.61%, 15.75% and 13.98 % for the three cases respectively. The value of j changes 24.97.5%, 25.47% and 24.86% respectively. While comparing laminar to transitional change, f decrease 18.86%, 20.04% and 17.93 % but the value of j changes 4.45%, 4.28% and 5.21% respectively. The effect of transverse pitch can be explained similar to longitudinal pitch effect but different value. With the increase of transverse pitch, the surface area is increased. So the flow area is expanded and hence the airflow becomes distributed corresponds to lower pressure drop. The same behavior as f is found for heat transfer. This can be explained that lower the transverse pitch, flow is more restricted and the airflow is dense due to close tube spacing and this enhance the heat transfer. Though similar performance is observed for longitudinal as well as transverse pitch in terms of heat transfer and pressure drop performance, but the efficiency shows a variation for respective cases. This justifies the significance of variation of different pitches.

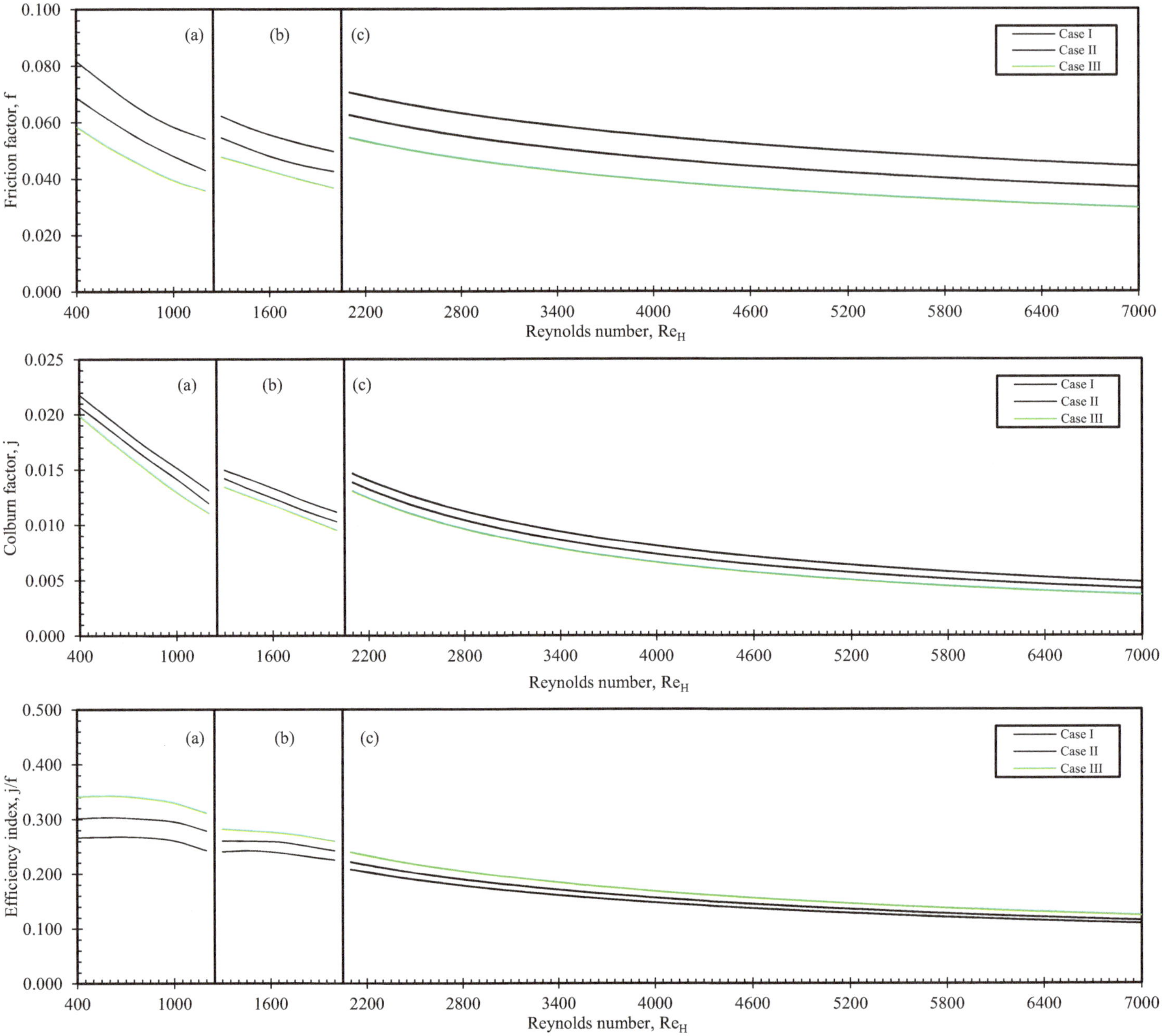

Fig. 12 Flow characteristics for the variation of three different fin pitch (Fp) for staggered tube arrangement in (c). Turbulent flow range ($2100 \leqslant Re_H \leqslant 7000$) compared with the published (a). Laminar flow range ($400 \leqslant Re_H \leqslant 1200$), (b). Transitional flow range ($1300 \leqslant Re_H \leqslant 2000$)

4.4. Effect of Fin Pitches (Fp)

Flow characteristics for the variation of different fin pitches (Fp) for staggered tube arrangement in turbulent flow range compared with the published laminar (Bhuiyan, Islam et al. 2011) and transitional flow range (Bhuiyan, Islam et al. 2012) is presented in figure 12. With the decrease of Fp, friction factor and Colburn factor both decreases but j/f increases. Heat transfer performance is quite similar as the pressure drop. This is because when Fp is reduced the flow becomes more streamlined resulting in better flow mixing. Also reduction in the Fp reduces the tube surface area which affects the pressure drop performance. This can also be explained on the basis of boundary layer concept. The boundary layer interruption could not have occurred at large fin pitches with faster inlet air velocities. Therefore, the heat transfers for high Reynolds Number were independent of the pitches. However for, smaller fin pitch range with lower inlet velocities, the heat transfer and pressure drop decreased with a reduction in fin pitches. This is because the interruption of the boundary layers between the fins resulted from an increase in the boundary layer thickness with a reduction in fin pitches. The enhancement of the j-factor with fin pitches may be the result of the delay of the boundary layer interruption to the next row. But the behavior is fully different with the characteristics found for Ll and Lt Cases. For the decrease of Fp from 3.53 mm to 2.53 mm, f decrease 17.07 %, 13.42 % and 12.17 % while j decrease 6.0 %, 6.59 % and 6.44 % for laminar, transitional and turbulent flow ranges respectively. For the decrease of Fp from 2.53 mm to 1.53 mm, f decrease16.50 %, 12.07 % and 13.86 % while j decrease 6.16%, 5.62% and 6.88% for laminar, transitional and turbulent flow ranges respectively. For laminar to transitional change, f increase 16.55% and 12.87% and 8.20% for the three fin pitch cases respectively. The value of j changes 24.5%, 25.23% and 24.79% respectively. For transitional to turbulent shift, f decrease 18.86%, 20.57% and 18.05% for the three fin pitch cases respectively. The value of j changes 4.88%, 4.72 % and 5.99% respectively.

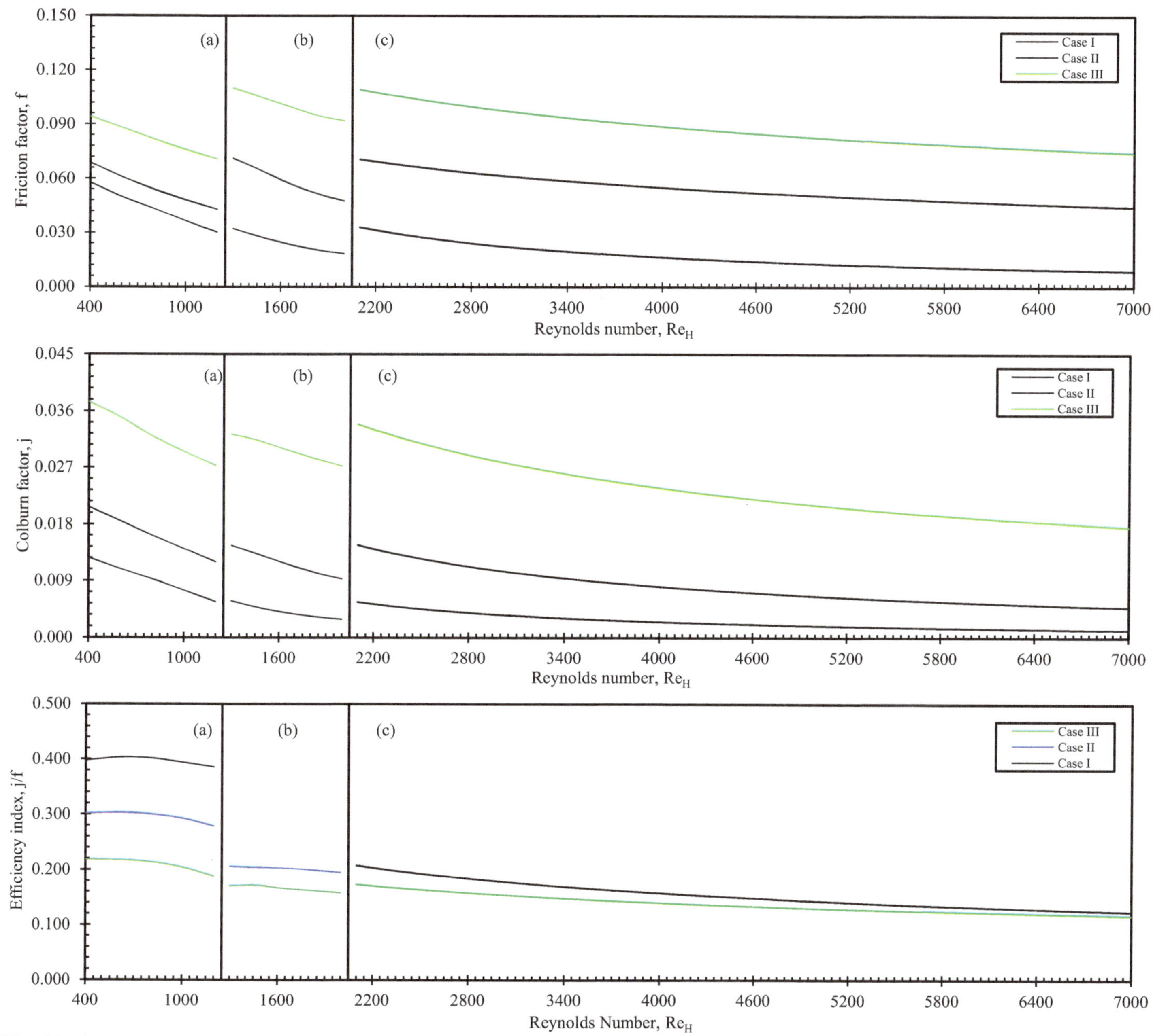

Fig. 13 Flow characteristics for the variation of three different wavy angles (Wa) for staggered tube arrangement in (c). Turbulent flow range ($2100 \leqslant Re_H \leqslant 7000$) compared with the published (a). Laminar flow range ($400 \leqslant Re_H \leqslant 1200$), (b). Transitional flow range ($1300 \leqslant Re_H \leqslant 2000$).

4.5. Effect of Wavy Angles (Wa)

Wavy angle (Wa) is one of the important parameters in the design of heat exchanger of wavy type. Little attention has given on the performance of different wavy angle in the literature. Figure 13 shows the flow characteristics for the variation of three different wavy angles (Wa) for staggered tube arrangement in turbulent flow range ($2100< Re_H<7000$) compared with the published laminar flow range ($400< Re_H <1200$) (Bhuiyan, Islam et al. 2011) and transitional flow range ($1300< Re_H<2000$) (Bhuiyan, Islam et al. 2012). It is found that with the increase of wavy angle, heat transfer and pressure drop increases drastically. But efficiency curves shows opposite trends. For the increase of Wa from 8° to 17.5° and 17.5° to 35°, heat transfer and pressure drop increases more than 50% in each cases. While comparing the flow model, for laminar to transitional change, f decreases 43.92% for wavy angle 8°, while for transitional to turbulent case, it increases 11.0%.

5. CONCLUSION

This numerical study has been conducted as a continuation of author's previous work considering wavy fin and tube heat exchanger. Commercial CFD Code is employed, coupled with user defined program. Code was validated with experimental results from literature and a reasonable matching has been observed. Simulated results were presented for turbulent flow range comparing previously published laminar and transitional flow ranges. There is clear difference in the performance of staggered and in-lined tube arrangement. Comparatively better performance is observed in staggered case. Detailed parametric study is carried out considering important pitches and wavy angles. The trend observed in turbulent flow is in line with the laminar and transitional flows. This parametric study provides a clear understanding and guideline for the design of similar type of heat exchanger depending upon requirements.

NOMENCLATURES

C_P	[J/kg K]	Specific heat
D	[m]	Tube diameter
F_p	[m]	Fin pitch
F_t	[m]	Fin thickness
H	[m]	Fin spacing
h	[W/m^{2}K]	Heat transfer coefficient
k	[m^2/s^2]	Kinetic energy
L	[m]	Flow length
Ll	[m]	longitudinal tube pitch
Lt	[m]	Transverse tube pitch
m	[kg/s]	Mass flow rate
P	[Pa]	Local pressure
P_{in}	[Pa]	Inlet pressure
P_k	[kg/m.s^3]	Share production
T	[°C]	Temperature
T_{in}	[°C]	Inlet temperature
T_{Wall}	[°C]	Wall temperature
U	[m/s]	Velocity
u_{in}	[m/s]	Inlet(frontal) velocity
ε	[m^2/s^3]	Turbulence dissipation
λ	[W/m.K]	Thermal conductivity
μ	[Ns/m^2]	Dynamic viscosity
μ_T	[Ns/m^2]	Turbulent viscosity
ρ	[kg/m^3]	Fluid density
ω	[s^{-1}]	Turbulent frequency
f		Friction factor
j		Colburn factor
Pr		Prandtl Number
Re		Reynolds number

6. REFERENCES

Amin, M. R., and Ramachandran, S. K., 2011, "Numerical Analysis of Airside Characteristics in Plain and Wavy Heat Exchangers in the Turbulent Flow Regime", *ASME 2011 International Mechanical Engineering Congress and Exposition, American Society of Mechanical Engineers*, **10** (Heat and Mass Transport Processes, Parts A and B): 97-107.
http://dx.doi.org/10.1115/IMECE2011-62330

Bhuiyan, A. A., Islam, A. K. M. S., and Amin, M. R., 2012, "Numerical Study of 3D Thermal and Hydraulic Characteristics of Wavy Fin-and-Tube Heat Exchanger," *Frontiers in Heat and Mass Transfer (FHMT)*, **3**(3).
http://dx.doi.org/10.5098/hmt.v3.3.3006

Bhuiyan, A. A., Amin, M. R., and Islam, A. K. M. S., 2012, "Three-Dimensional Performance Analysis of Plain Fin Tube Heat Exchangers in Transitional Regime," *Applied Thermal Engineering*, **50**(1): 445-454.
http://dx.doi.org/10.1016/j.applthermaleng.2012.07.034

Bhuiyan, A. A., Amin, M. R., Karim, M. R., and Islam, A. K. M. S., 2014, "Plate Fin and Tube Heat Exchanger Modeling: Effects of Performance Parameters for Turbulent Flow Regime," *International Journal of Automotive and Mechanical Engineering* **9**(1): 1768-1781.
http://dx.doi.org/10.15282/ijame.9.2013.25.0147

Bhuiyan, A. A., Islam, A. K. M. S., and Amin, M. R., 2011, "Numerical Prediction of Laminar Characteristics of Fluid Flow and Heat Transfer in Finned-Tube Heat Exchangers," *Innovative Systems Design and Engineering* **2**(6): 1-12.

Bhuiyan, A. A., Islam, A. K. M. S., 2010, "A Numerical Investigation on Heat Transfer and Pressure Drop Performance of Typical Wavy Fin-and-Tube Heat Exchanger," *The 7th Jordanian International Mechanical Engineering Conference (JIMEC'7)*, Jordan.

Bhuiyan, A. A., and Naser, J., 2014, Effect of Recycled Ratio on Heat Transfer Performance of Coal Combustion in a 0.5MWth Combustion Test Facility, *19th Australasian Fluid Mechanics Conference*, Melbourne, Austraia.

Bhuiyan, A. A., and Naser, J., 2015, "Computational Modelling of Co-firing of Biomass with Coal under Oxy-fuel Condition in a Small Scale Furnace," *Fuel*, **143**(0): 455-466.
http://dx.doi.org/10.1016/j.fuel.2014.11.089

Bhuiyan, A. A., and Naser, J., 2015, "Numerical Modelling of Oxy Fuel Combustion, the Effect of Radiative and Convective Heat Transfer and Burnout," *Fuel*, **139**(0): 268-284.
http://dx.doi.org/10.1016/j.fuel.2014.08.034

Bhuiyan, A. A., and Naser, J., 2015, "Numerical Modeling of Biomass Co-combustion with Pulverized Coal in a Small Scale Furnace," *Procedia Engineering*, **105**(C): 509-516.
http://dx.doi.org/10.1016/j.proeng.2015.05.083

Bhuiyan, A. A., and Naser, J., 2015, "Modeling of Slagging in Industrial Furnace: A Comprehensive Review," *Procedia Engineering*, **105**(C): 517-524.
http://dx.doi.org/10.1016/j.proeng.2015.05.084

Bhuiyan, A. A., Zaman, R. I., and Islam, A. K. M. S., 2010, "Numerical Analysis of Thermal and Hydraulic Performance of Fin and Tube Heat Exchangers," *13th Asian Congress of Fluid Mechanics (13ACFM)*, Bangladesh Society of Mechanical Engineers(BSME).

Chu, P., He, Y. L., Lei, Y. G., Tian, L. T., and Li, R., 2009, "Three-Dimensional Numerical Study on Fin-and-oval-tube Heat Exchanger with Longitudinal Vortex Generators," *Applied Thermal Engineering*, **29**(5–6): 859-876.
http://dx.doi.org/10.1016/j.applthermaleng.2008.04.021

Dong, J., Chen, Q., and Wei, W., 2012, "Experimental Investigation and Comparison on Fin Surfaces Performance of Flat Tube Heat Exchangers," *Advanced Materials Research*, **354-355:** 389-393.
http://dx.doi.org/10.4028/www.scientific.net/AMR.354-355.389

Dong, J., Su, L., Chen, Q., and Xu, W., 2013, "Experimental Study on Thermal–Hydraulic Performance of a Wavy Fin-and-Flat Tube Aluminum Heat Exchanger," *Applied Thermal Engineering*, **51**(1–2): 32-39.
http://dx.doi.org/10.1016/j.applthermaleng.2012.09.018

Du, X., Feng, L., Yang, Y., and Yang, L., 2013, "Experimental Study on Heat Transfer Enhancement of Wavy Finned Flat Tube with Longitudinal Vortex Generators," *Applied Thermal Engineering*, **50**(1): 55-62.
http://dx.doi.org/10.1016/j.applthermaleng.2012.05.024

Glazar, V., Trp, A., and Lenic, K., 2012, "Numerical Study of Heat Transfer and Analysis of Optimal Fin Pitch in a Wavy Fin-and-Tube Heat Exchanger," *Heat Transfer Engineering*, **33**(2): 88-96.
http://dx.doi.org/10.1080/01457632.2011.589312

Gong, J., Min, C., Qi, C., Wang, E., and Tian, L., 2013, "Numerical Simulation of Flow and Heat Transfer Characteristics in Wavy Fin-and-Tube Heat Exchanger with Combined Longitudinal Vortex Generators," *International Communications in Heat and Mass Transfer*, **43**: 53-56.
http://dx.doi.org/10.1016/j.icheatmasstransfer.2013.01.004

He, Y. L., Chu, P., Tao, W. Q., Zhang, Y. W., and Xie, T., 2013, "Analysis of Heat Transfer and Pressure Drop for Fin-and-Tube Heat Exchangers with Rectangular Winglet-type Vortex Generators," *Applied Thermal Engineering,* **61**(2): 770-783. http://dx.doi.org/10.1016/j.applthermaleng.2012.02.040

Hossain, A., Naser, J., and Imteaz, H., 2011, "Extended analytical turbulent diffusion model for particle dispersion and deposition in a horizontal pipe: comparison with CFD simulation," *Environmental Modeling & Assessment*, **16**(3): 295-311. http://dx.doi.org/10.1007/s10666-010-9248-9

Islam, A. K. M. S., Amin, M. R., Barna, S. F., Bhuiyan, A. A., and Banna, M. H., 2009, "Mixed Convection and Entropy Generation Characteristics Inside a Porous Cavity With Viscous Dissipation Effect," *ASME 2009 International Mechanical Engineering Congress and Exposition, American Society of Mechanical Engineers,* **9**(Heat Transfer, Fluid Flows, and Thermal Systems, Parts A, B and C): 533-542. http://dx.doi.org/10.1115/IMECE2009-10883

Jang, J. Y., and Chen, L. K., 1997, "Numerical Analysis of Heat Transfer and Fluid Flow in a Three-Dimensional Wavy-Fin and Tube Heat Exchanger," *International Journal of Heat and Mass Transfer*, **40**(16): 3981-3990. http://dx.doi.org/10.1016/S0017-9310(97)00047-1

Junqi, D., Jiangping, C., Zhijiu, C., Yimin, Z., and Wenfeng, Z., 2007, "Heat Transfer and Pressure Drop Correlations for the Wavy Fin and Flat Tube Heat Exchangers," *Applied Thermal Engineering*, **27**(11–12): 2066-2073. http://dx.doi.org/10.1016/j.applthermaleng.2006.11.012

Kim, N. H., Ham, J. H., and Cho, J. P., 2008, "Experimental Investigation on the Airside Performance of Fin-and-Tube Heat Exchangers having Herringbone Wave Fins and Proposal of a New Heat Transfer and Pressure Drop Correlation," *Journal of Mechanical Science and Technology*, **22**(3): 545-555. http://dx.doi.org/10.1007/s12206-007-1116-4

Kim, N. H., Youn, J. H., and Webb, R. L., 1996, "Heat Transfer and Friction Correlations for Wavy Plate Fin-and-Tube Heat Exchangers," *American Society of Mechanical Engineers, Heat Transfer Division, (Publication) HTD*, **330**: 43-52. http://dx.doi.org/10.1115/1.2824141

Lin, Z. M., Wang, L. B., and Zhang, Y. H., 2014, "Numerical Study on Heat Transfer Enhancement of Circular Tube Bank Fin Heat Exchanger with Interrupted Annular Groove Fin," *Applied Thermal Engineering,* **73**(2): 1465-1476.. http://dx.doi.org/10.1016/j.applthermaleng.2014.05.073

Lu, C. W., Huang, J. M., Nien, W., and Wang, C. C., 2011, "A Numerical Investigation of the Geometric Effects on the Performance of Plate Finned-Tube Heat Exchanger," *Energy Conversion and Management*, **52**(3): 1638-1643. http://dx.doi.org/10.1016/j.enconman.2010.10.026

Menter, F. R., 1994, "Two-equation eddy-viscosity turbulence models for engineering applications," *AIAA journal,* **32**(8): 1598-1605. http://dx.doi.org/10.2514/3.12149

Panse, S., 2005, "A Numerical Investigation of Thermal and Hydraulic Characteristics in 3D Plate and Wavy Fin-Tube Heat Exchangers for Laminar and Transitional Flow Regimes," *M. Sc. Thesis, Montana State University*, Bozeman, Montana, USA.

Pirompugd, W., Wongwises, S., and Wang, C. C., 2006, "Simultaneous Heat and Mass Transfer Characteristics for Wavy Fin-and-Tube Heat Exchangers under Dehumidifying Conditions," *International Journal of Heat and Mass Transfer*, **49**(1-2): 132-143. http://dx.doi.org/10.1016/j.ijheatmasstransfer.2005.05.043

Reddy, J. N., and Gartling, D. K., 2010, "The finite element method in heat transfer and fluid dynamics," *CRC press.*

Sohr, H., 2012, "The Navier-Stokes equations: An elementary functional analytic approach," *Springer.*

Tao, Y. B., He, Y. L., Huang, J., Wu, Z. G., and Tao, W. Q., 2007, "Three-Dimensional Numerical Study of Wavy Fin-and-Tube Heat Exchangers and Field Synergy Principle Analysis," *International Journal of Heat and Mass Transfer,* **50**(5-6): 1163-1175. http://dx.doi.org/10.1016/j.ijheatmasstransfer.2006.03.019

Tao, Y. B., He, Y. L., Wu, Z. G., and Tao, W. Q., 2007, "Numerical Design of an Efficient Wavy Fin Surface Based on the Local Heat Transfer Coefficient Study," *Journal of Enhanced Heat Transfer,* **14**(4): 315-332. http://dx.doi.org/10.1615/JEnhHeatTransf.v14.i4.50

Wang, C. C., 1999, "Investigation of Wavy Fin-and-Tube Heat Exchangers: A Contribution to Databank," *Experimental Heat Transfer,* **12**(1): 73-89. http://dx.doi.org/10.1080/089161599269825

Wang, C., Fu, W., and Chang, C., 1997, "Heat Transfer and Friction Characteristics of Typical Wavy Fin-and-Tube Heat Exchangers," *Experimental Thermal and Fluid Science*, **14**(2): 174-186. http://dx.doi.org/10.1016/S0894-1777(96)00056-8

Wang, C. C., Hwang, Y. M., and Lin, Y. T., 2002, "Empirical Correlations for Heat Transfer and Flow Friction Characteristics of Herringbone Wavy Fin-and-Tube Heat Exchangers," *International Journal of Refrigeration,* **25**(5): 673-680. http://dx.doi.org/10.1016/S0140-7007(01)00049-4

Wang, C. C., Liaw, J. S., and Yang, B. C., 2011, "Airside Performance of Herringbone Wavy Fin-and-Tube Heat Exchangers - Data with Larger Diameter Tube," *International Journal of Heat and Mass Transfer,* **54**(5-6): 1024-1029. http://dx.doi.org/10.1016/j.ijheatmasstransfer.2010.11.038

Wu, Y., Wu, M., Zhang, Y., and Wang, L., 2014, "Experimental Study of Heat and Mass Transfer of a Rolling Wheel," *Heat and Mass Transfer/Waerme- und Stoffuebertragung,* **50**(2): 151-159. http://dx.doi.org/10.1007/s00231-013-1228-6

Yan, W. M., and Sheen, P. J., 2000, "Heat Transfer and Friction Characteristics of Fin-and-Tube Heat Exchangers," *International Journal of Heat and Mass Transfer,* **43**(9): 1651-1659. http://dx.doi.org/10.1016/S0017-9310(99)00229-X

Yang, L. C., Asako, Y., Yamaguchi, Y., and Faghri, M., 1997, "Numerical Prediction of Transitional Characteristics of Flow and Heat Transfer in a Corrugated Duct," *Journal of Heat Transfer,* **119**(1): 62-69. http://dx.doi.org/10.1115/1.2824101

Yuan, Z., 2000, "Numerical Study of Periodically Turbulent Flow and Heat Transfer in a Channel with Transverse Fin Arrays," *International Journal of Numerical Methods for Heat & Fluid Flow,* **10**(8): 842-861. http://dx.doi.org/10.1108/09615530010359139

10

MIXED CONVECTION BOUNDARY LAYER FLOW OVER A VERTICALLY STRETCHING SHEET WITH CONVECTIVE BOUNDARY CONDITION AND EFFECT OF PARTIAL SLIP

Mitiku Daba*, P. Devaraj, S. V. Subhashini

Department of Mathematics, College of Engineering Guindy Campus, Anna University, Chennai, Tamil Nadu, Chennai-600 025, India

ABSTRACT

In the present study, we investigated a problem of steady laminar mixed convection flow over a vertically stretching sheet with partial slip under convective surface boundary condition. The governing partial differential equations of the boundary layer flow are reduced into a set of nonlinear ordinary differential equations using a suitable similarity transformations. The system of non linear ordinary differential equations are solved by the Keller box method. Velocity, temperature and heat transfer rate are analyzed by considering the important parameters: Prandtl number Pr, convective parameter ϵ, slip parameter K and mixed convection parameter λ on the fluid flow and the corresponding results are presented graphically and discussed in detail. The investigation shows that surface heat transfer rate is affected by Prandtl number Pr, slip parameter K and mixed convection parameter λ. Increasing Prandtl number and mixed convection parameter λ increases surface heat transfer rate but decreasing mixed convection parameter λ and increasing the slip parametr K reduces it. The numerical results obtained are compared with previously reported cases available from literature and they are found to be in very good agreement.
Keywords: *Boundary layer. Mixed convection. Convective boundary condition. Keller box method. Partial slip.*

1. INTRODUCTION

Mixed convection flow over a surface is a phenomenon having wider importance from both theoretical and practical situations due to its numerous applications in many engineering and geophysical areas. The processes occur when the effects of buoyancy forces in forced convection or the effects of forced flow in free convection become significant at the same time.

A free convection heat transfer due to the action of buoyancy and induced magnetic forces was investigated by Sparrow and Cess (1961) pointing out that presence of a magnetic field significantly affects the free convection heat transfer. Patil *et al.* (2009) considered a mixed convection flow over a continuously moving semi-infinite vertical plate due to the combined effects of thermal and mass diffusion by solving the coupled nonlinear partial differential equations numerically using implicit finite difference scheme in combination with quasilinearization. Boundary layer flow past a stretching plate have been studied by Ahmad *et al.* (2010) by using variable thermal conductivity to the flow problem and Aydin and Kaya (2009) analyzed the mixed convection flow of a viscous dissipating fluid about a vertical flat plate taking into account the effects of Ohmic heating and viscous dissipation. Bhattacharyya *et al.* (2013) examined effects of slip at the boundary on the mixed convective boundary layer flow over a flat plate. A problem of boundary layer flow over a stretching sheet in the presence of magnetic field has been solved by Pop and Tsung (1998). A similar work was done by Kumaran *et al.* (2010), who studied the transition effect of boundary layer flow due to an imposed and withdrawal of magnetic field over a viscous flow past a stretching sheet and have analyzed the problem numerically using one of the implicit finite difference methods called Crank-Nicholson. Later, Khan and Pop (2010) analyzed numerically the problem of laminar fluid flow past a stretching sheet in a nanofluid including the effects of Brownian motion and thermophoresis properties. A natural convection boundary-layer flow of nanofluid in a porous medium along a vertical plate be examined by Hady *et al.* (2011) wherein the problem has been solved numerically and analytically. Free convection and mass transfer flow over a vertical surface with the fluid considered to be electric conducting through a pours medium with constant heat flux in the presence of magnetic field is analyzed by Acharya, Dash, and Singh (2000) and Mansur and Ishak (2013) also studied the boundary layer flow of a nanofluid past a stretching/shrinking sheet with a convective boundary condition . Further, Bakar *et al.* (2012) extended the steady laminar flow over a stretching sheet with a convective boundary condition and considering the effect of partial slip. Most recently, Ali *et al.* (2014) discussed the problem of steady laminar magnetohydrodynamic (MHD) mixed convection stagnation-point flow of an incompressible viscous fluid over a vertical stretching sheet.

The flow over a stretching sheet has a wider applications in many industrial manufacturing processes such as the aerodynamic extrusion of plastic sheets and polymer extrusion, liquid film, hot rolling, wire drawing, paper production, drawing of plastic films, metal spinning, cooling

**Email: mitbru2007@yahoo.com*

of nuclear reactors during emergency shut down conditions, glass – fiber production and so on (Ferdows *et al.*, 2012; Khan and Pop, 2010; Mahdy, 2012; Nadeem *et al.*, 2013). This motivates us to study the problem of mixed convection flow over a stretching sheet. The aim of the present study is to investigate mixed convection boundary layer flow over a vertically stretching sheet with convective boundary condition and effect of partial slip. It is a new insight into such flow problems over a stretching sheet and also identifies the parameters that will affect the surface heat transfer rate. The boundary layer flow governed by the partial differential equations are first transformed into a system of nonlinear ordinary differential equations, before being solved numerically using Keller box method (Cebeci and Bradshaw, 1984; Na, 1979).

2. MATHEMATICAL FORMULATION

Consider a steady laminar mixed convection flow over a vertically stretching sheet of uniform surface temperature T_w immersed in an incompressible viscous fluid and a Cartesian coordinate system is considered in which the x-axis is measured along the plate in the upward direction and the y-axis is measured in the direction normal to the stretching sheet.

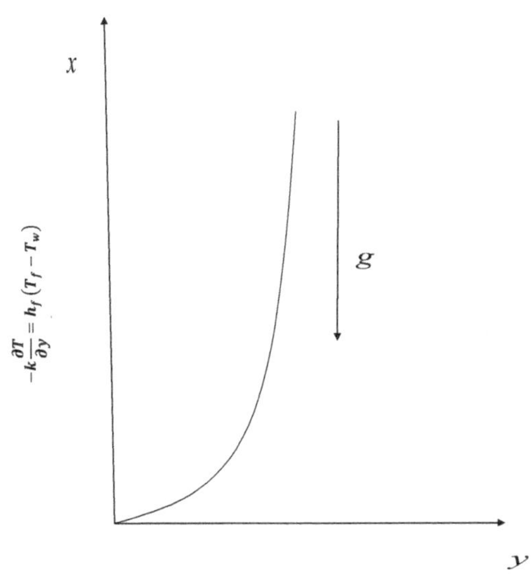

Fig. 1 Physical model and coordinate system

The sheet is stretched and its velocity is assumed to be $u_w = ax$, where a is a positive constant. The boundary-layer equations, under the Boussinesq approximation and for steady state flow conditions in the absence of heat generation and viscous dissipation, are given by:

$$\frac{\partial u}{\partial x} + \frac{\partial v}{\partial y} = 0 \quad (1)$$

$$u\frac{\partial u}{\partial x} + v\frac{\partial u}{\partial y} = \nu\frac{\partial^2 u}{\partial y^2} + g\beta(T - T_\infty) \quad (2)$$

$$u\frac{\partial T}{\partial x} + v\frac{\partial T}{\partial y} = \alpha\frac{\partial^2 T}{\partial y^2} \quad (3)$$

with the boundary conditions:

$u = u_w + L\frac{\partial u}{\partial y}$ and $v = 0$ at $y = 0$ and $u \to 0$ as $y \to \infty$ for the velocity field

$-k\frac{\partial T}{\partial y} = h_f(T_f - T_w)$ at $y = 0$ and $T \to T_\infty$ *as* $y \to \infty$ for the thermal field, where L is the proportionality constant of the velocity slip, u and v are the velocity components in the x and y directions respectively. We assume that the left surface of the sheet is heated by convection. We apply the following transformations to transform the nonlinear partial differential equations into a system of non linear ordinary differential equations.

$\eta = y\left(\frac{a}{\nu}\right)^{\frac{1}{2}}, \quad \psi(x,y) = (\nu a)^{\frac{1}{2}} f(\eta), \quad G(\eta) = \frac{T - T_\infty}{T_f - T_\infty}.$

The stream function $\psi(x,y)$ is defined as: $u = \frac{\partial \Psi}{\partial y}, v = -\frac{\partial \Psi}{\partial x}$. Continuity equation (1) is satisfied with the variable η and the dimensionless stream function $f(\eta)$. Using the similarity transformations and the function $f(\eta)$, equation (2) and (3) respectively reduce to the following ordinary differential equations:

$$f''' + \frac{f}{2}f'' - f'^2 + \lambda G = 0 \quad (4)$$

$$G'' + PrfG' = 0 \quad (5)$$

In order to get similarity solution we take $h_f = \frac{c}{\sqrt{x}}$, where c is constant. The transformed boundary conditions can be written as :

$$f'(0) = 1 + Kf''(0), \; f(0) = 0, \; G'(0) = -\epsilon(1 - G(0)), \quad (6)$$
$$f' \to 0, \; G \to 0 \text{ as } \eta \to \infty,$$

where $K = L\sqrt{\frac{a}{\nu}}$, $\epsilon = \frac{c}{k}\sqrt{\frac{\nu}{u_\infty}}$ and $\lambda = \frac{g\beta(T_f - T_\infty)}{a^2 L}$.

The physical quantities of interest are the local skin friction C_f and the local Nusselt number Nu_x which are defined as:

$$C_f = \frac{2\tau_w}{(\rho u_w^2)} \quad (7)$$

$$Nu_x = \frac{xq_w}{k(T_f - T_\infty)}, \quad (8)$$

where the surface shear stress τ_w and the surface heat flux q_w are given by

$$\tau_w = \mu\frac{\partial u}{\partial y}|_{y=0}, \text{ and } q_w = -k\frac{\partial T}{\partial y}|_{y=0}.$$

Using the similarity variable, equation (7) and (8) respectively becomes

$$\tfrac{1}{2}C_f\sqrt{Re_x} = f''(0), \quad \frac{Nu_x}{\sqrt{Re_x}} = -G'(0),$$

where $Re_x = \frac{U_w x}{\nu}$ is the Reynolds number.

3. RESULTS AND DISCUSSION

The non linear ordinary differential equations (4) and (5) with the boundary condition (6) were solved numerically by Keller box method. The computations have been carried out for different values of Prandtl number Pr, convective parameter ϵ, slip parameter K and mixed convection parameter λ. Figures 2 - 9 show the effect of these parameters on velocity & temperature profiles and Figures 10 - 12 discuss the effect of the parameters on surface heat transfer rate. In order to validate our method, we have compared steady state results of heat transfer coefficient $G'(0)$, η_δ and η_∞ with those of the works by Cortell (2008) and Anilkumar (2011). Where, η_δ is non-dimensional thickness of the boundary layer and η_∞ is the value of η at the free stream when the flow is the boundary layer flow on a flat plate with no effect of convective parameter ϵ, slip parameter K and mixed convection parameter λ. The comparison is given in Table 1.

Figures 2 and 3 depict the effect of mixed convection parameter λ and Prandtl number Pr on velocity profile keeping other parameters constant. We see from these figures that there is no significant effects of λ & Pr on velocity of the fluid. The velocity profile decreases as the slip parameter K increases as it is interpreted by Figure 4. From this we can deduce that as K increases the skin coefficient friction $f''(0)$ decreases in magnitude. Figure 5 explains the relationship between convective parameter ϵ and the non-dimensional temperature. As indicated in the figure, when the value of convective parameter ϵ getting large the temperature also rises. This is to mean that temperature of the hot fluid which is found to the left of the sheet is convicted to the right fluid. This causes the surface temperature to rise and consequently increases the temperature of the fluid which is found to the right of the stretching sheet. The reverse

Table 1 Comparison of $G'(0)$, η_δ and η_∞ with those of the works by Cortell (2008) and Anilkumar (2011) with no effect of ϵ, K and λ.

	Cortell (2008)			Anilkumar (2011)		
Pr	$G'(0)$	η_δ	η_∞	$G'(0)$	η_δ	η_∞
0.7	0.29268	5.64	8.36	0.29268	5.64	8.36
5	0.57669	2.75	4.07	0.57668	2.74	4.07
10	0.72814	2.17	3.22	0.72812	2.16	3.20
30	1.05173	1.50	2.24	1.05171	1.50	2.22
50	1.24729	1.27	1.90	1.24729	1.26	1.91
100	1.57183	1.01	1.52	1.57181	1.00	1.52
1000	3.38709	0.46	0.72	-	-	-

	Present		
Pr	$G'(0)$	η_δ	η_∞
0.7	0.29268	5.64	8.36
5	0.57670	2.75	4.07
10	0.72816	2.17	3.22
30	1.05173	1.50	2.24
50	1.24736	1.27	1.90
100	1.57197	1.00	1.51
1000	3.38843	0.47	0.72

situation also happen when the right hand side is assumed to be hot as it is shown by Figure 6.

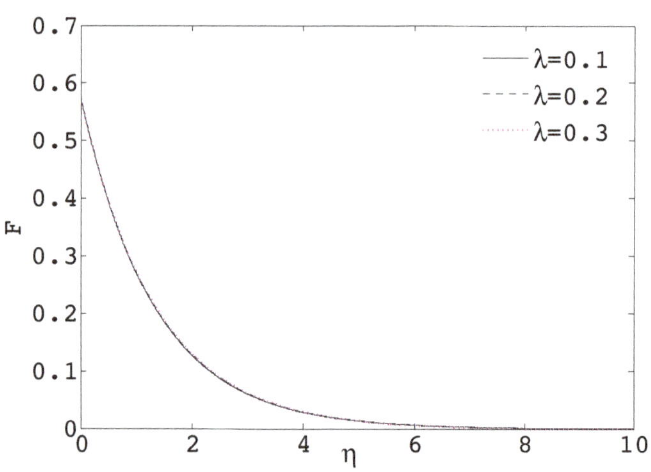

Fig. 2 Velocity profile for some values of λ

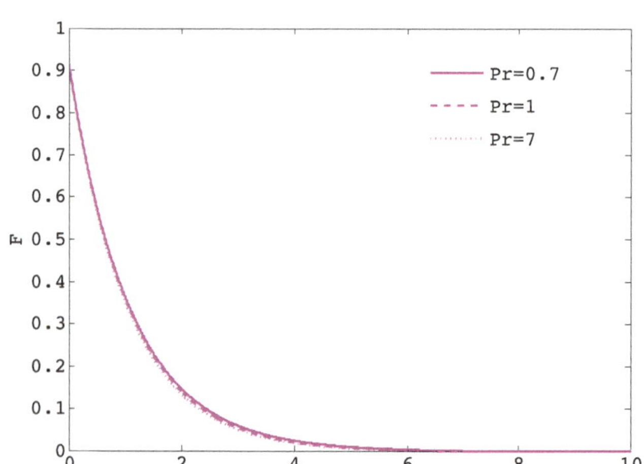

Fig. 3 Velocity profile for some values of Prandtl number

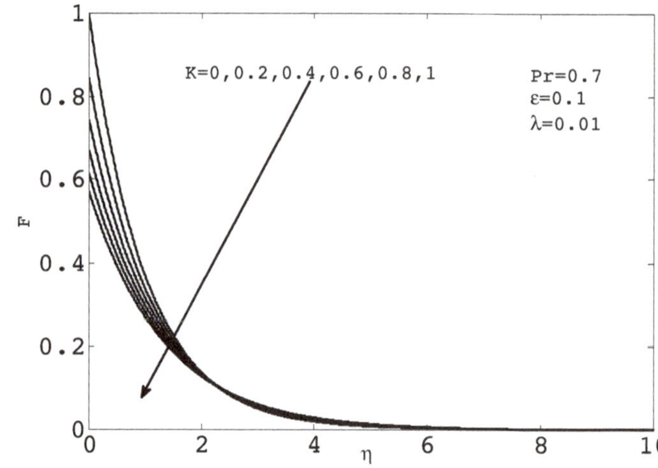

Fig. 4 Velocity profile for different values of slip parameter K

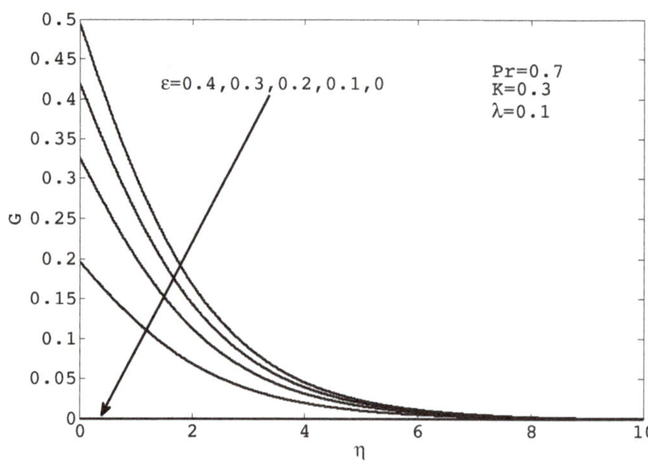

Fig. 5 Effect of convective parameter ϵ on temperature profile

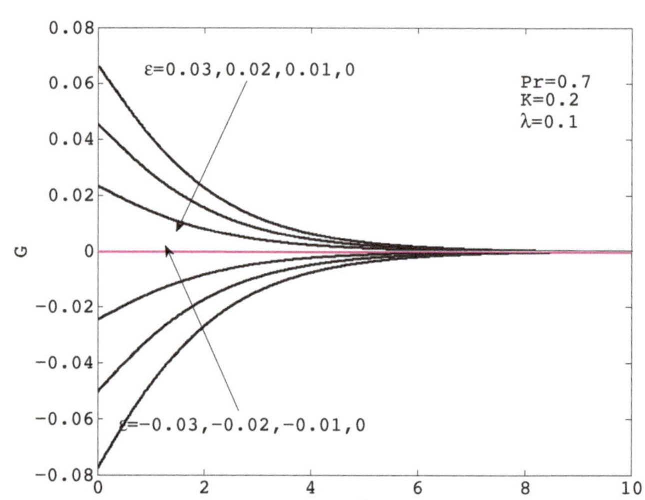

Fig. 6 Effect of convective parameter ϵ on temperature profile

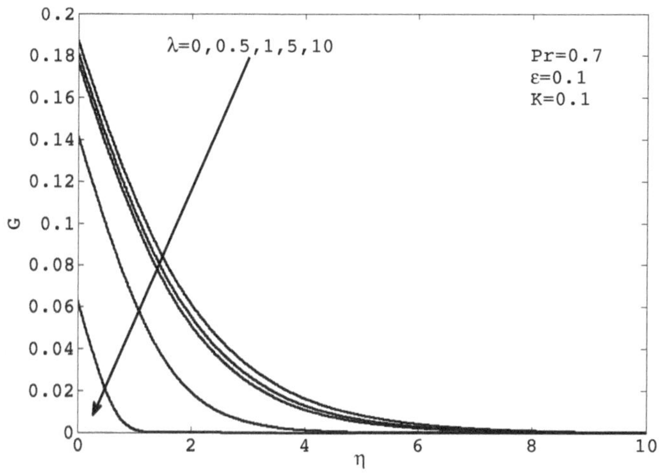

Fig. 7 Effect of mixed convection parameter λ on temperature profile

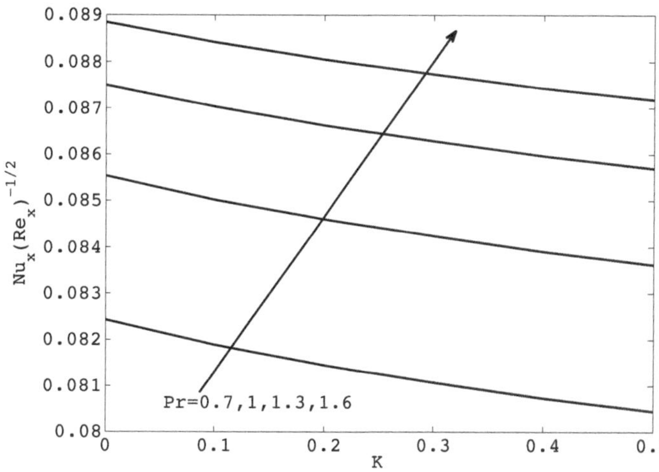

Fig. 10 Effect of Prandtl number Pr and slip parameter K on surface heat transfer rate when $\epsilon = 0.1$ and $\lambda = 0.5$

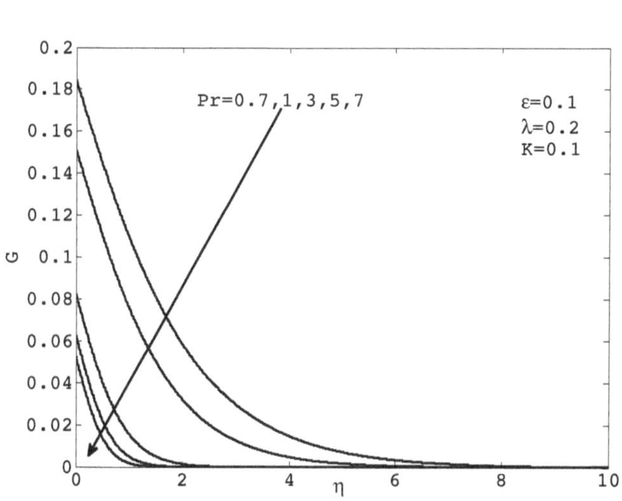

Fig. 8 Effect of Prandtl number on temperature profile

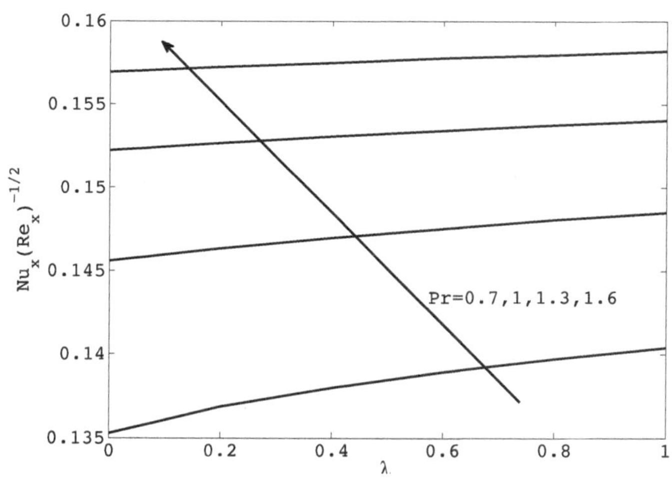

Fig. 11 Effect of Prandtl number Pr and mixed convection parameter λ on surface heat transfer rate when $\epsilon = K = 0.2$

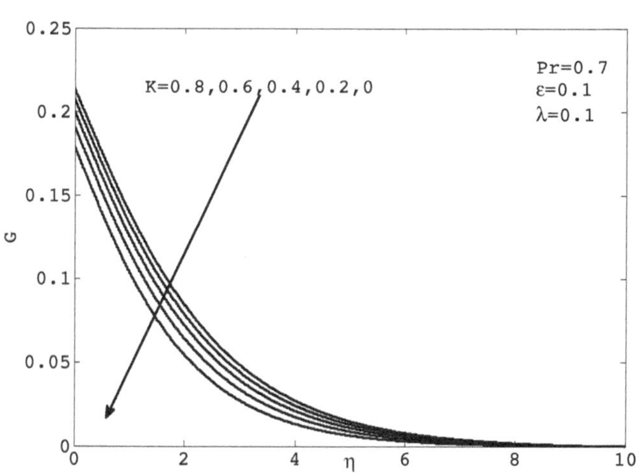

Fig. 9 Temperature profile for different values of K

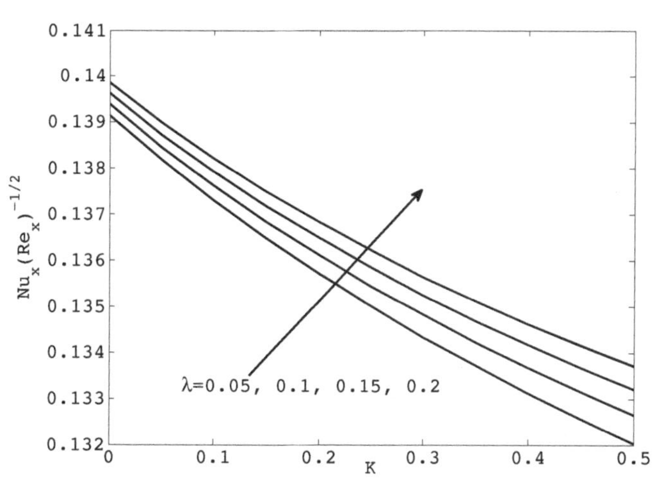

Fig. 12 Effect of mixed convection parameter λ and slip parameter K on surface heat transfer rate when $\epsilon = 0.2, Pr = 0.7$

Figure 7 illustrates that when the mixed convection parameter λ increases the thermal boundary layer getting thick. Consequently this reveals that the surface temperature gradient increases in magnitude in which this produces an increase in the surface heat transfer rate. Physically $\lambda > 0$ means that it is assisting flow and which means heating of the fluid or cooling of the surface of the stretching sheet. Effect of Prandtl number on temperature profile is depicted in Figure 8. According to the figure, as the Prandtl number Pr increases the thermal boundary layer is getting thin. This is because a higher Prandtl number fluid has relatively low thermal conductivity which reduces conduction. It is also evident from the figure that temperature gradient at the surface increases in magnitude as the Prandtl number increases. This indicates that there is an increase in the surface heat transfer rate. A correlation between the dimensionless temperature G and slip parameter K is presented on Figure 9. The figure tells us that as the slip parameter K increases the surface heat transfer rate decreases. Which implies that the dimensionless temperature tends to rise up. Influence of Prandtl number and slip paramter K on a local Nusselt number named as surface heat transfer rate is displayed in Figure 10. It is seen from the figure that decreasing the Prandtl number and increasing the slip parameter K will result in a decrease of the surface heat transfer rate. But increasing the Prandtl number and the mixed convection parameter λ increases the surface heat transfer rate as it is shown by Figure 11. Impact of mixed convection parametr λ and the slip parameter K on surface heat transfer rate is described by Figure 12. It is revealed by the figure that decreasing the mixed convection parameter λ and increasing the slip parameter K is to decrease the surface heat transfer rate.

4. CONCLUSIONS

We studied the problem of steady mixed convection boundary layer flow over a vertically stretching sheet in a viscous incompressible fluid with convective surface boundary condition and partial slip. The governing partial differential equations are transformed into ordinary differential equations using similarity transformation, which are more convenient for numerical computation. The transformed ordinary differential equations are then solved numerically by the Keller box method. The numerical results obtained are compared with previously reported cases available from literature and they are found to be in very good agreement. The effect of Prandtl number Pr, convective parameter ϵ, slip parameter K and mixed convection parameter λ on velocity, temperature and surface heat transfer rate are presented. From the present investigation, we found that:

- Effects of Prandtl number Pr and mixed convection parameter λ on velocity profiles are insignificant.
- Raising the Prandtl number or mixed convection parameter λ tends to increase the surface heat transfer rate.
- The velocity profile decreases as the slip parameter K increases.
- Rise of the temperature can be observed in an increase of convective parameter ϵ.
- A decrease of Prandtl number Pr and an increase of slip parameter K will result in a decrease of surface heat transfer rate.
- Increasing Prandtl number and mixed convection parameter λ increases the surface heat transfer rate but decreasing the mixed convection parameter λ and increasing the slip parametr K reduces it.

ACKNOWLEDGEMENTS

The first author would like to thank Indian Council for Cultural Relations, Ministry of External Affairs, Govt. of India for providing the research fellowship to carry out the research work.

NOMENCLATURE

C_f	local skin friction coefficient
c_p	specific heat capacity
F	dimensionless velocity
f	dimensionless stream function
G	dimensionless temperature
h_f	heat transfer coefficient
k	thermal conductivity
K	slip parameter
L	proportionality constant of the velocity slip
Nu_x	local Nusselt number
Pr	Prandtl number
q_w	surface heat flux
Re_x	local Reynolds number
T	temperature of the fluid
T_f	temperature of the hot fluid
T_∞	temperature of the fluid far away from the wall
T_w	temperature at the wall
u	velocity component in x-direction
U_w	stretching sheet wall velocity
v	velocity component in y-direction
Greek Symbols	
α	thermal diffusivity
ν	kinematic viscosity
μ	dynamic viscosity
ρ	density of the fluid
η	transformed variable
ψ	stream function
τ_w	wall shear stress
λ	mixed convection parameter
ϵ	convective parameter

REFERENCES

Acharya, M., Dash, G., and Singh, L., 2000, "Magnetic Field Effects On The Free Convection And Mass Transfer Flow Through Porous Medium With Constant Suction And Constant Heat Flux," *Indian Journal of Pure and Applied Mathematics*, **31**(4), 1–18.

Ahmad, N., Siddiqui, Z., and Mishra, M., 2010, "Boundary Layer Flow And Heat Transfer Past A Stretching Plate With Variable Thermal Conductivity," *International Journal of Non-Linear Mechanics*, **45**, 306–309. http://dx.doi.org/10.1016/j.ijnonlinmec.2009.12.006.

Ali, F., Nazar, R., Arifin, N., and POP, I., 2014, "Mixed Convection Stagnation-Point Flow on Vertical Stretching Sheet with External Magnetic field," *Applied Mathematics and Mechanics (English Edition)*, **35**(2), 155–166.
http://dx.doi.org/10.1007/s10483-014-1780-8.

Anilkumar, D., 2011, "Nonsimilar Solutions for Unsteady Mixed Convection From a Moving Vertical Plate," *Communications in Nonlinear Science and Numerical Simulation*, **16**, 3147–3157.
http://dx.doi.org/10.1016/j.cnsns.2010.11.017.

Aydin, O., and Kaya, A., 2009, "MHD Mixed Convection of a Viscous Dissipating Fluid about a Permeable Vertical Flat Plate," *Mathematical Modelling*, **33**, 4086–4096.
http://dx.doi.org/10.1016/j.apm.2009.02.002.

Bakar, N., Zaimi, W., Hamid, R., Bidin, B., and Ishak, A., 2012, "Boundary Layer Flow over q Stretching Sheet with a Convective Boundary Condition and Slip Effect," *World Applied Sciences Journal*, **17**, 49–53.

Bhattacharyya, K., Mukhopadhyay, S., and Layek, G., 2013, "Similarity Solution Of Mixed Convective Boundary Layer Slip Flow Over A Vertical Plate," *Ain Shams Engineering Journal*, **4**, 299–305.
http://dx.doi.org/10.1016/j.asej.2012.09.003.

Cebeci, T., and Bradshaw, P., 1984, *Physical and Computational Aspects of Convective Heat Transfer*, Springer-Verlag, New York.

Cortell, R., 2008, "Radiation effects in the Blasius flow," *Applied Mathematics And Computation*, **198**, 333–338.
http://dx.doi.org/10.1016/j.amc.2007.08.037.

Ferdows, M., Khan, M., Mahmud, M., and Shuyu, S., 2012, "MHD mixed convective boundary layer flow of a nanofluid through a porous medium due to an exponentially stretching sheet," *Mathematical Problems in Engineering*.
http://dx.doi.org/10.1155/2012/408528.

Hady, F., Ibrahim, F., M.R., S.A.G., and Eid, 2011, "Boundary-layer Non-Newtonian Flow Over Vertical Plate In Porous medium saturated with nanofluid," *Applied Mathematics and Mechanics (English Edition)*, **32**(12), 1577–1586.
http://dx.doi.org/10.1007/s10483-011-1524-7.

Khan, W., and Pop, I., 2010, "Boundary-layer Flow of a Nanofluid Past a Stretching Sheet," *International Journal of Heat and Mass Transfer*, **53**, 2477–2483.
http://dx.doi.org/10.1016/j.ijheatmasstransfer.2010.01.032.

Kumaran, V., Kumar, A., and Pop, I., 2010, "Transition Of MHD Boundary Layer Flow Past P Stretching Sheet," *Commun Nonlinear Sci Numer Simulat*, **15**, 300–311.
http://dx.doi.org/10.1016/j.cnsns.2009.03.027.

Mahdy, A., 2012, "Unsteady mixed convection boundary layer flow and heat transfer of nanofluids due to stretching sheet," *Nuclear Engineering and Design*, **249**, 248âĂŞ255.
http://dx.doi.org/10.1016/j.nucengdes.2012.03.025.

Mansur, S., and Ishak, A., 2013, "The Flow and Heat Transfer af a Nanofluid Past a Stretching or Shrinking Sheet with a Convective Boundary Condition," *Abstract and Applied Analysis*.
http://dx.doi.org/10.1155/2013/350647.

Na, T., 1979, *Computational methods in engineering boundary balue problems*, Academic Press, New York.

Nadeem, S., Haq, R., Akbar, N., Lee, C., and Khan, Z., 2013, "Numerical study of boundary layer flow and heat transfer of Oldroyd -B nanofluid towards a stretching sheet," *PLOS ONE*, **8**.
http://dx.doi.org/10.1371/journal.pone.0069811.

Patil, P., Roy, S., and Chamkha, A., 2009, "Double Diffusive Mixed Convection Flow over a Moving Vertical Plate in The Presence of Internal Heat Generation and a Chemical Reaction," *Turkish Journal of Engineering and Environmental Sciences*, **33**, 193–205.
http://dx.doi.org/10.3906/muh-0905-21.

Pop, I., and Tsung, Y., 1998, "A Note on MHD Flow over a Stretching Permeable Surface," *Mechanics Research Communications*, **25**, 263–269.

Sparrow, E., and Cess, R., 1961, "The Effect of a Magnetic Field on Free Convection Heat Transfer," *International Journal of Heat and Mass Transfer*, **3**, 267–274.

11

EXPERIMENTAL INVESTIGATION ON HEAT TRANSFER AND FRICTION FACTOR CHARACTERISTICS OF A STATIONARY SQUARE DUCT ROUGHENED BY V AND Λ-SHAPED RIBS

Anand Shukla[a], Alok Chaube[b], Shailesh Gupta[c†] and Arvind Sirsath[c]

[a]*Faculty of Engineering & Technology, Mahatma Gandhi Chitrakoot Gramodaya, Vishwavidyalaya, Satna-485331,India*
[b]*Department of Mechanical Engineering, Jabalpur Engineering College, Jabalpur-482011, India*
[c]*Department of Mechanical Engineering, Hitkarini College of Engineering & Technology, Jabalpur-482005, India*

ABSTRACT

One of the traditional methods used to improve the efficiency of a gas turbine is to increases the inlet temperature; thereby increasing the power output and in turn, the efficiency. The internal cooling passages of blades are roughened by artificial roughness to improve the cooling performance. The present study investigates the convective heat transfer and friction factor (pressure drop) characteristics of a rib-roughened square duct. The test section of the duct is roughened on its top and bottom wall with V and Λ- shaped square ribs. In the study, the Reynolds number (*Re*) varied from 10,000 to 40,000, the relative roughness height (e/D_h) is 0.060, relative roughness pitch (*p/e*) is 10 and rib attack angle (α) is taken as 45^0 and 60^0. The comparison of heat transfer and pressure drop for V and Λ- shaped ribs is presented in the form of Nusselt Number and friction factor. The results show that the Nusselt number enhancement decreases when the Reynolds number increases. The friction factor ratio is found to increases as Reynolds number increases. The thermal performance decreases when the Reynolds number increases. It seems that the ribs disturb the main flow resulting in the recirculation and secondary flow near the ribbed wall. The heat transfer coefficient and friction factor for all the cases are higher than that of the smooth duct. The V-shaped ribs have higher Nusselt number and friction factor than the Λ - shaped ribs. It is also concluded that the V -shaped ribs perform better than the Λ- shaped ribs. Ribs with $\alpha = 60^0$ produces higher heat transfer and friction factor than $\alpha = 45^0$. The results obtained in the study will help the turbine blade designer to design the effective cooling methods to cool the blades.

Keywords: Friction factor; Nusselt number; Reynolds number; thermal performance, V-shaped ribs.

1. INTRODUCTION

It has been made to apply different active and passive methods of heat transfer enhancement in rectangular / square ducts which can be used in compact heat exchangers for the automotive industry, air-conditioning and refrigerant applications, internal cooling for gas turbine blades, electrical circuits in electronic chipsets, etc. Achieving higher heat transfer rates through various augmentation techniques can result in substantial energy savings, more compact and less expensive apparatus with higher thermal efficiency. One of the methods to enhance the heat transfer in a flow passage is to roughen the surfaces with ribs which is popularly used in the design of internal cooling passage of turbine blade. The surfaces roughened with the ribs cause extra flow resistances. Hence to optimize the shape and orientation of rib-roughened surface, it is required to investigate the heat transfer and friction characteristics of various roughened surfaces.

The performance of the heat transfer surface with ribs depends significantly on the parameters of the flow structure, such as reattachment length of the separated streamline and turbulence intensities, as well as the area of the surface and Reynolds number. The ribs increase the level of mixing by turbulence and disturb the laminar sub-layer, also increase the surface area for convective heat transfer, thereby enhance the cooling capacity of the passage.

In general, mixing the main flow, reducing the flow boundary layer, raising the turbulent intensity, creating rotating and secondary flows are the main reasons for the increase of the heat transfer. Vortex generation is a promising technique for heat transfer augmentation. Geometrical parameters such as channel aspect ratio (*AR*), rib height-to-passage hydraulic diameter (e/D_h), rib attack angle (α), rib pitch-to-height ratio (*p/e*), rib shape, discretization of ribs and the manner in which the ribs are positioned relative to one another have considerable effects on heat transfer coefficient and friction factor of the passage (Han *et al.*, 2001). Han *et al.* (1978) observed the effect of rib shape, angle of attack and pitch to rib height ratio on heat transfer and friction factor characteristics of a rectangular duct with two opposite side roughened walls. They observed that the maximum value of heat transfer and friction factor occurs for square ribs, at $p/e = 10$ and $\alpha = 45^0$. Kiml *et al.* (2001) reported that the thermal performance of rib arrangements with $\alpha = 60^0$ is better than that with 45^0. Han and Park (1988) studied heat transfer and pressure losses with 90°, 60°, 45°, and 30° angle ribs in square and rectangular channels. The Reynolds number was from 10000 to 60000. They concluded that the higher thermal performance in the square channel was 30° rib angle and the higher thermal performance in the rectangular channel was 45° rib angle. The higher heat transfer with higher pressure drop in the square channel was 60° rib angle. The square channel in their study showed a larger increase in heat transfer performance than the wide aspect ratio channels.

Lau *et al.* (1991) observed that the replacement of continuous transverse ribs by inclined ribs in a square duct results in higher turbulence at the ribbed wall due to interaction of the primary and secondary flows. Metzger *et al.* (1990) investigated the effects of rib

† Corresponding author: E-mail address: gupta_shailu@rediffmail.com

angle and orientation on local heat transfer in a square channel. It was found that 60° angled ribs giving a two-cell pattern provided the best heat transfer performance. Lau *et al.* (1991) examined the turbulent heat transfer and frictional characteristics in a square channel with V-shaped ribs. The configuration investigated included V-shaped ribs angled at 45°, 60°, 90°, 120° and 135° for a P/e ratio of 10.The results indicated that the V-shaped ribs at (α = 46° and 60°) produced a 38%-46% and 47%-66% increase in thermal performance when compared to the 90° full rib. Similarly, when the following ribs were reversed, the heat transfer performance was enhanced by 26%-32% and 39%-48% respectively. This enhancement in thermal performance was accompanied by a 55%-79% increase in the pressure drop for various V-shaped configurations in comparison to the 90° rib. The V-shaped ribs pointing downwards (Λ-shaped ribs) was found to result in poor heat transfer performance. The crossed ribs whose performance was also investigated resulted in poor heat removal. It was also found that doubling the p/e ratio lowers the heat transfer augmentation and friction. Liou and Hwang (1992) studied the heat transfer and flow field in the ribbed channels. They found that the (p/e) of 10 resulted in the best heat transfer; the heat transfer showed a periodic behavior between consecutive ribs; and both heat transfer and friction factor increased with decreasing rib spacing. Liou and Hwang (1993) conducted experiments to measure the local as well as average heat transfer coefficients to compare the performance of square, triangular and semi-circular ribs and found that the square ribs give best heat transfer performance.

It is revealed from the literature that discrete ribs were performed outstanding as compared to continuous angled or V-shaped ribs (Cho *et al.*, 1992). It is also reviewed that V-shaped ribs pointing upwards (V-shaped) and pointing downwards (Λ-shaped) will give different Heat transfer and friction characteristics. Han and Zhang (2000) carried out experiments to study the heat transfer and pressure drop characteristics of a roughened square channel with various angled and V-shaped broken rib arrangement with rib attack angle of 45^0 and 60^0, the Reynolds number was from 15000 to 90000. They were reported that 60^0 V-shaped broken rib arrangements give better performance than 45^0 V-shaped broken ribs. They also concluded that broken ribs create heat transfer enhancement level of 2.5 to 4, while the enhancement created by the continuous ribs is only 2 to 3. Tanda (2004) investigated the heat transfer enhancement for one wall-ribbed rectangular channel of AR = 5:1 with continuous, 90° and V-broken ribs and found that the enhancement of the 90° broken ribs is around 1.8 times higher than that of continuous ribs. Han *et al.* (1991) presented experimental results on pressure drop and heat transfer in a square channel with ribs on two walls for nine different rib configurations. Regionally averaged heat transfer and friction factor were reported for rib pitch-to-height ratio (p/e) of 10 and the rib height to hydraulic diameter (e/D_h) of 0.0625. They concluded that the angled ribs and 'V' ribs provide higher heat transfer augmentation compared to continuous ribs. It was observed that the heat transfer augmentations and the friction factor were highest for the 60^0 orientation compared to 45^0 and 90^0 orientation amongst the angled ribs. The 60^0 'V' and 60^0 'Λ' offered higher heat transfer augmentation compared to the corresponding 45^0 orientation. The crossed rib configuration offered least heat transfer augmentation. The friction factor was found to be highest for 60^0 angled ribs compared to 45^0 and 90^0 configuration. Continuous rib configuration offered the least resistance and the 60^0 'Λ' case offered highest flow resistance among all configurations.

A very narrow channel (AR=8:1) with V-shaped, ʌ -shaped, and angled ribs was studied by Gao and Suden (2001). Using a liquid crystal technique, they too confirmed that V-shaped ribs result in the highest heat transfer enhancement and the highest frictional losses. They concluded that the V-shaped ribs yield the best overall thermal performance. Kim *et al.* (2001) also investigated heat transfer enhancement mechanisms in rectangular channels with V- and ʌ - shaped ribs using a flow visualization technique to examine the secondary flow behaviours created by the V-shaped ribs. Taslim *et al.* (1996) experimentally investigated the heat transfer and friction in channel roughened with angled, V-shaped and discrete ribs on two opposite walls for Reynolds number ranging from 5,000 to 30,000. The results showed that the 90° transverse ribs produced the lowest heat transfer performance. The 45° angled V-shaped ribs produced the highest heat transfer performance in comparison to other rib configurations. For V-shaped ribs facing downstream of flow, the one with lowest blockage ratio had better heat removal rate. The discrete ribs also produced better performance in comparison to the transverse ribs.

Muluwork *et al.* (1998) have investigated the effect of a staggered discrete V- apex up and down on the thermal performance. The Stanton number for V-down discrete ribs was higher than the corresponding V-upwards and transverse discrete roughened surfaces. The Stanton number ratio enhancement was found to be 1.32 to 2.47 in the range of parameters covered in the investigation. Further for the Stanton number, it was seen that the ribbed surface friction factor for V-downwards discrete ribs was highest among the three configurations investigated. Karwa (2003) has investigated and revealed the effect of transverse, inclined, V-continuous and V-discrete patterns on heat transfer and the friction factor in a rectangular duct. The ribs in the V-pattern were tested for both pointing upstream (V-upwards) and downstream (V-downwards) to the flow. The angle of inclination of the ribs in inclined and V-pattern was 60°. The enhancement in the Stanton number over the smooth duct was up to 137%, 147%, 134% and 142% for the V-up continuous, V-down continuous, V-up discrete and V-down discrete rib arrangement respectively. The friction factor ratio for these arrangements was up to 3.92, 3.65, 2.47 and 2.58, respectively. Based on the equal pumping power, V-down discrete roughness gave the best heat transfer performance.

From above, it can be stated that inclined and V-shaped ribs when arranged in a systematic manner produces excessive enhancement in convective heat transfer, however the friction factor also increases. Some of the researchers investigated that V-shaped ribs give higher heat transfer than ʌ-shaped ribs, whereas, some observed the reverse results.

In the present work, experimental investigations have been carried out on an opposite side roughened square duct having artificial roughness in the form of V and Λ-shaped rib, to determine the enhancement in heat transfer and friction factor to find out which one give the highest heat transfer performance. The optimum value of rib attack angle (α) for both types (V and Λ- shaped) of ribs have been obtained and discussed.

2. EXPERIMENTAL PROGRAM AND ROUGHNESS GEOMETRY

A schematic diagram of the test set up is shown in Fig.1. which was used by Gupta *et al.* (2013) and Chaube *et al.* (2014). The wooden square duct has an internal size of 3750 mm x 75 mm x 75 mm, which consists of an entrance section, a test section and an exit section of length 1500 mm ($20D_h$), 1500 mm ($20D_h$) and 750 mm ($10D_h$) respectively. The exit end of the duct is connected to 81 mm internal diameter G. I. (galvanized iron) pipe provided with a calibrated orifice plate through a square to circular transition piece. The outside of entire set-up from inlet to the orifice plate, were covered with 25 mm thick thermocole (foamed polystyrene) sheet, so that the heat losses from the test section can be minimized.

The aluminium plates are used as test plates which are made by gluing square aluminium ribs (4.5 mm x 4.5 mm) in a required distribution to serve as top and bottom ribbed walls of the test section. The plates are heated by means of separate heaters assembly, thus subjected to uniform heat flux of 0-1500 W/m^2. To measure mass flow rate of air through the duct, a calibrated orifice-meter is used which is connected with an inclined U-tube manometer with a control valve in the pipeline. Twenty thermocouples are affixed on backside of the test plate as shown in Fig. 2 to record the plate temperature at different locations. The bulk air temperature at the entrance section is measured by providing thermocouple at the entrance section, whereas bulk temperature of the air at the exit is recorded by three thermocouples mounted spanwise at the exit section after mixing (depthwise). To measure the pressure drop across the test section length a digital micro-manometer (Fluke-922) is connected between the required points. The airflow rate was varied to give the flow having Reynolds number in the range of 10000 to 40,000. Relevant data were noted under the steady-state condition for constant surface heat flux, which was assumed to have reached when the plate and air temperatures shows negligible variation for about 10- minute duration. The steady state for each test run was obtained in about 1.5 to 2 hours. In order to minimize the percentage error in measurement of temperatures, minimum heat flux value is so selected as to maintain the temperature of roughened plate around 20^{0}C - 30^{0}C above to that of mean bulk air temperature.

The V and Λ-shaped ribs used in the investigation are shown in Fig. 3. The relative roughness height (e/D_h) and relative roughness pitch (p/e) are kept constant as 0.060 and 10 respectively. Reynolds number was varied in the range of 5000 to 40000. Experimental data for both types of ribs were collected for attack angle (α) of 45^0 and 60^0.

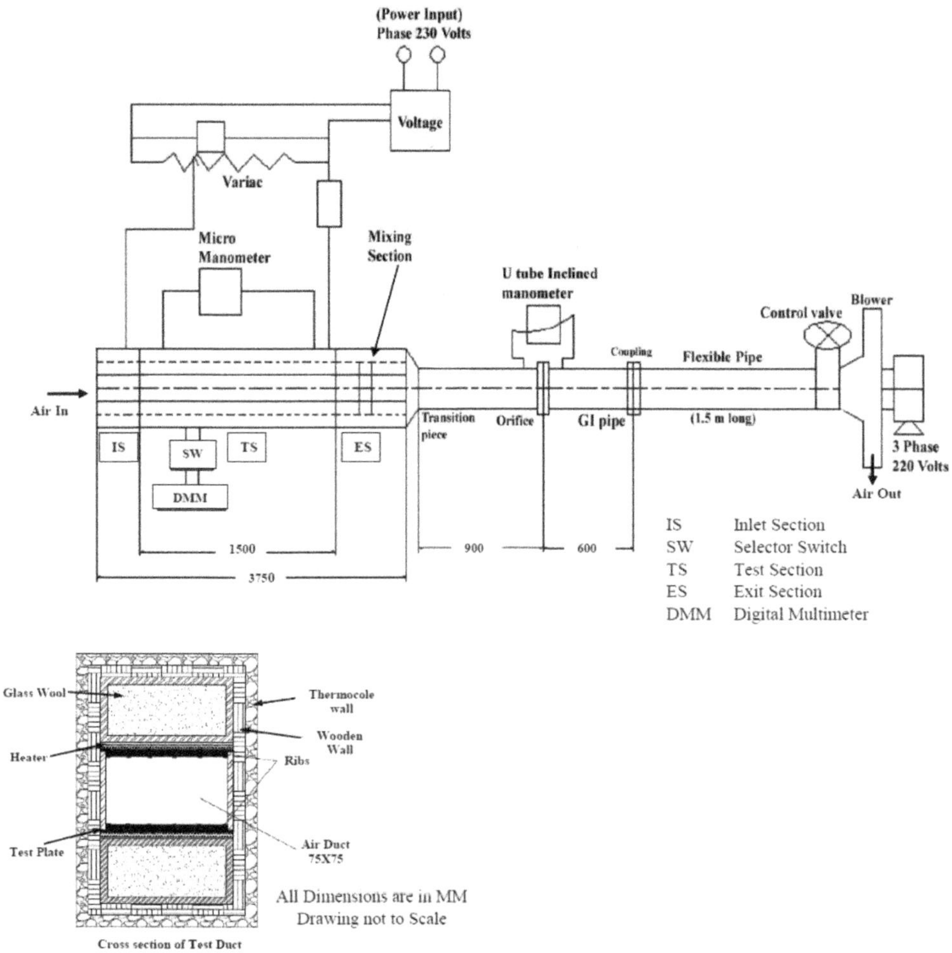

Fig. 1. Schematic diagram of experimental set-up

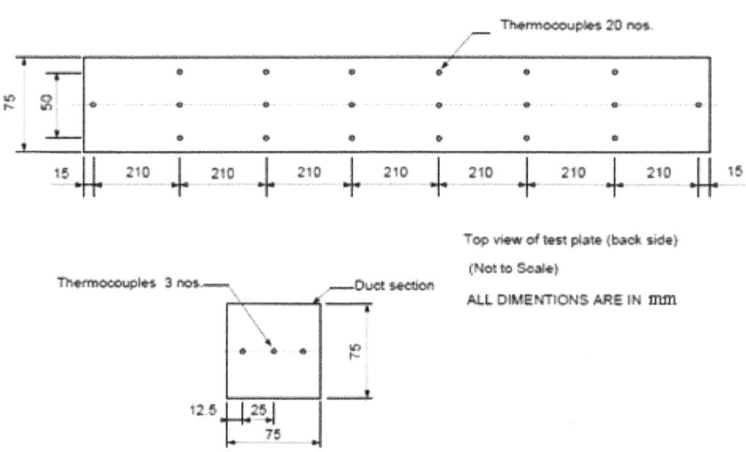

Fig. 2. Detailed locations of thermocouples

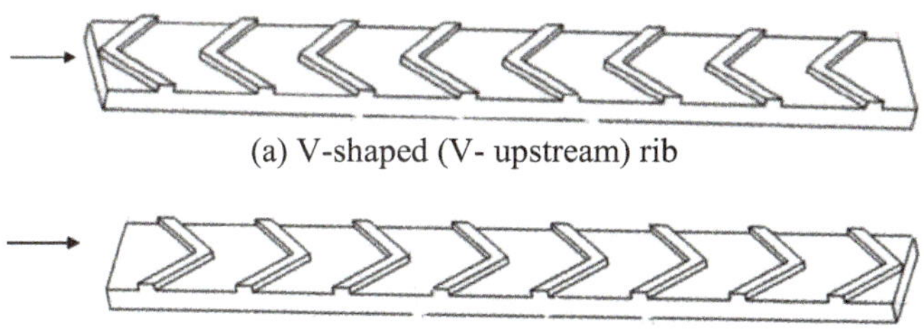

(a) V-shaped (V- upstream) rib

(b) Λ-shaped (V- downstream) rib

Fig. 3. Schematic of roughness geometry

3. DATA REDUCTION

The objective of this experiment is to investigate the Nusselt number and friction factor in the channel. The Reynolds number based on the channel hydraulic diameter, D_h, is given by

$$Re = \frac{\rho_a V D_h}{\mu_a} \tag{1}$$

where,
$$V = \frac{m}{\rho_a . W . H} \tag{2}$$

The mass flow rate, m, of air through the duct has been calculated from pressure drop measurement across the orifice plate.

$$m = C_d . A_o \left[\frac{2 . \rho_a . (\Delta P)_o}{1 - \beta^4} \right]^{0.5} \tag{3}$$

The pressure drop $(\Delta P)_o$ across the orifice plate is given by

$$(\Delta P)_o = 9.81 \text{ x } (\Delta h)_o \text{ x } \rho_m \text{ x } \sin\theta \tag{4}$$

The average heat transfer coefficient, h, is evaluated from the measured temperatures and heat inputs. With heat added uniformly to fluid (Q_u) and the temperature difference of plate and fluid ($T_p - T_f$), the average heat transfer coefficient will be evaluated from the experimental data via the following equations:

$$Q_u = m\, C_p\, (T_o - T_i) \tag{5}$$

$$h = \frac{Q_u}{A_p (T_p - T_f)} \tag{6}$$

The average value of plate temperature (T_p) is calculated as a weighted mean of the plate temperature measured at different locations and mean bulk air temperature (T_f) is calculated from the following equations.

$$T_f = \frac{T_i + T_o}{2} \tag{7}$$

$$T_p = \sum_{n=1}^{20} \frac{T_n}{20} \tag{8}$$

Then, average Nusselt number, Nu, is written as

$$Nu = \frac{h D_h}{k_a} \tag{9}$$

The friction factor is determined from the measured values of pressure drop, $(\Delta P)_d$ across the test section length, between the two points located 1.2 m apart.

$$f = \frac{2 (\Delta P)_d\, D_h}{4\, \rho_a\, L_f V^2} \tag{10}$$

where, $(\Delta P)_d$ is the pressure drop across the duct and is given by

$$(\Delta P)_d = 9.81 \text{ x } (\Delta h)_d \text{ x } \rho_m \tag{11}$$

A parameter known as thermo-hydraulic performance parameter 'η' is used to evaluate the effectiveness of artificially roughened surfaces accounting for the enhancement of Nusselt number and friction factor [Lewis, M. J., 1975]. It can be calculated by

$$\eta = [(Nu_r/Nu_o)/(f_r/f_o)^{1/3}] \tag{12}$$

Based on the calculations of the errors in the experimental measurements by various instruments used, the uncertainties in the calculated values of Reynolds number, Nusselt number, friction factor and thermo-hydraulic performance parameter are estimated as ± 1.65%, 1.94 %, ± 3.22 % and 2.25 % respectively (Holman, 2004).

4. VALIDATION OF EXPERIMENTAL DATA

Before taking the experimental data of artificially roughened ducts, the experimental set up was validated by keeping all surface as smooth surfaces and the results obtained in the form of Nusselt number and friction factor were compared with the standard correlations for Nusselt number and friction factor obtained with the standard Dittus-Boelter equation ($Nu_o = 0.023\, Re^{0.8}\, Pr^{0.4}$) for the Nusselt number and Blasius equation ($f_o = 0.079\, Re^{-0.25}$) for friction factor (Bhatti and Shah 1987).

The comparison of the experimental and predicted values of Nusselt number and friction factor as a function of Reynolds number is shown in Fig.4 and in Fig.5 respectively. It is observed that the average deviation between the predicted and experimental values has been ±5.9 % and ±7.3 % for Nusselt number and friction factor respectively. Thus the results obtained and the predicted values are in good agreement, which ensures the accuracy of the experimental data with the present experimental set-up.

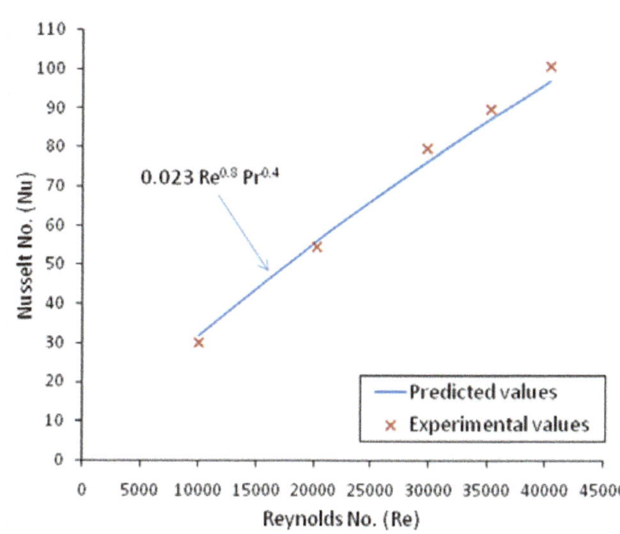

Fig. 4. Comparison of experimental and predicted values of Nusselt number for smooth duct

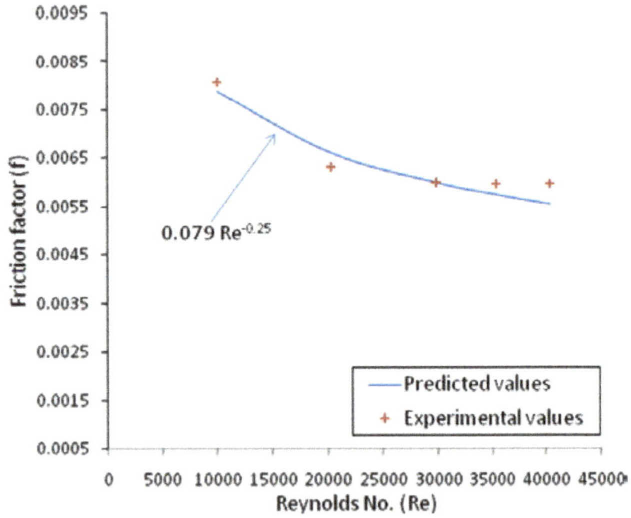

Fig. 5. Comparison of experimental and predicted values of friction factor for smooth duct

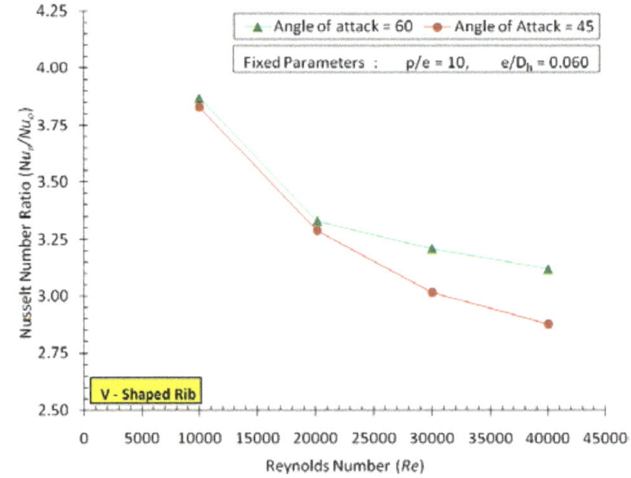

Fig. 7. Variation of Nusselt number ratio with Reynolds number at α=60^0 and 45^0 for V-shaped ribs.

5. RESULTS AND DISCUSSION

The values of Nusselt number and friction factor for ʌ-shaped and V-shaped rib geometry were computed on the basis of experimental data collected for various flow and roughness parameters. The effects of various parameters on Nusselt number and friction factor are presented in this section.

Figure 6 and Fig. 7 shows the values of Nusselt number, and Nusselt number ratio for V-shaped ribs respectively. It is clear from the figures that the Nusselt number increases as Reynolds number increases and the heat transfer enhancement is about 2.75 to 3.8 times than that of smooth duct. The formation of secondary flow by the ribs is mainly responsible for the increase in heat transfer coefficient.

The V-shaped ribs are responsible for the spanwise heat transfer coefficient. Gao and Sunden [23] suggested the four cell pattern of secondary flow. The swirling motion created by the secondary flow helped to generate a high overall average heat transfer coefficient. The heat transfer enhancement is higher for rib attack angle α= 60^0 than α= 45^0. The flow travels along the ribs from side wall and meets at the apex of the rib whereas for V ribs the flow deflects from the apex and then travels along the ribs to generate secondary flows. (as shown in Figure 8).

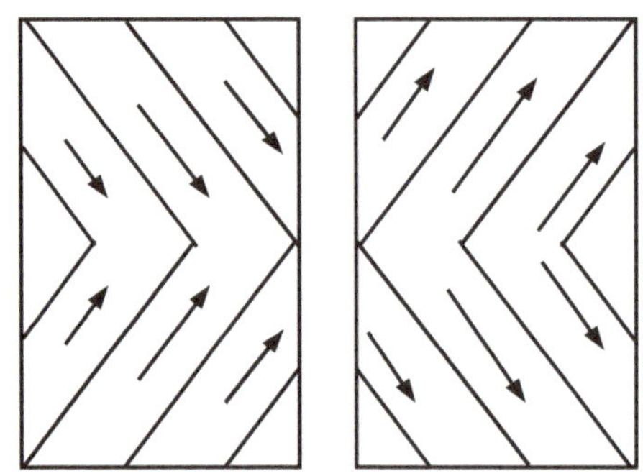

Fig. 8. Secondary flow generated in ʌ-shaped and V-shaped ribs

Figure 9 and 10 illustrates the variation of friction factor with Reynolds number for different rib attack angles of V-shaped ribs. It is observed from Figures 9 and 10 that the friction factor decreases as Reynolds number increases and it remains almost constant at higher Reynolds numbers. The increase in friction factor ranges from 7 to 8.8. The friction factor enhancement is more for 60^0 ribs as compared to 45^0 ribs.

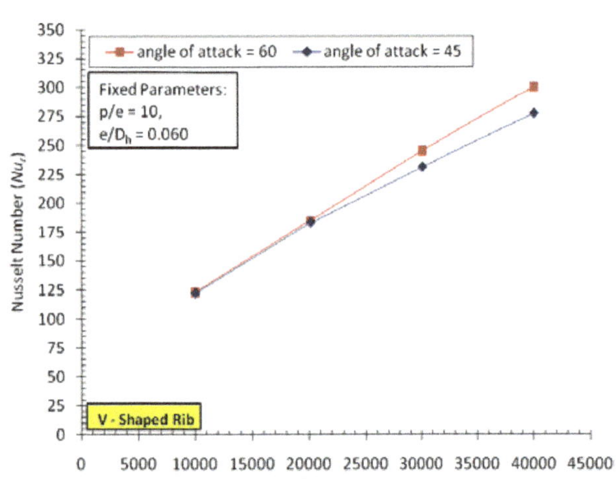

Fig. 6. Variation of Nusselt number with Reynolds number at α=60^0 and 45^0 for V-shaped ribs.

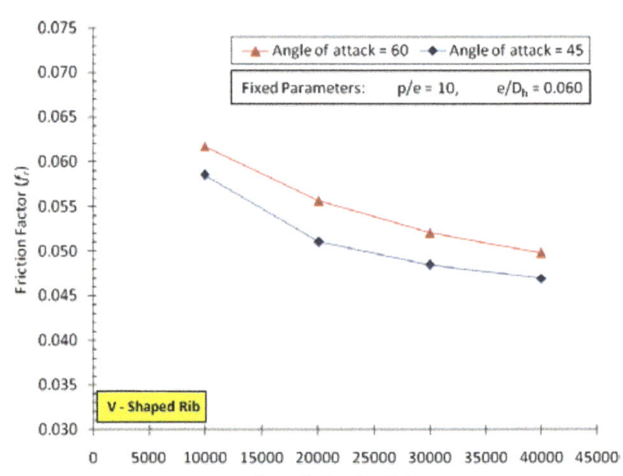

Fig. 9. Variation of friction factor with Reynolds number at α=60^0 and 45^0 for V-shaped ribs.

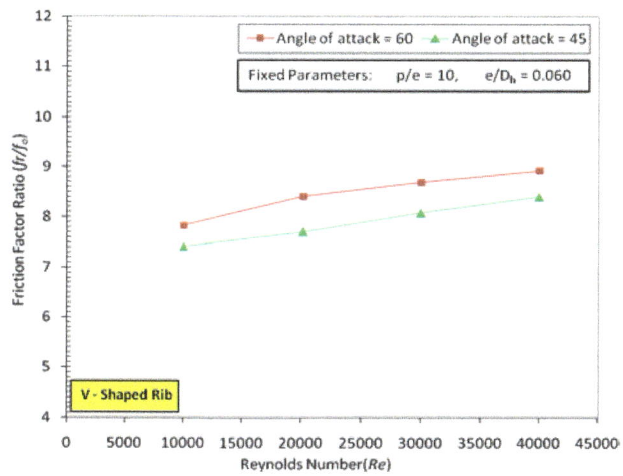

Fig. 10. Variation of friction factor ratio with Reynolds number at α=60^0 and 45^0 for V-shaped ribs.

Figures 11 to 14 show the Nusselt number and friction factor results for V – ribs pointing downstream (Λ- shaped ribs). In contrast to the V the Λ- shaped ribs produces less heat transfer and more friction factor. It is also observed that at low Reynolds numbers, the Λ-ribs produces lower heat transfer coefficient. It seems that a much weaker secondary flow is generated than the case of V-shaped ribs. Gao and sunden (2001) shows that in addition to the main flow V ribs also created two double cell counter rotating vortices but the direction is reversed as in the case of Λ-shaped ribs. Liou *et al.* (2000) investigated the secondary flow patterns for 45^0 V-shaped ribs pointing upstream and downstream in a square duct. The results conforms the results of present investigation and but opposite results are observed as compared to the results of Gao and Sunden (2001). The reason is that the aspect ratio in present case is 1 which produces the different kind of secondary flow as compared to the rectangular ducts. The wall confinement reduces the regions where secondary flows are directed towards the bottom wall and therefore curtails the heat transfer enhancement for the case of V- ribs pointing downstream. The friction factor enhancement is also somewhat reduces as compared to the case of V ribs pointing upstream.

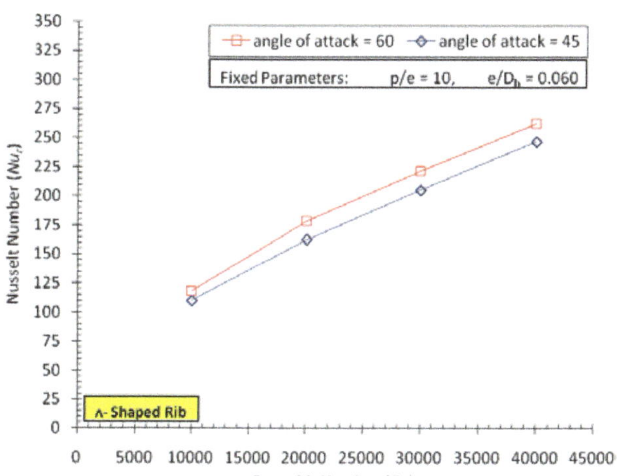

Fig. 11. Variation of Nusselt number with Reynolds number at α=60^0 and 45^0 for Λ-shaped ribs.

The Λ-shaped ribs produce less heat transfer whereas pressure penalty is greater than the V-shaped ribs. Thus the best rib geometry can be decided by considering the heat transfer and friction characteristics simultaneously. A parameter that facilitates the simultaneous consideration of thermal and hydraulic performance known as thermo-hydraulic performance (Lewis, 1975) is evaluated for both the cases and plotted as in Figure 15. It is observed from the figure that the thermo hydraulic performance for V-shaped ribs is higher than the Λ-shaped ribs . For all the cases studied the performance is decreases as Reynolds number increases. V-shaped ribs with α=45^{0}give comparable results as compared to the results of V-shaped ribs with α=60^0.

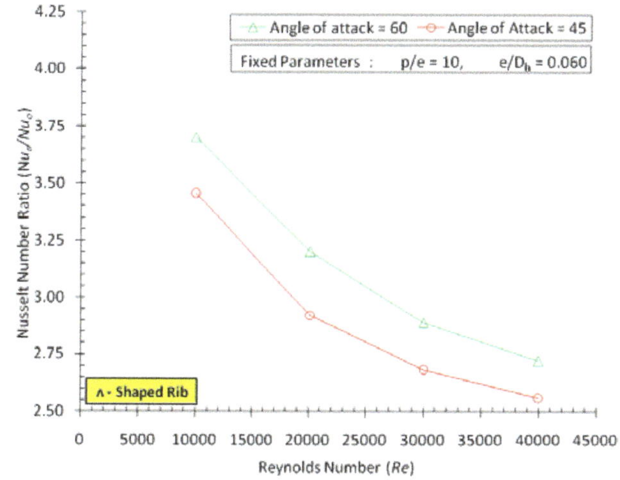

Fig. 12. Variation of Nusselt number ratio with Reynolds number at α=60^0 and 45^0 for Λ-shaped ribs.

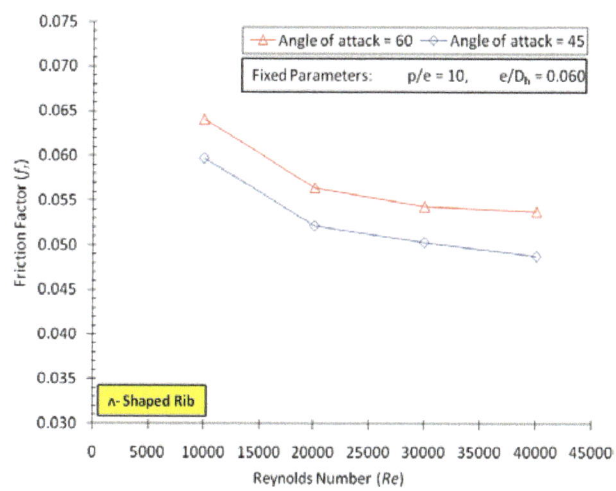

Fig. 13. Variation of friction factor with Reynolds number at α=60^0 and 45^0 for Λ-shaped ribs.

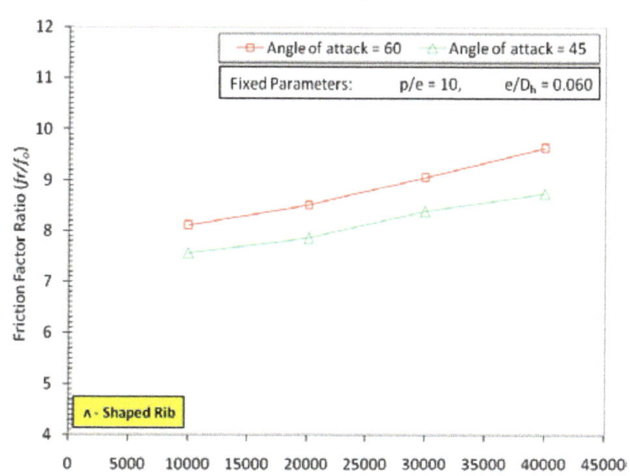

Fig. 14. Variation of friction factor ratio with Reynolds number at α=60^0 and 45^0 for Λ-shaped ribs.

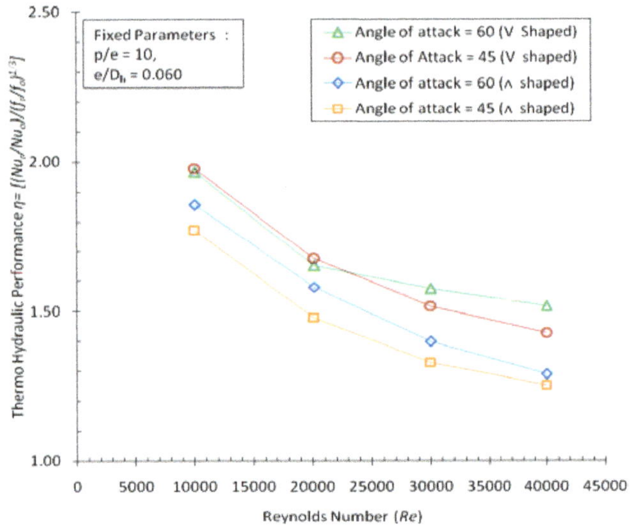

Fig. 15. Variation of Thermo hydraulic performance with Reynolds number at $\alpha=60^0$ and 45^0 for V and Λ-shaped ribs.

6. CONCLUSIONS

An experimental investigation has been carried out with the aim to compare V and Λ-shaped rib geometry for a square duct to found out rib geometry that gives maximum heat transfer with high value of thermo-hydraulic performance.

The following conclusions can be drawn from this work:

(1) In general, Nusselt number increases and friction factor decreases with an increase of Reynolds number. Nusselt number and friction factor are significantly higher as compared to those obtained for smooth duct. This is due to the distinct change in the fluid flow characteristics as a result of V and Λ–shaped roughness that causes flow separations, reattachments and the generation of secondary flows.

(2) The Nusselt number enhancement decreases when the Reynolds number increases. The friction factor ratio is found to increases as Reynolds number increases and becomes constant at high Reynolds numbers.

(3) The maximum value of Nusselt number has been found for V-shaped ribs with $\alpha=60^0$ corresponds to relative roughness height (e/D_h) of 0.06 and relative roughness pitch (p/e) of 10.

(4) The rib attack angle plays an important role for desirable fluid flow characteristics. The maximum value of Nusselt number for both the cases is obtained for 60° due to stronger secondary flow.

(5) The maximum enhancement in Nusselt number and friction factor values compared to smooth duct are of the order of 3.8 and 8.8 respectively.

(6) The thermal performance decreases when the Reynolds number increases. The V shaped ribs with $\alpha = 60^0$ give the highest thermal hydraulic performance for most of the Reynolds numbers. It produces somewhat less thermal hydraulic performance as compared to $\alpha = 45^0$ for a narrow range of Reynolds number due to less value of friction factor at these Reynolds numbers.

(7) The V shaped ribs with $\alpha = 45^0$ shows the comparable results as compared to the results of $\alpha = 60^0$.

NOMENCLATURE

A	Area of duct cross section, m^2
A_p	Area of roughened plate, m^2
A_{pipe}	cross section area of pipe, m^2
AR	Channel aspect ratio *(W/H)*
C_d	Coefficient of discharge for orifice meter
C_p	Specific heat of air at constant pressure, J/kg K
D_h	Channel hydraulic diameter of duct, m
D_p	Inside diameter of pipe, m
d/W	Relative gap position
e	Rib height, m
e/D_h	Relative roughness height
f	Fanning friction factor
f_o	Friction factor for smooth circular duct
f_r	Friction factor for roughened duct
f_r/f_o	Friction factor ratio
g/e	Relative gap width
H	Depth of duct, m
h	Convective heat transfer coefficient, W/m^2 K
k_a	Thermal conductivity of air, W/m K
L	Test section length, m
L_f	Duct length for calculation of friction factor, m
L/D_h	Test length to hydraulic diameter ratio
m	mass flow rate, kg/s
Nu	Nusselt number
Nu_o	Nusselt number for smooth circular duct
Nu_r	Nusselt number for roughened duct
Nu_r/Nu_o	Nusselt number ratio
p	Rib pitch, m
Pr	Prandtl number
$(\Delta P)_d$	Pressure drop in the test channel, N/m^2
$(\Delta P)_o$	Pressure drop across the orifice plate, N/m^2
p/e	Relative roughness pitch
Q_u	Useful Heat gain, W
Re	Reynolds number
T_f	Bulk mean air temperature, ^{0}C or K
T_i	Air inlet temperature, ^{0}C or K
T_o	Air outlet temperature, ^{0}C or K
T_p	Mean plate temperature, ^{0}C or K
V	Velocity of air, m/s
W	Width of duct, m

Greek Symbols

α	Rib angle of attack, degree
β	Ratio of orifice diameter to pipe diameter
θ	Angle of inclination of U-tube manometer, degree
μ_a	Dynamic viscosity of air, N s m^{-2}
ρ_a	Density of air, kg m^{-3}
ρ_m	Density of manometer fluid, kg m^{-3}

REFERENCES

Bhatti, M. S., Shah, R. K., 1987, "Turbulent and Transition Flow Convective Heat Transfer," *Handbook of Single-phase Convective Heat Transfer*, Wiley, New York.

Chaube, A., Gupta, S. and Verma, P., 2014, "Heat Transfer and Friction Factor Enhancement in a Square Channel Having Integral Inclined Discrete Ribs on Two Opposite Walls," *Journal of Mechanical Science and Technology*, **28** (5), 1927-1937. http://dx.doi.org/10.1007/s12206-014-0143-1

Cho, H. H., Wu, S. J., and Kwon H. J., 2000, "Local Heat / Mass Transfer Measurements in a Rectangular Duct with Discrete Ribs," *Journal of Turbomachinery*, **122**, 579-586. http://dx.doi.org/10.1115/1.1303049

Gao X., and Sunden, B., 2001, "Heat Transfer and Pressure Drop Measurement in Rib-Roughened Rectangular Ducts," *Journal of experimental thermal and fluid science*, **24**, 25-34. http://dx.doi.org/10.1016/S0894-1777(00)00054-6

Gupta, S., Chaube, A. and Verma, P., 2013, "Augmented Heat Transfer in Square ducts with Transverse and Inclined Ribs with and without a gap," *International Journal of Current Engineering and Technology*, **3**(2), 688-694.

Han J. C. and Park, J. S., 1988, "Developing Heat Transfer in Rectangular Channels with Rib Turbulators," *International Journal of Heat and Mass Transfer,* **31**, 183-195. http://dx.doi.org/10.1016/0017-9310(88)90235-9

Han, J. C. and Zhang, Y. M., 1992, "High Performance Heat Transfer Ducts with Parallel and V-Shaped Broken Ribs," *International Journal of Heat and Mass Transfer* **35**, 513-523. http://dx.doi.org/10.1016/0017-9310(92)90286-2

Han, J. C. Glicksman, L. R. and Rohsenow, W. M., 1978, "An Investigation of Heat Transfer and Friction for Rib-Roughened Surfaces," *International Journal of Heat and Mass Transfer,* **21**, 1143-1156. http://dx.doi.org/10.1016/0017-9310(78)90113-8

Han, J. C., Dutta, S. and Ekkad, S. V., 2001, *Gas Turbine Heat Transfer and Cooling Technology*, Taylor & Francis, Inc., New York.

Han, J. C., Zhang, Y. M., Lee, C. P., 1991, "Augmented Heat Transfer in Square Channels with Parallel, Crossed and V-shaped Angled Ribs," *Journal of Heat Transfer*, **113**, 590–596. http://dx.doi.org/10.1115/1.2910606

Holman, J. P., 2004, *Experimental Methods for Engineers*, Tata McGraw Hills, New Delhi.

Karwa, R., 2003, "Experimental Studies of Augmented Heat Transfer and Friction in Asymmetrically Heated Rectangular Ducts with Ribs on the Heated Wall in Transverse, Inclined, V-continuous And V-discrete Pattern," *Int. Comm. Heat Mass Transfer*, **30**, 241-250. http://dx.doi.org/10.1016/S0735-1933(03)00035-6

Kiml, R., Mochizuki, S, and Murata, A., 2001, "Heat Transfer Enhancement Mechanism in a Rectangular Passage with V and Λ Shaped Ribs," *J. Flow Visualization Image Process*, **8**, 51–68.

Kiml, R., Mochizuki, S., Murata, A., 2001 "Effects of Rib Arrangements on Heat Transfer and Flow Behavior in a Rectangular Rib Roughened Passage," *Journal of heat transfer*, **123** (4), 675-681. http://dx.doi.org/10.1115/1.1378019

Lau, S. C., Kukreja, R. T., McMillin, R.D., 1991, "Effects of V-shaped Rib Arrays on Turbulent Heat Transfer and Friction of Fully Developed Flow in a Square Channel," *Int. Journal of Heat and Mass Transfer*, **34** (7), 1605–1616. http://dx.doi.org/10.1016/0017-9310(91)90140-A

Lau, S. C., McMillin R. D., and Han, J. C., 1991, "Heat Transfer characteristics of turbulent flow in a square channel with angled rib", *Journal of Turbomachinery*, **113**, 367-374. http://dx.doi.org/10.1115/1.2927885

Lewis, M. J., 1975, "Optimizing the thermo-hydraulic performance of rough surfaces," *international Journal of Heat and Mass transfer*, **18**, 1243-1248. http://dx.doi.org/10.1016/0017-9310(75)90232-X

Liou T., and Hwang, D., 1992, "Turbulent Heat Transfer Augmentation and Friction in Periodic Fully Developed Channel Flows," *Journal of Heat Transfer*, **114**, 56-64. http://dx.doi.org/10.1115/1.2911267

Liou, T. M., Chen, C. C., Tsai, T. W., 2000, "Heat transfer and fluid flow in a square duct with different shaped vortex generators," *Journal of Heat Transfer,* **122** (2), 327-335. http://dx.doi.org/10.1115/1.521487

Liou, T. M., Hwang, J .J., 1993, "Effect of ridge shapes on turbulent heat transfer and friction in a rectangular channel," *International Journal of Heat and Mass Transfer*, **36**, 931–940. http://dx.doi.org/10.1016/S0017-9310(05)80277-7

Metzger, D. E. Fan, C. S., Yu, Y., 1990, "Effects of Rib Angle and Orientation on Local Heat Transfer in Square Channels With Angled Roughness Ribs," *Compact Heat Exchangers*, Hemisphere, Washington.

Muluwork, K. B., Saini J. S., and Solanki, S. C., 1998, "Studies on discrete RIB roughened solar air heater, In: *Proceedings of National Solar Energy Convention, Roorkee*, 75 -84.

Tanda, G., 2004, "Heat Transfer in Rectangular Channel with Transverse and V-shaped Broken Ribs," *International Journal of Heat and Mass Transfer*, **47**, 229–243. http://dx.doi.org/10.1016/S0017-9310(03)00414-9

Taslim, M. E., Li, T. and Kercher, T. M., 1996, "Experimental Heat Transfer and Friction in Channel Roughened with Angled, V-Shaped and Discrete Ribs on Two Opposite Walls," *ASME Journal of Turbomachinery*, **118**, 20-28. http://dx.doi.org/10.1115/1.2836602

HEAT TRANSFER ENHANCEMENT OF HEAT EXCHANGER INSERTED TWISTED PLATES FOR CARBON DIOXIDE GAS

Makoto Shibahara*

Graduate School of Maritime Sciences, Kobe University, Kobe, 658-0022, Japan

ABSTRACT

Overall heat transfer coefficients of the heat exchanger inserted twisted plates for CO_2 were measured. The overall heat transfer coefficients increased with mass flow rates of water at the same Reynolds number in the experiment. It is considered that the helically twisting fluid motions in the twisted heat exchanger were contributed to the heat transfer enhancements.

Keywords: *Heat transfer enhancement, Heat Exchanger, Twisted Plates*

1. INTRODUCTION

It is important to reduce greenhouse gases in atmosphere for a control of global warming. In order to decrease the concentration of carbon dioxide gas (CO_2) in the atmosphere, the renewable energy technologies have been developed such as the biological gasification with biomass fuel and the direct combustion. Recently, the isolation technology of the emitted CO_2 from the atmosphere is also considered to be effective. Liu *et al.* (2006) experimentally studied a solution process of CO_2 in seawater and pure water under various pressures and temperatures. On the other hand, Willcox *et al.* (1982) proposed the concept of a marine biomass plantation (ocean farm) to create biomass fuel such as a kelp, then they gave the economic analyses three decades ago. Based on the concept of the ocean farm, Shibahara *et al.* (2008) have suggested on the marine renewable energy power plant combined the ocean farms. Figure 1 shows the basic concept of marine renewable energy power plant system. This system consists of micro gas turbine, heat exchangers, pumps, cooling water, liquid oxygen, and vessel for alkali solution such as amine adsorption. When the kelp is used as the marine biomass fuel with the high concentration of oxygen in the combustion chamber, the exhaust gas is composed of carbon dioxide gas and water as shown in following chemical equation.

$$C + 2H_2 + 2O_2 \rightarrow CO_2 + 2H_2O \qquad (1)$$

Then, CO_2 captured by the alkali solution ships to the ocean farm. After distilling the adsorption solution with boilers, CO_2 is disposed into the ocean farm by the pipeline and promotes growth of kelp (Zou, 2005). In this system, high-efficiency compact heat exchanger becomes necessary for the limit of the marine plant space, therefore, it is important to investigate the heat transfer enhancement of the heat exchanger through CO_2 for the proposed power plant system.

Shibahara *et al.* (2008, 2009) have already reported about a transient heat transfer process for CO_2 flowing over a horizontal plate under wide experimental conditions assuming the plate-type heat exchanger. Transient heat transfer coefficients for CO_2 were measured to construct the fundamental database for the proposed power plant system. Moreover, the heat transfer coefficients for various gases flowing over a twisted plate were measured and obtained the empirical correlations in order to improve the heat transfer performance (Shibahara *et al.*, 2010; Shibahara *et al.*, 2013). As a result of experiment, the heat transfer coefficients of the twisted plate were 13~28% higher than that of the plate one.

The purpose of this study is to apply the heat transfer enhancement of twisted plates to a heat exchanger and to obtain the fundamental data of the heat exchanger for CO_2. In the experiment, the overall heat transfer coefficients of twisted heat exchanger were measured.

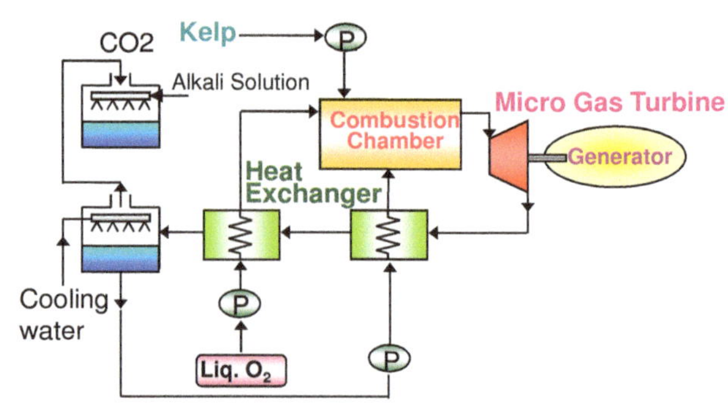

Fig.1 Schematic diagram of the marine renewable energy plant system (Shibahara et al. (2008)).

2. EXPERIMENTAL APPARATUS

Figure 2 shows the schematic diagram of the experimental apparatus. The experiment apparatus consists of a gas cylinder, mass flow meter, test section, flexible ribbon heater, isothermal bath, thermocouples and the data logger. The vacuum pump (ULVAC, DA-60D) was used to degas the tube. Flowing rate was measured with the mass flow meter (Azbil, CMS0050). And also, the pressure was measured with the pressure transducer (KEYENCE, AP-53A). The gas temperature in the

* E mail: sibahara@maritime.kobe-u.ac.jp

tube was heated to the desired temperature level by a flexible ribbon heater. The temperature near test section heater was measured by K-type thermocouples with a precision of ±1K. CO_2 with a high purity of 99.9% was used as the test fluid.

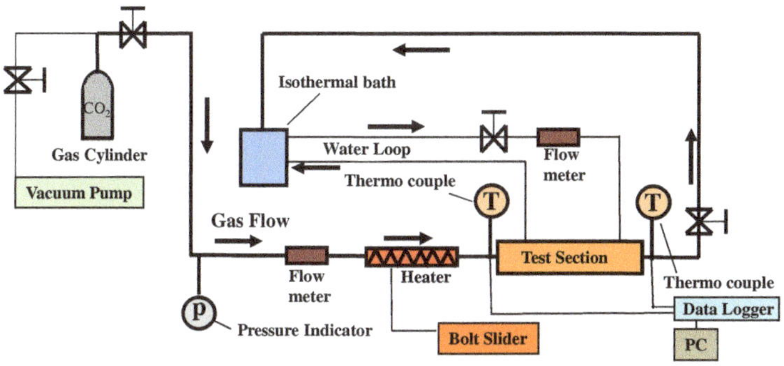

Fig.2 Schematic diagram of the experimental apparatus

Figure 3 shows the overview of the twisted heat exchanger. It consisted of a copper tube (ϕ8 x 150), twisted copper plates (10 x 150 x 0.3) in parallel copper plates (30 x 130 x 0.3). The swirl fluid motion was generated by the twisted plates outside the copper tube. Moreover, the twisted plate was also inserted in the copper tube to enhance heat transfer of CO_2. Figure 4 shows the detail of the heat exchanger. The width of the *i*nserted plate was 8 mm with the thickness of 0.3 mm and the pitch of the tape was 50 mm in the copper tube. The twist ratio is defined by the following equation (Fujita and Lopez, 1995).

$$\gamma = \frac{H}{W} \tag{2}$$

where, H [m] and W [m] are 180 degree twist-pitch of the twisted plate and width of the twisted plate. The twisted ratio was 9.0 in the copper tube. On the other hand, since the pitch of the external twisted plates was 130 mm as shown in Fig.4, the twisted ratio was 13.0.

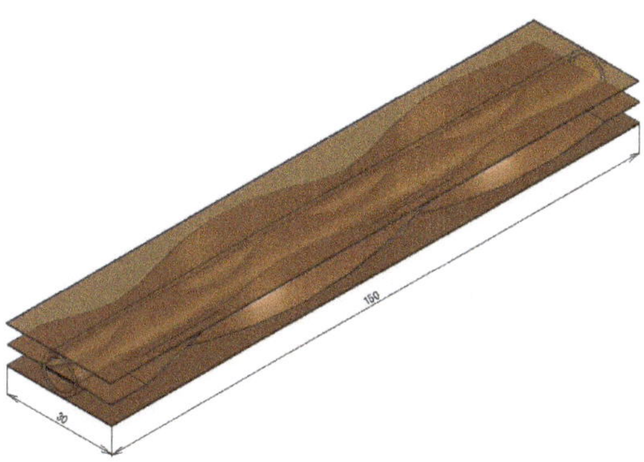

Fig.3 Overview of the twisted heat exchanger

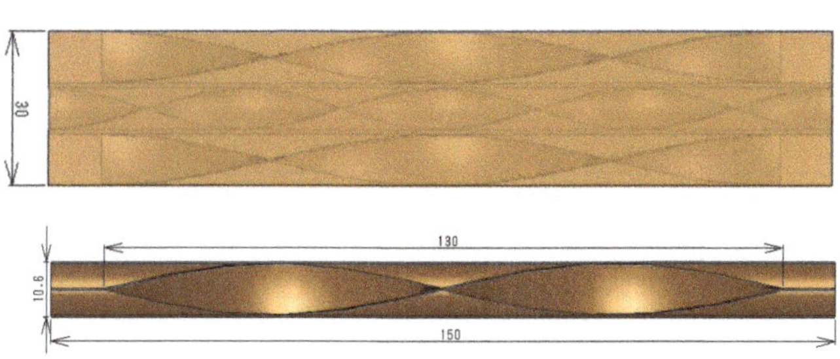

Fig.4 The detail of the twisted heat exchanger

2.1 Experimental Method

The mass flow rate was controlled using the regulator of gas cylinder at the experiment, and the gas temperature was heated through the flexible ribbon heater with a bolt slider. The gas temperatures were measured by the thermocouples connected to the data logger (KEYENCE GR3500). The gas flow velocity was ranged from 2.5 m/s to 7.2 m/s. The heat transfer performance of the twisted heat exchanger was evaluated by means of a wind-tunnel experiment and water test in an acryl tube with the inner diameter of 46 mm and the length of 190 mm as shown in Fig.5. The countercurrent flow was ranged from 2.0 m/s to 4.0 m/s measured by hot-wire flow meter (Testo435) in the wind-tunnel experiment. On the other hand, the mass flow rate of the countercurrent flow was ranged from 0 to 6.7 g/s in the water test.

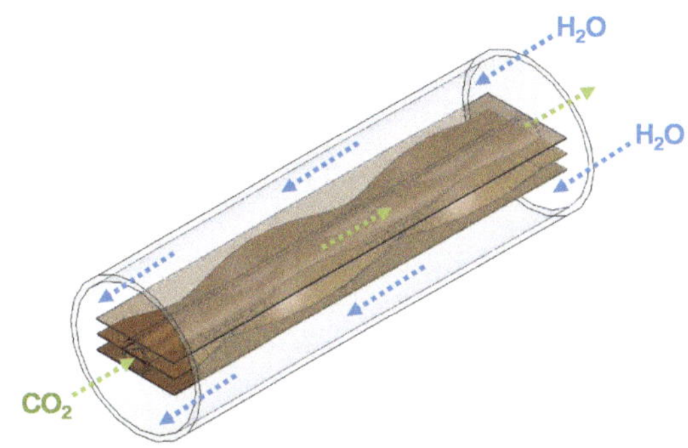

Fig.5 Water test of the heat exchanger

After these experiments, the heat transfer rate was calculated by the following equations as the data reduction.

$$Q = V\rho c_p (T_{in} - T_{out}) \tag{3}$$

where, V [m^3/s], ρ [kg/m^3], C_p [J/kg K], T_{in} [K], T_{out} [K] were volume of CO_2, density of CO_2, specific heat of CO_2, inlet gas temperature and outlet gas temperature, respectively. And also, overall heat transfer coefficient was defined as follow:

$$U = \frac{Q}{A\Delta T_{in}} \tag{4}$$

where, A [m^2] was estimated as the area of inner diameter of the copper tube since the inserted tape was lose fit in the copper tube. ΔT_{in} [K] was log mean temperature difference. On the other hand, the Reynolds number was defined as follows:

$$Re_h = \frac{uD_h}{\nu} \tag{5}$$

where, u [m/s], ν [m^2/s], D_h [m] were the flow velocity, kinematic viscosity and hydraulic diameter, respectively. Since the twisted plate was inserted in the tube, the hydraulic diameter (Fujita and Lopez, 1995) can be expressed as

$$D_h = D\frac{1-(4\delta/\pi D)}{1+(2/\pi)(1-\delta/D)} \tag{6}$$

where, D [m] and δ[m] were the inner diameter and thickness of plate, respectively.

2.2 Experimental Result

Figure 6 shows the time-dependence of the inlet gas temperature and outlet gas temperature under the gas pressure of 104.8 kPa. The gas flow velocity was 2.5 m/s and the countercurrent flow of air was 2.0 m/s. As shown in Fig.6, a rise of the inlet gas temperature was observed after turning on the flexible ribbon heater. The outlet gas temperature also ascends gradually depending on the increase of the inlet gas temperature. Since these gas temperature approaches to asymptotic value after 400 s, it is understood that the heat transfer process becomes steady-state condition.

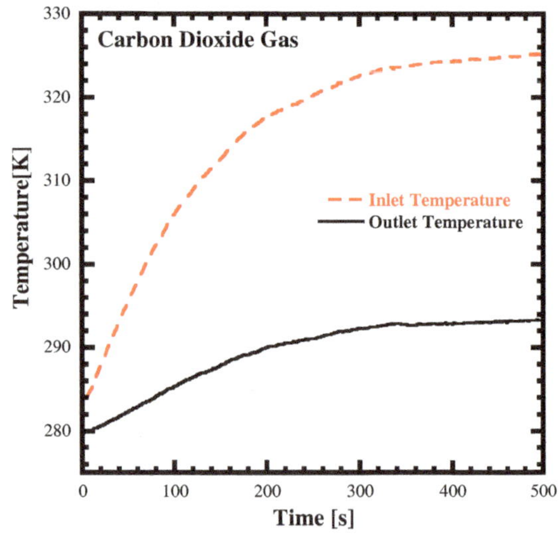

Fig.6 The time-dependence of the inlet and outlet temperature.

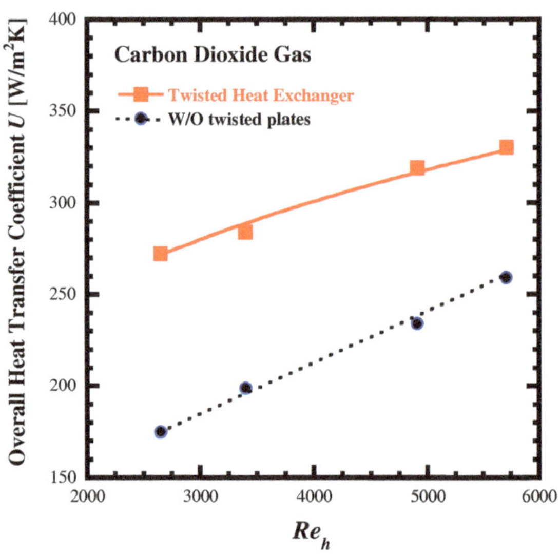

Fig.7 The relation of the overall heat transfer coefficients and Reynolds number at the countercurrent flow of 3m/s in air.

Figure 7 shows the overall heat transfer coefficient for the inlet gas temperature of 323.15 K at the various Reynolds number. The countercurrent flow of air was 3.0 m/s in the acryl tube. Since the overall heat transfer coefficient increased with Reynolds number, it is considered that convective heat transfer comes to govern in the acryl tube. For comparison, the experimental result of a heat exchanger without twisted plates was plotted in Fig.7. The overall heat transfer coefficients of the twisted heat exchanger were higher than that of a heat exchanger without the twisted plates. It is understood that they were also affected by the Reynolds number as well as usual convective heat transfer.

Figure 8 shows the relation of the overall heat transfer coefficients and Reynolds number at the various countercurrent flows in water. The inlet gas temperature was 323.15 K. The overall heat transfer coefficients increased with mass flow rates of water at the same Reynolds number. Comparing with the experimental result of the heat exchanger without the twisted plates, the overall heat transfer coefficients of the twisted heat exchanger were higher than that without the twisted plates at the mass flow rate of 1.7 g/s. It is considered that the swirl effect of the twisted plates contributed to the heat transfer enhancements. Thus, the fluid structure changed to the swirl flow caused by the twisted plate. In addition, the heat transfer enhancement in the each copper tube was contributed by the twisted tape. It was considered that the flow velocity near the twisted tape in the each copper tube increased due to the blockage of the flow path as well as a single tube inserted a twisted tape (Manglik and Bergles, 1993).

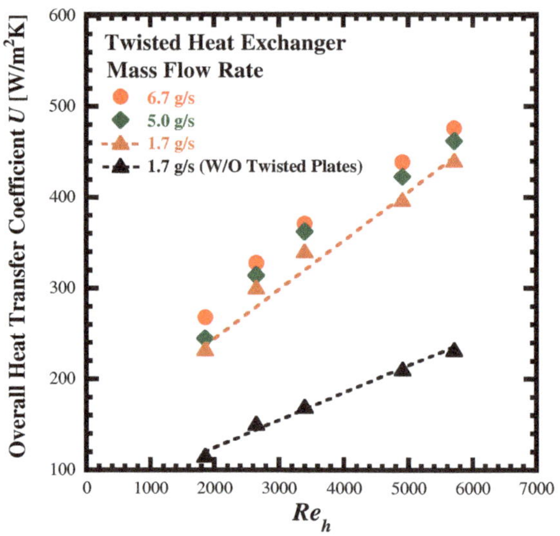

Fig.8 The overall heat transfer coefficients at various Reynolds number in water.

3. CONCLUSION

The heat transfer performance of the twisted heat exchanger was investigated experimentally. The overall heat transfer coefficients increased with mass flow rates of water at the same Reynolds number in the experiment. It is considered that the helically twisting fluid motions in the twisted heat exchanger were contributed to the heat transfer enhancements.

NOMENCLATURE

A	area of inner diameter, m^2
c_p	specific heat of CO_2, J/(KgK)
D	inner diameter, m
D_h	hydraulic diameter, m
H	180 degree twist-pitch
Q	heat transfer rate, W
Re	Reynolds number, uD_h/ν
T	temperature, K
ΔT	log mean temperature difference, K
U	overall heat transfer coefficient, W/m^2K
V	volume flow of CO_2, m^3/s
W	width of twisted plate, m
u	gas flow velocity, m/s
δ	plate thickness, m
γ	twist ratio
ρ	density of CO_2, kg/m^3
ν	kinematic viscosity, m^2/s

Subscripts

in	inlet
out	outlet
h	hydraulic diameter

ACKNOWLEDGMENTS

This work was partially supported by JSPS KAKENHI Grant Number 26820381.

REFERENCES

CFD2000, 2005, "Theoretical Background," Adaptive research.

Howard, A.W, 1982, "The Ocean as a Supplier of Food and Energy", Cellular and Molecular Life Sciences, **38(1)**, 31-35.
http://dx.doi.org/10.1007/BF01944522

Shibahara, M., Liu, Q., Fukuda, K., 2008, "Heat Transfer Performance of Plate Heater for Carbon Dioxide Gas in a Dispersed Power Plant System Using Marine Biomass Resource," *3rd Pan Asian Association of Maritime Engineering Societies and Advanced Maritime Engineering Conference*, 863-869.

Shibahara, M., Liu, Q., Fukuda, K., "Transient Forced Convection Heat Transfer for Carbon Dioxide Flowing over a Horizontal Plate with Exponentially Increasing Heat Input," IMECE2008-68603, *Proceedings of 2008 ASME International Mechanical Engineering Congress and Exposition*, Boston, MA.
http://dx.doi.org/10.1115/IMECE2008-68603

Shibahara, M., Liu, Q., Fukuda, K., 2009, "Transient Forced Convection Heat Transfer Due to Exponentially Increasing Heat Input for Helium Gas Flowing on a Narrow Plate," *Journal of Power and Energy Systems,* **3(1)**, 272-288.
http://dx.doi.org/10.1299/jpes.3.272

of the Japan Institute of Marine Engineering, **45**, 99-101.
http://dx.doi.org/10.5988/jime.45.1033

Shibahara, M., Liu, Q., Fukuda, K., 2013, "Enhancement of Forced Convection Heat Transfer Using Twisted Heater with Exponential Heat Inputs," HT2013-17227, *Proceedings of ASME 2013 Summer Heat Transfer Conference*, Minneapolis, MN.
http://dx.doi.org/10.1115/HT2013-17227

Manglik, R.M., Bergles, A.E., 1993, "Heat Transfer and Pressure Drop Correlations for Twisted – Tape Inserts in Isothermal Tubes: Part I – Laminar Flows," *Journal of Heat Transfer*, **115(4)**, 881-889.
http://dx.doi.org/10.1115/1.2911383

Manglik, R.M., Bergles, A.E., 1993, "Heat Transfer and Pressure Drop Correlations for Twisted – Tape Inserts in Isothermal Tubes: Part II – Transition and Turbulent Flows," *Journal of Heat Transfer*, **115**(4), 890-896.
http://dx.doi.org/10.1115/1.2911384

Liu, Q., Fukuda, K., Matsuda, T., 2006, "Study of Solution Process of Carbon Dioxide in Sea Water", *Journal of the Japan Institution of Marine Engineering (Special Issue)*, **41**, 144-149.
http://dx.doi.org/10.5988/jime.41.SI_144

Fujita, Y., and Lopez, A.M., 1995, "Heat Transfer Enhancement of Twisted Tape Inserts in Turbulent Pipe Flows," *Transactions of JSME Series B*, **61**(588), 280-287.
http://dx.doi.org/10.1299/kikaib.61.3044

Zou, D., 2005, "Effects of Elevated Atmospheric CO_2 on Growth, Photosynthesis and Nitrogen Metabolism in the Economic Brown Seaweed, Hizikia, Fusiforme (Sargassaceae, Phaeophyta)," Aquaculture, **250**, 726-735.
http://dx.doi.org/10.1016/j.aquaculture.2005.05.014

13

RECENT PROGRESS ON EXPERIMENTAL RESEARCH OF CRYOGENIC TRANSPORT LINE CHILLDOWN PROCESS

J. N. Chung*, Kun Yuan

Department of Mechanical and Aerospace Engineering, University of Florida
Gainesville, FL. 32611-6300, USA

ABSTRACT

Chilldown or quenching is a complicated process that initiates the cryogenic fluid line transport, and it involves unsteady two-phase heat and mass transfer. To advance our understanding of this process, we have reviewed recent experimental investigations. The chilldown process can be generally divided into three regimes: film boiling, transition boiling and nucleate boiling, and each regime is associated with a different flow pattern and heat transfer mechanism. Under low flow rate conditions, it is concluded that the two-phase flow regime is dispersed flow in the film boiling regime. The dispersed liquid phase is in the form of long filaments as the transport line is chilled down, and the vapor phase is generally superheated. Microgravity experiment for line chilldown was conducted basically in the film boiling region. Bottom wall heat flux was found to decrease under microgravity condition. Under the experimental condition, the gravity effect does not show a strong dependence on wall temperature and inlet flow rate.

Keywords: *Cryogenic chilldown, transport line, space applications.*

1. INTRODUCTION

Cryogenic fluids are widely used in industrial applications, space exploration, and cryosurgery systems and so on. In these systems, proper transport, handling and storage of cryogenic fluids are of great importance. When a cryogenic system is first started up, its walls and hardware must go through a transient chill down period prior to reaching a steady operation. Therefore, chilldown is the process of keeping the system adjusted to the low temperature scale which is usually several hundred degrees below the room temperature. The chilldown or quenching process is complicated, involving unsteady two-phase heat and mass transfer, and has not been fully understood.

Cryogens are usually used as a liquid fuels such as liquid hydrogen and oxygen in the rocket industry, liquid nitrogen and helium are frequently used to cool superconducting magnetic device for medical applications. In some special cases, for example, cryogenic fluids are used in spacecrafts not only at low temperatures but also under reduced gravity. Liquid oxygen and hydrogen are used to power launch vehicles and helium is used for pressurizing fuel tanks. For these applications, the transport and storage of cryogenic fluids under microgravity is regarded as an important element of the space mission.

Furthermore, on space exploration, efficient and safe use of cryogenic fluids in thermal management, power and propulsion, and life-support systems of a spacecraft during space missions involves transport, handling, and storage of these fluids under both terrestrial and microgravity conditions. The uncertainties about the flow regime and heat transfer characteristics pose severe design concerns. Moreover, the thermo-fluid dynamics of two-phase systems in microgravity encompass a wide range of complex phenomena that are not understood sufficiently for complete and optimal engineering design of the entire system.

Cryogenic fluids are also widely used in industrial systems. Until the early 1970s, liquid hydrogen was mainly used by NASA as a rocket fuel; however, development and growth of commercial markets have since outpaced this use. For example, liquid hydrogen is used in industrial applications such as metal processing, plate glass production, fat and oil hardening, semiconductor manufacturing, and pharmaceutical and chemical manufacturing. Today, the commercial market is many times larger than the government needs.

1.1 Role of Cryogenics in Space Exploration

The extension of human space exploration from a low earth orbit to a high earth orbit, then to Moon, Mars, and possibly asteroids and moons of other planets is one of NASA's biggest challenges for the new millennium. Integral to this is the effective, affordable, and reliable supply of cryogenic fluids. The efficient and safe utilization of cryogenic fluids in thermal management, power and propulsion, and life support systems of a spacecraft during space missions involves the transport, handling, and storage of these fluids in terrestrial and reduced gravity. The uncertainties about the flow pattern and heat transfer characteristics pose a severe design concern. Therefore, the design of cryogenic fluid storage and transfer system is very important and spawns research in several areas: for example, design of the vessel, the piping and draining system, insulation and safety devices. Moreover, the thermo-fluid dynamics of two-phase systems in reduced gravity encompasses a wide range of complex phenomena that are not understood sufficiently for engineering design to proceed.

Chilldown process is the inevitable initial stage during cryogenic fluid transport. Due to their low boiling points, boiling and two-phase flows

* *Corresponding author. Email: jnchung@ufl.edu*

are encountered in most of the cryogenic operations. The complexity of the problem results from the intricate interaction of the fluid dynamics and heat transfer, especially when phase-change (boiling and condensation) is involved. Because of the large stratification in densities between the liquid and vapor phases, the reduced gravity condition in space would strongly change the terrestrial flow patterns and accordingly affect the momentum and energy transport characteristics. Therefore, boiling and two-phase flow behave quite differently when the gravity levels are varied. The uncertainties about the flow pattern, pressure drop and heat transfer characteristics pose a severe design concern. For example, there is considerable disagreement over the chilldown heat fluxes and whether a unique rewetting temperature exists (Dhir et al. 1981; Piggott and Porthouse 1975). Different definitions of rewetting temperatures are reported in chilldown researches: Leidenfrost temperature, minimum film boiling temperature, quenching temperature from thermocouple observations and the temperature at which critical heat flux occurs (Dhir et al. 1981). For similar experimental observations, quiet different explanations were also suggested by different researchers. For example, it was reported that the rewetting velocity increased with increasing inlet flow rate, given the same initial wall temperature (Yamanouchi 1968, Duffey and Porthouse 1973, Thompson 1974), Duffey and Porthouse (1973) suggested that the flow rate effect resulted from increasing the wet side heat transfer coefficient with higher inlet flow. This improves the rate of axial heat conduction and hence leads to a faster rewetting rate. Thompson (1974), however, argued that the inlet flow rate affects precooling on the dry side rather than the heat transfer in the wet side. Additionally, because of the experimental difficulties, in general, there is very little heat transfer data for cryogenic flow boiling in reduced gravity. A representative work is reported by Adham-Khodaparast et al. (1995).

Another driving force of recent investigations, for example, comes from the need for further understanding of the cryogenic chilldown process under low mass flux in a thermodynamic vent system (TVS) on spacecrafts. A TVS is a system where a small amount of liquid is withdrawn from a cryogenic propellant tank and vented to remove heat from the bulk liquid cryogen in the tank and thus lower the tank pressure (Lin et al. 1991; Van Dresar et al. 2001, 2002). The mass flux in TVS system is generally very low. Systematic experiments for steady state low mass flux cryogenic two-phase flow were conducted by Van Dresar et al. (2001, 2002), the highly transient chilldown process was, however, not included.

1.2 Background and Literature Review

Cryogenic chilldown involves complex interaction of energy and momentum transfer among the two phases and the solid wall. Understanding of the boiling phenomenon, flow regime and heat transfer regime provides foundation for further insight into this dynamic process. The following section gives background information of boiling, two-phase flow regime and heat transfer regime. Previous experimental research that is related to cryogenic chilldown and microgravity boiling are reviewed and qualitatively assessed in the section below.

Boiling Curve

A boiling curve shows the relationship between the heat flux that the heater supplies to the boiling fluid and the heater surface temperature. According to the typical boiling curve (Figure 1), a chilldown (quenching) process usually starts from point E, and then goes towards point D in the film boiling regime as the wall temperature decreases. Point D is called the Leidenfrost point which signifies the minimum heater temperature required for the film boiling. For the film boiling process, the wall is so hot that liquid will vaporize before reaching the heater surface which causes the heater to be always in contact with vapor. When cooling beyond the Leidenfrost point, if a constant heat flux heater was used, then the boiling would shift from film to nucleate boiling (somewhere between points A and B) directly with a substantial decrease in the wall temperature because the transition boiling is an unstable process.

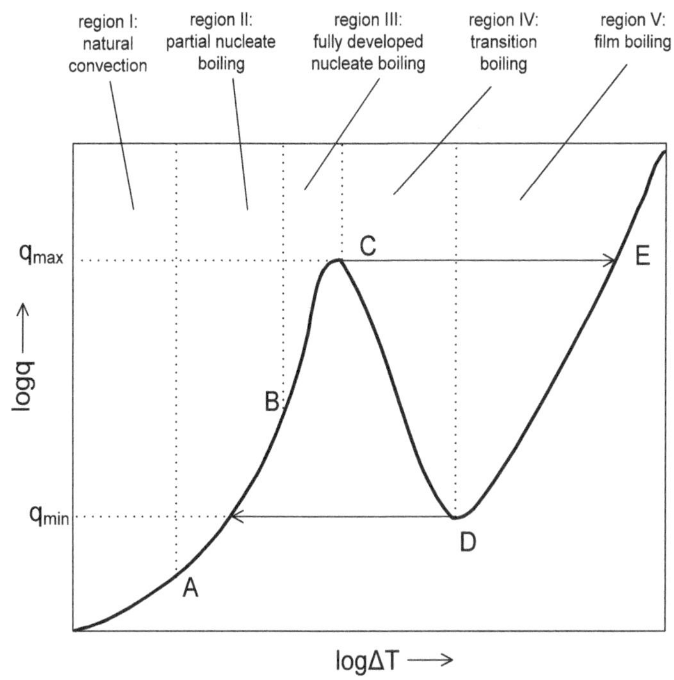

Fig. 1 Typical boiling curve.

Two-Phase Flow Regimes and Heat Transfer Regimes

The flow in cryogenic chilldown process is typically two-phase flow, because the wall temperature usually exceeds the liquid boiling temperature up to several hundred Celsius in the beginning. The topology of two-phase flow has an important effect on heat transfer and pressure drop in the flow channel. Therefore, generally the first step in two-phase flow experiment or modeling is to determine the two-phase flow regime. Commonly observed flow regimes in horizontal tubes are shown in Figure 2.

General descriptions of the two-phase flow regimes can be referred from Carey (1992) and Van Dresar and Siegwarth (2001). Flow regimes of common two-phase flow such as air-water have been extensively mapped from experiments. However, the published data for cryogens are limited.

It is also noted that definitions of the flow regimes are somewhat arbitrary. Qualitative assessment has not been done yet, and transition criteria between different flow regimes are not fully understood.

Different flow regimes are often associated with different heat transfer regimes. When phase change occurs as two-phase mixture flows along the channel, as that encountered in chilldown process, the situation is even more complicated (Carey 1992): different flow regimes are generally observed at different positions along the channel length. The sequence of flow regimes will primarily depend on the flow rate, channel orientation, fluid properties, and wall heat flux.

Some general information could be drawn from the findings of reflooding experiments designed for hypothetical loss-of-coolant accident (LOCA) in nuclear reactors, for example, the results of Chan and Banerjee (1981a, b, c) for horizontal tube and work of Cheng et al. (1978) for vertical tube.

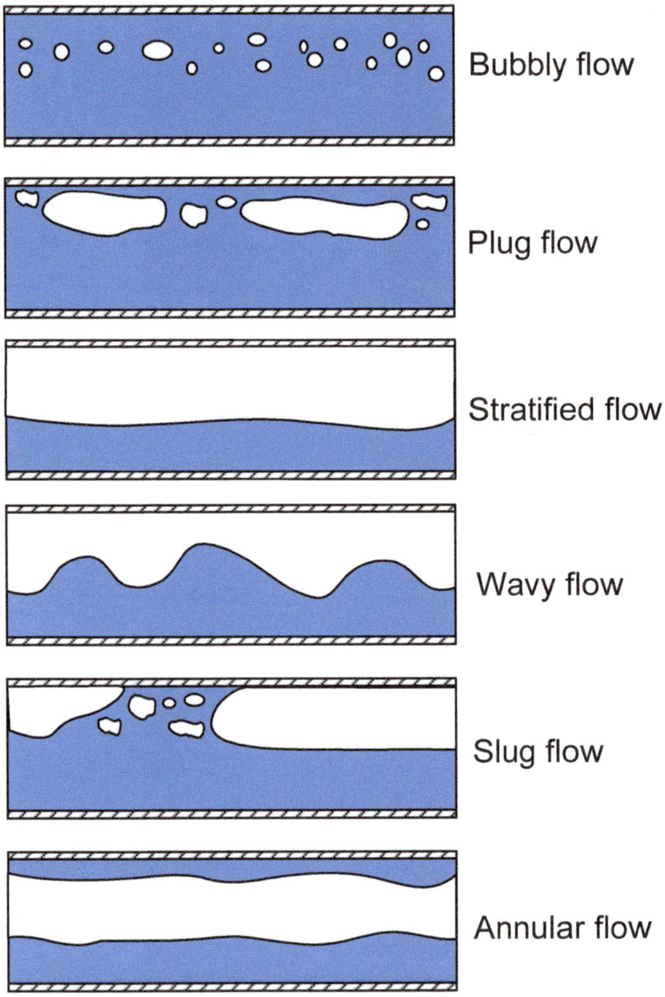

Fig. 2 Two-phase flow regimes inside a horizontal tube.

When the two-phase flow first enters the hot tube, the liquid phase evaporates very quickly and forms a vapor film that separates the liquid phase from touching the tube wall, and the two-phase flow is in film boiling state. Depends on the local quality and other thermo-hydraulic parameters, the flow regime can be dispersed flow or inverted annular flow. The corresponding heat transfer regime will be dispersed flow forced convection, which is also called dispersed flow film boiling (DFFB) in literature (Yadigaroglu et al. 1993; Andreani and Yadigaroglu 1996; Hammouda et al. 1997; Shah and Siddiqui 2000), or inverted annular film boiling (IAFB).

As the wall temperature decreases under certain degree, the liquid phase is able to contact the tube wall. The liquid-wall contacting front, which is often referred as quenching front (QF) or sputtering region, is characterized by violent boiling associated with significant wall temperature decrease, and propagates downstream with the flow. The heat transfer mechanism at the QF is transition boiling, which is more effective than the film boiling heat transfer. This establishment of liquid-wall contact is called rewetting phenomenon and has been a research interest for several decades.

After the QF, nucleate boiling heat transfer dominates. For a vertical tube, the flow regime can be annular flow, slug flow or bubbly flow; for horizontal tube, the flow regime is generally stratified flow. With further wall temperature decrease, the nucleate boiling stage gradually changes to pure convection until the wall temperature reaches the steady state, which denotes the end of the chilldown process.

Gravity Effect

Because of the differences in density and inertia, the two phases in two-phase flow are usually non-uniformly distributed across the pipe under terrestrial condition. The absence of gravity has important effects on flow regimes, pressure drop, and heat transfer of the two-phase flow. Surface-tension-induced forces and surface phenomena are likely to be much more important in space than they are on earth. Actually, all flow-regime-specific phenomena will be influenced by gravity level. As an example, Figure 3 compares the flow regime under both terrestrial and microgravity conditions; the difference is obvious.

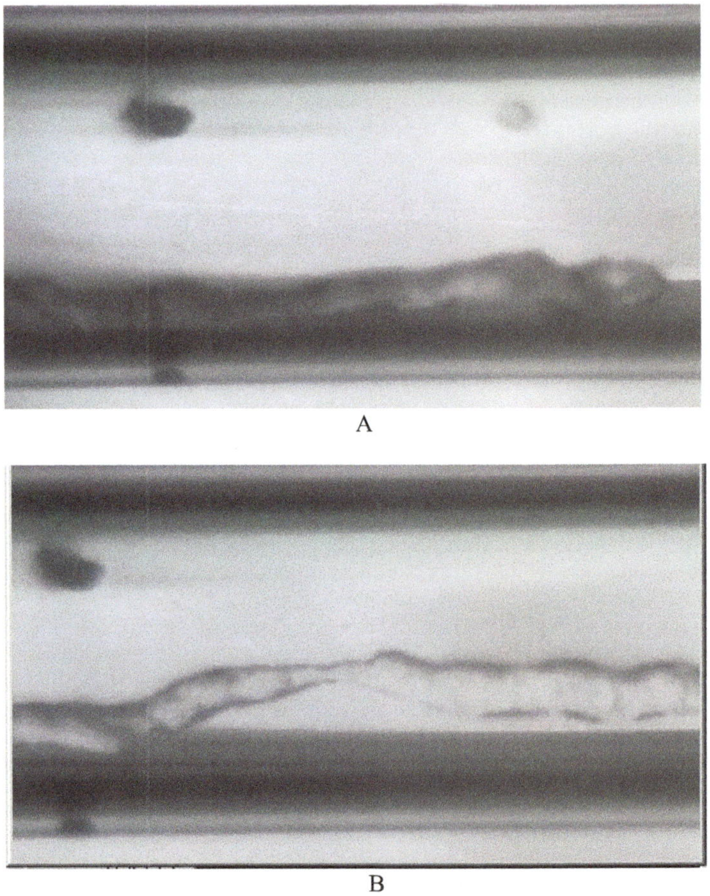

Fig. 3 Gravity effect on flow regimes. A) Flow regime in 1-g test. B) Flow regime in microgravity test.

Following is a simple scaling analysis that examines the gravity effects. For annular flow film boiling in a horizontal tube, the effect of gravity is assessed based on the ratio of Gr/Re^3, where Gr is the Grashof number and Re is the Reynolds number. The gravity effect is measured by the natural convection contribution characterized by the Gr, while the forced convective film boiling is scaled by the Reynolds number. According to Gebhart et al. (1988), Re^3 is used in the denominator when the flow is perpendicular to gravity for a horizontal tube. Re^2 is used when the flow is in the same direction of gravity. All the thermal properties are those of vapor because of film boiling. Figure 4 shows this ratio with the vapor flow velocity range of 0-0.5 m/s and a ΔT of 100 °C for the Gr estimation.

Based on Figure 4, if the vapor velocity is greater than 10 cm/s, then the Gr/Re^3 is less than 0.2. The natural convection is negligible for Gr/Re^3 less than 0.225 according to Gebhart et al. (1988). Therefore, a terrestrial gravity experiment with the vapor velocity greater than 10 cm/s would provide results that mimic the microgravity phenomenon.

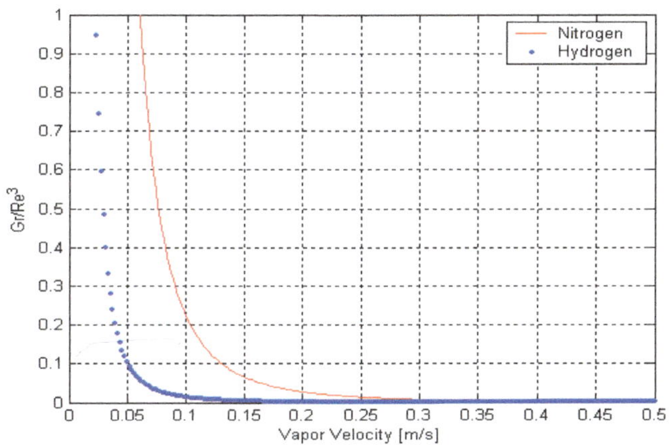

Fig. 4 Scaling analysis of gravity effects on two-phase flow.

Cryogenic two-phase flow and chilldown process is a complex problem for the scientific community to solve. The following summarizes the previous accomplishments from the experimental research.

Terrestrial cryogenic boiling and two-phase flow experiments

Numerous studies of cryogenic boiling in 1-g environment were conducted in the 1950s and 1960s. Brentari et al. (1965) gave a comprehensive review of the experimental studies and heat transfer correlations. For the fluids of oxygen, nitrogen, hydrogen and helium, it was found that for pool boiling, the Kutateladze (1952) correlation had the greatest reliability for nucleate boiling, while the Breen and Westwater (1962) correlation was best for film boiling. Maximum nucleate flux data were reasonably well predicted by the Kutateladze (1952) correlation. Although these correlations were selected as the best available, neither has particularly good agreement with experimental data. For the case of forced convection boiling, Brentari et al. (1965) reported that no correlation was found to be distinctly better. Some simple predictive methods were found to work as well as more complex schemes. In all boiling cases, it was questioned as to whether or not the predictive correlations include all of the significant variables that influence the boiling process. In particular, it was suggested that more detailed and better controlled experiments are needed and that more attention to surface and geometry effects is required.

Another comprehensive review of cryogenic boiling heat transfer addressing hydrogen, nitrogen and oxygen is given by Seader et al. (1965). It was reported that nucleate pool boiling results cannot be correlated by a single line but cover a range of temperature difference for a given heat flux. The spread is attributed to surface condition and geometry, and orientation. Maximum heat flux can be reduced by about 50% when going from 1-g to near 0-g. Seader et al. (1965) reported a fair amount of data for film pool boiling. Film boiling heat flux is reduced considerably at near 0-g conditions. Only a very limited amount of data is available for subcooled or saturated forced convective boiling and few conclusions were drawn. The lack of data for cryogenic forced convective boiling was also reported by Brentari and Smith (1965).

Relatively recent correlations have been published for 1-g saturated flow boiling of cryogens (Shah 1984; Gungor and Winterton 1987; Klimenko et al. 1989; Kandlikar 1990; Van Dresar et al. 2002) using the Convection number Co, Boiling number Bo and Froude number Fr as correlating parameters. Klimenko et al. (1989) investigated the effects of tube diameter and orientation on two-phase nitrogen flow and concluded that in vertical channels diameter effect was revealed in a transition from convective to less intensive nucleate boiling when the Froude number of a mixture Fr_m decreases from 40 to 10. On the contrary, in horizontal non-stratified flow, the reduction of the Fr_m number was accompanied by cross-section averaged heat transfer coefficient incensement of 20-30% in the nucleate boiling region. With $Fr_m \geq 40$, the geometry and orientation did not affect the heat transfer coefficient. Van Dresar et al. (2001) experimentally studied the near-horizontal two-phase flow of nitrogen and hydrogen. Unlike most of the other works which based on turbulent liquid flow, their work focused on laminar liquid flow and the results for low mass and heat flux flow were correlated with Froude number.

Terrestrial chilldown experiments

Research on cryogenic chilldown began in the 1960s with the development of rocket launching systems. Burke et al. (1960) studied the chilldown process of stainless-steel transfer lines of 60, 100 and 175 ft long with a 2.0 in. OD. The transfer lines were quenched by flowing liquid nitrogen. A sight glass was located near the discharge end for flow phenomena observation. Based on the wall temperature, liquid flow rate, and the observation, the chilldown process was simply divided into three stages: gas flow, two-phase flow and liquid flow. However, the flow regime information was lack in the experiments; moreover, the averaged wall temperature was used in their study. While other researchers (Bronson et al. 1962) pointed out that circumferential temperature gradient could be very large in cryogenic chilldown process. Early visualized study of flow regimes in a horizontal pipe during chilldown can be retrospect to Bronson et al. (1962). A 50 ft long with 1.375 in. ID test section was quenched by liquid hydrogen in this work. Results showed that the stratified flow is prevalent in the cryogenic chilldown process. Based on their experiments, rudimental models (Burke et al. 1960; Bronson et al. 1962) were also suggested to calculate chilldown time. Differences in flow regimes were not considered in these models; instead a gross effect was used. The work of both Burke et al. (1960) and Bronson et al. (1962) based on extremely long transfer lines, it is doubtful that their results can be applied over a short tube.

Chi and Vetere (1964) studied the chilldown process of hydrogen flowing through a 2-ft long thin wall copper tube. Thermocouples were installed on the outside wall and in the center of the tubing to measure the wall and stream temperatures at both the inlet and the outlet of the test section. The thermocouple beads in the center of the tubing were treated as control volume and their responses were used to identify different flow regimes. Chi and Vetere (1964) found that the void fractions were much larger, more than five times in some cases, than those given by previous investigators (Wright and Walters 1959; Hsu and Graham 1963). They attributed this difference to the effect of non-equilibrium nature of the chilldown process and concluded that thermodynamic equilibrium cannot exist in film boiling and transition regimes during chilldown. Another research work by Chi (1965) used several 26-in long aluminum tubes with $3/16$-in ID and ODs from $1/2$ to 2 inch. Unlike the thin wall copper tubing experiments (Chi and Vetere 1964), the temperature responses showed that slug flow was not observed until the aluminum test sections were almost cooled down, and the dominant flow regime was mist flow.

As mentioned before, chilldown of a hot surface or tube is of fundamental importance for the re-establishment of normal and safe temperature level following dryout in a LOCA in nuclear reactors. Liquid water or common refrigerant are usually used in this type of experiments. For example, Chan and Banerjee (1981a, b, c) and later Abdul-Razzak et al. (1992) used water to chilldown a preheated horizontal tubes. In their experiments, the chilldown process was divided into three regions, namely film boiling region, partially quenched region and totally quenched region. Different heat transfer mechanisms were involved in different regions.

Kawaji et al. (1985) experimentally investigated the chilldown process inside a vertical tube with different flow rates. Their results showed that for high flow rate, the entering liquid will initially boil through film boiling mechanism and then develop into inverted annular flow, dispersed flow and fully vapor flow, for low flow rate, the corresponding sequence was saturated boiling, annular flow, dispersed flow and fully vapor flow.

Recently, Velat (2004) experimentally studied the cryogenic chilldown in a horizontal pipe using nitrogen as working fluid. Pyrex glass tube with vacuum jacket was used for visualization. The flow regimes were recorded by high speed camera. Wall temperature histories and pressure drop along the pipe were recorded and associated with the visualized images.

Reduced gravity boiling and two-phase flow experiments

Because of the experimental difficulties there are very little heat transfer data for cryogenic flow boiling in reduced gravity. We were able to find just one report done by Antar and Collins (1997) investigated the cryogenic flow boiling in low gravity condition. The experimental results of two-phase flow under reduced gravity conditions using regular working fluids, such as R113, are also summarized here.

Adham-Khodaparast et al. (1995) investigated the flow film boiling during quenching of a hot flat surface with R-113. Micro-sensors were used to record instantaneous heat flux and heater surface temperature. They reported lower heat transfer rates during microgravity as compared to normal gravity and contributed that to thickening of the vapor layer. The wall superheat and the surface heat flux at the onset of rewetting and the maximum heat flux were found to increase with the inlet liquid subcooling, mass flux and gravity level. The effect of gravity was determined to be more important for low flow rates and less relevant for high flow rates. The two-phase flow regimes were not reported in their work.

Another quenching test under microgravity was done by Westbye et al. (1995). A hot thin-walled stainless steel tube was quenched by injection of subcooled R113 into the tube under both 1-g and microgravity conditions. The injection mass flux was 160-850 kg/m^2s. It was found that the rewetting temperatures were 15^0C-25^0C lower in microgravity than those obtained in 1-g, and the film boiling heat transfer coefficients in microgravity were less than those in 1-g tests. This resulted in much longer cooling periods in microgravity. It was also reported that once the tube was cooled sufficiently to allow axial propagation of the QF, the rewetting velocity was slightly greater in microgravity. The nucleate and transition boiling curves under microgravity were reported to be shifted to lower wall superheats as compared to 1-g results.

Antar and Collins (1997) reported cryogenic chilldown process under 1-g condition and on board a KC-135 aircraft. They observed that a sputtering leading core followed by a liquid filament annular flow regime. This flow regime is composed of a long and connected liquid column that is flowing in the center of the tube and is surround by a thick vapor layer. They attributed the filamentary flow to the lack of difference in the speed of vapor and liquid phases. On the heat transfer side, they reported that the quench process was delayed in low gravity and the tube wall cooling rate was diminished under microgravity conditions. The QF speed was found to be slower under the low gravity conditions.

2. CRYOGENIC LINE CILLDOWN RESEARCH

2.1 Terrestrial Chilldown Experiment

Progress of cryogenic two-phase chilldown under terrestrial condition is discussed in this section. We focus on gravity-driven flow experiments as it enhances the basic understanding of the chilldown process. The experimental system, procedure and measurement methods are provided in Yuan (2006).

In gravity-driven flow experiments, an insulated reservoir is used to generate the flow with an estimated mass flux range of 18-23 kg/m^2s by reviewing the recorded flow images. The results show that the cryogenic chilldown process can be divided into three stages that are associated with different heat transfer mechanisms.

Heat Transfer Study

For the horizontal tube orientation, the two-phase flow is generally stratified because of the gravitational force. Therefore the bottom side of the tube is chilled down first. The heat transfer mechanism at the bottom of the tube includes film boiling, transition boiling, and nucleate boiling, while the heat transfer mechanism at the top of the tube is mainly due to convection by the superheated vapor.

Wall temperature profiles

Wall temperatures are measured by 16 thermocouples, 9 are embedded very close to the inner surface of the tube wall; the other 7 thermocouples measure the outside wall temperature of the test section. The thermocouple locations are shown in Figure 5. The unit for the length dimension is mm.

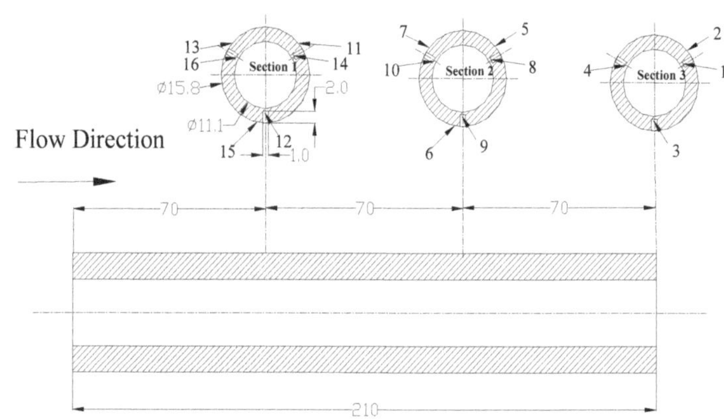

Fig. 5 Sketch of the test section and thermocouple locations for gravity-driven test.

Figure 6 gives the temperature profiles at the inlet and middle sections. It is found that a large temperature difference exists between the top and bottom of the test section, which was also reported by Bronson (1962) in the experiments of chilldown in heavy-wall transfer lines. This temperature difference suggests that different heat transfer mechanisms are encountered at the top and the bottom of the test section, respectively. Since the two phases are separated by the gravitational force, the heat transfer mechanism at the bottom is boiling heat transfer, while the heat transfer at the top is forced convection to the vapor phase.

Data reduction

In boiling heat transfer experiments, the surface heat flux as a function of the wall super heat is often of particular interest, because it denotes different heat transfer mechanisms and is crucial for engineering applications. However, in chilldown or quenching experiments, the surface heat flux can not be controlled independently. The surface heat flux can only be inferred from the wall temperature history by solving the inverse heat conduction problem (IHCP). For quenching of a vertical tube, in which the axisymmetrical assumption is generally held, the situation is much simpler.

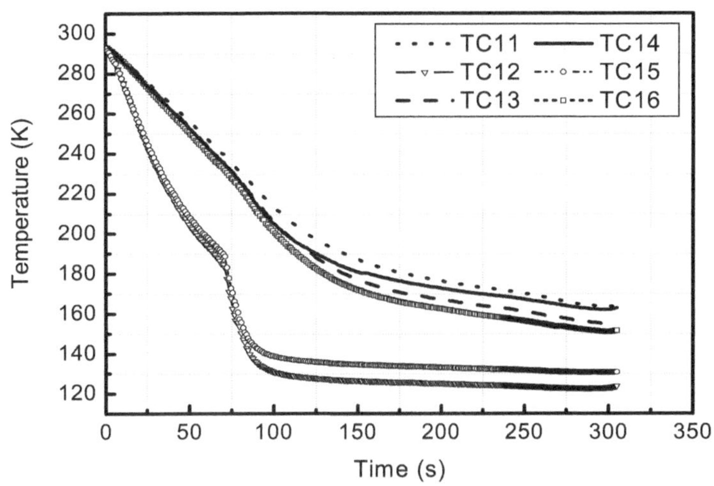

A

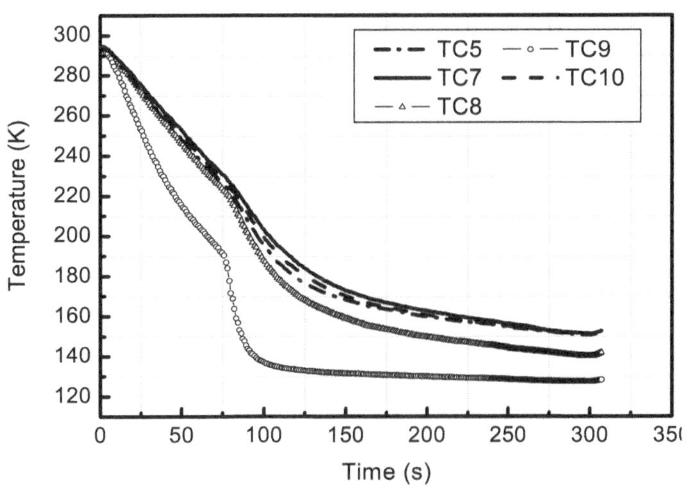

B

Fig. 6 Temperature profiles of gravity-driven test at different cross-sections. A) Inlet section. B) Middle section

Based on the exact solution of the IHCP, Burggraf (1964) developed a method to obtain the temperatures and heat fluxes at the inside tube wall from the temperature history data of the thermocouple welded on the outside of the test section. The inside wall temperature is given as follows:

$$T_i = T + \frac{r_o^2}{4\alpha}\left(\left(\frac{r_i}{r_o}\right)^2 - 1 - 2\ln\frac{r_i}{r_o}\right)\frac{dT}{dt} + \left(\frac{1}{64\alpha^2}\left(r_i^4 - 5r_o^4\right) - \frac{r_o^2 r_i^2}{8\alpha^2}\ln\frac{r_i}{r_o} - \frac{r_o^4}{16\alpha^2}\ln\frac{r_i}{r_o} + \frac{r_o^2 r_i^2}{16\alpha^2}\right)\frac{d^2T}{dt^2} + \ldots \quad (1)$$

where r_i and r_o denote the inner and outer radius, respectively; T_i and T are the inside wall temperature and temperature measured by thermocouple, respectively; α is the thermal diffusivity of the wall. Then the inside wall heat flux with the first three leading terms can be written as (Iloeje et al. 1975):

$$q''_w = \rho c\left(\frac{r_i^2 - r_o^2}{2r_i}\right)\frac{dT}{dt} + \frac{(\rho c)^2}{k}\left(\frac{r_i^3}{16} - \frac{r_o^4}{16r_i} - \frac{r_o^2 r_i}{4}\ln\frac{r_i}{r_o}\right)\frac{d^2T}{dt^2} + \frac{(\rho c)^3}{k^2}\left(\frac{r_i^5}{384} - \frac{3r_o^4 r_i}{128} + \frac{3r_o^2 r_i^3}{128} - \frac{r_o^6}{384r_i} - \frac{r_o^2 r_i^3}{32}\ln\frac{r_i}{r_o} - \frac{r_o^4 r_i}{32}\ln\frac{r_i}{r_o}\right)\frac{d^3T}{dt^3} \quad (2)$$

For horizontal chilldown experiments, the axisymmetrical assumption is not valid, and the solution of the IHCP suffers from the ill-posedness. In order to calculate heat transfer data from the transient temperature profiles, an energy balance is performed locally on a control volume of the tube wall at the thermocouple location. The change in the stored heat in the control volume is equated to the heat transported to the fluid and that transferred by conduction, minus the losses to the environment. This method was used by many researchers in analyzing the data from quenching tests (Abdul-Razzak et. al 1992; Westbye et al. 1995; Chen et al. 1979). The inside wall heat flux is then given as :

$$q''_w = -\rho c\left(\frac{r_o^2 - r_i^2}{2r_i}\right)\frac{dT}{dt} + \frac{k}{U^2}\left(\frac{r_o^2 - r_i^2}{2r_i}\right)\frac{d^2T}{dt^2} - \left(\frac{r_o}{r_i}\right)\left(q''_{conv} + q''_{rad}\right) + \frac{(r_o - r_i)}{r_o d\varphi}q''_{cond} \quad (3)$$

In the above, U is the rewetting velocity. The first term in the right-hand-side (RHS) of Equation (3) comes from the change of the stored heat in the control volume and is the dominant term in chilldown experiments. It is also interesting to note that this term equals the first term from the IHCP solution given in Equation (2).

To estimate the axial heat conduction term in rewetting experiments, Chen et al. (1979) assumed a constant rewetting velocity U, and the axial temperature gradient (dT/dz) was represented by $(1/U)(dT/dt)$. Then the effect of the axial conduction on the inside wall heat flux can be evaluated from the second term in the RHS of Equation (3). The axial conduction term is generally very small, and it is only important at the location near the quenching front, where the maximum axial temperature gradient exists between the dry side and the wet side of the wall. A comparison between the heat fluxes based on a 2-D heat transfer model (Cheng 1978), which considers the axial heat conduction, with those obtained by neglecting the axial conduction was given by Cheng et al. (1979). The comparison for the copper test section showed that the two mid-plane boiling curves were approximately the same and the net axial conduction at the mid-section is negligibly small. In our experiments, the temperature measurements show that the axial temperature gradient is relatively small, and additionally the heat conductivity of the glass is much smaller than that of the copper used by Cheng et al.(1979), therefore the axial heat conduction along the test section is neglected in the heat flux calculation. It is left in Equation (3) only for completeness.

The third term in Equation (3) accounts for the contribution from the radiation and natural convection inside the vacuum jacket. The last term in Equation (3) comes from the circumferential heat transfer. Here, $d\varphi$ is the differential azimuthal angle of the control volume and depends on the size of the control volume. In our calculation, the control volume is approximately assumed to be in thermal equilibrium with the embedded thermocouple. Based on this assumption, the control volume size is chosen with an average arc length of 2.0mm. The circumferential temperature gradient is calculated by a linear interpolation between the thermocouple measuring points.

Derivative of the temperature history is needed in calculating the first term in Equation (3). Generally, the finite difference method is not suitable to obtain the temperature derivatives, because it is very susceptive to small fluctuations, which are inevitable in measurements. Therefore, a least square technique, proposed by Iloeje et al. (1975) in a similar situation, is used to get smooth profiles. To increase accuracy, the temperature data are divided into segments and the order of the curve fit used for each segment is made as high as possible (but less than 6, otherwise it will be subject to temperature fluctuations) without reintroducing irregularities inherent in the raw data. Figure 7 shows a typical result obtained with the least square fit procedure, and Figure 8 compares the first term in Equation (3) calculated from the least square fit line with that from the finite difference method. It is clear that the finite difference method is not suitable for calculating the derivatives of measured temperature profiles.

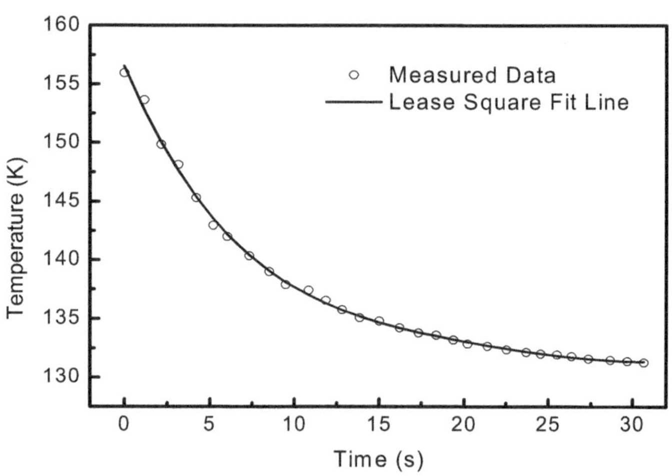

Fig. 7 Typical curve fit line of the experimental data.

We assume the vacuum jacket and the test section as long concentric cylinders, the radiation contribution to the heat flux, q''_{rad}, is then obtained by (Incropera and Dewitt 2002; Liao 2005):

$$q''_{rad} = \frac{\sigma_B \left(T_j^4 - T_o^4\right)}{\dfrac{1}{\varepsilon_t} + \dfrac{1-\varepsilon_j}{\varepsilon_j}\left(\dfrac{r_o}{r_j}\right)} \tag{4}$$

where σ_B is Stefen-Boltzmann constant, T_j is the temperature of the vacuum jacket, and is assumed to equal the room temperature, T_o is the outer wall temperature of the test section, ε_j and ε_t are the emissivity of the vacuum jacket and the test section, respectively, and r_o and r_j are the outer radius of the test section and the inner radius of the vacuum jacket, respectively.

To evaluate the natural convection heat transfer due to the residual air inside the vacuum jacket, Raithby and Hollands' correlation (Raithby and Hollands 1975; Liao 2005) is used:

$$q''_{conv} = \frac{k_{eff}}{r_o \ln\left(r_j/r_o\right)}\left(T_j - T_o\right) \tag{5}$$

where k_{eff} is the effective thermal conductivity given by Raithby and Hollands (1975).

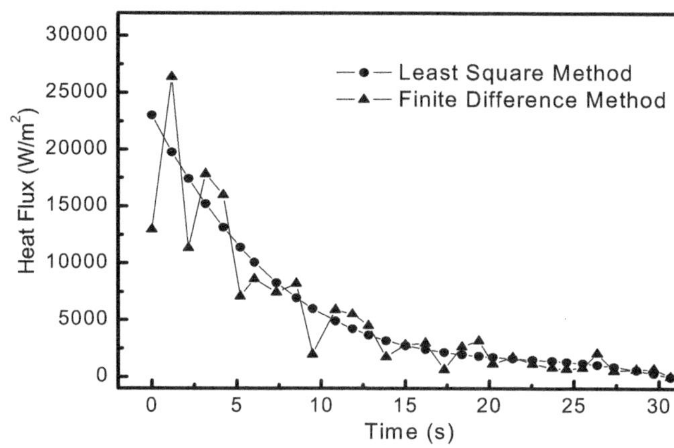

Fig. 8 Temperature derivatives calculated by least square fit method and by finite difference method

Heat transfer mechanisms

The data reduction method given above is used to analyze the transient temperature measurements. Left part of Figure 9 shows the calculated bottom wall heat flux as a function of the bottom wall temperature at the outlet location, while the right part represents the corresponding temperature profiles.

The shape of bottom heat flux is similar to the boiling curve from steady-state pool boiling experiments. This suggests that the chilldown process may share many common features with pool boiling experiments. Following the method to characterize different heat transfer mechanisms in pool boiling experiments, a maximum or critical heat flux (CHF) q''_{CHF} and a minimum heat flux $q''_{\min}$ are used to divide the chilldown heat transfer into three stages, which are film boiling, transition boiling and nucleate boiling, as shown in Figure 9.

Initially, the wall temperature is very high, liquid nitrogen evaporates voraciously when enters the test section; a vapor film will form and separate the liquid from contacting the wall, the two-phase flow is therefore in the film boiling stage. At decreased wall temperatures, the liquid begins to contact the wall and the heat transfer mechanism is transition boiling, which is characterized by increasing wall heat flux with decreasing wall superheat that is contrary to what in the film boiling region and nucleate boiling region. After passing the CHF point the heat transfer mechanism then changes to nucleate boiling.

The calculated bottom wall heat flux as a function of time is shown in Figure 10. It is obvious that the time of transition boiling is very short compared with the other two boiling stages.

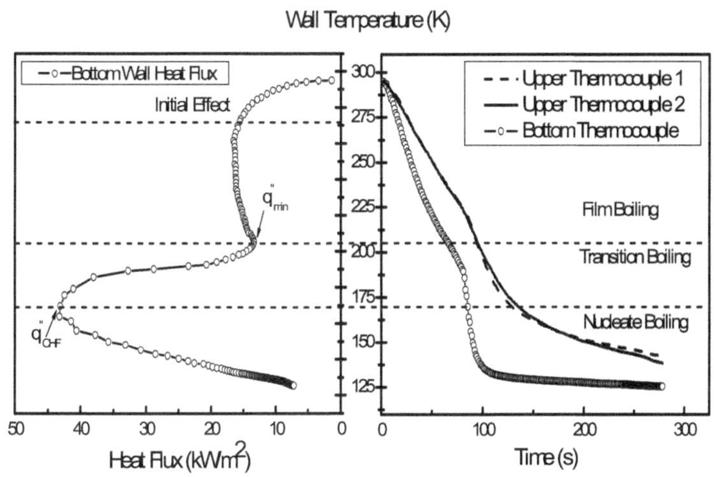

Fig. 9 Bottom wall heat flux and transient wall temperature profiles at the outlet cross-section.

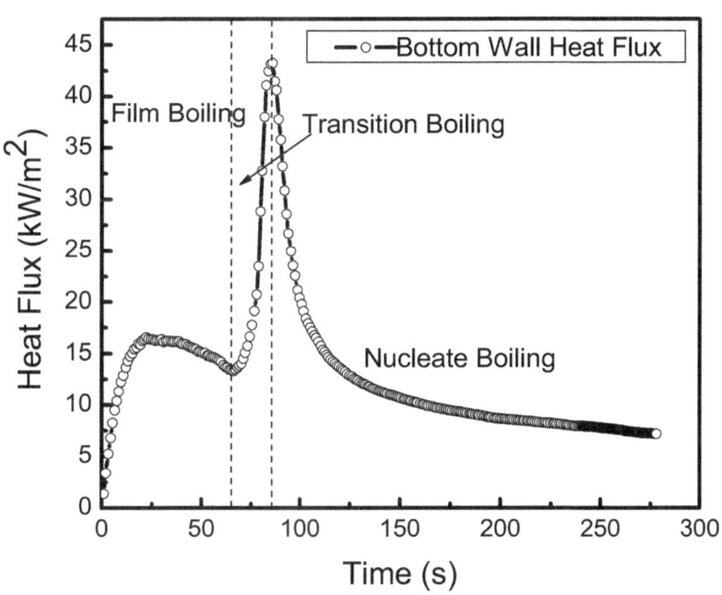

Fig. 10 Bottom wall heat flux at the outlet cross section as a function of time.

Similarity between the chilldown boiling curve and the pool boiling curve naturally leads one to compare the chilldown data with pool boiling correlations. The comparisons between the two turning points, namely the minimum heat flux and the CHF, are given below.

For the steady state film boiling, the correlation developed by Zuber (1959) is widely used to predict the minimum heat flux:

$$q''_{\min} = C h_{lv} \rho_v \left[\frac{\sigma g(\rho_l - \rho_v)}{(\rho_l - \rho_v)^2} \right]^{1/4} \tag{6}$$

here, σ is surface tension; g is gravitational acceleration; h_{lv} is the latent heat of vaporization, while C was suggested differently as 0.177 (Zuber 1959), 0.13 (Zuber 1958), or 0.09 (Berenson 1961). The resulting $q''_{\min}$ is then $13.0\,\mathrm{kW/m^2}$, $9.6\,\mathrm{kW/m^2}$, and $6.6\,\mathrm{kW/m^2}$, respectively. In Figure 9, the $q''_{\min}$ for chilldown is calculated as $13.3\,\mathrm{kW/m^2}$, which is a slightly larger than the steady state prediction with $C = 0.177$.

Based on the similarity between the CHF condition and column flooding, Kutateladze (1948) derived the following relation for the pool boiling CHF:

$$q''_{CHF} = 0.131 h_{lv} \rho_v \left[\frac{\sigma g(\rho_l - \rho_v)}{\rho_v^2} \right]^{1/4} \tag{7}$$

Zuber (1959) proposwed the identical correlation based on the analysis of Taylor and Helmholtz instability. For liquid nitrogen under atmospheric pressure, Equation (7) gives a CHF value of $160.7\,\mathrm{kW/m^2}$. The chilldown measurement in Figure 10 is only about 27% of this value. This big discrepancy is believed to come from the different experimental conditions between the chilldown and the pool boiling tests. In pool boiling experiments, the heat supplied to the fluid is maintained by a heater, while in chilldown tests it comes from the stored heat in the tube wall. In the film boiling region, the heat flux is generally small, therefore the tube wall can maintain a near constant heat flux condition and functions like the heater used in the pool boiling tests. However, in the transition boiling region, the stored energy in the tube wall is depleted so quickly that the experimental condition is very different from that of the pool boiling tests. The limited energy stored in the tube wall put a restriction on the value of CHF, which, therefore, is much less than that in the pool boiling data. Previous work by Bergles and Thompson (1970) also concluded that quantitively the differences between chilldown and steady-state boiling curves can be very large.

In Equation (6) and Equation (7), the minimum heat flux and CHF are not correlated with the thermal properties of the wall for pool boiling tests. However, as seen from above, the available heat flux to the flow is closely related to the energy stored in the wall. Therefore, the thermal properties, e.g. thermal conductivity, heat capacity, of the wall are expected to play a role in the chilldown process. For example, for the wall with a higher thermal conductivity, the energy transferred to the flow can be more quickly supplied by the surroundings, and therefore this will result in a higher value of CHF.

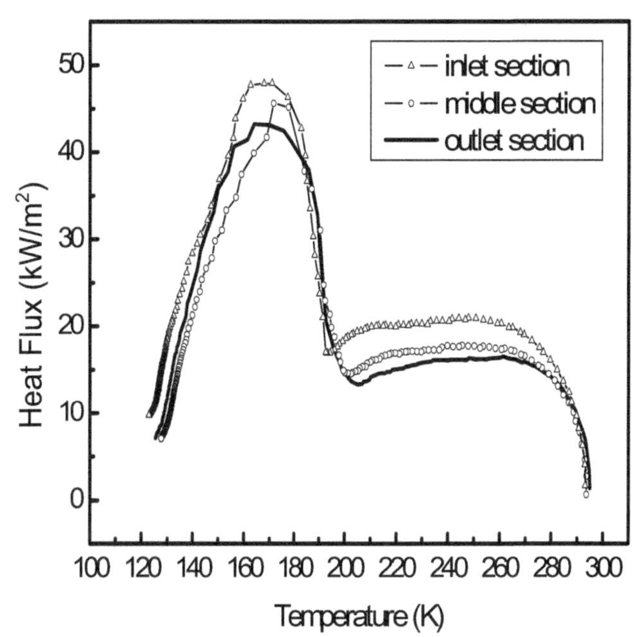

Fig. 11 Bottom wall heat fluxes at different axial locations.

Figure 11 shows the calculated heat flux in different axial locations. It is found that both the CHF and the minimum heat flux decrease with increasing axial distance from the inlet. This is also associated with an increase in the rewetting temperature.

In most of the previous quenching experiments, the film boiling heat flux was reported as either keeping a relatively constant value (Westbye et al. 1995; Cheng and Ragheb 1979) or decreasing as the test section is chilled down (Ganić and Rohsenow 1977; Bergles and Thompson 1970). In our experiments, however, the local heat flux first increases in a short period and then decreases gradually. This is consistent with the transient nature of the experiments. Generally the increasing period is expected to be shorter at higher mass fluxes if the other conditions are kept the same. The mass fluxes in previous investigations were much larger than that in current experiments, and therefore associated with a very short increasing time of the heat flux in the film boiling regime. This might be the reason that the increases of the heat flux were not recorded before.

Visualization Study

Heat transfer analysis shows the chilldown process can be divided into three stages with each stage associated with different heat transfer mechanisms. The visualization study casts more light on the characteristic of flow regimes during the chilldown process.

Figure 12 illustrates the overall chilldown sequence in our experiments, while some selected flow images in different region of chilldown are shown in Figure 13. In the beginning of the film boiling region, the flow pattern is basically dispersed flow, in which the liquid phase is dispersed as near spherical drops within a continuous vapor phase. The void fraction of the two-phase flow decreases as the tube is chilled down, and long liquid filaments, separated from wall by a thin vapor film, are observed to flow along the tube bottom. The length of the liquid filaments generally increases with decreasing wall temperature. For short liquid filaments, the flow regime is close to dispersed flow, while for very long filaments, the flow regime can be classified as highly skewed inverted annular flow. In the film boiling region, heat is primarily transferred from the wall by conduction through the vapor film and thus evaporate the liquid filaments, and by convection to the vapor phase.

Once the bottom wall temperature has been reduced low enough, transient boiling, characterized by intermittent liquid-wall contact and violent bubble generation, is observed. Shortly after the transient boiling, a continuous liquid-wall contact is established and the liquid nitrogen begins to pile up on the bottom wall. Many nucleation sites are observed to be suppressed as the wall temperature keeps decreasing. The prevailing boiling regime is nucleate boiling and the flow pattern is stratified flow or wavy flow.

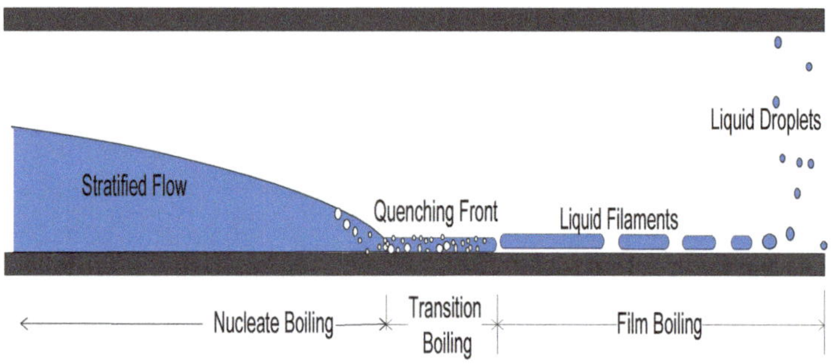

Fig. 12 Characteristics of horizontal chilldown under low flow rate.

Previous horizontal quenching tests (Chan and Banerjee1981a; Abdul-Razzak et al. 1992) showed the same sequence as our experiments, however, with different flow regimes mainly in the film boiling region. In their experiments, the flow regime in film boiling was inverted annular flow. The thickness of the liquid core increases with decreasing wall temperature, and then follows the stratified flow regime. While in our experiments, the thickness of the liquid filament is generally a constant, instead the liquid thickness increases in the nucleate boiling region as shown in Figure 12.

The difference in flow regimes is believed mainly due to different flow rates. The flow rate in the test of Chan and Banerjee (1981a) was about $150 \sim 450\,\mathrm{kg/m^2 s}$, while it was even higher ($400 \sim 1300\,\mathrm{kg/m^2 s}$) in the experiments by Abdul-Razzak et al. (1992). In current chilldown test, the flow rate is much lower. The heat stored in the tube wall causes dry-out of the liquid phase in the film boiling region and thus leads to the dispersed flow.

The visualization study also gives the detail of propagation the quenching front and how the liquid rewets the tube wall (Figure 14). The arrow in Figure 14 indicates the limit between the quenching front and the nucleate boiling region.

As shown in Figure 14, the quenching front is characterized by vehement bubble generation and intermittent liquid-wall contact. Following the quenching front, a continuous liquid-wall contact is established, the flow regime changes to stratified flow, and many nucleation sites are suppressed.

2.2 Cryogenic Line Chilldown under Microgravity Condition

Efficient and safe utilization of cryogenic fluids in a spacecraft during space missions demands thorough study of cryogenic two-phase flow and heat transfer under microgravity condition. However, researches on this field are very limited due to the experimental difficulties. In present study, experimental study of cryogenic chilldown under microgravity condition is conducted. The investigation focuses on the film boiling region. The microgravity experiments were performed in the 1.8-second drop tower at the University of Florida. The experimental system, procedure and measurement methods are provided in Yuan (2006).

Film Boiling under Microgravity

Among the various film boiling correlations for pool boiling, it is very interesting to note that the surface heat flux or heat transfer coefficient is usually correlated with $g^{1/4}$, in which g is the gravitational acceleration. For example, the highly cited Bromley's correlation (Bromley 1950) for horizontal tubes is:

$$h_{fb} = 0.62\left[\frac{k_v^3 h_{lv}' \rho_v g\left(\rho_l - \rho_v\right)}{\mu_v D \Delta T_w}\right]^{1/4} \tag{8}$$

For a horizontal surface, the mostly used correlation gives (Berenson 1961):

$$h_{fb} = 0.425\left[\left(\frac{k_v^3 h_{lv}' \rho_v g\left(\rho_l - \rho_v\right)}{\mu_v \Delta T_w}\right)\left(\frac{g\left(\rho_l - \rho_v\right)}{\sigma}\right)^{1/2}\right]^{1/4} \tag{9}$$

in which the $\left(g\left(\rho_l - \rho_v\right)/\sigma\right)^{1/2}$ term is actually a geometrical factor as pointed out by Berenson (1961), and has the dimension of [m^{-1}], thus in Equation (9) the film boiling heat transfer coefficient is still proportional to $g^{1/4}$. Liao (2005) and Chan (1995) also report the same relation. The list also includes the correlations given by Noyes (1963), Chang and Snyder (1960), etc. The only exception, to our knowledge, is the correlation proposed by Chang (1959), in which the film boiling heat transfer coefficient is suggested to be proportional to $g^{1/3}$.

It is also noted that some of the above correlations were derived based on quite different approaches, such as the analytical solution of flow in vapor film channel (Chan 1995; Liao 2005) and the vapor escape model based on Taylor-Helmholtz instability (Berenson 1961).

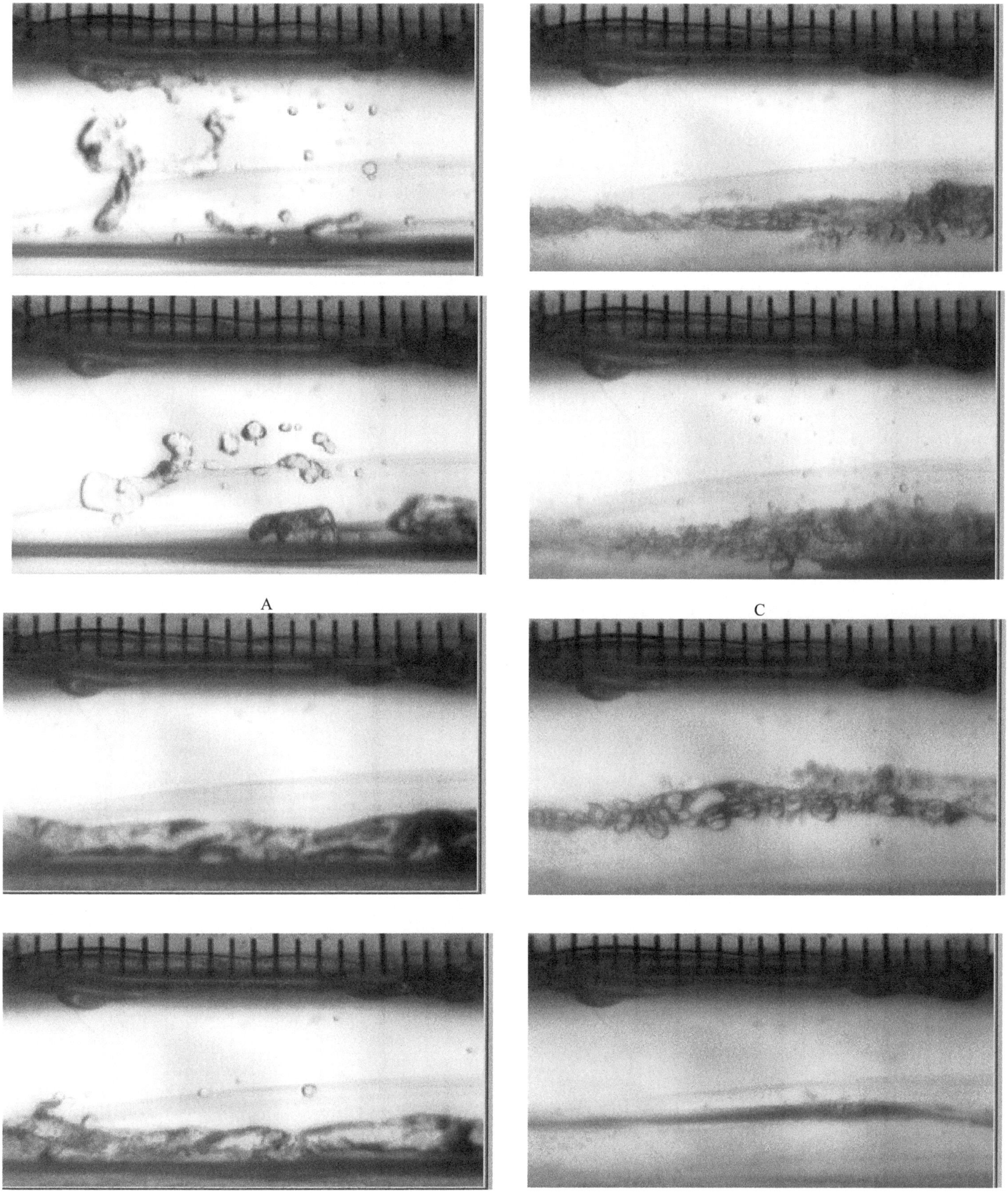

F

Fig. 13 Flow visualizations at different stages of chilldown. A) Initial stage. B) Film boiling stage. C) Transition boiling stage. D) Nucleate boiling stage.

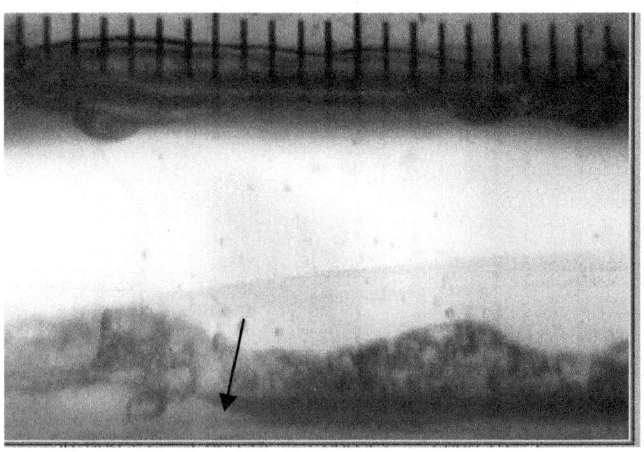

Fig. 14 Propagation of the quenching front.

Therefore, it is natural for one to think that the heat transfer characteristics in film boiling are related to $g^{1/4}$ and we can extend the above correlations to different g-levels. For example, the Bromley's correlation can be modified as:

$$h_{fb} = 0.62\left[\frac{k_v^3 h_{lv}' \rho_v g\left(\rho_l - \rho_v\right)}{\mu_v D \Delta T_w}\left(\frac{a}{g}\right)\right]^{1/4} \qquad (10)$$

in which, a is the specific local acceleration.

Table 1 Basic findings of flow film boiling in previous chilldown experiments.

	Adham-Khodaparast et al. (1995)	Westbye et al. (1995)	Antar and Collins (1997)
Experimental apparatus	horizontal surface	horizontal tube	vertical tube
Working fluid	R113	R113	LN_2
Flow pattern	NA	inverted annular flow	liquid filament annular flow
Heat transfer rate under microgravity	lower	lower	lower
Rewetting temperature under microgravity	lower	lower	lower
Rewetting speed under microgravity	NA	higher	lower

However, experiments by Merte and Clark (1964) showed that the heat flux and the Nusselt number were proportional to $(a/g)^{1/3}$ in the gravity range of $0.01 < a/g < 1$. On the other hand, the maximum and minimum heat fluxes were found to depend on gravity according to $(a/g)^{1/4}$. Unlike most of the other microgravity pool boiling research, which were performed under steady or quasi-steady conditions with a constant wall heat flux or a constant wall temperature, the experiments by Merte and Clark (1964) were highly transient. In their experiments, a small ball was quenched inside the saturated liquid nitrogen under different g-levels, the temperature of the ball kept decreasing, to this extent, their experimental condition was close to a cryogenic chilldown process. The drop tower used in their experiments can provide a free fall time of about 1.4 seconds.

For flow boiling experiments, the flow will play a role in both bulk convection and vapor removal mechanism, therefore the heat transfer coefficient will not have a simple relation with the local acceleration as that in pool boiling.

Chilldown investigations in reduced gravity is very limited, all of the previous experiments found in the open literature have been reviewed above. The basic findings for the flow film boiling are summarized in Table 1. All previous research reported lower heat transfer rates in reduced gravity, Adham-Khodaparast et al. (1995) attributed this reduction to the thickening of the vapor film.

Experimental Condition

Since the microgravity duration is relatively short, the experiment is designed to drop the apparatus at different wall temperatures with different mass fluxes. The wall temperatures are classified as three groups that are high wall temperature (about 270K), medium wall temperature (about 240K) and low wall temperature (about 210K). Different motor speeds are used to generate different mass fluxes.

The gravity level and the deceleration impact have been measured by an accelerometer (Omega ACC104A) mounted on an aluminum piece that is attached to the drag shield.

Flow Regime Visualization under Microgravity Condition

The flow regimes before and during the drop are compared in Figure 15 and Figure 16. The images on the left are taken before drop, while the microgravity images are on the right. With the experimental wall temperature range, the two-phase flow is in the dispersed flow film boiling (DFFB) state and the liquid phase is either in the form of small droplets or long filaments before drop. The characteristics of the 1-g flow regime are summarized in previous section.

Different flow behaviors have been recorded during the microgravity period. If the liquid phase before drop is in the form of dispersed droplets, these droplets will enter the central region still as droplets during the microgravity period (Figure 15B). For long liquid filaments, sometimes the filaments are lifted up and still maintain original shape during the drop (Figure 15D); in some other cases, the liquid filaments are broken and dispersed into the central region (Figure 16F) or the two-phase flow has a liquid-vapor core in the center with smaller liquid chunks at both top and bottom (Figure 16H).

Heat Transfer Study

Due to the thickening of the vapor film, the heat transfer rate in film boiling is generally lower under microgravity condition. However, this effect is expected to be relatively less important with increasing flow speeds. In this section, the heat transfer in the film boiling region under microgravity is discussed.

Wall Temperature Profiles

To get more data points during the microgravity period, the wall temperatures are measured only at one axial location. Figure 17 gives the typical temperature profiles with different mass fluxes.

The circled temperatures in Figure 17 approximately indicate the drop period, which includes release of the drag shield, microgravity period, impact on the air bag, and deceleration period.

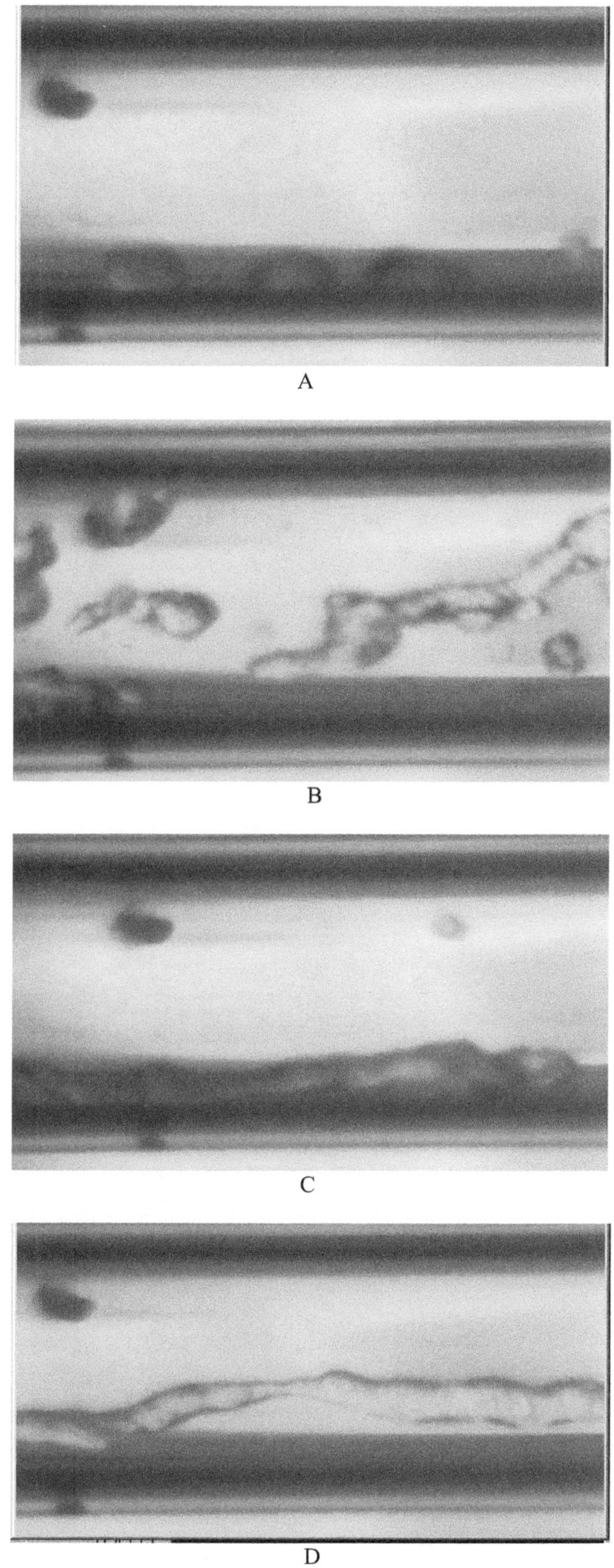

Fig. 15 Two-phase flow images under both 1-g and microgravity conditions. A) 1-g case 1. B) Microgravity case 1. C) 1-g case 2. D) Microgravity case 2.

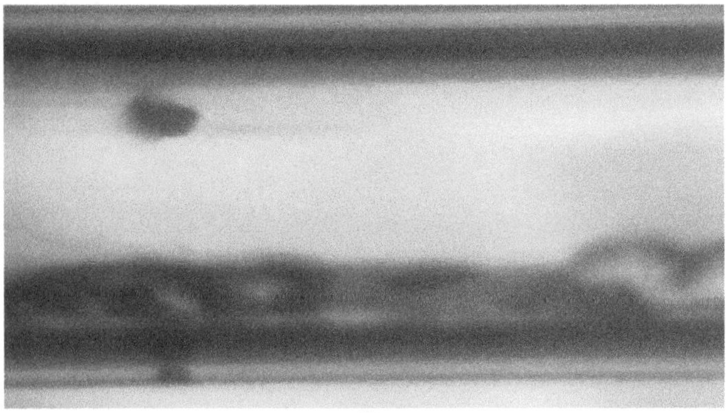

E

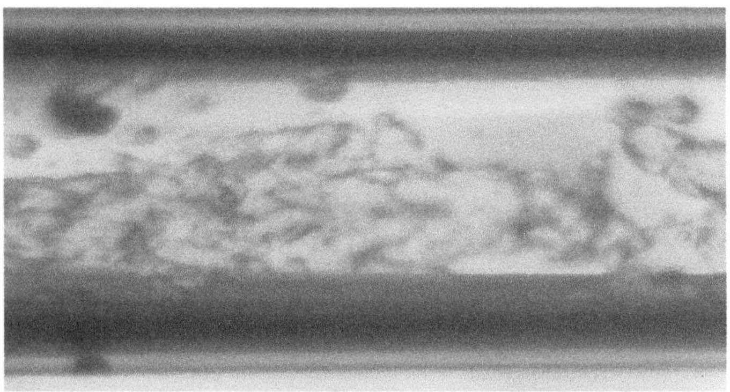

F

G

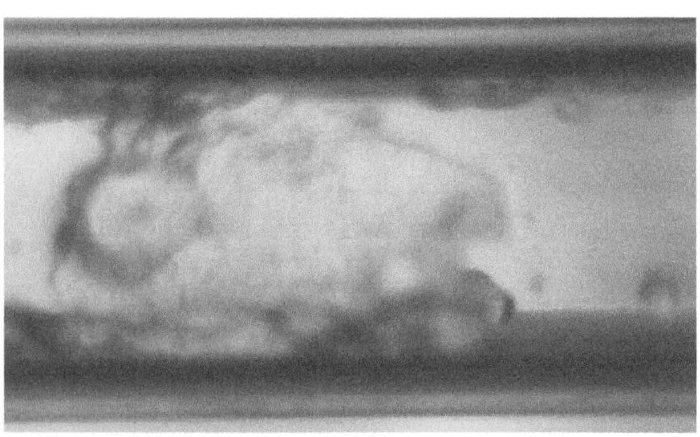

H

Fig. 16 Two-phase flow images under both 1-g and microgravity Conditions E) 1-g case 3. F) Microgravity case 3. G) 1-g case 4. H) Microgravity case 4.

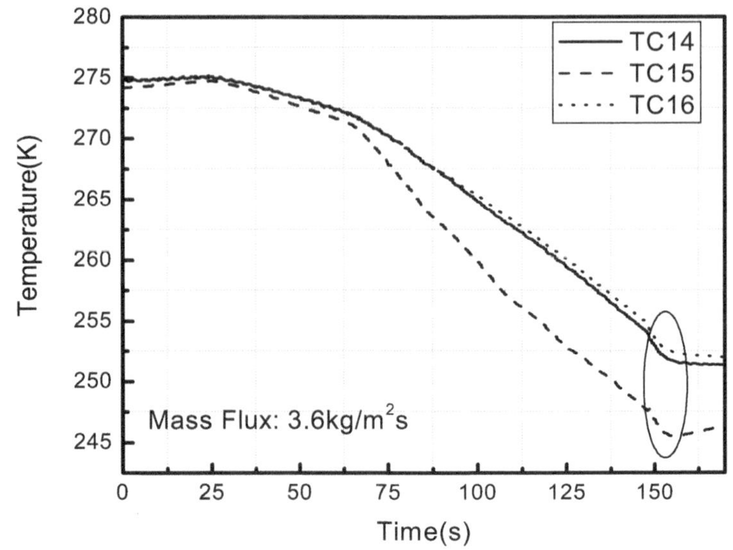

A

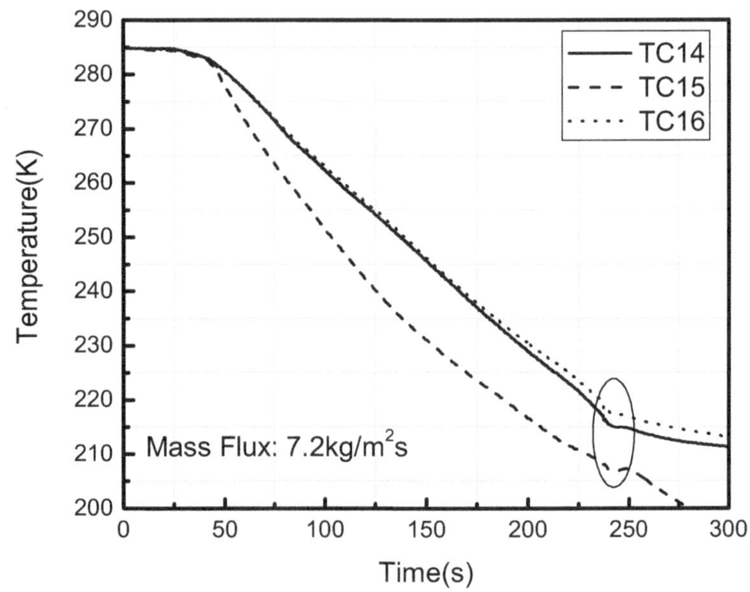

B

Fig. 17 Temperature profiles with different mass fluxes in microgravity test. A) Mass flux of 3.6 kg/m²s. B) Mass flux of 7.2 kg/m²s.

To examine the detail of the wall temperature response to the sudden removal of the gravity force, the temperature profiles are zoomed in. Figure 18 illustrates the typical results. It is observed that the temperature decreasing rate at the bottom wall is generally slower during the microgravity period, because the removal of the gravitational force will thicken the vapor film and reduce the film boiling heat flux. Since the top wall transfers heat mainly by convection, the gravity field is expected to have a less effect on the heat transfer at the top wall. However, as shown in Figure 15 and Figure 16, it is possible for the liquid phase to contact the top wall during the microgravity period and it results in higher heat removal rate. For the current transient experimental condition, the best way to represent the data is to calculate the averaged heat flux which is given in the following section.

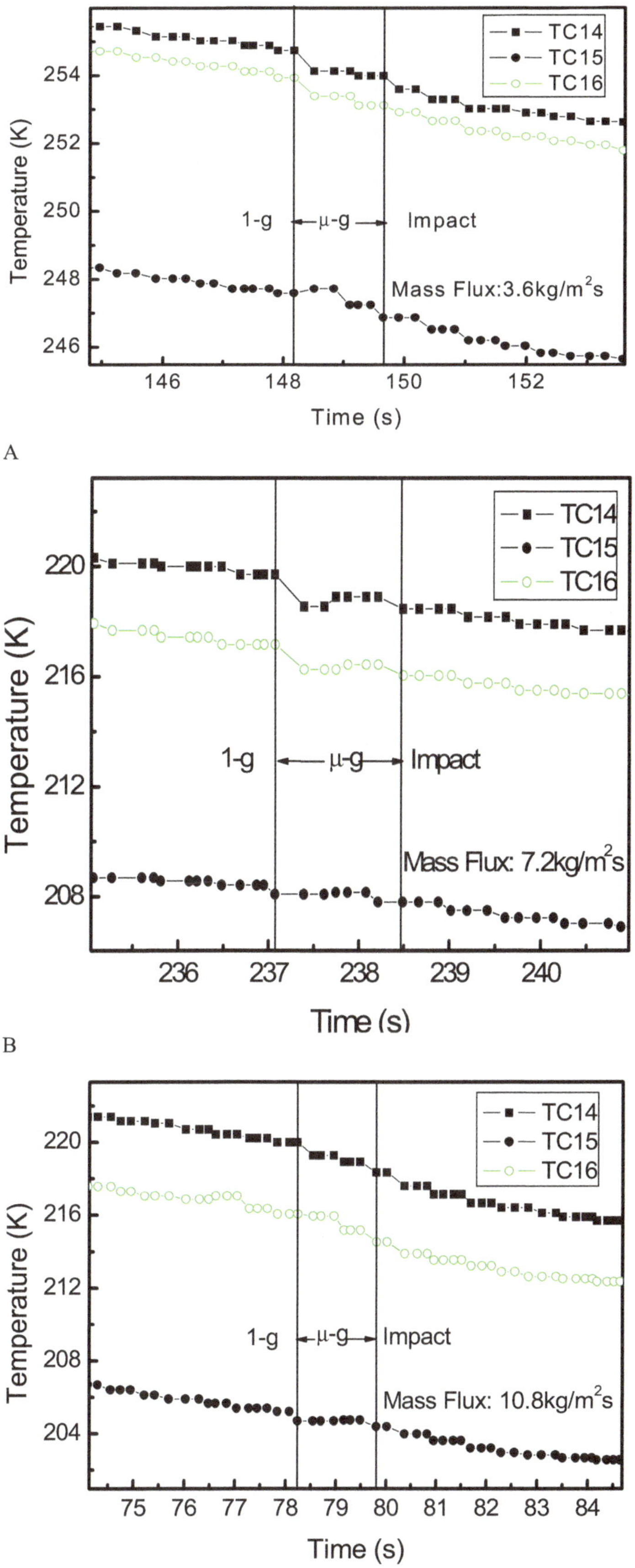

Fig. 18 Wall temperature response to microgravity. A) Mass flux of 3.6 kg/m^{2}s. B) Mass flux of 7.2 kg/m^{2}s. C) Mass flux of 10.8 kg/m^{2}s.

Wall Heat Flux

The gravity effect is shown in Figure 19, in which the ratio of the bottom heat flux before the drop to that during the drop is plotted for different flow rates. The heat flux at the bottom of the tube decreases under the microgravity condition and the ratio varies from the minimum of about 0.66 to about 0.90. The result does not show a strong dependence on the wall temperature and inlet flow rate. Two runs of the quenching test performed by Xu (1998) reported similar ratio of 0.7 and 0.8, however, in the work of Westbye et al. (1995), this ratio was found to be much less and ranged from 0.15 to 0.6.

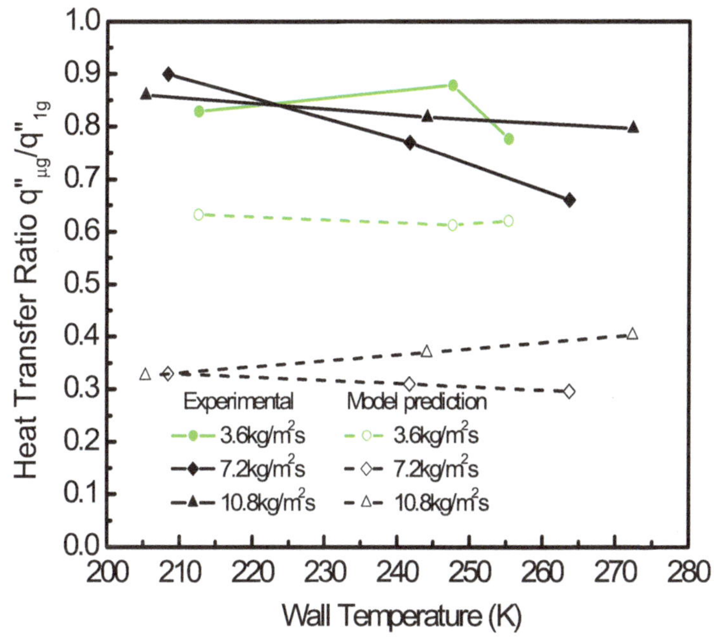

Fig. 19 Ratio of heat flux under microgravity to 1-g condition with different flow rates and comparison with model prediction.

The bottom wall is subject to film boiling with the liquid filaments in the central core and the convection to the super heated vapor phase. For the film boiling part, the heat flux and the heat transfer coefficient are proportional to $(a/g)^{1/3}$ as suggested by Merte and Clark (1964). With a simple model, if we assume that the convection is not affected by the microgravity condition, the model predictions are shown by the dash lines in Figure 15. The gravity level in the model calculation is 10^{-4}g. The model results are much scattered and significantly lower than the experimental values. This suggests that under the microgravity condition, the effect of convection may raise the heat flux, and therefore the total effect of the microgravity is less prominent than that in the pool boiling experiments.

3. CONCLUSIONS

The experiments show that the chilldown process can be generally divided into three regions: film boiling region, transition boiling region and nucleate boiling region, and each region is associated with different flow regime and heat transfer mechanism. After data reduction, chilldown boiling curve, which is similar to the pool boiling curve, is generated based on the wall temperature measurement. The limits between different regions of the chilldown process are clearly shown as the point of the critical heat flux (CHF) and the minimum heat flux on the chilldown boiling curve. It has been observed that the two-phase flow is dispersed in the film boiling region with liquid phase in the form of long filaments as the tube is chilled down, while the vapor phase is generally superheated. The heat transfer mechanism at the bottom wall is film boiling to the liquid filaments and convection to the superheated vapor phase, the upper wall transfers heat mainly by convection.

Cryogenic chilldown under microgravity condition was experimentally studied in the dispersed flow film boiling regime. During the film boiling, the bottom wall heat flux is found to decrease under microgravity condition. Under current experimental condition, the gravity effect does not show a strong dependence on wall temperature and inlet flow rate.

ACKNOWLEDGEMENTS

The support by the Andrew H. Hines, Jr./Progress Energy Endowment Fund is acknowledged.

NOMENCLATURE

a	Acceleration [ms^{-2}]
c	Specific heat [J $kg^{-1}K^{-1}$]
D	Diameter of the flow channel [m]
g	Gravitational acceleration constant 9.8[ms^{-2}]
h	Heat transfer coefficient [$Wm^{-2}K^{-1}$]; Enthalpy [Jkg^{-1}]; Water level [m]
h_{lv}	latent heat of evaporation[$Jkg^{-1}K^{-1}$]
h'_{lv}	Latent heat plus vapor sensible heat content [$Jkg^{-1}K^{-1}$]
k	Thermal conductivity [$Wm^{-1}K^{-1}$]
q''	Heat flux [Wm^{-2}]
r	Radius of the flow channel [m]
t	Time [s]
T	Temperature [K]
ΔT_w	wall superheat [K]

Greek letters

α	Thermal diffusivity [m^2s^{-1}]
ε	Emissivity
μ	Viscosity [$kgm^{-1}s^{-1}$]
ρ	Density [$kg\ m^{-3}$]
σ	Surface tension [$N\ m^{-1}$]
σ_B	Stefen Boltzmann constant 5.67×10^{-8} [$W\ m^{-2}K^{-4}$]

Subscripts

CHF	Critical heat flux
cond	Conduction
conv	Convection
fb	Film boiling
i	Interfacial; Inner
l	Liquid phase
min	Minimum
o	Outer
rad	Radiation
tc	Thermocouple
v	Vapor phase
w	Wall

REFERENCES

Abdul-Razzak, A., Shoukri, M., and Chan, A. M. C., 1992, "Rewetting of Horizontal Tubes," *Nuclear Engineering and Design*, **138**, 375-388. http://dx.doi.org/10.1016/0029-5493(92)90072-4

Adham-Khodaparast, K., Xu, J. J., and Kawaji, M., 1995, "Flow Film Boiling Collapse and Surface Rewetting in Normal and Reduced Gravity Conditions," *International Journal of Heat and Mass Transfer*, **38**, 2749-2760. http://dx.doi.org/10.1016/0017-9310(95)00026-6

Andreani, M., and Yadigaroglu, G., 1996, "A 3-d Eulerian-Lagrangian Model of Dispersed Flow Film Boiling Including a Mechanistic Description of the Droplet Spectrum Evolution I. The Thermal-Hydraulic Model," *International Journal of Heat and Mass Transfer*, **40**, 1753-1772. http://dx.doi.org/10.1016/S0017-9310(96)00253-0

Antar, B. N., and Collins, F. G., 1997, "Flow Boiling during Quenching in Low Gravity Environment," *International Journal of Microgravity Science and Technology*, **3**, 118-128. http://dx.doi.org/10.1016/0094-4548(78)90048-6

Berenson, P. J., 1961, "Film-Boiling Heat Transfer from a Horizontal Surface," *ASME Journal of Heat Transfer*, **83**, 351-358. http://dx.doi.org/10.1115/1.3682280

Bergles, A. E., and Thompson, W. G., 1970, "The Relationship of Quench Data to Steady-State Pool Boiling Data," *International Journal of Heat and Mass Transfer*, **13**, 55-68. http://dx.doi.org/10.1016/0017-9310(70)90024-4

Breen, B. P., and Westwater, J. W., 1962, "Effect of Diameter of Horizontal Tubes on Film Boiling Heat Transfer," *Chemical Engineering Progress,* **58**, 67-72.

Brentari, E. G., Giarratano, P. J., and Smith, R.V., 1965, *Boiling Heat Transfer for Oxygen, Nitrogen, Hydrogen, and Helium, NBS Technical Note No. 317*, National Bureau of Standards.

Brentari, E. G., and Smith, R. V., 1965, "Nucleate and Film Pool Boiling Design Correlations for O2, N2, H2 and He," *Advances in Cryogenic Engineering,* **10**, 325-341. http://dx.doi.org/10.1115/1.3244455

Bronson, J. C., Edeskuty, F.J., Fretwell, J.H., Hammel, E.F., Keller, W.E., Meier, K.L., Schuch, A.F., and Willis, W.L., 1962, "Problems in Cool-Down of Cryogenic Systems," *Advances in Cryogenic Engineering,* **7**, 198-205. http://dx.doi.org/10.1007/978-1-4757-0531-7_25

Burggraf, O. R., 1964, "An Exact Solution of Inverse Problem in Heat Conduction Theory and Application," *ASME Journal of Heat Transfer*, **86**, 373-382.
http://dx.doi.org/10.1115/1.3688700

Burke, J. C., Byrnes, W. R., Post, A. H., and Ruccia, F. E., 1960, "Pressure Cooldown of Cryogenic Transfer Lines," *Advances in Cryogenic Engineering*, **4**, 378-394.
http://dx.doi.org/10.1007/978-1-4757-0540-9_33

Carey, V. P., 1992, *Liquid-vapor phase-change phenomena,* Taylor & Francis Press, NewYork .

Chan, A. M. C., 1995, "Stratified Flow Film Boiling inside Horizontal Tubes," *ASME Journal of Heat Transfer,* **117**, 179-184.
http://dx.doi.org/10.1115/1.2822300

Chan, A. M. C., and Banerjee, S., 1981a, "Refilling and Rewetting of a Hot Horizontal Tube Part I: Experiment," *ASME Journal of Heat Transfer*, **103**, 281-286.
http://dx.doi.org/10.1115/1.3244454

Chan, A. M. C., and Banerjee, S., 1981b, "Refilling and Rewetting of a Hot Horizontal Tube Part II: Structure of a Two-Fluid Model," *ASME Journal of Heat Transfer,* **103**, 287-292.
http://dx.doi.org/10.1115/1.3244454

Chan, A. M. C., and Banerjee, S., 1981c, "Refilling and Rewetting of a Hot Horizontal Tube Part III: Application of a Two-Fluid Model to Analyze Rewetting," *ASME Journal of Heat Transfer*, **103**, 653-659.
http://dx.doi.org/10.1115/1.3244454

Chang, Y. P., 1959, "Wave Theory of Heat Transfer in Film Boiling," *ASME Journal of Heat Transfer,* **8**, 1-12.

Chang, Y. P., and Snyder, N. W., 1960, "Heat Transfer in Saturated Boiling," *Chemical Engineering Progress Symposium*, **56**, 25-38.

Chen, W. J., Lee, Y., and Groeneveld, D. C., 1979, "Measurement of Boiling Curves during Rewetting of a Hot Circular Duct," *International Journal of Heat and Mass Transfe,* **22**, 973-976.
http://dx.doi.org/10.1016/0017-9310(79)90039-5

Cheng, S. C., 1978, "Transition Boiling Curves Generated from Quenching Experiments using a Two Dimensional Model," *Letters in Heat and Mass Transfer,* **5**, 391-403.
http://dx.doi.org/10.1016/0735-1933(78)90019-2

Cheng, S. C., Ng, W. W. L., and Heng, K. T., 1978, "Measurements of Bboiling Curves of Subcooled Water under Forced Convective Conditions," *International Journal of Heat and Mass Transfer,* **21**, 1385-1392.
http://dx.doi.org/10.1016/0017-9310(78)90202-8

Cheng, S. C., and Ragheb, H., 1979, "Transition Boiling Data of Water on Inconel Surface under Forced Convective Conditions," *International Journal of Multiphase Flow,* **5**, 281-291.
http://dx.doi.org/10.1016/0301-9322(79)90026-0

Chi, J. W. H., 1965, "Cooldown Temperatures and Cooldown Time during Mist Flow," *Advances in Cryogenic Engineering*, **10**, 330-340.
http://dx.doi.org/10.1007/978-1-4684-3108-7_40

Chi, J. W. H., and Vetere, A.M., 1964, "Two-Phase Flow during Transient Boiling of Hydrogen and Determination of Nonequilibrium Vapor Fractions," *Advances in Cryogenic Engineering,* **9**, 243-253.
http://dx.doi.org/10.1007/978-1-4757-0525-6_29

Chung, J. N., and Olafsson, S. I., 1984, "Two-Phase Droplet Flow Convective and Radiative Heat Transfer," *International Journal of Heat and Mass Transfer*, **27**, 901-910.
http://dx.doi.org/10.1016/0017-9310(84)90011-5

Dhir, V. K., Duffey R. B. and Catton I., 1981, "Quenching Studies on a Zircaloy Rod Bundle," *ASME Journal of Heat Transfer*, **103**, 293-299.
http://dx.doi.org/10.1115/1.3244456

Duffey, R. B. and Porthouse, D. T. C., 1973, "The Physics of Rewetting in Water Reactor Emergency Core Cooling," *Nuclear Engineering and Design*, **25**, 379-394.
http://dx.doi.org/10.1016/0029-5493(73)90033-2

Ganić, E. N., Rohsenow, W. M., 1977, "Dispersed Flow Heat Transfer," *Intentional Journal of Heat and Mass Transfer,* **20**, 855-866.
http://dx.doi.org/10.1016/0017-9310(77)90115-6

Gebhart, B., Jaluria, Y., Mahajan, R.L., Sammakia, B., 1988, *Buoyancy-Induced Flow and Heat Transport*, Hemisphere Publishing Company.

Gungor, K.E. and Winterton, R.H.S., 1987, "Simplified General Correlation for Saturated Flow Boiling and Comparisons of Correlations with Data," *Chemical Engineering Research and Design*, **65**, 148-156.

Hammouda, N., Groeneveld, D. C., and Cheng, S. C., 1997, "Two-Fluid Modeling of Inverted Annular Film Boiling," *International Journal of Heat and Mass Transfer,* **40**, 2655-2670.
http://dx.doi.org/10.1016/S0017-9310(96)00278-5

Hsu, Y. Y., and Graham, R. W., 1963, *A Visual Study of Two-Phase Flow in a Vertical Tube with Heat Addition*, NASA-TN-1564.

Iloeje, O. C., Plummer, D. N., Rohsenow, W. M. and Griffith, P., 1975, "An Investigation of the Collapse and Surface Rewet in Film Boiling in Forced Vertical Flow," *ASME Journal of Heat Transfer*, **42**, 3387-3407.
http://dx.doi.org/10.1115/1.3450336

Incropera, F. P., and Dewitt, D. P., 2002, *Fundamentals of heat and mass transfer*, 5th edition. John Wiley & Sons, New York.

Kawaji, M., Ng, Y.S., Banerjee, S., and Yadigaroglu, G., 1985, "Reflooding with Steady and Oscillatory Injection: Part I-Flow Regimes, Void Fraction, and Heat Transfer," *ASME Journal of Heat Transfer*, **107**, 670-678.
http://dx.doi.org/10.1115/1.3247476

Klimenko, V.V., Fyodorov, M.V., and Fomichyov, Yu. A., 1989, "Channel Orientation and Geometry Influence on Heat Transfer with Two-Phase Forced Flow of Nitrogen," *Cryogenics*, **29**, 31-36.
http://dx.doi.org/10.1016/0011-2275(89)90008-8

Kutateladze, S.S., 1952, "Heat Transfer in Condensation and Boiling," *State Sci. and Tech. Pub. Of Lit. on Machinery*, Moscow (Atomic Energy Commission Translation 3770, Tech. Info Service, Oak Ridge, Tennessee.)

Liao, J., 2005, *Modeling Two-Phase Transport during Cryogenic Chilldown in a Pipeline*, Ph.D. dissertation, University of Florida.

Lin C-S, Van Dresar, N. T., and Hasan, M. M., 1991, *A Pressure Control Analysis of Cryogenic Storage Systems,* NASA TM 104409.

Merte, Jr., H., and Clark, J. A., 1964, "Boiling Heat Transfer with Cryogenic Fluids at Standard, Fractional, and Near-Zero Gravity," *ASME Journal of Heat Transfer*, **86**, 315-319.
http://dx.doi.org/10.1115/1.3688689

Noyes, R. C.,1963, "An Experimental Study of Sodium Pool Boiling Heat Transfer," *ASME Journal of Heat Transfer*, **85**, 125-132.
http://dx.doi.org/10.1115/1.3686029

Piggott, B. D. G., and Porthouse, D. T. C., 1975, "A Correlation of Rewetting Data," *Nuclear Engineering and Design*, **32**, 171-181.
http://dx.doi.org/10.1016/0029-5493(75)90128-4

Raithby, G. D., and Hollands, K. G. T., 1975, "A General Method of Obtaining Approximate Solutions to Laminar and Turbulent Free Convection Problems," *Advances in Heat Transfer*, **11**, 265-315, Academic Press, New York.
http://dx.doi.org/10.1016/S0065-2717(08)70076-5

Seader, J. D., Miller, W. S., and Kalvinskas, L.A., 1965, *Boiling heat transfer for cryogenics,* NASA Contractor Report CR-243.

Shah, M. M., 1984, "Prediction of Heat Transfer during Boiling of Cryogenic Fluids Flowing in Tubes," *Cryogenics,* **24**, 231-236.
http://dx.doi.org/10.1016/0011-2275(84)90148-6

Shah, M. M. and Siddiqui, M. A., 2000, "A General Correlation for Heat Transfer during Dispersed-Flow Film Boiling in Tubes," *Heat Transfer Engineering*, **21**, 18-32.
http://dx.doi.org/10.1080/01457630050144479

Thompson, T. S., 1974, "Process of Rewetting a Hot Surface by a Falling Liquid–Film," *Nuclear Engineering and Design*, **31**, 234-245.
http://dx.doi.org/10.1016/0029-5493(75)90144-2

Van Dresar, N. T. and Siegwarth, J. D., 2001, *Near-Horizontal, Two-Phase Flow Patterns of Nitrogen and Hydrogen at Low Mass and Heat Flux,* NASA TP 2001-210380.

Van Dresar, N.T., Siegwarth, J.D., and Hasan, M.M., 2002, "Convective Heat Transfer Coefficients for Near-Horizontal Two-Phase Flow of Nitrogen and Hydrogen at Low Mass and Heat Flux," *Cryogenics*, **41**, 805-811.
http://dx.doi.org/10.1016/S0011-2275(01)00173-4

Velat, C., 2004, *Experiments in cryogenic two phase flow,* Master thesis, University of Florida.

Westbye, C. J., Kawaji, M., and Antar B.N., 1995, "Boiling Heat Transfer in the Quenching of a Hot Tube under Microgravity," *AIAA Journal of Thermophysics and Heat Transfer,* **9**, 302-307.
http://dx.doi.org/10.2514/3.660

Wright, C. C., and Walters, H. H., 1959, *Single tube heat transfer tests, gaseous and liquid hydrogen*, WADC Technical Report 59, 423.

Xu, J. J., 1998, *Flow Boiling Heat Transfer in the Quenching of a Hot Surface under Reduced Gravity Conditions,* Ph.D. dissertation, University of Toronto.

Yadigaroglu, G., Nelson, R. A., Teschendorff, V., Murao, Y., Kelly, J., and Bestion, D., 1993, "Modeling of Reflooding," *Nuclear Engineering and Design,* **145**, 1-35.
http://dx.doi.org/10.1016/0029-5493(93)90056-F

Yamanouchi, A., 1968, "Effect of Core Spray Cooling in Transient State after Loss of Coolant Accident," *Journal of Nuclear Science and Technology*, **5**, 547-558.
http://dx.doi.org/10.1080/18811248.1968.9732504

Yuan, K., 2006, *Cryogenic Boiling and Two-Phase Flow Chilldown Process under Terrestrial and Microgravity Conditrions,* Ph.D. dissertation, University of Florida.

Zuber, N., 1958, "On Stability of Boiling Heat Transfer," *ASME Journal of Heat Transfer*, **80**, 711-720.
http://dx.doi.org/10.1115/1.3688707

Zuber, N., 1959, *Hydrodynamic aspects of boiling heat transfer,* Ph.D. dissertation, UCLA.

THE EFFECT OF MELTING ON MIXED CONVECTION HEAT AND MASS TRANSFER IN NON-NEWTONIAN NANOFLUID SATURATED IN POROUS MEDIUM

R.R. Kairi [a] and Ch. RamReddy[b, *]

[a] *Department of Mathematics, Islampur College, Islampur, West Bengal, India – 733 202*
[b] *Department of Mathematics, National Institute of Technology Warangal, India – 506 004*

ABSTRACT

In this paper, we investigated the influence of melting on mixed convection heat and mass transfer from the vertical flat plate in a non-Newtonian nanofluid saturated porous medium. The wall and the ambient medium are maintained at constant, but different, levels of temperature and concentration. The Ostwald–de Waele power-law model is used to characterize the non-Newtonian nanofluid behavior. A similarity solution for the transformed governing equations is obtained. The numerical computation is carried out for various values of the non-dimensional physical parameters. The variation of temperature, concentration, heat and mass transfer coefficients with the power-law index, mixed convection parameter, melting parameter, Brownian motion parameter, thermophoresis parameter, buoyancy ratio and Lewis number are discussed in a wide range of values of these parameters.

Keywords: *mixed convection, melting, non-Newtonian, nanafluid, Brownian motion, thermophoresis.*

1. INTRODUCTION

A nanofluid is a stable, uniform suspension of nanometer sized solid particles (< 100 nm) in conventional liquids such as water and ethylene glycol. An innovative procedure for improving heat transfer by using ultra fine solid particles in fluids has been used extensively during the last several years. For example, the thermal conductivity of copper oxide is about 100 times greater than that of water. Thus, the thermal conductivity of a suspension containing solid particles could be expected to be significantly greater than that of the base fluid. Xuan and and Roetzel (2000) proposed a homogeneous flow model where the convective transport equations of pure fluids are directly extended to nanofluids. The boundary layer flow in nanofluds has been analyzed recently by Nield and Kuznetsov (2009a, 2009b). Hojjat *et al.,* (2011) experimentally investigated the laminar convective heat transfer of non-Newtonian nanofluids with constant wall temperature. Gorla *et al.,* (2011) also analyzed mixed convective boundary layer flow over a vertical wedge embedded in a porous medium saturated with a nanofluid. They conclude that thermophoresis and buoyancy ratio enhanced the heat transfer rate while the same reduced the mass transfer rate. Moreover, the Brownian motion parameter enhanced mass transfer rate whereas the heat transfer rate decreased with the same. The effects of viscous dissipation on mixed convection heat and mass transfer along a vertical plate embedded in a nanofluid-saturated non-Darcy porous medium have been investigated by Rashad *et al.* (2014).

The study of melting phenomena in porous media has received much attention in recent years because of its important applications in casting, welding and magma solidification, permafrost melting and thawing of frozen ground etc. Epstein and Cho (1976) studied the melting heat transfer from a flat plate in a steady laminar case, while Kazmierczak *et al.,* (1986, 1987) analyzed melting from a vertical flat plate embedded in a porous medium in both free and forced convection processes. The heat transfer at the melting surface in the laminar boundary layer by using Karman-Pohlhausen method was discussed by Pozvonkov *et al.,* (1970). Bakier (1997) studied the melting effect on mixed convection from a vertical plate of arbitrary wall temperature both in aiding and opposing flows in a fluid saturated porous medium while Gorla *et al.,* (1999) considered a similar study with uniform wall temperature conditions.

Non-Newtonian power-law fluids are so widespread in industrial processes and in the environment that it would be no exaggeration to affirm that Newtonian shear flows are the exception rather than the rule. Natural convection in a non-Newtonian fluid about a vertical wall and that around horizontal cylinder and sphere in a porous medium was presented by Chen and Chen (1988(a),1988(b)), respectively. Nakayama and Koyama (1991) analyzed the more general case of free convection over a non-isothermal body of arbitrary shape embedded in a porous medium. Poulikakos and Spatz (1988) analyzed the melting phenomena on free convection from a vertical front in a non-Newtonian fluid saturated porous matrix. Recently Kairi and Murthy (2012) explored the melting effect of a non-Newtonian fluid on mixed convection with Soret effects in a non-Darcy porous medium. The influence of yield stress on free convective boundary layer flow of non-Newtonian nanofluids over a vertical plate in a porous medium was discussed by Hady *et al.,* (2011). Rashad *et al.,* (2011) investigated the boundary layer flow of a non-Newtonian nanofluid over a permeable vertical cone embedded in a porous medium. Since there are few investigations in the recent literature about the convective heat and mass transport in non-Newtonian nanofluids saturated with or without porous medium. This motivated us to investigate the significance of melting effect on mixed convection in a porous medium saturated with non-Newtonian nanofluids.

The main purpose of the present investigation is to analyze the effect of melting in non-Newtonian nanofluids on mixed convection in porous medium. The effects of Brownian motion and thermophoresis are included. Using Matlab BVP solver bvp4c, which is a finite difference code that implements the 3-stage Lobatto IIIa formula, a numerical solution of the boundary layers equations is obtained.

2. Mathematical Formulation

Consider the mixed convective heat and mass transfer from a vertical plate embedded in a Darcy porous medium saturated with a non-Newtonian nanofluid. It is assumed that the plate constitutes the interface between the liquid phase and the solid phase during melting inside the porous matrix. The co-ordinate system and flow model are shown in the fig. 1.

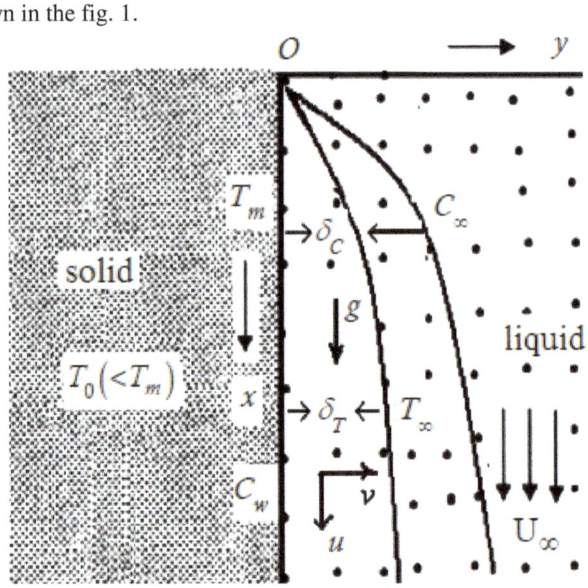

Fig. 1 Physical model and coordinate system (Kairi and Murthy 2012)

The x - coordinate is taken along the plate, the y - coordinate is measured normal to the plate, while the origin of the reference system is taken at the leading edge of the plate. The plate is at a constant temperature T_m at which the material of the porous matrix melts. The liquid phase temperature is $T_\infty(>T_m)$ and the temperature of the solid far from the interface is $T_0(<T_m)$. The concentration at the wall is C_w and the surrounding porous medium is maintained at constant concentration C_∞. The flow is steady, laminar and two dimensional. With the usual boundary layer and linear Boussinesq approximations, the governing equations, namely the equations of continuity, flow, energy and concentration for the isotropic and homogeneous porous medium may be written as

$$\frac{\partial u}{\partial x}+\frac{\partial v}{\partial y}=0 \tag{1}$$

$$\frac{\partial u^n}{\partial y}=-\frac{(1-C_\infty)\rho_{f_\infty}\beta gK}{\mu}\frac{\partial T}{\partial y}+\frac{(\rho_p-\rho_{f_\infty})gK}{\mu}\frac{\partial C}{\partial y} \tag{2}$$

$$u\frac{\partial T}{\partial x}+v\frac{\partial T}{\partial y}=\alpha\frac{\partial^2 T}{\partial y^2}+\tau\left[D_B\frac{\partial C}{\partial y}\frac{\partial T}{\partial y}+\frac{D_T}{T_\infty}\left(\frac{\partial T}{\partial y}\right)^2\right] \tag{3}$$

$$\frac{1}{\varepsilon}\left(u\frac{\partial C}{\partial x}+v\frac{\partial C}{\partial y}\right)=D_B\frac{\partial^2 C}{\partial y^2}+\frac{D_T}{T_\infty}\frac{\partial^2 T}{\partial y^2} \tag{4}$$

where $\alpha=\frac{k}{(\rho c)_f}$ and $\tau=\frac{\varepsilon(\rho c)_p}{(\rho c)_f}$, u and v are the Darcian velocity components along the x and y directions, T and C are the temperature and concentration, respectively, n is the power law index ($n<1$, $n=1$ and $n>1$, respectively, Pseudoplastic, Newtonian and Dilatant fluids). The parameters ρ_f, μ and β, respectively, are the density, viscosity and thermal expansion of the fluid while ρ_p is the density of the particle, g is the acceleration due to gravity, the effective heat capacity is $(\rho c)_f$ and effective thermal conductivity of the porous medium is k, D_B is the Brownian diffusion coefficient and D_T is thermophoretic diffusion coefficient.

The boundary conditions necessary to complete the problem formulation are written as

$$y=0:\quad k\frac{\partial T}{\partial y}=\rho_\infty\left[L+c_s(T_m-T_0)\right]v,\ T=T_m,\ C=C_w \tag{5}$$

$$y\to\infty:\quad u\to U_\infty,\ T\to T_\infty,\ C\to C_\infty \tag{6}$$

where L and c_s are latent heat of the solid and the specific heat capacity of the solid phase, respectively and U_∞ is the uniform free stream velocity.

The boundary condition (5) at the interface states that the temperature of the plate is equal to the melting temperature of the material saturating the porous matrix and the other condition means that the heat conducted to the melting surface is equal to the sum of heat of melting and the sensible heat required to raise the temperature of the solid, T_0 to its melting temperature T_m. It is important to note that eq. (5) is consistent with a coordinate system fixed to the melting surface, so that the interior of the solid appears to move towards the (stationary) melting surface with constant velocity equal to the melting velocity $v(x,0)$. In the present formulation, transient effects in the solid have been neglected. This assumption is valid as long as the melting solid is large compared to its thermal boundary layer thickness [see Epstein and Cho (1976)].

The continuity equation is automatically satisfied by defining a stream function $\psi(x,y)$ such that $u=\frac{\partial\psi}{\partial y}$ and $v=-\frac{\partial\psi}{\partial x}$. We introduce the following similarity transformation

$$\eta=\frac{y}{x}\chi^{-1}Pe_x^{1/2},\ \psi(\eta)=\alpha\chi^{-1}Pe_x^{1/2}f(\eta),\ \theta(\eta)=\frac{T-T_m}{T_\infty-T_m},$$

$$\phi(\eta)=\frac{C-C_w}{C_\infty-C_w},$$

where

$$\chi^{-1}=1+\sqrt{\frac{Ra_x}{Pe_x}},\ Pe_x=\frac{U_\infty x}{\alpha}\ \text{and}\ Ra_x=\frac{x}{\alpha}\left[\frac{(1-C_\infty)\rho_{f_\infty}Kg\beta(T_\infty-T_m)}{\mu}\right]^{1/n}$$

The above transformation reduces the system of partial differential equations into the following system of non-linear ordinary differential equations:

$$nf'^{\,n-1}f''=-(1-\chi)^{2n}(\theta'-N_r\phi') \tag{7}$$

$$\theta''+\frac{1}{2}f\theta'+N_b\theta'\phi'+N_t\theta'^2=0 \tag{8}$$

$$\phi''+\frac{Le}{2}f\phi'+\frac{N_t}{N_b}\theta''=0 \tag{9}$$

The boundary conditions are

$$\eta=0:\ f+2M\theta'=0,\ \theta=0,\ \phi=0 \tag{10}$$

$$\eta\to\infty:\ f'\to\chi^2,\theta\to 1,\phi\to 1 \tag{11}$$

In the above equations $M=\frac{c_f(T_\infty-T_m)}{L+c_s(T_m-T_0)}$ is the melting parameter, $Le=\alpha/D$ is the Lewis number, $N_r=\frac{(\rho_p-\rho_{f_\infty})(C_w-C_\infty)}{\rho_{f_\infty}\beta(T_m-T_\infty)(1-C_\infty)}$ is the buoyancy ratio, $N_t=\frac{\varepsilon(\rho c)_p D_T(T_m-T_\infty)}{(\rho c)_f\alpha T_\infty}$ is thermophoresis parameter and $N_b=\frac{\varepsilon(\rho c)_p D_B(C_w-C_\infty)}{(\rho c)_f\alpha}$ is Brownian motion parameter. We note that $\chi=0$ corresponds to pure free convection and $\chi=1$ corresponds to pure forced convection processes.

The local Nusselt number Nu_x and Sherwood number Sh_x are physical properties. The non-dimensional heat and mass transfer coefficients are defined as

$$Nu_x / \chi^{-1} Pe_x^{1/2} = \theta'(0) \quad (12)$$

$$Sh_x / \chi^{-1} Pe_x^{1/2} = \phi'(0) . \quad (13)$$

3. Results and discussion

The resulting ordinary differential equations (7)-(9) along with the boundary conditions (10)-(11) are solved using Matlab BVP solver bvp4c, which is a finite difference code that implements the 3-stage Lobatto IIIa formula. The integration length η_∞ varies with the parameter values and it has been suitably chosen each time such that the boundary conditions at the outer edge of the boundary layer are satisfied. The results obtained here are accurate up to the 4th decimal place. At first, the accuracy of the numerical solution was tasted by accurately producing the findings of Cheng and Minkowycz (1977) for the case of a Newtonian fluid without melting ($M = 0, n = 1$), then the findings of Chen and Chen (1988) for the case of natural convection of (i) a Pseudoplastic fluids without melting, (ii) a Newtonian fluids without melting, (iii) a Dilatant fluids without melting, and (iv) a Newtonian fluid with melting and they are shown in the Table (1). Also, the present results for the Nusselt and Sherwood numbers are compared with those obtained by Kairi and Murthy (2012) in the absence of Brownian motion and thermophoresis parameter and they are found to be good agreement, but the details are not shown here for brevity. The following values are considered for the parameters: $0.5 \le n \le 1.5$, $0 \le \chi \le 1$, $0 \le M \le 1$, $0.1 \le N_r \le 0.5$, $0.0 \le N_b \le 1.0$, $0.1 \le N_t \le 1$ and $0.1 \le Le \le 5$. It was reported that in case of non-Newtonian fluids, heat and mass transfer coefficients decreased with increases in the melting parameter (Kairi and Murthy 2012). The effect of these parameters in power law nanofluids on the velocity, temperature heat and transfer coefficient is presented through some selected values of these parameters. Their combined effects on the heat and mass transfer coefficients are discussed.

Table 1. *Comparison of values of $-\theta'(0)$ for free convection along a vertical plate in Newtonian and non-Newtonian fluid saturated porous medium in the absence of Brownian motion and thermophoresis parameter with $\chi = 0$, $N_r = 0$ and $Le = 0$.*

S.No.	Fluid type with/ without melting effects	Cheng and Minkowycz (1977)	Chen and Chen(1988)	Present
(i)	$M = 0, n = 0.5$	-----------	0.3768	0.37682
(ii)	$M = 0, n = 1$	0.4440	0.4437	0.44370
(iii)	$M = 0, n = 1.5$	-----------	0.4752	0.47525
(iv)	$M = 1, n = 1$	-----------	0.2910	0.29103

Figures 2 and 3 show that the variations of non-dimensional temperature θ and concentration ϕ against the similarity variable η for different values of thermophoresis parameter N_t and melting parameter M with fixed values of N_r, N_b, Le and χ. It is observed that as increasing in the thermopohoresis parameter, the temperature boundary layer thickness increases in the presence of melting parameter and it is also noted that with influence of thermopnoresis parameter an increment in the values of melting parameter the temperature profile decreases for both pseudoplastic and dilatant fluids. Further, the concentration decreases with increase of melting and thermophoresis parameter for both pseudoplastic and dilatant fluids.

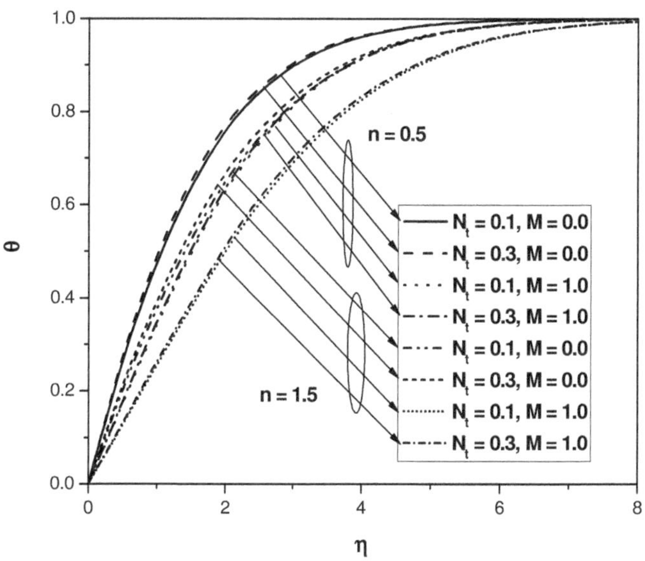

Fig. 2 Variation of temperature profiles for $n = 0.5$ and $n = 1.5$ with the similarity variable η varying N_t and M with fixed $\chi = 0.5$, $N_r = 1$, $N_b = 0.5$ and $Le = 1$.

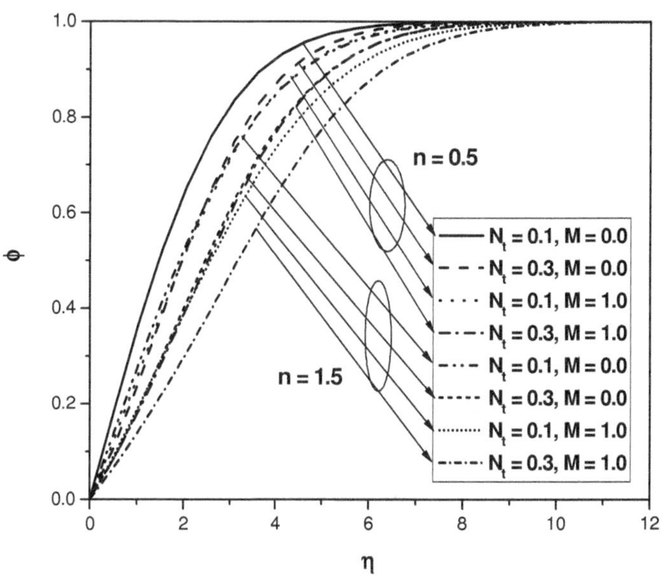

Fig. 3 Variation of concentration profiles for $n = 0.5$ and $n = 1.5$ with the similarity variable η varying N_t and M with fixed $\chi = 0.5$, $N_r = 1$, $N_b = 0.5$ and $Le = 1$.

The variations of the non-dimensional temperature θ and concentration distribution ϕ against the similarity variable η for different values of N_b and M are shown in Figs. 4 and 5, respectively. It is noted that for both pseudoplastic and dilatants fluids, the temperature and concentration boundary layer thicknesses increase as the value of N_b rises in the presence of melting parameter and the temperature and concentration decreases with increase of melting parameter with influence of Brownian motion.

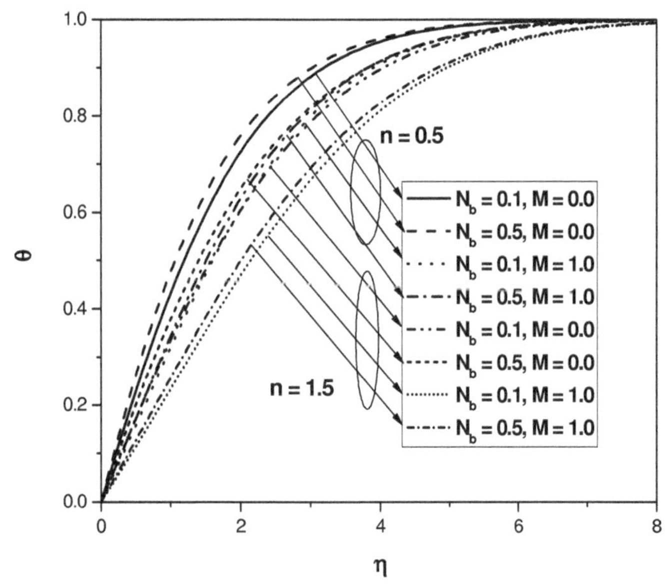

Fig. 4 Variation of temperature profiles for $n = 0.5$ and $n = 1.5$ with the similarity variable η varying N_b and M with fixed $\chi = 0.5$, $N_r = 1$, $N_t = 0.3$ and $Le = 1$.

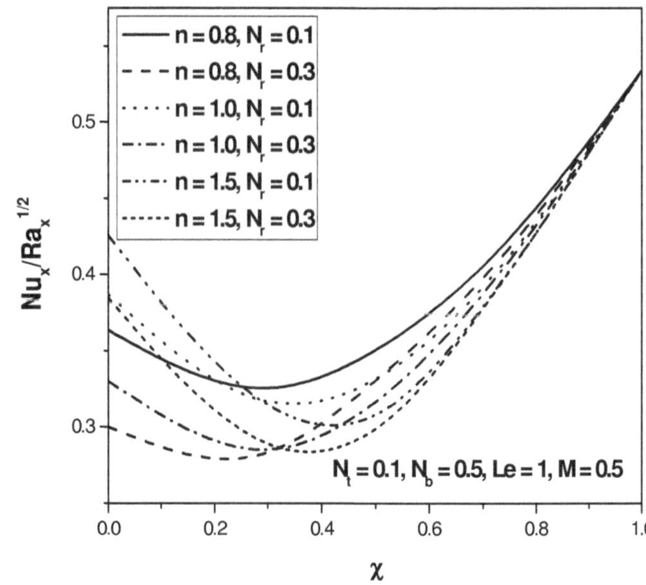

Fig. 6 Variation of heat transfer coefficient against χ varying n and N_r with fixed $Le = 1$, $N_b = 0.5$, $N_t = 0.1$ and $M = 0.5$.

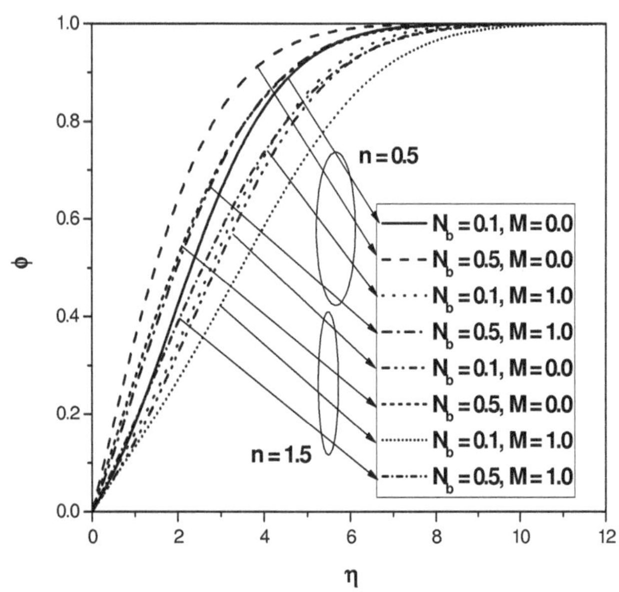

Fig. 5 Variation of concentration profiles for $n = 0.5$ and $n = 1.5$ with the similarity variable η varying N_b and M with fixed $\chi = 0.5$, $N_r = 1$, $N_t = 0.3$ and $Le = 1$.

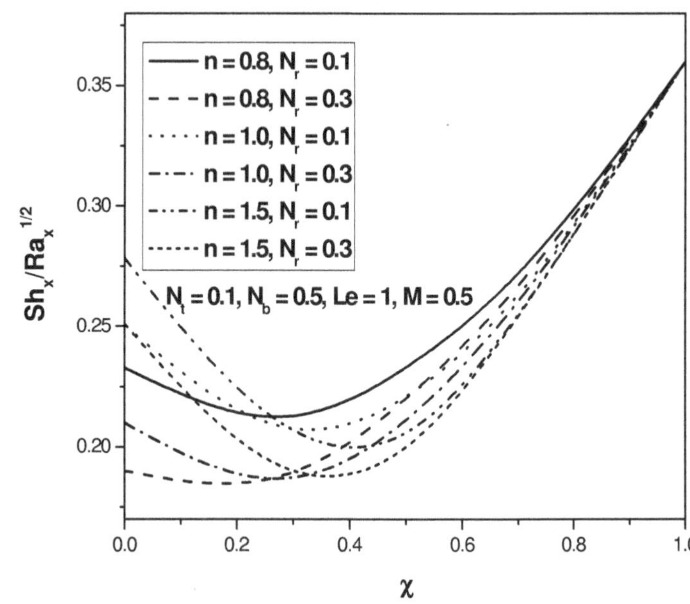

Fig. 7 Variation of mass transfer coefficient against χ varying n and N_r with fixed $Le = 1$, $N_b = 0.5$, $N_t = 0.1$ and $M = 0.5$.

In figures 6 and 7, the variations of heat and mass transfer coefficients are plotted against χ varying n and N_r with fixed values of other parameters. It is seen that heat and mass transfer coefficients are decreasing with increasing buoyancy ratio, N_r for pseudopastic, Newtonian and dilatants fluids. Similar results were reported by Kairi and Murthy (2012) while investigating the mixed convection heat and mass transfer of non-Newtonian fluids from a vertical surface embedded in porous medium.

Figures 8 and 9 prepared to illustrate the variations of the Nusselt and Sherwood numbers against Le for different values of M and N_t with fixed value of other parameters. It is observed that the with influence of melting parameter, the Nusselt number increases but, Sherwood number decreases with thermophoresis parameter, also it can be noted that the Nusselt and Sherwood number decrease with increase of melting parameter for both pseudoplastic and dilatant fluids. Moreover the thermophoresis effect on both heat and mass transfer rate is prominent for the pseudoplastics than that of dilatant nanofluids.

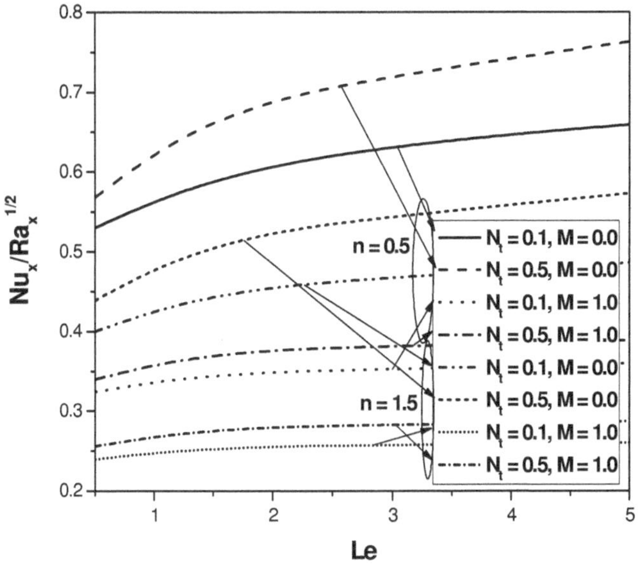

Fig. 8 Variation of heat transfer coefficient against Le varying n, M and N_t with fixed $N_b = 0.5$, $N_r = 0.1$ and $\chi = 0.5$.

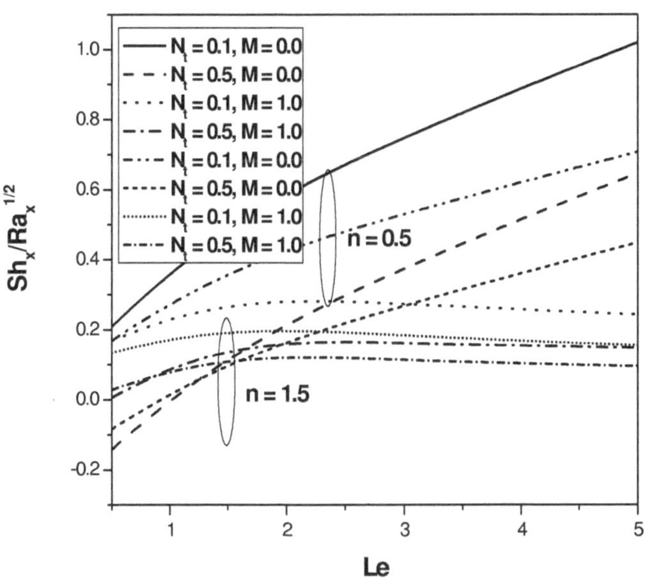

Fig. 9 Variation of mass transfer coefficient against Le varying n, M and N_t with fixed $N_b = 0.5$, $N_r = 0.1$ and $\chi = 0.5$.

Figures 10 and 11 prepared to show that the variation of heat and mass transfer coefficients for pseudoplastic and dilatant fluids against Le for varying M and N_b, and fixed value of other parameters. In the presence of melting parameter, the Nusselt and Sherwood number increases with increase of Brownian motion parameter N_b, and Lewis number Le for all types of non-Newtonian nanofluids. On the other hand, Nusselt and Sherwood number decrease with increase of melting parameter in the presence of thermophoresis for larger values of Le. The influences of the Brownian motion parameter on Nusselt and Sherwood numbers are superior for the pseudoplastic nanofluids with compared to dilatant nanofluids saturated porous medium.

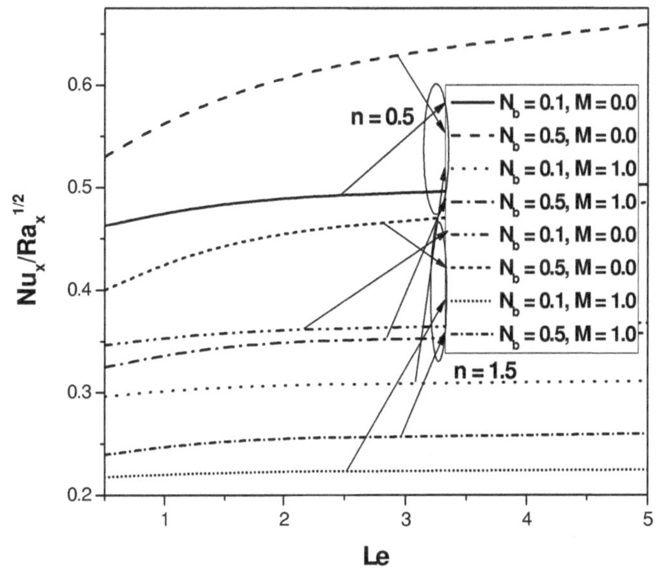

Fig. 10 Variation of heat transfer coefficient against Le varying n, M and N_b with fixed $N_t = 0.3$, $N_r = 0.1$ and $\chi = 0.5$.

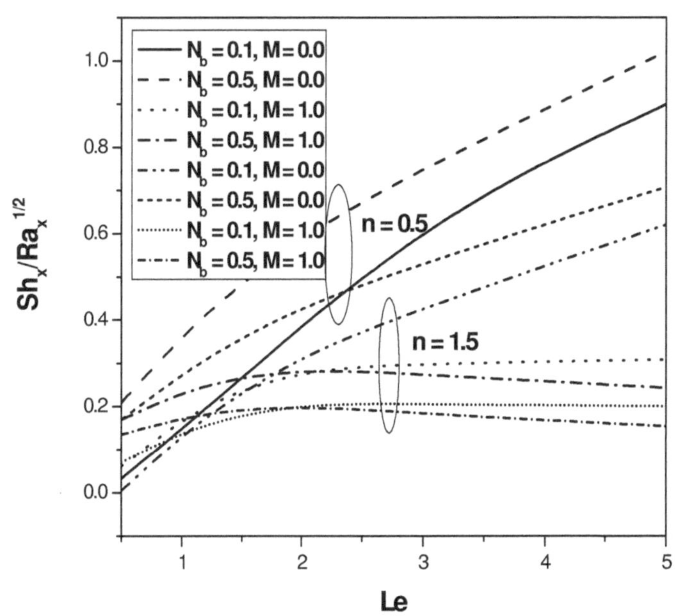

Fig. 11 Variation of mass transfer coefficient against Le varying n, M and N_b with fixed $N_t = 0.3$, $N_r = 0.1$ and $\chi = 0.5$.

4. CONCLUSIONS

In this study the effect of melting on mixed convection heat and mass transfer from a vertical flat plate in a porous medium saturated with non-Newtonian nanofluids is analyzed. It is noted that the temperature and concentration profiles as well as the heat and mass transfer coefficients are significantly affected by melting, thermophoresis and Brownian diffusion. The heat and mass transfer coefficients decrease with increasing M in the whole range of mixed convection for all types of non-Newtonian nanofluids. It is also noted that heat transfer coefficients increases but, mass transfer coefficient decrease with increasing buoyancy ratio N_r in the whole range of mixed convection for all types of non-Newtonian nanofluids. Finally, thermophoresis and Brownian motion effects on Nusselt and Sherwood numbers are prominent in pseudoplastic nanofluids when compared to dilatant nanofluids saturated porous medium.

ACKNOWLEDGEMENTS

The authors are thankful to the reviewers for their valuable suggestions and comments for improving the clarity of the manuscript. The financial supports received from the MHRD Research Seed Grant (RSG)Scheme 2014-2015 at NIT Warangal is gratefully acknowledged.

NOMENCLATURE

c_f	specific heat capacity of the convective fluid
c_s	specific heat capacity of the solid phase
C	concentration
C_w	concentration at the plate
C_∞	concentration at the ambient medium
D_B	Brownian diffusion coefficient
D_T	thermophoretic diffusion coefficient
f	dimensionless stream function
g	acceleration due to gravity
k	effective thermal conductivity
K	permeability of the porous medium
L	latent heat of the solid
Le	Lewis number
M	Melting parameter
n	power law index
N_r	buoyancy ratio
N_b	Brownian motion parameter
N_t	thermophoresis parameter
Nu_x	local Nusselt number
Pe_x	local Peclet number
Ra_x	local Rayleigh number
Sh_x	local Sherwood number
T	temperature
T_m	melting temperature
T_0	temperature at the solid phase
T_∞	temperature at the ambient media
u, v	average velocity components in x and y directions
U_∞	free stream velocity
x, y	coordinates along and perpendicular to the wall

Greek Symbols

α	effective thermal diffusivity
β	coefficient of thermal expansion
χ	mixed convection parameter
ε	porosity of the porous medium
ϕ	dimensionless concentration
η	similarity variable
μ	fluid consistency of the inelastic non-Newtonian power – law fluid
θ	dimensionless temperature
ρ_f	nanofluid density
ρ_p	nano-particle mass density
$(\rho c)_p$	heat capacity of nano-particle
$(\rho c)_f$	heat capacity of fluid
ψ	dimensionless stream function optical penetration depth

REFERENCES

Bakier, A.Y., 1997, "Aiding and Opposing Mixed Convection Flow in Melting from a Vertical Flat Plate Embedded in a Porous Medium," *Transport in Porous Media*, **29**, 127-139. http://dx.doi.org/10.1023/A:1006539027308

Cheng, P., and Minkowycz, W.J., 1977 "Free Convection about a Vertical Flat Plate Embedded in a Porous Medium with Application to Heat Transfer from a Dike," *Journal of Geophysical Research*, 82, 2040-2044. http://dx.doi.org/10.1029/JB082i014p02040

Chen, H. T., and Chen, C.K., 1988(a) "Natural Convection of Non-Newtonian Fluids along a Vertical Plate Embedded in a Porous Medium," *ASME Journal of Heat Transfer*, **110**, 257 -260. http://dx.doi.org/10.1115/1.3250462

Chen, H. T., and Chen, C. K., 1988(b), "Natural Convection of a Non-Newtonian Fluid about a Horizontal Cylinder and Sphere in a Porous Medium," *International Communications in Heat and Mass Transfer*, **15**, 605 - 614. http://article/pii/0735193388900516

Epstein, M., and Cho, D.H., 1976, "Melting Heat Transfer in Steady Laminar Flow over a Flat Plate," *ASME Journal of Heat Transfer*, **98**, 531–533. http://dx.doi.org/10.1115/1.3450595

Gorla, R.S.R., Mansour, M. A., Hassanien, I. A. and Bakier, A.Y., 1999, "Mixed Convection Effect on Melting from a Vertical Plate in a Porous Medium," *Transport in Porous Media*, **36**, 245-254. http://dx.doi.org/10.1023/A:1006566924390

Gorla, R.S.R., Chamkha, A.J., and Rashad, A.M., 2011,"Mixed Convective Boundary Layer Flow over a Vertical Wedge Embedded in a Porous Medium Saturated with a Nanofluid: Natural Convection Dominated Regime," *Nanoscale Research Letters*, **6**, 207-215. http://dx.doi.org/10.1186/1556-276X-6-207

Hady, F.M., Ibrahim, F.S., Abdel-Gaied, S.M., and Eid, R.M., 2011, "Influence of Yield Stress on Free Convective Boundary-Layer Flow of a Non-Newtonian Nanofluid Past a Vertical Plate in a Porous Medium," *The Journal of Mechanical Science and Technology*, **25(8)**, 2043-2050. http://dx.doi.org/10.1007/s12206-011-0628-0

Hojjat, M., Etemad, S. Gh., Bagheri, R., and Thibault, J., 2011, "Laminar Convective Heat Transfer of Non-Newtonian Nanofluids with Constant Wall Temperature," *Heat Mass Transfer*, **47**, 203-209. http://dx.doi.org/10.1007/s00231-010-0710-7

Kairi, R. R., and Murthy, P. V. S. N., 2012, "Effect of Melting on Mixed Convection Heat and Mass Transfer in a Non-Newtonian Fluid Saturated Non-Darcy Porous Medium," *ASME Journal of Heat Transfer*, **134**, 1-8. http://dx.doi.org/10.1115/1.4003899

Kazmierczak, M., Poulikakos, D., and Pop, I., 1986, "Melting from a Flat Plate Embedded in a Porous Medium in the Presence of Steady Natural Convection," *Numerical Heat Transfer*, **10**, 571–581. http://dx.doi.org/10.1080/10407788608913536

Kazmierczak, M., Poulikakos, D., and Sadowski, D., 1987, "Melting of a Vertical Plate in Porous Medium Controlled by Forced Convection of a Dissimilar Fluid," *International Communication in Heat and Mass Transfer*, **14**, 507-517.
http://dx.doi.org/10.1016/0735-1933(87)90015-7

Nakayama, A., and Koyama, H., 1991, "Buoyancy Induced Flow of Non-Newtonian Fluids over a Non-Isothermal Body of Arbitrary Shape in a Fluid-Saturated Porous Medium," *Applied Scientific Research,* **48**, 55 - 70.
http://dx.doi.org/10.1007/BF01998665

Nield, D.A., and Kuznetsov, A.V., 2009a, "The Cheng-Minkowycz Problem for Natural Convective Boundary-Layer Flow in a Porous Medium Saturated by a Nanofluid," *International Journal of Heat and Mass Transfer*, **52**, 5792-5795.
http://dx.doi.org/10.1016/j.ijheatmasstransfer.2009.07.024

Nield, D.A., and Kuznetsov, A.V., 2009b, "Thermal Instability in a Porous Medium Layer Saturated by a Nanofluid," *International Journal of Heat and Mass Transfer*, **52**, 5796-5801.
http://dx.doi.org/10.1016/j.ijheatmasstransfer.2009.07.023

Poulikakos, D., and Spatz, T.L., 1988, "Non-Newtonian Natural Convection at a Melting Front in a Permeable Solid Matrix," *International Communication in Heat and Mass Transfer*, **15**, 593-603.
http://dx.doi.org/10.1016/0735-1933(88)90050-4

Pozvonkov, F.M., Shurgalskii, E. F., and Akselrod, L.S., 1970, "Heat Transfer at a Melting Flat Surface under Conditions of Forced Convection and Laminar Boundary Layer," *International Journal of Heat and Mass Transfer*, **13**, 957-962.
http://dx.doi.org/10.1016/0017-9310(70)90162-6

Rashad, A., EL-Hakiem, M., and Abdou, M.M.M., 2011, "Natural Convection Boundary Layer of a Non-Newtonian Fluid about a Permeable Vertical Cone Embedded in a Porous Medium Saturated with a Nanofluid," *Computers and Mathematics with Applications*, **62**, 3140-3151.
http://dx.doi.org/10.1016/j.camwa.2011.08.027

Rashad, A.M., Chamkha, A.J., RamReddy, Ch. and Murthy, P.V.S.N. 2014, "Influence of Viscous Dissipation on Mixed Convection in a Non-Darcy Porous Medium Saturated with a Nanofluid," *Heat Transfer-Asian Research,* **43**, 397-411.

http://dx.doi.org/10.1002/htj.21083

Xuan, Y., and Roetzel, W., (2000), "Conceptions for Heat Transfer Correlation of Nanofluids," *International Journal of Heat and Mass Transfer*, **43**, 3701-3707.
http://dx.doi.org/10.1016/S00179310(99)00369-5

A REVIEW OF HOLE GEOMETRY AND COOLANT DENSITY EFFECT ON FILM COOLING

Srinath Ekkad[a,*] Je-Chin Han[b]

[a] *Dept. of Mechanical Engineering, Virginia Tech, Blacksburg, VA 24061, USA*
[b] *Dept. of Mechanical Engineering, Texas A&M University, College Station, TX 77843, USA*

ABSTRACT

Improved film cooling hole geometries and effect of coolant density on film cooling have been a focus since the 1970s. One of the first studies on modifying hole exit to improve film cooling effectiveness and quantifying coolant density effect was from Prof. Goldstein's group. This paper provides an overview of the development and implementation of hole exit geometries as well as coolant density study over the past few decades and the impact on future studies of advanced hole geometries under realistic engine-like coolant-to-mainstream density ratio conditions. This work is not intended to be a comprehensive review of the literature.

Keywords: *Film cooling, Hole geometry effect, Density ratio effect.*

1. FUNDAMENTALS OF FILM COOLING

"Film Cooling is the introduction of a secondary fluid (coolant or injected fluid) at one or more discrete locations along a surface exposed to a high temperature environment to protect that surface not only in the immediate region of injection but also in the downstream region" (Goldstein, 1971).

Turbine airfoil surfaces, shrouds, blade tips and endwalls are all cooled using discrete hole film cooling. A typical cooled airfoil is shown in Fig. 1. The figure shows the various locations where coolant is injected into the mainstream from inside the airfoil through discrete holes. Film cooling protects the airfoil surface directly compared to internal cooling techniques that remove heat from the inside surface. Film cooling also removes heat from blade surface through the film hole by internal convection. The thermal protection is expected to provide reduced heat load to the airfoil surface. Designers need to know the net heat load into the component surface when film is injected. Due to complex nature of discrete hole injection, there is a need to know the local wall temperature (T_w) under the film and the gas side heat transfer coefficient with film injection. Both these components are required to estimate reduced heat load to the surface.

Figure 2 shows the film injection geometry and heat transfer model associated with it. Typically, the heat load to surface without film cooling is represented as heat flux.

$$q''_0 = h_0\left(T_g - T_w\right) \tag{1}$$

where h_0 is the heat transfer coefficient on the surface with wall temperature T_w and oncoming gas temperature (T_g). When film is injected on the surface, the driving temperature is T_f, film temperature, which is a mixture of gas (T_g) and coolant temperature (T_c).

$$q'' = h\left(T_f - T_w\right) \tag{2}$$

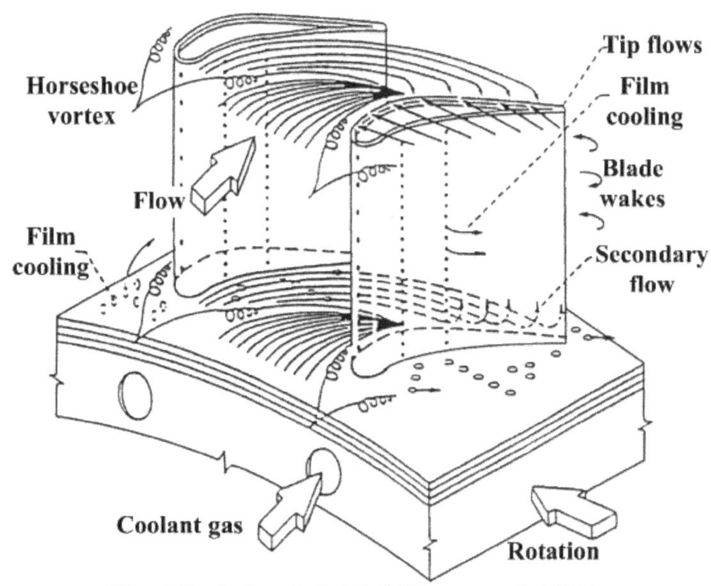

Fig. 1 Typical cooled airfoil (Hyams *et al.*, 1996).

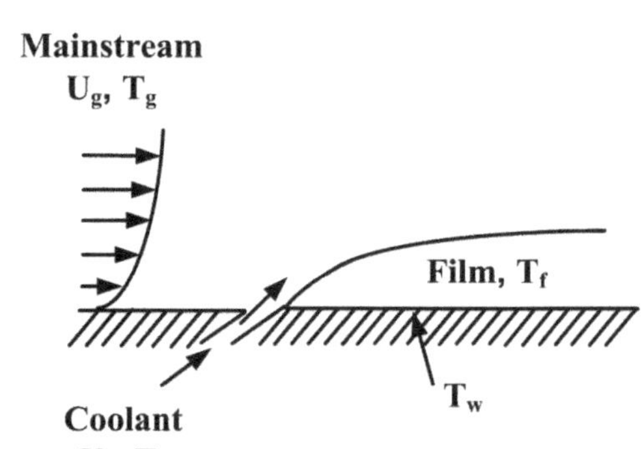

Fig. 2 Schematic of film cooling concept.

* *Corresponding authors. Email: sekkad@vt.ed*

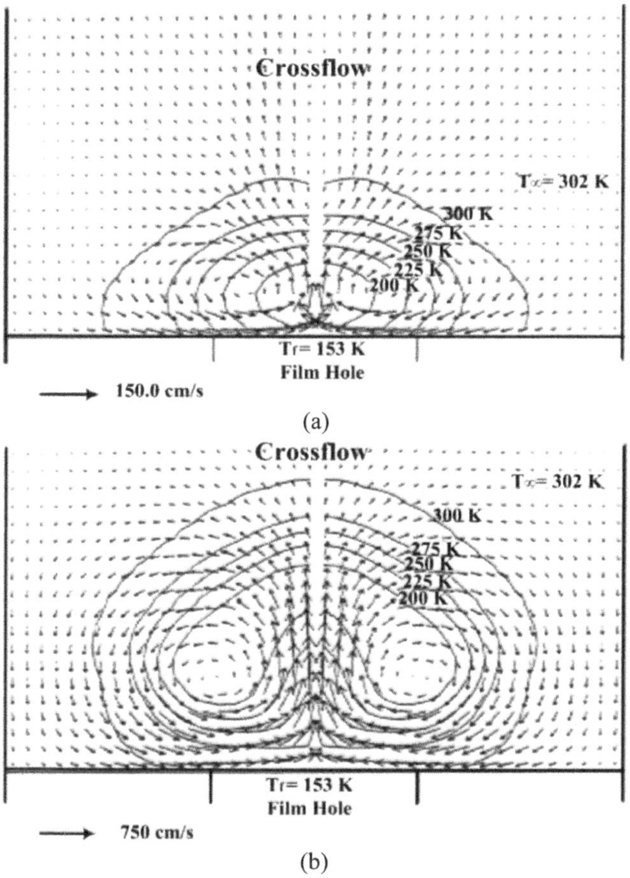

Fig. 3 Computed temperature distributions of injection through angled holes (Leylek and Zerkle, 1994).

where h is the heat transfer coefficient on the surface with film injection. Also, a new term film effectiveness (η) is introduced, where $\eta = (T_g - T_f)/(T_g - T_c)$. To obtain any benefit from film cooling, the heat load ratio, q''/q_0'', should be below 1.0.

$$\frac{q''}{q_0''} = \frac{h}{h_0}\frac{T_f - T_w}{T_g - T_w} = \frac{h}{h_0}\left\{1 - \eta\frac{T_g - T_c}{T_g - T_w}\right\} \quad (3)$$

Film cooled airfoils have leading edge film cooling holes called showerhead and downstream pressure and suction surface film cooling. The showerhead cooling has several film hole rows near the stagnation point on the blade leading edge. Pressure and suction surface cooling is achieved by single or multiple rows of injection holes at several locations. Fig. 1 is a schematic to show some of the important parameters affecting airfoil film cooling. The oncoming mainstream flow, blade wakes, rotation, secondary flow effects on endwalls, and tip flows are important external effects. The film hole geometry and configuration and coolant mass flow through the holes also affect film-cooling performance.

In general, heat transfer coefficients increase with coolant injection through cooling holes, however, the surface protection by film coolant should be greater than the penalty by the enhanced heat transfer coefficients. Therefore, the heat load ratio, Eq. (3), must be less than one for good film cooling designs.

Figure 3 shows computed temperature distributions downstream of injection through an angled hole (Leylek and Zerkle, 1994). The temperature distributions clearly show the counter-rotating vortex structure and kidney-shaped cross-section of the coolant jet. Two coolant-to-mainstream blowing rates of 0.5 and 2.0 are shown. At low blowing rate of 0.5, the jet velocity is lower than the mainstream flow.

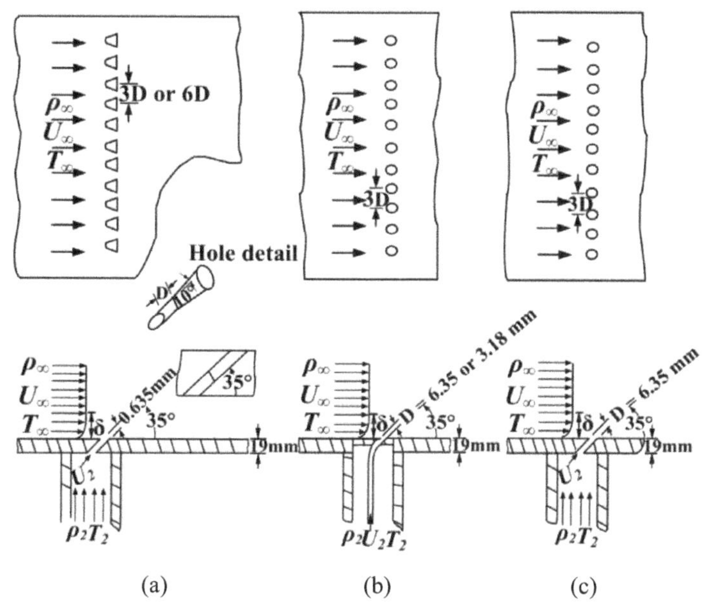

Fig. 4 Geometry of injection section: (a) shaped injection channels, (b) long cylindrical injection channels, (c) cylindrical injection channels through a thin wall (Eckert *et al.*, 1970).

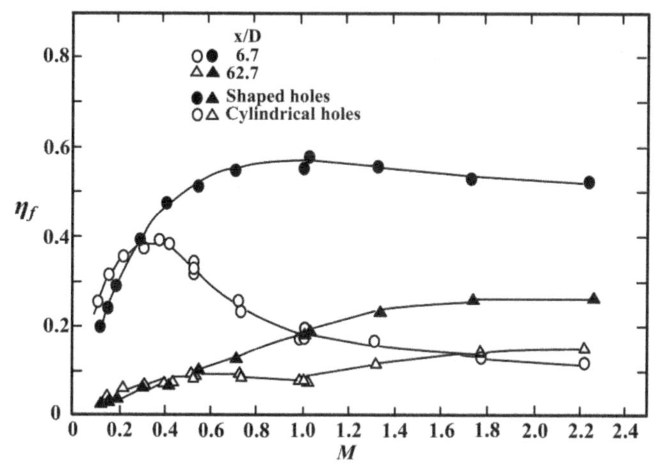

Fig. 5 Comparing diffuser hole performance with cylindrical hole (Eckert *et al.*, 1970).

The stronger mainstream forces the jet down toward the surface. However for a blowing rate of 2.0, the jet-mainstream interaction region is larger. The jet appears to lift-off farther from the surface. Also, more mainstream flow appears to entrain underneath the jet. Studies in redefining hole geometries have focused on this high lift-off effect for cylindrical hole exits at higher blowing ratios. Modifying the jet penetration into the mainstream has been the focus in improving the jet cooling effectiveness.

For better performance, cooling holes with compound angles, diffuser shaped expansions, and holes in slots are used. Increase in lateral momentum in compound angle holes, decrease in axial velocity in the shaped holes and reduced jet momentum and 2-D jet coverage in slots lead to an increase in cooling efficiency. This jet spreading provides a better lateral coverage thus a higher laterally averaged cooling efficiency is obtained.

There have been many film cooling studies in open literature, for example, Kercher (1998), documented more than a thousand paper's titles, Han *et al.* (2012) reviewed several hundreds of papers in their book's chapter 3. The present review paper focuses only on advanced film-hole geometry and coolant density effect on film cooling.

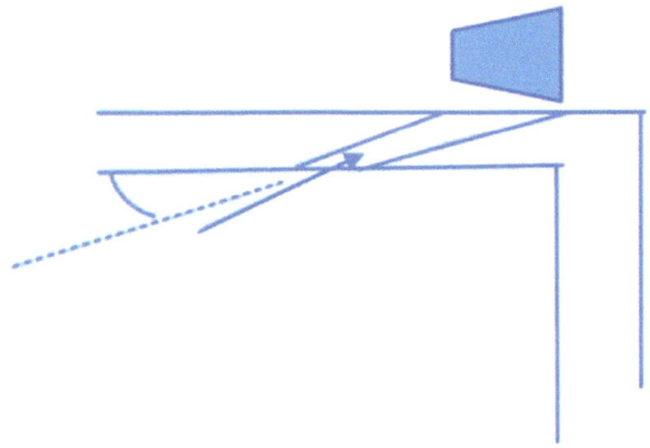

Fig. 6 Illustration of the geometry of trapezoidal holes as studied by Makki and Jakubowski (1986).

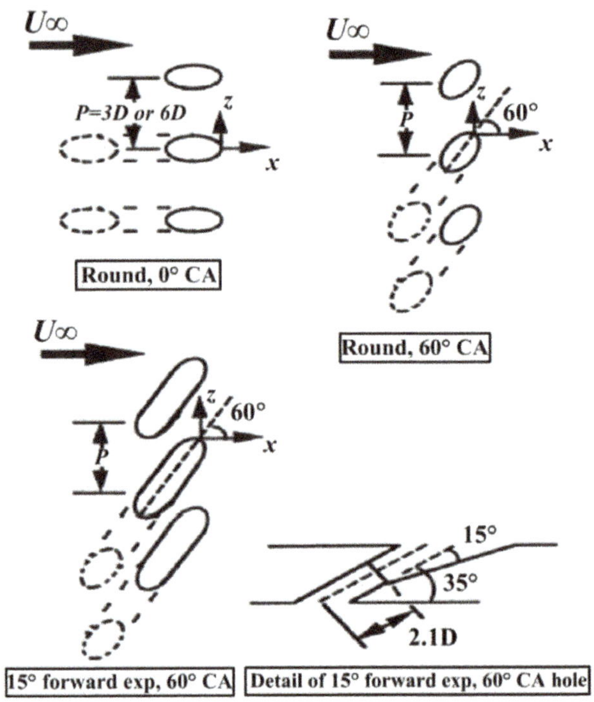

Fig. 7 Illustration of the hole configurations studied by Sen *et al.* (1996) and Schmidt *et al.* (1996).

2. FILM COOLING HOLE GEOMETRY DEVELOPMENT

2.1 Early Work (1970s)

Eckert *et al.* (1970) studied injection through circular tube shooting into a cross mainstream. They indicated that for the same flow of secondary fluid, considerably less cooling or protection is obtained with discrete holes than with injection through a continuous slot. This less efficient performance is apparently due to the penetration of the jets from the individual holes into the free stream, permitting the hot free stream gases to flow under the secondary fluid close to the surface to be protected. They suggested that the trajectory of a three-dimensional jet entering a cross-stream is primarily determined by the momentum flux (or dynamic head) ratio of the two fluids, namely $(\rho U^2)_c/(\rho U^2)_\infty$. This may be contrasted to two-dimensional film cooling in which the data can often be correlated using the blowing parameter (mass flux ratio), $(\rho U)_c/(\rho U)_\infty$. The last mentioned parameter is indicative of the enthalpy deficit of the entering secondary fluid and would be expected to play a role in predicting film cooling effectiveness. However, when flow conditions are such that the jet can separate from the surface, the momentum flux ratio cited above can also be of importance.

Goldstein *et al.* (1974) indicated that the passageway for the secondary fluid might be designed to produce a Coanda effect at the exit so that injected fluid would follow the surface rather than penetrate into the free stream. They studied three geometries, one with a "shaped" exit, the other with a long tube supplying the hole exit to compare with the baseline short cylindrical hole. Fig. 4 shows the three geometries studied.

Their experiments indicated the improvement of three-dimensional film cooling which can be obtained with shaped channels for the secondary flow. The use of a passage with an initially round cross section widened to each side by an angle of about 10 degrees significantly increased the film cooling effectiveness immediately downstream of either a single hole or a hole in a row of holes. In addition, the shaped hole appreciably increased the spreading of the secondary flow so that the film cooling effectiveness at positions laterally displaced from the centerline of a hole is also much greater than that found downstream of a cylindrical hole. One explanation of the increased effectiveness with shaped holes is that the mean velocity of the secondary flow is decreased with the larger exit area. This lower effective blowing rate causes the jet to stay closer to the wall rather than penetrating into the mainstream and accounts for the higher-film cooling effectiveness, particularly at high blowing rates.

Figure 5 summarizes the essential effect of exit hole shaping on centerline film effectiveness at two distances downstream of the holes (Eckert *et al.*, 1970). The significant decline in film cooling effectiveness with increased blowing through cylindrical secondary flow channels has been attributed to the penetration of the secondary fluid jets into the mainstream away from the adiabatic surface. Up to the blowing rate that produces separation from the wall (0.4-0.6), the two geometries give very similar results. For high blowing rates, the shaped hole geometry shows increasing effectiveness and flattens out.

2.2 Early Work (1980s)

Haller and Camus (1984) performed aerodynamic loss measurements on a two-dimensional transonic cascade with holes that had a spanwise flare angle of 25 deg and found significant improvements in film-cooling effectiveness without any additional loss penalty. Papell (1984) experimentally studied a novel cusp shape hole exit fand found that shaping the hole with a cusp (similar in appearance to a kidney bean) induced strong longitudinal vortex structures within the film hole. With the aid of flow visualizations, he summarized that creating these vortical structures enabled the crossflow to use its energy to force the jet down to the surface, rather than itself creating the counter-rotating vortices. Further, he placed the cusp on the leeward side (TE) to force the film hole secondary flow to rotate in a direction opposite of that traditionally observed in cylindrical film holes. Makki and Jakubowski (1986) studied trapezoidal exit holes and found that forward-and-laterally positioned film holes provided lower heat transfer coefficients and higher film cooling effectiveness when compared to round holes leading overall heat flux reduction to the test surface as shown in Fig. 6. Their conclusion was based on comparison of the cooling performance for forward-and-laterally expended holes with circular film hole geometries using a 35° inclination angle and a pitch to diameter ratio of 3. Note that more studies exist and only a handful could be provided as pertinent examples.

2.3 Work in 1990s

Sen *et al.* (1996) and Schmidt *et al.* (1996) studied the effect of compound-angle film-cooling on heat-transfer coefficients and film effectiveness. They studied three different hole geometries and measured heat-transfer coefficient enhancement and film effectiveness over a surface without film cooling. Compound-angle holes are holes that are inclined along the mainstream direction with an additional angle of inclination in the lateral or span-wise direction. Fig. 7

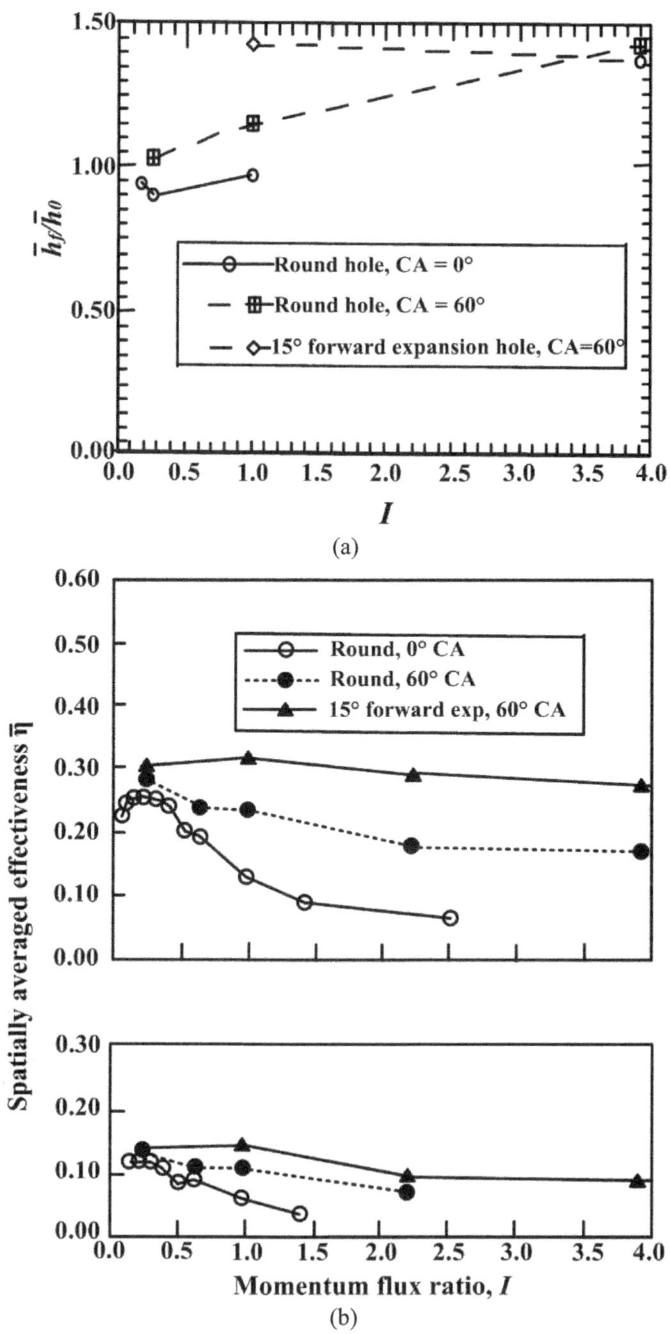

Fig. 8 Variation of spatially averaged: (a) heat transfer coefficient ratios, (b) film cooling effectiveness for configurations studied by Sen *et al.* (1996) and Schmidt *et al.* (1996).

illustrates the three-hole geometries studied by Sen *et al.* (1996). The first case was a round of holes and called the simple-angle hole, where the holes are inclined only along the mainstream, with no lateral angle. The second case was a compound angle hole that had an additional angle of 60° away from the mainstream in the lateral direction. The third case was the shaped hole with the 15° forward expansion and the 60° compound angle. As shown in the figure, these holes had the diffusing exit inclined further along the hole exit, at 15°.

Figure 8(a) presents the variation of spatially averaged heat-transfer coefficient with increasing momentum flux ratio (I). The heat-transfer coefficient ratio increases with increasing I for the round hole (CA = 0) and the round hole (CA = 60). The heat-transfer coefficient ratio is very high for the forward expansion hole. This may be due to increased lateral

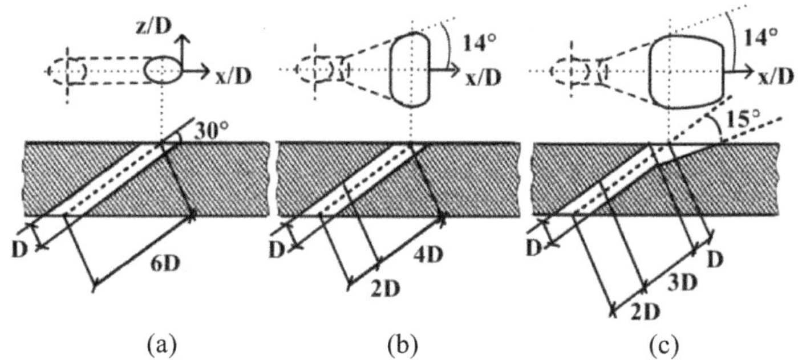

Fig. 9 Hole configuration studied by Gritsch *et al.* (1998b).

mixing of jets with the mainstream. However, the ratio decreases with increased forward-expansion hole at large I. The ratio values for the round hole, CA = 60, continually increase with increasing I. Fig. 8(b) compared the spatially averaged film effectiveness for the three geometries against the momentum flux ratio (I). The results are shown at two hole-to-hole, pitch-to-diameter (P/D) ratios of 3.0 and 6.0. Again, it is evident that hole geometry does not provide significant effect at low momentum flux ratio (or blowing ratio). But with compound-angle hole injection (CA = 60°), the film effectiveness is significantly improved at higher momentum flux ratios. With forward expanded exits and CA = 60°, film effectiveness improves further at higher I values. The effect is reduced for the larger P/D ratio of 6.0. Overall effectiveness values are also lower for each case with an increased P/D ratio.

Thole *et al.* (1998) showed that jet penetration as well as local velocity gradient in the coolant-mainstream mixing region was significantly reduced for holes with expanded exits as compared to a cylindrical hole at the same blowing rate. Gritsch *et al.* (1998b) investigated the same hole geometries as that of Thole *et al.* (1998) that included a cylindrical hole and compared with a fan-shaped diffuser hole and a laid-back fan shaped diffuser hole. Fig. 9 presents the three geometries studied by Thole *et al.* (1998) and Gritsch *et al.* (1998b). The diameter of the cylindrical hole and the diameter of the cylindrical inlet section of the expanded holes was 10 mm. The blowing ratio calculation was based on this inlet area for all three holes. The length-to-diameter ratio (L/D) was 6 for all hole geometries. The lateral expansion angle of both expanded holes was 14 deg, resulting in a hole width of 30 mm at the hole exit. The forward expansion angle of the laid-back fan-shaped hole was 15 deg, resulting in a hole length of 40 mm at the hole exit. The exit-to-entry area ratio of the fan-shaped and laid-back fan-shaped hole were 3.0 and 3.1, respectively (areas perpendicular to hole axis).

They indicated that, for the simple cylindrical hole, the film effectiveness decreases significantly with increasing blowing ratio. The jets penetrate the mainstream with increasing blowing ratio, resulting in lower film effectiveness. The distribution clearly shows jet separation at a higher blowing ratio. For the fan-shaped hole, the film effectiveness is not as significantly reduced with increasing blowing ratio. The jets spread better over a greater spanwise area due to the flared exit. However, at higher blowing ratio, the effect of jet separation is still evident just downstream of injection. For the laidback fan-shaped hole, the local film effectiveness downstream of the film hole is lower than that for the fan-shaped hole. However, the jet spreads more in the spanwise direction, providing higher film effectiveness laterally. Gritsch *et al.* (1998b) also presented the heat transfer coefficients for the same holes. They indicated that heat transfer coefficients for fan shaped holes were much lower due to increased cross-sectional area of the hole exits. The reduced momentum produces lower jet penetration and mixing thus reduced local turbulent production. Gritsch *et al.* (1998a) presented discharge coefficients for the holes studied in Gritsch *et al.* (1998a and 1998b). Laidback fan shaped and fan shaped holes produced similar discharge coefficients which were typically higher than for cylindrical holes especially for lower pressure ratios between coolant total pressure and mainstream pressure.

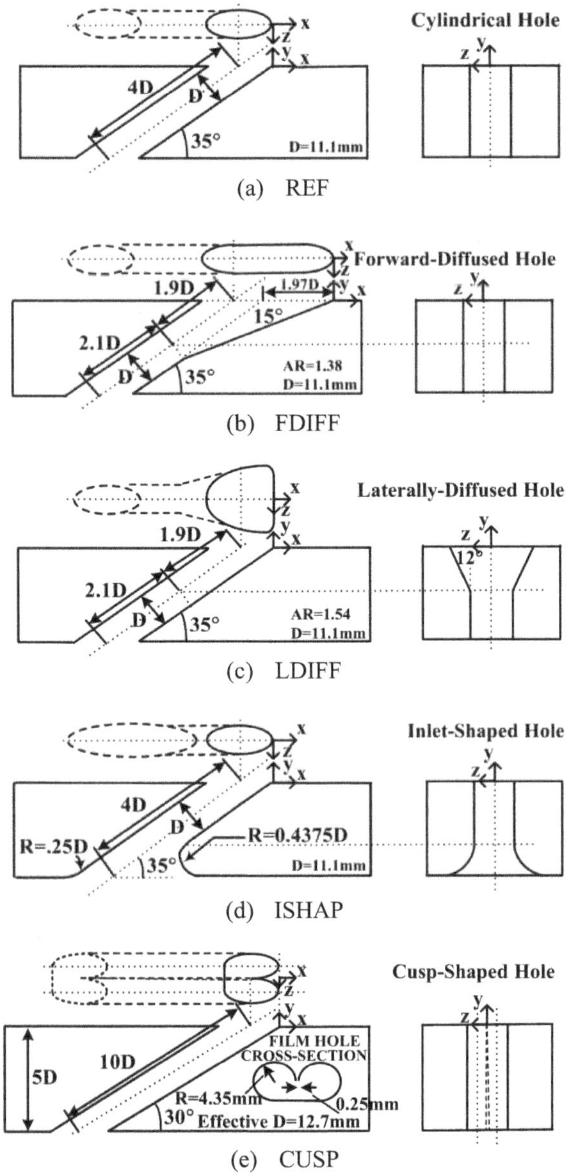

Fig. 10 Schematic of the five hole geometries studied by Hyams and Leylek (2000).

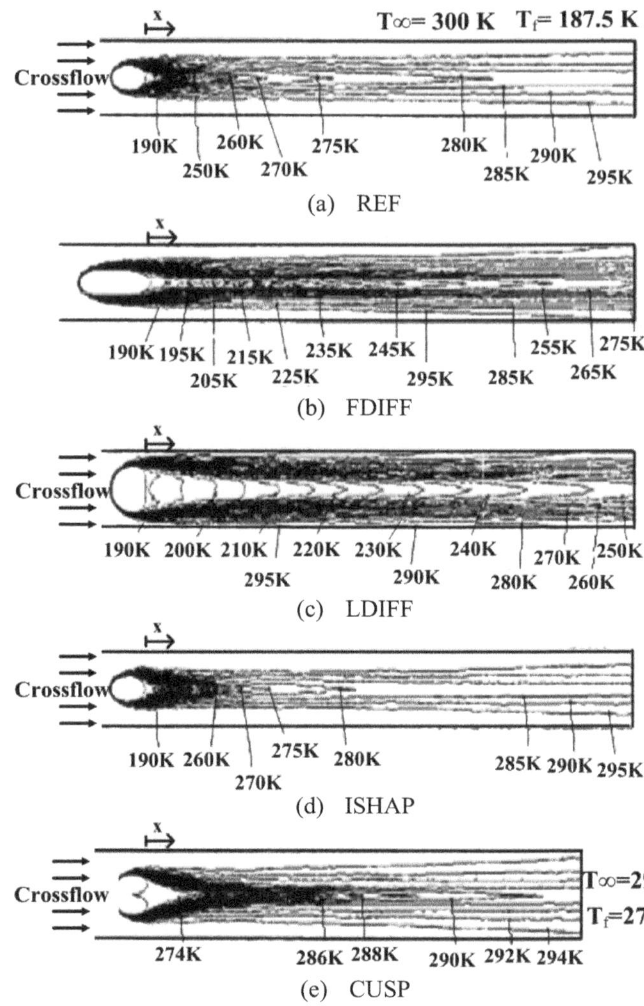

Fig. 11 Comparison of temperature footprints from the five geometries in Figure 9 (Hyams and Leylek, 2000).

Kohli and Thole (1997) studied the effect of coolant entrance condition for both cylindrical and diffuser shaped holes computationally. The coolant channel orientation played a major role in the effectiveness distributions on the test surface, based on their results. Hyams *et al.* (1996) studied the effects of slot jet shaping on the heat transfer downstream of a slot jet. It was found that shaping of the slot inlet and exit provided significant gains in the film cooling performance.

2.4 2000-2014

Hyams and Leylek (2000) computationally studied five distinct film cooling configurations in Fig. 10: (a) cylindrical film hole (reference case), (b) forward-diffused film hole, (c) laterally diffused film hole, (d) inlet shaped film hole, and (e) cusp-shaped film hole. Fig. 11 shows the five basic film hole shapes investigated in this work. The configurations were: (a) cylindrical (REF), (b) forward diffused (FDIFF), (c) laterally diffused (LDIFF), (d) inlet shaped (ISHAP), and (e) cusp shaped (CUSP). The fifth hole indicated as CUSP is the same geometry proposed by Papell (1984) in the 1980s. They found that the laterally diffused film hole provides the best coverage and highest effectiveness. The forward diffused film hole performs well along the centerline, but does not spread laterally. The cylindrical and inlet-shaped film holes perform poorly at both blowing ratios; the coolant lifts off from the test surface in these cases. Fig. 11 shows the computed temperature distributions for all five cases indicating the jet spreading and coverage.

Sargison *et al.* (2001a and 2001b) presented experimental measurements of the performance of a new film cooling-hole geometry – the Converging Slot-Hole or Console as shown in Fig. 12. This novel, patented geometry has been designed to improve the heat transfer and aerodynamic loss performance of turbine vane and rotor blade cooling systems. The laterally averaged adiabatic effectiveness results demonstrated that the console cooling effectiveness approached that of the slot, as does the fan shaped-hole effectiveness. The console laterally averaged heat transfer coefficient was similar to the slot, and higher than the cylindrical and fan shaped-hole results. The slot and console does not significantly change the boundary layer flow compared with the case of no film cooling, and hence the heat transfer coefficient is similar. The aerodynamic performance of the console was also investigated, and it was shown that the aerodynamic loss due to a console is significantly less than for fan shaped or cylindrical film cooling-holes.

Teng *et al.* (2001a and 2001b) reported the effect of unsteady wake on gill-hole film cooling with shaped holes. Results showed the film cooling effectiveness for laidback fan-shaped hole increases but heat transfer coefficients decreases as compared with the cylindrical holes. Saumweber and Schulz (2003) investigated the effect of elevated free-

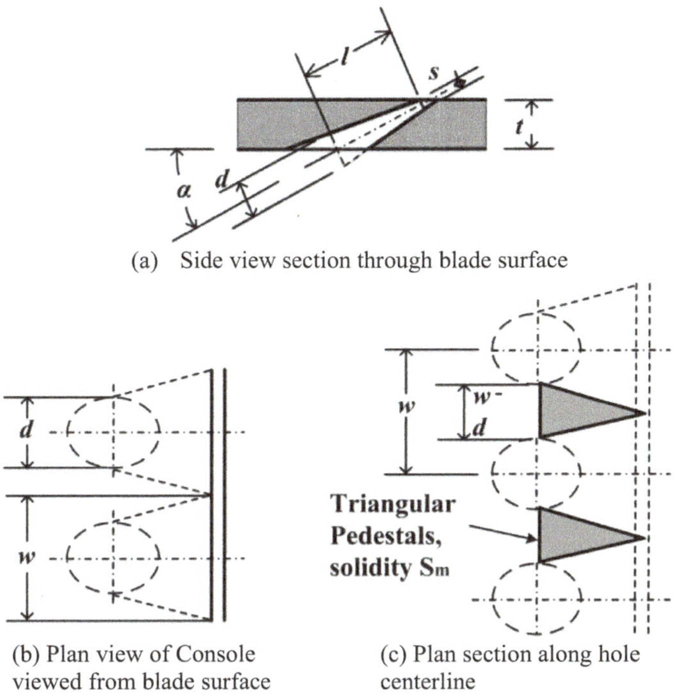

Fig. 12 Basic console or converging slot geometry proposed by Sargison *et al.* (2001a and 2001b).

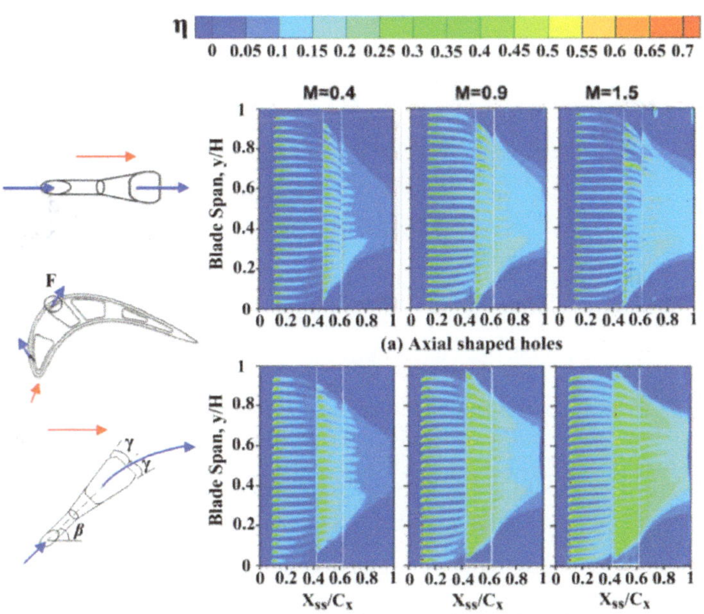

Fig. 13 Film cooling over a blade surface comparing axial and compound angled shaped holes by Gao *et al.* (2008 and 2009).

stream turbulence on film cooling performance of shaped holes. For cylindrical holes, elevated free-stream turbulence intensity lowers adiabatic film-cooling effectiveness at small to moderate blowing ratios. Reductions up to 40% for local effectiveness and 25% for laterally averaged effectiveness are found when the turbulence intensity is raised from 3.6 to 11%. They found that the effect of increased turbulence level is detrimental at all blowing ratios for the shaped holes as the lateral spreading of coolant could not be improved by elevated free-stream turbulence intensity. Bunker (2005) provided a review of shaped-hole technology both in terms of cooling effectiveness as well as manufacturing. Bunker quoted, "The current understanding of shaped film parametric effects and physics is relatively sparse, however, the acceptance of shaped film technology is based largely on the foundation of round film-hole data. The benefits of shaped-hole film cooling are real and substantial, so much so that these types of film holes are used whenever possible in the practice of cooling gas turbines. No single shaping of film hole stands out as an optimal geometry, as long as the general guidelines of diffusion angles and sufficient depth/length of transition are followed. As manufacturing methods progress further, additional unconventional shapes of film holes will become viable, possibly leading to even higher performance and the ability to cool components with less air." Bunker (2009) reviewed thirty discrete hole film cooling geometries and provided eight key factors to judge the potential performance for each film hole geometry including adiabatic effectiveness, manufacturing cost, repair, geometry sensitivity, flow sensitivity, operational tolerance, and strength/durability. Bogard and Thole (2006) provided a review of the literature including effects of free-stream turbulence, blowing ratio, coolant-to-mainstream density ratio, surface curvature, and hole shape on the performance of film cooling.

Saumweber and Schulz (2008a) studied the effects of free-stream Mach number and free-stream turbulence, including turbulence intensity, integral length scale, and periodic unsteady wake flow on a row of cylindrical and shaped film holes. The results clearly show that the performance gain with fan-shaped holes as compared to cylindrical holes is clearly overestimated at low laboratory type conditions. Saumweber and Schulz (2008b) studied the effect of coolant flow direction prior to entrance into the film hole for both cylindrical and shaped holes. With a coolant supply channel orientation perpendicular to the cooling hole axis, they found that the counter rotating vortex pair inside the hole - typically found for plenum conditions at the inlet - is suppressed. Instead, they saw a single helical motion that is induced inside the cylindrical part of the hole. The resulting flow effects downstream of the ejection location were very different for cylindrical and fan-shaped holes. Saumweber and Schulz (2008c) studied the effect of the expansion angle of the diffuser, the inclination angle of the hole, and the length of the cylindrical part at the hole entrance on cooling performance of diffuser holes. They indicated that a combination of parameters affect overall cooling performance and different situations requires different hole geometries.

Gao *et al.* (2008) and Gao and Han (2009) compared the effects of axial shaped hole and compound angled shaped hole on suction side and pressure side film cooling distributions by using pressure sensitive paint technique. The compound angled shaped hole provides higher film effectiveness than the axial angled shaped hole as shown in Fig. 13. The film traces were converged toward the suction side mid-span region, with no film effectiveness found near 25% of hub and 25% of tip, respectively. Note that the rotation effect is not considered in these papers. Gao and Han (2009) reported the influence of hole shape and angle on showerhead film cooling effectiveness by using pressure sensitive paint technique. Results showed the radial angle shaped hole.

Lu *et al.* (2005) investigated the film cooling performance for a row of cylindrical holes by embedding the row in transverse slots. Transverse slots could be formed around the holes in the event that the blade surface is sprayed with TBC leaving the holes embedded in trenches. These trenches could be 2-D slots or 3-D craters depending on the type of masking used during the TBC spray process. The geometry of the transverse slot greatly affected the cooling performance downstream of injection. They investigated the effect of the slot exit area and edge shape. They indicated that embedding the holes in a slot reduced the jet momentum at exit and also spread the jets and provided two-dimensional slot jet coverage compared to the three-dimensional nature of individual jets. The wider spread of jets provided a better over film effectiveness with slight increases in heat transfer coefficients. Waye and Bogard (2006) investigated adiabatic film cooling effectiveness of axial holes embedded within a transverse trench on the suction side of a turbine vane. The narrow trench configuration provided the best adiabatic effectiveness performance with increasing adiabatic effectiveness levels with increasing blowing due to the trench suppressed coolant jet

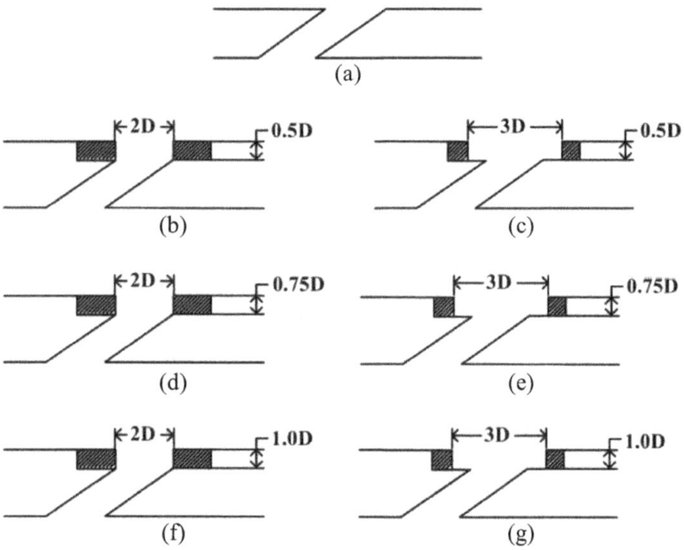

Fig. 14 Trench geometries studied by Lu *et al.* (2007a).

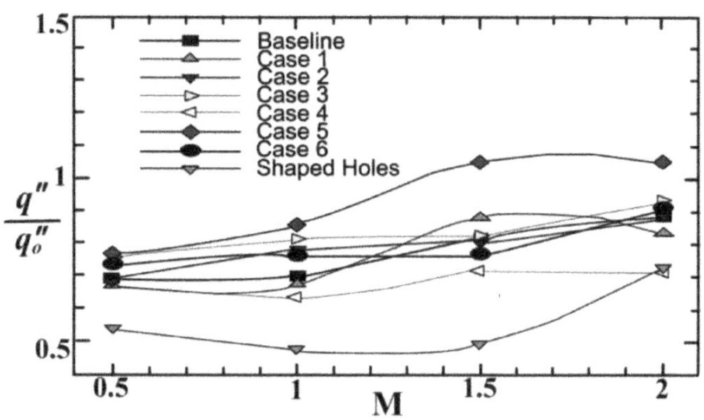

Fig. 15 Overall heat flux ratio comparisons for different trench geometries (Lu *et al.*, 2007a).

separation. Lu *et al.* (2007a) presented an experimental investigation of film cooling from cylindrical holes embedded in transverse trenches where different trench depths were studied with two different trench widths. Fig. 14 shows the various trench geometries studied. They presented overall heat flux ratio comparisons for different trenches and compared the performance of the trenched holes with a shaped hole configuration. Fig. 15 shows the heat flux ratio comparison. Baseline and cases 1-6 are as shown in Fig. 14. An optimum trench depth at 0.75 D was identified as shallow and deep trenches showed worse performance. This was confirmed by the flow distributions obtained from CFD predictions. Shaped diffuser hole reduced the jet momentum at hole exit and outperformed the trench cases.

Lu *et al.* (2007b) also presented results for cylindrical holes embedded in craters. Different crater geometries were considered for a typical crater depth which was identified by Lu *et al.* (2007a) as 0.75D. Fig. 16 shows the different crater geometries investigated in this study. The location of the hole inside the crater will have significant effect on the downstream cooling behavior of these holes.

Hassison *et al.* (2007) studied film cooling adiabatic effectiveness on the suction side of a simulated vane for various configurations of coolant holes embedded in shallow transverse trenches or circular and elliptical shaped depressions. A shaped hole configuration was also tested. They compared the best trench configurations with a typical shaped hole configuration and showed similar film effectiveness performance over the range of blowing ratios. Hassison *et al.* (2007)

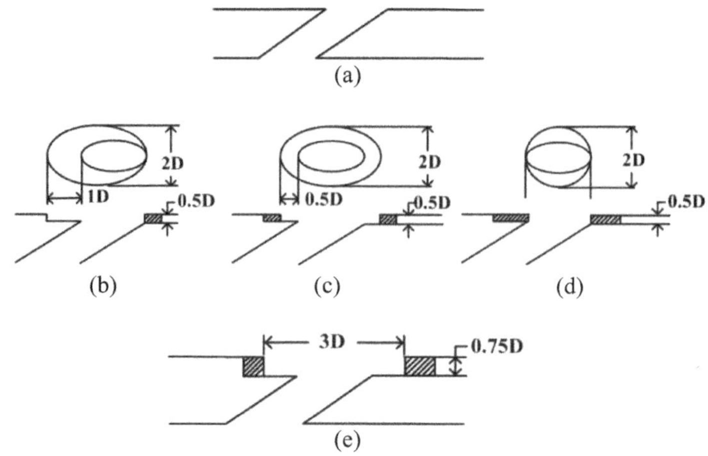

Fig. 16 Crater geometries studied by Lu et *al.* (2007b).

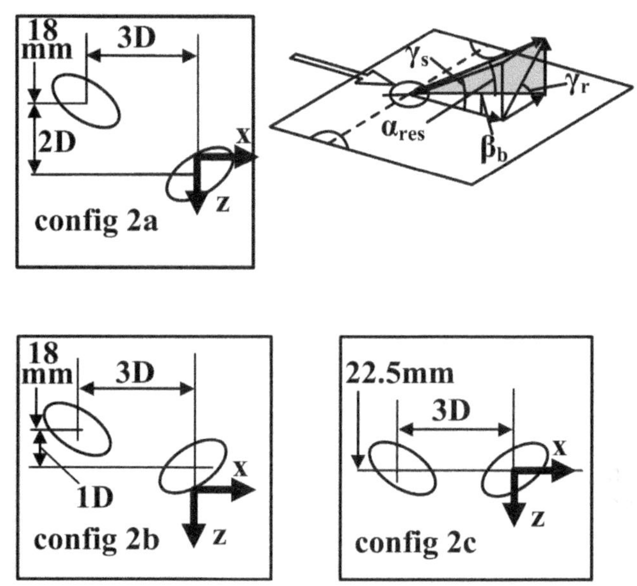

Fig. 17 Double Jet Concept presented by Kusterer *et al.* (2006).

presented both film cooling adiabatic effectiveness and heat transfer coefficients for cylindrical holes embedded in a 1d transverse trench on the suction side of a simulated turbine vane to determine the net heat flux reduction. Rallabandi *et al.* (2011) reported effect of upstream and downstream step on flat plate film cooling effectiveness on four separate hole geometries: simple angled (axial angle 30-deg) and compound angled (axial angle 30-deg, compound angle 45-deg) with cylindrical and fan-shaped film cooling holes. Results indicated an increase in film-cooling effectiveness in the region near the hole due to the upstream step for all the cases considered. This increase due to the step was found to be most significant in the case of compound angled cylindrical holes and least significant in the case of simple angled fan-shaped holes.

Kusterer *et al.* (2006) introduced the Double-jet Film-cooling (DJFC) as an alternative film-cooling technology to conventional film-cooling design. They showed improvement of the film-cooling effectiveness by application of the DJFC. They also showed basic applicability of the DJFC to a realistic blade cooling configuration and presented first test results under machine operating conditions. From their results, the DJFC has been able to replace a row of shaped holes on the suction side of the blade without negative effects on the suction side thermal load level. Fig. 17 shows the concepts presented by Kusterer *et al.* (2006).

Heidmann and Ekkad (2008) introduced a novel turbine film cooling hole shape that was conceived and designed at NASA Glenn Research Center. The primary focus of this paper is to study the film cooling performance for a row of cylindrical holes each supplemented with two symmetrical anti vortex holes which branch out from the main holes. This "anti-vortex" design is unique in that it requires only easily machinable round holes, unlike shaped film cooling holes and other advanced concepts. The hole design is intended to counteract the detrimental vorticity associated with standard circular cross-section film cooling holes. This vorticity typically entrains hot free-stream gas and is associated with jet separation from the turbine blade surface. They indicated that possible permutation would be to study the feasibility of increasing the pitch to diameter ratio of the main holes. This would reduce the total coolant flow rate for a given plenum pressure while maintaining or improving cooling effectiveness. This would be of great benefit to the engine cycle performance. Another design modification to consider would be increasing the size of the side holes to be the same as the main holes. This would simplify manufacturing and ease concerns about drilling very small holes.

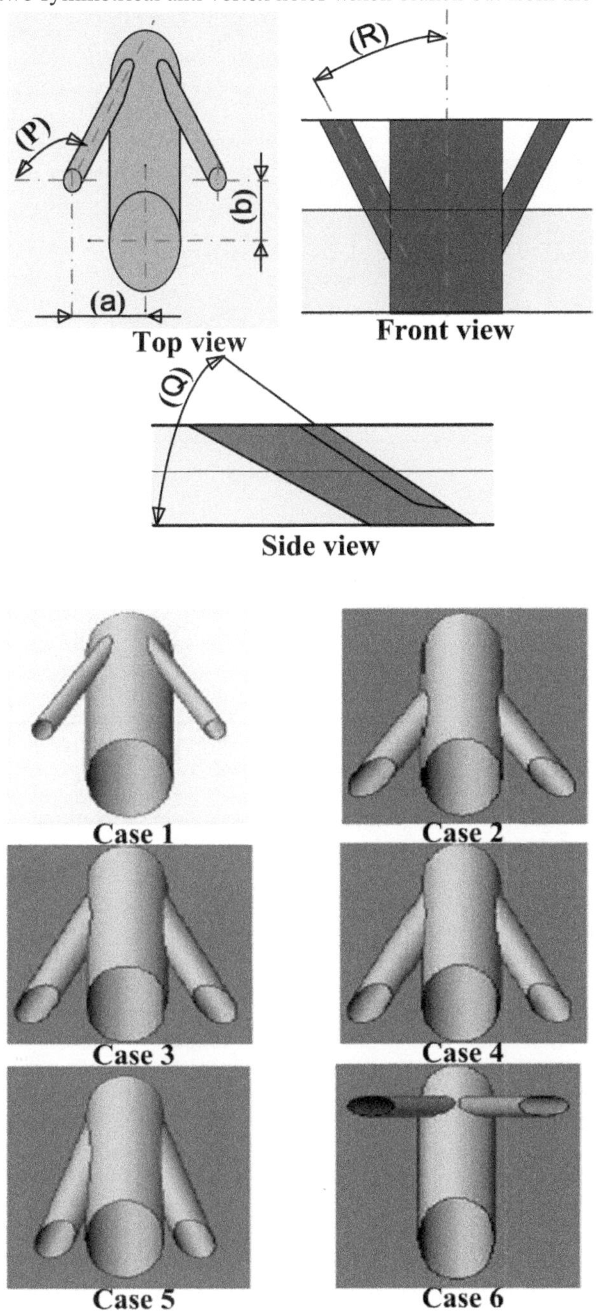

Fig. 18 Design of anti-vortex holes and different configurations by Dhungel *et al.* (2009).

Dhungel *et al.* (2009) presented a detailed study on the effects of geometry and orientations of anti-vortex holes branching out from the primary holes on film cooling have been presented. For all the cases, both heat transfer coefficients and film effectiveness were measured using a transient infrared thermography technique. It appears that the presence of anti vortex holes kills the kidney pair vortices and also reduces the momentum of the main jet hence bettering the film coverage in both downstream and lateral direction. When the anti vortex holes are nearer to the primary film cooling holes and are developing from the base of the primary holes, better film cooling was accomplished as compared to other anti vortex hole orientations. Narzary *et al.* (2011) presented a new anti-vortex design geometry where the two side holes, also of the same diameter, branch out from the root at 15° angle based on Dhungel *et al.* (2009). The pitch-to-diameter ratio was increased to 6.0 between the main holes. Results indicated significant improvement in film effectiveness with anti-vortex holes compared to cylindrical holes at all the blowing ratios studied. At any given blowing ratio, the anti-vortex hole design used 50% less coolant and provides at least 30-40% higher cooling effectiveness. Fig. 18 shows all the anti-vortex hole designs studied by Dhungel *et al.* (2009).

Chen *et al.* (2014) investigated the combined effects of hole geometry, blowing ratio, density ratio and free-stream turbulence intensity on flat plate film cooling using the steady state pressure sensitive paint (PSP) technique. They studied four common film hole geometries with forward injection and backward injection (by reversing the injection direction from forward to backward to the mainstream): simple angled cylindrical holes and fan-shaped holes, and compound angled ($\beta = 45°$) cylindrical holes and fan-shaped holes. They found out that the film cooling effectiveness with backward injection was greatly reduced for shaped holes as compared with the forward injection. However, significant improvements were observed in both simple angled and compound angled cylindrical holes at higher blowing ratios and density ratio (DR = 2).

Rezasoltani *et al.* (2015) conducted an experimental and numerical study to understand the effect of four different blade tip ejection configurations (hole arrangements) on the effectiveness. The holes were installed on the blade tips of the first rotor row of a 3-stage research turbine and hence the effect of rotation was also evaluated. The four hole arrangements included (a) plane tip with tip hole cooling, (b) squealer tip with tip hole cooling, (c) plane tip with pressure side edge compound angle hole cooling and (d) squealer tip with pressure side edge compound angle hole cooling. It was concluded that for all the configurations the overall effectiveness was increasing with blowing ratio and found to dramatically affect the flow behaviors at the tip region.

3. EFFECT OF COOLANT DENSITY STUDY

Experimental Methods: Both Thermal Methods and Mass Transfer Analogy Methods have been used for film cooling measurements. Thermal methods involve creating a temperature differential between the mainstream and the coolant in the laboratory in order to record the film cooling effectiveness. Measurement surfaces (blades, vanes, flat plates etc.) are typically made of a very low conductivity material to ensure that the wall surface temperature is almost equal to the film temperature. The surface temperatures are measured; knowing the mainstream and coolant temperatures, film cooling effectiveness can be estimated.

Surface temperatures can be measured by placing thermocouples at discrete locations; by using IR cameras, Thermochromic Liquid Crystals or Temperature Sensitive Paint to map the temperature profile. Transient thermal methods (such as the Transient Liquid Crystal Method) making use of a semi-infinite solid assumption have to measure the film cooling effectiveness and the heat transfer coefficient simultaneously have been

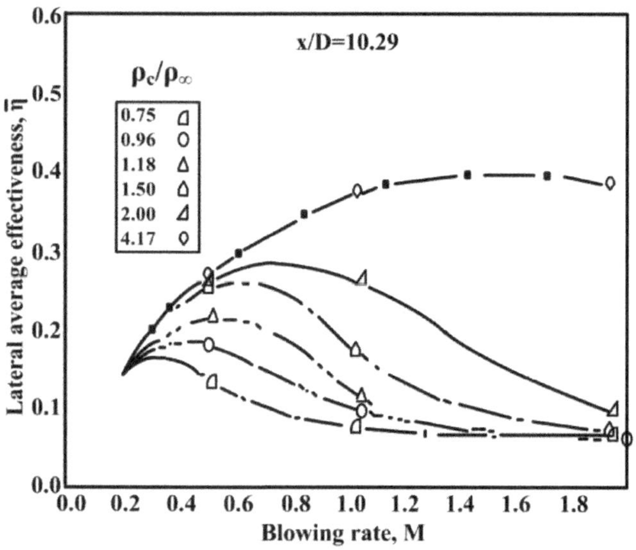

Fig. 19 Effect of blowing ratio on laterally averaged film effectiveness for different coolant density ratios Pedersen *et al.* (1977).

used. In order to generate reliable data using thermal methods, it is imperative that the measurement surfaces be comprised of a material with very low thermal conductivity. In regions proximate to holes, or in regions of high film cooling effectiveness, where thermal gradient within the measurement surface is significant, a dispersion of the effectiveness data might occur and the data corrections might need.

Mass Transfer Analogy Methods: The mass transfer analogy has been used in literature to avoid the conduction related issues by using a gas-chromatography based mass transfer analogy (sampling foreign gas at various discrete locations and estimating film cooling effectiveness); using the reaction of a diazo-coated polyester film with ammonia and water-vapor seeded coolant air to qualitatively and quantitatively arrive at film cooling effectiveness; using the naphthalene sublimation mass transfer analogy (with the coolant air saturated with naphthalene) to arrive at adiabatic effectiveness; by using the pressure sensitive paint (PSP) technique.

It is not easy to conduct heat transfer tests under real engine-like coolant to mainstream temperature conditions. Many researchers have employed the heat and mass transfer analogy method in order to simulate engine-like coolant to mainstream temperature conditions. Therefore, the detailed heat transfer coefficient as well as adiabatic film cooling effectiveness distributions can be determined at simulated engine coolant-to-mainstream density ratio conditions.

The underlying assumption governing the heat and mass transfer analogy requires that the flow-field be highly turbulent, i.e., turbulent Lewis number $Le_T \sim 1$. Turbulent Lewis number = (turbulent thermal diffusivity + thermal diffusivity) / (turbulent mass diffusivity + mass diffusivity). This assumption is usually valid over the surface of the gas turbine vane/blade/end-wall due to the high Reynolds numbers involved as well as various secondary mechanisms further inducing turbulence in the flow-field (such as high turbulence, surface roughness, leakage vortices, horse-shoe vortices, film cooling jets and periodic rotor/stator wakes). However, measurements have shown that the flow proximate to the leading edge portion of the blade is usually either laminar or intermittent, even with film cooling. This invalidates the governing assumption in the leading edge region, and more investigation is required to quantify this effect as reported by Han and Rallabandi (2010).

3.1 Early Work (1970s)

The coolant to mainstream density ratio (DR) in modern gas-turbine engines is typically around 2.0, due to the coolant temperature significantly lower than hot mainstream. Scaled down laboratory tests

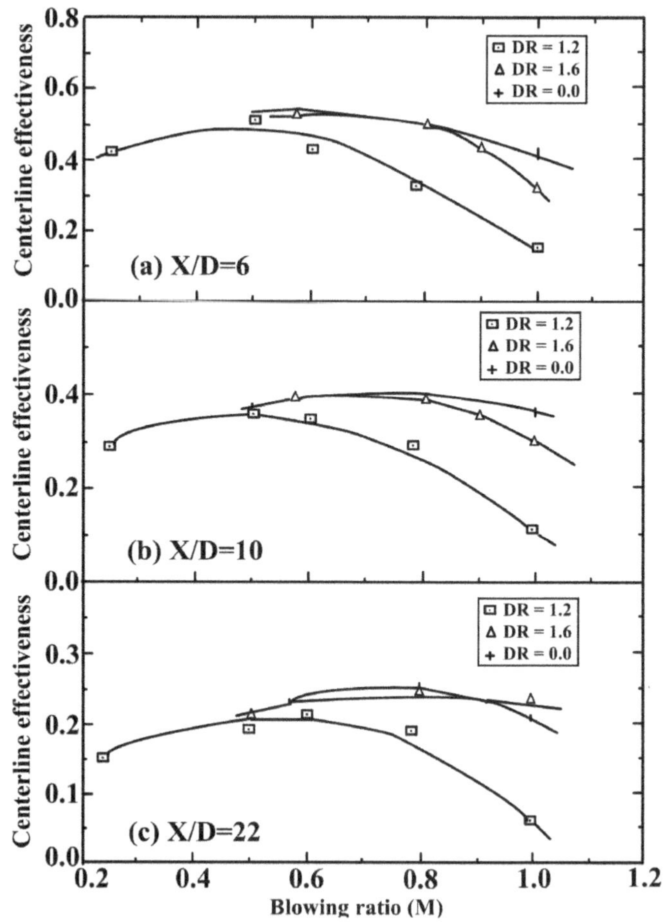

Fig. 20 Effect of blowing ratio and density ratio on film effectiveness (a) $X/D = 6$ (b) $X/D = 10$ (c) $X/D = 22$ (Sinha *et al.*, 1991).

(to simulate engine DR conditions) usually involve using a foreign gas with a higher density by Goldstein *et al.* (1974), and Pedersen *et al.* (1977) (mass transfer analogy method). In general, increasing DR at a given M results in a higher effectiveness, especially at higher blowing ratios, since the momentum of a high density coolant is lower at a given blowing ratio, there is a lower tendency to lift-off.

Pedersen *et al.* (1977) presented the effect of coolant-to-mainstream density ratio on film-cooling effectiveness. They studied density ratios varying from 0.75 to 4.17. Fig. 19 presents the effect of blowing ratio on lateral averaged film-cooling effectiveness as a function of coolant density for a single row of injection holes inclined 35° in the streamwise direction. Density ratios below 1.0 show a decrease in effectiveness with increasing blowing ratio. As density ratio increases, the peak effectiveness value moves toward higher blowing ratio. Higher-density coolant tends to stay closer to the surface compared to the lower-density coolant at the same blowing ratio. The jet momentum is higher for a lower-density coolant, and this may cause jets to lift off and produce lower effectiveness. At very high-density ratios of 4.17, effectiveness values are significantly higher and appear to increase even at high blowing ratios up to 2.0.

3.2 Work in 1980s

Most experimental results for turbine blades were obtained on simulated cascades under simulated engine flow conditions. Most of these studies used the adiabatic wall effectiveness technique. However, a few studies have also reported results with the superposition approach. Camci and Arts (1985a and 1085b) made high-pressure rotor blade heat-transfer measurements in a short-duration wind tunnel cascade facility. Blades with leading-edge film-cooling array and suction-side film cooling were

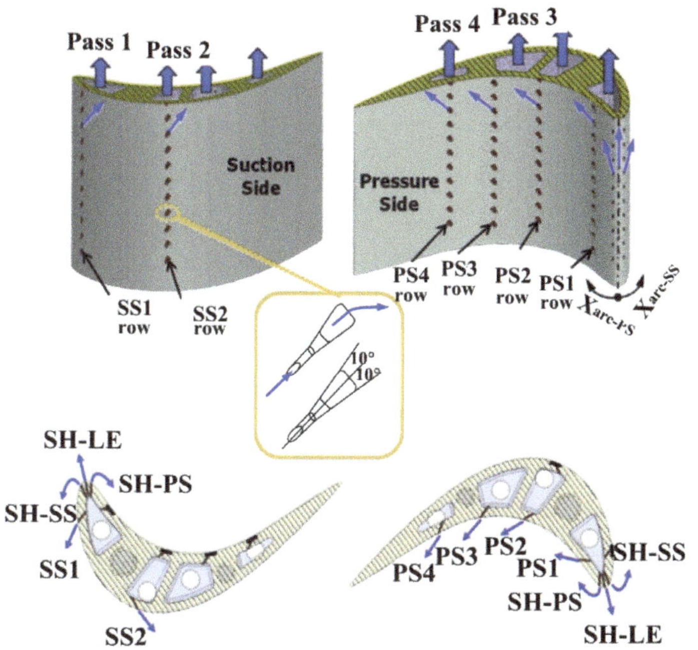

Fig. 21 Details of the test film-cooled blade with shaped cooling holes on the pressure and suction surfaces studied by studied by Liu *et al.* (2012).

tested. The coolant-to-mainstream temperature ratio T_c/T_g, was varied from 0.4 to 0.7. Coolant-to-mainstream mass flow ratio was also varied. Results showed that heat load reduction values increase with decreasing T_c/T_g, which is as expected. When cooler flow is injected, the blade receives more protection downstream of injection. In that case, lower heat-transfer levels are expected. Hylton *et al.* (1988) and Nirmalan and Hylton (1990) studied the effects of parameters such as Mach number, Reynolds number, coolant-to-gas temperature ratio, and coolant-to-gas pressure ratio on the C3X vane film cooling. The leading edge has a showerhead array of five equally spaced rows. Two rows each on the pressure and suction surfaces are located downstream. In general, results showed the Stanton number reduction values increase with decreasing T_c/T_g values from 0.85 to 0.65.

3.3 Work in 1990s

The effect of blowing ratio and density ratio on film effectiveness is clearly evident reported by Sinha *et al.* (1991) using chilling the coolant to very low temperatures (thermal method). For a constant-density ratio, the film effectiveness value increases with increasing blowing ratio and peaks and starts to drop off at very high blowing ratios. The blowing ratio at which peak effectiveness occurs is sometimes referred to as the optimum blowing ratio. It is typically in the range of 0.5 to 0.8.

From Fig. 20, it is evident that the effectiveness values for different coolant density ratios collapse into a single curve at low M values. However, the curves move away at higher blowing ratios with higher-density coolant providing higher effectiveness. This may be because higher-density coolant stays closer to the surface due to less mixing and lower momentum. It is important to note that there is no consistent trend in the way the film effectiveness values react to increasing blowing ratio for different coolant densities. Salcudean *et al.* (1994) studied the effects of coolant density on a symmetrical airfoil model in a cross-flow. They used the heat/mass transfer analogy to measure adiabatic wall effectiveness. They studied both single-row and two-row injection on the leading edge. The hole rows are located at ±15° and ±44° from stagnation. They used air and CO_2 as injectant to simulate different density ratios with respect to the mainstream. Ekkad *et al.* (1997) studied film cooling effectiveness over a flat surface with air and CO_2 injection through

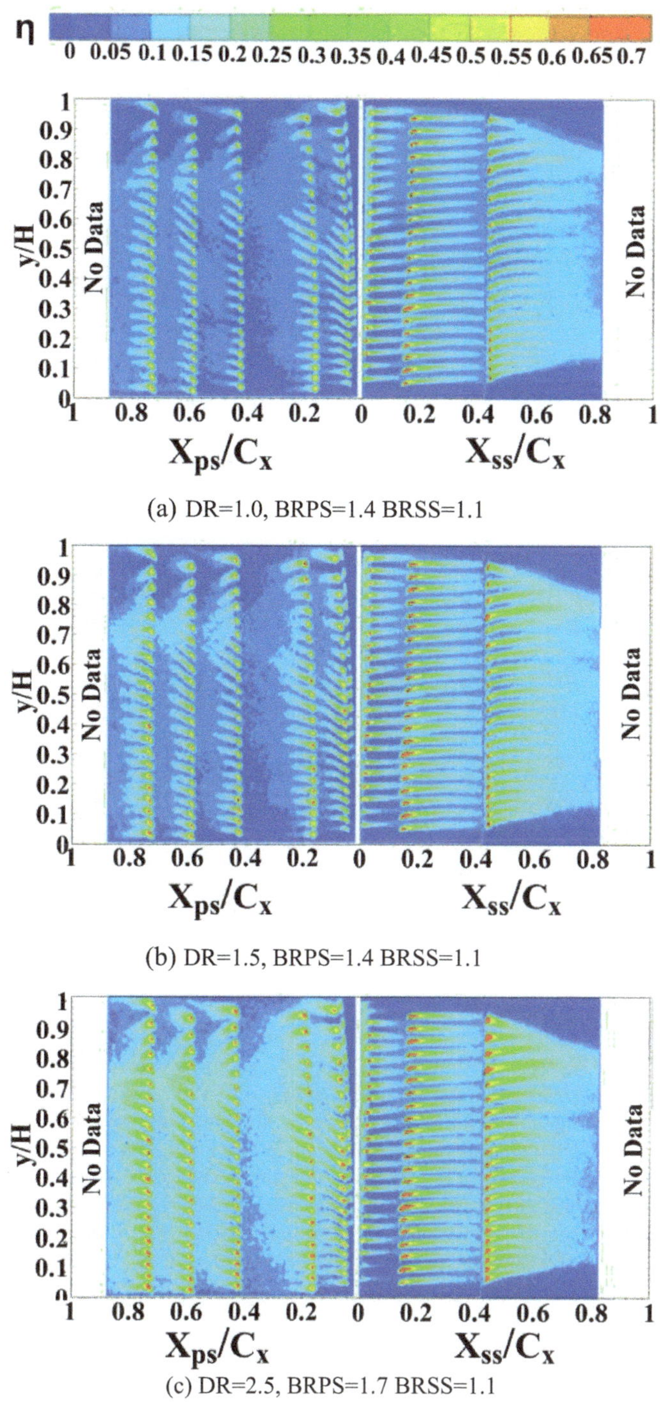

(a) DR=1.0, BRPS=1.4 BRSS=1.1

(b) DR=1.5, BRPS=1.4 BRSS=1.1

(c) DR=2.5, BRPS=1.7 BRSS=1.1

Fig. 22 Adiabatic effectiveness distribution on turbine blade pressure and suction surface with compound angled cylindrical cooling holes at three different density ratio (at M=1.5, Tu=10.5%) by Narzary *et al.* (2012).

compound angle holes using a transient liquid crystal image method. Ekkad *et al.* (1998) presented the effect of coolant density and free-stream turbulence on a cylindrical leading-edge model with one row of film-cooling holes on both sides of stagnation location. They used a transient liquid crystal technique to obtain detailed heat-transfer coefficient and film effectiveness distributions. To obtain a higher coolant-to-mainstream density ratio of 1.5, CO_2 was used as coolant, using regular compressed air for a density ratio of 1.0 simulated lower-density coolant. They found the effectiveness values for air as coolant are highest at a low blowing ratio of 0.4

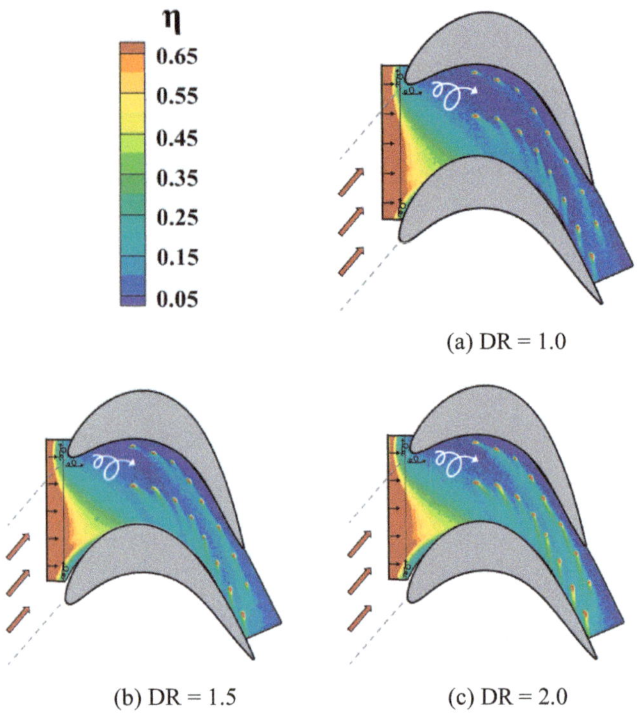

Fig. 23 Density ratio effect on adiabatic film-cooling effectiveness on the platform with compound angled cylindrical cooling holes (at M=1.5, Tu=10.5%, MFR=0.5%) by Liu *et al.* (2013).

and decrease with increase in blowing ratio. However, for CO_2, the highest film effectiveness is obtained at a blowing ratio of 0.8. The spanwise variations indicate jet-like streaks of high film effectiveness along the bottom of the film hole.

3.4 2000-2015

Ethridge *et al.* (2001) studied the effect of coolant-to-mainstream density ratio on a vane with high curvature. Dittmar *et al.* (2002) studied different film cooling hole configurations on the suction (convex) surface and concluded that shaped holes provide better coverage at higher blowing ratios by resisting jet penetration into the mainstream. The effect of coolant to mainstream density ratio and an unsteady wake was studied by Rallabandi *et al.* (2012) and Narzary *et al.* (2012) or Liu *et al.* (2012) using the pressure sensitive paint measurement method. Foreign gases with variable density (Nitrogen for DR = 1.0, CO_2 for DR = 1.5 and a mixture of Ar + SF6 for DR = 2.5) were used to simulate realistic engine density ratios. Due to the concave geometry of the pressure side, at higher blowing ratios, results show a reattachment of the lifted off jet; deterioration in film cooling effectiveness due to the average effect of the unsteady wake (Rallabandi *et al.*, 2012). Fig. 22 shows the test film-cooled blade with compound angled cylindrical holes on the pressure and suction surfaces reported by Narzary *et al.* (2012). Results show a longer coolant trace on the suction side compared with the pressure side; an increase in effectiveness at higher density ratios for a given blowing ratio (as seen from Fig. 22). However, Liu *et al.* (2012) reported that the density ratio effect on turbine blade film cooling with compound angled shaped-hole (see Fig. 21) is not as significant as that with compound angled cylindrical hole reported by Narzary *et al.* (2012). This is because the shaped-hole has already provided higher film cooling so that the density effect is reduced. The effect of coolant density ratio on end-wall film cooling effectiveness was studied by Liu *et al.* (2013) and Rezasoltani *et al.* (2013) using the pressure sensitive paint measurement method. Liu *et al.* (2013) studied platform film cooling with compound angled cylindrical holes and concluded that higher density coolants are more resilient to lift-off and result in higher film cooling effectiveness (as shown in Fig. 23). Results also indicate that shaped holes offer significantly better coverage than cylindrical holes, but the density effect is not as significant as that with the cylindrical holes as mentioned (not shown here). The reason is same as indicated in the previous blade film cooling study. The effect of hub secondary flows (horseshoe vortices etc.) on film cooling is evident from the film cooling effectiveness contours.

4. CONCLUSIONS

A review of advanced hole exit geometry and coolant density effect on film cooling has been provided with an intent to show the development of analysis and design of complex hole exit shapes and the influence of coolant density on performance. Currently, the shaped hole envisioned in the first paper (Goldstein *et al.*, 1974) is an industry standard. The benefits of complex hole exits to provide better cooling effectiveness and overall heat flux reduction has been clearly shown over the past 4 decades. However, the basic challenge has been that these conceptual hole designs need to be manufactured economically without compromising mechanical integrity of the hot gas path components. As researchers and designers develop these geometries and show significant improvements, manufacturing techniques will be developed to make these holes on actual geometries just as was envisioned in the research papers. Similarly, studying the coolant density effect on film cooling has led the way for many researchers to fully investigate its effect on turbine blade with shaped-hole film cooling under realistic engine-like coolant-to-mainstream density ratio conditions.

NOMENCLATURE

BRM	blowing ratio
C	blade chord length
CY	cylindrical hole
DR	coolant-to-mainstream density ratio
d	hole diameter
h	convective heat transfer coefficient (W/m^2-K)
I	momentum flux ratio (=BR^2/DR)
l	hole length
p	hole pitch
q"	heat flux
R	gas constant
T	temperature, K
Tu	turbulence intensity
V	velocity, m/s
X	axial distance from hole leading edge

Greek Symbols

ρ	fluid density
η	adiabatic film-cooling effectiveness

Subscripts

c	coolant
f	film
g	mainstream air
w	wall
0	no film case

REFERENCES

Bogard, D.G., and Thole, K.A., 2006, "Gas Turbine Film Cooling," *Journal of Propulsion and Power*, **22**(2), 249-270. http://dx.doi.org/10.2514/1.18034

Bunker, R.S., 2005, "A Review of Turbine Shaped Film Cooling Technology," *Journal of Heat Transfer*, **127**(4), 441-453. http://dx.doi.org/10.1115/1.1860562

Bunker, R.S., 2009, "Film Cooling: Breaking the Limits of Diffusion Shaped Holes," *Turbine 2009 Symposium on Heat Transfer in Gas Turbine Systems*, Antalya, Turkey. http://dx.doi.org/10.1615/ICHMT.10.1615/ICHMT.2009.HeatTransfGasTurbSyst.160

Camci, C., and Arts, T., 1985a, "Short-Duration Measurements and Numerical Simulation of Heat Transfer along the Suction Side of a Gas Turbine Blade," *Journal of Engineering for Gas Turbines and Power*, **107**(4),991-997. http://dx.doi.org/10.1115/1.3239846

Camci, C., and Arts, T., 1985b, "Experimental Heat Transfer Investigation around the Film Cooled Leading Edge of a High Pressure Turbine Rotor Blade," *Journal of Engineering for Gas Turbines and Power*, **107**(4), 1016-1021. http://dx.doi.org/10.1115/1.3239805

Chen, A.F, Li, S.J., and Han, J.C., 2014, "Film Cooling with Forward and Backward Injection for Cylindrical and Fan-Shaped Holes Using PSP Measurement Technique," *Proceedings of ASME Turbo Expo*, GT2014-26232, Dusseldorf, Germany. http://dx.doi.org/10.1115/GT2014-26232

Dhungel, A., Lu, Y., Philips, A., Ekkad, S.V., and Heidmann, J., 2009, "Film Cooling from a Row of Holes Supplemented with Anti Vortex Holes," *Journal of Turbomachinery*, **131**(2), 021007. http://dx.doi.org/10.1115/1.2950059

Dittmar, J., Schulz, A., and Wittig, S., 2002, "Assessment of Various Film Cooling Configurations using Shaped and Compound Angle Holes Based on Large Scale Experiments," *Proceedings of ASME Turbo Expo*, GT2002-30176, Amsterdam, The Netherlands. http://dx.doi.org/10.1115/GT2002-30176

Dorrington, R., Bogard, D.G., and Bunker, R.S., 2007, "Film Effectiveness Performance for Coolant Holes Imbedded in Various Shallow Trench and Crater Depressions," *Proceedings of ASME Turbo Expo*, GT2007-27992, Montreal, Canada. http://dx.doi.org/10.1115/GT2007-27992

Eckert, E.R.G., Eriksen, V.L., Goldstein, R.J. and Ramsey, J.W., 1970, "Film Cooling Following Injection Through Inclined Circular Tubes," *Israel J. of Technology*, **8**, 145-154.

Ekkad, S.V., Zapata, D., and Han, J.C., 1997, "Film Cooling Effectiveness Over a Flat Surface with Air and CO2 Injection Through Compound Angle Holes Using a Transient Liquid Crystal Image Method," *Journal of Turbomachinery*, **119**(3), 587-593. http://dx.doi.org/10.1115/1.2841161

Ekkad, S.V., Han, J.C., and Du, H., 1998, "Detailed Film Cooling Measurements on a Cylindrical Leading Edge model: Effect of Freestream Turbulence and Coolant Density," *Journal of turbomachinery*, **120**(4), 799-807. http://dx.doi.org/10.1115/1.2841792

Ethridge, M.I., Cutbirth, J.M., and Bogard, D.G., 2001, "Scaling of Performance for Varying Density Ratio Coolants on an Airfoil With Strong Curvature and Pressure Gradient Effects," *Journal of Turbomachinery*, **123**(2), 231-237. http://dx.doi.org/10.1115/1.1343457

Gao, Z., Narzary, D.P., and Han, J.C., 2008, "Film Cooling on a Gas Turbine Blade Pressure side or Suction Side with Axial Shaped Holes," *International Journal of Heat and Mass Transfer*, **51**(9-10), 2139-2152. http://dx.doi.org/10.1016/j.ijheatmasstransfer.2007.11.010

Gao, Z., Narzary, D.P., and Han, J.C., 2009, "Film Cooling on a Gas Turbine Blade Pressure Side or Suction Side with Compound Angle Shaped Holes," *Journal of Turbomachinery*, **131**(1), 011019. http://dx.doi.org/10.1115/1.2813012

Gao, Z., and Han, J.C., 2009, "Influence of Hole Shape and Angle on Showerhead Film Cooling using PSP Technique," *Journal of Heat Transfer*, **131**(6), 061701. http://dx.doi.org/10.1115/1.3082413

Goldstein, R.J., 1971, "Film Cooling," *Advances in Heat Transfer*, Academic Press, San Diego, USA, **7**, 321-379.

Goldstein, R.J., Eckert, E.R.G., and Burggraf, F., 1974, "Effects of Hole Geometry and Density on Three-Dimensional Film Cooling," *International Journal of Heat and Mass Transfer*, **17**(5), 595-607. http://dx.doi.org/10.1016/0017-9310(74)90007-6

Gritsch, M., Schulz, A., and Wittig, S., 1998a, "Heat Transfer Coefficient Measurement of Film Cooling Holes with Expanded Exits," *Proceedings of ASME Turbo Expo*, 98-GT-28, Stockholm, Sweden. http://dx.doi.org/10.1115/98-GT-028

Gritsch, M., Schulz, A., and Wittig, S., 1998b, "Adiabatic Wall Effectiveness Measurements of Film-Cooling Holes with Expanded Exits," *Journal of Turbomachinery*, **120**(3), 549-556. http://dx.doi.org/10.1115/1.2841752

Gritsch, M., Schulz, A., and Wittig, S., 1998c, "Discharge Coefficient Measurements of Film Cooling Holes with Expanded Exits," *Journal of Turbomachinery*, **120**(3), 557-563. http://dx.doi.org/10.1115/1.2841753

Haller, B.R., and Camus, J-J, 1984, "Aerodynamic Loss Penalty Produced by Film Cooling Transonic Turbine Blades," *Journal of Engineering for Gas Turbines and Power,* **106**(1), 198-205. http://dx.doi.org/10.1115/1.3239535

Han, J.C. and Rallabandi, A.P., 2010, "Turbine Blade Film Cooling Using PSP Technique," *Frontiers in Heat and Mass Transfer*, **1**(1), 1-21. http://dx.doi.org/10.5098/hmt.v1.1.3001

Han, J.C., Dutta, S., and Ekkad, S.V., 2012, "Chapter 3 Turbine Film Cooling," *Gas Turbine Heat Transfer and Cooling Technology*, Second Edition, CRC Press-Taylor & Francis Group, Boca Raton, Florida, ISBN# 978-1-4398-5568-3.

Harrison, K.L., Dorrington, J.R., Dees, J.E., Bogard, D.G., and Bunker, R.S., 2007, "Turbine Airfoil Net Heat Flux Reduction with Cylindrical Holes Embedded in a Transverse Trench," *Proceedings of ASME Turbo Expo*, GT2007-27996, Montreal, Canada. http://dx.doi.org/10.1115/GT2007-27996

Heidmann, J., and Ekkad, S.V., 2008, "A Novel Anti-Vortex Turbine Film Cooling Hole Concept," *Journal of Turbomachinery*, **130**(3), 031020. http://dx.doi.org/10.1115/1.2777194

Hyams, D., McGovern, K., and Leylek, J., 1996, "Effects of Geometry on Slot-Jet Film Cooling Performance," *Proceedings of ASME Turbo Expo*, 96-GT-187, Birmingham, UK. http://dx.doi.org/10.1115/96-GT-187

Hyams, D.G., and Leylek, J.H., 2000, "A Detailed Analysis of Film Cooling Physics: Part III – Streamwise Injection with Shaped Holes," *Journal of Turbomachinery*, **122**(1), 122-132. http://dx.doi.org/10.1115/1.555435

Hylton, L.D., Nirmalan, V., Sultanian, B.K., and Kaufmann, R.M., 1988, "The Effects of Leading Edge and Downstream Film Cooling on Turbine Vane Heat Transfer", NASA CR-182133.

Kercher, D.M., 1998, "A Film-Cooling CFD Bibliography: 1971-1996," *International Journal of Rotating Machinery*, **4**(1), 61-72. http://dx.doi.org/10.1155/S1023621X98000062

Kohli, A., Thole, K.A., 1997, "A CFD Investigation on the Effect of Entrance Flow Conditions in Discrete Film Holes," *Proceedings of ASME* 32nd National Heat Transfer, **12**, 223-232.

Kusterer, K., Bohn, D., Sugimoto, T., and Tanaka, R., 2006, "Double-jet Ejection of Cooling Air for Improved Film-cooling," *Proceedings of ASME Turbo Expo*, GT2006-90854, Barcelona, Spain. http://dx.doi.org/10.1115/GT2006-90854

Leylek, J.H., and Zerkle, R.D., 1994, "Discrete-Jet Film Cooling: A Comparison of Computational Results with Experiments," *Journal of Turbomachinery*, **116**(3), 358-368. http://dx.doi.org/10.1115/1.2929422

Li, S.J., Yang, S.F., and Han, J.C., 2014, "Effect of Coolant Density on Leading Edge Showerhead Film Cooling Using the Pressure Sensitive Paint Measurement Technique," *Journal of Turbomachinery*, **136**(5), 051011. http://dx.doi.org/10.1115/1.4025225

Liu, K., Yang, S.F., and Han, J.C., 2012, "Influence of Coolant Density on Turbine Blade Film-Cooling with Compound-Angle Shaped Holes," *Proceedings of ASME Turbo Expo*, GT2012-69117, Copenhagen, Denmark. http://dx.doi.org/10.1115/GT2012-69117

Liu, K., Yang, S.F., and Han, J.C., 2013, "Influence of Coolant Density on Turbine Platform Film Cooling with Stator-Rotor Purge Flow and Compound-Angle Holes," *Proceedings of ASME Turbo Expo*, GT2013-94155, San Antonio, Texas, USA. http://dx.doi.org/10.1115/GT2013-94155

Lu, Y., Nasir, H., and Ekkad, S.V., 2005, "Film Cooling from a Row of Holes Embedded in Transverse Slots," *Proceedings of ASME Turbo Expo*, GT2005-68598, Reno, Nevada, USA. http://dx.doi.org/10.1115/GT2005-68598

Lu, Y., Dhungel, A., Ekkad, S.V., and Bunker, R.S., 2007a, "Effect of Trench Width and Depth on Film Cooling from Cylindrical Holes Embedded in Trenches," *Proceedings of ASME Turbo Expo*, GT2007-27388, Montreal, Canada. http://dx.doi.org/10.1115/GT2007-27388

Lu, Y., Dhungel, A., Ekkad, S.V., and Bunker, R.S., 2007b, "Film Cooling Measurements for Cratered Cylindrical Inclined Holes," *Proceedings of ASME Turbo Expo*, GT2007-27386, Montreal, Canada. http://dx.doi.org/10.1115/GT2007-27386

Makki, Y.H., and Jakubowski, G.S., 1986, "An Experimental Study of Film Cooling from Diffused Trapezoidal Shaped Holes," *Proceedings of AIAA and ASME Joint Thermophysics and Heat Transfer Conference*, 86-1326, Boston, Massachusetts, USA.

Narzary, D.P., LeBlanc, C., and Ekkad, S.V., 2011, "Film Cooling Performance of an Anti-Vortex Hole on a Flat Plate," *Proceedings of ASME/JSME 2011 8th Thermal Engineering Joint Conference*, AJTEC2011-44161, Honolulu, Hawaii. http://dx.doi.org/10.1115/AJTEC2011-44161

Narzary, D.P., Liu, K.C., Rallabandi, A.P., and Han, J.C., 2012, "Influence of Coolant Density on Turbine Blade Film-Cooling using Pressure Sensitive Paint Technique," *Journal of Turbomachinery*, **134**(3), 031006. http://dx.doi.org/10.1115/1.4003025

Nirmalan, N.V., and Hylton, L.D., 1990, "An Experimental Study of Turbine Vane Heat Transfer with Leading Edge and Downstream Film Cooling," *Journal of Turbomachinery*, **112**(3), 477-487. http://dx.doi.org/10.1115/1.2927683

Papell, S.S., 1984, "Vortex Generating Flow Passage Design for Increased Film Cooling Effectiveness and Surface Coverage," *Proceedings of 22nd National Heat Transfer Conference,* 84-HT-22, Niagara Falls, New York, USA.

Pedersen, D.R., Eckert, E.R.G., and Goldstein, R.J., 1977, "Film Cooling with Large Density Diffferences between the Mainstream and the Secondary Fluid Measured by the Heat-Mass Transfer Analogy," *Journal of Heat Transfer*, **99**(4), 620-627. http://dx.doi.org/10.1115/1.3450752

Rallabandi, A.P., Grizzle, J., and Han, J.C., "Effect of Upstream Step on Flat Plate Film Cooling Effectiveness Using PSP," *Journal of Turbomachinery*, **133**(4), 041024. http://dx.doi.org/10.1115/1.4002422

Rallabandi, A.P., Li, S.J., and Han, J.C., 2012, "Unsteady Wake and Coolant Density Effects on Turbine Blade Film Cooling using Pressure Sensitive Paint Technique," *Journal of Heat Transfer*, **134**(8), 081701. http://dx.doi.org/10.1115/1.4006748

Rezasoltani, M., Schobeiri, M.T., and Han, J.C., 2013, "Experimental Investigation of the Effect of Purge Flow on Film Cooling Effectiveness on a Rotating Turbine with Non-Axisymmetric Endwall Contouring," *Proceedings of ASME Turbo Expo*, GT2013-94807, San Antonio, Texas, USA. http://dx.doi.org/10.1115/GT2013-94807

Rezasoltani, M., Lu, K., Schobeiri, M.T., and Han, J.C., 2015, "A Combined Experimental and Numerical Study of the Turbine Blade Tip Film Cooling Effectiveness under Rotation Condition," *Journal of Turbomachinery*, **137**(5), 051009. http://dx.doi.org/10.1115/1.4028745

Salcudean, M., Gartshore, I., Zhang, K., and Barnea, Y., 1994. "Leading Edge Film Cooling of a Turbine Blade Model Through Single and Double Row Injection: Effects of Coolant Density," *Proceedings of ASME Turbo Expo*, 94-GT-002, The Hague, Netherlands. http://dx.doi.org/10.1115/94-GT-002

Sargison, J.E., Guo, S.M., Oldfield, M.L.G., Lock, G.D., and Rawlinson, A.J., 2001a, "A Converging Slot-Hole Film-Cooling Geometry Part 1: Low-Speed Flat-Plate Heat Transfer and Loss," *Proceedings of ASME Turbo Expo*, 2001-GT-0126, New Orleans, Louisiana, USA. http://dx.doi.org/10.1115/2001-GT-0126

Sargison, J.E., Guo, S.M., Oldfield, M.L.G., Lock, G.D., and Rawlinson, A.J., 2001b, "A Converging Slot-Hole Film-Cooling Geometry Part 2: Transonic Nozzle Guide Vane Heat Transfer and Loss," *Proceedings of ASME Turbo Expo*, 2001-GT-0127, New Orleans, Louisiana, USA. http://dx.doi.org/10.1115/2001-GT-0127

Saumweber, C., and Schulz, A., 2003, "Interactions of Film Cooling Rows: Effects of Hole Geometry and Row Spacing on the Cooling Performance Downstream of the Second Row of Holes," *Proceedings of ASME Turbo Expo*, GT2003-38195, Atlanta, Georgia, USA. http://dx.doi.org/10.1115/GT2003-38195

Saumweber, C., and Schulz, A., 2008a, "Free-Stream Effects on the Cooling Performance of Cylindrical and Fan-Shaped Cooling Holes," *Proceedings of ASME Turbo Expo*, GT2008-51030, Berlin, Germany. http://dx.doi.org/10.1115/GT2008-51030

Saumweber, C., and Schulz, A., 2008b, "Comparison the Cooling Performance of Cylindrical and Fan-Shaped Cooling Holes with Special Emphasis on the Effect of Internal Coolant Cross-Flow," *Proceedings of ASME Turbo Expo*, GT2008-51036, Berlin, Germany. http://dx.doi.org/10.1115/GT2008-51036

Saumweber, C., and Schulz, A., 2008c, "Effect of Geometry Variations on the Cooling Performance of Fan-Shaped Cooling Holes," *Proceedings of ASME Turbo Expo*, GT2008-51038, Berlin, Germany. http://dx.doi.org/10.1115/GT2008-51038

Schmidt, D.L., Sen, B., and Bogard, D.G., 1996, "Film Cooling with Compound Angle Holes: Adiabatic Effectiveness," *Journal of Turbomachinery*, **118**(4), 807-813. http://dx.doi.org/10.1115/1.2840938

Sen, B., Schmidt, D.L., and Bogard, D.G., 1996, "Film Cooling with Compound Angle Holes: Heat Transfer," *Journal of Turbomachinery*, **118**(4), 800-806. http://dx.doi.org/10.1115/1.2840937

Sinha, A.K., Bogard, D.G., and Crawford, M.E., 1991, "Film-Cooling Effectiveness Downstream of a Single Row of Holes with Variable Density Ratio," *Journal of Turbomachinery*, **113**(3), 442-449. http://dx.doi.org/10.1115/1.2927894

Teng, S., Han, J.C., and Poinsatte, P.E., 2001a, "Effect of Film-Hole Shape on Turbine-Blade Heat-Transfer Coefficient Distribution," *Journal of Thermophysics and Heat Transfer*, **15**(3), 249-256. http://dx.doi.org/10.2514/2.6610

Teng, S., Han, J.C., and Poinsatte, P.E., 2001b, "Effect of Film-Hole Shape on Turbine-Blade Film Cooling-Performance," *Journal of Thermophysics and Heat Transfer*, **15**(3), 257-265. http://dx.doi.org/10.2514/2.6621

Thole, K., Gritsch, M., Schulz, A., and Wittig, S., 1998, "Flowfield Measurements for Film-Cooling Holes with Expanded Exits," *Journal of Turbomachinery*, **120**(2), 327-336. http://dx.doi.org/10.1115/1.2841410

Waye, S., and Bogard, D.G., 2006, "High Resolution Film Cooling Effectiveness Measurements of Axial Holes Embedded in a Transverse Trench with Various Trench Configurations," *Proceedings of ASME Turbo Expo*, GT2006-90226, Barcelona, Spain. http://dx.doi.org/10.1115/GT2006-90226

16

PERFORMANCE ASSESSMENT AND EMPIRICAL CORRELATION IN A HEAT EXCHANGER SQUARE DUCT WITH DIAGONAL INSERTED GENERATORS

Amnart Boonloi*

Department of Mechanical Engineering Technology, College of Industrial Technology, King Mongkut's University of Technology North Bangkok, Bangkok, 10800, Thailand

ABSTRACT

A mathematical analysis of the heat transfer enhancement, thermal performance and flow configurations in a heat exchanger square duct with diagonal inserted plate vortex generators is presented. The 30° V–shaped baffles are modified and placed on the double sides of the thin plate or frame (with no plate) which inserted diagonally in the square duct. The effects of blockage ratio (b/H, BR), the pitch ratio (p/H, PR), flow direction (V–Downstream and V–Upstream) and configuration of inserting plate are investigated for Reynolds number based on the hydraulic diameter of the square duct, D_h, Re = 100 – 2000. The finite volume method applies for the computational domains. The numerical results show that the use of the diagonal inserted generators can help to increase heat transfer rates and thermal performance in the square duct higher than the smooth duct with no generators. The presences of the diagonal inserted generators not only increase heat transfer rates but also increase very enlarges pressure loss. The maximum thermal enhancement factor is around 3.25 at the highest Reynolds number. In addition, the use of the diagonal inserted generators can help to install and also comfortable to maintenance in the heat exchange duct.

Keywords: *Diagonal inserted plate; Heat exchanger; Square duct; Thermal performance; Vortex generators*

1. INTRODUCTION

Various vortex generators; rib, baffle, winglet, etc., normally uses in the heating system. The vortex generators can help to increase heat transfer rates and the thermal performance lead to a compact heat exchanger and save more cost for the operating system. Many investigators had been studying the methods to augmented heat transfer rate in the heat exchanger with using vortex generators on both experimental and numerical. The numerical method can help to describe the flow configurations and heat transfer characteristics that the ways to improve the thermal performance and develop the design of the compact heat exchanger. The previous works for the investigations of the heat transfer augmentation by using a numerical method are as follows table 1.

As the previous works, there are found that the uses of the vortex generators were almost placed on the channel wall or on the tube wall. The vortex generators which placed on the tested walls were difficult to forming and installing. It has always performed a little gap between tested wall and the vortex generators. Therefore, the modified vortex generators for suitable and easy to install are important to investigate. Except from the installation vortex generators, the V–shaped vortex generators which provide a higher heat transfer rate and the thermal performance were found. The V–baffle vortex generators perform higher heat transfer rate and the thermal performance in comparison with other shapes such as inclined baffle. Therefore, this work will be focused on the installation method for the V–shaped baffle vortex generators in the heat exchanger square duct.

The investigations of various turbulators in flat plate-fin heat exchanger on both numerically and experimentally were reported by Joardar and Jacobi, 2005; Gentry and Jacobi, 1997; Chen and Shu, 2004; Wu and Tao, 2012; and Li-Ting *et al.*, 2009. They concluded that the uses of turbulators in the compact heat exchanger lead to enhance heat transfer and thermal performance in the heat system.

The V–shaped baffle vortex generators are placed on both sides of a thin plate and insert diagonally in the square duct with V–tip pointing downstream called "V–Downstream" and V–tip pointing upstream called "V–Upstream". Moreover, to reduce the pressure loss that is done by inserted thin plate, the design of the wire frame for installing V–shaped baffle diagonally is studied. The effects of the blockage ratio, pitch ratio, flow direction and Reynolds number are presented numerically in three dimensional.

2. V–SHAPED BAFFLES CONFIGURATIONS AND BOUNDARY CONDITION

The square duct with double sides of the V–baffle inserted diagonally is presented in Fig.1a, while the computational domain is presented in Fig. 1b. The modified V–baffle vortex generators in a square duct are referred from Refs. (Promvonge *et al.*, 2012). The periodic boundaries (Promvonge *et al.*, 2012) which the flow structure and heat transfer behavior profiles repeat itself from one to another module is applied for inlet and outlet of the computation domain. The tested fluid is air enters to the square duct with constant mass flow rate at an inlet temperature, T_{in} and flow over the V–baffle turbulators. The baffle height, b where b/H is identified as the blockage ratio, BR. The longitudinal pitch, L or the space between the baffle positioning is set to L = H, L/H is known as the pitch ratio, PR. As the literature reviews above, the flow attack angle of the V–baffle of 30° is used. The use of 30°V–baffle can be optimized between the augmenting heat transfer (45°V–baffle) and the reducing of the pressure loss (20°V–baffle). The case studies and the boundary conditions for this work are as follows table 2 and 3, respectively.

* *Corresponding author. Email: amnartb@kmutnb.ac.th*

Table 1 The investigations of the vortex generators with numerical method.

Authors	*Studied cases*	Nu/Nu_0	f/f_0	η
Jedsadaratanachai *et al.*, 2011	30° inclined baffle Inline, two opposite walls, square channel BR = 0.2 PR = 0.5 – 2.5 Re = 100 – 2000	1 – 9.2	1 – 21.5	3.78
Kwankaomeng and Promvonge, 2010	30° inclined baffle One side, square channel BR = 0.1 – 0.5 PR = 1.0 – 2.0 Re = 100 – 1000	1 – 9.23	1.09 – 45.31	3.1
Promvonge *et al.*, 2010	30° inclined baffle Inline, two opposite walls, square channel BR = 0.1 – 0.3 PR = 1.0 – 2.0 Re = 100 – 2000	1.2 – 11.0	2 – 54	4
Promvonge and Kwankaomeng, 2010	45° V–baffle Staggered, two opposite walls, AR=2 channel BR = 0.05 – 0.3 PR = 1.0 Re = 100 – 1200	1 – 11	2 – 90	2.75
Promvonge *et al.*, 2010	45° inclined baffle Inline–staggered, two opposite walls, square channel BR = 0.05 – 0.3 PR = 1.0 Re = 100 – 1000	1.5 – 8.5	2 – 70	2.6
Promvonge *et al.*, 2012	45° V–baffle Inline Downstream, two opposite walls, square channel BR = 0.1 – 0.3 PR = 1.0 – 2.0 Re = 100 – 2000	1 – 21	1.1 – 225	3.8
Boonloi, 2014	20° V–baffle Inline Downstream–Upstream, two opposite walls, square channel BR = 0.1 – 0.3 PR = 1.0 Re = 100 – 2000	1 – 13	1 – 52	4.2
Boonloi and Jedsadaratanachai, 2013	30° V–baffle Downstream, One side, square channel BR = 0.1 – 0.5 PR = 1.0 – 2.0 Re = 100 – 1200	1 – 14.49	2.18 – 313.24	2.44

Table 2 Case studies

Configuration	BR	PR	*Flow direction*	*Reynolds number*
V–baffle placed on thin plate (with plate)	0.1 – 0.3	1 – 2	V–Downstream V–Upstream	100 – 2000
V–baffle fixed with frame (no plate)	0.1 – 0.3	1 – 2	V–Downstream V–Upstream	100 – 2000

Table 3 Boundary conditions

Zones	*Boundary condition*
Inlet	Periodic boundary
Outlet	Periodic boundary
All of the square duct walls	- Constant temperature 310K - No slip wall condition
V–baffle vortex generators	Adiabatic wall condition
- Thin plate - Frame	Adiabatic wall condition
Tested fluid	Air at constant temperature 300K (Pr = 0.7)

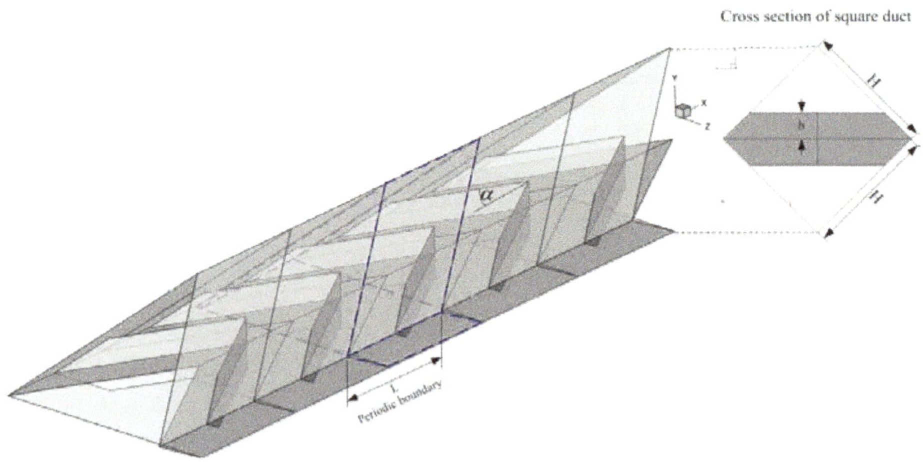

(a)

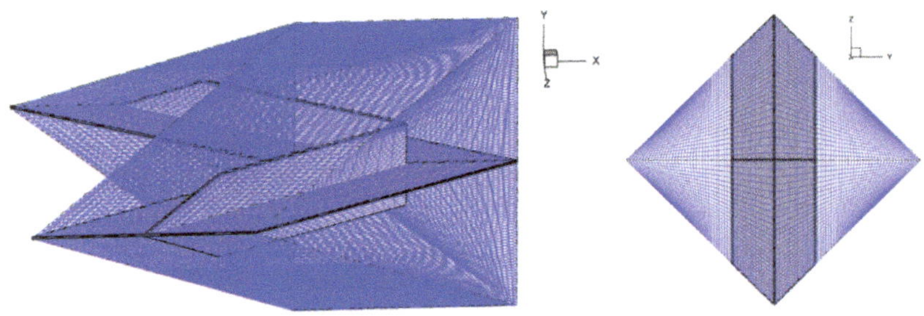

Computational domain

(b)

Fig. 1 (a) Square duct with double sides V–baffles taped inserted diagonally and (b) Computational domain.

3. MATHEMATICAL FOUNDATION AND GRID SYSTEM

The parameters of interest in the current work are the Reynolds number (Re), friction factor (f), Nusselt number (Nu) and Thermal Enhancement Factor (η). The Reynolds number is defined as:

$$\mathrm{Re} = \rho \bar{u} D / \mu \tag{1}$$

The friction factor, f is calculated by pressure drop, Δp across the periodic module of the square duct, L as:

$$\mathrm{f} = \frac{(\Delta p / L)}{\frac{1}{2}\rho \bar{u}^2} \tag{2}$$

The heat transfer is measured by the local Nusselt number which can be written as:

$$\mathrm{Nu}_x = \frac{h_x D}{k} \tag{3}$$

The average Nusselt number can be obtained by:

$$\mathrm{Nu} = \frac{1}{A}\int \mathrm{Nu}_x \partial A \tag{4}$$

The Thermal Enhancement Factor (η) is defined as the ratio of the heat transfer coefficient of an augmented surface, h to that of a smooth surface, h_0, at an equal pumping power and given by:

$$\eta = \left.\frac{h}{h_0}\right|_{pp} = \left.\frac{\mathrm{Nu}}{\mathrm{Nu}_0}\right|_{pp} = (\mathrm{Nu}/\mathrm{Nu}_0)/(\mathrm{f}/\mathrm{f}_0)^{1/3} \tag{5}$$

where, Nu_0 and f_0 stand for Nusselt number and friction factor for the smooth duct, respectively.

The mathematical foundations are denoted from Refs. (Promvonge *et al.*, 2012). The mathematical model for fluid flow and heat transfer in a square channel was developed under the following assumptions:

- Steady three–dimensional fluid flow and heat transfer
- The flow is laminar and incompressible
- Constant fluid properties
- Body forces and viscous dissipation are ignored
- Negligible radiation heat transfer

Based on the above assumptions, the tube flow is governed by the continuity, the Navier–Stokes equations and the energy equation. In the Cartesian tensor system these equations can be written as follows:

Continuity equation:

$$\frac{\partial}{\partial x_i}(\rho u_i) = 0 \tag{6}$$

Momentum equation:

$$\frac{\partial(\rho u_i u_j)}{\partial x_j} = -\frac{\partial p}{\partial x_i} + \frac{\partial}{\partial x_j}\left[\mu\left(\frac{\partial u_i}{\partial x_j} + \frac{\partial u_j}{\partial x_i}\right)\right] \tag{7}$$

Energy equation:

$$\frac{\partial}{\partial x_i}(\rho u_i T) = \frac{\partial}{\partial x_j}\left(\Gamma \frac{\partial T}{\partial x_j}\right) \tag{8}$$

where, Γ is the thermal diffusivity and is given by:

$$\Gamma = \frac{\mu}{Pr} \tag{9}$$

Apart from the energy equation discretized by the QUICK scheme, the governing equations were discretized by the second order upwind scheme, decoupling with the SIMPLE algorithm and solved using a finite volume approach (Patankar,1980; Roache, 1998). The solutions were considered to be converged when the normalized residual values were less than 10^{-5} for all variables, but less than 10^{-9} only for the energy equation.

The grid system of 180,400 cells was adopted for the current computational model due to the increasing the number of cells result in the different values on both heat transfer and friction factor less than ±0.25% as presented in Table 4 for BR = 0.20, Re = 2000, PR = 1, V–downstream case and with plate

Table 4 Grid system

Grid cell	Nu	f	% error Nu	% error f
81,000	40.01121	2.14385	-0.86460	6.53127
120,100	39.99520	2.14511	-0.82424	6.47646
180,400	39.76985	2.28956	-0.25615	0.17876
201,000	39.65485	2.28546	0.03376	0.35751
250,000	39.66824	2.29366	0	0

4. NUMERICAL RESULTS AND DISCUSSION

4.1 Validation of smooth duct

The validation of the computational domain is necessary for the investigation with the numerical methods. The verifications for both heat transfer and friction factor are studied by comparison with the previous values (Incropera and Dewitt, 2006) under a similar operating condition. The results shows agree well within ±0.15% on both heat transfer and friction factor in terms of Nu and f, respectively.

4.2 Flow topology and heat transfer characteristic of V–baffle with thin plate

The flow structures and heat transfer characteristics in a square duct with 30° V–baffle inserted diagonally (with plate), are present in Figs. 2 – 4. The flow configurations are displayed in term of streamlines in transverse planes while the heat transfer is presented in forms of temperature contours in transverse planes and the Nu_x contour.

Figs. 2a and b show streamlines in transverse planes (module/4) for V–Downstream and V–Upstream, respectively, for BR = 0.2, PR = 1 and Re = 800. As seen, the flow structure is appeared which consist four main vortex flows and small vortices at the corner of the square duct on both V–Downstream and V–Upstream cases. Considering at the lower part of the streamlines plane, the counter rotating vortex flows with common-flow-down and common-flow-up are induced for V–Downstream and V–Upstream, respectively. The core of the vortex flows is changed depending on the position of the V–baffle. The vortex flows recurrence as one to another module, so, the first plane and the fifth plane are similar flow profiles for both cases.

Figs. 3a and b display the temperature contours in transverse planes (module/4) for V–Downstream and V–Upstream, respectively, for BR = 0.2, PR = 1 and Re = 800. There are found that the use of vortex generators can help with mixing the temperature between the core of the duct and near the wall regimes. The V–Downstream case provides better mixing than the V–Upstream case. The blue contours, the temperature ≅ 300K, are clearly seen at the upper and lower corners of the V–Upstream case due to the jet flows of the V–Upstream case can't induce in these regimes. The heat transfer phenomena's are related to the flow configuration part

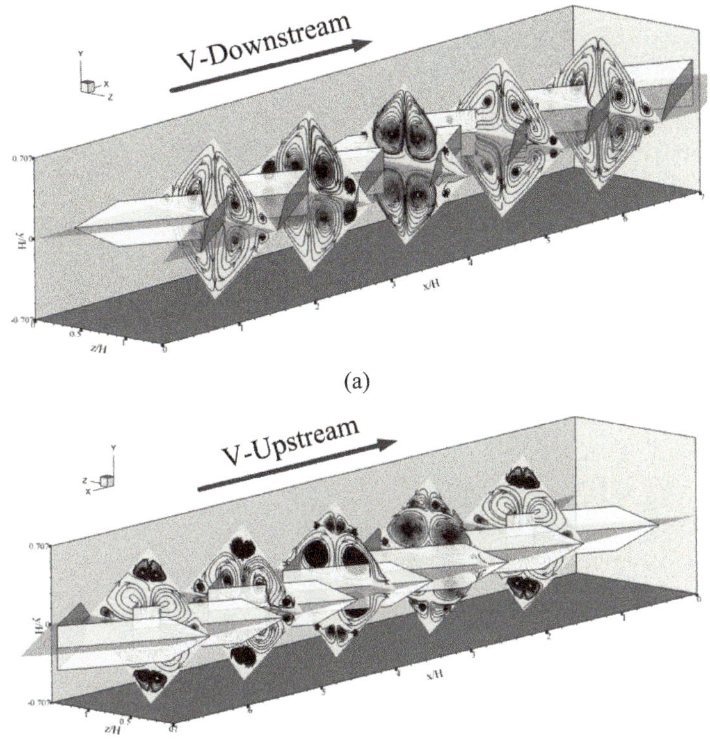

(a)

(b)

Fig. 2 Streamlines in transverse planes for double sides of 30° V–baffle inserted diagonally (with plate) (a) V–Downstream and (b) V–Upstream at BR = 0.2, PR = 1 and Re = 800.

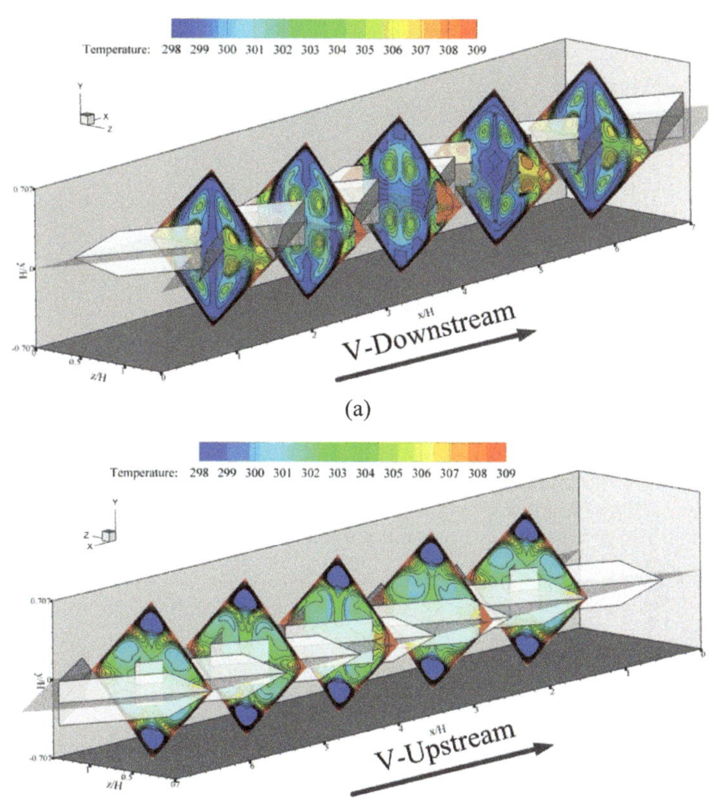

(a)

(b)

Fig. 3 Temperature contours in transverse planes for double sides of 30° V–baffle inserted diagonally (with plate) (a) V–Downstream and (b) V–Upstream at BR = 0.2, PR = 1 and Re = 800.

The heat transfer characteristics are also presented in Figs. 4a and b, in term of the Nusselt number contours on the duct walls for V–Downstream and V–Upstream, respectively, at BR = 0.2, PR = 1 and Re = 800. There are found that the use of V–baffle give higher heat transfer rate than the smooth duct for all cases. The V–Downstream performs higher heat transfer rate than the V–Upstream. The peaks of heat transfer areas are appearing, except for the small regimes at the duct corners.

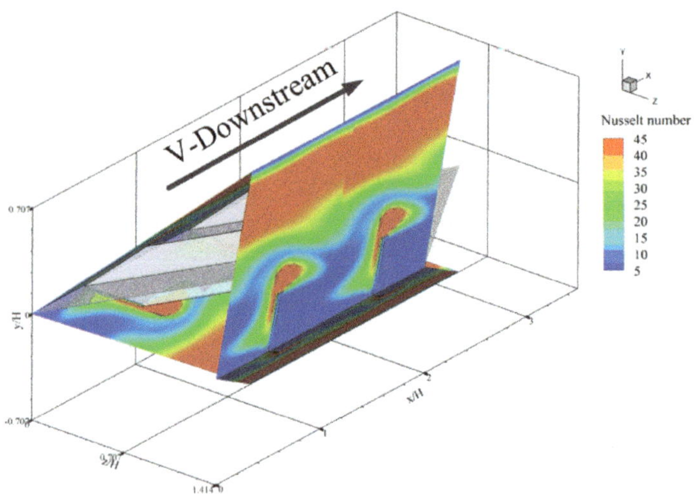

(a)

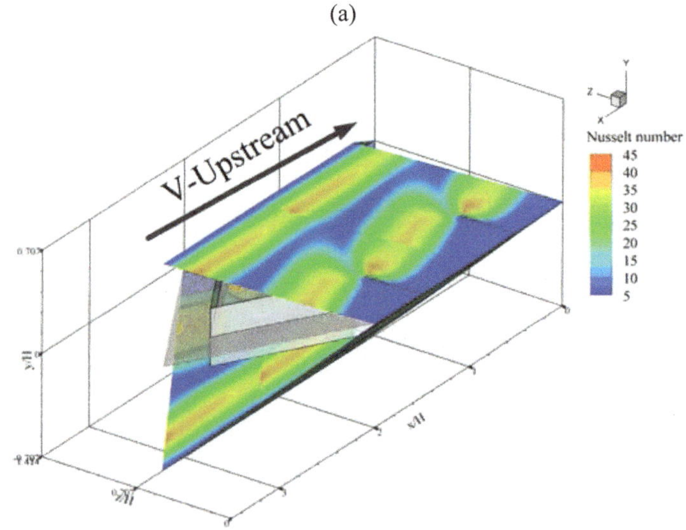

(b)

Fig. 4 Nu_x contours in transverse planes for double sides of 30° V–baffle inserted diagonally (with plate) (a) V–Downstream and (b) V–Upstream at BR = 0.2, PR = 1 and Re = 800.

4.3 Flow topology and heat transfer characteristic of V–baffle with no plate

As the pre-results, it is found that the use of V–baffle inserted diagonally in a square channel gives very enlarge pressure in comparison with the vortex generators which placed on the walls and also higher than the smooth square duct with no vortex generators. The new design for reducing the pressure loss in the heat exchange duct system is presented with changed the thin plate as a frame (no plate). The streamlines in transverse planes, temperature contours and Nux contours are presented as Figs. 5, 6 and 7, respectively.

As seen, the flow configurations and heat transfer characteristics are similar profile as V–baffle with plate case, but the V–Upstream case of no plate seems to be the higher heat transfer rate than with a plate when considering at the Nu_x contours in Fig. 7b

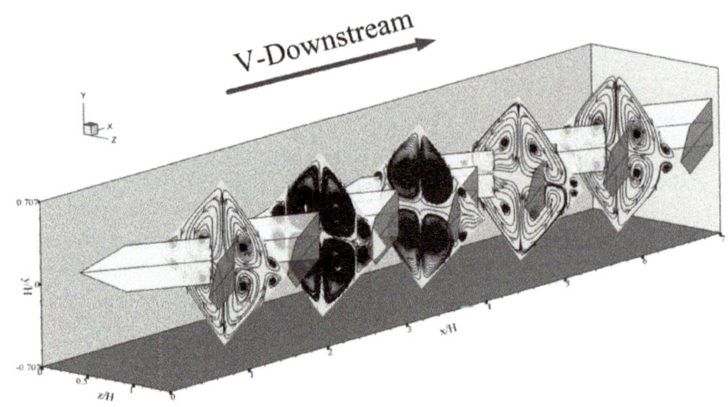

(a)

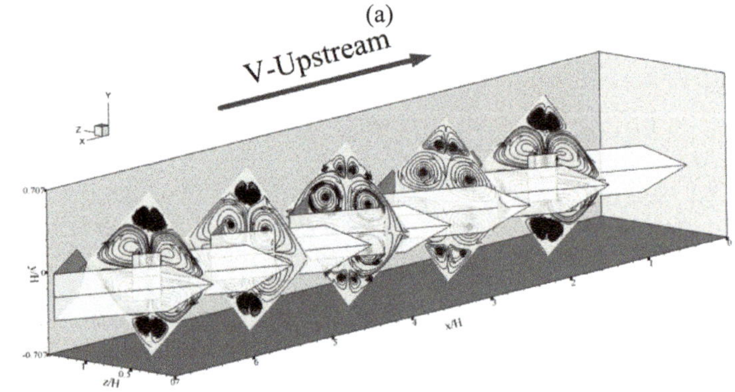

(b)

Fig. 5 Streamlines in transverse planes for double sides of 30° V–baffle inserted diagonally (no plate) (a) V–Downstream and (b) V–Upstream at BR = 0.2, PR = 1 and Re = 800.

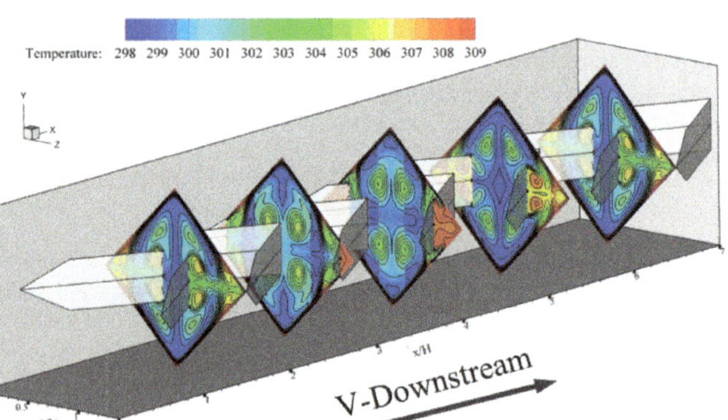

(a)

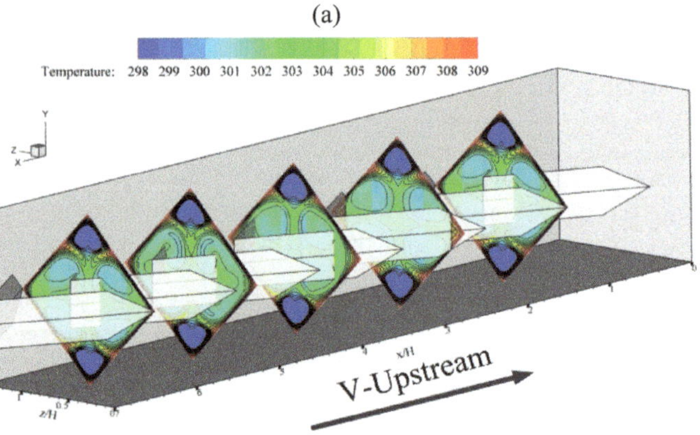

(b)

Fig. 6 Temperature contours in transverse planes for double sides of 30° V–baffle inserted diagonally (no plate) (a) V–Downstream and (b) V–Upstream at BR = 0.2, PR = 1 and Re = 800.

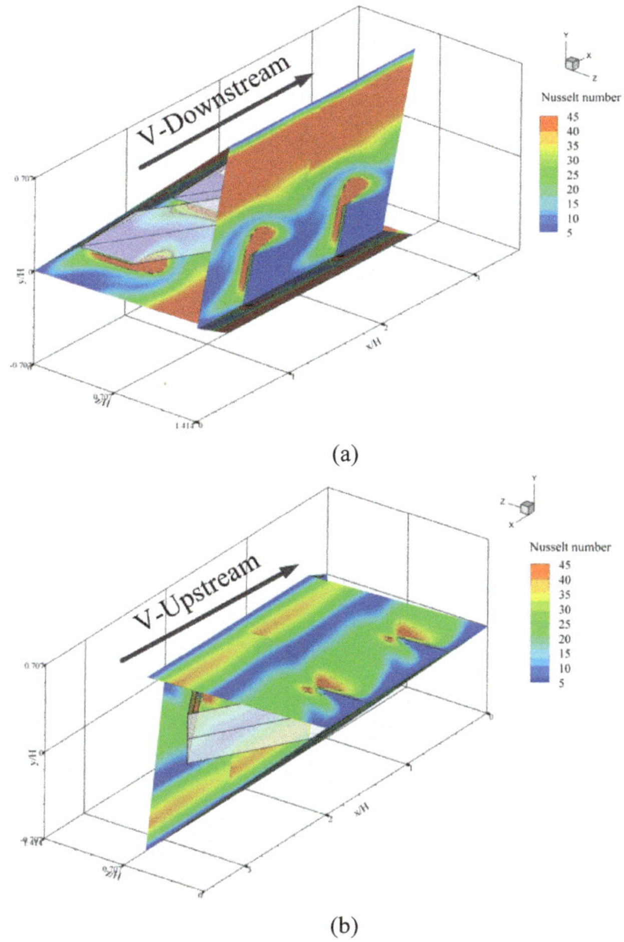

Fig. 7 Nu_x contours in transverse planes for double sides of 30° V–baffle inserted diagonally (no plate) (a) V–Downstream and (b) V–Upstream at BR = 0.2, PR = 1 and Re = 800.

4.4 Performance Evaluation

The performance evaluations of the 30° V–baffle vortex generators inserted diagonally are presented in terms of Nu/Nu_0, f/f_0 and η as Figs. 8 – 13. The effects of Reynolds number, blockage ratio and the pitch ratio on both V–Downstream and V–Upstream are investigated. In general, the rising BR and Reynolds number result in the increasing heat transfer rate and the friction factor while the increasing PR leads to the reducing trends on both heat transfer and friction factor values.

For 30° V–baffle with plate, there are found that the increasing rate of the heat transfer for V–Downstream provides higher than the V–Upstream, especially, the BR > 0.1. The Nu/Nu_0 values are around 2 – 20 and 2 – 11 times over the smooth square duct for V–Downstream and V–Upstream, respectively. This means that the maximum heat transfer for V–Downstream is higher than the V–Upstream around 2 times. The f/f_0 value, the V–Downstream performs the highest friction factor than the V–Upstream around 2 times at BR = 0.3, Re = 2000 and PR = 1. The f/f_0 values are around 1 – 320 and 1 – 150 times above the smooth duct with no vortex generators for V–Downstream and V–Upstream, respectively. The peak range for the increasing friction factor is found at 0.25 < BR < 0.30. The thermal enhancement factor, η, is found to be maximum at BR = 0.15 and PR = 1 around 3.2 on the highest Reynolds number for V–Downstream while around 2.75 at the lowest BR, PR = 1 for V–Upstream.

For 30° V-baffle no plate, the Nu/Nu_0 is around 2 – 21 and 2 – 12.5 times higher than the smooth duct with no vortex generators for V–Downstream and V–Upstream, respectively, while the f/f_0 is about 1 – 320 and 1 – 150 times over the smooth duct. The optimum thermal enhancement factor is about 3.25 in both cases with the differential case. The V–Downstream performs the maximum thermal enhancement factor at BR = 0.2 and PR = 1, while the V–Upstream performs maximum point at BR = 0.1 and PR = 1.5, at the highest Reynolds number.

Figs. 14 – 15 show a comparison between with plate and no plate cases for Nu/Nu_0, f/f_0 and η, respectively, at PR = 1. As seen, there are notices that the heat transfer and friction factor values are nearly values on both cases. This means that the aim to reduce the pressure loss by changing thin plate as a wire frame to installing the V–baffle is not beneficial, but can help to increase the thermal enhancement factor for V–Upstream cases, from 2.75 to 3.25.

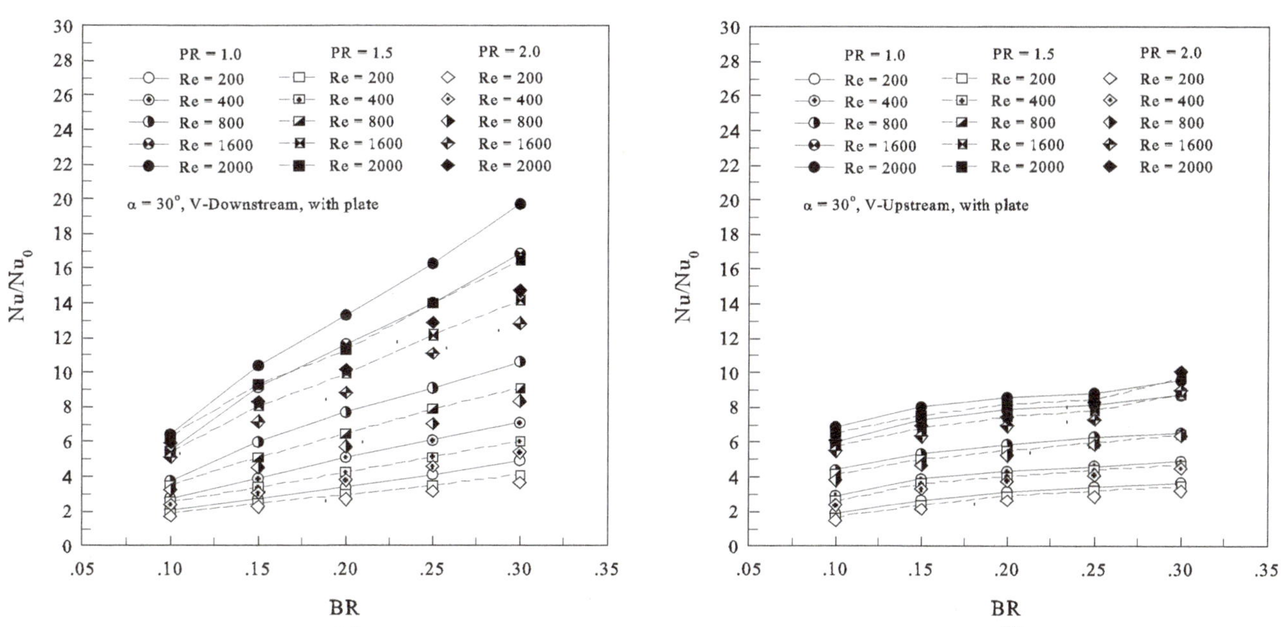

Fig. 8 The variations of Nu/Nu_0 with BR values at various PR and Re values for 30° V–baffle inserted diagonally (with plate) for (a) V–Downstream and (b) V–Upstream.

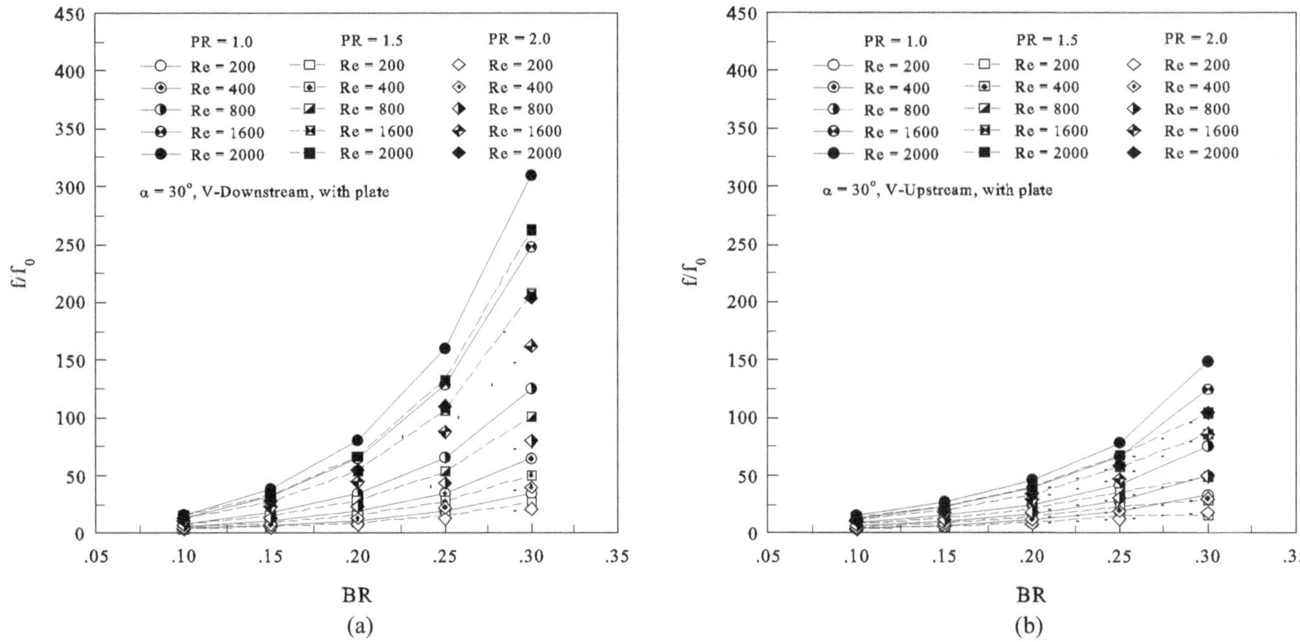

Fig. 9 The variations of f/f0 with BR values at various PR and Re values for 30° V–baffle inserted diagonally (with plate) for (a) V–Downstream and (b) V–Upstream.

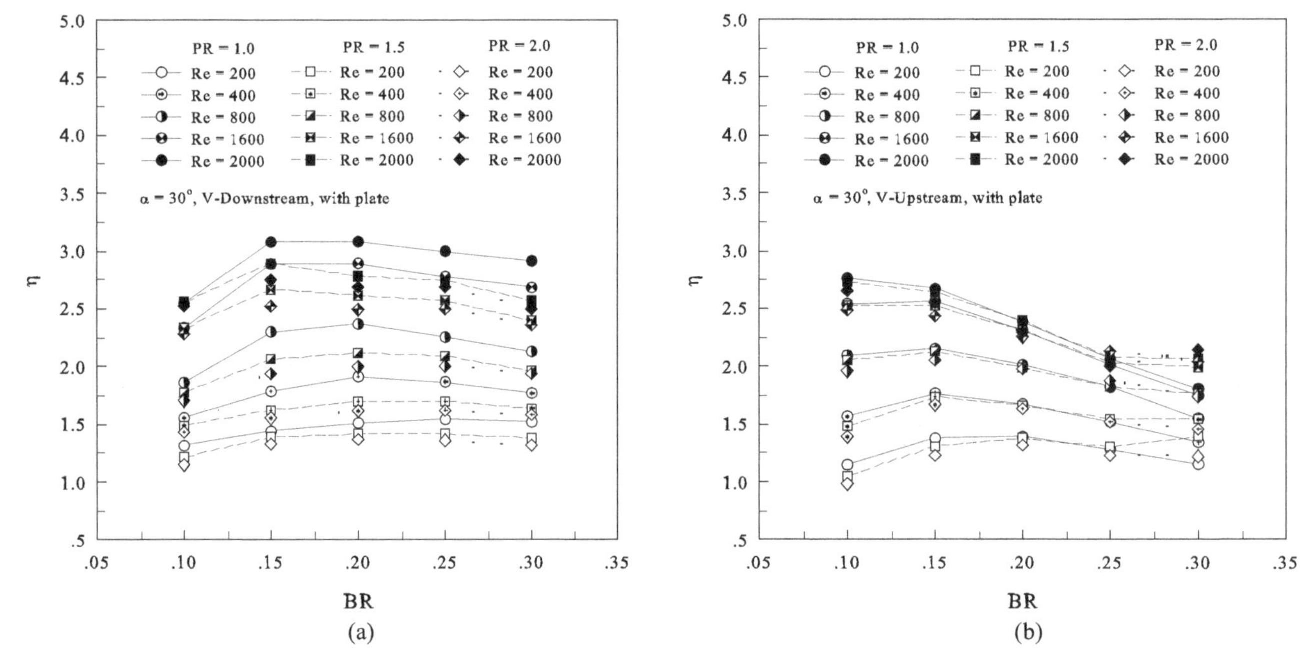

Fig. 10 The variations of η with BR values at various PR and Re values for 30° V–baffle inserted diagonally (with plate) for (a) V–Downstream and (b) V–Upstream.

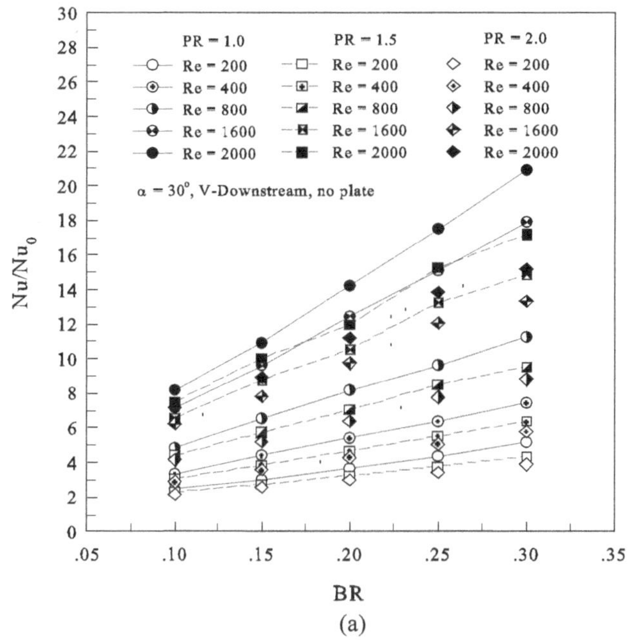

Fig. 11 The variations of Nu/Nu0 with BR values at various PR and Re values for 30° V–baffle inserted diagonally (no plate) for (a) V–Downstream and (b) V–Upstream.

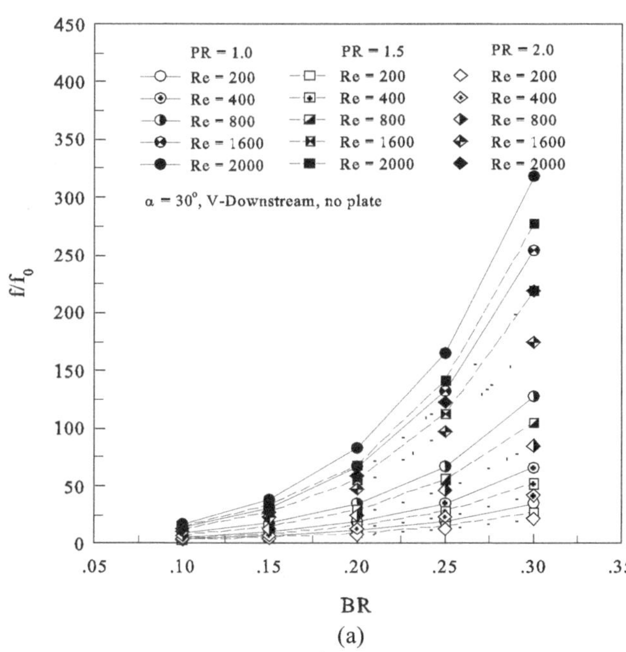

Fig. 12 The variations of f/f0 with BR values at various PR and Re values for 30° V–baffle inserted diagonally (no plate) for (a) V–Downstream and (b) V–Upstream.

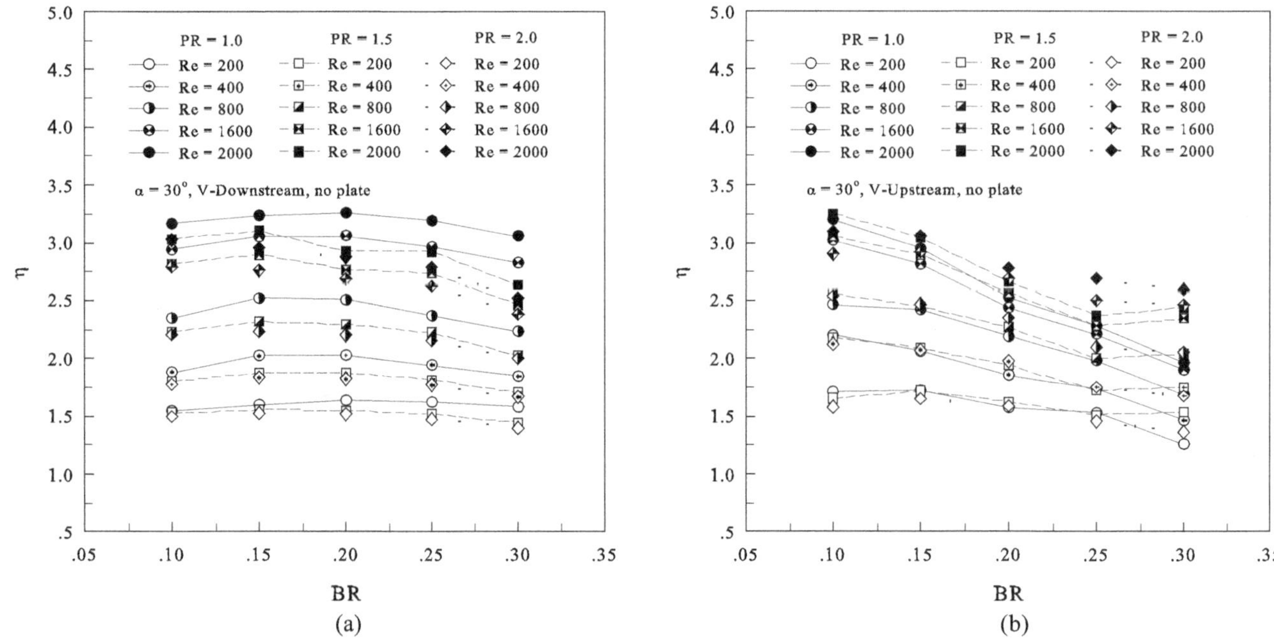

Fig. 13 The variations of f/f_0 with BR values at various PR and Re values for 30° V–baffle inserted diagonally (no plate) for (a) V–Downstream and (b) V–Upstream.

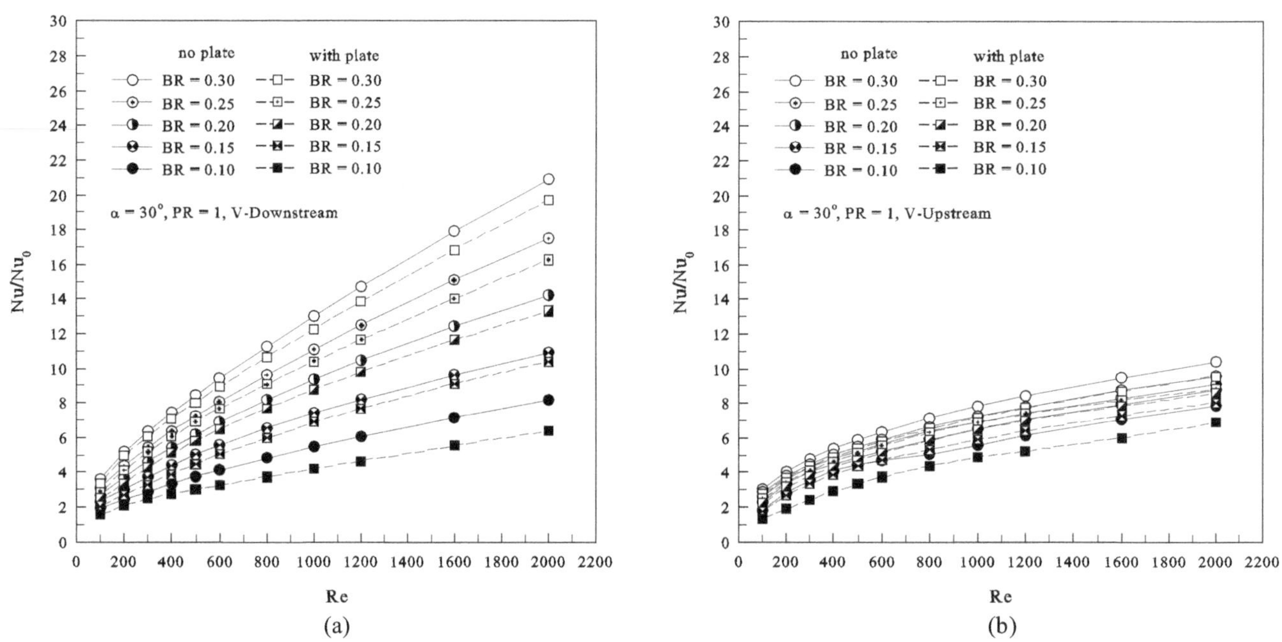

Fig. 14 Comparisons of with plate and no plate cases in term of Nu/Nu_0 with Re values (a) V–Downstream and (b) V–Upstream.

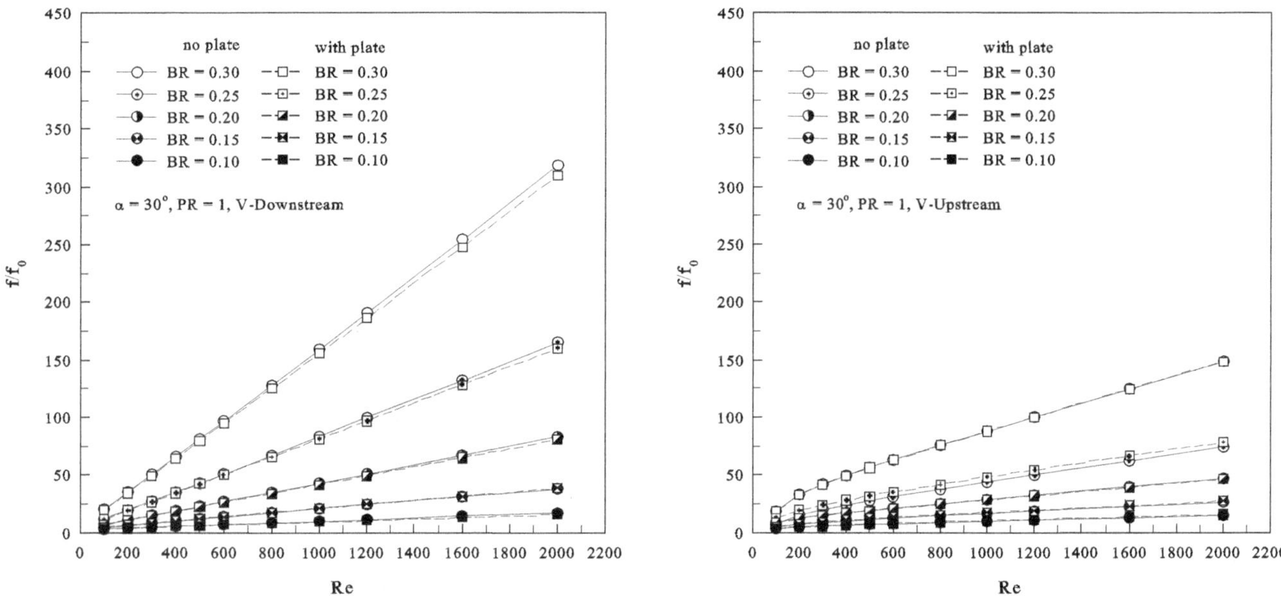

(a) (b)

Fig. 15 Comparisons of with plate and no plate cases in term of f/f_0 with Re values (a) V–Downstream and (b) V–Upstream.

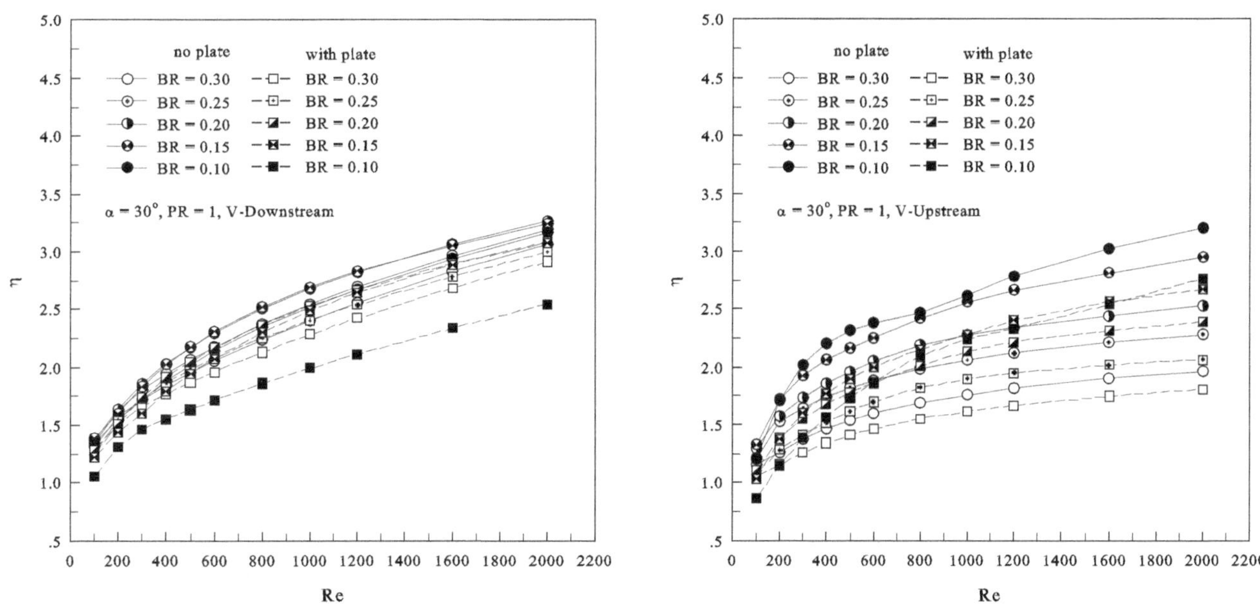

(a) (b)

Fig. 16 Comparisons of with plate and no plate cases in term of η with Re values (a) V–Downstream and (b) V–Upstream.

Table 5 Empirical correlations for 30° V–baffle inserted diagonally

Case	*Empirical correlation*	*Eq. no.*	*Using range*
V–Downstream with plate	$Nu/Nu_0 = 0.117\,Re^{0.551}\,Pr^{0.4}(BR+1)^{5.347}(PR+1)^{-0.576}$	(10)	$\alpha = 30°$ BR = 0.1 – 0.3 PR = 1 – 2 Re = 100 – 2000
	$f/f_0 = 0.026\,Re^{0.777}(BR+1)^{14.344}(PR+1)^{-0.841}$	(11)	
V–Upstream with plate	$Nu/Nu_0 = 0.185\,Re^{0.484}\,Pr^{0.4}(BR+1)^{3.012}(PR+1)^{-0.329}$	(12)	
	$f/f_0 = 0.092\,Re^{0.625}(BR+1)^{11.011}(PR+1)^{-0.834}$	(13)	
V–Downstream with no plate	$Nu/Nu_0 = 0.147\,Re^{0.550}\,Pr^{0.4}(BR+1)^{4.513}(PR+1)^{-0.538}$	(14)	
	$f/f_0 = 0.020\,Re^{0.806}(BR+1)^{14.816}(PR+1)^{-0.875}$	(15)	
V–Upstream with no plate	$Nu/Nu_0 = 0.243\,Re^{0.462}\,Pr^{0.4}(BR+1)^{2.319}(PR+1)^{-0.148}$	(16)	
	$f/f_0 = 0.073\,Re^{0.647}(BR+1)^{11.457}(PR+1)^{-0.779}$	(17)	

5. CONCLUSIONS

The fully developed periodic laminar forced convection in a heat exchanger square channel with 30° V–baffles inserted diagonally is investigated numerically in three dimensional. The use of the inserted vortex generators types can help to install and maintenance in the heat exchanger square duct system. The effects of blockage ratio, pitch ratio, Reynolds number, flow direction and the vortex generators configurations are presented. The conclusions of this work are as follows:

- The use of the diagonally inserted vortex generators can help to increase heat transfer rates and the thermal performance but also increase the pressure loss in the heat transfer system.
- The increasing BR, Reynolds number with reducing PR, produce a higher heat transfer rate and friction factor for all cases. The V–Downstream case performs higher heat transfer rate and friction factor than the V–Upstream case.
- In the cases studied, the augmentations are found in ranges 2 – 21 and 1 – 320 times for heat transfer and friction factor, respectively, for 30° V–baffle inserted diagonally in the square channel. The optimum thermal enhancement factor is around 3.25 for V–Downstream case.
- The different installation methods, placing on the thin plate and inserted with a wire frame, it is found that the wire frame (no plate) is not beneficially for reducing the pressure loss. In both cases, the heat transfer rate and pressure loss values are very nearly values, but the use of the wire frame can help to improve the thermal performance for V–Upstream case.
- The empirical correlations for diagonally inserted vortex generators are presented as table. 5.

ACKNOWLEDGEMENTS

The funding of this study is supported by King Mongkut's University of Technology North Bangkok, Thailand. The author would like to thanks Dr.Withada Jedsadaratanachai and Assoc. Prof. Dr.Pongjet Promvonge, KMITL for the suggestions

NOMENCLATURE

A	heat transfer area, m^2
BR	blockage ratio, (b/H)
b	baffle height, m
H	hydraulic diameter of square duct
f	friction factor
GCI	grid convergence index
h	convective heat transfer coefficient, W m^{-2} K^{-1}
k	thermal conductivity, W m^{-1} K^{-1}
L	cyclic length of one cell (or axial pitch length, H), m
Nu	Nusselt number
p	static pressure, Pa
Pr	Prandtl number
PR	pitch ratio, L/H
Re	Reynolds number
T	temperature, K
u_i	velocity in x_i-direction, m s^{-1}
$\bar{u}$	mean velocity in channel, m s^{-1}
Greek letter	
μ	dynamic viscosity, kg s^{-1}m^{-1}
Γ	thermal diffusivity
α	rib inclination angle or angle of attack, degree
η	thermal enhancement factor
ρ	density, kg m^{-3}
Subscript	
in	inlet
0	smooth duct
w	wall
pp	pumping power
ref	reference

REFERENCES

Boonloi, A., and Jedsadaratanachai, W., 2013, "3D Numerical study on laminar forced convection in V–baffled square channel," *American Journal of Applied Sciences,* 10, 1287 – 1297. http://dx.doi.org/10.3844/ajassp.2013.1287.1297

Boonloi, A., 2014, "Effect of Flow Attack Angle of V–Ribs Vortex Generators in a Square Duct on Flow Structure, Heat Transfer, and Performance Improvement," *Modelling and Simulation in Engineering*, Article ID 985612, 11 pages. http://dx.doi.org/10.1155/2014/985612

Chen, TY., and Shu, HT., 2004, "Flow Structures and Heat Transfer Characteristics in Fan Flows with and without Delta-wing Vortex Generators," *Experimental Thermal and Fluid Science,* 28, 273 – 282. http://dx.doi.org/10.1016/S0894-1777(03)00107-9

Gentry, MC., and Jacobi, AM., 1997, "Heat Transfer Enhancement by Delta–wing Vortex Generators on a Flat Plate: Vortex Interactions with the Boundary Layer," *Journal of Experimental Thermal and Fluid Science*, 14, 231 – 242. http://dx.doi.org/10.1016/S0894-1777(96)00067-2

Incropera, F., and Dewitt, P.D., 2006, *Introduction to heat transfer*, 5rd ed., John Wiley & Sons, Hoboken, NJ.

Jedsadaratanachai, W., Suwannapan, S., and Promvonge, P., 2011, "Numerical study of laminar heat transfer in baffled square channel with various pitches," *Energy Procedia*, 9, 630 – 642. http://dx.doi.org/10.1016/j.egypro.2011.09.073

Joardar, A., and Jacobi, AM., 2005, "Impact of Leading Edge Delta-wing Vortex Generators on the Thermal Performance of the Flat Tube, Louvered-fin Compact Heat Exchanger," *International Journal of Heat and Mass Transfer*, 48, 1480 – 1493. http://dx.doi.org/10.1016/j.ijheatmasstransfer.2004.10.018

Kwankaomeng, S., and Promvonge, P., 2010, "Numerical prediction on laminar heat transfer in square duct with 30° angled baffle on one wall," *International Communication in Heat and Mass Transfer*, 857 – 866. http://dx.doi.org/10.1016/j.icheatmasstransfer.2010.05.005

Li-Ting, T., Ya-Ling, H., Yong-Gang, L., and Wen-Quan, T., 2009, "Numerical Study of Fluid Flow and Heat Transfer in a Flat-plate Channel with Longitudinal Vortex Generators by Applying Field Synergy Principle Analysis," *International Communications in Heat and Mass Transfer*, 36, 111 – 20. http://dx.doi.org/10.1016/j.icheatmasstransfer.2008.10.018

Patankar, S.V., 1980, *Numerical heat transfer and fluid flow*, McGraw–Hill, New York.

Promvonge, P., Jedsadaratanachai, W., and Kwankaomeng, S., 2010, "Numerical study of laminar flow and heat transfer in square channel with 30° inline angled baffle turbulators," *Applied Thermal Engineering*, 30, 1292 – 1303. http://dx.doi.org/10.1016/j.applthermaleng.2010.02.014

Promvonge, P., and Kwankaomeng, S., 2010, "Periodic laminar flow and heat transfer in a channel with 45° staggered V–baffles," *International Communication in Heat and Mass Transfer*, 37, 841 – 849. http://dx.doi.org/10.1016/j.icheatmasstransfer.2010.04.002

Promvonge, P., Sripattanapipat, S., and Kwankaomeng, S., 2010, "Laminar periodic flow and heat transfer in square channel with 45° inline baffles on two opposite walls," *International Journal of Thermal Sciences*, 49, 963 – 975.
http://dx.doi.org/10.1016/j.ijthermalsci.2010.01.005

Promvonge, P., Jedsadaratanachai, W., Kwankaomeng, S., and Thianpong, C., 2012, "3D simulation of laminar flow and heat transfer in V–baffled square channel," *International Communication in Heat and Mass Transfer*, 39, 85 – 93.
http://dx.doi.org/10.1016/j.icheatmasstransfer.2011.09.004

Roache, P.J., 1998, *Verification and Validation in Computational Science and Engineering*, Hermosa Publishers, Albuquerque, NM, ISBN 0913478083.

Wu, JM., and Tao, WQ., 2012, "Effect of Longitudinal Vortex Generator on Heat Transfer in Rectangular Channels," *Applied Thermal Engineering*, 37, 67 – 72.
http://dx.doi.org/10.1016/j.applthermaleng.2012.01.002

17

EFFECT OF WALL THERMAL CONDUCTIVITY ON HYDROGEN-ASSISTED CATALYTIC IGNITION CHARACTERISTICS OF PROPANE-AIR AT MICRO-SCALES IN DIFFERENT FEEDING MODES

Junjie Chen*, Xuhui Gao, Deguang Xu

School of Mechanical and Power Engineering, Henan Polytechnic University, Jiaozuo, Henan, 454000, China

ABSTRACT

Effect of wall thermal conductivity on hydrogen self-ignition and hydrogen-assisted ignition of propane-air mixtures in different feeding modes from ambient cold-start conditions were investigated numerically with chemical kinetic model in Pt/γ-Al_2O_3 catalytic micro-combustors. For the steady and transient state, effect of wall thermal conductivity on self-ignition characteristics of lean hydrogen-air mixtures was presented, and hydrogen-assisted combustion of propane-air mixtures was investigated numerically in the co-feed mode and the sequential feed mode. The computational results indicate the large thermal inertia of the micro-combustor solid structure leads to slow temperature dynamics, and transient response is dominated by the thermal inertia. The heat localization in poorly conducting walls leads to fast ignition and shorter steady state time. In general, the concentration of hydrogen required for propane ignition increased with increasing wall thermal conductivity, decreasing inlet velocity, and decreasing inlet equivalence ratio of propane-air mixtures. In the co-feed mode, the combustion characteristics of hydrogen-assisted propane qualitatively resemble the selectively preheating initial portion of the combustion chamber wall. In the sequential feed mode, the time taken to reach steady state, the hydrogen cut-off time, the propane ignition time and the cumulative propane emissions increased with increasing wall thermal conductivity; the ignition characteristics are similar to partially preheating the initial segment for low and moderate wall thermal conductivity values (0.5 and 20 W/m·K); however, the ignition characteristics are close to completely heating the micro-combustor wall for high wall thermal conductivity values (200 W/m·K).

Keywords: *Hydrogen-assisted ignition; Self-ignition; Thermal conductivity; Micro-combustion; Catalytic combustion*

1. INTRODUCTION

In recent years, hydrocarbon-fueled catalytic micro-combustors have been the focus of intense efforts for a variety of portable power production systems with demonstrated energy densities considerably higher than those obtained with state-of-the-art Li-ion batteries (Ju and Maruta, 2011). Applications range from catalytic micro-thrusters for space applications (Volchko *et al.*, 2006), to micro-reactors used for fuel reforming in micro solid oxide fuel cells (Cheekatamarla *et al.*, 2008), and to scaled-down thermal engines, wherein catalytic micro-combustors are used for direct chemical-to-thermal energy conversion (Gomez *et al.*, 2007; Karim *et al.*, 2008). Noble metal catalysts (such as Pt, Pd, Rh, etc.) are employed in most cases, because of their superior activity for the total oxidation and reforming of hydrocarbons at modest reactor temperatures (Aryafar and Zaera, 1997).

Various configurations of catalytic micro-combustors have so far been introduced. For instance, heat recirculating Swiss-roll burners make use of the excess exhaust gas enthalpy for increased combustion stability (Ahn *et al.*, 2005), while conventional straight catalytic channels (Federici *et al.*, 2009) and catalytic foam structures (Jin and Kwon, 2009) have also been employed in micro-combustors study. The aforementioned studies have focused on thermal management and combustion efficiency. On the other hand, numerical simulations of catalytic micro-combustors encompass investigations of combustion stability and performance using 1-D models with lumped heat and mass transport coefficients (Kaisare *et al.*, 2008; Ronney, 2003), 2-D models with detailed chemistry kinetics but without heat conduction in the solid wall (Maruta *et al.*, 2002), 2-D models with simplified chemistry kinetics accounting for heat conduction in the solid (Deshmukh *et al.*, 2004; Stefanidis and Vlachos, 2009), and finally full 2-D CFD models where all relevant heat transfer mechanisms in the solid (including heat conduction and surface radiation heat transfer) along with detailed homogeneous and heterogeneous chemistry kinetics are accounted (Karagiannidis *et al.*, 2007). It is emphasized that although Sherwood and Nusselt number empirical correlations have been proposed for laminar channel flows with surface catalytic reactions (Groppi *et al.*, 1995), their universal applicability under different reacting conditions is not warranted. Moreover, 1-D models fail to describe homogeneous combustion due to the strong dependence of homogeneous reactions on the boundary layer profiles of temperature and species (Mantzaras and Benz, 1999).

Many experimental and numerical investigations which have been conducted are concentrated on the effects of some crucial options such as micro-tube/channel geometry (Kaisare and Vlachos, 2007; Kim and Maruta, 2006; Norton and Vlachos, 2004), heat loss (Hua *et al.*, 2005; Norton and Vlachos, 2004), thermal and radical quenching (Kim *et al.*, 2006), fluid flow characteristic and equivalence ratio (Hua *et al.*, 2005; Zhang *et al.*, 2007), and pre-heating of fresh entrance reactive mixture (Hua *et al.*, 2005; Ju and Choi, 2003; Kaisare and Vlachos, 2007; Leach and Cadou, 2005; Norton and Vlachos, 2003; Raimondeau *et al.*, 2002) on combustion phenomena in micro-combustors. As results of many investigations, it can be inferred that the micro-combustor

* *Corresponding author. Email: comcjj@163.com*

geometry such as wall thickness (Hua *et al.*, 2005), combustor diameter (axisymmetric models) or distance between two planar plates (2D models), heat transfer surface (Kaisare and Vlachos, 2007) and inserting backward facing-step (Ju and Choi, 2003) play crucial roles in establishing a stable and efficient combustion in a micro-tube/channel. This phenomena may occur due to several reasons such as heat loss variations, heat transfer from post-flame to pre-flame zone (Hua *et al.*, 2005; Kaisare and Vlachos, 2007; Leach and Cadou, 2005), the increment of reactive mixture residence time (Norton and Vlachos, 2003) and enhancement of mixing process, respectively. Also, some chemical and physical properties of the micro-combustor wall and the inlet mixture such as fuel type (Norton and Vlachos, 2003), wall thermal conductivity (Baigmohammadi *et al.*, 2013; Norton and Vlachos, 2003, 2004; Raimondeau *et al.*, 2002; Rana *et al.*, 2014; Veeraragavan and Cadou, 2011; Zarvandi *et al.*, 2012; Zhou *et al.*, 2009), equivalence ratio and inlet velocity can influence combustion process in micro-combustors. Therefore, variation of the wall thermal conductivity and its physical properties can govern the combustion by proper pre-heating of fresh incoming mixture (Raimondeau *et al.*, 2002) or thermal and radical species quenching processes (Kim *et al.*, 2006).

The wall thermal conductivity plays a vital role in combustion characteristics (Baigmohammadi *et al.*, 2013; Norton and Vlachos, 2003, 2004; Rana *et al.*, 2014; Zhou *et al.*, 2009). Low wall thermal conductivity limits the upstream heat transfer through the wall, which limits the preheating of the feed, inhibiting the onset of combustion, and causes blowout. Low wall thermal conductivity also causes hot spots of high temperatures within the wall, which can lead to mechanical failure. High thermal conductivity walls are essentially isothermal and have lower temperatures. However, they offer a larger hot area for external heat transfer and become susceptible to spatially global-like extinction (Norton and Vlachos, 2003). When a low-power device is being designed, more insulating materials should be favored to minimize external heat losses. Conversely, a high-power device would favor more conductive materials (Rana *et al.*, 2014; Zhou *et al.*, 2009). Furthermore, the wall material thermal conductivity plays a competing role in flame stability (Norton and Vlachos, 2003, 2004; Raimondeau *et al.*, 2002). Walls transfer heat upstream for ignition of the cold incoming gases but at the same time are responsible for heat losses. Consequently, there is an optimum wall thermal conductivity in terms of flame stability, which appears to be that of common ceramics such as alumina and silica (Norton and Vlachos, 2005; Norton *et al.*, 2004). Regarding material lifetimes, higher wall thermal conductivities reduce the wall temperature gradients and hotspots and should be preferred (Kaisare and Vlachos, 2007).

Until recently, most of the numerical and experimental investigations have focused on the steady-state performance of catalytic micro-combustors and micro-reformers (Fanaee and Esfahani, 2014; Hsueh *et al.*, 2011; Leu *et al.*, 2013; Sari and Zohrabian, 2014). On the other hand, a limited number of investigates investigated the transient response and particularly the light-off (ignition process). Ignition at problem-specific operating conditions is key to the overall acceptance of given micro-combustors design. Furthermore, ignition determines the time length required for start-up and the associated pollutant emissions during this period. The start-up of a propane-fueled, platinum-coated micro-combustor, with added hydrogen to facilitate light-off, has been demonstrated experimentally (Norton and Vlachos, 2005), while the self-ignition of methanol-air mixtures in similar micro-reactors coupled to thermo-electrical devices have been investigated (Karim *et al.*, 2008). The former work mainly focused on developing a start-up strategy for the specific micro-combustor and micro-reactor, while the latter demonstrated the feasibility of integrating a self-igniting catalytic micro-combustor to a power generating system. Similarly, catalytic ignition and extinction in partial oxidation of methane-air mixtures, diluted with exhaust gas recycle, have been investigated numerically and experimentally in rhodium-coated honeycomb reactors for gas-turbine applications (Schneider *et al.*, 2008). Although the last investigation did not pertain to micro-reactors, it nevertheless addressed crucial catalytic ignition requirements in accordance to stringent residence time limitations and reactor inlet temperature.

The study of hydrogen-assisted catalytic ignition necessarily starts with the fuels used. The hydrogen-assisted catalytic ignition characteristics of methane in a Pt-coated monolith reactor have been investigated experimentally (Deutschmann *et al.*, 2000), achieving ignition with a minimum hydrogen concentration of 4.75% (on a molar basis). The experimental results showed that the amount of hydrogen required for ignition increases as methane concentration is decreased. Also, during ignition, the main effect of hydrogen is thermal. The hydrogen ignited first on catalyst surface, and then the temperature of mixtures increased from room temperature to ignition point. A kind of catalytic micro-channel burner have been fabricated to investigate hydrogen catalytic combustion and the characteristics of propane in hydrogen-assisted catalytic ignition (Norton and Vlachos, 2005; Norton *et al.*, 2004). A minimum hydrogen concentration of 3.6% mole fraction was found to be required. They found that the ignition process had two stages: hydrogen ignition, and then propane ignition. At low hydrogen concentrations, propane kinetically inhibits hydrogen catalytic combustion.

Most current prototypes of micro-combustors depend on exterior heating to generate energy for ignition, and the additional equipment necessary to power the exterior heaters can negate mass advantages of micro-combustors. In this work, effect of wall thermal conductivity on hydrogen self-ignition and hydrogen-assisted ignition of propane-air mixtures in different feeding modes from ambient cold-start conditions are investigated numerically in catalytic micro-combustors. Moreover, for the steady and transient state, a reduced-order reaction model is employed to examine the ignition characteristics of hydrogen. Finally, the hydrogen-assisted combustion of propane-air mixtures is investigated numerically in different operation modes of fuel feed.

2. NUMERICAL MODELS AND SIMULATION APPROACH

2.1 Model Geometry and Mesh

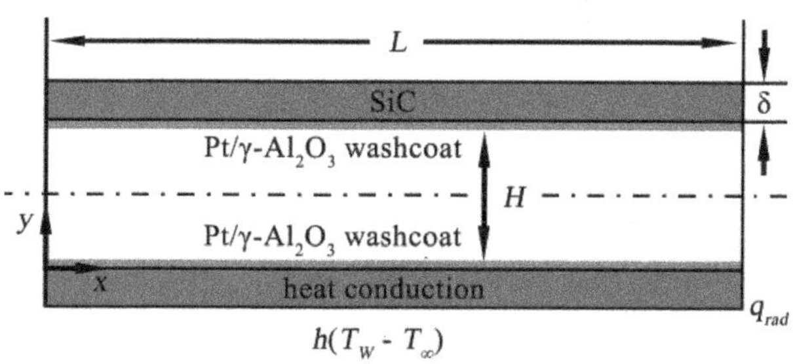

Fig. 1 Schematic diagram of the micro-channel geometry.

A schematic diagram of the catalytic micro-combustor modeled is shown in Fig. 1. The kinetics module plug-in for FLUENT® Release 6.3 was employed to simulate the flow and combustion of hydrogen-propane-air mixtures in the plane micro-channel of length $L = 20.0$ mm, height $H = 0.2$ mm, and solid wall thickness $\delta = 0.2$ mm. Inner horizontal surfaces contained Pt/γ-Al_2O_3 catalyst washcoat. The properties of Pt/γ-Al_2O_3 supported noble metal catalysts are shown in Table 1. In the present work, in order to couple heat transfer, fluid dynamics, and hetero-/homogeneous chemical kinetics, FLUENT-KINETICS was employed to simulate the catalytic combustion process.

Table 1 The properties of Pt/γ-Al_2O_3 catalyst washcoat.

Property	*Value*
Catalyst surface site density Γ (mol/cm^2)	2.7×10^{-9}
Average pore diameter d_{pore} (m)	2.08×10^{-8}
Catalyst porosity ε_{cat}	0.4
Catalyst tortuosity τ_{cat}	8.0

2.2 Chemical Kinetics for Hydrogen Combustion

The chemical kinetics (Bui *et al.*, 1997) for the combustion of hydrogen-air is adopted, which developed the model via the hierarchical model reduction of 13-step surface catalytic reaction mechanism. The resulting reaction rate of combustion is given by the following reduced-order reaction rate kinetics:

$$-r_{\text{cat, H}_2} = \eta A_0 T^{\beta} e^{\frac{-E_{a,H_2}^{ads}}{RT}} C_{H_2} \quad (1)$$

Where the effectiveness factor η is 1, the pre-exponential factor A_0 is 1280 cm $K^{-0.5} \cdot s^{-1}$, the temperature exponent β is 0.5, and the activation energy $E_{a,H2}$(ads) is 0 kJ·mol^{-1} because of hydrogen adsorption is the non-activated process (Bui *et al.*, 1997).

Norton et al. (2004) reported that homogeneous combustion of near-stoichiometric hydrogen-air mixtures occurs in the larger gap size of 1.0 mm, but not for the smaller gap size of 0.25 mm. Even in the 1.0 mm micro-combustor, hydrogen light-off in gas phase was observed beyond the equivalence ratio of 0.33. Moreover, Norton et al. (2004) also reported that this kinetic model is valid for the concentration of hydrogen in the range of the equivalence ratio $0.008<\varphi_{H2}<0.424$. Mantzaras (2008) observed hydrogen light-off in gas phase for the equivalence ratio of 0.2 in 1.2 mm diameter micro-channels. Therefore, in this work, homogeneous combustion of hydrogen-air mixtures is neglected and numerical simulations are restricted for hydrogen concentrations below the equivalence ratio of 0.2.

2.3 Chemical Kinetics for Propane Combustion

Pt/γ-Al_2O_3 catalysts were adopted for catalytic micro-combustion of propane-air mixtures. The reason is that the propane conversion of noble metal catalysts decreases in the order Pt/γ-Al_2O_3 > Pd/γ-Al_2O_3 > Rh/γ-Al_2O_3 at the stoichiometric propane-oxygen ratio (Men *et al.*, 2009). The reduced-order reaction rate kinetics for catalytic combustion of propane-air are as follows:

$$r_{\text{cat, C}_3\text{H}_8} = \frac{k_{C_3H_8}^{ads} C_{s, C_3H_8}}{\left(1+\sqrt{\frac{k_{O_2}^{ads} C_{s, O_2}}{k_{O_2}^{des}}}\right)^2} \quad (2)$$

Where $r_{cat,C3H8}$ is the surface catalytic reaction rate of propane. $C_{s,i}$ is concentration of adsorbed species *i*. k_i is adsorption or desorption rate constant of species *i*, and determined as follows:

$$k_i^{ads} = \frac{s_0}{\Gamma}\sqrt{\frac{RT}{2\pi M_i}}\left(\frac{T}{T_{ref}}\right)^{\beta_i^{ads}} e^{\frac{-E_{a,i}^{ads}}{RT}} \quad (3)$$

$$k_i^{des} = A_0\left(\frac{T}{T_{ref}}\right)^{\beta_i^{des}} e^{\frac{-E_{a,i}^{des}}{RT}} \quad (4)$$

Where s_0 is the sticking coefficient. Γ is Pt/γ-Al_2O_3 catalyst surface site density. M_i is the molecular weight of species *i*. T_{ref} is the temperature of reference conditions. The values of kinetic parameters for the catalytic combustion of lean propane are shown in Table 2.

Table 2 Kinetic parameters for the catalytic combustion of lean propane over Pt/γ-Al_2O_3.

	A_0 (s^{-1}) or s_0	β	E_a (kcal·mol^{-1})
C_3H_8 adsorption	0.06	0.154	4
O_2 adsorption	0.0542	0.766	0
O_2 desorption	8.41×10^{12}	-0.796	*

* $E_{a,O_2}^{des} = 52.8 - \frac{2.3T}{300} - 32.0\theta_{O\cdot}$

Surface area factor η = 1.7, temperature of reference conditions T_{ref} = 300 K.

The activation energy $E_{a,O2}$(des) of oxygen desorption depends on the coverage $\theta_{O\cdot}$ of oxygen radical, which is calculated as follows:

$$\theta_{O\cdot} = 1 - \frac{1}{\left(1+\sqrt{\frac{k_{O_2}^{ads} C_{s,O_2}}{k_{O_2}^{des}}}\right)} \quad (5)$$

2.4 Boundary Conditions

The incoming flow of hydrogen-air or hydrogen-propane-air mixtures was fully premixed and have uniform inlet temperature of 300 K. The thermal boundary condition on the wall is the heat loss to the ambient air. The heat losses to the surroundings are calculated using the following equation:

$$q = h(T_{w,o} - T_{\infty}) + \varepsilon\delta(T_{w,o}^4 - T_{\infty}^4) \quad (6)$$

Where the exterior convective heat transfer coefficient h is 20 W/m^2·K in this study. $T_{w,o}$ is the temperature at the exterior wall surface. The ambient temperature T_∞ is 300K. The solid wall emissivity ε is taken to be 0.8 and δ is the Stefan-Boltzmann constant.

2.5 Computation scheme

For the steady state simulation, the mass, momentum, energy and species conservation equations are solved implicitly with the 2D steady-state double-precision segregated solver using the under-relaxation method. The aforementioned conservation equations are discretized using second-order upwind scheme. The "SIMPLE" algorithm is used to couple the pressure and velocity. The fluid density is calculated using the ideal gas law. The fluid thermal conductivity, specific heat, and viscosity are calculated using the mass fraction weighted average of species properties. The species specific heat is calculated using the piecewise polynomial fit of temperature. The uniform grid size of 0.005 mm was used to mesh the computational domain. Grid-independence of the solution was tested. The convergence of the CFD simulation was judged based on the residuals of all governing equations. The simulation results are achieved with residuals smaller than 1.0×10^{-6}.

For the transient state simulation, the "PISO" algorithm is used to couple the pressure and velocity. The transient solid equation was solved with a second order accurate, fully implicit scheme by using a quadratic backward time discretization. The coupled flow and solid phases were solved iteratively and convergence was achieved at each time step when the solid temperature did not vary at any position along the wall by more than 10^{-5} K (Karagiannidis and Mantzaras, 2010). Catalytic chemical reaction times are assumed to be shorter than the heat conduction times in the solid, so as to ensure chemical equilibration at the local wall temperature during an integration time step of the solid. The time step Δt = 50.0 ms used in this work had to be longer than the chemical time scales present during catalytic ignition in micro-combustors.

The simulations were performed on a high-performance cluster consisting of 4 × E5-2600 processors and 48 GB of RAM. Parallel processing was used, and the message passing interface (MPI) was used to transmit information between nodes. For the transient state simulation, the calculation time of each simulation varied between 40.0 minutes and 8 hours, depending on the difficulty of the problem.

3. EFFECT OF WALL THERMAL CONDUCTIVITY ON CATALYTIC IGNITION CHARACTERISTICS OF HYDROGEN

3.1 Steady State Behavior

Despite the high diffusivity of hydrogen, the catalytic combustion of hydrogen-air at micro-scales is diffusion limited (Kamada *et al.*, 2014; Schultze *et al.*, 2013; Sen and Avci, 2014; Yang *et al.*, 2014). The incoming hydrogen-air mixtures was fully premixed with the equivalence ratio of 0.2. The effect of inlet velocity on the temperature along central axis and interior wall as well as the hydrogen conversion for adiabatic and non-adiabatic micro-combustors at steady state are shown in Fig. 2-4. The micro-combustion occurs close to the inlet section and only a fraction of micro-combustor length is required to attain complete hydrogen conversion, even for higher velocity case. The maximum temperature is attained at the beginning of the interior wall, by reason of the surface catalytic reaction. The maximum temperature on the interior wall increases as the inlet velocity is increased or heat transfer coefficient is decreased. In the case of non-adiabatic micro-combustors, the upstream interior wall gets heated up because of the surface catalytic reaction, whereas the net heat loss to the ambience occurs in the downstream section. Therefore, for the moderate to high heat losses case, the direction of heat transfer is from the solid wall to the gas phase in upstream section because of the surface catalytic reaction, whereas the direction reverses in the downstream section by reason of the heat losses. In adiabatic micro-combustors or at higher inlet velocity, the aforementioned reversal is not observed.

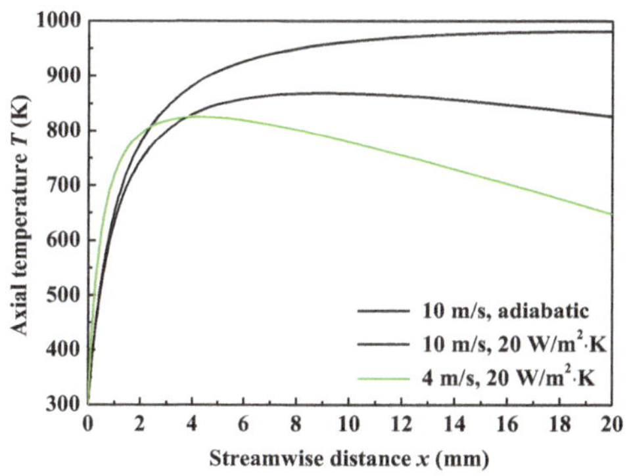

Fig. 2 Temperature along central axis for adiabatic and non-adiabatic micro-combustors at steady state.

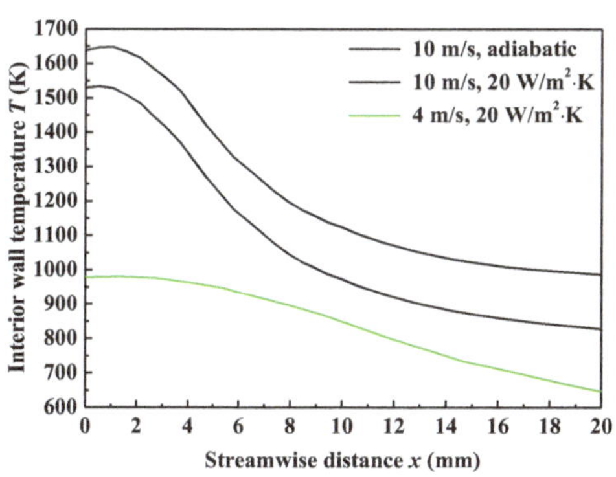

Fig. 3 Temperature along interior wall for adiabatic and non-adiabatic micro-combustors at steady state.

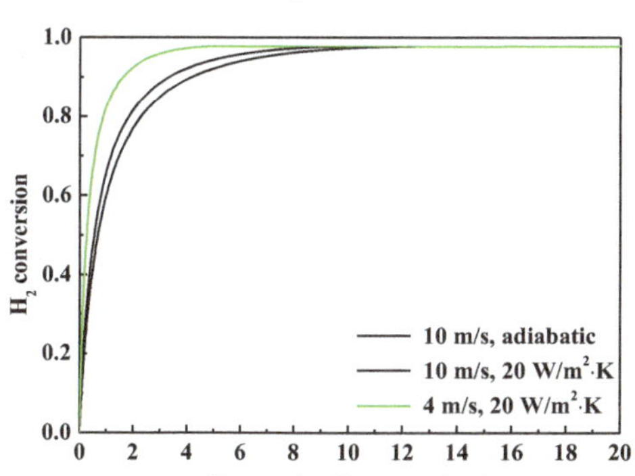

Fig. 4 Hydrogen conversion for adiabatic and non-adiabatic micro-combustors at steady state.

The wall thermal conductivity also affects the overall temperature distribution in micro-combustors. For lower wall thermal conductivities, the significant temperature gradients exist as observed, whereas the interior wall temperature profiles are nearly flat for the highly conducting walls in Fig. 5-6.

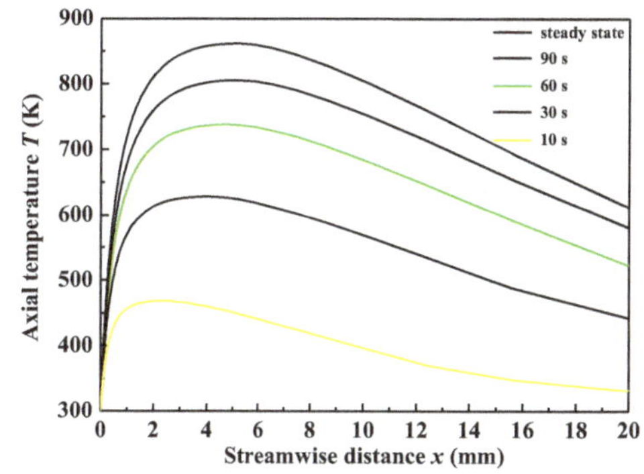

(a) $k_s = 0.5$ W/m·K

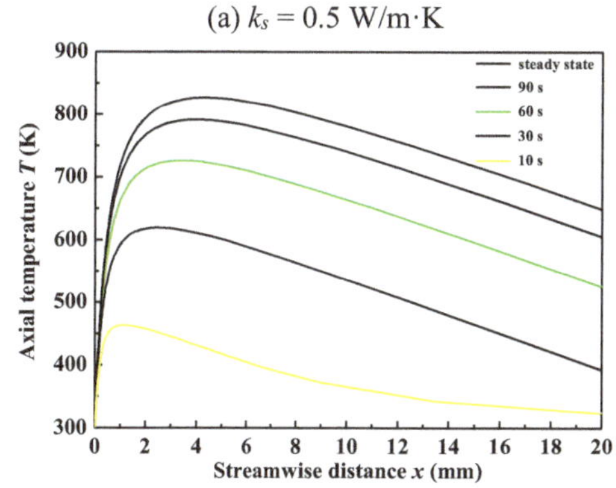

(b) $k_s = 20$ W/m·K

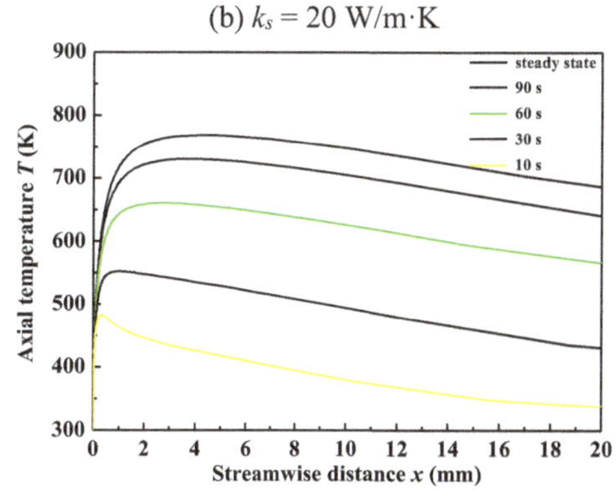

(c) $k_s = 200$ W/m·K

Fig. 5 Transient evolution of the central axial temperature profiles in the gas phase for hydrogen catalytic ignition for different wall thermal conductivities. φ (hydrogen) = 0.7, $u_0 = 2$ m/s, $h = 20$ W/m²·K, and $k_s = 20$ W/m·K.

3.2 Transient Behavior

In the present work, the primary purpose is to analyze the transient response of the micro-combustor. The micro-combustor initially contains only air and starting at ambient conditions of 300 K. The premixed hydrogen-air mixtures is fed starting at the time $t = 0$ and the desired inlet velocity of 4 m/s and equivalence ratio of 0.2. Numerical simulations are run for enough time to ensure the reaction system reaches steady state. Norton and Vlachos (2004) investigated numerically the effects of dimensions, external heat losses, wall thermal conductivity, and operating conditions on combustion characteristics and the steady-state, self-sustained flame stability of propane-air. The result shown that the wall thermal conductivity is vital in determining the flame stability, as the walls are responsible for the majority of the external heat losses as well as the upstream heat transfer. Therefore, the effects of the wall thermal conductivity on the ignition characteristics of propane-air are investigated in this work.

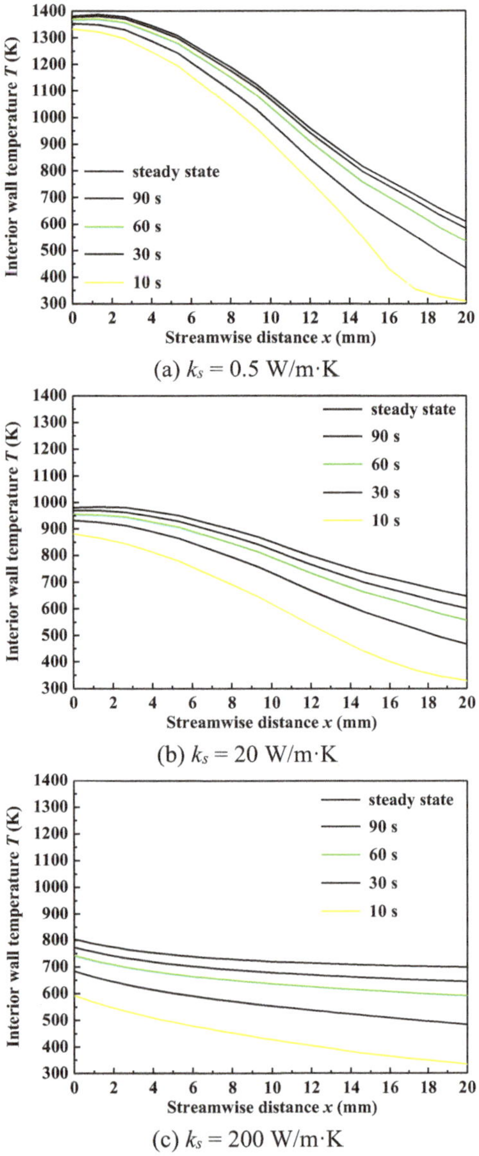

Fig. 6 Transient evolution of the interior wall temperature profiles on the wall for hydrogen catalytic ignition for different wall thermal conductivities. φ (hydrogen) = 0.7, u_0 = 2 m/s, h = 20 W/m^2·K, and k_s = 20 W/m·K.

For low (0.5 W/m·K), moderate (20 W/m·K) and high (200 W/m·K) thermal conductivities of wall material, the temporal responses of the central axial and the interior wall temperature profiles from ambient cold-start conditions are shown in Fig. 5 and 6, respectively. The thermal conductivities of wall material have a significant effect on the temperature distribution; in the bulk gas and the wall, the large temperature gradients exist in low thermal conductivity walls whereas the temperature profiles are nearly flat for the highly conducting walls. However, the hydrogen conversion is not affected by the thermal conductivity. In all cases, front-end ignition is observed and hydrogen conversion attain the steady state profile in Fig. 4 in less than 1.0 s. Due to the hydrogen conversion profile does not change for most of the simulation time after about 1.0 s, the aforementioned behavior is often referred to as pseudo-steady-state. Compared with the reaction and diffusion processes, the large thermal inertia of the micro-combustor solid structure leads to slow temperature dynamics, and transient response is dominated by the thermal inertia (Kaisare *et al.*, 2005; Kikas *et al.*, 2003).

The variations of the outlet temperature in gas phase for different wall thermal conductivities are shown in Fig. 7. The time when the micro-combustor reaches steady state is denoted by symbols. The symbols time is defined at which the maximum deviation of the outlet temperature is less than 0.5 K from the steady state values. The steady state time increases as the wall thermal conductivity is increased. For low wall thermal conductivity, a high temperature gradient exists on the wall, which will make the homogeneous combustion shift upstream and the micro-combustor will have a higher peak temperature (Chen *et al.*, 2007). On the contrary, for high wall thermal conductivity, a low temperature gradient on the wall will make the homogeneous combustion shift downstream and the micro-combustor will have a higher outlet temperature. For different inlet velocities and wall thermal conductivities, the time taken to reach steady state are shown in Fig. 8. As the inlet velocity is increased, the steady state time decreases due to the maximum wall temperature increase and the net energy supplied. As observed in Fig. 7-8, for lower wall thermal conductivity, the steady state time is shorter. The reason is that the heat localization in poorly conducting walls leads to fast ignition. At low inlet velocity, the effect of wall thermal conductivity on the steady state time is especially significant; as the inlet velocity is increased, the effect diminishes as a result of the increased convective heat transfer.

In general, the catalytic combustion reaches pseudo-steady state instantaneously; the combustion region is a short zone near the inlet section; and the higher inlet velocities lead to faster transient response.

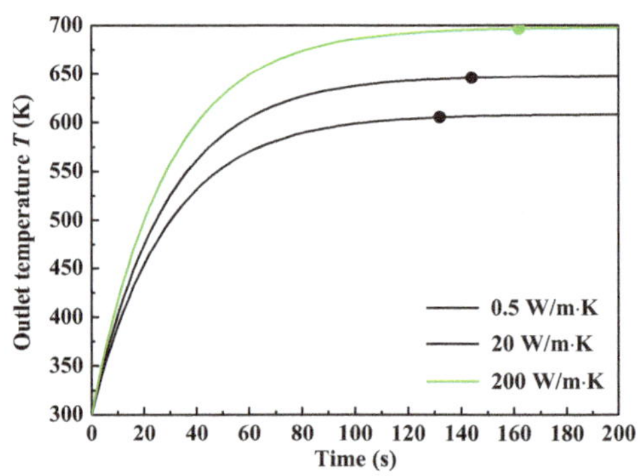

Fig. 7 Outlet temperature in gas phase versus time for different wall thermal conductivities. The symbols represent the time when the micro-combustor reaches steady state.

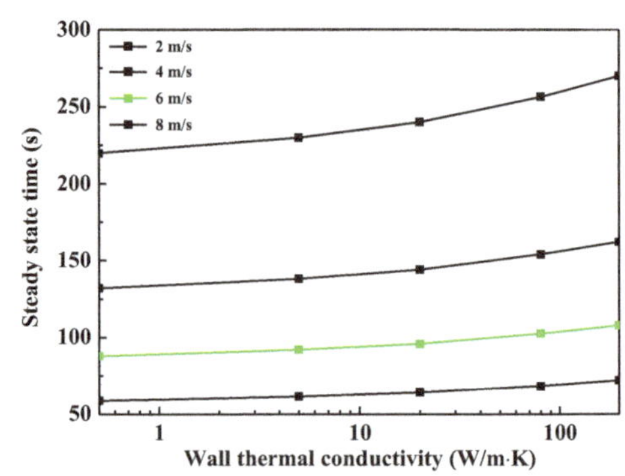

Fig. 8 The entire time required to reach steady state from initial conditions for different inlet velocities and wall thermal conductivities.

4. EFFECT OF WALL THERMAL CONDUCTIVITY ON CATALYTIC IGNITION CHARACTERISTICS OF HYDROGEN-ASSISTED PROPANE

The self-ignition nature of hydrogen-air mixtures at micro-scales of ceramic burners offers an opportunity to self-ignite hydrocarbons. This concept may be a way toward elimination of ignition sources from micro-scale devices leading to further reduction of system size. Furthermore, since hydrogen is a main target for fuel cell applications, one can envision storage of small amounts of hydrogen during device operation from reforming of hydrocarbons that is subsequently used for startup. Here, we explore the feasibility of this idea, calculate the necessary hydrogen content needed to self-ignite propane-air mixtures, and finally study the synergism of binary fuels.

Ignition of binary mixtures is not a new concept. For example, Deutschmann *et al.* (2000) computationally and experimentally studied the use of hydrogen combustion to ignite methane-air mixtures in a catalytic monolith. They found that the thermal release from the combustion of sufficiently fuel-rich hydrogen-air mixtures causes methane ignition. However, high temperatures limited their experimental study to initial transients only.

Hydrogen combustion can be employed to ignite propane-air, because the wall temperatures reached in hydrogen-air combustion are higher than the ignition temperature of propane. Which is referred to as hydrogen-assisted ignition of propane. In the present work, the thermal effect of hydrogen on hydrogen-assisted catalytic ignition is the main influence factor, because the concentration of hydrogen added is low. Therefore, the chemical effect is not consider in this work. Due to absence of detailed reaction mechanism of hydrogen-assisted catalytic ignition of propane-air mixtures, the variation of free radicals and the effect of hydrogen on the ignition temperature of propane are also not investigated in this work.

Here, we capitalize on the self-ignition of hydrogen at micro-scales and focus on the self-ignition of propane by adding hydrogen. There are different ways to carry out such simulation. One is to simultaneously start flowing all reagents. Another is to first start the hydrogen-air mixture followed by introduction of the hydrocarbon. In the present work, transient response of different feed methods (two startup methods of propane) is analyzed:

(a) Co-feed mode: premixed hydrogen-propane-air mixtures;

(b) Sequential feed mode: switched from hydrogen-air to propane-air mixtures.

The first method entails simultaneously turning on the flow of hydrogen, propane, and air until the propane ignites (co-feed mode), and then turning off the hydrogen flow and increasing the flow rates of propane and air so that the total flow rate remains constant throughout the simulation.

In the second method, hydrogen-air mixtures at the desired equivalence ratio are fed into the micro-combustor firstly. Once the sufficiently high temperature (ignition temperature of propane) has been reached, the hydrogen feed is cut off and the feed of propane-air mixtures is started. Simultaneously, keep total flow rate constant.

Norton and Vlachos (2005) performed experimental studies on hydrogen assisted self-ignition of propane-air mixtures in catalytic micro-combustors. They demonstrated that co-feed mode results in a faster heatup of the system as compared with sequential feed mode; co-feed mode is simple and has the additional benefit of heat release from both fuels; co-feeding hydrogen with propane appears as a good startup strategy. Therefore, comparison of the advantage and disadvantage between the co-feed and sequential feed modes is not taken into consideration. In the present work, a detailed study of hydrogen-assisted propane catalytic combustion is performed for two modes of operation: co-feed and sequential feed modes.

4.1 Co-Feed Mode

The bifurcation behavior of propane catalytic ignition in hydrogen-assisted co-feed mode are shown in Fig. 9. The inlet velocity is set to 10.0 m/s. The equivalence ratio (0.6 and 0.7) of premixed propane-air mixtures is kept fixed. The flow rates of propane and air are decreased as the hydrogen concentration is increased starting from a zero value, while the hydrogen flow rate increases to maintain the total volumetric flow rate constant. As hydrogen concentration is increased, the maximum temperature increases gradually because of hydrogen combustion in the feed. Then, after a particular concentration of hydrogen, the heat released on hydrogen combustion is sufficient to ignite propane combustion and the turning point bifurcation is observed in Fig. 9. For comparison, the ignition temperatures in inlet feed preheating mode are also shown in Fig. 9. For the hydrogen-assisted ignition case, the maximum wall temperatures at the ignition bifurcation are slightly higher than the inlet preheating case.

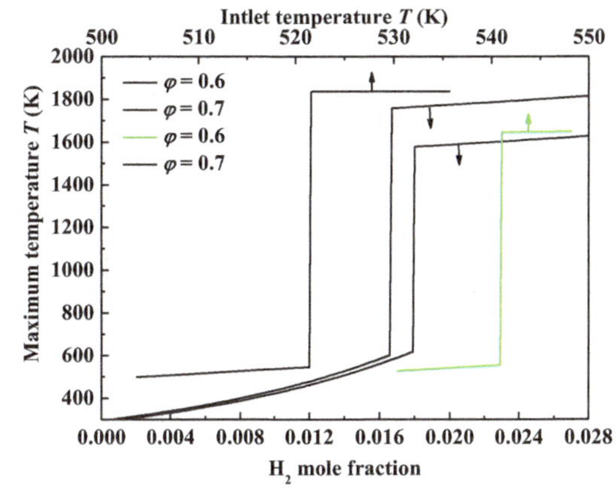

Fig. 9 The ignition turning point bifurcation in micro-combustors for catalytic propane-air ignition in hydrogen-assisted co-feed mode (black and red lines, bottom abscissa) and inlet preheating mode (green and blue lines, top abscissa).

The effect of inlet equivalence ratio on the hydrogen requirement for self-ignition is shown in Fig. 10. The inlet velocity is set to 10 m/s. For comparison, the ignition temperatures of propane without hydrogen addition in inlet feed preheating mode are also shown in Fig. 10. The hydrogen requirement in the feed reduces as the inlet equivalence ratio of propane-air is increased. For the hydrogen-assisted ignition case, the aforementioned trend is qualitatively similar to the inlet preheating case.

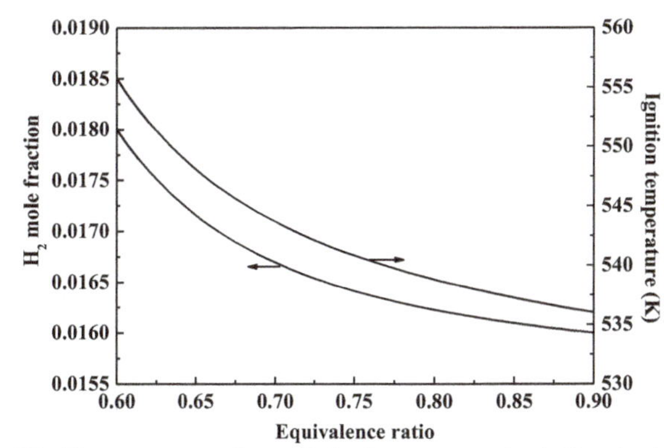

Fig. 10 The concentration of hydrogen required for ignition in hydrogen-assisted co-feed mode (black line, left ordinate) and the ignition temperature (red line, right ordinate) for different equivalence ratios of propane-air mixtures.

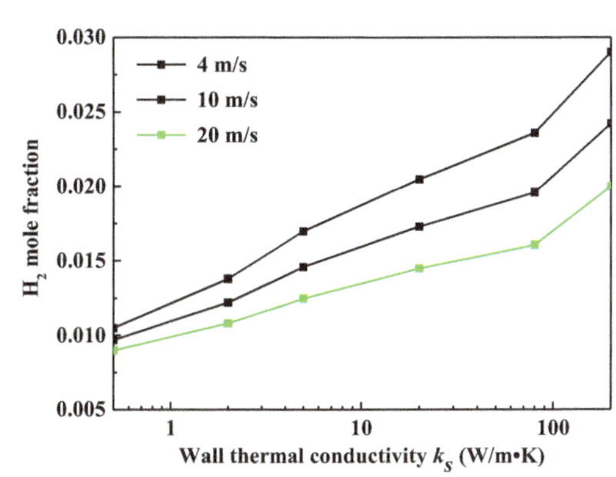

Fig. 11 The concentration of hydrogen required for ignition in hydrogen-assisted co-feed mode for different inlet velocities and wall thermal conductivities. φ (propane) = 0.7, h = 20 W/m^2·K, and k_s = 20 W/m·K.

As observed in Fig. 11, inlet velocity and wall thermal conductivity both have strong effects on the concentration of hydrogen required for propane ignition. The wall temperatures increase as the inlet velocity is increased. Therefore, the concentration of hydrogen required for propane ignition decreases as the velocity is increased. Likewise, the concentration of hydrogen required for propane ignition decreases as the wall thermal conductivity is decreased. The reason is that the hot spots created near the inlet section for lower thermal conductivity material as observed in Fig. 2 help the hydrogen-propane-air mixtures get ignite with lesser hydrogen.

In the present work, the transient response of ignition in the co-feed mode is analyzed as follows.

The equivalence ratio φ = 0.7 of premixed propane-air is kept fixed and the concentration of hydrogen required is 0.0005 excess of the minimum hydrogen concentration required for ignition. In the co-feed ignition mode, the transient responses of bulk gas phase temperature, interior wall temperature and propane conversion for three different values of wall thermal conductivity are shown in Fig. 12, 13 and 14, respectively.

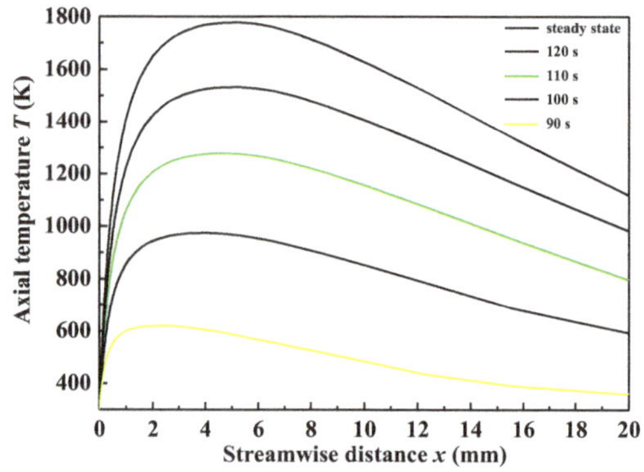

(a) k_s = 0.5 W/m·K

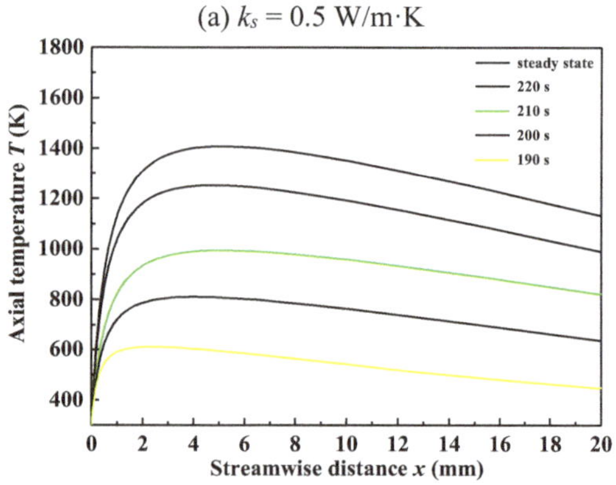

(b) k_s = 20 W/m·K

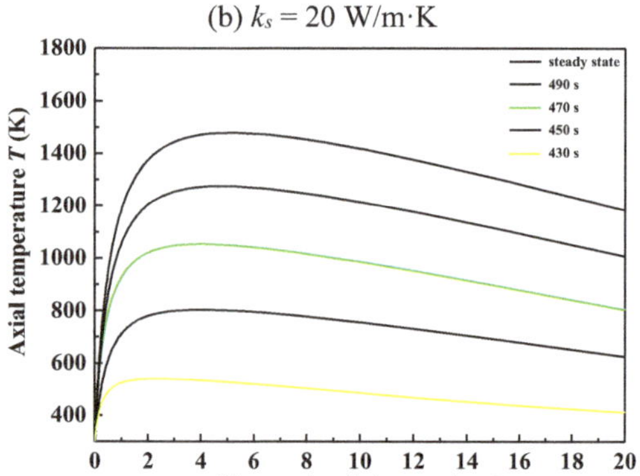

(c) k_s = 200 W/m·K

Fig. 12 Transient response of bulk gas phase temperature profiles for ignition in hydrogen-assisted co-feed mode for different wall thermal conductivities. φ (propane) = 0.7, u_0 = 2 m/s, h = 20 W/m^2·K, and k_s = 20 W/m·K.

As before, hydrogen combustion ignites in approximately 1.0-2.0 s and directly reaches its steady state profile (magenta line: hydrogen in Fig. 14). For most of the simulation time, note that the hydrogen conversion profiles do not show any noticeable change, and we solve the fully transient issue without making any pseudo-steady assumptions. At various different times, the axial variation in propane conversion is plotted in Fig. 14. The ignition time (t_{ign}) is defined as the time taken for propane conversion rate to reach 50% at the exit in catalytic micro-combustors.

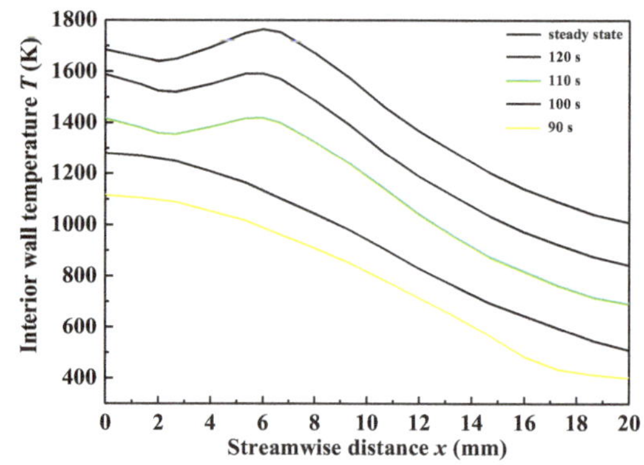

(a) k_s = 0.5 W/m·K

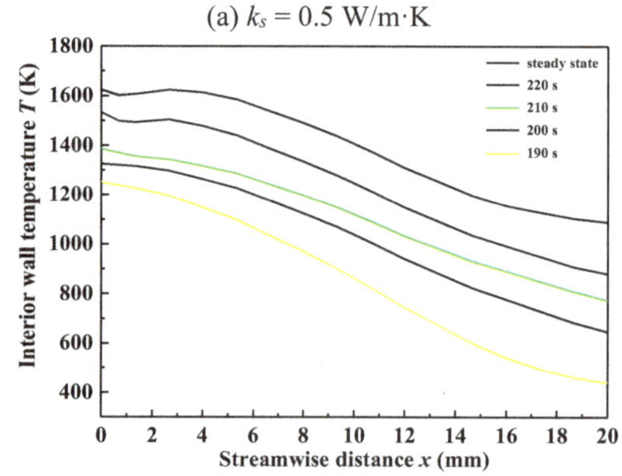

(b) k_s = 20 W/m·K

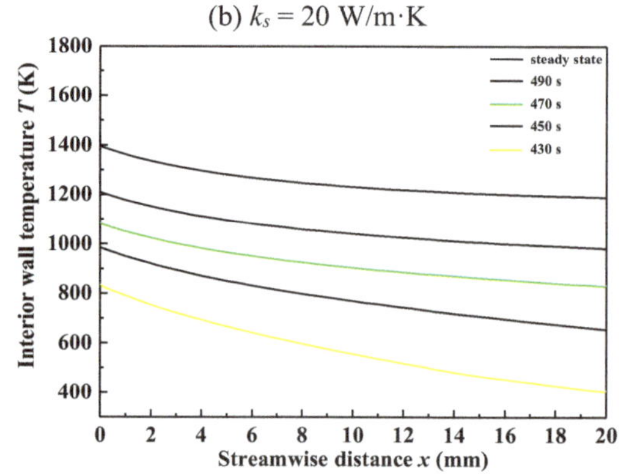

(c) k_s = 200 W/m·K

Fig. 13 Transient response of interior wall temperature profiles for ignition in hydrogen-assisted co-feed mode for different wall thermal conductivities. φ (propane) = 0.7, u_0 = 2 m/s, h = 20 W/m^2·K, and k_s = 20 W/m·K.

As the wall thermal conductivity is increased, the catalytic ignition time increases as observed in Fig. 14: from 98 s for insulating walls (0.5 W/m·K), to 202 s for moderately thermal conducting wall (20 W/m·K), and 448 s for highly thermal conducting walls (200 W/m·K). The aforementioned trend can be attributed to heat localization in the wall materials with low thermal conductivity, which leads to hot spot formation and faster ignition. In addition, in the co-feed ignition mode, the front-end ignition in catalytic micro-combustors is observed in all cases. Therefore, the bulk gas phase and interior wall temperatures (Fig. 12 and 13) reach their steady state values quickly once propane is ignited. Therefore, in the co-feed ignition mode, the catalytic combustion characteristics of hydrogen-assisted propane qualitatively resemble the selectively preheating initial portion of the combustion chamber wall.

As observed in Fig. 13, for low thermal conductivity walls (0.5 W/m·K), the interior wall temperatures first drop, then increase and then drop again at steady state. This trend is mostly because hydrogen combustion zone is located significantly upstream compared to propane. In the region between them, the interior wall temperatures drop slightly by reason of heat losses to the surroundings.

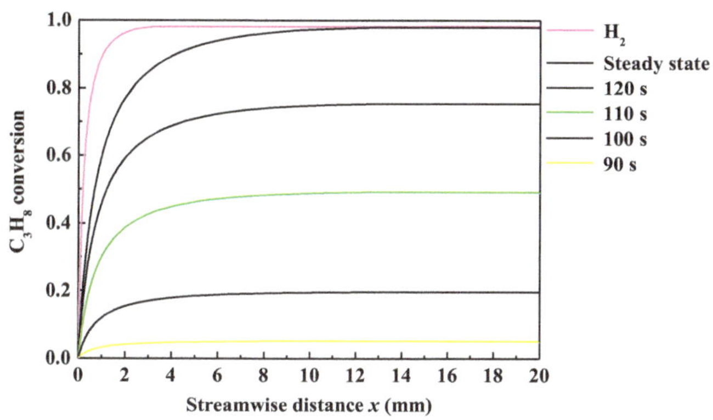

(a) k_s = 0.5 W/m·K

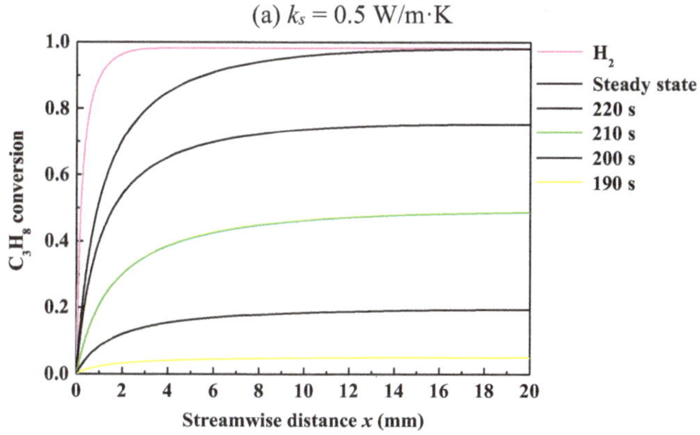

(b) k_s = 20 W/m·K

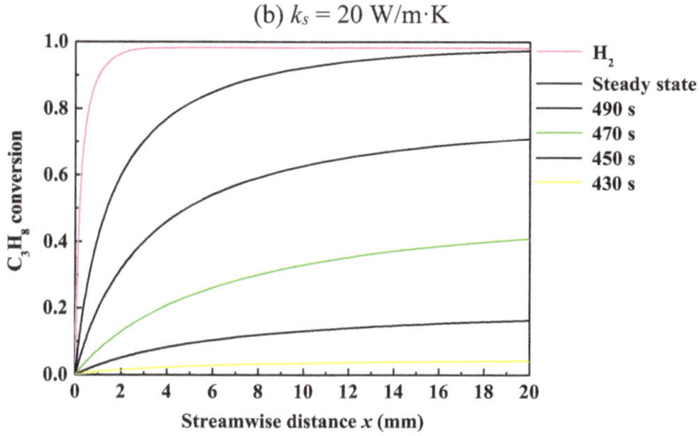

(c) k_s = 200 W/m·K

Fig. 14 Transient response of propane conversion profiles for ignition in hydrogen-assisted co-feed mode for different wall thermal conductivities. φ (propane) = 0.7, u_0 = 2 m/s, h = 20 W/m²·K, and k_s = 20 W/m·K.

The gas phase temperature and propane conversion vs. the cumulative time at the exit for different wall thermal conductivities are shown in Fig. 15. As observed, the triangles represent the ignition time t_{ign} at which propane is ignited, and the circles represent the steady state time t_{ss} at which propane catalytic combustion taken to reach steady state.

In general, under lean combustion conditions, the premixed hydrogen-propane-air mixtures in co-feed mode can lead to ignition without the need for external heating. Hydrogen ignites instantaneously, and followed by propane ignition after propane ignition temperature is reached. These results are consistent with the conclusions from the previous works of Deutschmann *et al.* (2000), who experimentally investigated the hydrogen-assisted catalytic ignition characteristics of methane in a Pt-coated monolith reactor, and found that the main effect of hydrogen is thermal during ignition; the hydrogen ignited first on the surface of catalysts, and then the temperature of mixtures increased from room temperature to ignition point.

The concentration (on a molar basis) of hydrogen required for propane catalytic ignition varies in the range of 0.008-0.028, and decreases for: lower wall thermal conductivity, higher inlet velocity, and higher equivalence ratio of propane-air mixtures.

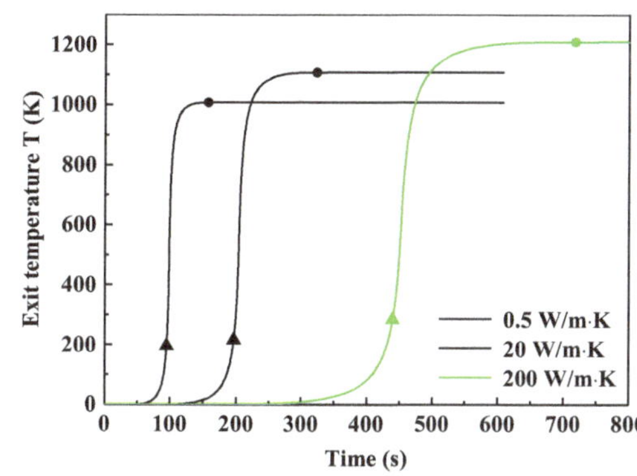

(a) Exit temperature

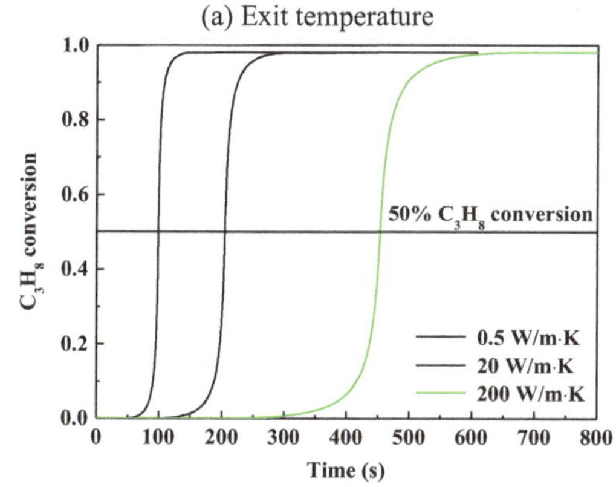

(b) Propane conversion

Fig. 15 Exit temperature and propane conversion profiles for ignition in hydrogen-assisted co-feed mode for different wall thermal conductivities. φ (propane) = 0.7, u_0 = 2 m/s, h = 20 W/m²·K, and k_s = 20 W/m·K. The triangles represent the ignition time t_{ign}, and the circles represent the steady state time t_{ss}.

4.2 Sequential Feed Mode

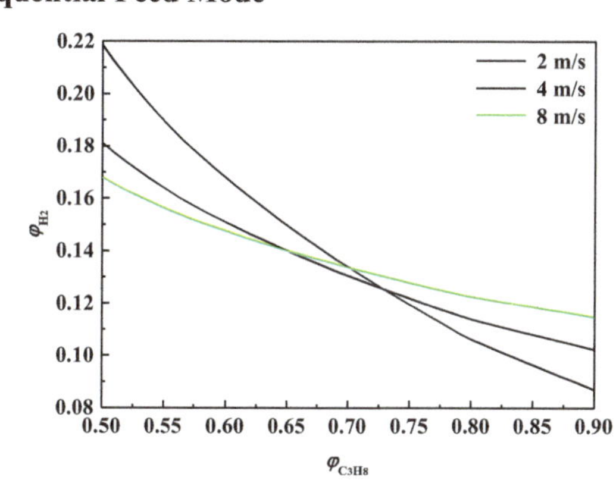

Fig. 16 The minimum equivalence ratio of hydrogen-air required for the propane ignition of a particular equivalence ratio of propane-air for different inlet velocities.

In the sequential feed mode, effect of equivalence ratio of propane-air on the minimum equivalence ratio of hydrogen required for propane ignition for different inlet velocities is shown in Fig. 16. The minimum concentration of hydrogen required for propane ignition decreases as the equivalence ratio of propane-air is increased. For higher inlet velocity, the concentration of hydrogen required for propane ignition is expected to be lower in agreement with the previous section. The region below each curve represents that the hydrogen concentration is insufficient for propane ignition. While this behavior is existent at higher equivalence ratio of propane-air ($\varphi_{C3H8} > 0.73$), the situation is entirely different at lower equivalence ratio of propane-air ($\varphi_{C3H8} < 0.64$). Note that the wall temperature is still higher as the velocity is increased, keeping equivalence ratio constant. However, the location of the combustion zone is pushed downstream at higher inlet velocities. This prevents stabilization of propane combustion when we switch over from hydrogen-air to propane-air inlet. Hence, we see a reversal in the trend at lower propane equivalence ratio.

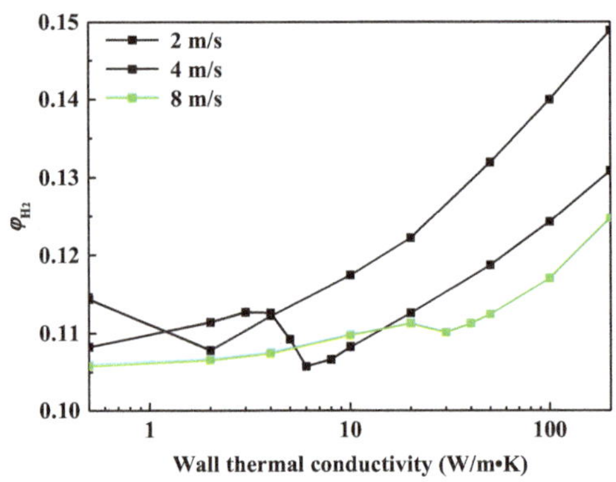

(a) Concentration of hydrogen required

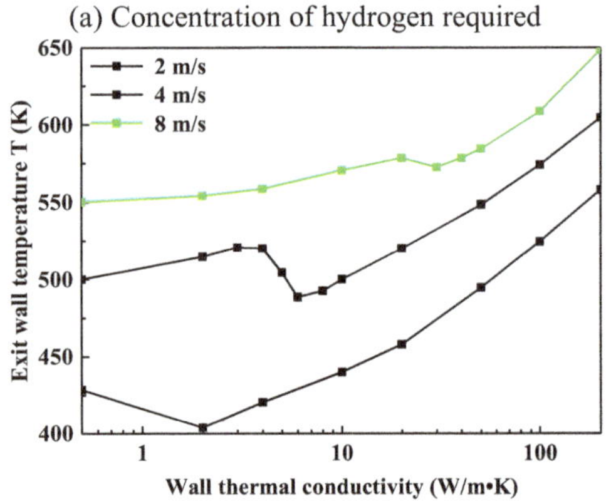

(b) Exit wall temperature

Fig. 17 Effect of wall thermal conductivity on exit wall temperature and the minimum concentration of hydrogen required for propane ignition for different inlet velocities.

Effect of wall thermal conductivity on exit wall temperature and the minimum concentration of hydrogen required for propane catalytic ignition for different inlet velocities are shown in Fig. 17. The exit wall temperature is the minimum wall temperature that should be attained before switching from hydrogen-air to propane-air mixtures to guarantee stable propane catalytic ignition through the parametric study.

In the sequential feed mode, as wall thermal conductivity is decreased, the concentration of hydrogen required decreases consistently with the co-feed mode results. However, this variation trend is not monotonic unlike the co-feed mode. Such as inlet velocity of 4 m/s, as the wall thermal conductivity is decreased until 6 W/m·K, the minimum concentration of hydrogen required decreases monotonically. However, as the wall thermal conductivity is further decreased to 0.5 W/m·K, the minimum concentration of hydrogen required increases slightly and then continue to decrease. Similar variation trends can be observed for the inlet velocities of 2 and 8 m/s as well.

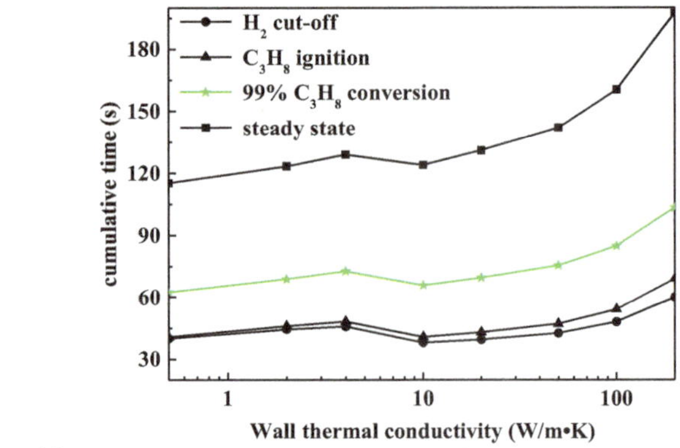

Fig. 18 Effect of wall thermal conductivity on hydrogen cut-off time, propane ignition time, 99% propane conversion and steady state time in the sequential feed mode. φ (hydrogen) = 0.148, φ (propane, after hydrogen cut-off) = 0.7, u_0 = 2 m/s, h = 20 W/m²·K, and k_s = 20 W/m·K.

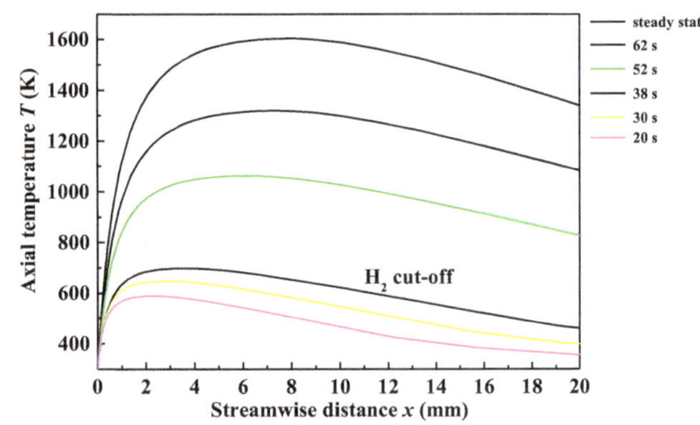

(a) k_s = 0.5 W/m·K

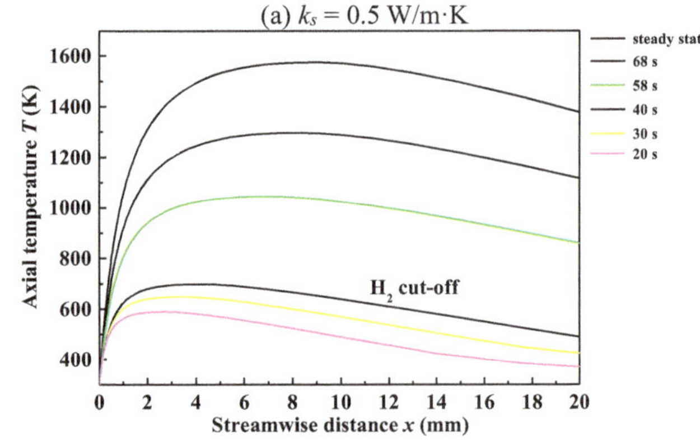

(b) k_s = 20 W/m·K

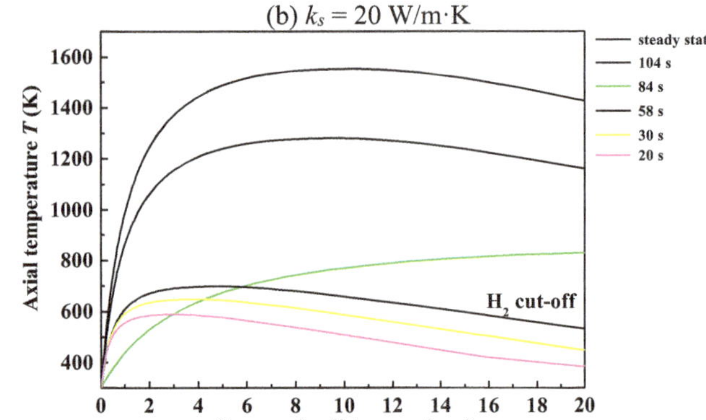

(c) k_s = 200 W/m·K

Fig. 19 Transient response of bulk gas phase temperature profiles in the sequential feed mode for different wall thermal conductivities. φ (hydrogen) = 0.148, φ (propane, after hydrogen cut-off) = 0.7, u_0 = 2 m/s, h = 20 W/m²·K, and k_s = 20 W/m·K.

Effect of wall thermal conductivity on the hydrogen cut-off time, propane ignition time, 99% propane conversion and steady state time are shown in Fig. 18. For three different wall thermal conductivities (0.5, 20 and 200 W/m·K), the temporal variations in bulk gas phase temperature, interior wall temperature and propane conversion profiles are shown in Fig. 19, 20 and 21. The flow is switched from hydrogen-air to propane-air mixtures as the exit wall temperatures exceed the corresponding temperatures shown in Fig. 17 (b). The switch is represented as circles in Fig. 18 and by blue lines in Fig. 19.

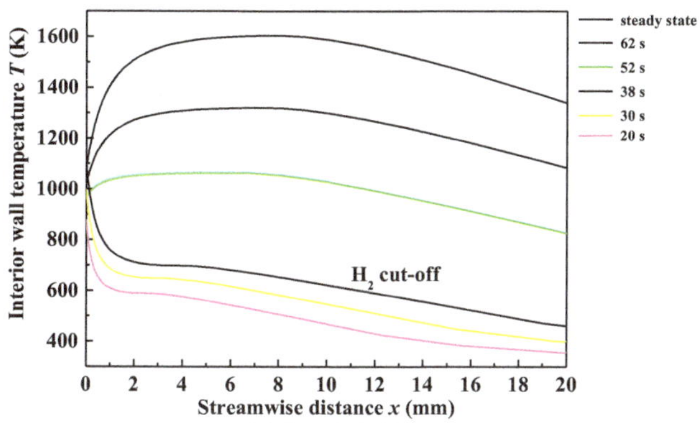

(a) $k_s = 0.5$ W/m·K

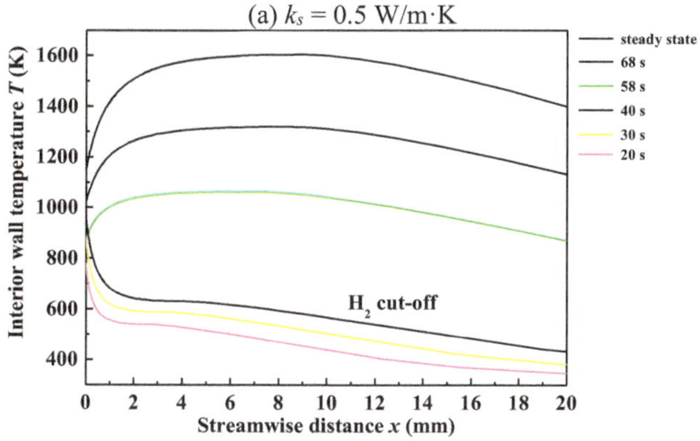

(b) $k_s = 20$ W/m·K

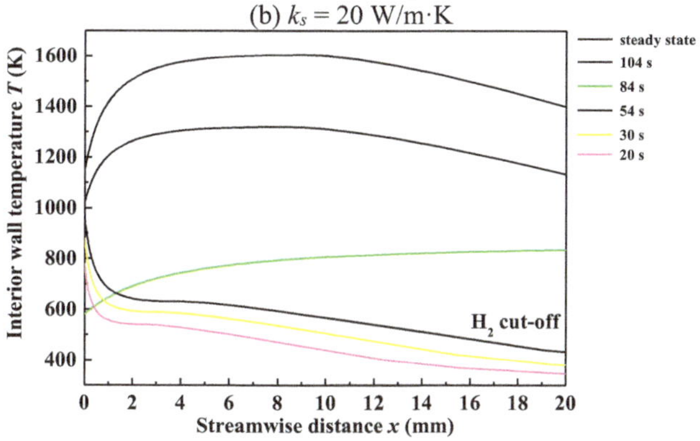

(c) $k_s = 200$ W/m·K

Fig. 20 Transient response of interior wall temperature profiles in the sequential feed mode for different wall thermal conductivities. φ (hydrogen) = 0.148, φ (propane, after hydrogen cut-off) = 0.7, $u_0 = 2$ m/s, $h = 20$ W/m^2·K, and $k_s = 20$ W/m·K.

The hydrogen cut-off times do not vary significantly ($40s \leq t_{cut\text{-}off} \leq 60s$) as the wall thermal conductivity is increased. For all cases, propane ignition times require less than 10 s and 99% propane conversion is reached (as observed in Fig. 18) in a short time thereafter once the flow of propane-air mixtures is started. However, compared with lower thermal conductivity walls, the times taken to reach steady state are significantly higher for highly thermal conductivity walls. For low and moderate wall thermal conductivity values (0.5 and 20 W/m·K) as observed in Fig. 19, 20 and 21, the catalytic ignition characteristics are similar to partially preheating the initial segment of the micro-combustor. However, for high wall thermal conductivity values (200 W/m·K), the ignition characteristics are close to completely heating the catalytic micro-combustor wall, and the back-end ignition can be observed.

The cumulative propane emissions and exit gas temperature for different wall thermal conductivities are shown in Fig. 22. Initially, there are no propane emissions until propane flow is started. The propane emissions with sharp jump occurs between the switch time and 99% propane conversion time (represented as circles in Fig. 22 (a)). Moreover, the exit gas temperature increases rapidly as the inlet flow is switched from lean hydrogen-air ($\varphi_{H2} = 0.148$) to propane-air ($\varphi_{C3H8} = 0.7$) mixtures. For high thermal conductivity walls (200 W/m·K), Net propane emissions (22.84 mg) at the end of 200 s increased by 3 times for low thermal conductivity walls (5.54 mg (0.5 W/m·K) and 5.58 mg (20 W/m·K)).

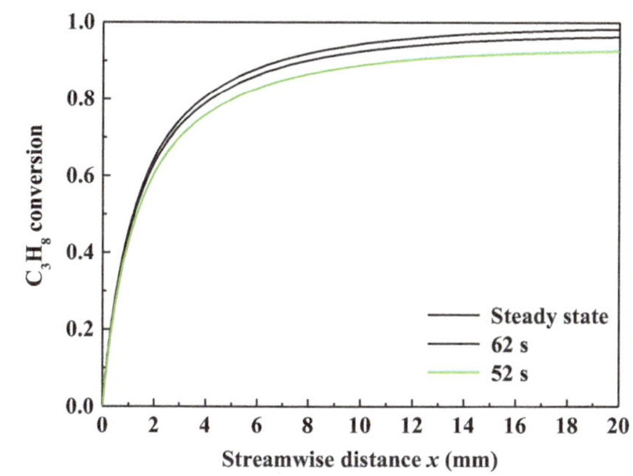

(a) $k_s = 0.5$ W/m·K

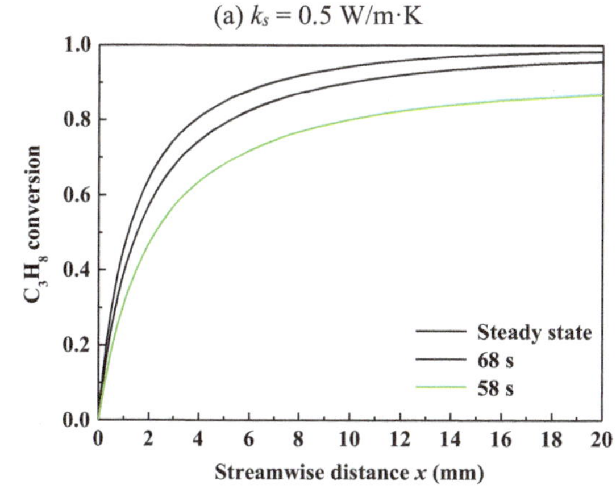

(b) $k_s = 20$ W/m·K

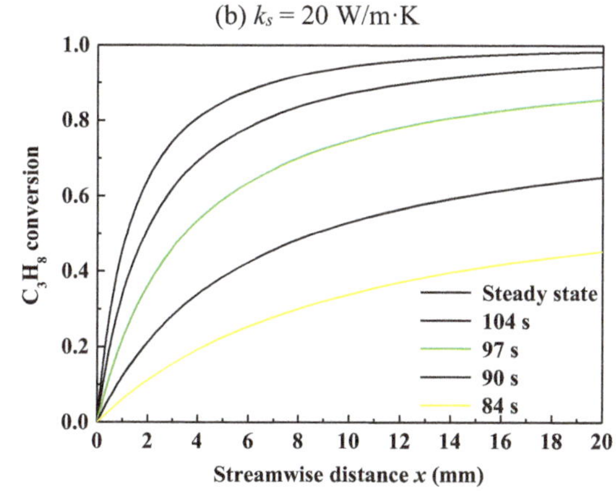

(c) $k_s = 200$ W/m·K

Fig. 21 Transient response of propane conversion profiles in the sequential feed mode for different wall thermal conductivities. φ (hydrogen) = 0.148, φ (propane, after hydrogen cut-off) = 0.7, $u_0 = 2$ m/s, $h = 20$ W/m^2·K, and $k_s = 20$ W/m·K.

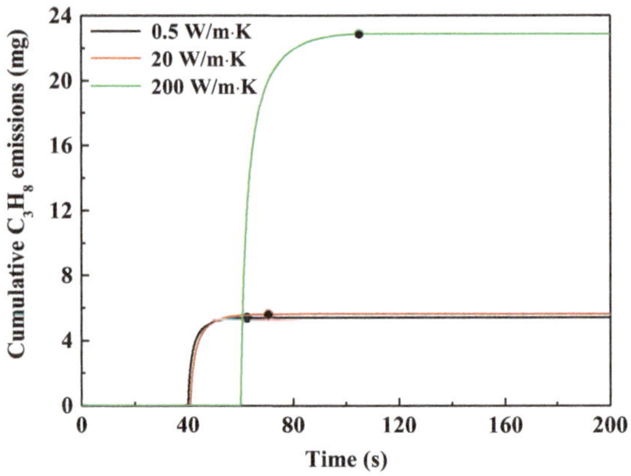

(a) Cumulative propane emissions

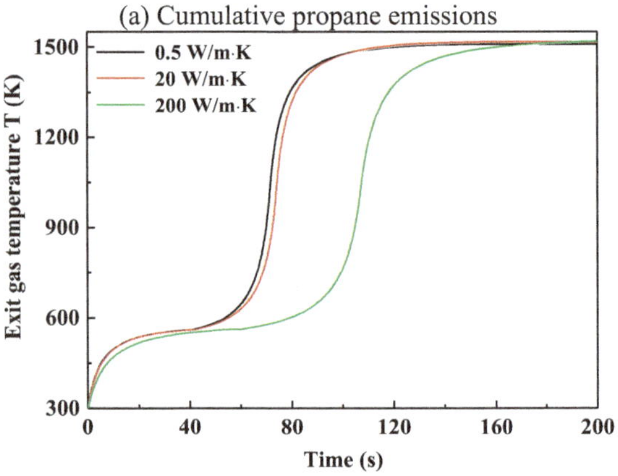

(b) Exit gas temperature

Fig. 22 Cumulative propane emissions and exit gas temperature in the sequential feed mode for different wall thermal conductivities. φ (hydrogen) = 0.148, φ (propane, after hydrogen cut-off) = 0.7, u_0 = 2 m/s, h = 20 W/m^2·K, and k_s = 20 W/m·K. Circles represent the time taken to reach 99% propane conversion in panel (a).

5. CONCLUSIONS

The addition of hydrogen to the initial mixture offers a way to light-off catalytic combustion of propane on Pt/γ-Al_2O_3. In this study, for the steady and transient state, effect of wall thermal conductivity on hydrogen self-ignition and hydrogen-assisted ignition of propane-air mixtures in different feeding modes from ambient cold-start conditions were investigated numerically with chemical kinetic model in Pt/γ-Al_2O_3 catalytic micro-combustors. The following conclusions were obtained from this micro-scale combustion characteristics study.

(1) Effect of wall thermal conductivity on the combustion and self-ignition characteristics of hydrogen. Hydrogen conversion reaches the steady state in less than 1.0 s and front-end ignition is observed. Compared with the diffusion and reaction processes, the large thermal inertia of the micro-combustor solid structure leads to slow temperature dynamics, and transient response is dominated by the thermal inertia. The time required to reach steady state decreases with increasing inlet velocity and decreasing wall thermal conductivity. The heat localization in poorly conducting walls leads to fast ignition and shorter steady state time.

(2) Effect of wall thermal conductivity on catalytic ignition characteristics of hydrogen-assisted propane in the co-feed mode. The concentration (on a molar basis) of hydrogen required for propane ignition varies from 0.008 to 0.028. The concentration of hydrogen required for propane ignition increased with increasing wall thermal conductivity, decreasing inlet velocity, and decreasing inlet equivalence ratio of propane-air mixtures. The combustion characteristics of hydrogen-assisted propane qualitatively resemble the selectively preheating initial portion of the combustion chamber wall. Hydrogen ignites instantaneously, and followed by propane ignition after propane ignition temperature is reached.

(3) Effect of wall thermal conductivity on catalytic ignition characteristics of hydrogen-assisted propane in the sequential feed mode. The wall thermal conductivity, inlet velocity, and inlet equivalence ratio of propane-air mixtures have significant effect on the concentration of hydrogen required for propane ignition. In general, the concentration of hydrogen required for propane ignition increased with increasing wall thermal conductivity, decreasing inlet velocity, and decreasing inlet equivalence ratio of propane-air mixtures. While this behavior is existent at higher wall thermal conductivity (k_s > 30 W/m·K) for wall thermal conductivity and inlet velocity. The time taken to reach steady state, the hydrogen cut-off time, the propane ignition time and the cumulative propane emissions increased with increasing wall thermal conductivity. For low and moderate wall thermal conductivity values (0.5 and 20 W/m·K), the ignition characteristics are similar to partially preheating the initial segment of the micro-combustor. However, for high wall thermal conductivity values (200 W/m·K), the ignition characteristics are close to completely heating the micro-combustor wall, and the back-end ignition can be observed.

REFERENCES

Ahn, J., Eastwood, C., Sitzki, L., and Ronney, P.D., 2005, "Gas-phase and catalytic combustion in heat-recirculating burners," *Proceedings of the Combustion Institute,* **30**(2), 2463-2472.
http://dx.doi.org/10.1016/j.proci.2004.08.265

Aryafar, M., and Zaera, F., 1997, "Kinetic study of the catalytic oxidation of alkanes over nickel, palladium, and platinum foils," *Catalysis Letters,* **48**(3-4), 173-183.
http://dx.doi.org/10.1023/a:1019055810760

Baigmohammadi, M., Sadeghi, S.S., Tabejamaat, S., and Zarvandi, J., 2013, "Numerical study of the effects of wire insertion on CH_4(methane)/air pre-mixed flame in a micro combustor," *Energy,* **54**, 271-284.
http://dx.doi.org/10.1016/j.energy.2013.01.047

Bui, P.-A., Wilder, E., Vlachos, D., and Westmoreland, P., 1997, "Hierarchical reduced models for catalytic combustion: H_2/air mixtures near platinum surfaces," *Combustion Science and Technology,* **129**(1-6), 243-275.
http://dx.doi.org/10.1080/00102209708935728

Cheekatamarla, P.K., Finnerty, C.M., and Cai, J., 2008, "Internal reforming of hydrocarbon fuels in tubular solid oxide fuel cells," *International Journal of Hydrogen Energy,* **33**(7), 1853-1858.
http://dx.doi.org/10.1016/j.ijhydene.2008.02.004

Chen, G.-B., Chen, C.-P., Wu, C.-Y., and Chao, Y.-C., 2007, "Effects of catalytic walls on hydrogen/air combustion inside a micro-tube," Applied Catalysis a-General, 332(1), 89-97.
http://dx.doi.org/10.1016/j.apcata.2007.08.011

Deshmukh, S.R., Mhadeshwar, A.B., and Vlachos, D.G., 2004, "Microreactor modeling for hydrogen production from ammonia decomposition on ruthenium," *Industrial & Engineering Chemistry Research,* **43**(12), 2986-2999.
http://dx.doi.org/10.1021/ie030557y

Deutschmann, O., Maier, L.I., Riedel, U., Stroemman, A.H., and Dibble, R.W., 2000, "Hydrogen assisted catalytic combustion of methane on platinum," *Catalysis Today,* **59**(1-2), 141-150.
http://dx.doi.org/10.1016/s0920-5861(00)00279-0

Fanaee, S.A., and Esfahani, J.A., 2014, "Two-dimensional analytical model of flame characteristic in catalytic micro-combustors for a

hydrogen-air mixture," *International Journal of Hydrogen Energy,* **39**(9), 4600-4610.
http://dx.doi.org/10.1016/j.ijhydene.2013.12.191

Federici, J.A., Wetzel, E.D., Geil, B.R., and Vlachos, D.G., 2009, "Single channel and heat recirculation catalytic microburners: An experimental and computational fluid dynamics study," *Proceedings of the Combustion Institute,* **32**(2), 3011-3018.
http://dx.doi.org/10.1016/j.proci.2008.07.005

Gomez, A., Berry, J.J., Roychoudhury, S., Coriton, B., and Huth, J., 2007, "From jet fuel to electric power using a mesoscale, efficient Stirling cycle," *Proceedings of the Combustion Institute,* **31**(2), 3251-3259.
http://dx.doi.org/10.1016/j.proci.2006.07.203

Groppi, G., Belloli, A., Tronconi, E., and Forzatti, P., 1995, "A comparison of lumped and distributed models of monolith catalytic combustors," *Chemical Engineering Science,* **50**(17), 2705-2715.
http://dx.doi.org/10.1016/0009-2509(95)00099-q

Hsueh, C.-Y., Chu, H.-S., Yan, W.-M., Leu, G.-C., and Tsai, J.-I., 2011, "Three-dimensional analysis of a plate methanol steam micro-reformer and a methanol catalytic combustor with different flow channel designs," *International Journal of Hydrogen Energy,* **36**(21), 13575-13586.
http://dx.doi.org/10.1016/j.ijhydene.2011.07.099

Hua, J., Wu, M., and Kumar, K., 2005, "Numerical simulation of the combustion of hydrogen-air mixture in micro-scaled chambers. Part I: fundamental study," *Chemical Engineering Science,* **60**(13), 3497-3506.
http://dx.doi.org/10.1016/j.ces.2005.01.041

Jin, J., and Kwon, S., 2009, "Microcatalytic combustion of H_2 on Pt/Al_2O_3-coated nickel foam," *Combustion Science and Technology,* **181**(2), 211-225.
http://dx.doi.org/10.1080/00102200802424526

Ju, Y., and Choi, C.W., 2003, "An analysis of sub-limit flame dynamics using opposite propagating flames in mesoscale channels," *Combustion and Flame,* **133**(4), 483-493.
http://dx.doi.org/10.1016/s0010-2180(03)00058-0

Ju, Y., and Maruta, K., 2011, "Microscale combustion: technology development and fundamental research," *Progress in Energy and Combustion Science,* **37**(6), 669-715.
http://dx.doi.org/10.1016/j.pecs.2011.03.001

Kaisare, N.S., Deshmukh, S.R., and Vlachos, D.G., 2008, "Stability and performance of catalytic microreactors: Simulations of propane catalytic combustion on Pt," *Chemical Engineering Science,* **63**(4), 1098-1116.
http://dx.doi.org/10.1016/j.ces.2007.11.014

Kaisare, N. S., Lee, J. H., and Fedorov, A. G., 2005, "Hydrogen generation in a reverse-flow microreactor: 1. Model formulation and scaling," *AICHE Journal,* **51**(8), 2254-2264.
http://dx.doi.org/10.1002/aic.10492

Kaisare, N.S., and Vlachos, D.G., 2007, "Extending the region of stable homogeneous micro-combustion through forced unsteady operation," *Proceedings of the Combustion Institute,* **31**(2), 3293-3300.
http://dx.doi.org/10.1016/j.proci.2006.07.031

Kaisare, N.S., and Vlachos, D.G., 2007, "Optimal reactor dimensions for homogeneous combustion in small channels," *Catalysis Today,* **120**(1), 96-106.
http://dx.doi.org/10.1016/j.cattod.2006.07.036

Kamada, T., Nakamura, H., Tezuka, T., Hasegawa, S., and Maruta, K., 2014, "Study on combustion and ignition characteristics of natural gas components in a micro flow reactor with a controlled temperature profile," *Combustion and Flame,* **161**(1), 37-48.
http://dx.doi.org/10.1016/j.combustflame.2013.08.013

Karagiannidis, S., and Mantzaras, J., 2010, "Numerical investigation on the start-up of methane-fueled catalytic microreactors," *Combustion and Flame,* **157**(7), 1400-1413.
http://dx.doi.org/10.1016/j.combustflame.2010.01.008

Karagiannidis, S., Mantzaras, J., Jackson, G., and Boulouchos, K., 2007, "Hetero-/homogeneous combustion and stability maps in methane-fueled catalytic microreactors," *Proceedings of the Combustion Institute,* **31**(2), 3309-3317.
http://dx.doi.org/10.1016/j.proci.2006.07.121

Karim, A.M., Federici, J.A., and Vlachos, D.G., 2008, "Portable power production from methanol in an integrated thermoeletric/microreactor system," *Journal of Power Sources,* **179**(1), 113-120.
http://dx.doi.org/10.1016/j.jpowsour.2007.12.119

Kikas, T., Bardenshteyn, I., Williamson, C., Ejimofor, C., Puri, P., and Fedorov, A. G., 2003, "Hydrogen production in a reverse-flow autothermal catalytic microreactor: From evidence of performance enhancement to innovative reactor design," *Industrial & Engineering Chemistry Research,* **42**(25), 6273-6279.
http://dx.doi.org/10.1021/ie030021c

Kim, K.T., Lee, D.H., and Kwon, S., 2006, "Effects of thermal and chemical surface-flame interaction on flame quenching," *Combustion and Flame,* **146**(1-2), 19-28.
http://dx.doi.org/10.1016/j.combustflame.2006.04.012

Kim, N.I., and Maruta, K., 2006, "A numerical study on propagation of premixed flames in small tubes," *Combustion and Flame,* **146**(1-2), 283-301.
http://dx.doi.org/10.1016/j.combustflame.2006.03.004

Leach, T.T., and Cadou, C.P., 2005, "The role of structural heat exchange and heat loss in the design of efficient silicon micro-combustors," *Proceedings of the Combustion Institute,* **30**(2), 2437-2444.
http://dx.doi.org/10.1016/j.proci.2004.08.229

Leu, C.-H., King, S.-C., Huang, J.-M., Chen, C.-C., Tzeng, S.-S., Lee, C.-I., Chang, W.-C., and Yang, C.-C., 2013, "Visible images of the catalytic combustion of methanol in a micro-channel reactor," *Chemical Engineering Journal,* **226**(0), 201-208.
http://dx.doi.org/10.1016/j.cej.2013.04.040

Mantzaras, J., 2008, "Catalytic combustion of syngas," *Combustion Science and Technology,* **180**(6), 1137-1168.
http://dx.doi.org/10.1080/00102200801963342

Mantzaras, J., and Benz, P., 1999, "An asymptotic and numerical investigation of homogeneous ignition in catalytically stabilized channel flow combustion," *Combustion and Flame,* **119**(4), 455-472.
http://dx.doi.org/10.1016/s0010-2180(99)00071-1

Maruta, K., Takeda, K., Ahn, J., Borer, K., Sitzki, L., Ronney, P.D., and Deutschmann, O., 2002, "Extinction limits of catalytic combustion in microchannels," *Proceedings of the Combustion Institute,* **29**(1), 957-963.
http://dx.doi.org/10.1016/s1540-7489(02)80121-3

Men, Y., Kolb, G., Zapf, R., Pennemann, H., and Hessel, V., 2009, "Total combustion of propane in a catalytic microchannel combustor," *Chemical Engineering Research & Design,* **87**(1A), 91-96.
http://dx.doi.org/10.1016/j.cherd.2008.07.010

Norton, D.G., and Vlachos, D.G., 2003, "Combustion characteristics and flame stability at the microscale: a CFD study of premixed methane/air mixtures," *Chemical Engineering Science,* **58**(21), 4871-4882.
http://dx.doi.org/10.1016/j.ces.2002.12.005

Norton, D.G., and Vlachos, D.G., 2004, "A CFD study of propane/air microflame stability," *Combustion and Flame,* **138**(1-2), 97-107. http://dx.doi.org/10.1016/j.combustflame.2004.04.004

Norton, D.G., and Vlachos, D.G., 2005, "Hydrogen assisted self-ignition of propane/air mixtures in catalytic microburners," *Proceedings of the Combustion Institute,* **30**(2), 2473-2480. http://dx.doi.org/10.1016/j.proci.2004.08.188

Norton, D.G., Wetzel, E.D., and Vlachos, D.G., 2004, "Fabrication of single-channel catalytic microburners: effect of confinement on the oxidation of hydrogen/air mixtures," *Industrial & Engineering Chemistry Research,* **43**(16), 4833-4840. http://dx.doi.org/10.1021/ie049798b

Raimondeau, S., Norton, D., Vlachos, D.G., and Masel, R. I., 2002, "Modeling of high-temperature microburners," *Proceedings of the Combustion Institute,* **29**(1), 901-907. http://dx.doi.org/10.1016/s1540-7489(02)80114-6

Rana, U., Chakraborty, S., and Som, S.K., 2014, "Thermodynamics of premixed combustion in a heat recirculating micro combustor," *Energy,* **68**, 510-518. http://dx.doi.org/10.1016/j.energy.2014.02.070

Ronney, P.D., 2003, "Analysis of non-adiabatic heat-recirculating combustors," *Combustion and Flame,* **135**(4), 421-439. http://dx.doi.org/10.1016/j.combustflame.2003.07.003

Sari, A., and Zohrabian, A., 2014, "Simulation study of the effect of feed moisture on autothermal reforming in short contact time catalytic micro channels," *International Journal of Hydrogen Energy,* **39**(7), 3269-3285. http://dx.doi.org/10.1016/j.ijhydene.2013.12.105

Schneider, A., Mantzaras, J., and Eriksson, S., 2008, "Ignition and extinction in catalytic partial oxidation of methane-oxygen mixtures with large H_2O and CO_2 dilution," *Combustion Science and Technology,* **180**(1), 89-126. http://dx.doi.org/10.1080/00102200701487087

Schultze, M., Mantzaras, J., Bombach, R., and Boulouchos, K., 2013, "An experimental and numerical investigation of the hetero-/homogeneous combustion of fuel-rich hydrogen/air mixtures over platinum," *Proceedings of the Combustion Institute*, **34**(2), 2269-2277. http://dx.doi.org/10.1016/j.proci.2012.05.029

Sen, I., and Avci, A. K., 2014, "Simulation of exhaust gas reforming of propane in a heat exchange integrated microchannel reactor," *International Journal of Hydrogen Energy*, **39**(2), 844-852. http://dx.doi.org/10.1016/j.ijhydene.2013.10.148

Stefanidis, G.D., and Vlachos, D.G., 2009, "Controlling homogeneous chemistry in homogeneous-heterogeneous reactors: application to propane combustion," *Industrial & Engineering Chemistry Research,* **48**(13), 5962-5968. http://dx.doi.org/10.1021/ie801480m

Veeraragavan, A., and Cadou, C.P., 2011, "Flame speed predictions in planar micro/mesoscale combustors with conjugate heat transfer," *Combustion and Flame,* **158**(11), 2178-2187. http://dx.doi.org/10.1016/j.combustflame.2011.04.006

Volchko, S.J., Sung, C.J., Huang, Y.M., and Schneider, S.J., 2006, "Catalytic combustion of rich methane/oxygen mixtures for micropropulsion applications," *Journal of Propulsion and Power,* **22**(3), 684-693. http://dx.doi.org/10.2514/1.19809

Yang, W., Deng, C., Zhou, J., Liu, J., Wang, Y., and Cen, K., 2014, "Experimental and numerical investigations of hydrogen-air premixed combustion in a converging-diverging micro tube," *International Journal of Hydrogen Energy*, **39**(7), 3469-3476. http://dx.doi.org/10.1016/j.ijhydene.2013.12.102

Zarvandi, J., Tabejamaat, S., and Baigmohammadi, M., 2012, "Numerical study of the effects of heat transfer methods on $CH_4/(CH_4 + H_2)$-air pre-mixed flames in a micro-stepped tube," *Energy,* **44**(1), 396-409. http://dx.doi.org/10.1016/j.energy.2012.06.015

Zhang, Y., Zhou, J., Yang, W., Liu, M., and Cen, K., 2007, "Effects of hydrogen addition on methane catalytic combustion in a microtube," *International Journal of Hydrogen Energy,* **32**(9), 1286-1293. http://dx.doi.org/10.1016/j.ijhydene.2006.07.023

Zhou, J., Wang, Y., Yang, W., Liu, J., Wang, Z., and Cen, K., 2009, "Combustion of hydrogen-air in catalytic micro-combustors made of different material," *International Journal of Hydrogen Energy,* **34**(8), 3535-3545. http://dx.doi.org/10.1016/j.ijhydene.2009.01.032

NATURAL CONVECTION ON A POROUS VERTICAL PLATE IN A DOUBLY STRATIFIED NON-DARCY POROUS MEDIUM

D. Srinivasacharya[a,*], A.J. Chamkha[b], O. Surender[a], A.M. Rashad[c d]

[a] *Department of Mathematics, National Institute of Technology, Warangal - 506004, India*
[b] *Manufacturing Engineering Department, The Public Authority for Applied Education and Training, Shuweikh 70654, Kuwait*
[c] *Department of Mathematics, Aswan University, Faculty of Science, 81528, Egypt*
[d] *Department of Mathematics, Salman bin Abdul Aziz University, College of Science and Humanity Studies, AL-Kharj, Saudi Arabia*

ABSTRACT

The aim of the present article is to analyze the influence of thermal and mass stratification on natural convection heat and mass transfer over a porous vertical plate with uniform and constant wall temperature and concentration in porous medium. The Brinkman-Forchheimer based model is employed to describe the flow in the porous medium. The nonlinear governing equations and their associated boundary conditions are initially cast into dimensionless forms by pseudo-similarity variables. The resulting system of nonlinear, coupled partial differential equations is then solved numerically. The influence of pertinent parameters on the dimensionless velocity, temperature, concentration, heat and mass transfer coefficients are presented graphically.

Keywords: *Thermal stratification, Solutal stratification, Suction/injection, Non-Darcy porous medium.*

1. INTRODUCTION

The analysis of free convection heat and mass transfer in porous media has been an active field of research. It plays a crucial role in diverse applications, such as thermal insulation, distributions of temperature and moisture over agricultural fields and groves of fruit trees and damage of crops due to freezing and pollution of the environment, extraction of crude oil and chemical catalytic reactors, etc. The influence of suction and injection on convective heat and mass transfer over a vertical porous plate embedded in a porous medium has been the subject of many investigations owing to its relevance in many engineering and geophysical applications. Several researchers have studied natural convection heat and mass transfer over different geometries in porous medium. The influence of suction and injection on natural convective flow has been first studied by Eichhorn (1960). Soundalgekar et al. (1983) numerically investigated the problem of MHD free convection past a semi-infinite vertical plate with suction and injection. Pop and Watanabe (1992) studied the influence of suction or injection on steady free convection from a vertical cone. Makinde (2005) considered the combined free convection boundary layer flow past a permeable vertical plate with thermal radiation when the plate moves in its own plane. Alam et al. (2006) numerically studied the Dufour and Soret effects on unsteady MHD free convection and mass transfer flow past an infinite vertical porous plate embedded in a porous medium. Mbeledogu and Ogulu (2007) presented analytical closed-form solution of the unsteady hydro-magnetic natural convection heat and mass transfer flow of a rotating, incompressible, viscous Boussinesq fluid in the presence of radiative heat transfer and a first order chemical reaction between the fluid and the diffusing species. Ibrahim et al. (2008) reported analytical solutions for heat and mass transfer by laminar flow of a Newtonian, viscous, electrically conducting and heat generation/absorbing fluid on a continuously vertical permeable surface in the presence of a radiation, a first-order homogeneous chemical reaction and the mass flux. Rajeshwari et al. (2009) considered the effect of chemical reaction on the forced and free convection boundary layer that flows in a semi-infinite expanse of an electrically conducting viscous incompressible fluid past vertical porous plate. Chamkha et al. (2010) investigated the effects of chemical reaction on unsteady free convective heat and mass transfer on a stretching surface in a porous medium. Singh et al. (2011) presented an analytical solution for three-dimensional free convection, laminar flow of an incompressible viscous fluid past an infinite, vertical, and porous plate embedded within a highly porous medium. Husnain et al. (2012) studied unsteady boundary layer flow and heat transfer characteristics of a viscous fluid through porous media in the case of variable viscosity and variable Prandtl number. Bhattacharyya et al. (2013) considered unsteady MHD boundary layer flow with diffusion of chemically reactive species undergoing first-order chemical reaction over a permeable stretching sheet with suction or blowing and also with power-law variation in wall concentration.

In all the above papers, the effect of stratification has been neglected. Stratification is a formation/ deposition of layers which occur due to temperature variations, concentration differences or the presence of different fluids. The analysis of natural convection in a doubly stratified medium is a fundamentally interesting and important problem because of its wide range of engineering applications. These applications include heat rejection into the environment such as lakes, rivers, and seas; thermal energy storage systems such as solar ponds; and heat transfer from thermal sources such as the condensers of power plants. It is important to examine the temperature stratification and concentration differences of hydrogen and oxygen in lakes and ponds as they may directly affect the growth rate of all cultured species. Also, the analysis of thermal stratification is important for solar engineering because higher energy efficiency can be achieved with better stratification. It has been shown by scientists that thermal stratification

* Corresponding author Email: dsc@nitw.ac.in, dsrinivasacharya@yahoo.com

in energy storage may considerably increase system performance. Although the effect of stratification of the medium on the heat removal process in a porous medium is important, very little work has been reported in literature.

Early studies on convection transport focused on seeking similarity solution because similar variables can give great physical insight with minimal efforts. However, the non-similarity boundary layer flows are more general in nature and in our everyday life, and thus are more important than the similarity ones. Hence, in the present study we have made an attempt to obtain non-similar solutions for the problem of natural convection on a vertical porous plate with constant suction/injection in a stable doubly stratified non-Darcian fluid saturated porous medium in which the ambient temperature and concentration varies linearly. The effects of pertinent parameters on dimensionless velocity, temperature, concentration, heat and mass transfer coefficients have been examined and presented graphically.

2. MATHEMATICAL FORMULATION

Consider the steady natural convection heat and mass transfer along a porous vertical plate embedded in a stable, doubly stratified viscous fluid saturated non-Darcy porous medium. Consider that the plate is subject to constant suction or injection v_0. The porous medium is considered to be homogeneous and isotropic (i.e., uniform with a constant porosity and permeability) and is saturated with a fluid which is in local thermodynamic equilibrium with the solid matrix. The fluid has constant properties except the density in the buoyancy term of the balance of momentum equation. The fluid flow is moderate, so the pressure drop is proportional to the linear combination of fluid velocity and the square of velocity (Forchheimer flow model is considered). Choose the coordinate system such that x-axis is along the vertical plate and y-axis normal to the plate. The plate is maintained at uniform wall temperature T_w and concentration C_w. The ambient medium is assumed to be vertically linearly stratified with respect to both temperature and concentration in the form $T_\infty(x) = T_{\infty,0} + A_1 x$ and $C_\infty(x) = C_{\infty,0} + B_1 x$ respectively, where A_1 and B_1 are constants varied to alter the intensity of stratification in the medium. The values of T_w and C_w are assumed to be greater than the ambient temperature $T_{\infty,0}$ and concentration $C_{\infty,0}$ at any arbitrary reference point in the medium (inside the boundary layer).

By employing laminar boundary layer flow assumptions, Boussinesq approximation, using the Brinkman-Forchheimer model and Dupuit Forchheimer relationship, the governing equations are given by.

$$\frac{\partial u}{\partial x} + \frac{\partial v}{\partial y} = 0 \tag{1}$$

$$\frac{1}{\varepsilon^2}\left(u\frac{\partial u}{\partial x} + v\frac{\partial u}{\partial y}\right) = \frac{\nu}{\varepsilon}\frac{\partial^2 u}{\partial y^2} + g\beta_T(T - T_\infty) + g\beta_C(C - C_\infty) - \frac{\mu}{\rho K_p}u - \frac{b}{K_p}u^2 \tag{2}$$

$$u\frac{\partial T}{\partial x} + v\frac{\partial T}{\partial y} = \alpha\frac{\partial^2 T}{\partial y^2} \tag{3}$$

$$u\frac{\partial C}{\partial x} + v\frac{\partial C}{\partial y} = D\frac{\partial^2 C}{\partial y^2} \tag{4}$$

where u and v are Darcy velocity components in x and y directions respectively, T is the temperature, C is the concentration, g is the acceleration due to gravity, ρ is the density, b is the Forchheimer constant, K_p is the permeability, ε is the porosity, μ is the dynamic coefficient of viscosity, β_T is the coefficient of thermal expansion, β_C is the coefficient of solutal expansions, α is the effective thermal diffusivity and D is the effective solutal diffusivity of the medium.

The boundary conditions are

$$u = 0,\quad v = v_0,\quad T = T_w,\quad C = C_w,\quad \text{at}\quad y = 0 \tag{5a}$$

$$u = 0,\quad T = T_\infty(x),\quad C = C_\infty(x)\quad \text{as}\quad y \to \infty \tag{5b}$$

In view of the continuity Eq. (1), we introduce the stream function ψ such that

$$u = \frac{\partial \psi}{\partial y},\quad v = -\frac{\partial \psi}{\partial x} \tag{6}$$

Substituting Eqn. (6) in Eqs. (1)-(4) and then invoking the following non-similar transformations

$$\left.\begin{array}{l} \xi = \dfrac{x}{L},\ \eta = \dfrac{Gr^{1/4}}{L\xi^{1/4}}y,\ \psi = \dfrac{\mu Gr^{1/4}\xi^{3/4}}{\rho}f(\xi,\eta) \\ T - T_\infty(x) = (T_w - T_{\infty,0})\theta(\xi,\eta),\ C - C_\infty(x) = (C_w - C_{\infty,0})\phi(\xi,\eta) \end{array}\right\} \tag{7}$$

where T_∞ and C_∞ are ambient temperature and concentration resp. and α , D are thermal and solutal diffusivity resp.

Substituting the above transformations in Eqs. (1) - (4), we obtain the following nonlinear system of partial differential equations

$$\left.\begin{array}{l} \dfrac{1}{\varepsilon}f''' + \dfrac{3}{4\varepsilon^2}ff'' - \dfrac{1}{2}f'^2 + \theta + B\phi - \dfrac{\xi^{1/2}}{DaGr^{1/2}}f' - \dfrac{F_c}{Da}\xi f'^2 \\ = \dfrac{\xi}{\varepsilon^2}\left[f'\dfrac{\partial f'}{\partial \xi} - f''\dfrac{\partial f}{\partial \xi}\right] \end{array}\right\} \tag{8}$$

$$\frac{1}{Pr}\theta'' - \varepsilon_1\xi f' + \frac{3}{4}f\theta' = \xi\left[f'\frac{\partial \theta}{\partial \xi} - \theta'\frac{\partial f}{\partial \xi}\right] \tag{9}$$

$$\frac{1}{Sc}\phi'' - \varepsilon_2\xi f' + \frac{3}{4}f\phi' = \xi\left[f'\frac{\partial \phi}{\partial \xi} - \phi'\frac{\partial f}{\partial \xi}\right] \tag{10}$$

The boundary conditions are transformed as follows

$$\left.\begin{array}{l} f'(\xi,0) = 0,\ 3f(\xi,0) + 4\xi\left(\dfrac{\partial f}{\partial \xi}\right)_{\eta=0} = \gamma\xi^{1/4}, \\ \theta(\xi,0) = 1 - \varepsilon_1\xi,\ \phi(\xi,0) = 1 - \varepsilon_2\xi \quad \text{at}\quad \eta = 0 \end{array}\right\} \tag{11a}$$

$$f'(\xi,\infty) = 0,\ \theta(\xi,\infty) = 0,\ \phi(\xi,\infty) = 0\ \ as\ \ \eta \to \infty \tag{11b}$$

where the prime denotes differentiation with respect to η, γ is suction/ injection parameter,

$Gr = \dfrac{g\beta_T(T_w - T_{\infty,0})L^3}{\nu^2}$ is the Grashof number, $Pr = \dfrac{\nu}{\alpha}$ is the Prandtl number, $Sc = \dfrac{\nu}{D}$ is the Schmidt number, $F_c = \dfrac{b}{L}$ is the Forchheimer number, $Da = \dfrac{K_p}{L^2}$ is the Darcy number, $B = \dfrac{\beta_C(C_w - C_{\infty,0})}{\beta_T(T_w - T_{\infty,0})}$ is the buoyancy ratio, $\varepsilon_1 = \dfrac{A_1 L}{(T_w - T_{\infty,0})}$ and $\varepsilon_2 = \dfrac{B_1 L}{(C_w - C_{\infty,0})}$ are the thermal and solutal stratification parameters.

Results of practical interest are skin friction coefficient, heat and mass transfer rates. The local skin friction coefficient C_f, local Nusselt number Nu_ξ and the local Sherwood number Sh_ξ, are, respectively given by

$$Gr^{1/4}C_f = 2\xi^{-3/4}f''(\xi,0),\quad \frac{Nu_\xi}{Gr^{1/4}} = -\xi^{3/4}\theta'(\xi,0),\quad \frac{Sh_\xi}{Gr^{1/4}} = -\xi^{3/4}\phi'(\xi,0) \tag{12}$$

3. NUMERICAL METHOD

Equations (8) to (10) represent an initial-value problem with ξ playing the role of time. This general non-linear problem cannot be solved in closed form and, therefore, a numerical solution is necessary to describe the physics of the problem. The implicit, tri-diagonal finite-difference method similar to that discussed by Blottner (1970) has proven to be adequate and sufficiently accurate for the solution of this kind of problems. Therefore, it is adopted in the present work. All first-order derivatives with respect to ξ are replaced by two-point backward-difference formulae when marching in the positive ξ direction. Then, all second-order differential equations in η are discretized using three-point central difference quotients. This discretization process produces a tri-diagonal set of algebraic equations at each line of constant ξ which is readily solved by the well known Thomas algorithm (see Blottner (1970)). During the solution, iteration is employed to deal with the nonlinearity aspect of the governing differential equations. The problem is solved line by line starting with line $\xi = 0$ where similarity equations are solved to obtain the initial profiles of velocity, temperature and concentration and marching forward in ξ until the desired line of constant ξ is reached. Variable step sizes in the η direction with $\Delta\eta_1 = 0.001$ and a growth factor $G = 1.035$ such that $\Delta\eta = G\Delta\eta_1$ and constant step sizes in the ξ direction $\Delta\xi = 0.01$ are employed. These step sizes are arrived at after many numerical experimentations performed to assess grid independence. The convergence criterion employed in the present work is based on the difference between the current and the previous iterations. When this difference reached 10^{-5} for all points in the ξ directions, the solution was assumed converged and the iteration process was terminated. All the computations were carried out for fixed values of $\varepsilon = 0.8, B = 1.0, Pr = 0.71, Gr = 5.0, Sc = 0.22, F_c = 1.0$ and $Da = 1.0$, $\varepsilon_1 = 0.1, \varepsilon_2 = 0.3, \gamma = 0.1$. Some of the effects of the parameters on the physical quantities are very low (almost negligible) and hence they are not presented graphically.

4. RESULTS AND DISCUSSION

The non-dimensional velocity of the fluid falls near the plate and increases away from the plate with the rise of Forchheimer number as shown in Fig. 1. Since F_c represents the inertial drag, an increase in the Forchheimer number enhances the resistance to the flow and so a decrease in the fluid velocity near the plate ensues. Here $F_c = 0$ represents the case where the flow is Darcian. The velocity is maximum in this case due to the total absence of inertial drag. We notice from Fig. 2 that the temperature of the fluid diminishes with the rise of thermal stratification parameter. When thermal stratification is taken into consideration, the effective temperature difference between the plate and the ambient fluid will decrease; therefore, the thermal boundary layer is thickened and the temperature is reduced. Fig. 3 reveals that the concentration of the fluid decreases with the increase of

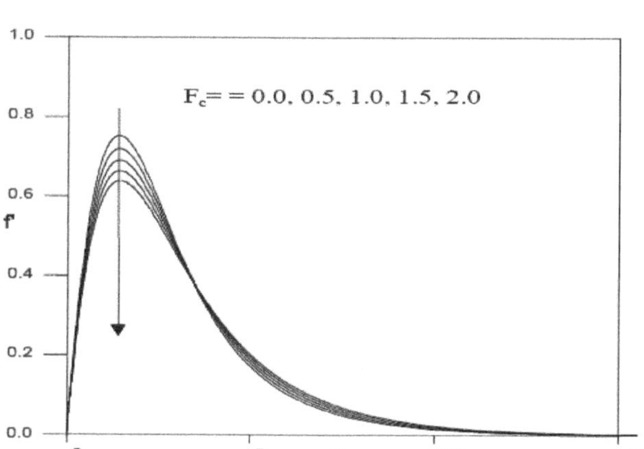

Fig. 1 Non-dimensional velocity profiles for different values of Forchheimer number

the mass stratification parameter. It is observed that the non-dimensional temperature and concentration values would become negative inside the boundary layer for different values of the stratification parameters depending on the values of other parameters.

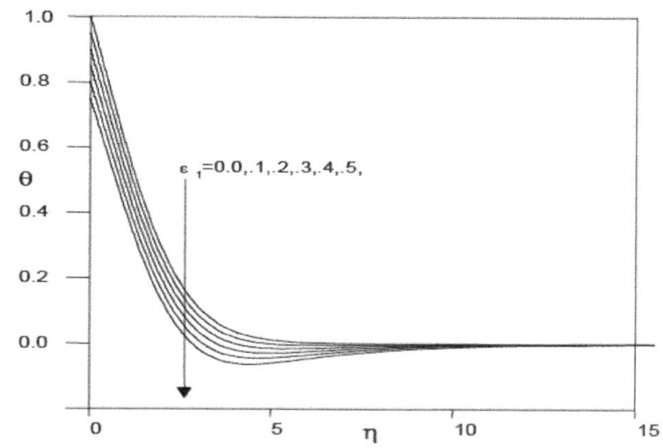

Fig. 2 Non-dimensional temperature profiles for different values of thermal stratification parameter

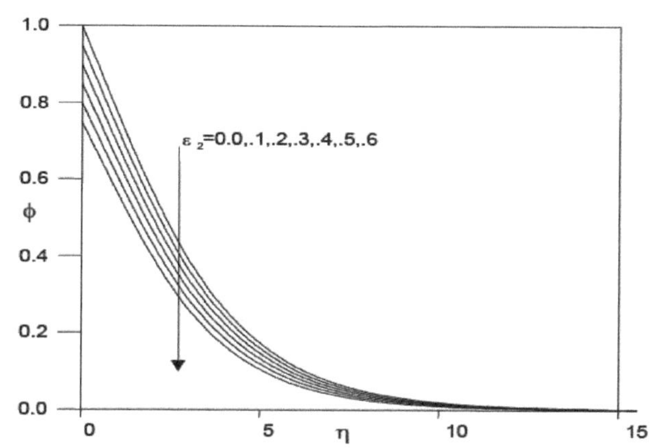

Fig. 3 Non-dimensional concentration profiles for different values of solutal stratification parameter

The influence of Darcy number Da on non-dimensional velocity, temperature and concentration profiles with η is presented in Figs. 4-6. Figure 4 shows that the velocity of the fluid is enhanced with the hike of Darcy number. The dimensionless temperature and concentration of the fluid decrease with the rise of Darcy number as shown in Figs. 5 and 6 respectively.

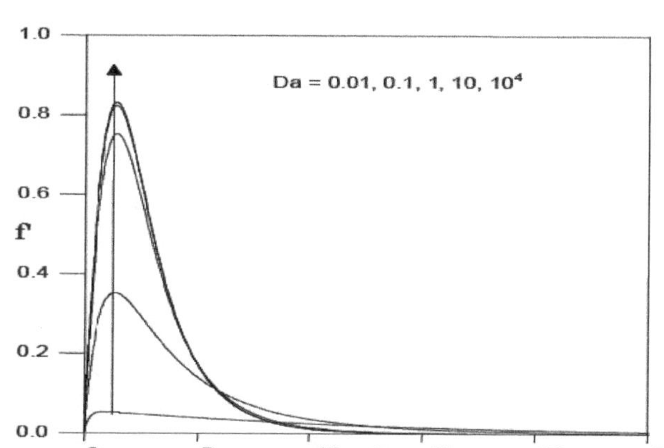

Fig. 4 Non-dimensional velocity profiles for different values of Darcy number

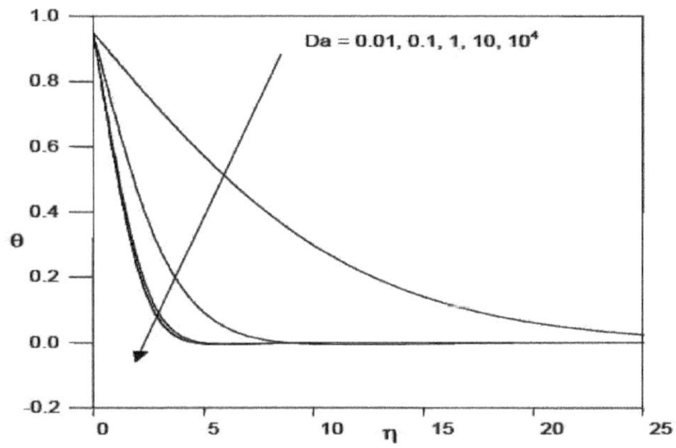

Fig. 5 Non-dimensional temperature profiles for different values of Darcy number

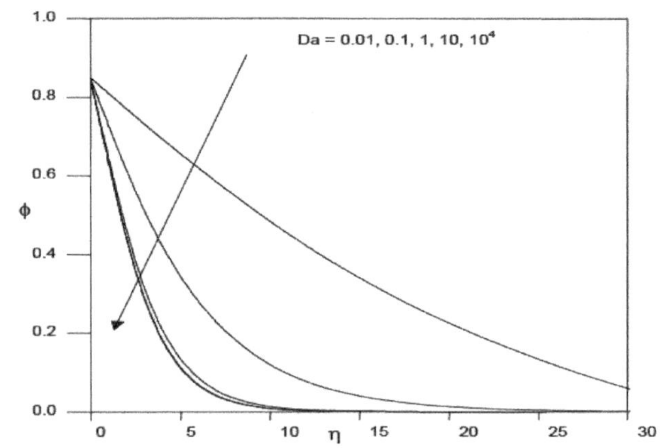

Fig. 6 Non-dimensional concentration profiles for different values of Darcy number

An increase in the intensity of suction/injection parameter causes a significant enhancement in the non-dimensional velocity, temperature and concentration in the boundary layer as shown in Figs. 7-9.

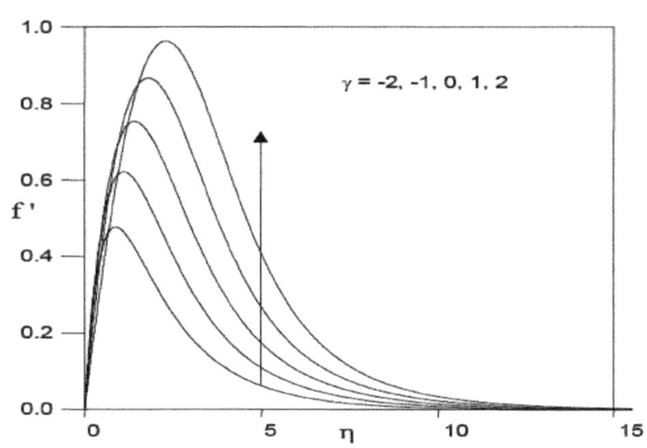

Fig. 7 Non-dimensional velocity profiles for different values of suction/injection parameter

The local heat and mass transfer coefficients (Nusselt number Nu_ξ and Sherwood number Sh_ξ) for different values of thermal and solutal stratification parameters are presented in Figs. 10-11. Figure 10 shows that the non-dimensional local heat transfer rate decreases with the increase in the value of thermal stratification parameter ε_1. Figure 11 illustrates that the local mass transfer coefficient decrease with the increase in the thermal stratification parameter. This is due to the effective mass transfer between the plate and the ambient medium decreases as the thermally stratified effect increases. The dimensionless local heat and mass transfer coefficients are reduced with the rise of solutal stratification parameter ε_2 as shown in Figs. 12-13.

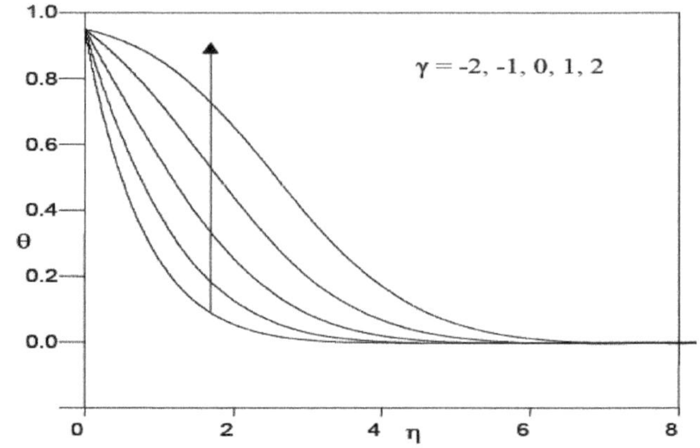

Fig. 8 Non-dimensional temperature profiles for different values of suction/injection parameter

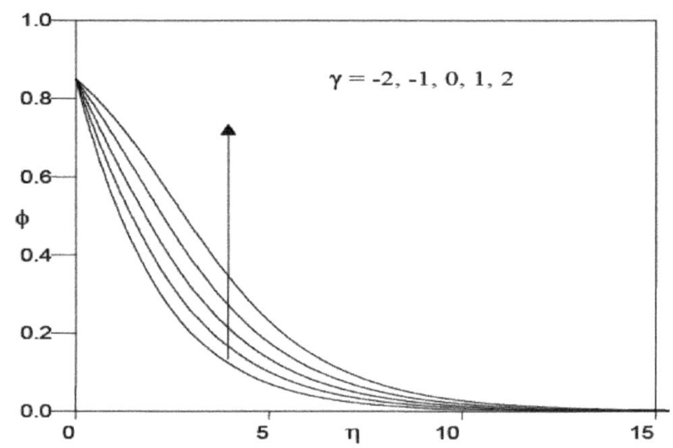

Fig. 9 Non-dimensional concentration profiles for different values of suction/injection parameter

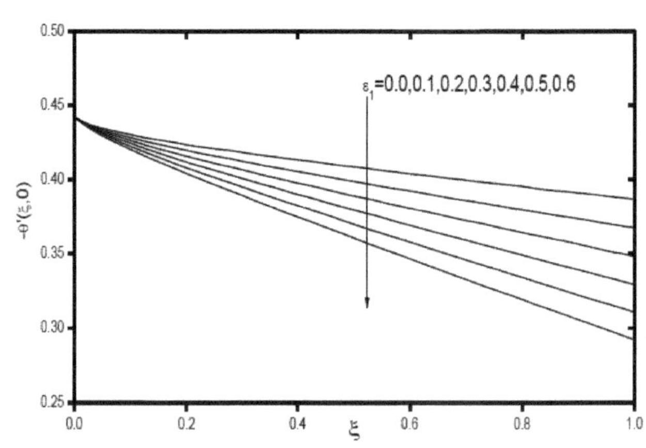

Fig. 10 Non-dimensional heat transfer rate for different values of thermal stratification parameter

Both local heat and mass transfer rates decreases with the increase in Forchheimer number F_c as shown in Figs. 14 and 15. Since F_c represent the inertial drag, thus an increase in the Forchheimer number retards the flow and hence diminishes the dimensionless heat and mass transfer rates in the boundary layer.

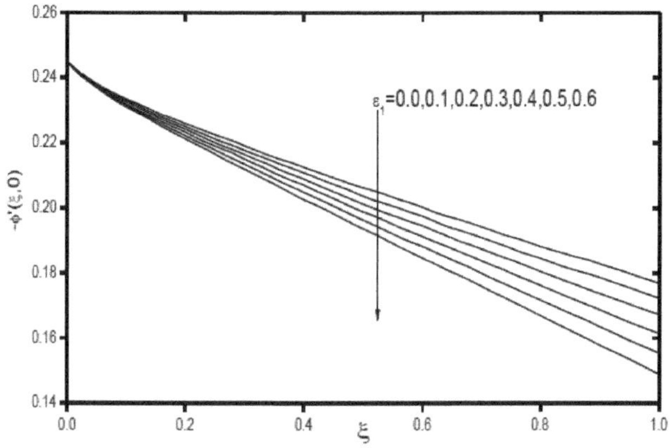

Fig. 11 Non-dimensional mass transfer rate for different values of thermal stratification parameter

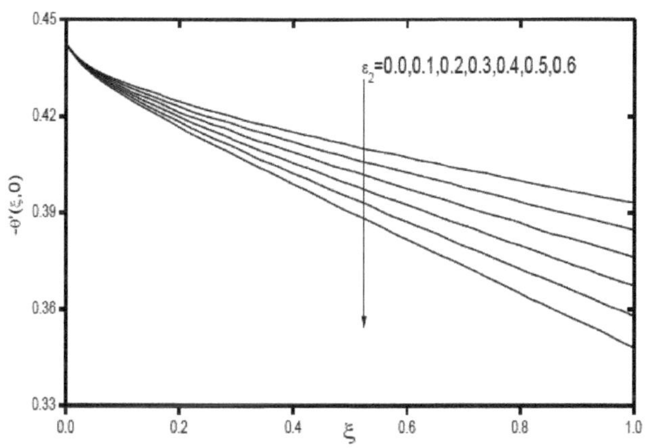

Fig. 12 Non-dimensional heat transfer rate for different values of solutal stratification parameter

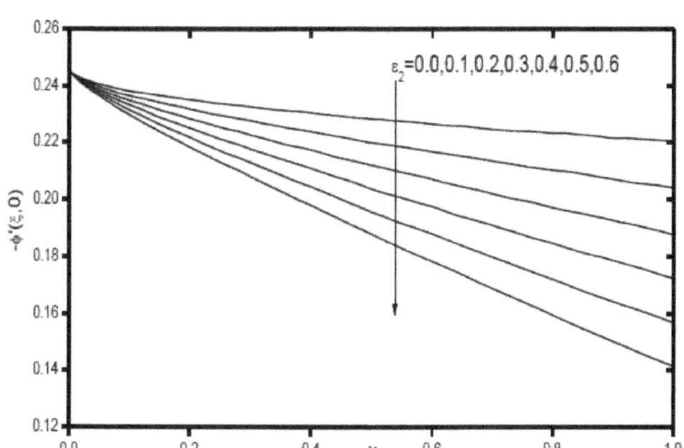

Fig. 13 Non-dimensional mass transfer rate for different values of solutal stratification parameter

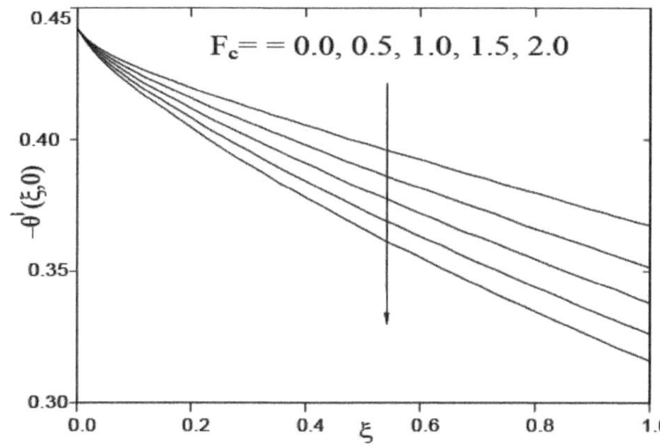

Fig. 14 Non-dimensional heat transfer rate for different values of Forchheimer number.

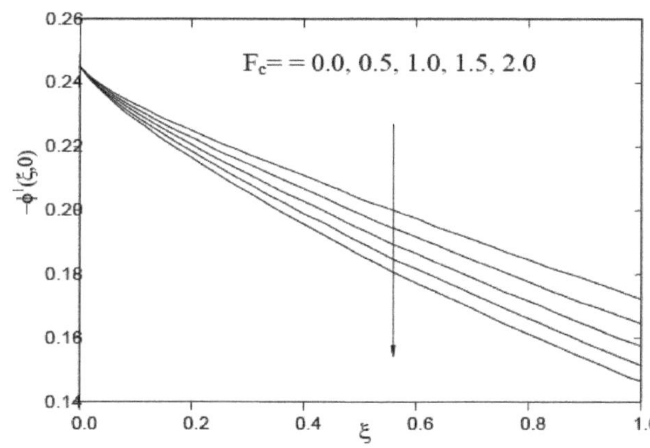

Fig. 15 Non-dimensional mass transfer rate for different values of Forchheimer number

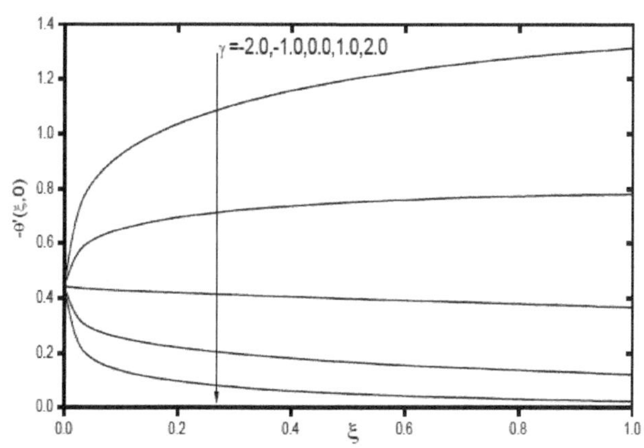

Fig. 16 Non-dimensional heat transfer rate for different values of suction/injection parameter.

The influence of suction/injection parameter γ on local heat and mass transfer coefficients is shown in Figs. 16-17. An increase in the value of suction/injection parameter causes a significant decrease in the local heat and mass transfer coefficients as depicted in Figs. 16 and 17.

The dimensionless skin friction coefficient is appreciably decreased with the increase of thermal and solutal stratification parameters, Forchheimer number and suction/injection parameter as shown in Figs. 18-21

5. CONCLUSIONS

In this article, the effect of double stratification on natural convection heat and mass transfer over a vertical porous plate in a fluid saturated non-Darcy porous medium with constant suction/injection has been analyzed using boundary layer approximation. The plate is maintained at uniform and constant wall temperature and concentration. The governing non-dimensional partial differential equations are solved numerically using Blottner method.

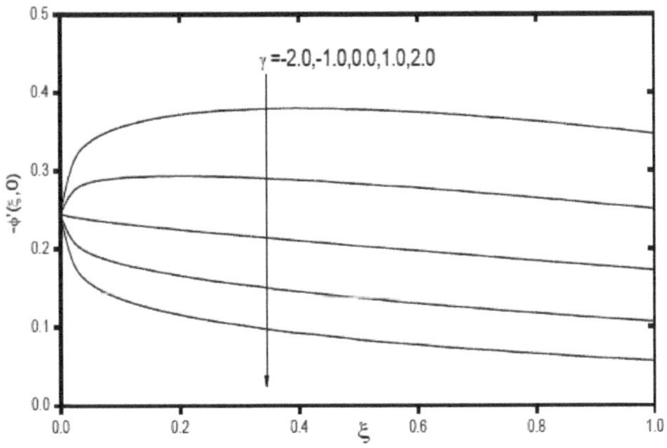

Fig. 17 Non-dimensional mass transfer rate for different values of suction/injection parameter.

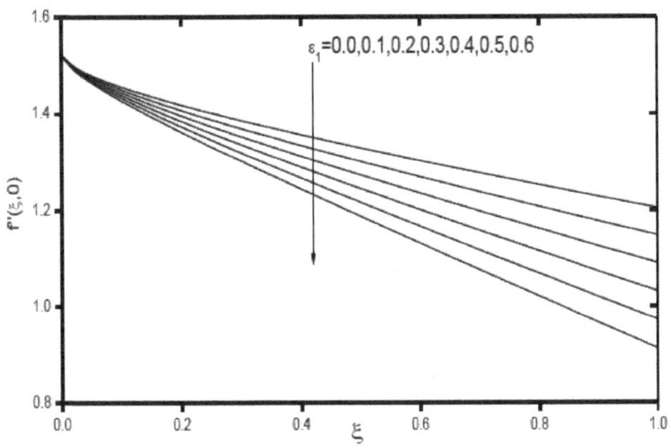

Fig. 18 Non-dimensional skin friction coefficient for different values of thermal stratification parameter

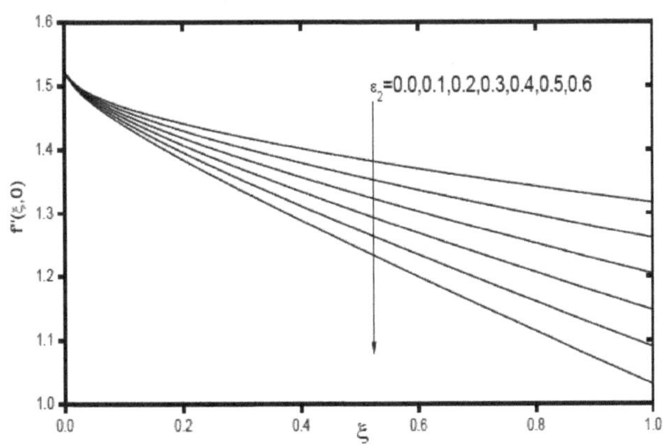

Fig. 19 Non-dimensional skin friction coefficient for different values of solutal stratification parameter

From the above study, we conclude that

1. An increase in thermal stratification parameter, decreases temperature distributions, non-dimensional heat and mass transfer coefficients, and skin friction coefficient.
2. An increase in mass stratification parameter reduces concentration distributions, non-dimensional heat and mass transfer coefficients in the boundary layer.
3. The higher value of Darcy number results in lower velocity but the higher temperature, concentration in the boundary layer.
4. The non-dimensional velocity of the fluid decreases near the plate and increases away from the plate with the increase of Forchheimer number. The higher value of Forchheimer number results in lower local heat and mass transfer rates.
5. An increase in the intensity of suction/injection parameter significantly enhanced the non-dimensional velocity, temperature and concentrations in the boundary layer but reduced both local heat and mass transfer rates.
6. The dimensionless skin friction coefficient is notably decreased with the increase of stratification parameters, Forchheimer number and suction/injection parameter.

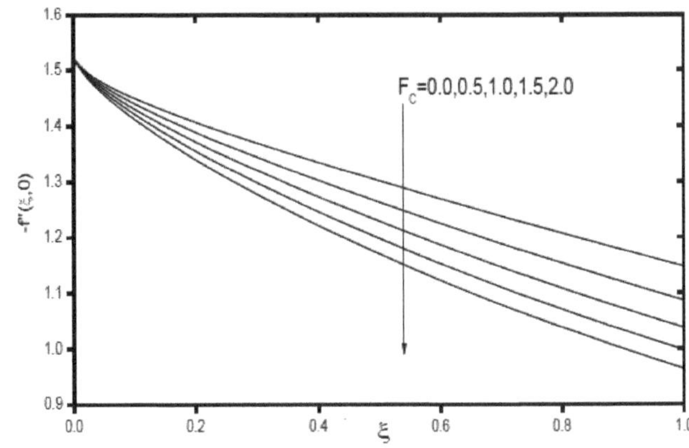

Fig. 20 Non-dimensional skin friction coefficient for different values of Forchheimer number

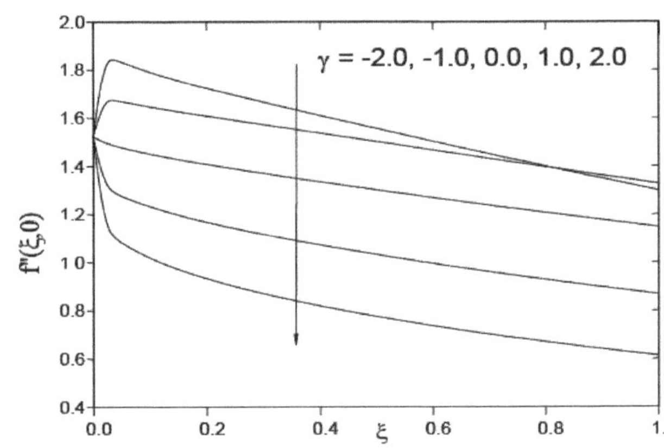

Fig. 21 Non-dimensional skin friction coefficient for different values of suction/injection parameter

REFERENCES

Alam, M., Rahman, M., and Samad, M., 2006, "Dufour and Soret Effects on Unsteady MHD Free Convection and Mass Transfer Flow Past a Vertical Porous Plate in a Porous Medium," *Nonlinear Analysis: Modelling and Control*, **11**(3), 217–226.

Bhattacharyya, K., Mukhopadhyay, S., and Layek, G., 2013, "Unsteady MHD Boundary Layer Flow with Diffusion and First-Order Chemical Reaction Over a Permeable Stretching Sheet With Suction or Blowing," *Chemical Engineering Communications*, **200**, 379–397. http://dx.doi.org/10.1080/00986445.2012.712577

Blottner, F., 1970, "Finite-difference Methods of Solution of the Boundary Layer Equations," *AIAA Journal*, **8**, 193–205. http://dx.doi.org/10.2514/3.5642

Chamkha, A., Aly, A., and Mansour, M., 2010, "Similarity Solution for Unsteady Heat and Mass Transfer From a Stretching Surface Embedded in a Porous Medium with Suction/ Injection and Chemical Reaction Effects," *Chemical Engineering Communications*, 197, 846–859. http://dx.doi.org/10.1080/00986440903359087

Husnain, S., Mehmood, A., Beg, O., and Ali, A., 2012, "Suction and Blowing Effects on Unsteady Flow and Heat Transfer Through Porous Media with Variable Viscosity," *Journal of Porous Media*, **15**(3), 293–202. http://dx.doi.org/10.1615/JPorMedia.v15.i3.70

Ibrahim, F., Elaiw, A., and Bakr, A., 2008, "Effect of the Chemical Reaction and Radiation Absorption on the Unsteady MHD Free Convection Flow Past a Semi Infinite Vertical Permeable Moving Plate with Heat Source and Suction," *Communications in Nonlinear Science and Numerical Simulation*, **13**, 1056–1066. http://dx.doi.org/10.1016/j.cnsns.2006.09.007

Makinde, O., 2005, "Free Convection Flow with Thermal Radiation and Mass Transfer Past a Moving Vertical Porous Plate," *International communications in Heat and Mass Transfer*, **32**, 1411–1419. http://dx.doi.org/10.1016/j.icheatmasstransfer.2005.07.005

Mbeledogu, I., and Ogulu, A., 2007, "Heat and Mass Transfer of an Unsteady MHD Natural Convection Flow of a Rotating Fluid Past a Vertical Porous Flat Plate in the Presence of Radiative Heat Transfer," *International Journal of Heat and Mass Transfer*, **50**, 1902–1908. http://dx.doi.org/10.1016/j.ijheatmasstransfer.2006.10.016

Pop, I., and Watanabe, T., 1992, "Free Convection with Uniform Suction or Injection From a Vertical Cone for Constant Wall Heat Flux, " *International Communications in Heat and Mass Transfer*, **19**, 275–283. http://dx.doi.org/10.1016/0735-1933(92)90038-J

Rajeshwari, R., Jothiram, B., and Nelson, V., 2009, "Chemical Reaction, Heat and Mass Transfer on Nonlinear MHD Boundary Layer Flow Through a Vertical Porous Surface in The Presence of Suction," *Applied Mathematical Sciences*, **3**, 2469–2480.

Singh, A., Singh, P., and Singh, N., 2011, "Effects of Periodic Permeability and Suction Velocity on Three-Dimensional Free Convection Flow Past a Vertical Porous Plate Embedded in Highly Porous Medium," *Journal of Porous Media*, **14**(5), 451–460. http://dx.doi.org/10.1615/JPorMedia.v14.i5.70

Soundalgekar, V., Singh, M., and Takhar, H., 1983, "MHD Free Convection Past a Semi-Infinite Vertical Plate with Suction and Injection," *Nonlinear Analysis, Theory, Methods & Applications*, **7**(9), 941–944.

NUMERICAL SIMULATION OF METAL-PLASTIC COMPOSITE HEAT RADIATOR WITH HEMISPHERICAL MICROSTRUCTURE ARRAY

Hui Jiang[a,b], Daming Wu[a,b,c], Jian Zhuang[a,b*], Ying Liu[a,b,c], Changqing Huang[a,b]

[a] *College of Mechanical and Electrical Engineering, Beijing University of Chemical Technology, Beijing, 100029, China*
[b] *Polymer Material Processing Equipment Engineering Research Center of The Ministry of Education, Beijing , 100029, China*
[c] *State Key Laboratory of Organic–Inorganic Composites, Beijing University of Chemical Technology, Beijing , 100029, China*

ABSTRACT

A new type of metal-plastic composite heat radiator with hemispherical microstructure array was proposed in this paper. The influence of the geometrical parameters of the microstructure array, including size of the hemisphere, configuration of hemisphere, tilt angle of the radiator, thermal conductivity and radiation emissivity of the plastic, on the process of heat transfer under natural convection were numerically simulated. It was concluded that the metal-plastic composite heat radiator with hemispherical microstructure array had comparable heat transfer behaviors with those of metal heat radiator. So it is possible to replace metal heat radiator by such a metal-plastic composite heat radiator in case of high moisture or corrosion applications. Under the same volume, the cooling capacity of the heat radiator made of metal-plastic was 20 times as the heat radiator made of metal. The heat flux of plastic plate with hemispherical microstructure was superior to the ordinary flat plastic plate. And the former heat flux was 2050 W/m^2, about two times as the latter, 1272W/m^2. The optimal inclination angle of the heat radiator with maximum heater flux was around 45°.

Keywords: *metal-plastic composite heat radiator; hemi-sphere microstructure array; microstructure; numerical simulation*

1. INTRODUCTION

With the development of Micro Electro Mechanical Systems (MEMS) and micro chemical mechanical system, the integration density of electronic device becomes higher. Power consumption and heat output was becoming more and more sophisticated, however, the size was becoming smaller and smaller and heat flux of the electronic device was increasing .In order to meet the requirements of the development of the electronics industry, including LED lamps and lanterns, compact, lightweight and highly efficient microstructure heat radiators are needed.

The cooling forms of microstructure heat radiator were summarized as natural convection and forced convection. Many people had done a lot of research on natural convection under different structure. Yu *et al.* (2010, 2011, 2012) investigated natural convection heat transfer around a radial heat radiator adapted for dissipating heat on a circular LED (light emitting diode) light and optimized the heat radiator. The numerical results were validated with experimental results and it showed a good agreement. Parametric studies were performed to compare the effects of the number of fins, long fin length, middle fin length and heat flux on the thermal resistance and average heat transfer coefficient. In addition, the effect of radiation on heat transfer was examined by varying emissivity, and it was found that the maximum radiation contribution was 27%. Xaman *et al.* (2008) presented a two-dimensional numerical study of combined heat transfer (laminar and turbulent natural convection, surface thermal radiation and conduction) in a square cavity with a glass wall. Results show that the flow pattern is not symmetric due to the combined effect of non-isothermal glass wall and radiative exchange inside the cavity. Tari *et al.* (2013), in order to obtain a validated model that is used for investigating inclined orientations of a heat sink, investigated the steady-state natural convection from heat radiators with parallel arrangement of rectangular cross section vertical plate fins on a vertical base numerically. Rao *et al.* (2006) solve the problem of natural convection heat transfer from a horizontal fin array by treating the adjacent internal fins as two-fin enclosures. The problem of natural convection heat transfer from a horizontal fin array is theoretically formulated by treating the adjacent internal fins as two-fin enclosures. Mueller *et al.* (2006) made the predictions and measurements of the temperature plate fins on a vertical base numerically. Rao *et al.* (2006) solved the problem of natural convection heat transfer from a horizontal fin array by treating the adjacent internal fins as two-fin enclosures. The comparison was made to experimental results, and the agreement between the model and experiment is very good. Results show that the heat loss due to radiation is typically 15–20% of the total. Lauriat (2006) studied numerically the heat transfer by natural convection and surface radiation in a two-dimensional vented enclosure in contact with a cold external ambient and a hot internal ambient by Lauriat (2006),and the interplay between convection and radiation was discussed and it was shown that radiation contribution in heat transfer dominated for all of the cases investigated. A comprehensive theoretical and experimental study was carried out on the thermal performance of a pin-fin heat sink by Kobus (2005). By the way of analyzing the theoretical model, the geometric factors had effects on the heat dissipation .And experiments were studied at this time, the results showed that the prediction model and experimental model had a good consistence. Kim *et al.* (2009) compare thermal performances of two types of heat sinks commonly used in the electronic equipment industry: plate-fin and pin-fin heat sink. Jang *et al.* (2012) optimized the cooling performance and mass of a pin-fin radial heat sink. The effects of various geometric parameters on the thermal resistance and mass of the heat sink were investigated; these indicated that the system was sensitive to the number of fin arrays, as well as the length of the long and middle fins. A systematic theoretical investigation of the effects of fin spacing, fin height, fin length and temperature difference between fin and surroundings on the free convection heat transfer from horizontal fin arrays was

*Corresponding author. Email: vipzhuangjian@163.com

carried out by Senol *et al.* (2000). It has been shown that it is not possible to obtain optimum performance in terms of overall heat transfer by only concentrating on one or two parameters. To sum up, through the numerical simulation, the model and the experimental study, in the case of natural convection, geometric parameters, inclination angle, and the effects of surface radiation on heat dissipation were achieved.

In this paper, effects of heat transfer of the metal-plastic composite heat radiator with hemispherical microstructure array was studied, which included comparison of the materials used, comparison of the size of the geometric size, comparison of arrangement way, comparison of radiator angle, comparison of thermal conductivity of the conductive plastic, contrast of radiant emittance of heat conduction plastic, contrast of different shape, and so on. According to the first type boundary condition, the heat flux was studied.

2. MATHEMATICAL MODE

The metal-plastic composite heat radiator consists of metal diffusion plate and plastic heat radiator with hemispherical microstructure array, as shown in Fig.1.

Considering the first type of boundary condition, the heat generated by the heat source through the metal diffusion plate and plastic heat radiator, and then by natural convection, forced convection, and radiation spreads to the surrounding environment. Keeping surface temperature of heat source a constant, environmental temperature also keeps constant. When heat flux of heat source on the surface reaches a heat balance, the greater the heat flux, the better the heat transfer effect of the heat radiator.

According to above proposed physical model, a mathematical model is made and a numerical simulation for different shapes of metal-plastic composite microstructure is carried out.

The following assumptions are proposed for the mathematical model:

1) The metal and plastic materials are isotropic.
2) The contact thermal resistance between metal and plastic layer is not considered.
3) The viscous dissipation of the air is ignored, and the air flow is laminar at steady state.
4) Only the changes in air density caused by floating force are considered.
5) The changes of heat transfer law caused by micro scale effect are not considered.

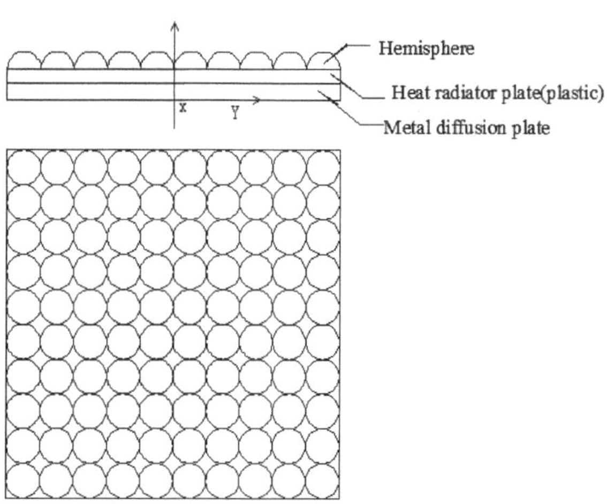

Fig. 1 Schematic of a metal-plastic composite heat radiator with hemisphere microstructure array

The heat, which is from the heat source, through the metal diffusion layer, plastic layer, finally the microstructure to strengthen heat transfer effect, is spread to the air.

3. GOVERNING EQUATIONS

3.1 In the air:

Continuity equation：

$$\frac{\partial \rho \mu}{\partial \mathrm{x}}+\frac{\partial \rho v}{\partial \mathrm{y}}+\frac{\partial \rho w}{\partial z}=0 \quad (1)$$

Momentum equation in x direction：

$$\mathrm{u}\frac{\partial(\rho\mu)}{\partial x}+v\frac{\partial(\rho\mu)}{\partial y}+w\frac{\partial(\rho\mu)}{\partial z}=-\frac{\partial p}{\partial x}+\mu(\frac{\partial^2 u}{\partial x^2}+\frac{\partial^2 u}{\partial y^2}+\frac{\partial^2 u}{\partial z^2})\text{-}\rho g \quad (2)$$

Momentum equation in y direction:

$$\mathrm{u}\frac{\partial(\rho v)}{\partial x}+v\frac{\partial(\rho v)}{\partial y}+w\frac{\partial(\rho v)}{\partial z}=-\frac{\partial p}{\partial y}+\mu(\frac{\partial^2 v}{\partial x^2}+\frac{\partial^2 v}{\partial y^2}+\frac{\partial^2 v}{\partial z^2}) \quad (3)$$

Momentum equation in z direction:

$$u\frac{\partial(\rho v)}{\partial x}+v\frac{\partial(\rho v)}{\partial y}+w\frac{\partial(\rho v)}{\partial z}=-\frac{\partial p}{\partial z}+\mu(\frac{\partial^2 v}{\partial x^2}+\frac{\partial^2 v}{\partial y^2}+\frac{\partial^2 v}{\partial z^2}) \quad (4)$$

Energy equation:

$$u\frac{\partial T_a}{\partial x}+v\frac{\partial T_a}{\partial y}+w\frac{\partial T_a}{\partial z}=\frac{\lambda}{\rho c_p}(\frac{\partial^2 T_a}{\partial x^2}+\frac{\partial^2 T_a}{\partial y^2}+\frac{\partial^2 T_a}{\partial z^2}) \quad (5)$$

where the air is incompressible, the density difference of the air is proportional to the temperature difference.

$$\rho=\rho_\infty[1-\beta(T-T_\infty)] \quad (6)$$

where ρ_∞ is the density which corresponds to T_∞, and β is the volume expansion factor.

3.2 In the solid

3.2.1 Energy equation:

Heat conduction equation from heat source to the metal diffusion plate is as follows:

$$Q=-\lambda_m A\frac{dT_m}{dx} \quad (7)$$

where Q is the heat that is generated by the heat source, λ_m is coefficient of thermal conductivity of the metal, A is the heat dissipation area, and T_m is the temperature of the metal along the x direction.

In the same way, heat conduction equation from the metal diffusion plate to the plastic heat base is as follows:

$$-\lambda_m A\frac{dT_m}{dx}=-\lambda_P A\frac{dT_P}{dx} \quad (8)$$

where λ_P is coefficient of thermal conductivity of the plastic,Tp is the temperature of the plastic along the x direction.

In addition, heat conduction equation from the plastic heat radiator to the air is as follows.

$$-\lambda_P A\frac{dT_P}{dx}=h_t A_e(T_e-T_f)+\sigma A_e\varepsilon_t(T_e^4-T_f^4) \quad (9)$$

where h_t is the total convection heat transfer coefficient, A_e is the expanded surface area, T_e is the temperature of the microstructure surface, and T_f is the temperature of the air.

3.2.2 The boundary conditions:

The fourth boundary condition between the surfaces of metal diffusion and plastic heat radiator are as follows:

$$T_p(x,y,z)|_{cs}=T_m(x,y,z)|_{cs},\ -\lambda_p\frac{\partial T_p(x,y,z)}{\partial n}|_{cs}=-\lambda_m\frac{\partial T_m(x,y,z)}{\partial n}|_{cs} \quad (10)$$

The boundary condition between the surface of the plastic microstructure and air is as follows:

$$-\lambda_p\frac{\partial T_p(x,y,z)}{\partial n}=h_t(T_e-T_f)+\sigma\varepsilon_t(T_e^4-T_a^4) \quad (11)$$

Equations (1-11) are the model that is used to solve the convective heat transfer of air flow and solid internal heat conduction. Based on the mathematical model, we can solve the heat dissipation problem of composite microstructure heat radiator using the FloEFD software.

4. NUMERICAL SIMULATIONS

In this paper, the finite element software FloEFD9.0 is used for numerical simulation. By inputting Pro/E drawings of heat sink of different size and structure into the FloEFD software, and then conducting numerical simulation, a variety of graphs and numerical values of different physical quantities are presented. For hemispherical metal - plastic composite microstructure heat radiator in this paper, there are four kinds of arrangement mode, which are regular hexagon, regular triangle, square, square closely (Fig.2). The numerical simulations of the four arrangement mode of heat radiator are done in this paper. The boundary condition is the first type of boundary condition, that is, the temperature of the heat source is a constant value. Making the heat flux of heat source as the goal, the greater its value is, the better the overall heat radiator cooling effect is. Physical parameters of the different materials are shown in table 1.The heat flux of heat radiator is obtained from equation 12.

$$q=h_t(T_e-T_f)+\sigma\varepsilon_t(T_e^4-T_a^4) \quad (12)$$

where q is heat flux through surface of the heat source, h_t is the equivalent for the surface heat transfer coefficient of the overall heat radiator, ε_t is equivalent for the radiation emissivity of the overall heat radiator, T_f is the temperature of the air on the surface of the heat radiator, and T_a is environment temperature. Generally in this paper: $T_f= T_a =25$℃.

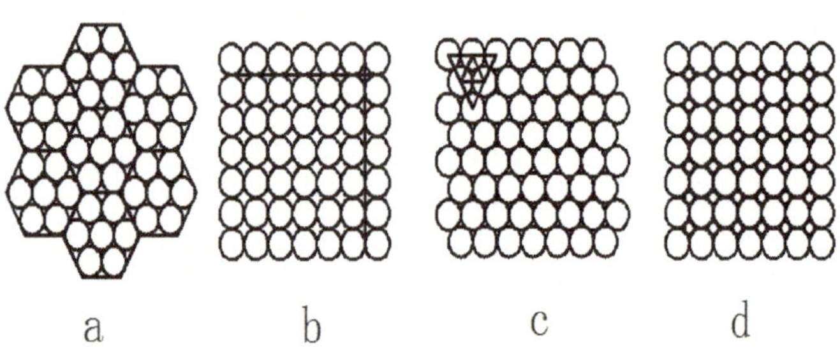

Fig. 2 Four kinds of arrangement of hemisphere (a) hexagonal (b) square(c) triangle (d) square closely

Table1 Overall heat radiator size parameters (mm)

No.	parameters	size
1	Thickness of metal diffusion plate	0.5
2	Thickness of the heat radiator plate	0.5
3	Metal and the heat radiator plate size	10×10
4	Geometric size of the hemisphere	0.5

Table 2 Physical parameters of different materials

Material Parameters	Aluminum 6061	PPS	Air
Density (Kg/m³)	2688.9	1800	1.205
Specific heat (J/Kg·K)	902	1100	1006.43
Thermal conductivity (W/(m·K))	237	20	0.02637

4.1 The influence on the whole flux of changing material

The material of heat radiators is Al and PPS, respectively. The overall heat radiator size parameters are in Fig. 3. The volume of metal heat radiator with rectangular fins is five times as metal-plastic composite heat radiator with hemispherical microstructure array. In this paper, two kinds of heat radiator are studied together. Under the same condition (the first type of boundary condition), we could have the results in Fig.4.The X axis is the temperature of surface of the heat source. The Y axis is the overall surface heat flux. The red dot curve is on behalf of the heat radiator Al-Al (The material of heat radiator is Al). And the black square dot curve is on behalf of the heat radiator is Al-PPS (The material of heat radiator is PPS). With the surface temperature rising, the heat flux is also a linear upward trend from the Fig.4 .In contrast, the heat flux in the form of Al - PPS is higher than that in the form of Al - Al. The reason is that radiation emission rate (0.98) of the PPS is higher than that of Al (0.1).The heat dissipation of the former by radiation is higher than the latter, thus the heat flux is higher than the latter. In addition, you can see that PPS could take the place of Al as the heat radiator. And we could conclude, under the same volume, the cooling capacity of the metal-plastic composite heat radiator with hemispherical microstructure array is about 20 times as metal heat radiator with rectangular fins. Because the plastics have large surface area, corrosion resistance, light weight, low manufacturing cost and easy processing, etc. So the metal - plastic composite microstructure heat radiator has a much broader use space.

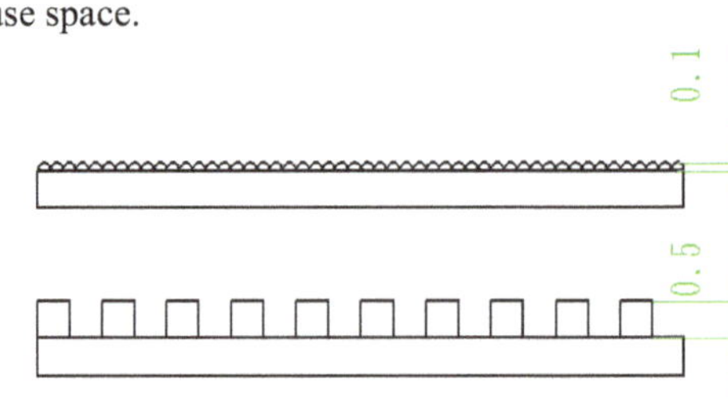

Fig. 3 The structure of two kinds of heat radiator

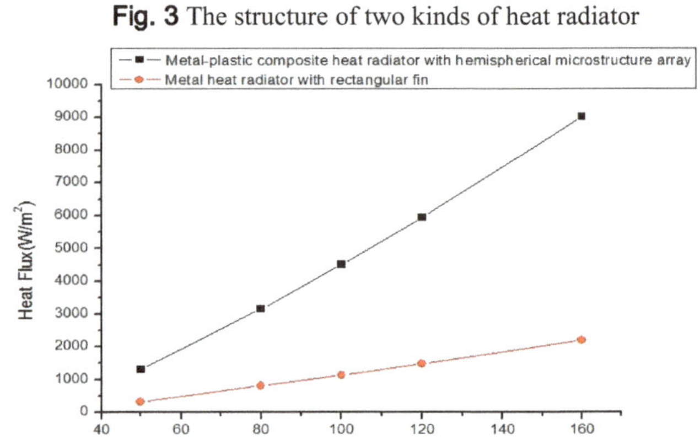

Fig. 4 The contrast of material of heat radiator between Al and PPS

4.2 The influence on the overall heat flux of changing hemisphere arrangement mode

Hemispherical arrangement influences the surface area and the thermal performance of the heat radiator further. Hemispherical arrangement mode includes regular triangle, square, square closely and regular hexagon. From the Fig.5, it can be seen that as temperatures rise, the trend of the heat flux is also rising and the higher the temperature, the faster the change of heat flux. Since the arrangement of hemisphere is compact, it could be seen that the heat flux changes among them are very small. When the temperature is 150° C, heat flux approached 4000W/m^2. And at this time the square closely arrangement mode has the highest heat flux. The reason is that the square closely arrangement mode is the most compact. So it has the highest surface heat flux.

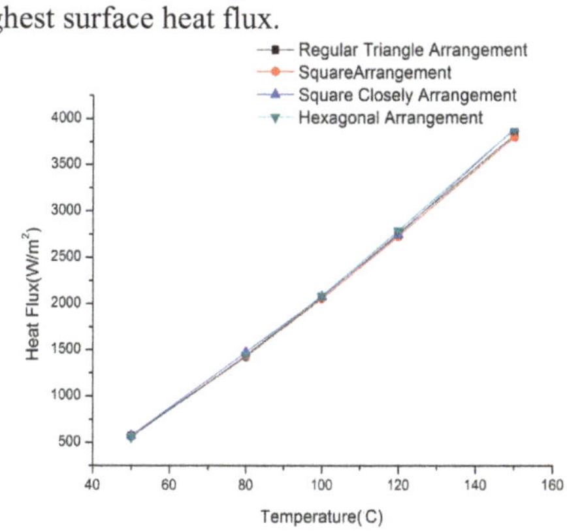

Fig. 5 The influence of four kinds of arrangement for heat flux

4.3The influence on the overall heat flux of adding different microstructure on the surface

From Fig. 3, by adding the microstructure on the surface, the surface area of the heat radiator is increased, which has influences on the heat dissipation of the heat radiator. As shown in Fig.6, the black square represents heat flux curve without adding microstructure. Red circle represents heat flux curve with the rectangular groove microstructure. Blue upward triangle represents heat flux curve with adding a regular triangle groove microstructure. Green downward triangle represents heat flux curve with adding hemispherical microstructure. As shown clearly in Fig.6, the heat flux of the heat radiator by adding the microstructure is greater than heat radiator without adding the microstructure. The reason is that increasing the surface area of heat radiator by adding microstructure, which makes the heat radiator more compact, makes the cooling capacity greater. So the performance of heat dissipation is increased. Among the rectangular groove, triangular groove and hemispherical microstructure, although their differences of numerical results are very small, it increases with the rising of the temperature.

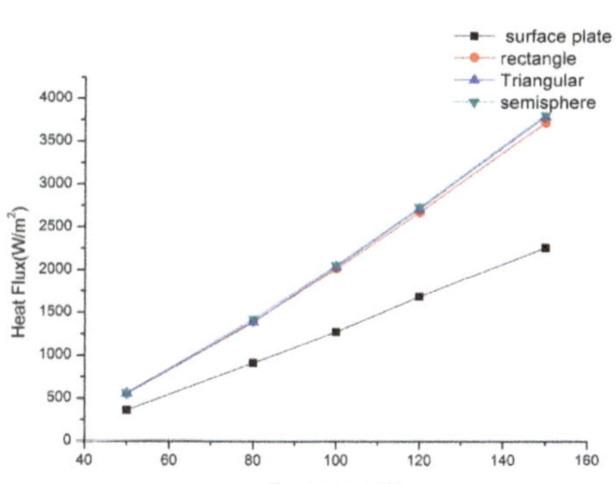

Fig. 6 Contrast among surface plate, triangular groove, rectangular groove, and hemispherical microstructure

4.4 The influence on the overall heat flux of changing thermal conductivity of the plastic

According to the first type of boundary condition ,the temperature of the heat source is 100℃, and the temperature of environment temperature is 25℃.The heat flux of the heat radiator changing with the thermal conductivity of plastic is presented in Fig.7.From 5W/m·K-50W/m·K, the heat flux is linear to the thermal conductivity of plastic. And when the thermal conductivity is greater than 50 W/m·K, the heat flux variation is small and heat flux basically reached its maximum 2930W/m^2. That is to say, when the thermal conductivity reaches 50W/m·K, the modification by adding filler material to increase the heat flux is meaningless. We need to seek other ways to make heat radiator improved and enhanced.

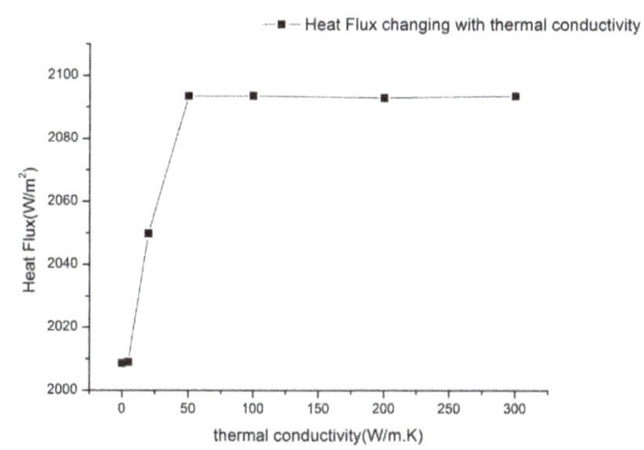

Fig. 7 The heat flux with the change of thermal conductivity

4.5The influence on the overall heat flux of different radiation emissivity

According to the first kind of boundary condition, the temperature of the heat source is 100℃, and the temperature of environment temperature is 25℃. We can see heat flux curve with the change of the different radiation emissivity in Fig.8. In general, the bigger the radiation emissivity is, the greater the heat flux is. The overall heat flux increases with the radiation emissivity rising. And at the initial stage (0-0.25), heat flux increases with the the radiation emissivity linearly. Afterwards it reaches a stable phase (0.25- 0.75), although the radiation emission rate increases, the heat flux dose not increase. Finally with the increasing of the radiation emissivity (0.75- 0.98), heat flux increases linearly again.

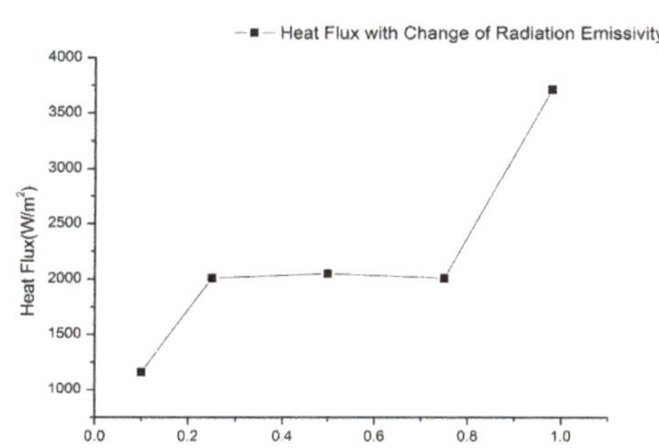

Fig. 8 The heat flux with change of radiation emissivity

4.6 Influence on the overal heat fiux of geometric size of hemisphere

Particle size of the hemisphere has certain effects on heat flux in the Fig.9. When the particle size from 0.5 mm to 2 mm, the change of heat flux is not obvious, that is to say, the size of the particle size has no obvious efffects on heat flux. But from 2 mm to 5 mm, the heat flux has apparent change. And as temperature rises, the difference is more and more obvious.The reason is due to the decrease of the particle size and increase of thc surface area of the heat dissipation, and then it has influence on heat flux. Reducing the size of the particle size further can make the heat flux bigger. But it is limited to computer level, so further simulation is not going on in this paper.

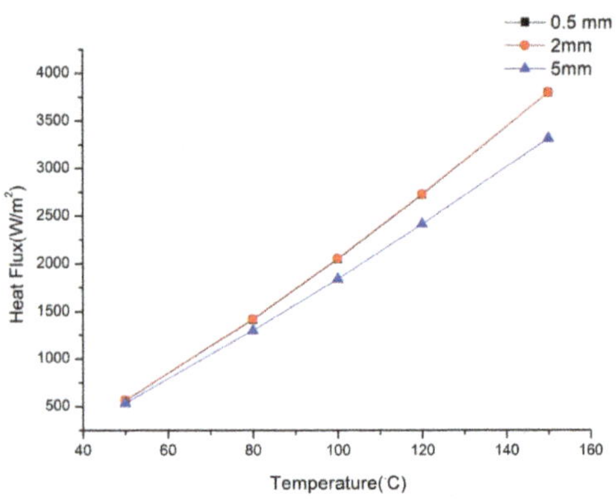

Fig. 9 The heat flux with changing the particle size

4.7 Influence on the overall heat flux of inclination angle

Because of the natural phenomena of the natural convection, hot air is always to rise. So the inclination angle of the heat radiator has certain effects on heat flux. In this paper,we get five points from 0 - 180°, that is 0°, 45°, 90°, 135°, 180°. From Fig. 10, you can see from the angle of 0° to 45°, the trend of the heat flux is slowly rising. And there is a peak around 45°, that is, 2059 W/m^2. From 45° – 180°, the heat flux has a tendency to fall sharply. And from high to low, heat flux changes is about half of their original value. Therefore in the process of actual situations, we should pay attention to inclination angle. Angle should be around 45° , and the greatest cooling effect can be achieved.

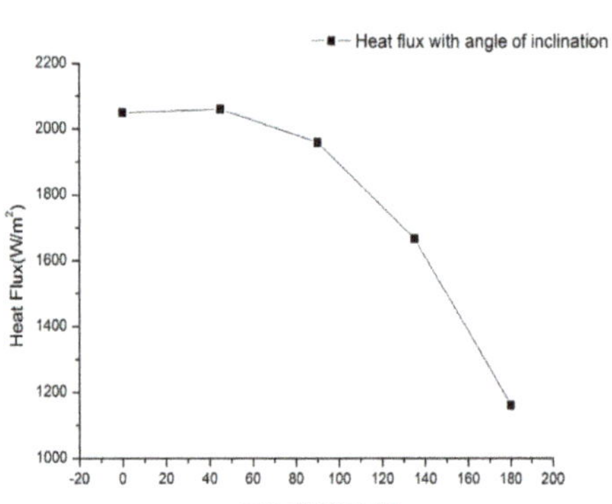

Fig. 10 The heat flux with the angle of inclination

Figures 11 shows velocity vector diagrams of angle of 0°, 90°, and 180°, respectively. As shown in the figure, with the changes of Angle, the velocity vector changes as well. Around the radiator, the size , direction of the velocity vector and the compact degree will also be affected. The compact degree of angle of 0° is higher than 90° and 180°. Therefore, the heat flux of the former is higher than the latter.

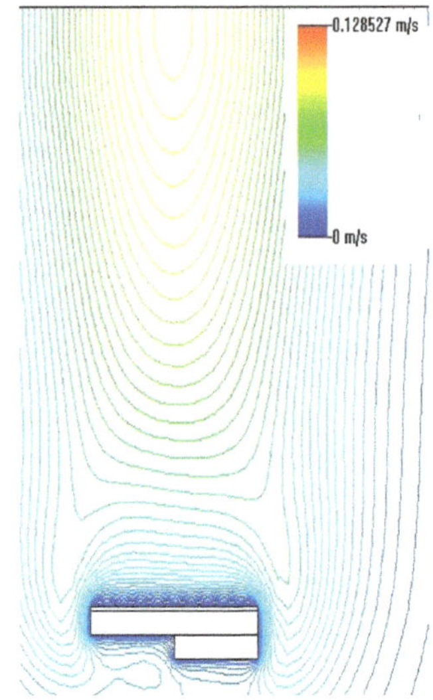

a (0°)

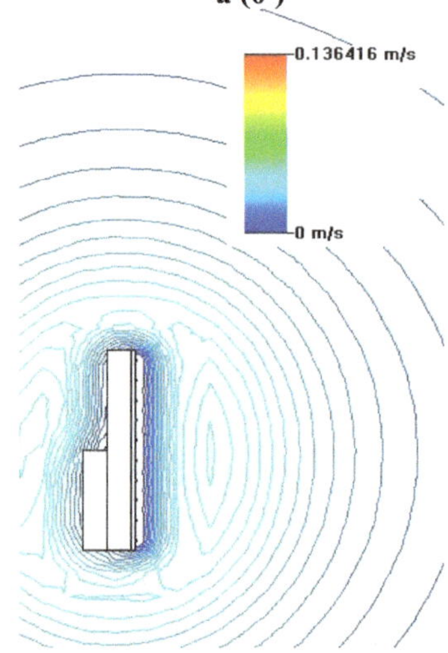

b (90°)

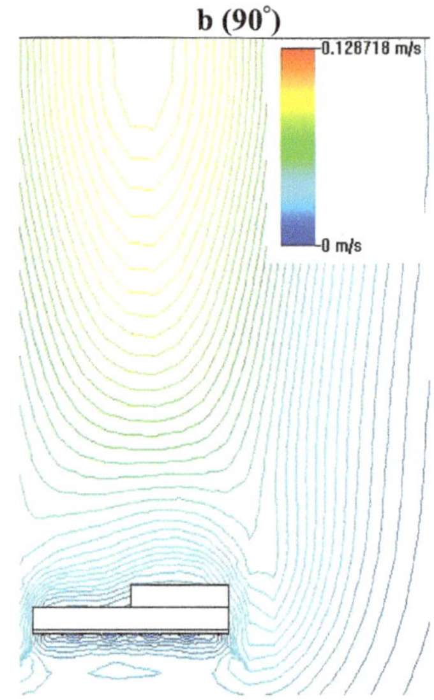

c (180°)

Fig. 11 Velocity vector diagrams

5. CONCLUSIONS

A new type of metal-plastic composite heat radiator with hemisphere microstructure array is put forward. And the different structure parameters and physical parameters of heat radiator are simulated. We get the following conclusion: The metal-plastic composite heat radiator with hemispherical microstructure array can completely replace the metal heat radiator with rectangular fins. The cooling capacity of the metal-plastic composite heat radiator with hemispherical microstructure array is about 20 times as metal heat radiator with rectangular fins. In some situations, and the former was easier to be machined, cheaper, anticorrosion and lighter than the latter. By adding the hemisphere microstructure on the surface, it could increase the surface area. Compared with the ordinary surface plate heat radiator, when to achieve heat balance, the former can achieve greater heat flux. Compared with micro triangle groove and micro rectangular groove, the heat flux of the hemispherical microstructure heat radiator is slightly larger. With the increasing of thermal conductivity of the heat radiator, heat flux grows lineally at the initial stage. There is a value around 50 W/m·K, after which heat flux is almost the same. With the changes of radiation emissivity of the plastic, heat flux grows lineally at the initial stage. There is a value around 0.25, after which, heat flux remains almost the same. Again there is a value around 0.75, after which, heat flux continues to grow linearly. The geometric size of hemisphere has influences on heat flux. And with the increase of geometric size of the hemisphere (0.5mm-2mm-5mm), the overall heat flux of heat radiator becomes smaller. The inclination of heat radiator has influences on heat flux. There is a peak around 45°, before which, heat flux increases linearly, and after which, heat flux decreases sharply.

ACKNOWLEDGEMENTS

The authors would like to express gratitude to the National Natural Science Foundation of China Youth Science Fund (No. 51203010)

NOMENCLATURE

T	temperature (K)
c_p	coefficient of heat capacity (J/Kg·K)
u	x-component of velocity (m/s)
v	y-component of velocity (m/s)
w	z-component of velocity (m/s)
g	acceleration of gravity (m/s^2)
h	convection coefficient (W/m^2·K)
q	heat flux (W/m^2)
Greek symbols	
λ	thermal conductivity (W/m·K)
ρ	air density (Kg/m^3)
μ	the dynamic viscosity of air (N/m^2·s)
ε	radiation emissivity
Subscripts	
Cs	contact surface
n	along the gradient direction
m	metal
p	plastic
e	expanded
∞	at infinity
0	initial condition
a	ambient
f	fluid
t	total

REFERENCES

Jang, D., S.H. Yu and K.S. Lee, 2012, "Multidisciplinary Optimization of a Pin-Fin Radial Heat Sink for Led Lighting Applications," *International Journal of Heat and Mass Transfer,* **55**, 515-521.
http://dx.doi.org/10.1016/j.ijheatmasstransfer.2011.02.012

Kobus, C. J., and T. Oshio,2005, "Development of a Theoretical Model for Predicting the Thermal Performance Characteristics of a Vertical Pin-Fin Array Heat Sink under Combined Forced and Natural Convection with Impinging Flow, "*International Journal of Heat and Mass Transfer*, **48**, 1053-1063.
http://dx.doi.org/10.1016/j.ijheatmasstransfer.2004.09.042

Kim, D.K., S, J.K., and J.K. Bae. 2009, "Comparison of Thermal Performances of Plate-Fin and Pin-Fin Heat Sinks Subject to an Impinging Flow," *International Journal of Heat and Mass Transfer*, **52**, 3510-3517.
http://dx.doi.org/10.1016/j.ijheatmasstransfer.2009.02.041

Lauriat, G., and G. Desrayaud, 2006, "Effect of Surface Radiation on Conjugate Natural Convection in Partially Open Enclosures," *International Journal of Thermal Sciences*, **45**, 335-346.
http://dx.doi.org/10.1016/j.ijthermalsci.2005.07.002

Mueller, D.W., and H.I. Abu-Mulaweh, 2006, "Prediction of the Temperature in a Fin Cooled by Natural Convection and Radiation," *Applied Thermal Engineering*, **26**, 1662-1668.
http://dx.doi.org/10.1016/j.applthermaleng.2005.11.014

Rao, V., Dharma, S. V., Naidu, B., Govinda, R.,and K. V. Sharma, 2006, "Heat Transfer from a Horizontal Fin Array by Natural Convection and Radiation—a Conjugate Analysis," *International Journal of Heat and Mass Transfer*, **49**, 3379-3391.
http://dx.doi.org/10.1016/j.ijheatmasstransfer.2006.03.010

Senol Baskaya., M, S., Murat, O., 2000, "Parametric Study of Natural Convection Heat Transfer from Horizontal Rectangular Fin Arrays," *International Journal of Thermal Sciences*, **39**, 797–805
http://dx.doi.org/10.1016/S1290-0729(00)00271-4

Tari, I., and Mehdi, M., 2013, "Natural Convection Heat Transfer from Inclined Plate-Fin Heat Sinks," *International Journal of Heat and Mass Transfer*, **56**, 574-593.
http://dx.doi.org/10.1016/j.ijheatmasstransfer.2012.08.050

Xamán, J., J. Arce, Álvarez, G. and Chávez, Y., 2008, "Laminar and Turbulent Natural Convection Combined with Surface Thermal Radiation in a Square Cavity with a Glass Wall," *International Journal of Thermal Sciences*, **47**, 1630-1638.
http://dx.doi.org/10.1016/j.ijthermalsci.2008.01.012

Yu, S.H., Lee, K.S., and Yook, S.J., 2010, "Natural Convection around a Radial Heat Sink," *International Journal of Heat and Mass Transfer,* **53**, 2935-2938.
http://dx.doi.org/10.1016/j.ijheatmasstransfer.2010.02.032

Yu, S.H., Lee, K.S., and Yook, S.J., 2011, "Optimum Design of a Radial Heat Sink under Natural Convection,"*International Journal of Heat and Mass Transfer*, **54**, 2499-2505.
http://dx.doi.org/10.1016/j.ijheatmasstransfer.2011.02.012

Yu, S.H., Jang, D., and Lee, K.S., 2012, "Effect of Radiation in a Radial Heat Sink under Natural Convection," *International Journal of Heat and Mass Transfer*, **55**, 505-509.
http://dx.doi.org/10.1016/j.ijheatmasstransfer.2011.09.026

HEAT TRANSFER BASED NUMERICAL INVESTIGATION OF AIRCRAFT CABIN ENVIRONMENT FROM VARIOUS INLET CONDITIONS

Zhuohuan Hu[a,*] Lulu Wang[a], Hui Wang[b], Mo Yang[a]

[a] *School of Energy and Power Engineering, University of Shanghai for Science and Technology, Shanghai, 200093, China*
[b] *National Computer Network and Information Safety Administration Center Shanghai Network Access Point, Shanghai, 201315, China*

ABSTRACT

A numerical investigation was conducted to study the heat transfer in an aircraft cabin and the effects of air distribution under different angle and inlet velocity conditions. The Reynolds-averaged Navier–Stokes equations and the low Reynolds number turbulence model were used to simulate the airflow in the cabin. Mathematical statistics was used to process the relevant data, and statistical results revealed that different inlet angles and velocities significantly affect air temperature and flow field. The study also determined a set of optimum matching inlet vane angles and inlet velocities that result in an environment which meets standard requirements and is energy saving. Additional studies were also conducted on the thermal comfort in cabin environments under optimal conditions. The conclusions of this thesis are scientific and feasible, and they can be used to improve the comfortableness of cabin environments.

Keywords: *airliner cabin, numerical simulation, heat transfer, PMV–PPD.*

1. INTRODUCTION

Ensuring flight security, increasing energy conservation, and improving the thermal comfort of passengers and pilots during flight are vital considerations in aircraft design (Shou et al., 2004; Zhang et al., 2007; Zhang et al., 2009). Pressure, temperature, velocity, and humidity are the main factors that influence comfortableness in aircraft cabins. Flow field, temperature field, and air quality in the cabin are related to relevant boundary conditions, such as inlet temperature, inlet velocity, inlet angle and position, and air humidity. As such, a comfortable cabin environment is realized under reasonable boundary conditions.

Previous studies have conducted numerical analyses on the heat and mass transfer involved in the internal fluid flow in aircraft cabins. J Günther et al. (2006) studied the airflow in an aircraft cabin using a combination of numerical simulation and experimentation by particle image velocimetry (PIV). Previous studies (Bosbach et al., 2006) have proven that the higher-order low Reynolds number turbulence model precisely fits the actual situation in simulating the flow in an aircraft cabin. Literature (Günther et al., 2006) also studied the flow of fresh air in the vicinity of a luggage compartment by building a cabin mock-up. Wang et al. (2008) and Yan et al. (2009) studied the unsteady airborne pollutant transport mechanism in an aircraft cabin mock-up without considering the heat source. Nevertheless, the results of their numerical simulation were in good agreement with those of the experiment. Bianco et al. (2009) established two- and three-dimensional simulation models of an executive aircraft cabin and investigated the thermal and velocity fields from the perspective of the external environment and fuselage material. The two-dimensional results were in agreement with the three-dimensional results. Kühn et al. (2009) studied the forced and mixed convection and received temperature field in the cabin mock-up of a passenger aircraft using PIV. Proper parameters are crucial in designing the air-conditioning of a passenger aircraft to ensure the thermal comfort of cabin passengers.

At the other end of the research spectrum, literatures (Poussou et al., 2010; Mazumdar et al., 2011; Mazumdar and Chen, 2007) investigated the influences of a moving human body on the airflow and contaminant transport in an aircraft cabin and measured the flow field and contaminant transport using planar laser-induced fluorescence and the techniques of particle image velocimetry. These experimental data were used to validate a Computational Fluid Dynamic (CFD) model. The results demonstrated that the CFD model effectively captures the characteristic flow features and contaminant transport observed in a small-scale model. Aakash and Chen (2012) developed a CFD model to predict the ozone distribution in an aircraft cabin. The agreement between the CFD results and available experimental data was generally good. The ozone concentration at the breathing zone in the cabin environment can effectively be used to assess the health risk to passengers and develop strategies that ensure a healthy cabin environment. Liu et al. (2012) studied pesticide concentrations and depositions after spraying part of an aircraft cabin. Furthermore, literatures (Isukapalli et al., 2013; You and Zhao, 2013) proposed a simplified method of quickly assessing the particle deposition rate in aircraft cabins and found that the influence of occupancy on the particle deposition rate of the cabin may be negligible.

The thermal comfort of the environment for passenger movement has also been increasingly investigated (Zhang, 2003; Maidment et al., 2004; Wang, 2009). Thermal comfort is a neither hot nor cold state to the environment, which is used to describe the degree of satisfaction on thermal environment. The referred research in the field of thermal comfort was carried out by Fanger in the 1970s, in which he developed an empirical model named predicted mean vote - predicted percentage dissatisfied (PMV-PPD) and the equations. He also pointed out that six

* *Corresponding author. Email: huzhuohuan@hotmail.com*

factors affected thermal sensation: four physical variables (air temperature, air velocity, relative humidity, mean radiant temperature), and two personal variables (clothing insulation and activity level) (Fanger, 1967). The PMV index represents the overwhelming number of people who feel hot or cold in the same environment. As such, the PMV index can be used to predict the thermal response of the human body to a thermal environment. And then, the results of the PMV equation are divided into seven levels to obtain PMV thermal sensation scale, from cold (-3) to hot (+3) in Table 1 (Charles, 2003). Although most people are satisfied with the thermal environment, some remain dissatisfied because of physiological differences. The PPD index represents the percentage of people who are dissatisfied with the thermal environment. The PMV-PPD standards were prescribed by the International Standardization Organization (ISO 7730) in1984 (Fanger, 1984).

Table 1 PMV thermal sensation scale.

PMV value	-3	-2	-1	0	1	2	3
Thermal sensation	Cold	Cool	Slightly cool	Neutral	Slightly warm	Warm	Hot

Several studies have also been conducted on the thermal comfort of indoor environments. Yao and Chen (1997) simulated and analyzed indoor climates and the thermal comfort in rooms with natural ventilation and buried pipe ventilation systems using thermal dynamic simulation and CFD techniques in conjunction with the thermal comfort PMV/PPD model. The buried pipe ventilation system can improve indoor thermal comfort in hot climates. Yu and Zhang (2008) simulated flow and humidity fields at the peak of the indoor cooling load in isothermal and nonisothermal air supply with different floor temperatures. The parametric change in the air-conditioning system that satisfies the thermal comfort of the human body can be achieved using the PMV–PPD thermal comfort index. Schellen et al. (2013) proposed a new indoor thermophysiological model and simulation of the indoor environment around the human body's thermal sensation. The results show that the thermophysiological model combined with the thermal comfort model serves an important role in the effective evaluation of complex daily thermal environments. Buratti et al. (2013) studied the effect of glazing systems on the thermal comfort of a lecture room and determined the optimum glazing system that results in a comfortable environment.

The engineering context of the present study is composed of the thermal environment of a double-service aircraft that carries passengers and goods. This study focuses on the interior heat source and mixture of airflow; it considers the thermal radiation along the solid walls in the cabin as negligible. The landing stage is the most difficult task because it is not an independent stage in the actual processes of a flight. This study employs FLUENT to analyze the temperature and flow fields in an airliner cabin during the landing process. A set of optimum matching inlet vane angles and velocities that generate a comfortable environment are determined. Additional studies on thermal comfort using the PMV–PPD index verify that the conclusions of the present study are scientific and feasible, two qualities that have been typically disregarded in previous studies. In conclusion, the study of heat and mass transfer in aircraft cabins improves not only flight safety but also the energy-saving equipment and the comfort of passengers.

2. MATHEMATICAL MODELS

2.1 Turbulence Models

The actual situation of the heat and mass transfer in the workplace in an aircraft is quite complex. Thus, heat and mass transfer models should be simplified to obtain the gas state–space mathematical models in the cabin. This study is premised on the following assumptions:

- Air is an incompressible fluid that conforms to the Boussinesq hypothesis.
- The turbulent flow of the working medium in the cabin is steady.
- The thermal radiation around the solid walls and human body in the cabin can be neglected.
- The heat dissipation caused by the viscous force can be neglected because of the incompressible airflow at low speed.

The speed of airflow in the aircraft cabin is quite low, and the viscosity of the turbulence near the wall is slight. This study adopts the k–ε turbulence model that belongs to the Launder–Sharma low Reynolds number flow in the aircraft cabin (Zhai et al., 2007; Zhang et al., 2007).

2.2 Governing Equations

The continuity, Navier–Stokes, energy, and k–ε turbulence model equations were used to calculate the airflow. In the Cartesian coordinate space, three-dimensional governing equations (Shen and Yuan, 2009; Fan, 2008) in tensor forms were used to solve the problems of steady-state and incompressible turbulent flow. These equations are as follows:

$$\frac{\partial}{\partial x_j}(\rho u_i) = 0 \tag{1}$$

$$\frac{\partial(\rho u_i u_j)}{\partial x_j} = -\frac{\partial p}{\partial x_i} + \frac{\partial}{\partial x_j}\left[\mu\left(\frac{\partial u_i}{\partial x_j} + \frac{\partial u_j}{\partial x_i} - \frac{2}{3}\delta_{ij}\frac{\partial u_k}{\partial x_k}\right) - \rho\overline{u_i'u_j'}\right] \tag{2}$$

$$\frac{\partial(\rho u_i T)}{\partial x_i} = \frac{\partial}{\partial x_i}\left(\frac{\mu}{P_r}\frac{\partial T}{\partial x_i} - \rho\overline{u_i'T'}\right) + S_T \tag{3}$$

The turbulent kinetic energy k equation and the dissipation rate ε equation in the k–ε model are calculated as follows:

$$\frac{\partial}{\partial x_i}(\rho u_i k) = \frac{\partial}{\partial x_i}\left[\left(\mu + \frac{\mu_t}{\sigma_k}\right)\frac{\partial k}{\partial x_i}\right] + G_k + G_b - \rho\varepsilon - Y_M \tag{4}$$

$$\frac{\partial}{\partial x_i}(\rho u_i \varepsilon) = \frac{\partial}{\partial x_i}\left[\left(\mu + \frac{\mu_t}{\sigma_\varepsilon}\right)\frac{\partial \varepsilon}{\partial x_i}\right] + \frac{C_{1\varepsilon}\varepsilon}{k}(G_k + C_{3\varepsilon}G_b) - C_{2\varepsilon}\rho\frac{\varepsilon^2}{k} \tag{5}$$

where ρ, p and T are the air density (kg/m^3), static pressure (P_a), and air temperature (K), respectively; u_i is the velocity along the x_i direction (m/s); P_r is the Prandtl number; μ and μ_t are the laminar and turbulent dissipation rates, respectively; S_T is the source term; G_k is the turbulent kinetic energy caused by the average flow velocity gradient; G_b is the turbulent kinetic energy caused by buoyancy; and Y_M is the effect of the compressible turbulent flow pulsation expansion on the total dissipation rate. The viscosities of the turbulence coefficient are as follows: $\mu_t = \rho C_\mu k^2/\varepsilon$, $C_{1\varepsilon} = 1.44$, $C_{2\varepsilon} = 1.92$, and $C_{1\varepsilon} = 0.09$.

2.3 Thermal Comfort Index

The PMV–PPD method is based on the heat balance equation for human body and the PMV–PPD index (Yang and Chen, 2002; Liu, 2007; Xu, 2005), calculated as follows:

$$\begin{aligned} PMV = {} & [0.303\exp(-0.036M) + 0.028]\{M - W \\ & -3.05[5.73 - 0.07(M - W) - P_a] - 0.0173M(5.87 - P_a) \\ & -0.0014M(34 - t_a) - 0.42(M - W - 58) \\ & -3.96*10^{-8} f_d[(t_d + 273)^4 - (t_r + 273)^4] - f_d a_c(t_d - t_a)\} \end{aligned} \tag{6}$$

$$PPD = 100 - 95\exp[-(0.03353PMV^4 + 0.2179PMV^2)] \tag{7}$$

where M is the body's metabolic heat production rate (W/m^2), W is the body's external power (W/m^2), P_a is the partial pressure of the water vapor in the air (kP_a), t_a is the air temperature of the surroundings (°C), t_r is the mean radiant temperature of the surroundings (°C), t_d is the surface temperature of the clothes (°C), f_d is the clothing area coefficient, and a_c is the convective heat transfer coefficient.

The following assumptions are taken: $t_a = t_r$, $M = 70W/m^2$, $W = 0$, and $a_c = 12.1\sqrt{v}$. Fanger pointed out that the climate formula should be multiplied by 0.5 to 1.0 in various countries and 0.7 in China as correction coefficients because the PMV test subjects are westerners (Ji et al., 2003).

3. NUMERICAL ANALYSIS

The velocity field, temperature field, and thermal comfort of the internal environment in the passenger and cargo aircraft cabins are analyzed below. Figure 1 shows the structure sketch of the aircraft cabin.

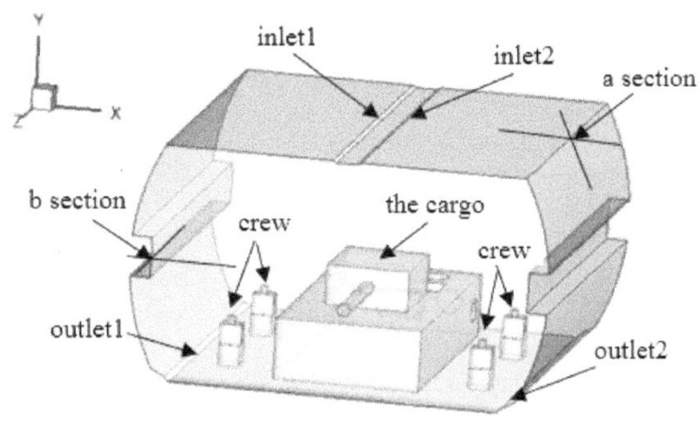

Fig. 1 The structure sketch of the aircraft cabin.

3.1 Boundary Conditions

The air inlet is based on boundary conditions, which assume an air inlet speed of 1 m/s to 9 m/s (Zhang and Ang, 2002). The magnitude of Reynolds numbers is between 10^5 and 10^6. And the inlet temperature is 20°C; this model sets two symmetrical air inlets. The flow field in the cabin is examined under inlet angle conditions of 30°, 45°, 60°, 75°, and 90°. Figure 2 shows various inlet angles of inlet1 and inlet2.

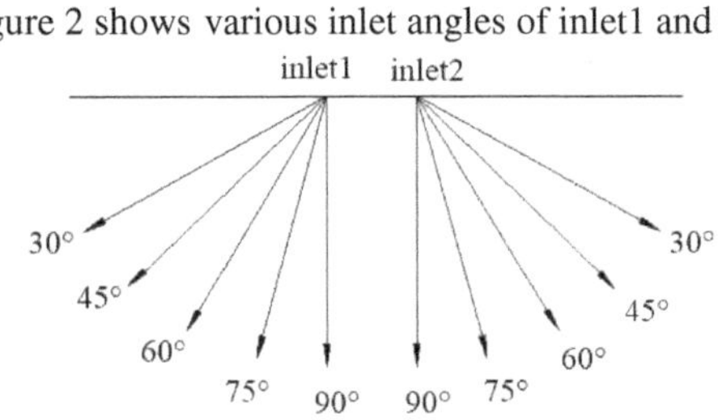

Fig. 2 Various inlet angles of inlet1 and inlet2.

The solid walls in the aircraft cabin are all set under non-slip conditions, and the speed is assumed as negligible. In this model, the internal walls of the cabin release heat outward. The human body is regarded as a source of heat, and the front and rear boundaries are set as planes of symmetry.

3.2 Analysis of Flow and Temperature Fields

The flow of air admission is controlled by changing the air inlet velocity. The final data are analyzed using the method of mathematical statistics. The influence of air distribution under different inlet air angles and velocities is also investigated. The inlet velocities of 3 and 6 m/s are selected as the examples here because of the large number of images. For being convenient to study and analyze, cutting a center longitudinal section is a section, at the same time cutting a center horizontal section is b section, as shown in Figure 1. Figure 3 shows the velocity field of the vertical section under different inlet angles. Figure 4 shows the velocity field of the horizontal section under different inlet angles. Figure 5 shows the velocity variance of different inlet velocities under different inlet angles (z section for the vertical section and y section for the horizontal section). Figures 3, 4, and 5 show the relatively obvious effects of the inlet angle and velocity on the flow field in the cabin.

Figure 6 shows the temperature field of the vertical section under different inlet angles. Figure 7 shows the temperature field of the horizontal section close to the walls under different inlet angles. Figure 8 shows the temperature variance of the different inlet velocities under different inlet angles. Figures 6, 7, and 8 show the relatively obvious effects of the inlet angle and velocity on the temperature field in the cabin.

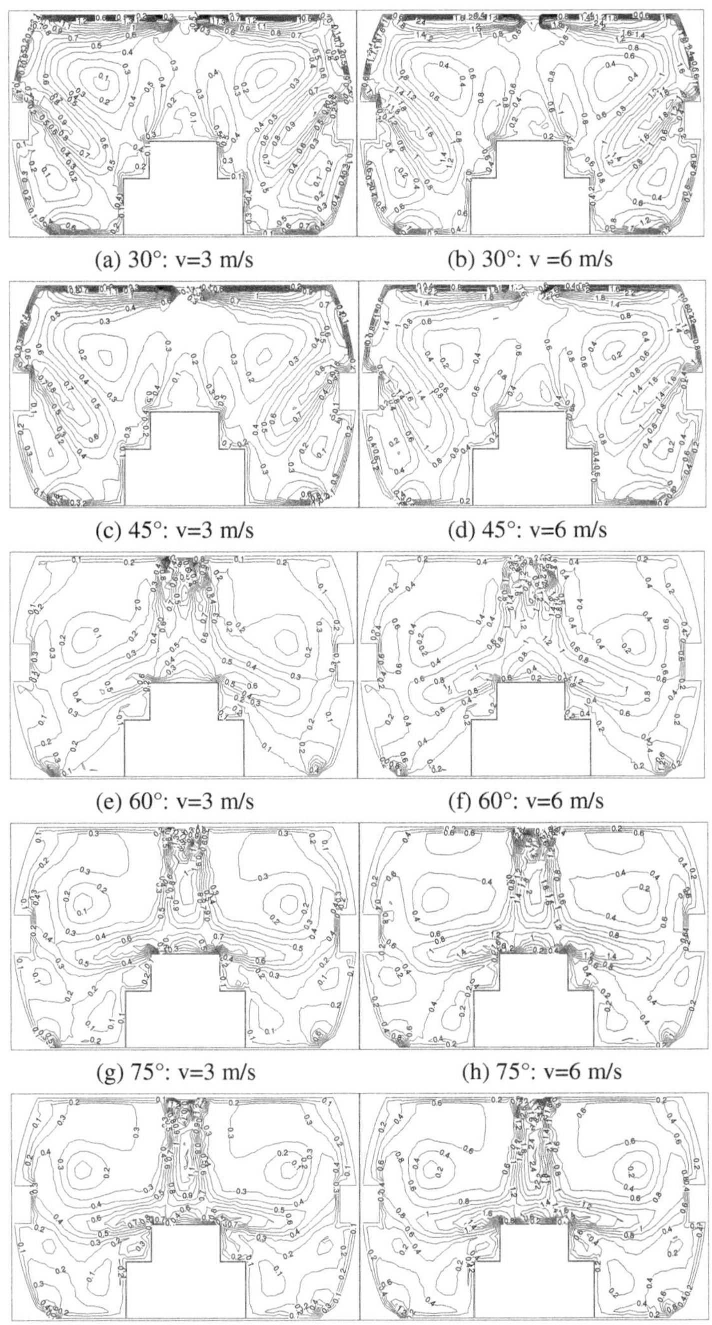

(a) 30°: v=3 m/s (b) 30°: v =6 m/s

(c) 45°: v=3 m/s (d) 45°: v=6 m/s

(e) 60°: v=3 m/s (f) 60°: v=6 m/s

(g) 75°: v=3 m/s (h) 75°: v=6 m/s

(i) 90°: v=3 m/s (j) 90°: v=6 m/s

Fig. 3 Flow field of the center longitudinal section under different air inlet velocities.

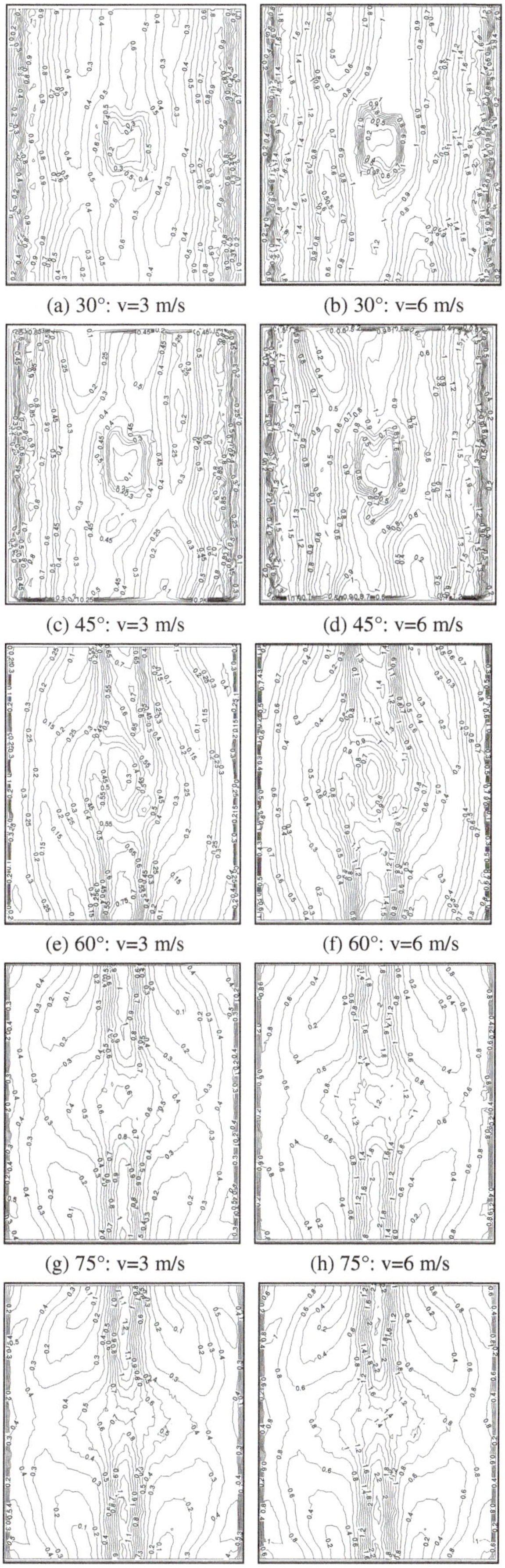

(a) 30°: v=3 m/s (b) 30°: v=6 m/s

(c) 45°: v=3 m/s (d) 45°: v=6 m/s

(e) 60°: v=3 m/s (f) 60°: v=6 m/s

(g) 75°: v=3 m/s (h) 75°: v=6 m/s

(i) 90°: v=3 m/s (j) 90°: v=6 m/s

Fig. 4 Flow field of the center horizontal section under different air inlet velocities.

Several swirls occur in Figures 3 and 4 because of the blocking effect of the walls and objects in the cabin, which can lead to an increasingly balanced flow field. However, these swirls negatively influence the renewal of air from the perspective of mass transfer. In addition, the original inlet velocity at the inlet port rapidly decreases because of the effects of cabin air resistance and wall frictions. The velocity gradient of most parts of the cabin, excluding the inlet port, is lower than 2 m/s when the inlet velocity increases from 1 m/s to 9 m/s. Although the air velocity in the cabin increases along with the inlet velocity, the former increases only by approximately 0.2 m/s to 0.4 m/s, whereas the latter increases by 3 m/s.

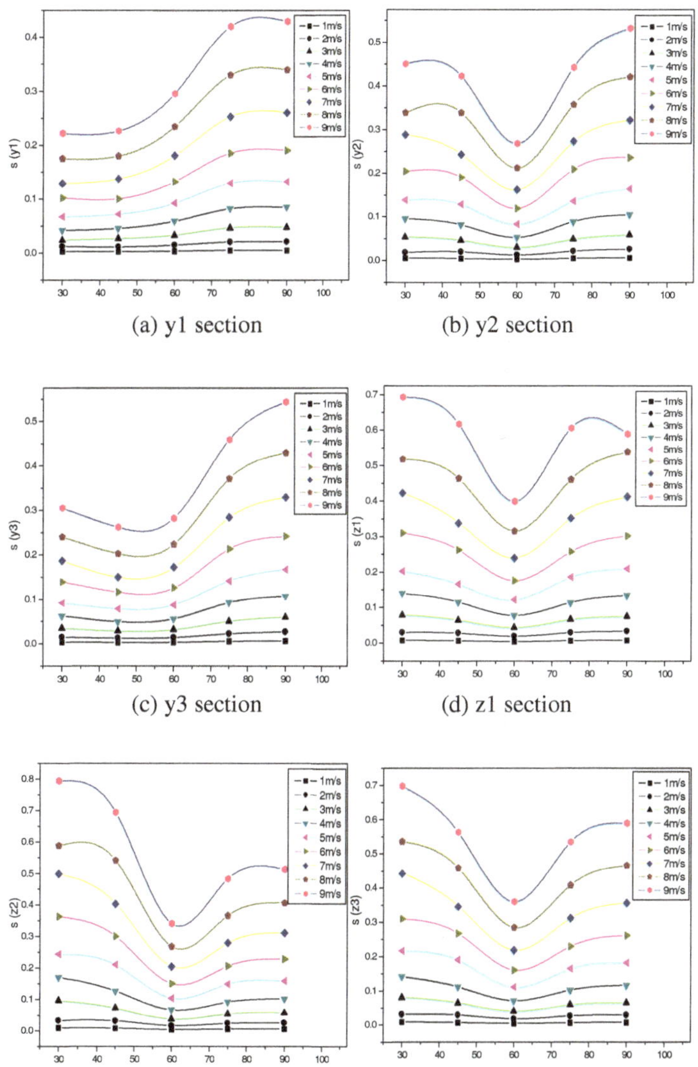

(a) y1 section (b) y2 section

(c) y3 section (d) z1 section

(e) z2 section (f) z3 section

Fig. 5 Variance of velocity under different velocities according to the air inlet angles.

Figure 5 shows that noticeable changes occur in velocity variances under different velocities and similar variation rules in each section. In addition, the magnitude of the overall variance is small, and a relatively small variance around the 60° inlet angle is detected. The velocity variance increases slightly as the inlet velocity increases. The slower the inlet velocity, the smaller the percentage of speed variation, and the more uniform the flow field, from the perspective of flow field uniformity. Based on mass transfer, an increase in inlet velocity improves the gas flow inside and outside, which in turn positively influences the renewal of the working medium. Hence, a smaller inlet velocity is not always better.

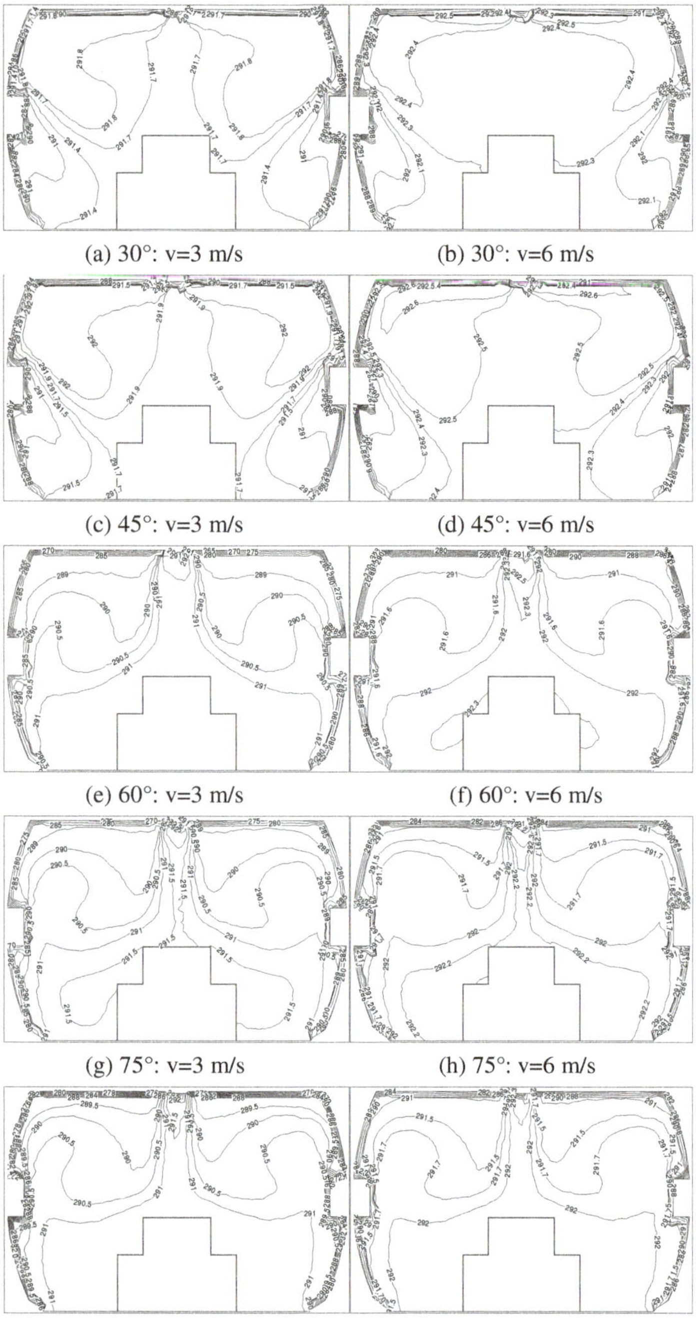

(a) 30°: v=3 m/s (b) 30°: v=6 m/s

(c) 45°: v=3 m/s (d) 45°: v=6 m/s

(e) 60°: v=3 m/s (f) 60°: v=6 m/s

(g) 75°: v=3 m/s (h) 75°: v=6 m/s

(j) 90°: v=3 m/s (i) 90°: v=6 m/s

Fig. 6 Temperature field of the center longitudinal section under different air inlet velocities.

Figures 6 and 7 show that the temperature within a small distance from the wall changes dramatically because of heat dissipation, whereas the center temperature tends to be uniform. The temperature field in the entire cabin is inferior to the preceding temperature field with larger inlet flow rate. The effect of improving temperature comfort by increasing the inlet airflow is not significant when the temperature in the cabin is increased to a certain degree. In addition, the wall temperature improves and the temperature gradient slows down with the inlet velocity increasing.

Figure 8 shows that the temperature variance in the cabin becomes small because of the increase in the inlet velocity. The trend of the decrease in temperature variance is significant when the velocity increases from 1 m/s to 3 m/s, but relatively insignificant when the velocity increases from 3 m/s to 9 m/s. The order of magnitude of the temperature variance is relatively larger than that of the velocity variance. The previous analysis suggests that the temperature

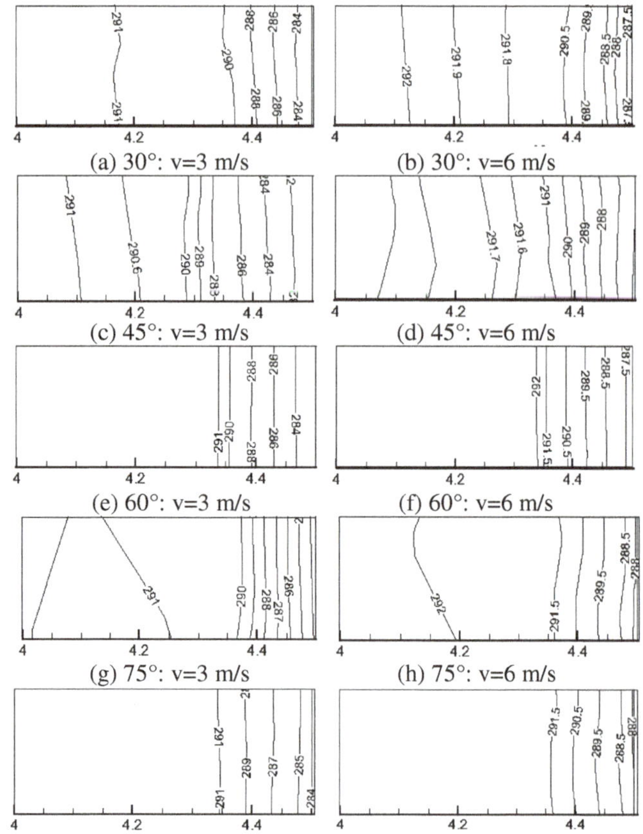

(a) 30°: v=3 m/s (b) 30°: v=6 m/s

(c) 45°: v=3 m/s (d) 45°: v=6 m/s

(e) 60°: v=3 m/s (f) 60°: v=6 m/s

(g) 75°: v=3 m/s (h) 75°: v=6 m/s

(i) 90°: v=3 m/s (j) 90°: v=6 m/s

Fig. 7 Isothermal diagram of the center cross section near the wall under different inlet velocities.

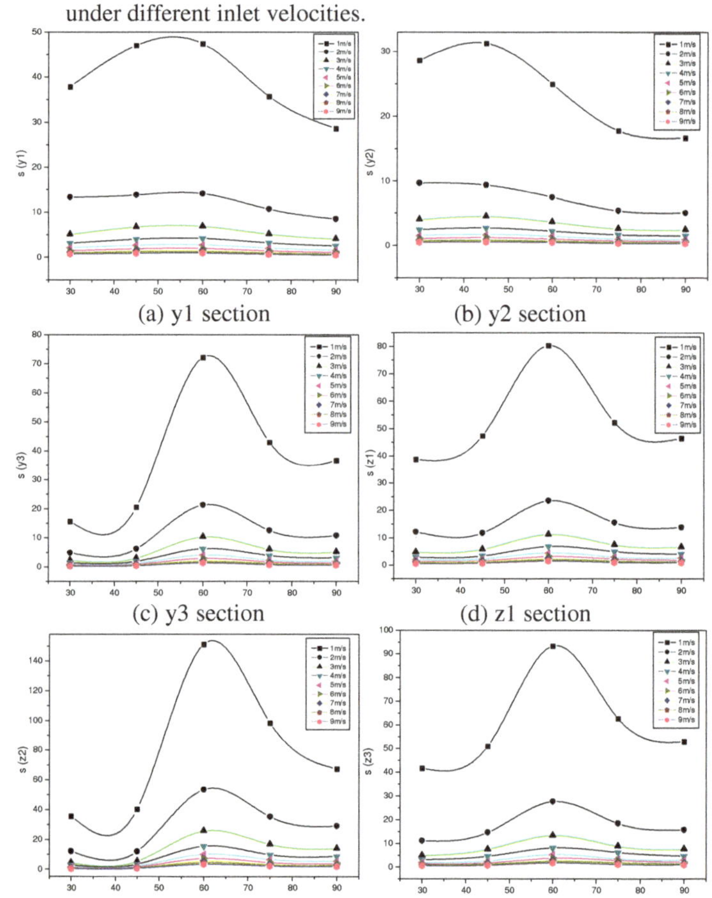

(a) y1 section (b) y2 section

(c) y3 section (d) z1 section

(e) z2 section (f) z3 section

Fig. 8 Variance of temperature under different velocities according to the air inlet angles.

fluctuation in the cabin does not conform to the principle of comfort when velocity is relatively small. Moreover, the effect of an improvement in temperature comfort caused by an increase in velocity is not obvious when the velocity of the cabin is relatively large. Figure 5 shows that the velocity fluctuation is fortified by an increase in the inlet velocity. Therefore, an inlet velocity with a mid-range value is the optimal velocity.

Table 2 Inlet mass flow corresponding to different inlet velocities.

v(m/s)	*1*	*2*	*3*	*4*	*5*	*6*	*7*	*8*	*9*
S(kg/s)	2.3	4.7	7.1	9.4	11.8	14.2	16.6	18.9	21.3

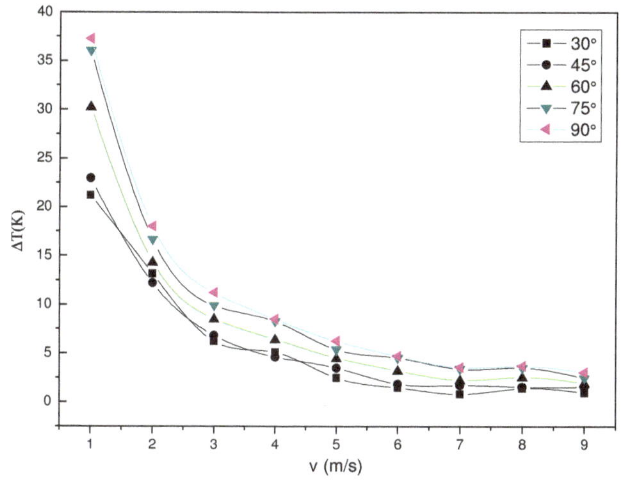

Fig. 9 The difference between maximum temperature and minimum temperature corresponding to different inlet velocities.

Table 2 shows inlet flow rate increase with the inlet velocity increasing and more energy will be consumed to produce the required flow rate. However, energy consumption increases and temperature difference decreases as the inlet flow increases, as shown in Figure 9. On the premise of satisfying occupant comfort, It is hope to save energy. According to aviation medicine, the normal comfort temperature region should be between 15 and 26°C and the temperature field should be uniform. Furthermore, The deviation of the cabin temperature along the vertical and horizontal directions from regulations should not be too large. The difference between maximum temperature and minimum temperature should be within the range of 6°C. The temperature of the cabin wall, floor, and top wall should be consistent with the cabin temperature. Otherwise, the passengers will feel uncomfortable because of the influence of thermal radiation and convection. Consequently, the comfort demand can be satisfied basically, when the air inlet velocity is not less than 3 m/s ; and the energy consumption can be reduced to the extent possible, when the air inlet velocity is between 3 and 6 m/s.

3.3 Analysis of PMV–PPD

The air thermal comfort in the cabin is discussed here based on the previous results. The PMV field of air comfort in the cabin and the PPD were obtained through a simulation under inlet angle conditions of 30° and 45° and inlet velocity of 1 m/s to 9 m/s. The inlet velocities of 3 and 6 m/s are selected as the examples here because of the large number of images. Figure 9 shows the PMV and PPD fields of a cross section.

The PMV in the cabin is a function of air temperature and air speed (other parameters have constant values), and PPD is a function of PMV. Figure 10 shows that the PMV field is similar to the velocity distribution law in the cabin, given that the temperature gradient in the cabin is small and the speed rate is large. Moreover, the PMV gradient is large near the cabin walls and the objects in the cabin because of the fast speed rate and the intense temperature changes. PPD field changes depending only on PMV.

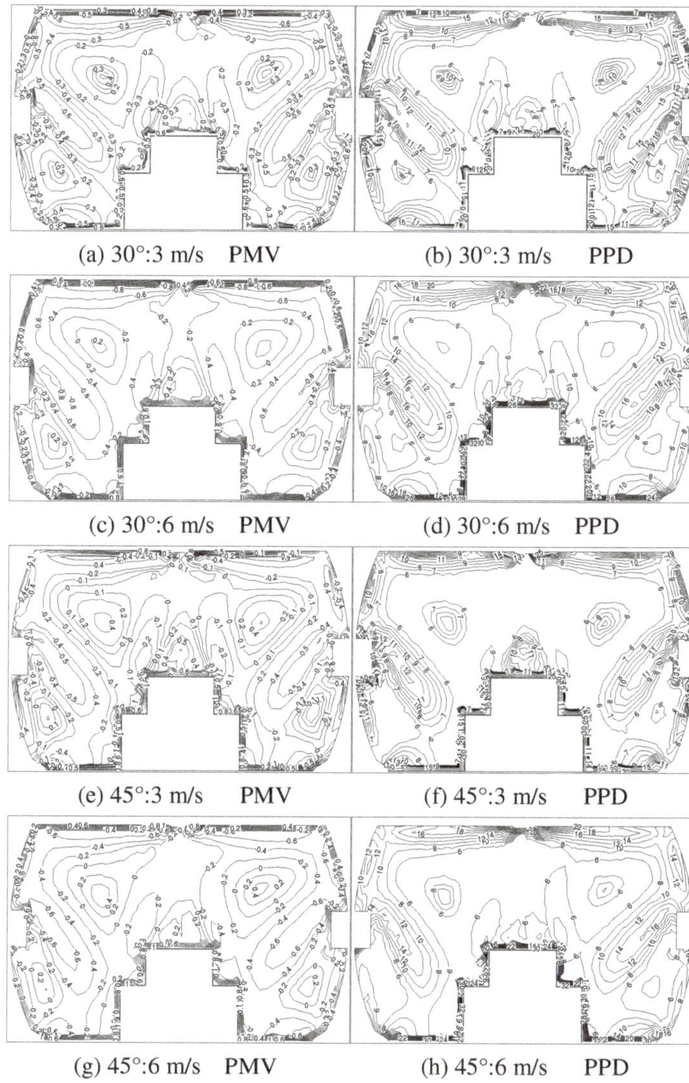

Fig. 10 PMV and PPD fields of a cross section.

Table 3 PMV range under different inlet velocities in the cabin.

PMV	*30°(min)*	*30°(max)*	*45°(min)*	*45°(max)*
1m/s	-2.31	0.75	-2.51	0.82
2m/s	-1.63	1.03	-1.76	1.07
3m/s	-1.24	1.14	-1.3	1.17
4m/s	-1.14	1.2	-1.17	1.22
5m/s	-1.1	1.22	-1.02	1.25
6m/s	-0.94	1.23	-0.99	1.27
7m/s	-0.93	1.25	-0.98	1.28
8m/s	-0.92	1.27	-0.96	1.29
9m/s	-0.9	1.28	-0.92	1.29

According to the Code for Design of Heating Ventilation and Air Conditioning, the thermal comfort of heating and air-conditioning in a room should follow the PMV and PPD index determination of moderate thermal environment and the regulations of thermal comfort conditions (Hua et al., 2011). It must also have the average PMV and PPD. The ideal PMV and PPD index values are as follows: the PMV index is between -1 and + 1, and the PPD index is not greater than 27%. Figure 10 and Table 3 show that the PMV index is between -1.3 and + 1.3 with a velocity not lower than 3 m/s and that the PMV in most cabin regions is not beyond the range of -1 to 1. The PPD in most cabin regions is not greater than 20%, and the PPD index near the wall surface exceeds 20%. Hence, the conclusions meet the criteria of thermal comfort.

4. CONCLUSIONS

Pressure, temperature, velocity, and humidity are the main factors that influence comfortableness in an aircraft cabin. Flow field, temperature field and air quality in the cabin are related to relevant boundary conditions, such as inlet temperature, inlet velocity, inlet angle and position, and air humidity. A comfortable cabin environment is realized under reasonable boundary conditions. The average Reynolds equation and the low Reynolds number turbulence model were used to simulate the cabin airflow. A numerical investigation was conducted on the heat transfer in a cabin to investigate the effects of air distribution under different angle and inlet velocity conditions. Additional studies were also conducted on the thermal comfort of the cabin environment under these inlet conditions.

The results of the analysis are as follows:

- The velocity variance is at the minimum level, and the temperature variance is at the maximum level, when the air inlet angle is 60°. The temperature difference between the wall and human body is too large and does not satisfy the comfort requirement. As such, the comfort demand of passengers is satisfied when the air inlet angle in the region is between 35° and 45°.
- The velocity variance in the cabin increases (i.e., the increase in velocity fluctuation) as the inlet flow increases, and the order of magnitude for the variance is relatively small.
- The temperature variance in the cabin decreases (i.e., an increasingly uniform temperature field) as the inlet flow increases. The trend of the decrease in temperature variance is significant when the velocity increases from 1 m/s to 3 m/s, but relatively insignificant when the velocity increases from 3 m/s to 9 m/s.
- Both energy consumption and velocity fluctuation increase as the inlet flow increases and the temperature distribution becomes increasingly uniform. Consequently, the comfort demand can be satisfied, and the energy consumption can be reduced to the extent possible, when the air inlet velocity is between 3 and 6 m/s.
- These conclusions prove that the PMV–PPD index of the cabin environment of comfort can satisfy the human body's basic thermal comfort demand using small inlet angles and reasonable inlet velocities

This study provides a mathematical model that reflects the actual situation for future research and optimization design of control system for comfortable cabin environments.

ACKNOWLEDGEMENTS

This work is supported in part by National Science Foundation of China (51276118).

REFERENCES

Bosbach, J., Pennecot, J., Wagner, C., Raffel, M., Lerche, T., and Repp, S., 2006, "Experimental and Numerical Simulations of Turbulent Ventilation in Aircraft Cabins," *Energy*, **31**(5), 694–705. http://dx.doi.org/10.1016/j.energy.2005.04.015

Bianco, V., Manca, O., Nardini, S., and Roma, M., 2009, "Numerical Investigation of Transient Thermal and Fluidynamic Fields in an Executive Aircraft Cabin," *Applied Thermal Engineering*, **29**(16), 3418–3425. http://dx.doi.org/10.1016/j.applthermaleng.2009.05.020

Buratti, C., Moretti, E., Belloni, E., and Cotana, F., 2013, "Unsteady Simulation of Energy Performance and Thermal Comfort in Non-residential Buildings," *Building & Environment*, **59**(328), 482–491. http://dx.doi.org/10.1016/j.buildenv.2012.09.015

Charles, K. E., 2003, "Fanger's Thermal Comfort and Draught Models," *Institute for Research in Construction National Research Council of Canada*, Ottawa, K1A 0R6, Canada IRC Research Report RR-162 October, 10. http://dx.doi.org/10.4224/20378865

Fanger, P. O., 1967, "Calculation of Thermal Comfort, Introduction of a Basic Comfort Equation," *ASHRAE transactions*, **73**(2), III-4.

Fanger, P. O., 1984, "Moderate Thermal Environments Determination of the PMV and PPD Indices and Specification of the Conditions for Thermal Comfort," *ISO 7730*.

Fan J., 2008, "FLUENT Advanced Application and Example Analysis," *Tsinghua University Press.*

Günther, G., Bosbach, J., Pennecot, J., Wagner, C., Lerche, T., and Gores, I., 2006, "Experimental and Numerical Simulations of Idealized Aircraft Cabin Flows," *Aerospace Science & Technology*, **10**(7), 563–573. http://dx.doi.org/10.1016/j.ast.2006.02.003

Hua, D., Li, T., Liu, W., Jin, S., and Q., 2011, "Moderate Thermal Environment—determination of the PMV and PPD Indices and Specification of the Conditions for Thermal Comfort," *Beijing, National Standards Publishing House.*

Isukapalli, S. S., Mazumdar, S., George, P., Wei, B., Jones, B., and Weisel, C. P., 2013, "Computational Fluid Dynamics Modeling of Transport and Deposition of Pesticides in an Aircraft Cabin," *Atmospheric Environment*, **68**(1), 198–207. http://dx.doi.org/10.1016/j.atmosenv.2012.11.019

Ji, X., Li, G., and Dai, Z., 2003, "Influencing Factors and the Research Progress on Forecasting and Evaluating Indoor Thermal Comfort," *Journal of Hygiene Research,* **32**(3), 295-299.

Kühn, M., Bosbach, J., and Wagner, C., 2009, "Experimental Parametric Study of Forced and Mixed Convection in a Passenger Aircraft Cabin Mock-up," *Building & Environment*, **44**(5), 961–970. http://dx.doi.org/10.1016/j.buildenv.2008.06.020

Liu, W., Mazumdar, S., Zhang, Z., Poussou, S. B., Liu, J., and Lin, C. H., 2012, "State-of-the-art Methods for Studying Air Distributions in Commercial Airliner Cabins," *Building & Environment*, **47**: 5–12. http://dx.doi.org/10.1016/j.buildenv.2011.07.005

Liu, X., 2007, "The Study of Thermal Comfort and Energy Saving Schemes for Hospital," *Refrigeration and Air Conditioning*, **21**(4): 85-89.

Mazumdar, S., Poussou, S. B., Lin, C. H., Isukapalli, S. S., Plesniak, M. W., and Chen, Q., 2011, "Impact of Scaling and Body Movement on Contaminant Transport in Airliner Cabins," *Atmospheric Environment*, **45**(33), 6019–6028. http://dx.doi.org/10.1016/j.atmosenv.2011.07.049

Mazumdar, S., and Chen, Q., 2007, "Impact of Moving Bodies on Airflow and Contaminant Transport Inside Aircraft Cabins," *Proceedings of Roomvent*, 10th International Conference on Air Distribution in Rooms, Helsinki, Finland.

Maidment, G., Missenden, J., and Ampofo, F., 2004, "Underground Railway Environment in the UK part 1: Review of Thermal Comfort," *Applied Thermal Engineering*, **24**(5), 611-631. http://dx.doi.org/10.1016/j.applthermaleng.2003.10.017

Poussou, S. B., Mazumdar, S., Plesniak, M. W., Sojka, P. E., and Chen, Q., 2010, "Flow and Contaminant Transport in an Airliner Cabin Induced by a Moving Body: Model Experiments and CFD Predictions," *Atmospheric Environment*, **44**(24), 2830–2839. http://dx.doi.org/10.1016/j.atmosenv.2010.04.053

Rai, A. C., and Chen, Q., 2012, "Simulations of Ozone Distributions in an Aircraft Cabin Using Computational Fluid Dynamics," *Atmospheric Environment*, **54**(5), 348–357. http://dx.doi.org/10.1016/j.atmosenv.2012.02.010

Shou, R.Z., He, H.S., 2004, "Aircraft's Environment Control," Beijing University of Aeronautics and Astronautics Press, Beijing.

Schellen, L., Loomans, M. G. L. C., Kingma, B. R. M., Wit, M. H. D., Frijns, A. J. H., and Lichtenbelt, W. D. V. M., 2013, "The Use of a Thermophysiological Model in the Built Environment to Predict Thermal Sensation: Coupling with the Indoor Environment and Thermal Sensation," *Building & Environment*, **59**, 10–22. http://dx.doi.org/10.1016/j.buildenv.2012.07.010

Shen, H., and Yuan, X., 2009, "Numerical Simulation and Experiment on Air Flow and Heat Transfer in Fighter Plane Cockpit," *Acta Aeronautica Et Astronautica Sinica*, **30**(1), 30-39.

Wang, A., Zhang, Y., Sun, Y., and Wang, X., 2008, "Experimental Study of Ventilation Effectiveness and Air Velocity Distribution in an Aircraft Cabin Mockup," *Building & Environment*, **43**, 337–343. http://dx.doi.org/10.1016/j.buildenv.2006.02.024

Wang, Z., 2006, "A Field Study of the Thermal Comfort in Residential Buildings in Harbin," *Building & Environment*, **41**(8), 1034–1039. http://dx.doi.org/10.1016/j.buildenv.2005.04.020

Xu, J., 2005, "Discussion on the Application Range of the PMV Equation," *Journal of Chongqing Architecture University*, **27**(3), 13-18.

Yan, W., Zhang, Y., Sun, Y., and Li, D., 2009, "Experimental and CFD Study of Unsteady Airborne Pollutant Transport within an Aircraft Cabin Mock-up," *Building & Environment*, **44**(1), 34-43. http://dx.doi.org/10.1016/j.buildenv.2008.01.010

You, R., and Zhao, B., 2013, "A Simplified Method for Assessing Particle Deposition Rate in Aircraft Cabins," *Atmospheric Environment*, **67**(2), 80–84. http://dx.doi.org/10.1016/j.atmosenv.2012.10.046

Yao, R., Chen, Q., 1997, "Prediction of Indoor Climate and Thermal Comfort for Cooling in Ventilated Residence Rooms," *HV&AC*, **27** (6): 5-9.

Yu, Y., and Zhang, G. W., 2008, "Thermal Environment Simulation and Analysis of Radiant Floor Cooling Combining with Displacement Ventilation in the Air Conditioned Room," *Refrigeration Air Conditioning & Electric Power Machinery*, **29**(119): 11-16.

Yang, P., and Chen, H., 2002, "The Application of PMV-PPD in the Air-conditioning Train," *Construction Machinery for Hydraulic Engineering & Power Station*, **23**(86): 22-24.

Zhang, T., and Chen, Q., 2007, "Novel Air Distribution Systems for Commercial Aircraft Cabins," *Building & Environment*, **42**(4), 1675–1684. http://dx.doi.org/10.1016/j.buildenv.2006.02.014

Zhang, Z., Chen, X., Mazumdar, S., Zhang, T., and Chen, Q., 2009, "Experimental and Numerical Investigation of Airflow and Contaminant Transport in an Airliner Cabin Mockup," *Building & Environment*, **44**(1), 85–94. http://dx.doi.org/10.1016/j.buildenv.2008.01.012

Zhang H., 2003, "Human Thermal Sensation and Comfort in Transient and Non-uniform Thermal Environments. Berkeley," *USA: University of California*.

Zhai, J. Z., Zhang, Z., Zhang, W., and Chen, Q., 2007, "Evaluation of Various Turbulence Models in Predicting Airflow and Turbulence in Enclosed Environments by CFD: Part 1—Summary of Prevalent Turbulence Models," *HVAC & R Research*, **13**(6), 853-870. http://dx.doi.org/10.1080/10789669.2007.10391459

Zhang, Z., Zhang, W., Zhai, J. Z., and Chen, Q., 2007, "Evaluation of Various Turbulence Models in Predicting Airflow and Turbulence in Enclosed Environment by CFD. Part-2: Comparison with Experimental Data from Literature," *HVAC&R Research*, **13**(6), 871-886. http://dx.doi.org/10.1080/10789669.2007.10391460

Zhang, D., and Ang, H., 2002, "Numerical Simulation of Air Distribution in an Aircraft Cabin," *Journal of Nanjing University of Aeronautics & Astronautics,* **34**(5), 484-487.

ANALYSIS OF MHD TRANSIENT FREE CONVECTION FLOW OF A NEWTONIAN FLUID PAST AN INFINITE VERTICAL POROUS PLATE

M. Umamaheswar[a], M. C. Raju[a*], S. V. K. Varma[b]

[a] *Department of Humanities and Sciences, Annamacharya Institute of Technology and Sciences Rajampet (Autonomous), Rajampet, Andhra Pradesh, 516126 India*

[b] *Department of Mathematics, S. V. University, Tirupati, 517502, Andhra Pradesh, India*

ABSTRACT

An investigation is carried out to analyze the unsteady MHD free convection, heat and mass transfer flow of a Newtonian fluid past an infinite vertical porous plate with homogeneous chemical reaction and heat absorption/generation. A uniform magnetic field is applied perpendicular to the plate. The non-dimensional governing equations are solved numerically by using finite difference method. The effects of various parameters governing the flow on velocity, temperature, concentration, skin friction, Nusselt number and Sherwood number are studied through graphs. It is noticed that velocity decreases with an increase in Magnetic field while it increases with an increase in Grashof number, modified Grashof number& Permeability parameter. Temperature increases with an increase in heat absorption and it shows reverse effect in the case of heat generation and Prandtl number. Concentration decreases with increase in Schmidt number and chemical reaction parameter.

Keywords: *MHD, Free convection, Porous medium, Vertical plate, Newtonian fluid, Finite difference Method and Chemical reaction.*

1. INTRODUCTION

In recent years, the analysis of hydro magnetic convection flow involving heat and mass transfer in porous medium has attracted the attention of many scholars because of its possible applications in diverse fields of science and technology such as soil sciences, astrophysics, Geophysics, nuclear power reactors etc. Magneto hydrodynamic free convection flow of an electrically conducting fluid in different porous geometries is of considerable interest to the technical field due to its frequent occurrence in industrial, technological and geothermal applications. As an example, the geothermal region gases are electrically conducting and undergo the influence of magnetic field. Also, it has applications in nuclear engineering in connection with reactors cooling. The interest in this field is due to the wide range of applications in engineering and geophysics, such as the optimization of the solidification processes of metals and metal alloys, the study of geothermal sources, the treatment of nuclear fuel debris, the control of underground spreading of chemical wastes and pollutants and the design of MHD power generators. Many papers concerned with the problem of MHD free convection flow in porous media have been published in the literature.

Chamkha (2003) studied MHD flow of a uniformly stretched vertical permeable surface in the presence of heat generation/absorption and a chemical reaction. Kim (2000) analyzed unsteady MHD convective heat transfer past a semi-infinite vertical porous moving plate with variable suction. Chamkha (2004) has studied unsteady MHD convective heat and mass transfer past a semi-infinite vertical permeable moving plate with heat absorption. Pal (2013) analyzed Hall current and MHD effects on heat transfer over an unsteady stretching permeable surface with thermal radiation. Cortell (2007) investigated MHD flow and mass transfer of an electrically conducting fluid of second grade in a porous medium over a stretching sheet with chemically reactive species.

On the other hand, Seth et al. (2011) analyzed MHD natural convection flow with radiative heat transfer past an Impulsively Moving Plate with ramped wall temperature. Muthucumaraswamy and Janakiraman (2006) studied MHD and radiation effects on moving isothermal vertical plate with variable mass diffusion. Umamaheswar et al. (2013) studied Unsteady MHD free convective visco-elastic fluid flow bounded by an infinite inclined porous plate in the presence of heat source, viscous dissipation and ohmic heating. Singh (2001) has studied MHD free convection and mass transfer flow with heat source and thermal diffusion. Mishra et al. (2013) analyzed Mass and heat transfer effect on MHD flow of a visco-elastic fluid through porous medium with oscillatory suction and heat source. Vajravelu et al. (2013) studied unsteady convective boundary layer flow of a viscous fluid at a vertical surface with variable fluid properties. Hayat et al. (2011) studied MHD flow and heat transfers over permeable stretching sheet with slip conditions. Pal and Chatterjee (2010) analyzed Heat and Mass Transfer in MHD Non-Darcian Flow of a Micro polar Fluid over a Stretching Sheet Embedded in a Porous Media with Non-Uniform Heat Source and Thermal Radiation.

Muthucumaraswamy and Ganesan (2002) studied natural convection on a moving isothermal vertical plate with chemical reaction. Patil et al. (2013) analyzed Chemical Reaction Effects on Unsteady Mixed Convection Boundary Layer Flow Past a Permeable Slender Vertical Cylinder Due to a Nonlinearly Stretching Velocity. Kandasamy et al. (2005) studied Chemical reaction, heat and mass transfer on MHD flow over a vertical stretching surface with heat source and thermal stratification effects. Takhar et al. (2000) studied flow and mass transfer on a stretching sheet with a magnetic field and chemically reactive species. Raptis and Perdikis (2006) studied viscous flow over a non-linearly stretching sheet in the presence of a chemical reaction and magnetic field. Kandasamy et al. (2010) analyzed Lie group analysis for the effect of temperature dependent fluid viscosity with thermophoresis and chemical reaction on MHD free convective heat and mass transfer over a porous stretching surface in the presence

* Corresponding author email: mcrmaths@yahoo.co.in, umasvu8@gmail.com

of heat source / sink. Anjalidevi and Kandasamy (2000) studied effects of a chemical reaction heat and mass transfer on MHD flow past a semi-infinite plate. Sonth et al. (2002) analyzed heat and mass transfer in a visco-elastic fluid flow over an accelerating surface with heat source/sink and viscous dissipation. Hossain and Mandal (1985) investigated mass transfer effects on the unsteady hydromagnetic free convection flow past an accelerated vertical porous plate. Khan et al. (2014) studied MHD boundary layer flow of a nanofluid containing gyro tactic microorganisms past a vertical plate with Navier slip. Seini and Makinde (2014) studied boundary layer flow near stagnation-points on a vertical surface with slip in the presence of transverse magnetic field. Jha, (1991) investigated MHD free convection mass transform flow through a porous medium. Prakash et al. (2013) studied diffusion-thermo and radiation effects on unsteady MHD flow through porous medium past an impulsively started infinite vertical plate with variable temperature and mass diffusion. Seth and Ansari (2010) analyzed MHD natural convection flow past an impulsively Moving vertical plate with ramped wall temperature in the presence of thermal diffusion with heat absorption. Seth et al. (2013) studied effects of thermal radiation and rotation on unsteady hydromagnetic free convection flow past an impulsively moving vertical plate with ramped temperature in a porous medium. Seth et al. (2014a) analyzed hydromagnetic natural convection flow with radiative heat transfer past an accelerated moving vertical plate with ramped temperature through a porous medium. Seth et al. (2014c) considered Hall effects on unsteady MHD natural convection flow of a heat absorbing fluid past an accelerated moving vertical plate with ramped temperature. Seth et al. (2014b) investigated effects of Hall current and rotation on unsteady MHD natural convection flow with heat and mass transfer past an impulsively moving vertical plate in the presence of radiation and chemical reaction.

Based on the above studies we have considered the unsteady MHD free convection, heat and mass transfer flow of a Newtonian fluid past an infinite vertical porous plate with homogeneous chemical reaction and heat absorption/generation. We have extended the work of Kaprawi (2015) with the novelty of considering heat source, chemical reaction, porous permeability, magnetic field parameter and mass transfer.

2. FORMULATION OF THE PROBLEM:

We consider a viscous incompressible, electrically conducting, heat absorbing/generating and chemically reacting Newtonian fluid flow past an infinite vertical porous. A magnetic field of uniform strength is applied perpendicular to the plate. Let x^*-axis is taken along the plate in the vertically upward direction and the y^*-axis is taken perpendicular to the plate. At time $t \leq 0$, the plate is maintained at the temperature higher than ambient temperature T_∞ and the fluid is at rest. At time $t > 0$, the plate is linearly accelerated with increasing time in its own plane and the temperature decreases with temperature $T = 1/(1+at)$. Similarly the species concentration decreases with time t. It is assumed that the effect of viscous dissipation is negligible. By usual Boussineq's and boundary layer approximation, the unsteady flow is governed by the following equations:

$$\frac{\partial u^*}{\partial t^*} = \nu \frac{\partial^2 u^*}{\partial y^{*2}} + g\beta_T\left(T^* - T_\infty\right) + g\beta_C\left(C^* - C_\infty\right) - \frac{\sigma B_0^2 u^*}{\rho} - \frac{\nu}{k} u^* \quad (1)$$

$$\rho C_p \frac{\partial T^*}{\partial t^*} = k_T \frac{\partial^2 T^*}{\partial y^{*2}} - Q^*\left(T^* - T_\infty\right) \quad (2)$$

$$\frac{\partial C^*}{\partial t^*} = D\frac{\partial^2 C^*}{\partial y^{*2}} - K_r^*\left(C^* - C_\infty\right) \quad (3)$$

The corresponding initial and boundary conditions are

$$\left.\begin{aligned} & u^* = 0, T^* = T_\infty, C^* = C_\infty \quad \text{for all } y^*, t^* \leq 0 \\ & t^* > 0 : u^* = U_0 a^* t^*, T^* = T_\infty + \left(\frac{T_s^* - T_\infty}{1 + At^*}\right), \\ & C^* = C_\infty + \left(\frac{C_s^* - C_\infty}{1 + At^*}\right) \quad \text{at } y^* = 0 \\ & u^* \to 0, T^* \to T_\infty, C^* \to C_\infty \quad \text{as } y^* \to \infty \end{aligned}\right\} \quad (4)$$

where $A = \dfrac{U_0^2}{\nu}$

The non-dimensional quantities are as follows:

$$u = \frac{u^*}{U_0}, t = \frac{t^* U_0^2}{\nu}, y = \frac{y^* U_0}{\nu}, \theta = \frac{T^* - T_\infty}{T_s^* - T_\infty}, C = \frac{C^* - C_\infty}{C_s^* - C_\infty},$$

$$Gr = \frac{\nu g \beta_T \left(T_s^* - T_\infty\right)}{U_0^3}, \text{ (Grashof number)}$$

$$Gm = \frac{\nu g \beta_C \left(C_s^* - C_\infty\right)}{U_0^3}, \text{ (Modified Grashof number)}$$

$$M = \frac{\sigma B_0^2 \nu}{\rho U_0^2}, \text{ (Magnetic parameter)}$$

$$K = \frac{k U_0^2}{\nu^2}, \text{ (Porous permeability)}$$

$$\Pr = \frac{\rho \nu C_p}{k_T}, \text{ (Prandtl number)}$$

$$Q = \frac{Q^* \nu^2}{k_T U_0^2}, \text{ (Heat Source/Sink)}$$

$$Sc = \frac{\nu}{D}, \text{ (Schmidt number)}$$

$$Kr = \frac{K_r^* \nu}{U_0^2}, \text{ (Chemical reaction parameter)}$$

After introducing the non-dimensional quantities into the equations (1)-(3), these equations reduces to

$$\frac{\partial u}{\partial t} = \frac{\partial^2 u}{\partial y^2} + Gr\,\theta + Gm\,C - M\,u - \frac{1}{K}u \quad (5)$$

$$\frac{\partial \theta}{\partial t} = \frac{1}{\Pr}\frac{\partial^2 \theta}{\partial y^2} - \frac{Q}{\Pr}\theta \quad (6)$$

(1)

$$\frac{\partial C}{\partial t} = \frac{1}{Sc}\frac{\partial^2 C}{\partial y^2} - Kr\,C \quad (7)$$

The corresponding initial and boundary conditions are

$$\left.\begin{aligned} & u = 0, \theta = 0, C = 0 \quad \text{for all y, t} \leq 0 \\ & t > 0 : u = at, \theta = \frac{1}{1+t}, C == \frac{1}{1+t} \quad \text{at } y = 0 \\ & u \to 0, \theta \to 0, C \to 0 \quad \text{as } y \to \infty \end{aligned}\right\} \quad (8)$$

3. METHOD OF SOLUTION

Equations (5)-(7) are linear partial differential equations and are to be solved by using the initial and boundary conditions (8). However exact solution is not possible for this set of equations and hence we solve these equations by finite-difference method. The equivalent finite difference schemes of equations for (5)-(7) are as follows:

$$\frac{u_{i,j+1}-u_{i,j}}{\Delta t}=Gr\,\theta_{i,j}+Gm\,C_{i,j}+\frac{u_{i-1,j}-2u_{i,j}+u_{i+1,j}}{(\Delta y)^2}-M\,u_{i,j}-\frac{1}{K}u_{i,j} \quad (9)$$

$$\frac{\theta_{i,j+1}-\theta_{i,j}}{\Delta t}=\frac{1}{\Pr}\frac{\theta_{i-1,j}-2\theta_{i,j}+\theta_{i+1,j}}{(\Delta y)^2}-\frac{Q}{\Pr}\theta_{i,j} \quad (10)$$

$$\frac{C_{i,j+1}-C_{i,j}}{\Delta t}=\frac{1}{Sc}\frac{C_{i-1,j}-2C_{i,j}+C_{i+1,j}}{(\Delta y)^2}-Kr\,C_{i,j} \quad (11)$$

Here, the suffix i refer to y and j to time. The mesh system is divided by taking $\Delta y = 0.1$. From the initial condition in (8), we have the following equivalent:

$$u(i,0)=0,\ \theta(i,0)=0, C(i,0)=0 \text{ for all } i \quad (12)$$

The boundary conditions from (8) are expressed in finite-difference form as follows

$$u(0,j)=at, \theta(0,j)=\frac{1}{1+t}, C(0,j)=\frac{1}{1+t} \text{ for all } j$$
$$u(i_{max},j)=0, \theta(i_{max},j)=0, C(i_{max},j)=0 \text{ for all } j \quad (13)$$

(Here i_{max} was taken as 200)

First the velocity at the end of time step viz, u (i, j+1)(i=1,200) is computed from (9) in terms of velocity, temperature and concentration at points on the earlier time-step. Then θ (i, j +1) is computed from (10) and C (i, j +1) is computed from (11). The procedure is repeated until t = 0.5 (i.e. j = 500). During computation Δt was chosen as 0.001.

Skin-friction:

The skin-friction in non-dimensional form is given by

$$\tau=-\left(\frac{du}{dy}\right)_{y=0}, \text{where } \tau=\frac{\tau^1}{\rho U_0^2}$$

Rate of heat transfer:

The dimensionless rate of heat transfer is given by

$$Nu=-\left(\frac{d\theta}{dy}\right)_{y=0}$$

Rate of mass transfer:

The dimensionless rate of mass transfer is given by

$$Sh=-\left(\frac{dC}{dy}\right)_{y=0}$$

4. RESULT AND DISCUSSION:

The effects of various physical parameters viz., the Schmidt number (Sc), the thermal Grashof number (Gr), the modified Grashof number (Gm), magnetic parameter (M),Prandtl number (Pr) and chemical reaction parameter (Kr) are exhibited in the figures 1-15 and studied by choosing arbitrary values.

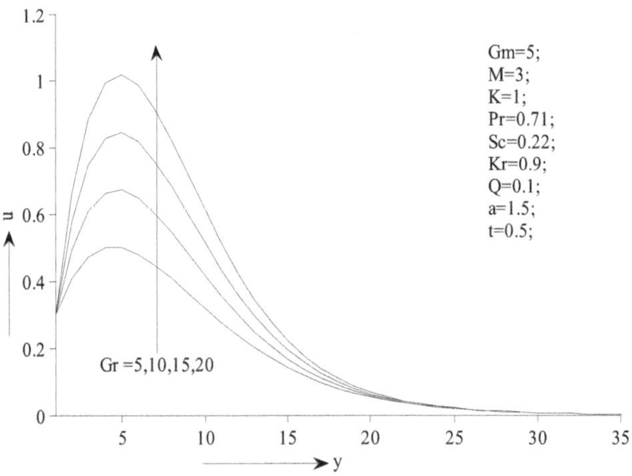

Fig. 1 Effect of Grashof number on velocity

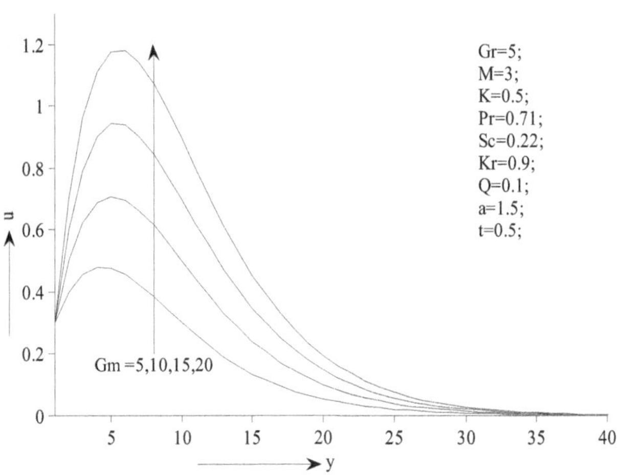

Fig. 2 Effect of Modified Grashof number on velocity

In Fig.1, effect of thermal Grashof number on velocity is presented. As Gr increases, velocity also increases. This is due to the buoyancy which is acting on the fluid particles due to gravitational force that enhances the fluid velocity. A similar effect is noticed from Fig.2, in the presence of modified Grashof number, which also increases fluid velocity. In figure 3, velocity profiles are displayed with the variation in magnetic parameter. From this figure it is noticed that velocity gets reduced by the increase of magnetic parameter. When an electrically conducting fluid moves in the presence of an applied magnetic field, a magnetic force, called Lorentz force, is generated in the flow field whose tendency is to resist the fluid motion. Due to this reason fluid velocity is getting retarded on increasing magnetic parameter (*M*). Fig.4 depicts the variations in velocity profiles for different values of Permeability parameter. From this figure it is noticed that, velocity increases as K increases. Fig.5 indicates that a rise in Pr substantially reduces the temperature in the viscous fluid. It can be found from Fig.5 that the thickness of thermal boundary layer decreases on increasing Pr. Fig.6 depicts the effect of heat absorption on temperature. It is noticed that the temperature decreases as an increase in the heat absorption parameter.

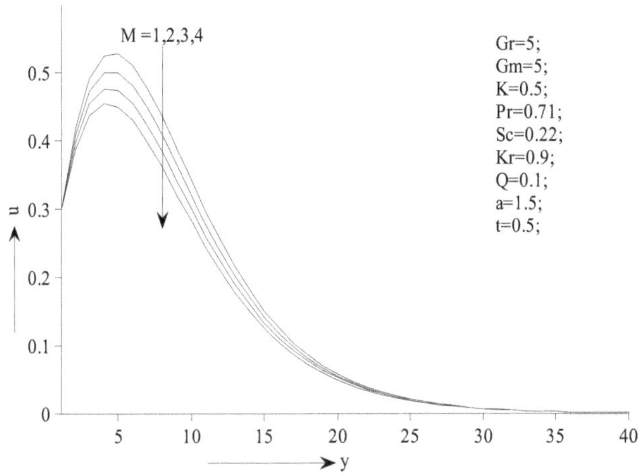

Fig. 3 Effect of Magnetic parameter (M) on velocity

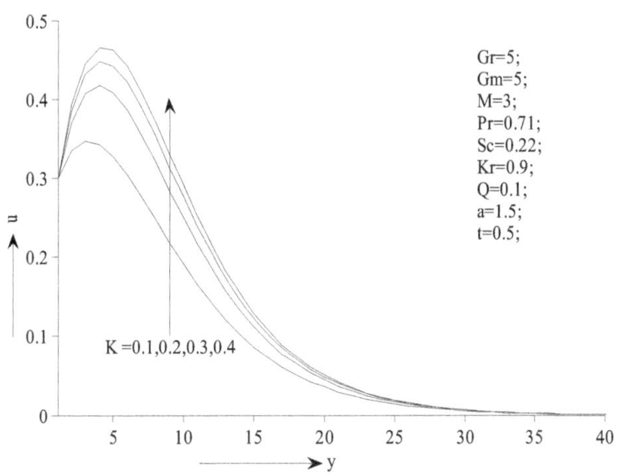

Fig. 4 Effect of Permeability parameter on velocity

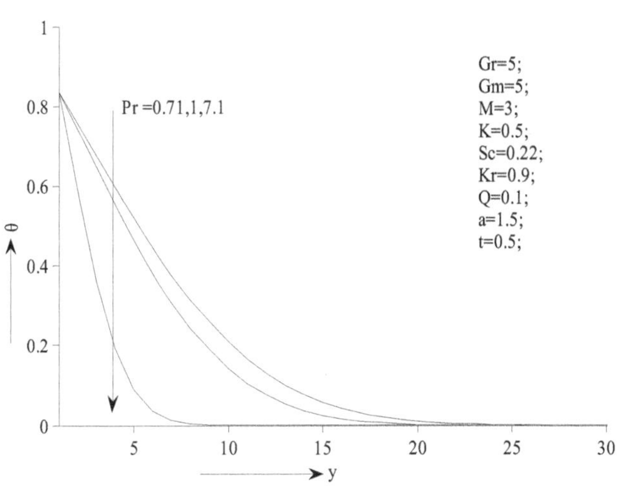

Fig. 5 Effect of Prandtl number (Pr) on temperature

The central reason behind this effect is that the heat absorption causes a decrease in the kinetic energy as well as thermal energy of the fluid. The momentum and thermal boundary layers get thinner in case of heat absorbing fluids. It shows reverse effect in the case of heat generation parameter. Influence of Schmidt number on concentration is shown in Fig.7, from this figure it is noticed that concentration decreases with an increase in Schmidt number. Because, Schmidt number is a dimensionless number defined as the ratio of momentum diffusivity and mass diffusivity, and is used to characterize fluid flows in which there are simultaneous momentum and mass diffusion convection processes. Therefore, concentration boundary layer decreases with an increase in Schmidt number. From Fig.8, we observe that the concentration(C) decreases as chemical reaction (Kr) increases.

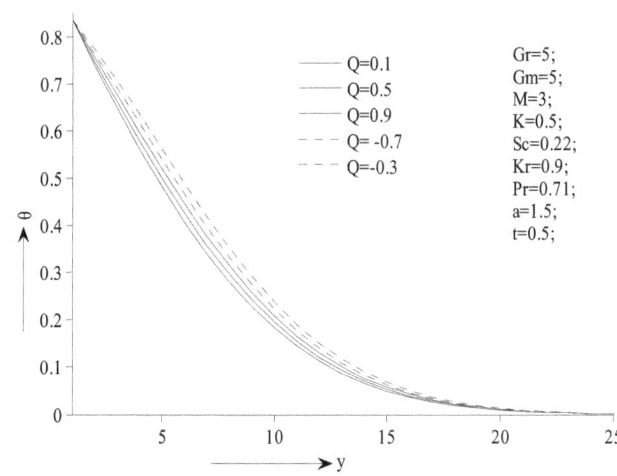

Fig. 6 Effect of Heat absorption/Generation (Q) on temperature

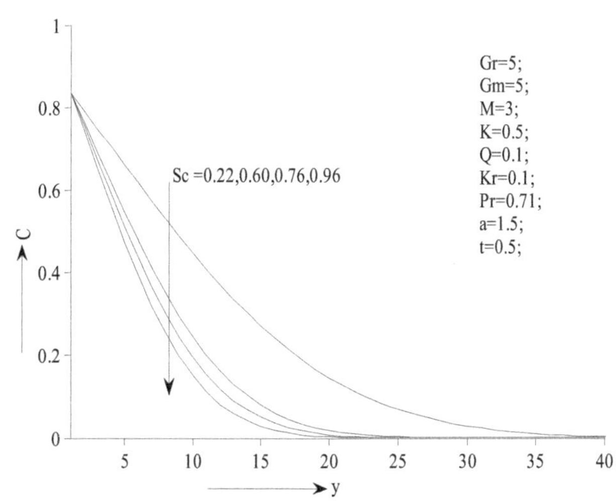

Fig. 7 Effect of Schmidt (Sc) on Concentration

The effects of some of the above parameters on skin friction, Nusselt number and Sherwood number are also analyzed with the help of graphs. The effects of Grashof number (G), modified Grashof number (Gm) and magnetic parameter (M) on skin friction are represented in figures 9-11. From these figures it is noticed that skin friction decreases as an increase in Grashof number, modified Grashof number but skin friction increases as an increase in Magnetic parameter (M). Figures 12 and 13 demonstrate the effect of Prandtl number (Pr) and heat absorption parameter (Q) on Nusselt number (Nu). It is observed that Nusselt number increase as Prandtl number and heat absorption parameter increase. Figures 14 and 15 show the effect of Schmidt number (Sc) and chemical reaction (Kr) on Sherwood number. It is noticed that Sherwood increases as increase in Schmidt number and chemical reaction.

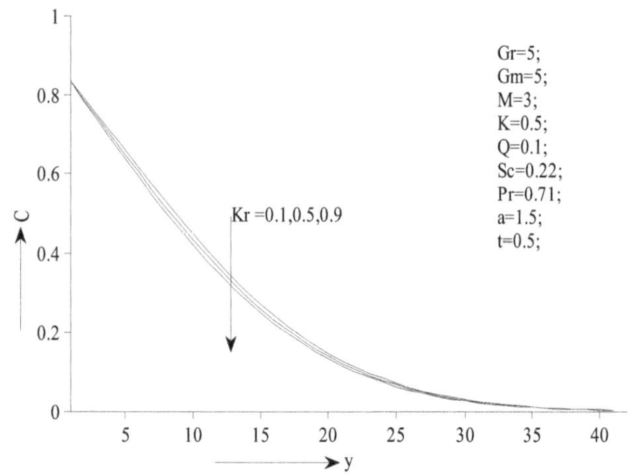

Fig. 8 Effect of Chemical reaction (Kr) on C

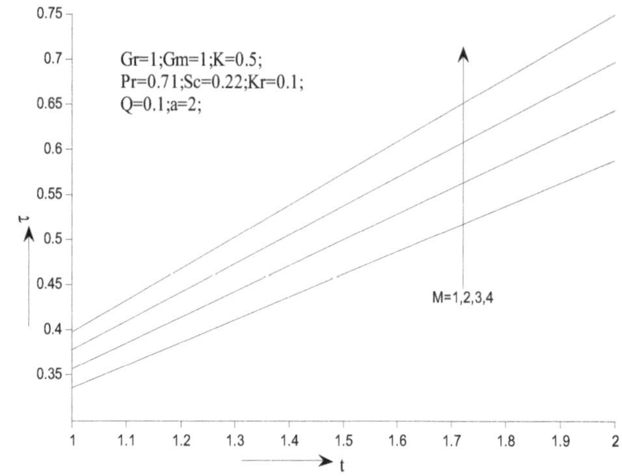

Fig. 11 Effect of Magnetic parameter (M) on Skin friction

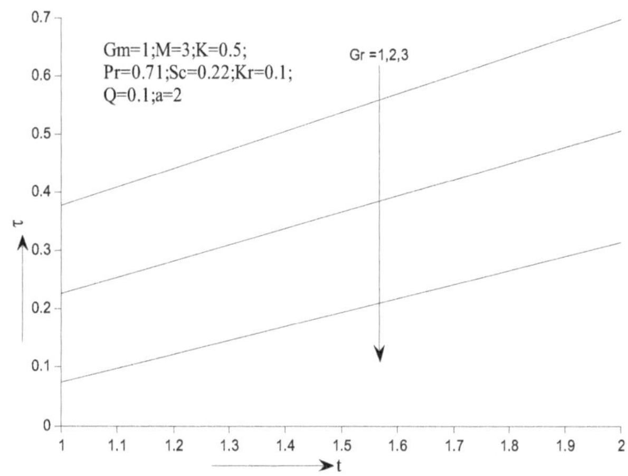

Fig. 9 Effect of Grashof number (Gr) on Skin friction

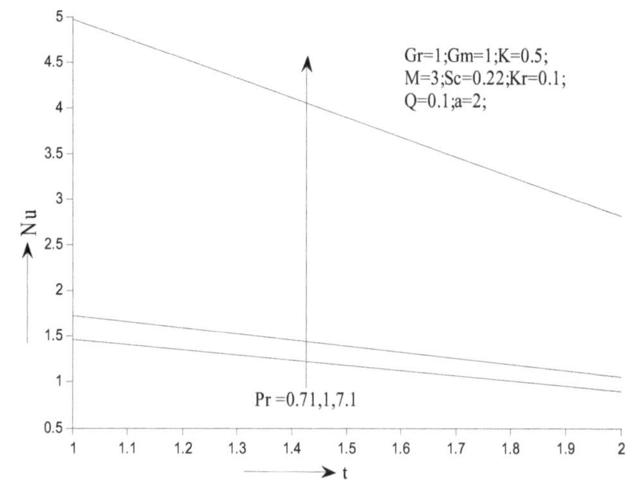

Fig. 12 Effect of Prandtl number (Pr) on Nusselt number

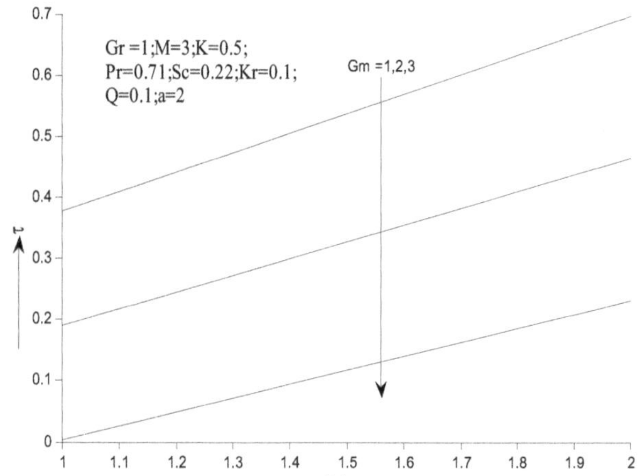

Fig. 10 Effect of modified Grashof number (Gm) on Skin friction

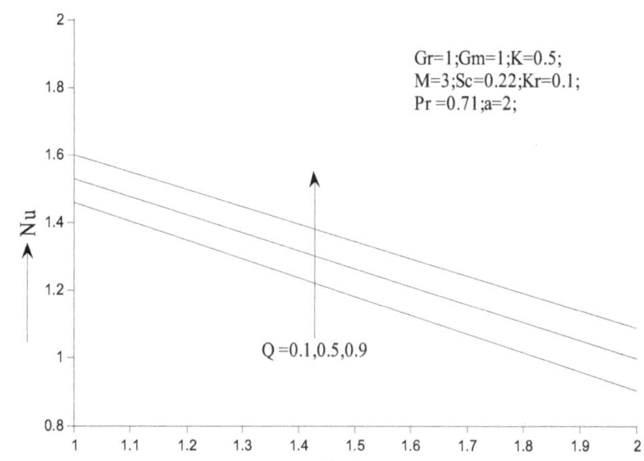

Fig. 13 Effect of heat absorption parameter (Q) on Nusselt number

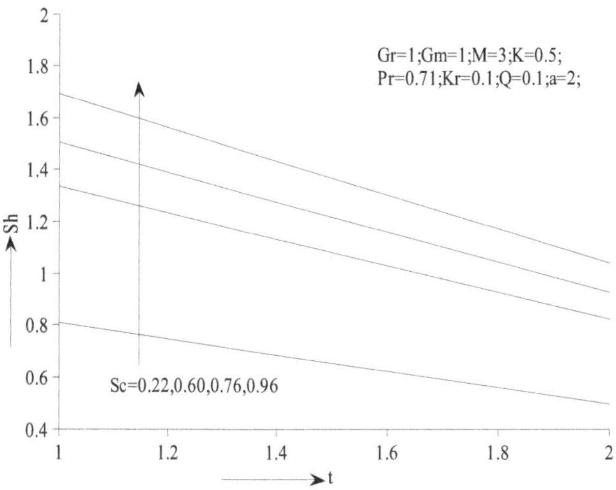

Fig. 14 Effect of Schmidt number (Sc) on Sherwood number

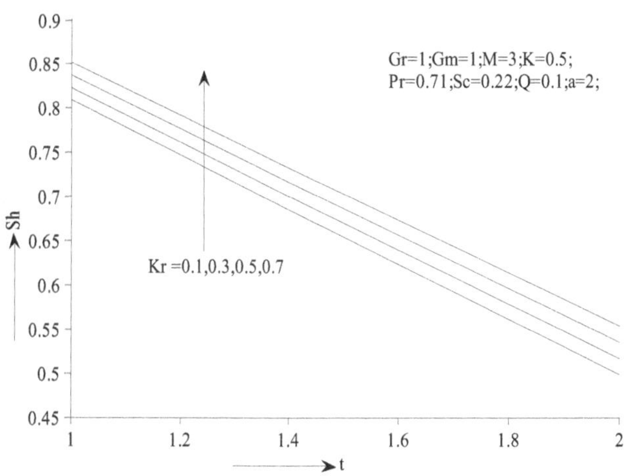

Fig. 15 Effect of Chemical reaction parameter (Kr) on Sherwood

5. CONCLUSION:

We investigated the analysis of MHD transient free convection flow of a Newtonian fluid past an infinite vertical porous plate. The governing boundary-layer equations are formulated with appropriate boundary conditions. The governing boundary layer equations are simplified and non-dimensionalized. The dimensionless equations are solved using the Finite difference method. The effects of various physical parameters which are: Grashof number (Gr), modified Grashof number (Gm), magnetic field (M), Schmidt number(Sc), Prandtl number (Pr) and chemical reaction parameter (Kr) are considered on the dimensionless velocity, temperature and concentration. Computations on the variation of local skin friction, Nusselt number and Sherwood number are also discussed through graphs. From the graphs plotted, we discover that:

1) Velocity decreases with an increase in Magnetic field while it increases with increase in Grashof number modified Grashof number, Permeability parameter.
2) Temperature increases with increase in heat absorption and it shows reverse effect in the case of heat generation and Prandtl number.
3) Concentration decreases with increase in Schmidt number and chemical reaction parameter.
4) Local skin friction decreases with increase in Grashof number, modified Grashof number where as it has reverse effect in the case of magnetic parameter.
5) Nusselt number increases with increase in Prandtl number and heat absorption parameter.
6) Sherwood number decreases with increase in Schmidt number and chemical reaction.

REFERENCES

Anjalidevi, S. P., and Kandasamy, R., 2000, "Effects of a Chemical Reaction Heat and Mass Transfer on MHD Flow Past a Semi-Infinite Plate," *Z. Angew.Math.Mech.*, **80**, 697-701.
http://dx.doi.org/10.1002/1521-4001(200010)80:10%3C697::aid-zamm697%3E3.0.co;2-f

Chamkha, A.J., 2003, "MHD Flow of a Uniformly Stretched Vertical Permeable Surface in the Presence of Heat Generation/Absorption and a Chemical Reaction," *Int. Commun. Heat Mass Transfer*, **30**(3), 413-422. http://dx.doi.org/10.1016/S0735-1933(03)00059-9

Chamkha, A.J., 2004, "Unsteady MHD Convective Heat and Mass Transfer Past A Semi-infinite Vertical Permeable Moving Plate with Heat Absorption," *Int. J. Eng. Sci.*, **42**, 217-230.
http://dx.doi.org/10.1016/S0020-7225(03)00285-4

Cortell, R., 2007, "MHD Flow And Mass Transfer Of An Electrically Conducting Fluid Of Second Grade In A Porous Medium Over A Stretching Sheet With Chemically Reactive Species," *Chemical Engineering and Processing*, **46**(8), 721–728.
http://dx.doi.org/10.1016/j.cep.2006.09.008

Hayat, T., Qasim, M., & Mesloub, S., 2011, "MHD Flow and Heat Transfer over Permeable Stretching Sheet with Slip Conditions," *International Journal for Numerical Methods in Fluids*, **66**(8), 963–975.
http://dx.doi.org/10.1002/fld.2294

Hossain, M. A., & Mandal, A. C., 1985, "Mass Transfer Effects On The Unsteady Hydromagnetic Free Convection Flow Past an Accelerated Vertical Porous Plate," *J. Phys. D: Appl. Phys.*, **18**, 63–169.
http://dx.doi.org/10.1088/0022-3727/18/7/003

Jha, B.K., 1991, "MHD Free Convection Mass Transform Flow through A Porous Medium," *Astrophysics and Space Science*, **175**(2), 283-289.
http://dx.doi.org/10.1007/BF00644290

Kandasamy, R., Muhaimin, I., and Salim, H. , 2010, "Lie Group Analysis for the Effect of Temperature – Dependent Fluid Viscosity with Thermophoresis and Chemical Reaction on MHD Free Convective Heat and Mass Transfer over a Porous Stretching Surface In The Presence Of Heat Source / Sink," *Comm. Nonlinear Sci. Num. Simul*, 15, P2109–2123.
http://dx.doi.org/10.1016/j.cnsns.2009.09.016

Kandasamy, R., Periasamy, K., and Prabhu, K. K. S., 2005, "Chemical Reaction, Heat And Mass Transfer on MHD Flow over a Vertical Stretching Surface with Heat Source and Thermal Stratification Effects," *Int. J. Heat Mass Transfer,* 48, pp.4557–4561.
http://dx.doi.org/10.1016/j.ijheatmasstransfer.2005.05.006

Kaprawi, S., 2015, "Analysis of Transient Natural Convection Flow Past an Accelerated Infinite Vertical Plate," *International Journal of Engineering Research,* **4**(2), 47-50.

Khan, W.A., Makinde, O.D., **&** Khan, Z.H.**,** 2014, "MHD Boundary Layer Flow of a Nanofluid Containing Gyro Tactic Microorganisms Past a Vertical Plate with Navier Slip," *International Journal of Heat and Mass Transfer*, 74, 285-291.
http://dx.doi.org/10.1016/j.ijheatmasstransfer.2014.03.026

Kim, J.Y., 2000, "Unsteady MHD Convective Heat Transfer Past A Semi-infinite Vertical Porous Moving Plate with Variable Suction," *Int. J. Eng. Sci.*, **38**, 833-845. http://dx.doi.org/10.1016/S0020-7225(99)00063-4

Mishra, S.R., Dash, G.C., and Acharya, M., 2013, "Mass and Heat Transfer Effect on MHD Flow of a Visco-elastic Fluid through Porous Medium with Oscillatory Suction and Heat Source," *International Journal of Heat and Mass Transfer*, **57**(2), 433-438. http://dx.doi.org/10.1016/j.ijheatmasstransfer.2012.10.053

Muthucumaraswamy, R., and Ganesan, P., 2002, "Natural Convection on a Moving Isothermal Vertical Plate with Chemical Reaction," *Journal of Engineering Physics and Thermophysics,* **75**(1), 51 – 75. http://dx.doi.org/10.1023/A:1014826924926

Muthucumaraswamy, R., Janakiraman, B., 2006, "MHD and Radiation Effects on Moving Isothermal Vertical Plate With Variable Mass Diffusion," *Journal of Theoretical and Applied Mechanics*, **33**, 17–29. http://dx.doi.org/1450-5584/2006/1450-55840601017M

Pal, D., 2013, "Hall Current and MHD Effects on Heat Transfer over an Unsteady Stretching Permeable Surface with Thermal Radiation," *Computers and Mathematics with Applications*, **66**, 1161-1180. http://dx.doi.org/10.1016/j.camwa.2013.07.010

Pal, D., Chatterjee, S. , 2010, "Heat and Mass Transfer In MHD Non-Darcian Flow of a Micro Polar Fluid over a Stretching Sheet Embedded in a Porous Media with Non-Uniform Heat Source and Thermal Radiation," *Communications in Nonlinear Science and Numerical Simulat*ion, **15**, 1843-1857. http://dx.doi.org/10.1016/j.cnsns.2009.07.024

Patil, PM., Roy, S., and Pop, I. , 2013, "Chemical Reaction Effects on Unsteady Mixed Convection Boundary Layer Flow Past a Permeable Slender Vertical Cylinder Due To A Nonlinearly Stretching Velocity," *Chemical Engineering Communications,* 200 (3), 398-417. http://dx.doi.org/10.1080/00986445.2012.712578

Prakash, J., Bhanumathi, D., Vijaya Kumar, A. G., and Varma, S. V. K., 2013, "Diffusion- Thermo And Radiation Effects On Unsteady MHD Flow Through Porous Medium Past An Impulsively Started Infinite Vertical Plate With Variable Temperature And Mass Diffusion," *Trans. Porous. Med.*, **96**, 135-151. http://dx.doi.org/10.1007/S11242-012-0078-x

Raptis, A., and Perdikis, C., 2006, "Viscous Flow over a Non-linearly Stretching Sheet In The Presence Of a Chemical Reaction and Magnetic Field," *International Journal of Non-Linear Mechanics*, **41**(4), pp.527–529. http://dx.doi.org/10.1016/j.ijnonlinmec.2005.12.003

Seini, Y.I., and Makinde, O.D., 2014, "Boundary Layer Flow Near Stagnation-points on a Vertical Surface with Slip in the Presence of Transverse Magnetic Field," *International Journal of Numerical Methods for Heat and Fluid Flow,* **24**(3), 643-653. http://dx.doi.org/10.1108/hff-04-2012-0094

Seth, G. S., and Ansari, Md. S., 2010, "MHD Natural Convection Flow Past an Impulsively Moving Vertical Plate with Ramped Wall Temperature in the Presence of Thermal Diffusion with Heat Absorption," *Int. J. Appl. Mech. Eng.*, 15(1), 199-215.

Seth, G.S. Ansari, Md. S., and Nandkeolyar, R., 2011, "MHD Natural Convection Flow with Radiative Heat Transfer Past an Impulsively Moving Plate with Ramped Wall Temperature," *Heat Mass Transfer*, **47**, 551–561. http://dx.doi.org/10.1007/S00231-010-0740-1

Seth, G. S., Hussain, S. M., and Sarkar, S., 2014a, "Hydromagnetic Natural Convection Flow With Radiative Heat Transfer Past an Acclerated Moving Vertical Plate With Ramped Temperature Through A Porous Medium," *J. Porous Media*, **17**(1), 67-79. http://dx.doi.org/10.1615/jpormedia.v17.i1.50

Seth, G. S., Hussain, S. M., and Sarkar, S., 2014b, "Effects of Hall Current and Rotation On Unsteady MHD Natural Convection Flow with Heat and Mass Transfer Past an Impulsively Moving Vertical Plate in the Presence of Radiation and Chemical Reaction," *Bulg. Chem. Comm.*, **46**(4), 704-718.

Seth, G.S., Nandkeolyar, R., and Ansari, Md. S., 2013, "Effects of Thermal Radiation and Rotation on Unsteady Hydromagnetic Free Convection Flow Past an Impulsively Moving Vertical Plate with Ramped Temperature in a Porous Medium," *J. Appl. Fluid Mech.*, **6**(1), 27-38.

Seth, G. S., Sharma, R., & Hussain S. M., 2014c, "Hall Effects on Unsteady MHD Natural Convection Flow of a Heat Absorbing Fluid Past an Accelerated Moving Vertical Plate with Ramped Temperature," *Emirates J. Eng. Res.*, 19(2), 19-32.

Singh, A.K., 2001, "MHD Free Convection and Mass Transfer Flow with Heat Source and Thermal Diffusion," *Journal of Energy Heat and Mass Transfer*, **23**, 167-178.

Sonth, R. M., Khan, S. K., Abel, M. S. , and Prasad, K. V. , 2002, "Heat and Mass Transfer in a Visco-elastic Fluid Flow over an Accelerating Surface With Heat Source/sink And Viscous Dissipation," *Heat and Mass Transfer*, **38**(3), 213–220. http://dx.doi.org/10.1007/S002310100271

Takhar, H. S, Chamkha, A. J and Nath, G., 2000, "Flow and Mass Transfer on a Stretching Sheet with a Magnetic Field and Chemically Reactive Species," Int. J. Eng. Sce. **38**, 13-16. http://dx.doi.org/10.1016/S0020-7225(99)00079-8

Umamaheswar, M., Varma, S. V. K., and Raju, M. C., 2013, "Unsteady MHD Free Convective Visco-elastic Fluid Flow Bounded by an Infinite Inclined Porous Plate in the Presence of Heat Source, Viscous Dissipation and Ohmic Heating," *International Journal of Advanced Science and Technology*, **61**, 39-52. http://dx.doi.org/10.14257/ijast.2013.61.05

Vajravelu, K., Prasad, K.V., and Chiu-On Ng, 2013, "Unsteady Convective Boundary Layer Flow of a Viscous Fluid at a Vertical Surface with Variable Fluid Properties," *Non-linear Analysis: Real World Applications,* **14**, 455-464. http://dx.doi.org/10.1016/j.nonrwa.2012.07.008

ENTROPY GENERATION IN BOUNDARY LAYER FLOW OF A MICRO POLAR FLUID OVER A STRETCHING SHEET EMBEDDED IN A HIGHLY ABSORBING MEDIUM

M. Y. Abdollahzadeh Jamalabadi*

Chabahar Maritime University, Chabahar 99717-56499, Iran

ABSTRACT

An analytical study of entropy generation in steady boundary layer flow, heat and mass transfer characteristic of 2D convective flow of a micro polar fluid over a stretching sheet embedded through a highly absorbing medium is performed. The governing equations are continuity, momentum boundary layer, micro rotation, and energy takes into account of Rosseland approximation for thermal radiation sources are solved analytically. The governing system of partial differential equations is first transformed into a system of non-linear ordinary differential equations using similarity transformation. The transformed equations are non-linear coupled differential equations which are then linearized by quasi-linearization method and solved very efficiently by the Homotopy analysis method. The special case of the first branch (K = 0, classical Newtonian fluid) is compared with the existing numerical results of stretching flow in good agreement. In addition, favorable comparisons with previously published work on various special cases of the problem are obtained. The effects of various physical parameters of entropy generation are presented graphically and in tabular form.

Keywords: *Thermal radiation, boundary layer flow, mixed convection, micro polar fluid, entropy*

1. INTRODUCTION

Most phenomena in heat and fluid flow are essentially nonlinear and are described by nonlinear equations. Heat transfer in free and mixed convection by simultaneous radiation take place in many applications such as the design of cooling systems for electronic devices (Barletta 1999), solar energy (Barletta 2002), combustion flows (Kandlikar 2005), furnaces (Zhao and Lu 2002), thermal radiative loading (Abdollahzadeh *et al.* 2012, Abdollahzadeh *et al.* 2013, Abdollahzadeh *et al.* 2014a), and many others (Abdollahzadeh *et al.* 2014b, Azizah *et al.* 2013,Shadloo *et al.* 2015). Furthermore, there is a considerable increase in the research interest in the flows of micro-polar fluids over the past few decades due to the occurrence of these fluids in industrial processes such as exotic lubricants (Khonsari 1990) the flow of colloidal suspensions or polymeric fluids and liquid crystals (Busuke 1969). The boundary layer flow over a continuously stretching surface finds many important applications in engineering processes, such as polymer extrusion and drawing of plastic films, and the applied magnetic field may play an important role in controlling momentum and heat transfers in the boundary layer flow of different fluids over a stretching sheet. Hoyt and Fabula (1964) have shown experimentally that fluids containing minute polymeric additives display a considerable reduction in the skin friction (about 30%). The theory of micro polar fluids, which display the effect of local rotary inertia and non-symmetrical stress tensor, was initially introduced and investigated by Eringen (2001), takes into account fluids consisting of randomly oriented particles suspended in a viscous medium. Extensive reviews of the theory and its applications could be found in the review article by Ariman *et al.* (1973). As well the flow induced by a stretching boundary is important in the extrusion processes in plastic and metal industries.

Recently, extensive attentions have been attracted to the flow induced by a shrinking sheet. Fang *et al.* (2011) investigated the steady momentum and heat transfer of a viscous fluid flow over a linearly stretching/shrinking sheet and then a convective boundary condition were further studied analytically. There are some analytic techniques for nonlinear problems, such as perturbation techniques (essentially based on the existence of small or large parameters) that are well known and widely applied. One of the semi-exact methods to solve PDE equations, called the Homotopy Analysis Method (HAM), first proposed by Liao in 1992 in his Ph.D thesis (Liao 1992,Liao 2003). Neither perturbation techniques, nor non-perturbation methods such as the artificial small parameter methods, the δ-expansion method, and Adomian's decomposition method can provide us with a convenient way to adjust and control convergence region and rate of approximation series. This method has already been applied successfully to solve many problems in fluid mechanics (Yun 2013, Rajeev 2014, Yan 2013). In this method, it is possible to adjust and control the convergent region, and this is the most important feature of this technique in comparison to other techniques. In all the previous investigations, the effects of radiation on the flow and heat transfer have not been provided analytically. In the last decade, many researchers have studied the entropy generation in fluid flow and heat transfer over surfaces.

Aiboud and Saouli (2010) presented the application of second law analysis of thermodynamics to viscoelastic flow over a stretching surface. They analytically obtained the velocity and temperature profiles using the Kummer's functions and computed the entropy generation number. Makinde (2011) analyzed the inherent irreversibility in the boundary layer flow of variable viscosity fluid over a semi-infinite flat plate under the influence of thermal radiation. Using local similarity solution technique and shooting quadrature, he numerically obtained the velocity, temperature and entropy generation number. Dehsara *et al.* (2012) numerically analyzed entropy generation for the mixed convection flow over a nonlinear stretching inclined transparent plate embedded in a porous medium due to solar radiation. Butt *et al.* (2012a)

* Email: my.abdollahzadeh@gnu.ac.kr

discussed the boundary layer flow and heat transfer analysis of a second grade fluid over a stretching sheet through a porous medium and investigated the effect of viscoelasticity on entropy generation using the Homotopy analysis method (HAM). Butt *et al.* (2012b) reported the effects of velocity slip on entropy generation in the boundary layer flow over a vertical surface with convective boundary condition. They numerically solved the governing equations using the shooting method and presented expressions for the entropy generation number and Bejan number. Malvandi *et al.* (2012) analytically studied the steady two-dimensional boundary layer flow over an isothermal flat plate by homotopy perturbation method (HPM) and analyzed the entropy generation inside the boundary layer. Galanis and Rashidi (2012) studied the entropy generation in non-Newtonian fluids due to heat transfer in entrance region of ducts. Butt *et al.* (2012c) made an investigation on entropy generation within a steady laminar mixed convective flow of an over a stretching sheet. They solved the governing equations using the homotopy analysis method and obtained expressions for local entropy generation number, Bejan number, and average Bejan number. Butt and Ali (2013) analyzed the effects of entropy generation in flow over a permeable stretching sheet embedded in a porous medium. They obtained analytical solutions of momentum and energy equations in terms of Kummer's functions and computed the entropy generation number and Bejan number. Noghrehabadi *et al.* (2013) analyzed the boundary layer heat transfer and entropy generation over an isothermal linearly stretching sheet with heat generation/absorption. They took into account the development of concentration gradient due to slip and the effects of Brownian motion and thermophoresis. Dehsara *et al.* (2014) investigated the entropy generation of mixed convection flow over a nonlinear stretching inclined transparent plate embedded in a porous medium. Using a numerical algorithm, they solved the governing equations, in the presence of the effects of viscous dissipation, variable magnetic field and solar radiation. More literature survey makes it clear that the entropy generation has not been investigated for the flow and heat transfer of a micropolar fluid over a stretching surface.

Motivated by the works mentioned above, in this paper the HAM is used to find an analytical solution to the velocity and temperature fields with the radiation effect in stretching sheet immersed in micro polar fluid. The purpose of this attempt is to analyze the entropy generation of micropolar flow over a flat plate. Therefore, the local similarity solutions will be found and then the series solution will be first computed using HAM and then its convergence will be discussed in detail. Afterward, the results for the entropy generation profiles will be calculated and discussed from the physical point of view. The entropy generation is calculated using the entropy relation by substituting the velocity and temperature fields obtained from the momentum and entropy equations.

2. GOVERNING EQUATIONS AND NUMERICAL METHOD

Consider a steady laminar two-dimensional mixed convection thermal boundary layer flow of an incompressible, electrically conducting, viscous, and incompressible micro polar fluid towards a stretching surface coinciding with the plane y=0, the flow region y>0. The origin (x=0) fixed as shown in Fig. 1. The x -axis is taken in the direction along the sheet and the y -axis is taken perpendicular to it. With these assumptions the governing equations are given by (Nazar 2004) as follow:

The equation of continuity

$$\frac{\partial u}{\partial x}+\frac{\partial \upsilon}{\partial y}=0, \tag{1}$$

The equation of momentum

$$u\frac{\partial u}{\partial x}+\upsilon\frac{\partial u}{\partial y}=\left(\nu+\frac{\kappa}{\rho}\right)\frac{\partial^2 u}{\partial y^2}+\frac{\kappa}{\rho}\frac{\partial N}{\partial y}, \tag{2}$$

The equation of angular momentum

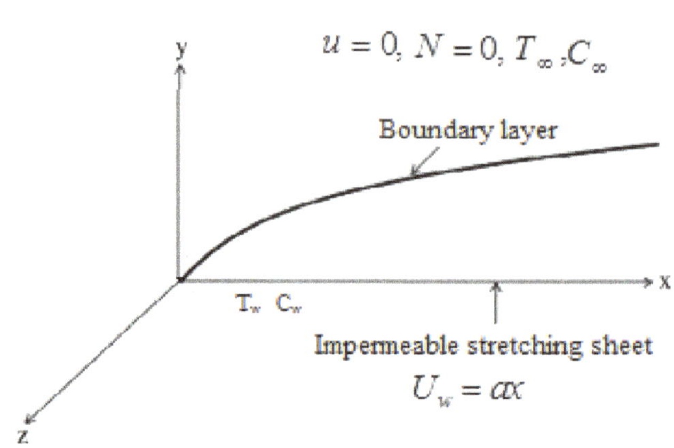

Fig. 1 Schematic of the boundary layer induced by stretching sheet.

$$u\frac{\partial N}{\partial x}+\upsilon\frac{\partial N}{\partial y}=\frac{\gamma}{\rho j}\frac{\partial^2 N}{\partial y^2}-\frac{\kappa}{\rho j}\left(2N+\frac{\partial u}{\partial y}\right), \tag{3}$$

The equation of energy

$$u\frac{\partial T}{\partial x}+\upsilon\frac{\partial T}{\partial y}=\frac{k}{\rho c_P}\frac{\partial^2 T}{\partial y^2}-\frac{1}{\rho c_P}\frac{\partial q_r}{\partial y}, \tag{4}$$

The equation of mass diffusion

$$u\frac{\partial C}{\partial x}+\upsilon\frac{\partial C}{\partial y}=D\frac{\partial^2 C}{\partial y^2}, \tag{5}$$

where u and v are the velocity components along the x and y directions, ρ is the density of the liquid, T is the temperature of the fluid, C_p is the specific heat at constant pressure, ν is the kinematic viscosity, σ is the electrical conductivity of the fluid, N is the components of micro-rotation or angular velocity whose rotation is in the direction of the $x-y$ plane and j,γ and κ are the micro inertia per unit mass, spin gradient viscosity and vortex viscosity, respectively. Furthermore, the spin gradient viscosity γ, which defines the relationship between the coefficient of viscosity and micro inertia as follows (Ahmadi 1976):

$$\gamma=(\mu+\kappa/2)j=\mu(1+K/2)j, \tag{6}$$

where μ is the dynamic viscosity and $K=\kappa/\mu$ is the dimensionless viscosity ratio and is called the material parameter. It is worth mentioning here that relation (6) is invoked to allow equations (1)–(4) to predict the correct behavior in the limiting case when microstructure effects become negligible, and the micro rotation, N, reduces to the angular velocity (Yücel 1989). It is also worth mentioning that the case K = 0 describes the classical Navier–Stokes equations for a viscous and incompressible Newtonian fluid. The flow is generated by the action of two equal and opposite forces along the x -axis and the sheet is stretched in such a way that the velocity at any instant is proportional to the distance from the origin (x-component of the velocity varies linearly along it).

$$u(y=0)=U_w=\alpha x \tag{7}$$

Furthermore sheet kept at constant temperature T_w.

$$T(y=0)=T_w \tag{8}$$

which is important parameter throughout the boundary layer (Hadjadj *et al.* 2015). The other physical boundary conditions for the problem under study are given by:

$$\upsilon(y=0)=0, \tag{9}$$

$$N(y=0)=-m\left.\frac{\partial u}{\partial y}\right|_{y=0} \quad (10)$$

$$C(y=0)=C_w, \quad (11)$$

$$u(y=\infty)=0, \quad (12)$$

$$N(y=\infty)=0, \quad (13)$$

$$T(y=\infty)=T_\infty, \quad (14)$$

$$C(y=\infty)=C_\infty, \quad (15)$$

where T_w is the wall temperature of the fluid and T_∞ is the temperature of the fluid far away from the sheet, C_w is the wall concentration of the solute and C_∞ is the concentration of the solute far away from the sheet and m is the boundary parameter. It should be mentioned that m is a constant such that $0 \leqslant m \leqslant 1$. The case when m=0, is called strong concentration, which indicates N=0 near the wall characterizes concentrated particle flows in which the micro-elements close to the wall surface are unable to rotate. The case when m=1/2 indicates the disappearing of anti-symmetric part of the stress tensor and means weak concentrations whereas m=1 is used for modelling of turbulent boundary layer flows. This assumption is entreated to permit the field of equations to predict the correct behavior in the limiting case when the microstructure effects become negligible, and the micro rotation *N*, reduces to the angular velocity. Following Rosseland approximation (Brewster 1992) with radiative heat flux q_r in equation (4) is modeled as:

$$q_r=-\frac{4\sigma^*}{3k^*}\frac{\partial T^4}{\partial y}, \quad (16)$$

where σ^* and k^* are the Stephan–Boltzman constant and the mean absorption coefficient, respectively. As the differences within the flow are such that T^4 can be expressed as a linear function of temperature, $T^4=T_\infty^4+4T_\infty^3(T-T_\infty)+...$, expanding T^4 in a Taylor series about T_∞ and neglecting higher order terms thus,

$$T^4 \approx 4T_\infty^3 T-3T_\infty^4. \quad (17)$$

By substituting equations (16) and (17) in equation (4), one obtains:

$$u\frac{\partial T}{\partial x}+\upsilon\frac{\partial T}{\partial y}=\alpha(1+N_R)\frac{\partial^2 T}{\partial y^2}, \quad (18)$$

where $\alpha=k/(\rho c_P)$ is the thermal diffusivity, $N_R=16\sigma^* T_\infty^3/(3kk^*)$ is the radiation parameter (Datti *et al.* 2004), and k is the thermal conductivity. Introducing the stream function $\psi(x,y)$ such that:

$$u=\frac{\partial\psi}{\partial y}, \quad (19)$$

$$\upsilon=-\frac{\partial\psi}{\partial x}, \quad (20)$$

The governing Eqs. (2) and (3) admit a self-similar solution of the form:

$$\psi(x,y)=(\nu x U_w)^{0.5} f(\eta), \quad (21)$$

where f is the dimensionless stream function and η is the similarity variable:

$$\eta=(\frac{U_w}{\nu x})^{0.5} y, \quad (22)$$

Note that the definitions of ψ, u and v in the above expressions satisfy the continuity equation (1) identically. Also for micro rotation:

$$N=U_w(\frac{U_w}{\nu x})^{0.5} h(\eta), \quad (23)$$

The thermal boundary conditions for solving Eq. (18) depend on the type of heating process considered. Now the non-dimensional temperature and concentration are defined as

$$\theta(\eta)=\frac{T-T_\infty}{T_w-T_\infty}. \quad (24)$$

$$\phi(\eta)=\frac{C-C_\infty}{C_w-C_\infty}. \quad (25)$$

Substituting equations (21) through (25) into the equations (2), (3), (5) and (18), the resulting non-linear third-order ordinary differential equations are:

$$(1+K)f'''+ff''-f'^2+Kh'=0, \quad (26)$$

$$(1+\frac{K}{2})h''+fh'-f'h-K(2h+f'')=0, \quad (27)$$

$$\frac{1}{\Pr}(1+N_R)\theta''+f\theta'-2f'\theta=0, \quad (28)$$

$$\phi''+Sc(\phi' f-2\phi f')=0, \quad (29)$$

where the prime denotes the derivative with respect to the similarity variable (defined in Eq . 22).The Equations (26)-(29) are subject to the following reduced boundary conditions at the wall:

$$f(0)=0, \quad (30)$$

$$f'(0)=1, \quad (31)$$

$$h(0)=-mf''(0), \quad (32)$$

$$\theta(0)=1, \quad (33)$$

$$\phi(0)=1, \quad (34)$$

and at the free stream :

$$f'(\eta=\infty)=0, \quad (35)$$

$$h(\eta=\infty)=0, \quad (36)$$

$$\theta(\eta=\infty)=0, \quad (37)$$

$$\phi(\eta=\infty)=0, \quad (38)$$

In the above equations, primes denote differentiation with respect to η and *Pr* is the dimensionless Prandtl number and is equal to:

$$\Pr=\frac{\mu C_p}{k_\infty}. \quad (39)$$

and also the Sc is the dimensionless Schmidt number is equal to:

$$Sc=\frac{\nu}{D}. \quad (40)$$

It is noted that *K=0* is corresponded to viscous fluid case, which is studied before without consideration of the thermal radiation effect

(Grubka and Bobba 1985). Furthermore It is worth mentioning that when m = 1/2, one can take g(η)=-1/2f'(η), so combining Eqs. (26) and (27) to reduce to a single non-linear ordinary differential equation as

$$(1+K/2)f'''+ff''-f'^2=0, \tag{41}$$

subject to the appropriate boundary conditions

$$f(0)=0, \tag{42}$$

$$f'(0)=1, \tag{43}$$

$$f'(\eta=\infty)=0, \tag{44}$$

The physical quantities of most interest are the skin friction coefficient C_f and the local Nusselt number Nu_x, which are defined respectively as:

$$C_f=\frac{\tau_w}{\rho U_w^2}, \tag{45}$$

$$Nu_x=\frac{xq_w}{k(T_w-T_\infty)}, \tag{46}$$

where the wall shear stress τ_w and the surface heat flux q_w for micro polar boundary layer flow are given by:

$$\tau_w=\left[(\mu+\kappa)\frac{\partial u}{\partial y}+\kappa N\right]_{y=0}, \tag{47}$$

$$q_w=-k(\frac{\partial T}{\partial y})_{y=0}, \tag{48}$$

Substituting (26) and (27) into (45) through (48), the following obtain:

$$C_f\,\mathrm{Re}_x^{0.5}=(1+K)f''(0)+Kh(0), \tag{49}$$

$$Nu_x/\mathrm{Re}_x^{0.5}=-\theta'(0), \tag{50}$$

where $\mathrm{Re}_x=U_w x/\nu$ is the local Reynolds number.

3. APPLICATION OF HOMOTOPY ANALYSIS METHOD

The governing equations for heat transfer of continuous stretching sheet immersed in a micro polar fluid are expressed by equations (26) through (29). For brevity, the complete operators are defined as follows:

$$N_f[f(\eta;q)]=(1+K)\frac{\partial^3 f(\eta;q)}{\partial\eta^3}+ K\frac{\partial h(\eta;q)}{\partial\eta}-\left(\frac{\partial f(\eta;q)}{\partial\eta}\right)^2+f(\eta;q)\frac{\partial^2 f(\eta;q)}{\partial\eta^2}, \tag{51}$$

$$N_h[h(\eta;q)]=(1+\frac{K}{2})\frac{\partial^2 h(\eta;q)}{\partial\eta^2}-K(\frac{\partial^2 f(\eta;q)}{\partial\eta^2}+2h(\eta;q)) -\frac{\partial f(\eta;q)}{\partial\eta}h(\eta;q)+f(\eta;q)\frac{\partial h(\eta;q)}{\partial\eta}, \tag{52}$$

$$N_\theta[\theta(\eta;q)]=\frac{1}{\mathrm{Pr}}(1+N_R)\frac{\partial^2\theta(\eta;q)}{\partial\eta^2} +f(\eta;q)\frac{\partial\theta(\eta;q)}{\partial\eta}, \tag{53}$$

$$N_\phi[\phi(\eta;q)]=\frac{1}{Sc}\frac{\partial^2\phi(\eta;q)}{\partial\eta^2} +f(\eta;q)\frac{\partial\phi(\eta;q)}{\partial\eta}-2\phi(\eta;q)\frac{\partial f(\eta;q)}{\partial\eta}, \tag{54}$$

where $q\in[0,1]$ is the embedding parameter. As the embedding parameter increases from *0* to *1*, the parameters expressed by $U(\eta;q)$, $V(\eta;q)$, $Y(\eta;q)$ and $Z(\eta;q)$ vary from the initial guess, $U_0(\eta)$, $V_0(\eta)$ $Y_0(\eta)$, and $Z_0(\eta)$, to the exact solution, $U(\eta)$, $V(\eta)$, $Y(\eta)$ and $Z(\eta)$. For q = 0

$$f(\eta;0)=U_0(\eta), \tag{55}$$

$$h(\eta;0)=V_0(\eta), \tag{56}$$

$$\theta(\eta;0)=Y_0(\eta), \tag{57}$$

$$\phi(\eta;0)=Z_0(\eta), \tag{58}$$

and at q = 1

$$f(\eta;1)=U(\eta), \tag{59}$$

$$h(\eta;1)=V(\eta), \tag{60}$$

$$\theta(\eta;1)=Y(\eta), \tag{61}$$

$$\phi(\eta;1)=Z(\eta). \tag{62}$$

As the embedding parameter q increases from 0 to 1, functions varies from the initial guess to the exact solution. Expanding $f(\eta;q)$, $h(\eta;q)$, $\theta(\eta;q)$ and $\phi(\eta;q)$ in a power series of q in by Taylor's theorem; Similar to the Blasius which provided a solution in power series, using Taylor series with respect to q results in:

$$f(\eta;q)=U_0(\eta)+\sum_{m=1}^{\infty}U_m(\eta)q^m, \tag{63}$$

$$h(\eta;q)=V_0(\eta)+\sum_{m=1}^{\infty}V_m(\eta)q^m, \tag{64}$$

$$\theta(\eta;q)=Y_0(\eta)+\sum_{m=1}^{\infty}Y_m(\eta)q^m, \tag{65}$$

$$\phi(\eta;q)=Z_0(\eta)+\sum_{m=1}^{\infty}Z_m(\eta)q^m, \tag{66}$$

where,

$$U_m(\eta)=\frac{1}{m!}\frac{\partial^m f(\eta;q)}{\partial q^m}\bigg|_{q=0}, \tag{67}$$

$$V_m(\eta)=\frac{1}{m!}\frac{\partial^m h(\eta;q)}{\partial q^m}\bigg|_{q=0}, \tag{68}$$

$$Y_m(\eta)=\frac{1}{m!}\frac{\partial^m \theta(\eta;q)}{\partial q^m}\bigg|_{q=0}, \tag{69}$$

$$Z_m(\eta)=\frac{1}{m!}\frac{\partial^m \phi(\eta;q)}{\partial q^m}\bigg|_{q=0}. \tag{70}$$

Note that the zero-order deformation equations (63-66) contain the auxiliary parameter and the auxiliary function. Assuming that both auxiliary parameter and the auxiliary function are properly chosen so that the series is convergent at q = 1.

It is known from Liao (2003) that if the solution series (63-66) converges, where the functions are governed by Equations (26-29) under the definitions (67-69), it must be the solution of Equations (26-29). Therefore According to Liao (2003), it is needed only to focus on the convergence of the solution series by properly choosing of q. Homotopy analysis method can be expressed by many different base functions (Liao

1992), according to the governing equations; it is straightforward to use a base function in the form of:

$$U(\eta)=\sum_{m=1}^{\infty}\sum_{p=1}^{\infty} b_{pm}\eta^{p}e^{-m\eta}, \quad (71)$$

$$V(\eta)=\sum_{m=1}^{\infty}\sum_{p=1}^{\infty} k_{pm}\eta^{p}e^{-m\eta}, \quad (72)$$

$$Y(\eta)=\sum_{m=1}^{\infty}\sum_{p=1}^{\infty} d_{pm}\eta^{p}e^{-m\eta}, \quad (73)$$

$$Z(\eta)=\sum_{m=1}^{\infty}\sum_{p=1}^{\infty} c_{pm}\eta^{p}e^{-m\eta}, \quad (74)$$

where b_{pm}, k_{pm}, d_{pm} and c_{pm} are the coefficients to be determined, and m is the so-called spatial-scale parameter. When the base function is selected, the auxiliary functions $H_f(\eta)$, $H_h(\eta)$, $H_\theta(\eta)$, and $H_\varphi(\eta)$, initial approximations $U_0(\eta)$, $V_0(\eta)$, $Y_0(\eta)$ and $Z_0(\eta)$ and the auxiliary linear operators L_f, L_h, L_θ and L_φ must be chosen in such a way that the corresponding high-order deformation equations have solutions with the functional form similar to the base functions. It is worth mentioning that the presence of expressions such as $\eta \sin(m\eta)$ prevents the convergence of the analytical solution. This method is referred to as the rule of solution expression (Liao 2003). In the first step the zeroth-order solutions deformation of equations of (26)–(29) are estimated by using HAM. Although the solution series (63-69) given by power functions may be valid in the whole region $\eta\in[0,+\infty)$, it is still an analytic-numerical solution, because the initial condition had to be given by numerical techniques. Their convergence regions are dependent on q, and when η is small, a large number of terms are needed to gain an accurate approximation for a large η. Therefore, they are not efficient solution expressions of Equations (26-29). This is mainly because the base functions defined by (63-69) do not automatically satisfy the boundary conditions at infinity. Choosing exponential function as the base functions to express, the auxiliary linear operators L_f, L_h, L_θ and $L\varphi$ are chosen through:

$$L_f[f(\eta;q)]=\frac{\partial^3 f(\eta;q)}{\partial\eta^3}+\frac{\partial f(\eta;q)}{\partial\eta}, \quad (75)$$

$$L_h[h(\eta;q)]=\frac{\partial^2 h(\eta;q)}{\partial\eta^2}+\frac{\partial h(\eta;q)}{\partial\eta}, \quad (76)$$

$$L_\theta[\theta(\eta;q)]=\frac{\partial^2 \theta(\eta;q)}{\partial\eta^2}+\frac{\partial \theta(\eta;q)}{\partial\eta}. \quad (77)$$

$$L_\phi[\phi(\eta;q)]=\frac{\partial^2 \phi(\eta;q)}{\partial\eta^2}+\frac{\partial \phi(\eta;q)}{\partial\eta}. \quad (78)$$

According to the rule of solution expression, Equations corresponding to the auxiliary linear operators in Equations (75) through (78) result in the following properties:

$$L_f[c_1+c_2e^{\eta}+c_3e^{-\eta}]=0, \quad (79)$$

$$L_h[c_4+c_5e^{-\eta}]=0, \quad (80)$$

$$L_\theta[c_6+c_7e^{-\eta}]=0, \quad (81)$$

$$L_\phi[c_8+c_9e^{-\eta}]=0, \quad (82)$$

According to the rule of solution expression and the initial conditions, the initial approximations U_0, V_0, Y_0 and Z_0, as well as the integral constants, c_1 through c_9, are expressed as:

$$U_0(\eta)=c_1+c_2e^{\eta}+c_3e^{-\eta}, c_1=0, c_2=0, c_3=\eta, \quad (83)$$

$$V_0(\eta)=c_4+c_5e^{-\eta},\ c_4=0, c_5=2m, \quad (84)$$

$$Y_0(\eta)=c_6+c_7e^{-\eta},\ c_6=0, c_7=1,, \quad (85)$$

$$Z_0(\eta)=c_8+c_9e^{-\eta},\ c_8=0, c_9=1, \quad (86)$$

Let us denote a nonzero auxiliary parameter and H(η) a nonzero auxiliary function. Then one can construct the so-called zero-order deformation equation for any constants C_1- C_9.The zero[th] order deformation equation and their boundary condition for $f(\eta)$, $h(\eta)$, $\theta(\eta)$, and $\varphi(\eta)$ are constructed as the following equations:

$$(1-q)L_f[f(\eta;q)-U_0(\eta)]=q\hbar_f H_f(\eta)N_f[f(\eta;q)], \quad (87)$$

$$(1-q)L_h[h(\eta;q)-V_0(\eta)]=q\hbar_h H_h(\eta)N_h[h(\eta;q)], \quad (88)$$

$$(1-q)L_\theta[\theta(\eta;q)-Y_0(\eta)]=q\hbar_\theta H_\theta(\eta)N_\theta[\theta(\eta;q)]. \quad (89)$$

$$(1-q)L_\phi[\phi(\eta;q)-Z_0(\eta)]=q\hbar_\phi H_\phi(\eta)N_\phi[\phi(\eta;q)]. \quad (90)$$

subject to the boundary conditions similar to (30)-(38).Those must be the exact solution of Equations (26-29) as long as it is convergent. Note that the solution (87-90) contains three auxiliary parameters q, η, and $\hbar$. It should be emphasized that, for any given values of q and η, we still have the freedom to choose a proper value of the auxiliary parameter $\hbar$ to control and adjust the convergence region and rate of the solution (87-90), when necessary. According to the rule of solution expression and from equations (87) through (90), the auxiliary functions $H_f(\eta)$, $H_h(\eta)$), $H_\theta(\eta)$ and $H_\varphi(\eta)$can be chosen as follows:

$$H_f(\eta)=\eta^{p}e^{-m\eta}, \quad (91)$$

$$H_h(\eta)=\eta^{p}e^{-m\eta}, \quad (92)$$

$$H_\theta(\eta)=\eta^{p}e^{-m\eta}. \quad (93)$$

$$H_\phi(\eta)=\eta^{p}e^{-m\eta}. \quad (94)$$

the nth-order deformation equations is obtained by differentiating equations (87) through (90), m times, with respect to the embedding parameter q and then setting $q=0$ in the final expression and dividing it by $m!$, are reduced to:

$$\begin{aligned}U_m(\eta)&=\chi_m U_{m-1}(\eta)+c_1+c_2e^{q}+c_3e^{-\eta}\\ &\hbar\left(\frac{1}{2}\int_0^{\eta} \begin{matrix} H_f(\eta)R_m(U_{m-1})d\eta- \\ \frac{1}{2}e^{-q}\int_0^{\eta}H_f(\eta)e^{q}R_m(U_{m-1})d\eta\end{matrix}\right.,\end{aligned} \quad (95)$$

$$\begin{aligned}V_m(\eta)&=\chi_m V_{m-1}(\eta)+\\ &\hbar\int_0^{\eta}\int_0^{\mu}H_\varphi(\eta)e^{\eta}R_m(V_{m-1})d\eta\, d\mu+c_4+c_5e^{-\eta},\end{aligned} \quad (96)$$

$$\begin{aligned}Y_m(\eta)&=\chi_m Y_{m-1}(\eta)+\\ &\hbar\int_0^{\eta}\int_0^{\mu}H_\theta(\eta)e^{\eta}R_m(Y_{m-1})d\eta\, d\mu+c_6+c_7e^{-\eta}.\end{aligned} \quad (97)$$

$$\begin{aligned}Z_m(\eta)&=\chi_m Z_{m-1}(\eta)+\\ &\hbar\int_0^{\eta}\int_0^{\mu}H\phi(\eta)e^{\eta}R_m(Z_{m-1})d\eta\, d\mu+c_8+c_9e^{-\eta}.\end{aligned} \quad (98)$$

Equations (95)-(98) is the mth order deformation equation for $f(\eta)$, $h(\eta)$, and $\theta(\eta)$, where

$$\chi_m=\begin{cases}0, & m\le 1\\ 1, & m>1\end{cases}. \quad (99)$$

and :

$$\begin{aligned}R_m(U_{m-1})&=(1+K)\frac{d^3U_{m-1}(\eta)}{d\eta^3}-K\frac{dV_{m-1}(\eta)}{d\eta}\\ &+\sum_{z=0}^{m-1}\frac{dU_z(\eta)}{d\eta}\frac{dU_{m-1-z}(\eta)}{d\eta}-\sum_{z=0}^{m-1}U_z(\eta)\frac{d^2U_{m-1-z}(\eta)}{d\eta^2},\end{aligned} \quad (100)$$

$$\begin{aligned}R_m(V_{m-1})&=(1+\frac{K}{2})\frac{d^2V_{m-1}(\eta)}{d\eta^2}+K(\frac{d^2U_{m-1}(\eta)}{d\eta^2}+2V_{m-1}(\eta))\\ &-\sum_{z=0}^{m-1}V_z(\eta)\frac{dU_{m-1-z}(\eta)}{d\eta}+\sum_{z=0}^{m-1}U_z(\eta)\frac{dV_{m-1-z}(\eta)}{d\eta}),\end{aligned} \quad (101)$$

$$R_m(Y_{m-1}) = \frac{1}{\Pr}(1+N_R)\frac{d^2Y_{m-1}(\eta)}{d\eta^2} - \sum_{z=0}^{m-1} U_z(\eta)\frac{dY_{m-1-z}(\eta)}{d\eta} + 2\sum_{z=0}^{m-1} Y_z(\eta)\frac{dU_{m-1-z}(\eta)}{d\eta}, \tag{102}$$

$$R_m(Z_{m-1}) = \frac{1}{Sc}\frac{d^2Z_{m-1}(\eta)}{d\eta^2} - \sum_{z=0}^{m-1} U_z(\eta)\frac{dZ_{m-1-z}(\eta)}{d\eta} + 2\sum_{z=0}^{m-1} Z_z(\eta)\frac{dU_{m-1-z}(\eta)}{d\eta}, \tag{103}$$

The rate of convergence can be increased when suitable values are selected for *m* and *p*. According to the rule of solution expression, the suitable values for *m* and *p* are {*p=0, m=1*}. Consequently, the corresponding auxiliary functions were determined as $H_f(\eta) = H_h(\eta) = H_\theta(\eta) = H_\varphi(\eta) = e^{-\eta}$. As a result of this selection, the first and second terms of the solution's series are as follows:

$$U_0(\eta) = \eta e^{-\eta}, \tag{103}$$

$$U_1(\eta) = \frac{h}{12}(9K - 12Km - 9\eta e^{-\eta} - 5e^{-\eta} + 3e^{-\eta}\eta^2 - 9e^{-\eta}K\eta + 3e^{-\eta}K\eta^2 + \cdots) \tag{104}$$

$$Y_0(\eta) = e^{-\eta}, \tag{105}$$

$$Y_1(\eta) = \frac{1}{\Pr}0.0104166he^{-\eta}(-32e^{-\eta} - 32N_Re^{-\eta} + 12e^{-2\eta}\eta\Pr + 32 + 32N_R + \cdots) \tag{106}$$

$$V_0(\eta) = 2me^{-\eta}, \tag{107}$$

$$V_1(\eta) = -\frac{1}{36}he^{-\eta}(-18m - 81Km + 8K - 24me^{-\eta} + 36Kme^{-\eta} + 12Ke^{-\eta}\eta + \cdots) \tag{108}$$

$$Z_0(\eta) = e^{-\eta}, \tag{109}$$

$$Z_1(\eta) = \frac{1}{Sc}0.0104166he^{-\eta}(12e^{-2\eta}\eta Sc + \cdots) \tag{110}$$

The above Equations are the semi-analytical results of the homotopy analysis method (HAM) to the nonlinear coupled partial differential equations of a boundary layer flow of a micro polar fluid over a stretching sheet embedded in a highly absorbing medium system. Here the concept of the homotopy from topology is employed to generate a convergent series solution for nonlinear systems. This was enabled by utilizing a homotopy-Mclaurin series to deal with the nonlinearities in the system.

4. ACCURACY AND CONVERGENCE

The system of coupled non-linear ordinary differential equations (26–29) together with the boundary conditions (30-38) is solved by converting it to an initial value problem. The analytical solution should converge. As pointed out by Liao (2003), the convergence and rate of approximation for the HAM solutions strongly depend on the value of auxiliary parameter h. For simplicity, first the convergence of that is dependent on $\hbar$ and q is considered. By means of the so-called h-curve, it is straightforward to choose a proper value of h which ensures that the solution series is convergent. It should be noted that the auxiliary parameter $\hbar$, as pointed out by Liao (2003) controls the convergence and accuracy of the solution series. In order to define a region such that the solution series is independent on $\hbar$, $\hbar$-curves are plotted as shown in Figure 1. The influence of _ on the convergence of *f*(0) can be investigated by plotting the so-called h-curves of *f*(0), as shown in Figure 1.The region where the distribution of f, f', f'', h, h', θ, θ' versus $\hbar$ is a horizontal line is known as the convergence region for the corresponding function. It is seen that the interval between -0.2 and -0.9 can be chosen as the convergence region. Since the interval for the admissible values of hg corresponds to the line segments nearly parallel to the horizontal axis. For example, it was known that the admissible values for the parameter hg are -1.7 < h< -0.5 in the first branch of solution for the stretching case from Fig. 1. In this paper $\hbar = -1.5$. According to Liao (2003), it is needed only to focus on correctly choosing the auxiliary parameter and the spatial-scale parameter so that the solution is convergent. The homotopy-Pad´e technique is applied to accelerate the convergence of f'(0). For given q and η, by means of plotting the corresponding h-curves of $f(0)$, a negative $\hbar$ with a small enough value of $|\hbar|$ may be chosen to ensure that the corresponding solution series converge.

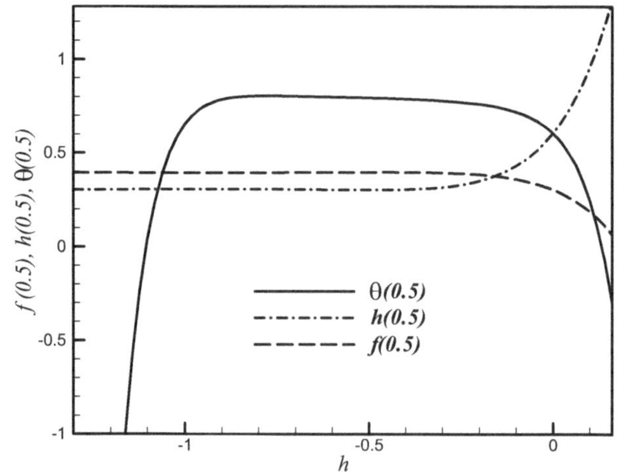

Fig. 2 $\hbar$-curves to show the convergence region.

Table 1. Values of $-f''(0)$ for various values of *m* when *K* = 0.

m	Cortell (2007)	Present with 15th order
0.2	0.766758	0.766991
0.5	0.889477	0.889634
1	1	1
1.5	1.061587	1.061643
3	1.148588	1.148635

The values of f''(0) for m <1 and m > 1 results in the classical Newtonian fluids case (K = 0) are compared with existing one branch of the numerical results of Cortell (2007) in the case of stretching flow. From the comparison listed in Tables 1, it is found that they are in good agreement. Physically, f'(0) is related to the friction of the fluid on the plate and therefore has important physical meanings. It is found that the [m,m] homotopy-Pad´e approximants of f'(0) do not depend upon the auxiliary parameter. Besides, the convergence rate of the [m,m] homotopy-Pad´e approximants of f'(0) is not sensitive to auxiliary parameter η, as shown in Table 1. Furthermore in Table 2, a comparison has been made with previous works to show the influence of the iterations order in accuracy. It can be easily seen that, by increasing the number of iterations in series solution, the accuracy increases, which are in very good agreement with other works. Note that the solution (103-110) is explicit, purely analytic, and uniformly valid in the whole region $\eta \in [0,+\infty)$. Thus, it can be regarded as a definition of the solution of Blasius' viscous flow problems governed by Equations (26-29). As shown using the homotopy analysis method, one can obtain many different solution expressions of a nonlinear problem, even if the solution is unique.

Table 2. Values of $-\theta'(0)$ for different *Pr* (m = 0.5, N_R = 0, *K* = 0)

Pr	Grubka and Bobba (1985)	Ali (1994)	Chen (1998)	Present with 6th order	Present with 15th order
0.01	0.0099	-	0.00991	0.0089	0.00991
0.72	0.4631	0.4617	0.46315	0.4595	0.04631
1	0.582	0.5801	0.58199	0.5793	0.0582
3	1.1652	1.1599	1.16523	1.16381	1.1652
10	2.308	2.296	2.30796	2.2998	2.30791
100	7.7657	-	7.76536	7.7532	7.76532

5. ENTROPY GENERATION

It is easy to show that the rate of a one-way destruction of useful work in an engineering system W_{lost} is directly proportional to the rate of entropy generation:

$$W_{lost} = T_0 \dot{S}_g''' \quad (111)$$

where T_0 is the absolute temperature of the ambient. Assuming a finite size control volume at an arbitrary point in a two dimensional convection flow field and applying the second law of Thermodynamics, as Bejan (1996) suggested entropy generation per unit time and per unit volume equation is:

$$\dot{S}_g''' = k\left(\frac{\nabla T}{T}\right)^2 + \frac{\mu}{T}\left(\left(\frac{\partial u}{\partial y}\right)^2 + \left(\frac{\partial v}{\partial x}\right)^2\right) \quad (112)$$

Substituting Eqs. (21), (22) and (24) into Eq. (111) a relation for entropy generation for the flow over a horizontal flat plate is obtained:

$$S_{total} = \frac{\dot{S}_g'''}{S} = \frac{1}{2\Pr\operatorname{Re}Ec}\left(\frac{\theta'}{\theta+\theta_r}\right)^2\left(\frac{\eta^2}{2\operatorname{Re}}+1\right) + \frac{1}{2\operatorname{Re}}\left(\frac{f''^2}{\theta+\theta_r}\right) \quad (113)$$

where S is

$$S = \frac{\rho^2 U_w^4}{\mu(T_w - T_\infty)} \quad (114)$$

Temperature ratio is

$$\theta_r = \frac{T_\infty}{T_w - T_\infty} \quad (115)$$

and the Eckert number (Ec) is a dimensionless number which characterize dissipation by express the relationship between a flow's kinetic energy and enthalpy as

$$Ec = \frac{U_w^4}{C_p(T_w - T_\infty)} \quad (116)$$

Equation (112) denotes the entropy generation in terms of similarity parameters as well as similarity functions. So, it is straightforward to calculate the entropy generation with the aid of similarity solution. It must be noted that in Eq. (112) the first term is because of heat transfer and the latter one is due to fluid friction; Bejan number is defined as follows:

$$Be = \frac{1}{2\Pr\operatorname{Re}EcS_{total}}\left(\frac{\theta'}{\theta+\theta_r}\right)^2\left(\frac{\eta^2}{2\operatorname{Re}}+1\right) \quad (117)$$

As is obvious, Be yields the share of S h and S f in total generated entropy, Be = 1 is the limit at which the heat transfer irreversibility dominates, Be = 0 is the opposite limit at which the irreversibility is dominated by fluid friction effects and Be=1/2 is the case that the heat transfer and fluid friction entropy generation rates are equal.

6. RESULTS AND DISCUSSION ERRORS

In order to study the effects of governing parameters on the flow and heat transfer characteristics, the graphical presentation of analytical results is presented in this section. Figure 3 presents the f profiles for various values of K (1, 10, 100, and 1000) when m = 0.9, Pr = 1 and NR = 1.From this Figure it is obvious that the f profiles increase as values of K increase. By observation of the increase in the curvature, it can also be noticed that the velocity gradient at the surface f″ (0) decreases (in an absolute sense) as K increases. The same trend in similar fixed values is seen in Figure 4 for the dimension-less velocity. From this, it can be concluded that micro polar fluids exhibit drag reduction compared to viscous fluids. The negative velocity gradient at the surface, f″ (0), as shown in Figure 4 reveals that a drag force is exerted by the stretching sheet on the fluid. This outcome is expected, because the development of the boundary layer is only induced by velocity profile.

Figure 5 shows the variations of the angular velocity profile with respect to K (5, 10, 30, 50, and 500) when m = 0.1, Pr = 1 and NR = 1 parameter. It can be seen that as K increases, the profiles decrease at the beginning (near the surface) and then (as η increases) increase smoothly. The effect of N_R parameter over the temperature distribution when m = 1, Pr = 0.7 and K = 100 is represented in Figure 6. The choice of such high K is to increase the convective effect and signify the thermal radiation heating. It is clear that the temperature is reduced due to the increase in K values. This claim is in agreement with the results presented in Table 3, which shows that as the value of K increases, the values of skin friction factors decreases (in an absolute sense) and subsequently a decrease in the temperature profiles occurs. Here the N_R values are 0, 1, 2, 5, and 10. As revealed in the Figure 6 by increase of N_R the dimensionless temperature increases. It can be observed that the temperature distribution increases in the flow region as values of N_R increase. This is due to the heat energy stored in the liquid because of frictional heating. It is also seen that the temperature gradient at the surface decreases (in an absolute sense) as N_R increases. This observation is evidence that values of being lower for $N_R > 0$ compared to $N_R = 0$ as it presented in Table 4. Thus, the heat transfer rate at the surface is lower in the presence of radiation. However, it can also expect that the heat transfer rate at the surface is increased as K increases. This observation is in agreement with the results of Table 4.

Figure 7 shows a different trend when the effect of Pr on the temperature distribution was explored as m and K values were kept constant at 0.5 and 100, respectively, and N_R = 0. Here the Pr values are 0, 0.05, 0.1, 0.2, 0.3, 0.5, 1, 2, and 10. From Figure 7 it can be seen that increasing Pr leads to decrease the temperature distribution which implies that thermal boundary layer thickness decreases and on the contrary, leads to an increase in the heat transfer rate at the surface.

Figure 8 illustrates the effect of Sc (0.01, 0.1, .3,and 1) on the dimension-less concentration distribution as m and K values were kept constant at 0.5 and 1, respectively, and N_R = 0. From Figure 8 it can be concluded that increasing Sc leads to decrease the dimension-less concentration distribution which point toward that concentration boundary layer thickness decreases and on the contrary, leads to an increase in the mass transfer rate at the surface.

Figure 9 presents the φ profiles for various values of K (1, 10, and 100) when m = 0.1, Sc = 0.1 and N_R = 1.From this Figure it is obvious that the φ profiles decrease as values of K increase. By observation of the increase in the curvature, it can also be noticed that the concentration gradient at the surface c'' (0) increases (in an absolute sense) as K increases. Such increase on mass convection is important in high Schmitt numbers.

Figure 10 depicted the φ profiles for various values of m (0,.5,0.8,0.9, and 1) when K = 1, Sc = 0.1 and N_R = 1.Most of the previous investigations have been done based on considering the boundary parameter m = 0 or m = 0.5 (m = 0 represents concentrated particle flow in which the elements close to the surface are unable to rotate and m = 0.5 indicates the vanishing of asymmetric part of the stress tensor). In the present work, more values of m have been considered. The concentration enhancement effect of boundary parameter m on concentration profiles is signified in Figure 10. It can also be found that the concentration gradients (mass transfer rate) at surface are larger for larger values of m. Figure 11 presents the f" profiles for various values of K (10, 50, 100, and 500) when m = 0.5.From this Figure it is obvious that the f" profiles increase as values of K increase. In addition the curvature of f" decreases (in an absolute sense) as K increases and it approaches the linear profile shape. The f" values are the significant part in the entropy generation rate that rules the velocity component of entropy (S_U).

Figure 12 show the θ" profiles for various values of K (10, 50, 100, and 500) when m = 0.5.From this Figure it is obvious that the θ" profiles increase smoothly as values of K increase. A the θ" value is the significant part in the entropy generation rate that rules the temperature component of entropy (S_θ).

Thermal entropy profiles for various values of Prandtl number (m=0.1, K=100, NR=1, Re =1, r=0.1, and Ec = 1) are shown in Figure 13. As shown most of the thermal component of entropy is occurred near the wall and after the η=1 this contribution is negligible. As well the velocity component of entropy profiles (m=0.1, K=100, NR=1, Re =1, and θ_r=0.1) for various values of Prandtl number are revealed in the Figure 14. As illustrated the maximum of the velocity component of entropy take place at the distance =1 from the wall and decrease by distance from this point.

The summation of this components (m=0.1, K=100, NR=1, Re =1, θr =0.1, and Ec = 1) are displayed in the Figure 15. Total entropy profiles for various values of Prandtl number have maximum near the wall and after the η=1 are diminishing. To have a better sense from contribution of entropy components the Figure 16 is plotted to show the Bejan number profiles for various values of Prandtl number. By increase of Prandtl number the Bejan number increased near the wall and decrease in the distances far from the wall. The neutral point is take place near the η=1 point. It must be mentioned that at the beginning of the plate, Be is approximately unity (Be ≅ 0.9), which means in the total entropy generation, heat transfer plays more important role than fluid friction but as we go along the surface, with decrease in temperature gradient, heat transfer share in total generated entropy decreases.

The thermal boundary layer thickness becomes thicker for stronger Prandtl number. Figure 17 exposed the variation of average Bejan number from η=0 to η=4 versus Prandtl number. By increase of c the average of Bejan number decreased smoothly. Variations of thermal boundary layer thickness and temperature gradients for various values of Prandtl number clearly lead to decrease in Bejan number.

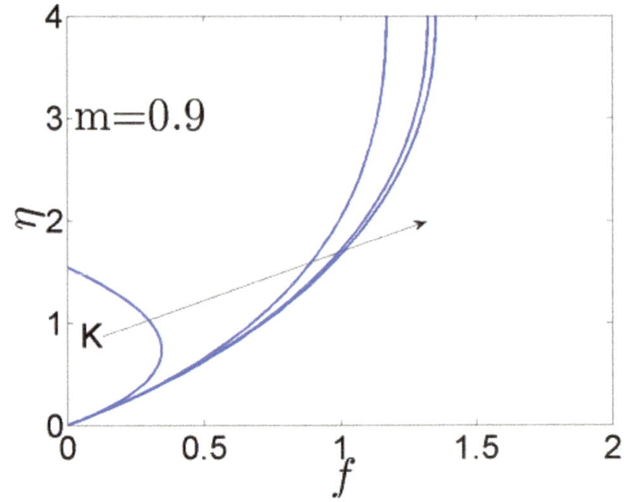

Fig. 3 Variation of $f\ (\eta)$ with η at different values of *K*.

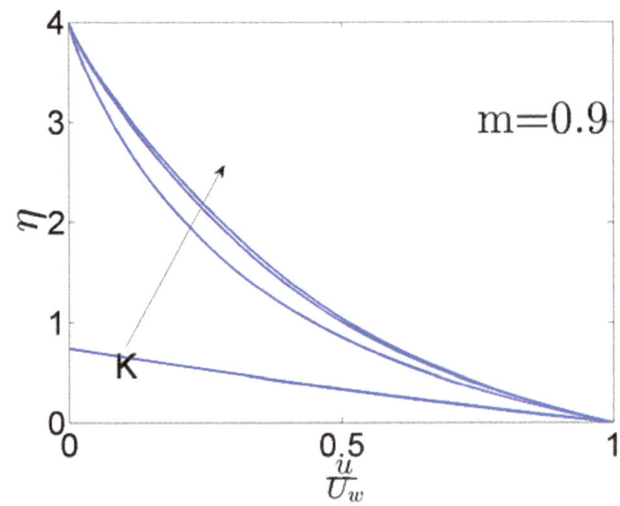

Fig. 4 Variation of velocity profile $f'\ (\eta)$ with η at different values of *K*.

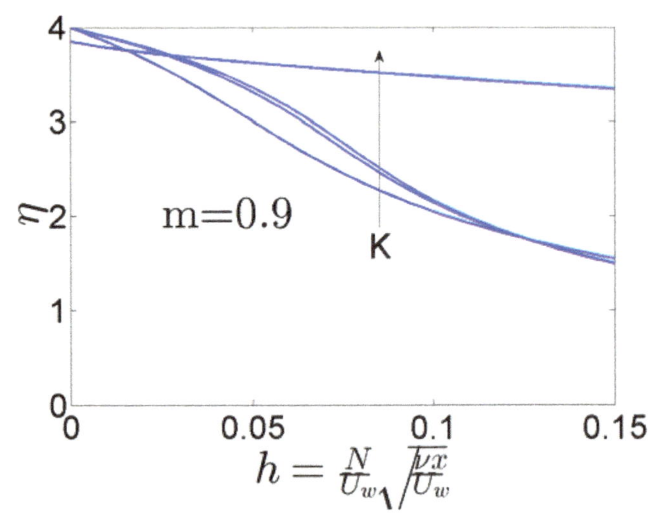

Fig. 5 Angular velocity *h (η)* for various values of *K*.

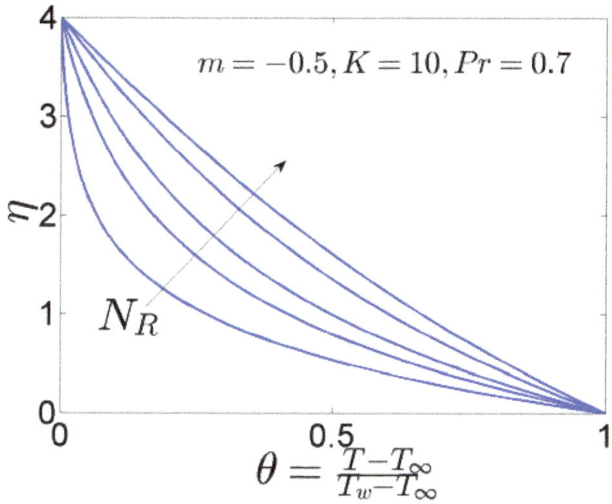

Fig. 6 Temperature profiles $\theta(\eta)$ for various values of N_R.

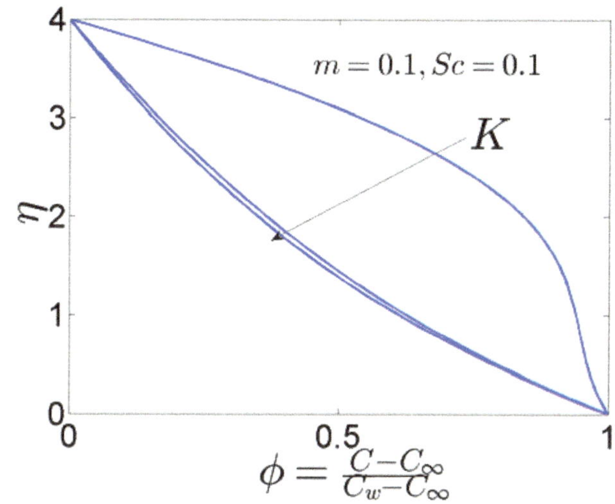

Fig. 9 Concentration profiles $\phi(\eta)$ for various values of K.

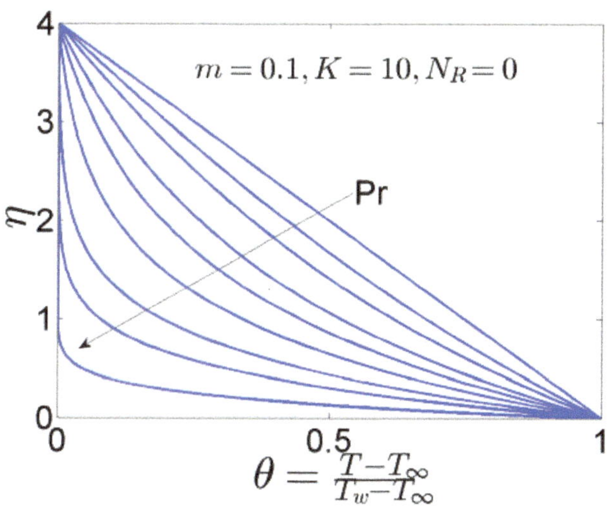

Fig. 7 Temperature profiles $\theta(\eta)$ for various values of Pr.

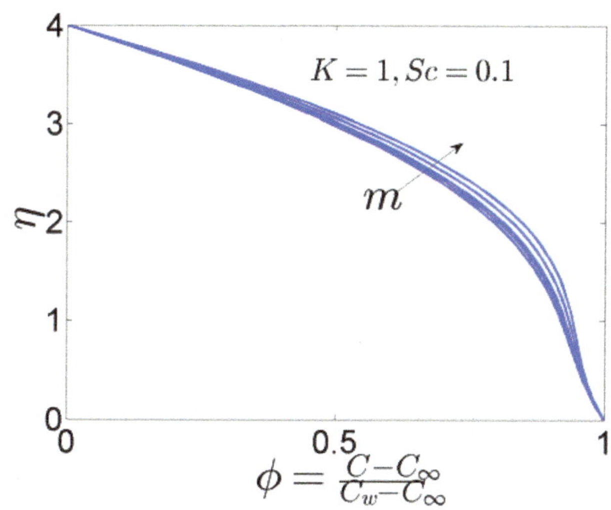

Fig. 10 Concentration profiles $\phi(\eta)$ for various values of m.

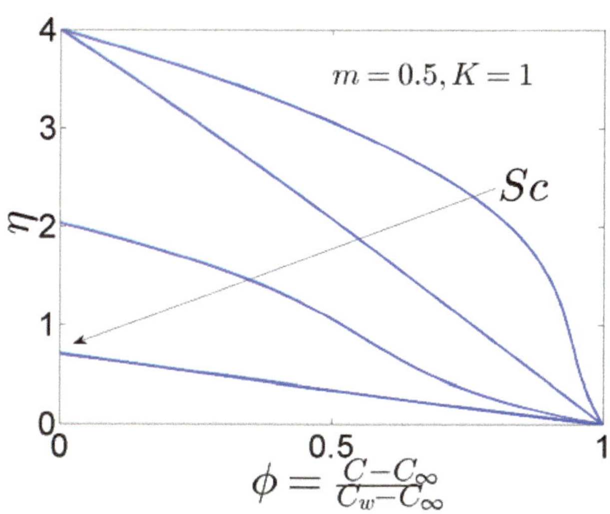

Fig. 8 Concentration profiles $\phi(\eta)$ for various values of Sc.

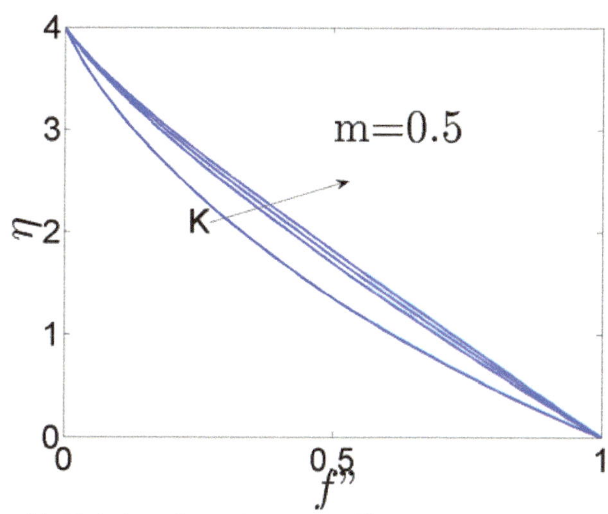

Fig. 11 Variation of velocity profile $f''(\eta)$ with η at different values of K.

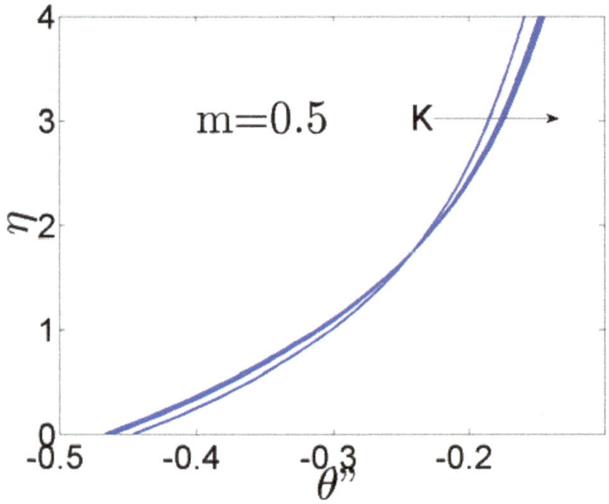

Fig. 12 Variation of velocity profile $\theta''(\eta)$ with η at different values of K.

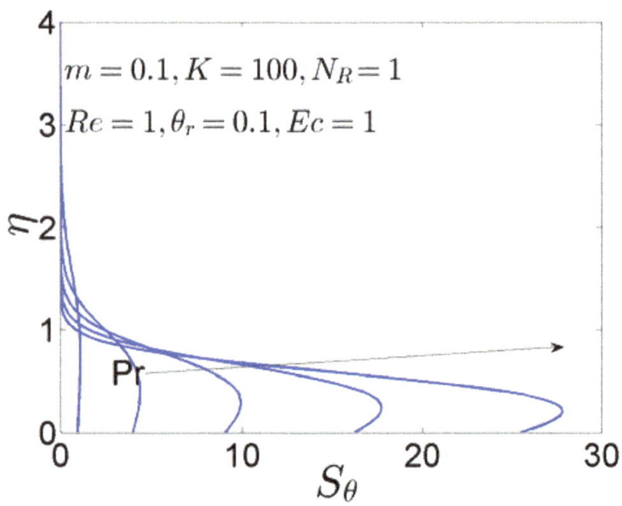

Fig. 13 Thermal entropy profiles $S_\theta(\eta)$ for various values of Pr.

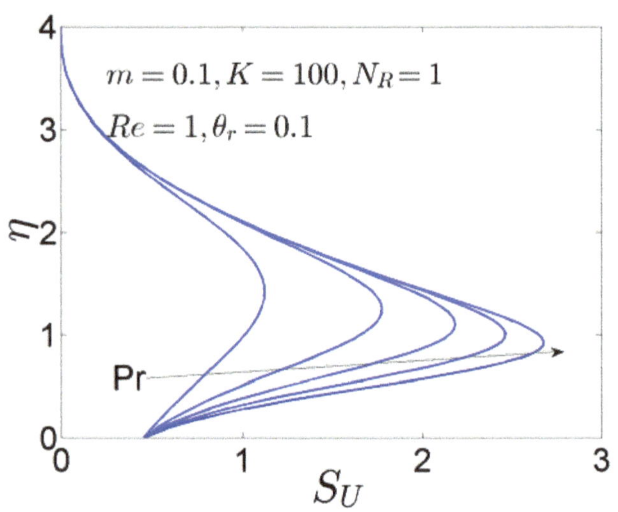

Fig. 14 Velocity entropy profiles $S_U(\eta)$ for various values of Pr.

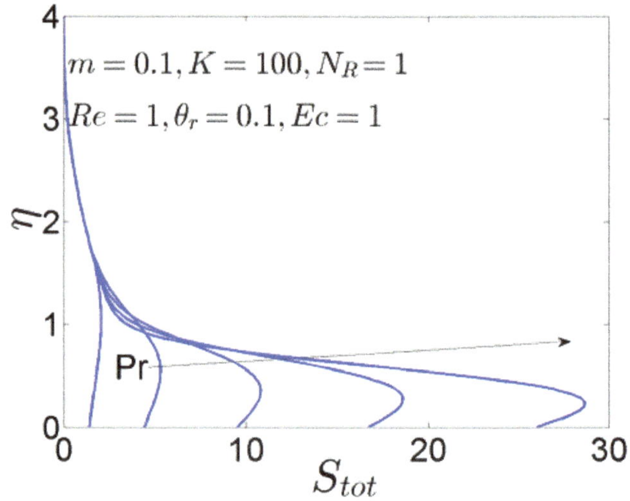

Fig. 15 Total entropy profiles $S_{tot}(\eta)$ for various values of Pr.

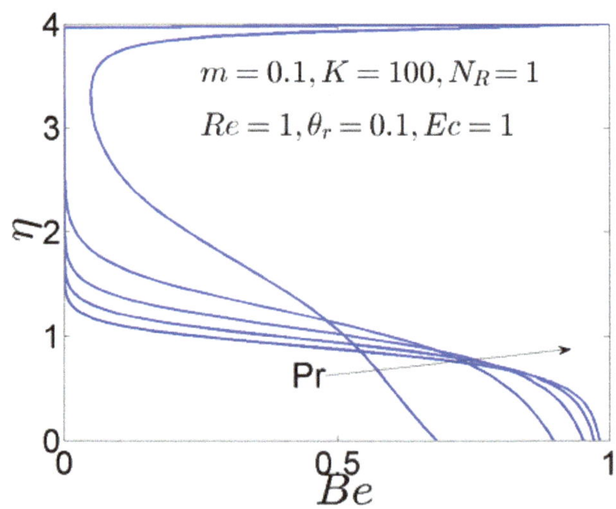

Fig. 16 Bejan number profiles $Be(\eta)$ for various values of Pr.

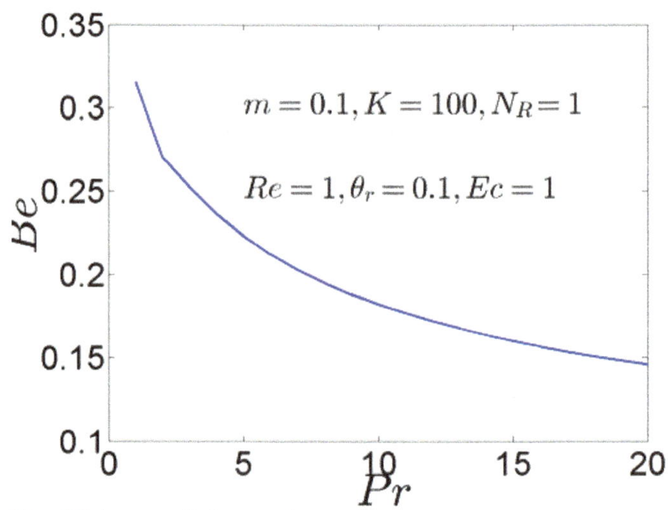

Fig. 17 Average Bejan number vs Pr.

Table 3. Values of $-\theta'(0)$ for various values of *Pr*, N_R, *m* and *K*.

Pr	N_R	*m*	*K*	$Nu_x/\mathrm{Re}_x^{0.5}$
			0	0.41181
			1	0.42723
1	1	0.5	2.5	0.47041
			5	0.52179
			10	0.57507
		0		0.45513
		0.2		0.45217
1	1	0.4	1	0.44869
		0.6		0.4446
		0.8		0.43975
		1		0.43378
	0			0.61332
	0.5			0.50777
1	1	0.5	1	0.42723
	5			0.34031
0.5				0.36031
1				0.42723
2.5	1	0.5	1	0.68679
5				0.90555
10				1.11815

Table 4. Values of Skin friction factor when *Pr=1* and N_R *=1* for various values of *m* and *K*.

m	K	$C_f\,\mathrm{Re}_x^{0.5}$
	0	-8.0252
	1	-6.5356
0.5	2.5	-5.1566
	5	-4.2871
	10	-3.3208
0		-10.264
0.2		-8.9227
0.4	1	-7.3884
0.6		-5.6153
0.8		-3.5205
1		-1.0118

7. CONCLUSIONS

In this paper the problem of steady two-dimensional heat transfer of continuous stretching sheet immersed in a micro polar fluid in the existence of radiation have studied. Using existence similarity transformations the governing equations have been transformed into non-linear ordinary differential equations and solved analytically by using HAM. The obtained results were validated against previous works. Effects of the various parameters such as Prandtl number, the material parameter, boundary parameter and radiation parameter on the temperature, velocity and angular velocity profiles, as well as the skin-friction coefficient and the local Nusselt number were examined. The following conclusions can be drawn as a result of the HAM calculations:

-An increase in the value of *K* leads to a decrease of the angular velocity profiles, the wall temperature, concentration and the skin-friction coefficient; however, the velocity profile and the local Nusselt number are increasing.

-An increase in the value of *m* leads to an increase of the angular velocity profiles, the wall temperature, concentration, and a decrease of the velocity profile, the local Nusselt number and the skin-friction coefficient.

-An increase in the value of N_R leads to a reduction of the local Nusselt number due to the wall temperature enhancement, while the trend is opposite for *Pr* number.

-An increase in the value of *Sc* leads to a reduction of the concentration.

-Be is plotted and its variation is studied as well. It is found that more entropy generates in boundary layer with increase the Re and Ec while decrease with the increase of Pr numbers.

REFERENCES

Abdollahzadeh Jamalabadi, M.Y., Ghassemi, M., Hamedi, M.H., 2012, "Two- Dimensional Simulation of Thermal Loading with Horizontal Heat Sources," *Proc. Inst. Mech. Eng. C J. Mech. Eng. Sci.*, **226**, 1302-1308
http://dx.doi.org/10.1177/0954406211419613

Abdollahzadeh Jamalabadi, M.Y., Ghassemi, M., Hamedi, M.H., 2013, "Numerical Investigation of Thermal Radiation Effects on Open Cavity with Discrete Heat Sources," *Int. J. Numer Method Heat Fluid Flow*, **23**, 649-661
http://dx.doi.org/10.1108/09615531311323791

Abdollahzadeh Jamalabadi, M.Y., 2014a, "Experimental Investigation of Thermal Loading of a Horizontal Thin Plate Using Infrared Camera," *J. King Saud Univ. – Eng. Sci*, **26,** 159-167.
http://dx.doi.org/10.1016/j.jksues.2013.03.002

Abdollahzadeh Jamalabadi, M.Y., Park, J.H., 2014b,"Thermal Radiation, Joule Heating, And Viscous Dissipation Effects on MHD Forced Convection Flow with Uniform Surface Temperature," *Open J. Fluid Dyn*, **4**, 125-132
http://dx.doi.org/10.4236/ojfd.2014.42011

Ahmadi, G., 1976, "Self- Similar Solution Of Incompressible Micropolar Boundary Layer Flow over a Semi-infinite Plate," *Int J Eng Sci*,**14**, 639-646.
http://dx.doi.org/10.1016/0020-7225(76)90006-9

Aiboud,S., Saouli,S., 2010, "Entropy Analysis For Viscoelastic Magnetohydrodynamic Flow over a Stretching Surface," Int. J. Nonlinear Mech., **45**, 482–489
http://dx.doi.org/10.1016/j.ijnonlinmec.2010.01.007

Ali, M.E., 1994, "Heat Transfer Characteristics of a Continuous Stretching Surface", Heat Mass Transfer, **29**, 227-234.
http://dx.doi.org/10.1007/BF01539754

Ariman, T., Turka, M.A., Sylvester,N.D., 1973, "Microcontinuum Fluid Mechanics– Review", *Int J Eng Sci*, **11**, 905-930.
http://dx.doi.org/10.1016/0020-7225(73)90038-4

Azizah,Y.N., Anuar,I., Ioan,P., 2013, "Hydromagnetic Flow and Heat Transfer Adjacent to a Stretching Vertical Sheet in a Micropolar Fluid," *Thermal Science*, **17**, 525-532
http://dx.doi.org/10.2298/TSCI100308198Y

Barletta, A., 1999, "Analysis of Combine Forced and Free Flow in a Vertical Channel with Viscous Dissipation and Isothermal-Isoflux Boundary Conditions," *ASME J. Heat Transfer*, **121**,349-356.
http://dx.doi.org/10.1115/1.2825987

Barletta, A., 2002, "Fully Developed Mixed Convection and Flow Reversal in a Vertical Rectangular Duct with Uniform Wall Heat Flux," *Int. J. Heat Mass Transfer*, **45**, 641-654.
http://dx.doi.org/10.1016/S0017-9310(01)00160-0

Bejan,A., 1996, Entropy Generation Minimization, (second ed.) CRC, Boca Raton

Brewster, M.Q., 1992, Thermal Radiative Transfer Properties, Wiley, New York

Busuke,H., Tatsuo, T., 1969, "Two-Dimensional Flows of Linear Micropolar Fluids," *Int J Eng Sci*, **7**, 515–522
http://dx.doi.org/10.1016/0020-7225(69)90036-6

Butt,A.S., Unawar,S.M. Mehmood, A., Ali, A. , 2012a, "Effect of viscoelasticity on Entropy Generation in a Porous Medium over a Stretching Plate," World Applied Sciences Journal, **17**, 516–523
http://dx.doi.org/10.1.1.390.2183

Butt,A.S., Munawar,S., Ali, A. Mehmood, A., 2012b, "Entropy Generation in Hydrodynamic Slip Flow over a Vertical Plate with Convective Boundary," *Journal of mechanical science and technology*, **26**, 2977–2984
http://dx.doi.org/10.1007/s12206-012-0701-3

Butt,A.S., Munawar,S. , Ali,A., Mehmood, A., 2012c, " Entropy Analysis of Mixed Convective Magnetohydrodynamic Flow of a Viscoelastic Fluid over a Stretching Sheet," *Z. Naturforsch. A – J. Phys. Sci.*, **67**, 451–459
http://dx.doi.org/10.5560/ZNA.2012-0055

Butt,A.S., Ali,A., 2013,"Entropy Generation in MHD Flow over a Permeable Stretching Sheet Embedded in a Porous Medium in the Presence of Viscous Dissipation," *Int. J. Exergy*, **13**, 85–101
http://dx.doi.org/10.1504/IJEX.2013.055779

Chen, C.H., 1998, "Laminar Mixed Convection Adjacent to Vertical, Continuously Stretching Sheets", *Heat Mass Transfer*, **33**, 471-476.
http://dx.doi.org/10.1007/s002310050217

Cortell, R., 2007, "Viscous Flow and Heat Transfer over a Nonlinearly Stretching Sheet," *Appl. Math. Comput.* 184, 864–873.
http://dx.doi.org/10.1016/j.amc.2006.06.077

Datti, P.S, Prasadb,K.V., Abelc,M. S., Joshic, A., 2004," MHD Viscoelastic Fluid Flow over a Non-Isothermal Stretching Sheet," *Int J Eng Sci*, **42**, 935-946.
http://dx.doi.org/10.1016/j.ijengsci.2003.09.008

Dehsara, M., Habibi-Matin, M., Dalir, N., 2012, "Entropy Analysis for MHD Flow over a Non-Linear Stretching Inclined Transparent Plate Embedded in a Porous Medium due to Solar Radiation," *Mechanika*, **18**, 524–533
http://dx.doi.org/10.5755/j01.mech.18.5.2694

Dehsara, M., Dalir,N., Nobari, M.R.H., 2014, "Numerical Analysis Of Entropy Generation in Nanofluid Flow over a Transparent Plate in Porous Medium in Presence of Solar Radiation, Viscous Dissipation and Variable Magnetic Field," *J. Mech. Sci. Technol.*, **28**, 1819–1831
http://dx.doi.org/10.1007/s12206-014-0329-6

Eringen, A. C., 2001,"Microcontinuum Field Theories. II: Fluent Media," Springer, New York, USA

Fang, T.G.,Yaoa,S., Pop,I., 2011, "Flow and Heat Transfer over a Generalized Stretching/shrinking Wall Problem—Exact Solutions of the Navier–Stokes Equations," *Int. J. Non-Linear Mech.*, **46**, 1116–1127
http://dx.doi.org/10.1016/j.ijnonlinmec.2011.04.014

Galanis,N., Rashidi, M.M., 2012, " Entropy Generation in Non-Newtonian Fluids due to Heat and Mass Transfer in the Entrance Region of Ducts," *Heat Mass Transfer*, **48**, 1647–1662
http://dx.doi.org/10.1007/s00231-012-1009-7

Grubka, L.J., Bobba, K.M., 1985, "Heat Transfer Characteristics of a Continuous, Stretching Surface with Variable Temperature," *ASME J. Heat Transfer*, **107**, 248–250.
http://dx.doi.org/10.1115/1.3247387

Hadjadj, A., Ben-Nasr, O., Shadloo, M.S., Chaudhuri, A., 2015, "Effect of Wall Temperature in Supersonic Turbulent Boundary Layers: A Numerical Study," *International Journal of Heat and Mass Transfer,* **81**, 426–438
http://dx.doi.org/10.1016/j.ijheatmasstransfer.2014.10.025

Hoyt, J.W., Fabula, A.G., 1964, "The Effect of Additives on Fluid Friction," U.S. Naval ordinance Test station Report, USA.

Kandlikar, S.G., 2005, "High Flux Heat Removal With Microchannels - A Roadmap of Challenges and Opportunities," *Heat Transfer Eng.*, **26**,5–14.
http://hdl.handle.net/1850/7413

Khonsari, M.M., 1990, "On the Self-exited Whirl Orbits of a Journal in a Sleeva Bearing Lubricated with Micropolar Fluids," *Acta Mechanica*, **81**, 235
http://dx.doi.org/10.1007/BF01176991

Liao, S.J., 1992, The Proposed Homotopy Analysis Technique for the Solution of Nonlinear Problems, Ph.D. Thesis, Shanghai Jiao Tong University

Liao, S.J., 2003, Beyond Perturbation: Introduction to Homotopy Analysis Method, Boca Raton: Chapman & Hall/CRC Press

Makinde,O.D., 2011, "Second Law Analysis for Variable Viscosity Hydromagnetic Boundary Layer Flow with Thermal Radiation and Newtonian heating Entropy," **13**, 1446–1464
http://dx.doi.org/10.3390/e13081446

Malvandi,A., Ganji, D.D., Hedayati, F., Kaffash, M.H., Jamshidi, M., 2012, "Series Solution of Entropy Generation Toward an Isothermal Flat Plate," *Therm. Sci.*, 16, 1289–1295
http://dx.doi.org/10.2298/TSCI1205289M

Nazar, R., Amina,N., Filipb, D., Pop,I. , 2004, "Stagnation Point Flow of a Micropolar Fluid Towards a Stretching Sheet," *Int J Nonlinear Mech*,**39**, 1227-1235.
http://dx.doi.org/10.1016/j.ijnonlinmec.2003.08.007

Noghrehabadi,A., Saffarian, M.R., Pourrajab,R., Ghalambaz,M.,2013, "Entropy Analysis for Nanofluid Flow over a Stretching Sheet in the Presence of Heat Generation/Absorption and Partial Slip," *J. Mech. Sci. Technol.*, **27**, 927–937
http://dx.doi.org/10.1007/s12206-013-0104-0

Rajeev, 2014, "Homotopy Perturbation Method for a Stefan Problem with Variable Latent Heat", *Thermal Science*, **18**, 391-398
http://dx.doi.org/10.2298/TSCI110627008R

Shadloo, M.S., Hadjadj, A., Hussain, F., 2015, "Statistical Behavior of Supersonic Turbulent Boundary Layers with Heat Transfer at $M_\infty = 2$," *International Journal of Heat and Fluid Flow*, **53**, 113-134 http://dx.doi.org/10.1016/j.ijheatfluidflow.2015.02.004

Yan, L.M., 2013, "Modified Homotopy Perturbation Method Coupled with Laplace Transform for fractional heat transfer and porous media equations," *Thermal Science*, **17**, 1409-1414 http://dx.doi.org/TSCI1305409Y

Yücel, A., 1989, "Mixed Convection in Micropolar Fluid Flow over a Horizontal Plate with Surface Mass Transfer," *Int J Eng Sci*, **27**, 1593-1602. http://dx.doi.org/10.1016/0020-7225(89)90153-5

Yun, Y.S.,Chaolu, T., 2013, "Homotopy Perturbation Method for Viscous Heating in Plane Couette Flow," *Thermal Science*, **17**,1355-1360 http://dx.doi.org/10.2298/TSCI1305355Y

Zhao, C.Y., Lu, T.J., 2002, "Analysis of Microchannel Heat Sinks for Electronics Cooling," *Int. J. Heat Mass Transfer* **45**, 4857–4869. http://dx.doi.org/10.1016/S0017-9310(02)00180-1

23

DEVELOPMENT OF A 10 KW MICROWAVE APPLICATOR FOR THERMAL CRACKING OF LIGNITE BRIQUETTES

Benjamin Lepers[a,*], Thomas Seitz[a], Guido Link[a], John Jelonnek[a,b], Mark Zink[c]

[a]*Karlsruhe Institute of Technology, Institute for Pulsed Power and Microwave Technology (IHM), Eggenstein Leopoldshafen, 76344, Germany*
[b]*Karlsruhe Institute of Technology, Institute for Highfrequency Techniques and Electronic (IHE), Karlsruhe, 71131, Germany*
[c]*TU Bergakademie Freiberg, Institute of Energy Process Engineering and Chemical Engineering (IEC), Fuchsmuehlenweg 9, 09599 Freiberg, Germany*

ABSTRACT

A compact 10 kW microwave applicator operating at 2.45 GHz for fast volumetric heating and thermal cracking of lignite briquettes has been successfully designed and tested. In this paper, the applicator design and construction are presented together with a sequentially coupled electromagnetic, thermal-fluid and mechanical Comsol model. In a first step, this model allows us to calculate the power density inside the lignite material and the temperature distribution in the applicator for different water flow rates. In a second step, the total stress due to the thermal dilatation, the internal pressure inside the ceramic and the contact pressure from the seals at both ends of the ceramic can be computed. In addition, preliminary experimental results are given and showed a significant mechanical strength weakening of the briquettes with microwave heating.

Keywords: *Microwave applicator, microwave heating, thermo elastic model, silicon nitride, lignite, thermal cracking*

1. INTRODUCTION

In the framework of a collaborative project between IHM from KIT and IEC from TU Bergakademie Freiberg for microwave heating of lignite briquettes, a 10 kW compact applicator operating at 2.45 GHz has been successfully designed and tested. This applicator delivers volumetric heating to the briquettes which are moving inside a microwave transparent ceramic silicon nitride tube with high mechanical strength. Upstream to this applicator, briquettes are produced from the lignite coal powder with a briquetting press machine Rosin *et al.* (2014). Downstream of this applicator, after being heated by the microwave field, the briquettes are moving toward the gasifier for syngas (mixture of hydrogen and carbon monoxide) production for example. This system allows the transfer of lignite material from the production machine to the gasifier with pressure up to 65 bars. Innovative feeding systems for pressurized gasification reactors are currently being developed by TU Freiberg Bergakademie. Important requirements for such systems are a simple design and high reliability. The briquetting press is a continuously operating machine with mass flow up to 25 t/h. The sealing between the briquetting machine and the gasifier is made by the briquettes plug in the feed channel. It makes the briquetting press a self sealing system Rosin *et al.* (2014).

With fast microwave volumetric heating (the moving briquettes crosses the ceramic tube in 9 s and passes in front the rectangular aperture in 2 s, at the production rate 75 briquettes per minute), thermal cracking was observed together with an increase of the briquette friability. Volumetric heating induces thermal cracking and can be used as a thermal pretreatment to reduce the energy requirement for grinding or can be used directly as a communition tool (i.e desctruction, breakage) for the lignite material Batar (2004). High power microwave heating of coal with exposure time of few seconds increases significantly the grindability Binner *et al.* (2014), and consequently is beneficial for the overall efficiency of the gasifier. Microwave heating of lignite is much faster than conventional heating of coal Pickles *et al.* (2014) (with the value $\epsilon_r = 5.48 - 0.73j$ for fresh lignite, the skin depth is about 6 cm). This work presents the design of a microwave applicator used for continuous volumetric heating of lignite briquettes together with a Comsol multiphysics model used to calculate the power density, temperature, displacement and stress field in the applicator parts. In addition, some first experimental results are given.

2. THE APPLICATOR DESIGN

The applicator is made of a silicium nitride (Frialit GP79, Friatec) cylinder inserted in an water cooled aluminum block (Alumec 100, Alcoa) of dimension $32 \times 26 \times 22$ cm as shown in Fig. 1. A rectangular wave guide crosses the aluminum bloc to deliver microwave energy to the lignite material moving inside the ceramic cylinder. Microwave radiation at 2.45 GHz is created in the magnetron and travels through the manual tuner and the wave guide. The applicator input power is up to 10 kW corresponding to $8 - 9$ kW of absorbed power by the lignite briquettes (depending of the lignite loss factor, ie water content). A three pin manual tuner element is installed after the magnetron and a variable short circuit wave guide element is installed at the end of the applicator wave guide as shown in Fig. 2. These two elements allow a good transmission of the microwave energy from the applicator to the lignite material load. The measured S_{11} magnitude at 2.45 GHz was below -15 dB. In operation,

* *Corresponding author. Email: benjamin.lepers@kit.edu*

Fig. 1 Aluminum block Alumec 100 with the ceramic silicon nitride and the graphite seal, the M22 thread insert are visible

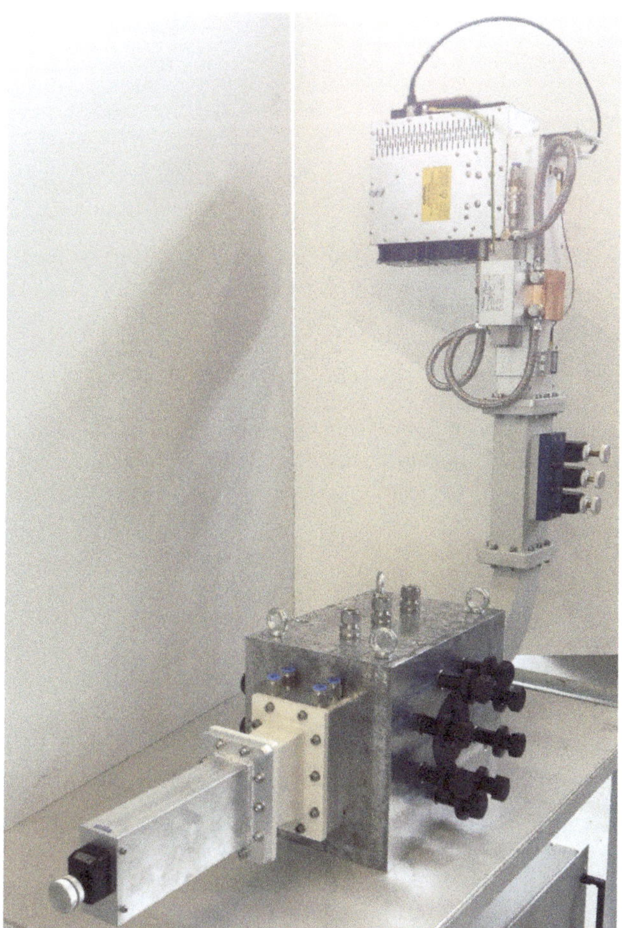

Fig. 2 Applicator with magnetron, manual tuner, aluminum bloc, silicon nitride ceramic, sliding short, wave guide transition (in white) and microwave windows

the measured reflected power was about $2-3\,\%$ of the input power. To provide significant heating to the lignite briquettes, the following engineering choices were made for this applicator:

- A high mechanical strength and low dielectric loss ceramic was used for the high pressure microwave transparent tube. Silicon nitride (GP79 Friatec) offers these two characteristics with a tensile strength above 400 GPa and a low relative dielectric loss factor around 0.02.

- For an effective sealing, the displacement due to thermal dilatation mismatch between the aluminum frame holder and the ceramic tube has to be minimized. A good solution was found with the aluminum alloy (Alumec 100, high strength aluminum alloy) together with inner cooling water channels. The high electrical conductivity of aluminum allows the reduction of electrical wall losses, and its high thermal conductivity renders the water cooling system much more efficient then a block made of stainless steel for example. With a water flow rate of 6 l/min the maximum temperature of the aluminum bloc in operation is kept below 45 °C and so the thermal expansion can be limited.

- A compact design which fits near the briquetting press machine installation where limited space is available.

3. CONSTRUCTION

The silicon nitride tube was made by the company Friatec (Silicon nitride GP 79 Hot pressed sintered). The length is 220 mm with inner and outer diameter of 54 and 105 mm respectively. Its allows the passage of lignites briquettes of 50 mm in diameter. The aluminum block is made of aluminium alloy Alumec 100 from the company Alcoa. The machining steps are detailed below:

- the raw aluminum block was cut to the specific dimensions $320 \times 260 \times 220$ mm

- the hole of diameter 105 mm through the depth of the aluminum block (220 mm length) was machined with wire electric discharge machining (EDM, electric discharge that remove the material).

- the rectangular channel of cross section 109×54 mm (WR 430 wave guide) was machined with EDM through the 32 cm length of the aluminum block, it intersects the hole of diameter 105 mm made for the ceramic.

- 8 holes of diameter 25 mm were drilled on the two opposite faces of the aluminum block for the thread insert as shown in Fig. 1.

- 3 safety holes of diameter 12 mm with a Swagelock connection type where drilled. In case of over pressure in the space between the outer ceramic tube and aluminum block, the gas will escape through these 3 holes (visible on Fig. 2).

- a cooling circuit inside the aluminium block in the vicinity of the ceramic tube was made by drilling holes of 6 mm in diameter. In total there are 7 straight cooling channel connected. Upstream and downstream, near the ceramic aperture, there are 2×3 channels encircling the ceramic tube. One channel is making the connection between the upstream and downstream part of the cooling circuit.

Once all parts were available, the applicator was built by assembling two set of parts. The first set is made of the magnetron head and the manual tuner which is used to minimize the reflected power. The second set is made of a 90 ° right angle wave guide (because of space limitation on the installation site), the transition wave guide and the microwave windows which are screwed on each side of the aluminum block at the rectangular channel location. In addition, a sliding short is screwed to the transition

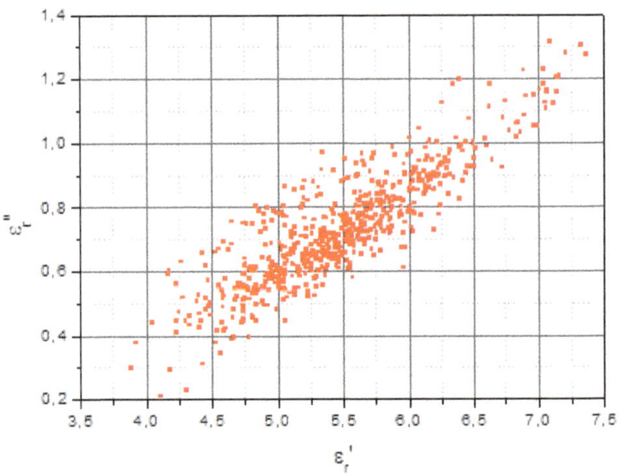

Fig. 3 Real and imaginary part of the relative complex permittivity of Braun coal for 100 briquettes with grain size smaller than 10 mm and water content of about 18 % in mass. The average value used for the electromagnetic simulation is $\epsilon_{rc} = 5.48 - 0.73j$

Properties	Lignite	Silicon nitride	Alumec 100
Relative complex dielectric constant ϵ_{rc} [-]	$5.48 - 0.73j$	$8 - 0.02j$	–
Density ρ [kg/m^3]	1200	3200	2850
Thermal conductivity k [W/(mK)]	0.3	21	160
Specific heat C_p [J/(kg.K)]	1200	700	880
Thermal expansion coeff α [10^{-6}/K]	–	2.3	23
Elastic young modulus E [GPa]	–	320	70
poisson coefficient ν [-]	–	0.28	0.33

Table 1 Dielectric, thermal and mechanical properties for the lignite coal, silicon nitride and Aluminum material used in the sequentially coupled model

wave guide. The sliding short is a tuning element and can minimize the reflected power by adjusting the location of the short circuit. This sliding short works in combination with the manual tuner. The transition wave guide parts have a conical geometry and make the transition from a WR 340 (86 × 43 mm) to a WR 430 (109 × 54 mm) rectangular standard wave guide.

4. MATERIAL PROPERTIES

The electromagnetic, thermal and mechanical properties of the lignite (Romonta), Silicon nitride GP 79 (Friatec) and Aluminum Alumec 100 (Alcoa) were assumed constant and are given in the Tab. 1 (Properties expressed as a function of the temperature will rend the model non linear, and do not bring much more information in the design phase of the applicator). Lignite is a semi mineral organic material consisting of Carbon (73 %), oxygen (20 %), hydrogen (5 %) and nitrogen (1.2 %). Since the briquettes are made from brown coal powder, the material is homogeneous and isotropic at a macroscopic scale and so the real and imaginary part of the complex permittivity are scalar number. For the simulation, the relative permittivity used is $\epsilon_r = 5.48 - 0.73j$ and is obtained from the averaging of the dielectric measurements of fresh lignite briquettes (with water content) performed in our laboratory (see Fig. 3). For the ceramic tube, a silicon nitride (GP 79, Friatech) was chosen because of a very high mechanical strength and low dielectric losses. For the aluminum block, the high strength aluminum Alumec 100 alloy was the best choice, because of a good electrical and thermal conductivity together with a high mechanical strength. In addition, Aluminum alloys are materials easy to be machined.

5. MICROWAVE AND THERMAL MODEL

A standard waveguide WR 430 intersecting with the silicon nitride tube was modelled with Comsol 4.3 to compute the S_{11} magnitude ($|S_{11}|dB = 10\,log(\frac{P_r}{P_i})$, with P_r and P_i the reflected and incident power), the electric field $\boldsymbol{E}$ and the power density q distribution by solving the wave equation

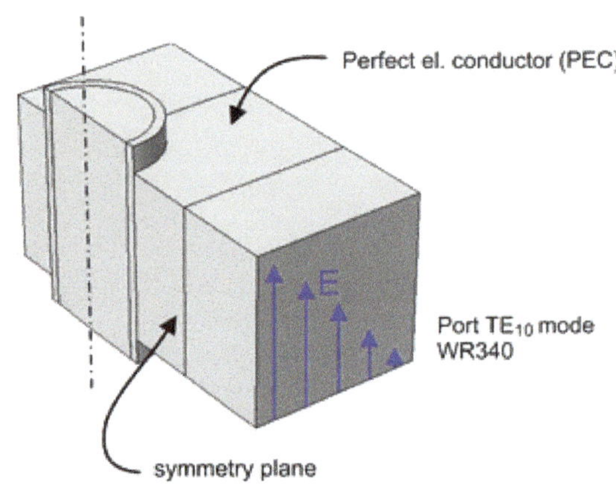

Fig. 4 Electromagnetic model with boundary conditions, from the geometry and field symmetry, only one half is modelled

in the frequency domain at 2.45 GHz:

$$\frac{1}{\mu_r}\nabla \times (\nabla \times \boldsymbol{E}) - k_0^2 \epsilon_{rc} \boldsymbol{E} = 0 \tag{1}$$

$$k_0^2 = \omega^2 \mu_0 \epsilon_0$$

$$\epsilon_{rc} = \epsilon_r^{'} - j\epsilon_r^{''}$$

$$q = \frac{1}{2}\omega \epsilon_0 \epsilon_r^{''} \boldsymbol{E}^2$$

with k_0 the wave number in free space, ϵ_{rc} the relative complex dielectric permittivity, $\omega = 2\pi f$ the pulsation and q the power density calculated from the electric field. Lignite is a non magnetic material so the relative magnetic permeability is $\mu_r = 1$. $\mu_0 = 4\pi \times 10^{-7}$ H/m and $\epsilon_0 = 8.854 \times 10^{-12}$ F/m are respectively the permeability and permittivity of free space. Using the symmetry of the applicator and of the fundamental mode TE_{10}, only half of the applicator geometry was modelled. On the symmetry plane, a perfect magnetic wall is used (tangential components of the magnetic field do not exist in this plane, $\boldsymbol{n} \times \boldsymbol{H} = 0$). On both ends of the ceramic a perfect electric boundary condition was used ($\boldsymbol{n} \times \boldsymbol{E} = 0$), because of the electrical conductive graphite seal. The aluminum walls were also assumed to be perfect electric conductors. A scattering boundary condition was used on both surfaces where the lignite enters and exits the ceramic tube. The boundary conditions for the electromagnetic model are shown in Fig. 4. With some quarter wave length matching elements inserted inside the wave guide, it was possible to achieve a good transmission of the energy from the wave guide to the lignite briquettes. Later, during the applicator development process, the matching elements inserted inside the wave guide were removed and an external manual three pin tuner was used instead. In a second step, the power density q obtained from the electromagnetic model was used to compute with the heat equation the temperature distribution inside the lignite material and the ceramic tube. The effect of the moving briquettes at constant velocity $\boldsymbol{v}$ inside the ceramic tube was included with the option "translational motion" in the Comsol thermal model. This effect is modelled by the transport term $\rho C_p \boldsymbol{v} \nabla T$ shown in Eq. 2. The temperature field in the lignite, and ceramic domain is obtained by solving the heat equation:

$$\rho C_p \frac{\partial T}{\partial t} + \rho C_p \boldsymbol{v} \cdot \nabla T = \nabla \cdot (k \nabla T) + q \qquad (2)$$

with ρ, C_p and k the density, specific heat and thermal conductivity of the material.

6. THERMO-MECHANICAL MODEL

A thermo mechanical model was developed in order to compute the von Mises stress in the two critical parts of the applicator: the ceramic tube and the aluminum bloc. The ceramic is submitted to three source of mechanical stress.

- The non uniform temperature distribution inside the ceramic wall leads to thermal stress.
- A graphite joint is used on both end face of the ceramic tube, with an axial stress of 50 MPa for a good sealing,
- The internal pressure during the process can reach 65 bars.

With a 3-D mechanical model, the total stress was computed with the thermal stress (due to the temperature gradient) and with two pressure boundary conditions on the inner wall of the ceramic and on both flat surfaces. The mechanical model is based on force equilibrium of solids (or Newton law extended to solids):

$$\rho \ddot{\mathbf{u}}(\mathbf{x}, t) - \nabla \cdot \boldsymbol{\sigma}(\mathbf{x}, t) - \boldsymbol{\kappa}(\mathbf{x}, t) = 0 \qquad (3)$$

with ρ the density of the solid, $\mathbf{u}$ the displacement vector, $\boldsymbol{\sigma}$ is the material stress tensor for small deformation. $\boldsymbol{\kappa}$ is the external volume forces such as gravity. In our case, the gravity can be neglected in comparison to the internal thermal stress. In addition, the inertial term $\rho\ddot{\mathbf{u}}$ is neglected in comparison to the thermal stress because the time of the duration of the "heat pulse" (microwave heating about$1-3$ s) is much larger than the characteristic time of propagation of stress waves inside the solid (speed of sound in the material over a characteristic length of the solid). The equation (3) without the inertial term is also called the quasi static approximation equilibrium of solids. The constitutive law between the stress and the strain used to model the behaviour of the ceramic and the metal parts is a linear elastic model or generalised 3D Hooke's law:

$$\boldsymbol{\sigma} = \frac{E}{1+\nu}\boldsymbol{\epsilon} + \frac{E\nu}{(1+\nu)(1-2\nu)} tr(\boldsymbol{\epsilon})\mathbf{I} \qquad (4)$$
$$-\alpha \frac{E}{1-2\nu} \Delta T \mathbf{I}$$

Three parameters are used to define the relationship between the strain $\boldsymbol{\epsilon}$ and the stress $\boldsymbol{\sigma}$ in the case of thermal expansion with a temperature difference ΔT. The Young modulus E, which represents the stiffness of the

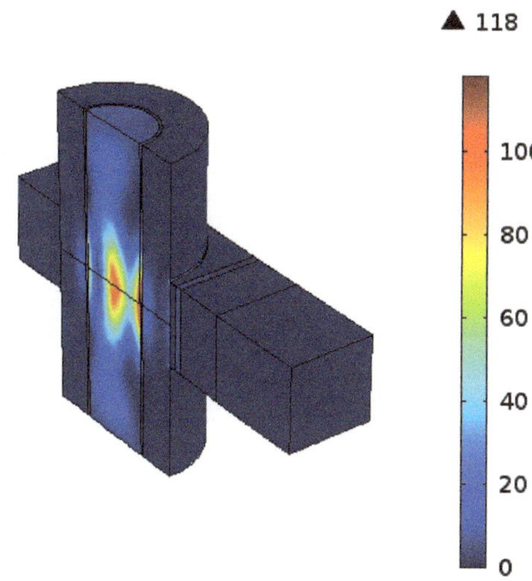

Fig. 5 Power density [$\mathrm{MW/m^3}$] for 10 kW input power

material, the poisson ratio ν which is the ratio between longitudinal and lateral displacement, and the linear thermal expansion coefficient α. The Von Mises stress $\sigma_m = \frac{1}{\sqrt{2}}\sqrt{(\sigma_1 - \sigma_2)^2 + (\sigma_2 - \sigma_3)^2 + (\sigma_3 - \sigma_1)^2}$, is a convenient scalar quantity used to evaluate the stress state inside the solid and to compare with the maximal material strength. σ_i are the main stress components of the stress tensor.

7. THERMAL AND FLUID MODEL

The velocity field inside the water channels was computed with the laminar flow physics of Comsol (Navier stokes equation for an incompressible fluid in laminar regime) in steady state. Then, the thermal model was solved in the solid and in the fluid domain. To take into account the heat transfer from the solid domain to the fluid domain, the advection term $\boldsymbol{v}\rho C_p \cdot \nabla T$ is used with $\boldsymbol{v}$ the velocity field obtained with the steady laminar incompressible fluid simulation.
Comment: The author had also the opportunity to test the pipe flow module in Comsol. The calculated temperature field in the solid domain (aluminum block) was similar for both cases, but the water channels are defined as lines and incorporate the analytic solution based on energy balance for a laminar flow in a tube (poiseuille flow, parabolic profile). The computation time is much shorter than the fluid simulation with Navier Stokes equations.

8. SIMULATION RESULTS AND DISCUSSION

The figure 5 shows the power density distribution inside the lignite and ceramic tube. There are 3 maximum located in the ceramic and lignite material (one in front of the aperture of the wave guide at the interface ceramic/lignite, one in the center, and one at the interface ceramic/lignite on the opposite side of the wave guide aperture). A maximum power density of about 100 $\mathrm{MW/m^3}$ was achieved inside the lignite material.

The temperature field inside the lignite material and ceramic is shown in Fig. 6 after one hour of continuous running with briquettes moving from bottom to top with an average velocity of 2.5 cm/s which correspond to 75 briquettes per minute. The initial lignite temperature material was set to 80° C which is the temperature of the briquettes leaving the briquette press machine and entering in the inner tube of the applicator. The temperature field follows the power density distribution and is not uniform. The maximum temperature is close to 300° C. In this model, the vaporisation of water was not taken into account, and consequently the temperature is overestimated in this current model. The implication is not problematic for the design phase of this applicator because the thermal stress in the ceramic and aluminum parts of the applicator will be also overestimated. Using the heat source from the electromagnetic model, the

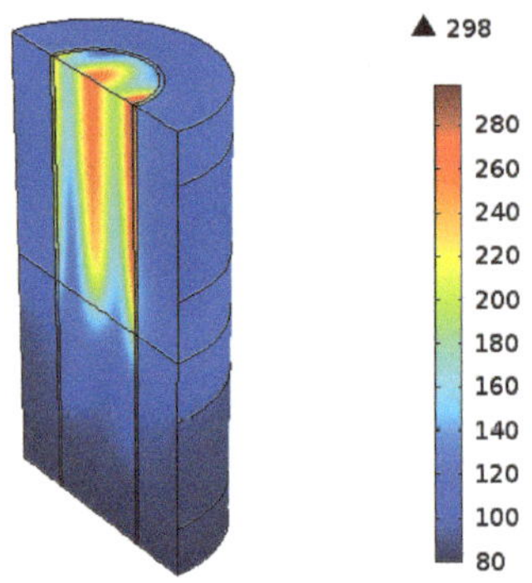

Fig. 6 Temperature [°C] distribution after 1 h of continuous running with a constant briquette translation velocity of $v = 2.5$ cm/s, corresponding to 75 briquettes per minute

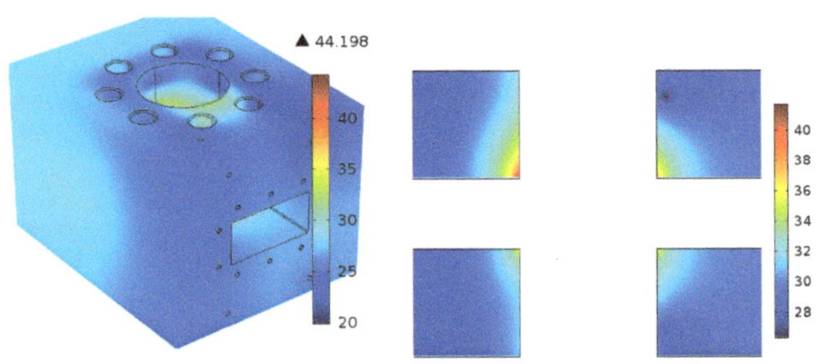

a) b)

Fig. 7 Temperature after 1 h of continuous running with a water flow rate of 6 l/min in the aluminium block a) and in the midle plane of the block b)

temperature field inside the ceramic and metal block was computed for 6 l/min water flow rate. The figure 7 shows the temperature distribution inside the aluminum block (a)) after 1 h of continuous running at 10 kW input power and in the middle plane of the aluminium block b). The white domain in the Fig. 7 is the wave guide filled with air. The ceramic and lignite material have also be hided to see the temperature in this plane. For a flow rate of 6 l/min, the maximum temperature inside the steel block is about 44 ° C in the aluminum block. For the same flow rate and input power, the maximal temperature calculated for a steel block is about 120 ° C . The non uniformity of the temperature distribution (or temperature gradients) is much higher for the steel block than the aluminum block because of its lower thermal conductivity. This shows that thermal expansion can be mitigate by using and aluminum alloy together with a sufficient water flow. For this application, the use of Stainless steel is not recommended because of lower electrical and thermal conductivity than aluminum.

The maximum total Von Mises stress calculated with the thermo mechanical model (temperature field as an input) was about 40 − 60 MPa in the middle of the ceramic as shown in Fig. 8 a) (green and yellow range on the figure, the maximum is not physical since its correspond to a corner singularity with fixed displacements). This stress value is well below the maximum tensile and compressive strength of the silicon nitride tube (more than 400 MPa for the tensile strength and 3000 MPa for the compressive strength). For certifications reasons, the ceramic cylinder was also tested with high pressure water at 5 times the nominal pressure, ie 325 bars.

A modified model with the full geometry was used to compute the stress and displacement distribution inside the aluminum block. This block experiences stress from the inhomogeneous temperature distribution and from the 2×8 M22 screws used to connect the applicator to the flange for

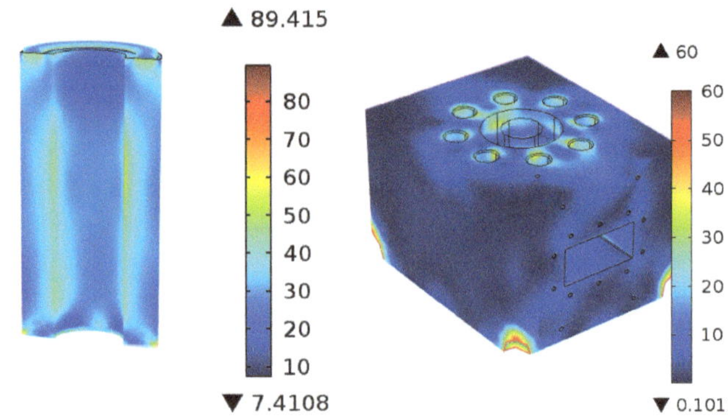

a) b)

Fig. 8 Von Mises stress [MPa] in the ceramic Silicium nitride tube and inside the Aluminum block after 1 h of continuous running. Thermal stress, internal pressure, pressure from the gasket and forces from the screws are included in this model, high stress at the 4 corners are due to the fixed displacement boundary condition. The scale is limited to 60 MPa

a good sealing. After 1 h of continuous running, the maximal value for the Von Mises stress in steady state conditions was about 30 − 50 MPa in the vicinity of the screw thread insert (Fig. 8 b)). This value was well below the maximum tensile strength of Alumec 100 which is about 500 MPa. The maximum displacement was below 0.15 mm for a force of 60 kN per screw and flow rate of 6 l/min.

Overall, these thermo mechanical simulations have shown a good design of the applicator with stress level well below the maximal strength of the materials. In addition, the cooling circuit is essential in order to limit the maximal temperature and consequently the thermal expansion of the block.

9. EXPERIMENTAL RESULTS

The initial tests were performed without the briquetting press machine and the ceramic cylinder was simply filled with liqnite briquettes. For these static conditions (no moving briquettes through the ceramic canal) few tests at a maximum power of 4 kW and for a maximum time duration of 10 s were performed. The temperature field just after microwave treatment was measured with a thermo camera as shown in Fig. 9. Surface temperature of about 100 ° C were measured.

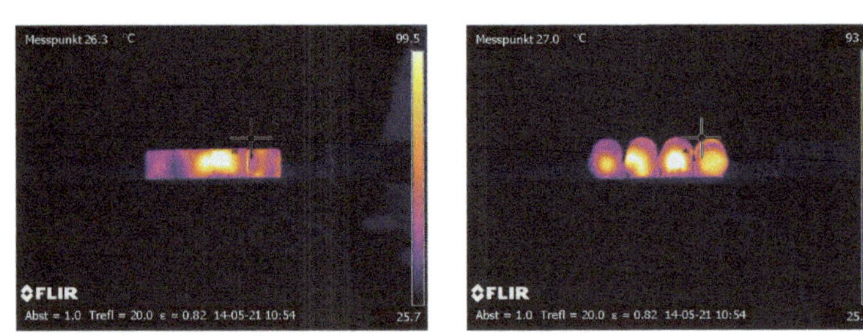

a) b)

Fig. 9 Surface temperature obtained with a thermo camera, lateral and front view of 4 lignite briquette after microwave treatment, max temperature 99 ° C

Additional tests were performed at atmospheric pressure with the applicator connected to the briquetting press and a maximal power of 10 kW could be used. The input power ranges from 0.5 − 10 kW with 0.5 kW step. The controlled software (Muegge MV220V) allows the control of the input power and the monitoring of the reflected power. Lignite briquettes were moving at various speed between 50 to 75 briquettes

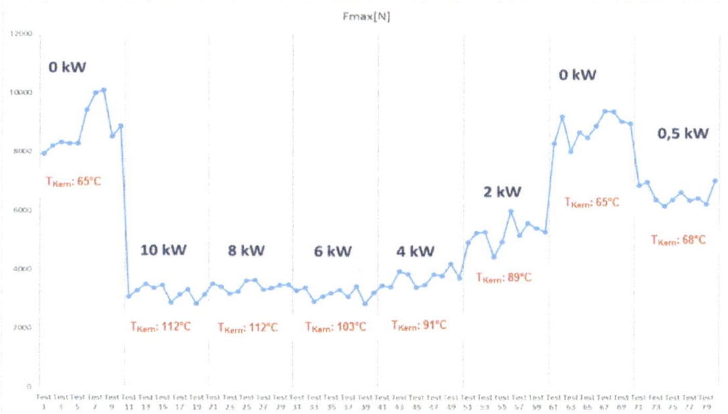

Fig. 10 Mean force [N] required to break the briquettes after microwave irradiation, for power ranging from 0 to 10 kW

per minute. Cracks formation with surface and core temperature of the lignite briquettes higher than 120 °C were measured.

A radial crack from the center to the outside was observed on all the microwave heated briquettes. Briquettes were heated at different power and were collected at the exit of the metallic tube (connected to the applicator) to measure the core temperature with a thermo coupled, after drilling a hole in the center. After that, they were put in a compressive stress machine, to measure the force required to induce breakage. The temperature was obtained from averaging 3 temperature measurements with 3 briquettes with the same input power and speed operating conditions. The figure 10 shows the force required to break the briquettes for different input power level together with the measured core temperature of the briquette (in red). It is interesting to see that even with a power of 500 W the briquettes are significantly weakened (9 kN for a briquette without microwave heating and 6 kN for a briquette heated with 500 W). The core temperature are all above $100°C$ for power ranging from 6 to 10 kW. The temperature values should be taken cautiously because many sources of uncertainty exist (time during manipulation, hole drilling, resident time of the briquette in the metallic tube). Overall the Fig. 10 shows that microwave heating significantly increase the friability of the lignite briquettes. It is believed that a combination of thermal stress and pore pressure from water vaporization is the cause of thermal cracking.

10. CONCLUSION

A compact 10 kW microwave applicator for heating and thermal cracking of lignite briquettes was successfully designed and tested. With 10 kW input power, this applicator allows the heating of wet lignite briquettes (water content in mass of about $\sim 15\%$) with temperature above $100 - 120$ °C and with average briquettes velocities in the range 1.6 to 2.5 cm/s. The utilization of a silicon nitride tube is the key of this high pressure applicator because this material combines both a high mechanical strength and very low dielectric losses. In operation, the aluminum block incorporating an efficient water cooling circuit remains below 45 °C as predicted by the thermal simulations.

The preliminary strength measurements have shown a significant mechanical weakening of the briquettes under microwave heating. Internal cracks are created during fast microwave heating, it is not clear yet what is the dominant phenomenon between pore pressure from water expansion/vaporization and thermal stress due to temperature gradients. Additional experimental tests and measurements are planned to study further the lignite coal material behaviour under fast microwave heating.

During the development phase of the applicator, the use of a sequentially coupled model between the electromagnetic, thermo fluid and mechanical physics allows the design of a reliable applicator with the desired specifications. It gives some good indications about the physical behaviour of the applicator in operation.

Finally, this applicator could be used for other lossy materials in liquid, solid, powder or granular form for continuous volumetric heating with internal pressure ranging from atmospheric pressure to 65 bars. Applications in chemistry and recycling are foreseen.

ACKNOWLEDGEMENTS

The authors would like to thank Dr. Heiner Gutte, TU-Bergakademie Freiberg, for the founding within the DER-project FE-35025755 "Studie zur berührungslosen Erwärmung und Schwächung von Kohlebriketts mit Mikrowellen für die Feststoffvergasung".

NOMENCLATURE

ρ	density (kg/m^3)
C_p	specific heat (J/kg · K)
k	thermal conductivity(W/(m.K))
$\boldsymbol{\sigma}$	stress tensor (Pa)
$\boldsymbol{\epsilon}$	strain tensor (Pa)
$\boldsymbol{I}$	identity tensor (Pa)
E	Young modulus (Pa)
ν	poisson ratio (-)
α	linear expansion coefficient (1/K)
ϵ_{rc}	relative complex permittivity (-)
μ_r	relative permeability (-)
$\boldsymbol{E}$	Electric field vector (V/m)
ω	pulsation (rad/s)
q	power density (W/m^3)
T	Temperature (K)
$\boldsymbol{v}$	average velocity of the briquette (m/s)
$\boldsymbol{n}$	normal vector to a given surface (-)
$\boldsymbol{H}$	magnetic field(H/m)
ϵ_0	permittivity of free space (F/m)
μ_0	permeability of free space(H/m)
∇	gradient operator
$\nabla\cdot$	divergent operator
$\nabla\times$	curl operator

REFERENCES

Batar, T., 2004, "Theory and Applications of Microwave Energy in Communition," *Euro Ceramics VIII*, 1399–1402. http://dx.doi.org/10.4028/www.scientific.net/KEM.264-268.1399.

Binner, E., Lester, E., Kingman, S., Dodds, C., Wu, J.R.T., Wardle, P., and Mathews, J., 2014, "A Review of Microwave Coal Processing," *Journal of Microwave Power and Electromagnetic Energy*, **48**, 35–60.

Pickles, C., Gao, F., and S.Kelebek, 2014, "Microwave Drying of a Low Rank Sub Bitumous Coal," *Minerals Engineering*, **62**, 31–42. http://dx.doi.org/10.1016/j.mineng.2013.10.011.

Rosin, A., Schroder, H., and Repke, J., 2014, "Briquetting Press as Lock Free Continuous Feeding System for Pressurized gasifiers," *Fuel*, **116**, 871–878. http://dx.doi.org/10.1016/j.fuel.2013.03.024.

ACCELERATING MHD FLOW OF A GENERALIZED OLDROYD-B FLUID WITH FRACTIONAL DERIVATIVE

Yaqing Liu[a, *], Jinyu Ma[b]

[a] *Gengdan Institute of Beijing University of Technology, Beijing 101301, China*
[b] *NCPC Hebei huamin pharmaceutical Co., Ltd, Shijiazhuang 052165,China*

ABSTRACT

This paper presents an exact solution for the magnetohydrodynamic (MHD) flow of an incompressible generalized Oldroyd-B fluid due to an infinite accelerating plate. The fractional calculus approach is introduced to establish the constitutive relationship of the Oldroyd-B fluid. The solutions in terms of Fox H-function are obtained by using the Laplace transform. When $N = 0$ the solutions corresponds to the generalized Oldroyd-B fluids, while $\theta \to 0$ and $\lambda \to 0$ describes the Maxwell fluid and the generalized second fluid, as limiting cases of our general results, respectively.

Keywords: *Generalized Oldroyd-B fluid; Exact solution; Laplace transform; Fox H-function.*

1. INTRODUCTION

Non-Newtonian fluids are now considered to play a more important and appropriate role in industrial and technological applications. An important class of non-Newtonian fluids is viscoelastic fluids which exhibit both elastic and viscous property and many models of constitutive equations have been proposed to describe the transport behavior of these fluids. The Oldroyd-B fluid is a special viscoelastic non-Newtonian fluids, it has been especially applied to the problems of having small dimensionless relaxation and retardation times.

Recently, the fractional derivatives (Podlubny, 1999) are found to be quite flexible for describing the behaviors of viscoelastic fluids. Many researchers have studied different problems related to such fluids. In their works, the constitutive equations for generalized non-Newtonian fluids are modified from the well known fluid models by replacing the time derivative of an integer order by the so-called Riemann-Liouville fractional calculus operators. Haitao and Xu (2007) investigated the Stokes' problem for a viscoelastic fluid with a generalized Oldroyd-B model. Khan et al. (2009) and Hyder et al. (2009) considered some accelerated flows and due to pressure gradient flows of generalized Oldroyd-B fluid, respectively. Hyder (2009) discussed the flows of generalized Oldroyd-B fluid between two side walls perpendicular to the plate. Fetecau, Prasad and Rajagopal (2007), Con. Fetecau et al. (2009) and Vieru et al. (2008) investigated some accelerated flows of a generalized Oldroyd-B fluid. Hayat et al. (2007, 2009) studied the flow of a Maxwell fluid between two side walls. Khan et al. (2006) studied the MHD flow of a generalized Oldroyd-B fluid in a circular pipe.

Analytic solutions are important than numerical solutions because these are valid on the whole domain of definition whereas the numerical solutions are only valid at chosen points in the domain of definition. In this paper, we consider the MHD flow of an incompressible generalized Oldroyd-B fluid due to an infinite accelerating plate. The exact solutions for the velocity and shear stress fields are obtained by using the discrete Laplace transform technique for the fractional calculus. Moreover, the solutions for generalized Oldroyd-B fluid, fractional second grade fluid and fractional Maxwell fluid can be recovered by the current analysis.

2. GOVERNING EQUATIONS

We consider the MHD flow of an incompressible generalized Oldroyd-B fluid due to an infinite accelerating plate. Assuming the velocity field and stress of the form

$$V = u(y,t)\mathrm{i}\,,\; S = S(y,t) \tag{1}$$

where u is the velocity and i is the unit vectors in the x-direction, taking account of the initial condition

$$S(y,0) = 0\,,\; y > 0 \tag{2}$$

the fluid being at rest up to the time $t = 0$, we get

$$(1 + \lambda D_t^{\alpha})S_{xy} = \mu(1 + \lambda_r D_t^{\beta})\partial_y u(y,t) \tag{3}$$

$S_{yy} = S_{zz} = S_{xz} = S_{yz} = 0$, $S_{xy} = S_{yx}$. Consider that the conducting fluid is permeated by an imposed magnetic field B_0 which acts in the positive y - coordinate. In the low- magnetic Reynolds number approximation, the magnetic body force is represented $\sigma B_0^2 u$. Then, in the absence of a pressure gradient in the x-direction, the equation of motion yields the following scalar equations:

$$\rho\frac{\partial u}{\partial t} = \frac{\partial}{\partial y}S_{xy} - \sigma B_0^2 u \tag{4}$$

* *Corresponding author. Email: liuyaqing1981@163.com*

where ρ is the constant density of the fluid. Eliminating S_{xy} between Eq. (3) and Eq. (4), we arrive at the following fractional differential equation

$$(1+\lambda D_t^{\alpha})\frac{\partial u(y,t)}{\partial t}=\nu(1+\lambda_r D_t^{\beta})\frac{\partial^2 u(y,t)}{\partial y^2}-M^2(1+\lambda D_t^{\alpha})u(y,t) \quad (5)$$

where D_t^{α} and D_t^{β} are based on Riemann- Liouville's definition is defined as (Podlubny, 1999), $\nu=\mu/\rho$ is the kinematic viscosity and $M^2=\frac{\sigma B_0^2}{\rho}$. The associate initial and boundary conditions as follow:

$$\text{Initial condition: } u(y,0)=\frac{\partial u(y,0)}{\partial t}=0\,,\quad y>0 \quad (6)$$

$$\text{Boundary conditions: } u(0,t)=At\,,\quad t>0 \quad (7)$$

$$u(y,t)\,,\ \frac{\partial u(y,t)}{\partial y}\to 0 \ \text{ as } y\to\infty\,,\quad t>0 \quad (8)$$

where u is velocity in the x -coordinate direction.

3. EXACT SOLUTIONS

Employing the non-dimensional quantities in Eqs. (5)-(8):

$$u^*=\frac{u}{A}\,,\ y^*=\frac{yA\rho}{\mu}\,,\ t^*=\frac{tA^2\rho}{\mu}\,,\ \lambda^*=\lambda\left(\frac{A^2\rho}{\mu}\right)^{\alpha},\ \lambda_r^{\ *}=\lambda_r\left(\frac{A^2\rho}{\mu}\right)^{\beta} \quad (9)$$

Dimensionless motion and equations can be given (for brevity the dimensionless mark "*" is omitted here)

$$(1+\lambda D_t^{\alpha})\frac{\partial u(y,t)}{\partial t}=(1+\lambda_r D_t^{\beta})\frac{\partial^2 u(y,t)}{\partial y^2}-M^*(1+\lambda D_t^{\alpha})u(y,t) \quad (10)$$

where $M^*=\frac{M^2}{\rho^2A^2}$. Using Laplace transforms principle of sequential fractional derivatives, yields

$$\frac{\partial^2 U(y,t)}{\partial y^2}-\frac{(p+M^*)(1+\lambda p^{\alpha})}{(1+\lambda_r p^{\beta})}U(y,t)=0 \quad (11)$$

$$\text{Boundary conditions: } U(0,t)=\frac{1}{p^2} \quad (12)$$

$$U(y,t)\,,\ \frac{\partial U(y,t)}{\partial y}\to 0 \ \text{ as } y\to\infty \quad (13)$$

Solving Eqs. (11)- (13), one obtains

$$U=\frac{1}{p^2}\exp[-(\frac{(p+M^*)(1+\lambda p^{\alpha})}{1+\lambda_r p^{\beta}})^{\frac{1}{2}}y] \quad (14)$$

The stress field can be calculated from Eq. (3). Taking the Laplace transform of Eq. (3) and introducing Eq. (14), obtain

$$\overline{\tau}(y,p)=-\frac{1}{p^2}(p+M^*)^{\frac{1}{2}}(\frac{1+\lambda p^{\alpha}}{1+\lambda_r p^{\beta}})^{-\frac{1}{2}}\exp[-(\frac{(p+M^*)(1+\lambda p^{\alpha})}{1+\lambda_r p^{\beta}})^{\frac{1}{2}}y] \quad (15)$$

where $\overline{\tau}(y,p)$ is the Laplace transform of $\tau(y,t)$ and $\tau(y,t)=S_{xy}/(A^2\rho)$.

In order to avoid the burdensome calculations of residues and contour integrals, we will apply discrete inverse Laplace transform to get to the velocity and the stress fields. Writing Eq. (14) as series forms

$$U(y,p)=\frac{1}{p^2}+\sum_{k=1}^{\infty}\frac{(-y)^k}{k!}\sum_{l=0}^{\infty}\frac{(-M^*)^l}{l!}\left(\frac{\lambda}{\lambda_r}\right)^{\frac{k}{2}}\sum_{m=0}^{\infty}\frac{(-\lambda)^m}{m!}\sum_{n=0}^{\infty}(-\lambda_r^{\ -1})^n$$
$$\times\frac{1}{n!}\frac{\Gamma(l-\frac{k}{2})}{\Gamma(-\frac{k}{2})}\frac{\Gamma(m-\frac{k}{2})}{\Gamma(-\frac{k}{2})}\frac{\Gamma(n+\frac{k}{2})}{\Gamma(\frac{k}{2})}\frac{1}{p^{\frac{k}{2}(\beta-\alpha-1)+\alpha m+\beta n+l+2}} \quad (16)$$

Then, applying the inversion formulae term by term for the Laplace transform, yields

$$u(y,t)=t+\sum_{k=1}^{\infty}\frac{(-y)^k}{k!}\sum_{l=0}^{\infty}\frac{(-M^*)^l}{l!}\left(\frac{\lambda}{\lambda_r}\right)^{\frac{k}{2}}\sum_{m=0}^{\infty}\frac{(-\lambda)^m}{m!}\sum_{n=0}^{\infty}(-\lambda_r^{\ -1}t^{\beta})^n\frac{1}{n!}$$
$$\times\frac{\Gamma(l-\frac{k}{2})}{\Gamma(-\frac{k}{2})}\frac{\Gamma(m-\frac{k}{2})}{\Gamma(-\frac{k}{2})}\frac{\Gamma(n+\frac{k}{2})}{\Gamma(\frac{k}{2})}\frac{t^{\frac{k}{2}(\beta-\alpha-1)+\alpha m+l+1}}{\Gamma(\frac{k}{2}(\beta-\alpha-1)+\alpha m+\beta n+l+2)}\,. \quad (17)$$

In terms of Fox H-function, we rewriting the above equation as the simpler form

$$u(y,t)=t+\sum_{k=1}^{\infty}\frac{(-y)^k}{k!}\sum_{l=0}^{\infty}\frac{(-M^*)^l}{l!}\left(\frac{\lambda}{\lambda_r}\right)^{\frac{k}{2}}\sum_{m=0}^{\infty}\frac{(-\lambda)^m}{m!}t^{\frac{k}{2}(\beta-\alpha-1)+\alpha m+l+1}$$
$$\times H_{3,5}^{1,3}\left[\lambda_r^{\ -1}t^{\beta}\left|\begin{matrix}(1-l+\frac{k}{2},0),(1-m+\frac{k}{2},0),(1-\frac{k}{2},1),\\(0,1),(1+\frac{k}{2},0),(1+\frac{k}{2},0),(1-\frac{k}{2},0),(\frac{k}{2}(\alpha+1-\beta)-\alpha m-l-1,\beta)\end{matrix}\right.\right] \quad (18)$$

where the property of the Fox H-function is

$$\sum_{n=0}^{\infty}\frac{(-z)^n\Pi_{j=1}^{p}\Gamma(a_j+A_jn)}{n!\Pi_{j=1}^{q}\Gamma(b_j+B_jn)}=H_{p,q+1}^{1,p}\left[z\left|\begin{matrix}(1-a_1,A_1),\cdots(1-a_p,A_p)\\(0,1),(1-b_1,B_1),\cdots(1-b_q,B_q)\end{matrix}\right.\right] \quad (19)$$

Similarly with the velocity field, we write Eq. (15) as the series form

$$\overline{\tau}(y,p)=-\sum_{k=0}^{\infty}\frac{(-y)^k}{k!}\sum_{l=0}^{\infty}(-M^*)^l\frac{1}{l!}\left(\frac{\lambda}{\lambda_r}\right)^{\frac{k-1}{2}}\sum_{m=0}^{\infty}\frac{(-\lambda)^{-m}}{m!}$$
$$\times\sum_{n=0}^{\infty}\frac{(-\lambda_r)^{-n}}{n!}\frac{\Gamma(l-\frac{k+1}{2})}{\Gamma(-\frac{k+1}{2})}\frac{\Gamma(m-\frac{k-1}{2})}{\Gamma(-\frac{k-1}{2})}\frac{\Gamma(n+\frac{k-1}{2})}{\Gamma(\frac{k-1}{2})}$$
$$\times\frac{1}{p^{\frac{k-1}{2}(\beta-\alpha-1)+\alpha m+\beta n+l+1}} \quad (20)$$

And using the discrete inverse Laplace transform, yields

$$\tau(y,t)=-\sum_{k=0}^{\infty}\frac{(-y)^k}{k!}\sum_{l=0}^{\infty}(-M^*)^l\frac{1}{l!}\left(\frac{\lambda}{\lambda_r}\right)^{\frac{k-1}{2}}\sum_{m=0}^{\infty}\frac{(-\lambda)^{-m}}{m!}$$

$$\times\sum_{n=0}^{\infty}\frac{(-\lambda_r)^{-n}}{n!}\frac{\Gamma(l-\frac{k+1}{2})}{\Gamma(-\frac{k+1}{2})}\frac{\Gamma(m-\frac{k-1}{2})}{\Gamma(-\frac{k-1}{2})}\frac{\Gamma(n+\frac{k-1}{2})}{\Gamma(\frac{k-1}{2})}$$

$$\times\frac{t^{\frac{k-1}{2}(\beta-\alpha-1)+\alpha m+\beta n+l}}{\Gamma(\frac{k-1}{2}(\beta-\alpha-1)+\alpha m+\beta n+l+1)} \tag{21}$$

In terms of the Fox H-function, we obtain

$$\tau(y,t)=-\sum_{k=0}^{\infty}\frac{(-y)^k}{k!}\sum_{l=0}^{\infty}(-M^*)^l\frac{1}{l!}\left(\frac{\lambda}{\lambda_r}\right)^{\frac{k-1}{2}}\sum_{m=0}^{\infty}\frac{(-\lambda)^{-m}}{m!}t^{\frac{k-1}{2}(\beta-\alpha-1)+\alpha m+l}$$
$$\times H^{1,3}_{3,5}\left[\lambda_r^{-1}t^{\beta}\left|\begin{matrix}(1-l+\frac{k+1}{2},0),(1-m+\frac{k-1}{2},0),(1-\frac{k-1}{2},1),\\(0,1),(1+\frac{k+1}{2},0),(1+\frac{k-1}{2},0),(1-\frac{k-1}{2},0),(\frac{k-1}{2}(\alpha+1-\beta)-\alpha m-l,\beta)\end{matrix}\right.\right] \tag{22}$$

4. SPECIAL CASES

(i) When $M^*=0$, our solution reduced to the generalized Oldroyd-B fluid, as obtained by Khan et al. (2009). We attain to the velocity field

$$u(y,t)=t+\sum_{k=1}^{\infty}\frac{(-y)^k}{k!}\left(\frac{\lambda}{\lambda_r}\right)^{\frac{k}{2}}\sum_{m=0}^{\infty}\frac{(-\lambda)^m}{m!}t^{\frac{k}{2}(\beta-\alpha-1)+\alpha m+1}$$
$$\times H^{1,2}_{2,4}\left[\lambda_r^{-1}t^{\beta}\left|\begin{matrix}(1-m+\frac{k}{2},0),(1-\frac{k}{2},1),\\(0,1),(1+\frac{k}{2},0),(1-\frac{k}{2},0),(\frac{k}{2}(\alpha+1-\beta)-\alpha m-1,\beta)\end{matrix}\right.\right] \tag{23}$$

and the associated shear stress

$$\tau(y,t)=-\sum_{k=0}^{\infty}\frac{(-y)^k}{k!}\left(\frac{\lambda}{\lambda_r}\right)^{\frac{k-1}{2}}\sum_{m=0}^{\infty}\frac{(-\lambda)^{-m}}{m!}t^{\frac{k-1}{2}(\beta-\alpha-1)+\alpha m}$$
$$\times H^{1,2}_{2,4}\left[\lambda_r^{-1}t^{\beta}\left|\begin{matrix}(1-m+\frac{k-1}{2},0),(1-\frac{k-1}{2},1),\\(0,1),(1+\frac{k-1}{2},0),(1-\frac{k-1}{2},0),(\frac{k-1}{2}(\alpha+1-\beta)-\alpha m,\beta)\end{matrix}\right.\right] \tag{24}$$

(ii) If $\alpha\neq 0$, $\lambda\to 0$, then Eqs.(18) and (22) can be simplified as

$$u(y,t)=t+\sum_{k=1}^{\infty}\frac{(-y)^k}{k!}\sum_{l=0}^{\infty}(-M^*)^l\frac{1}{l!}\left(\frac{1}{\lambda_r}\right)^{\frac{k}{2}}t^{\frac{k}{2}(\beta-1)+l+1}$$
$$\times H^{1,2}_{2,4}\left[\lambda_r^{-1}t^{\beta}\left|\begin{matrix}(1+\frac{k}{2}-l,0),(1-\frac{k}{2},1)\\(0,1),(1+\frac{k}{2},0),(1-\frac{k}{2},0),(\frac{k}{2}(1-\beta)-l-1,\beta)\end{matrix}\right.\right] \tag{25}$$

$$\tau(y,t)=-\sum_{k=0}^{\infty}\frac{(-y)^k}{k!}\sum_{l=0}^{\infty}(-M^*)^l\frac{1}{l!}\left(\frac{1}{\lambda_r}\right)^{\frac{k-1}{2}}t^{\frac{k-1}{2}(\beta-1)+l}$$
$$\times H^{1,2}_{2,4}\left[\lambda_r^{-1}t^{\beta}\left|\begin{matrix}(1-\frac{k+1}{2}+l,0),(1-\frac{k-1}{2},1)\\(0,1),(1+\frac{k+1}{2},0),(1-\frac{k-1}{2},0),(\frac{k-1}{2}(1-\beta)-l,\beta)\end{matrix}\right.\right] \tag{26}$$

which represent the velocity and the stress fields for a generalized second grade fluid.

(iii) If $\beta\neq 0$, $\lambda_r\to 0$, then Eqs. (19) and (22) can be simplified as

$$u(y,t)=t+\sum_{k=1}^{\infty}\frac{(-y)^k}{k!}\sum_{l=0}^{\infty}\frac{(-M^*)^l}{l!}\lambda^{\frac{k}{2}}t^{-\frac{k}{2}(\alpha+1)+l+1}$$
$$\times H^{1,2}_{2,4}\left[\lambda^{-1}t^{\alpha}\left|\begin{matrix}(1+\frac{k}{2}-l,0),(1+\frac{k}{2},1)\\(0,1),(1+\frac{k}{2},0),(1+\frac{k}{2},0),(\frac{k}{2}(\alpha+1)-l-1,\alpha)\end{matrix}\right.\right] \tag{27}$$

$$\tau(y,t)=-\sum_{k=0}^{\infty}\frac{(-y)^k}{k!}\sum_{l=0}^{\infty}(-M^*)^l\frac{1}{l!}\lambda^{\frac{k-1}{2}}t^{-\frac{k-1}{2}(\alpha+1)+l}$$
$$\times H^{1,2}_{2,4}\left[\lambda^{-1}t^{\alpha}\left|\begin{matrix}(1+\frac{k+1}{2}-l,0),(1+\frac{k-1}{2},1)\\(0,1),(1+\frac{k+1}{2},0),(1+\frac{k-1}{2},0),(\frac{k-1}{2}(\alpha+1)-l,\alpha)\end{matrix}\right.\right] \tag{28}$$

which correspond to the similar solutions for a generalized Maxwell fluid.

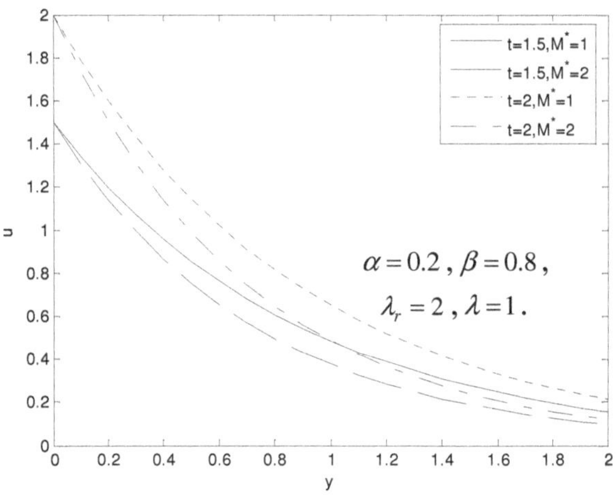

Fig. 1 Velocity fields for different values of t and M^* when keeping α, β, λ_r, λ fixed.

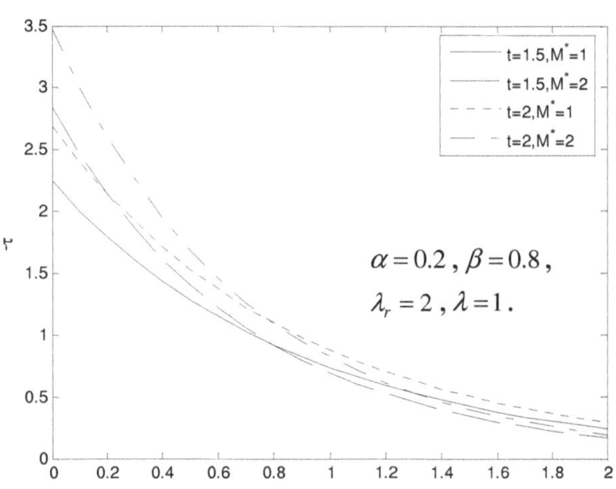

Fig. 2 Stress fields for different values of t and M^* when keeping α, β, λ_r, λ fixed.

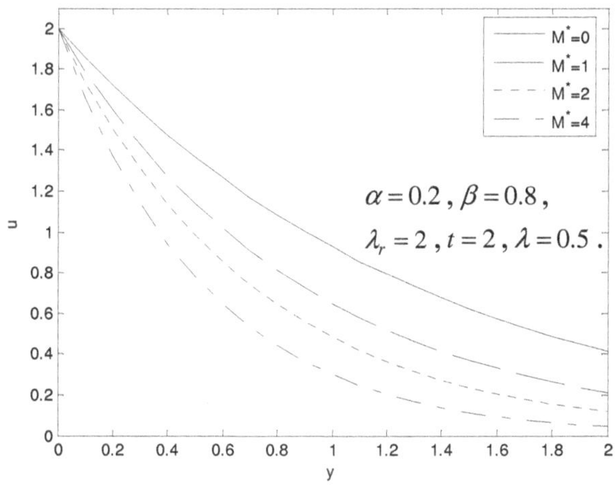

Fig. 3 Velocity fields for different values of M^* when keeping α, β, λ_r, λ, t fixed.

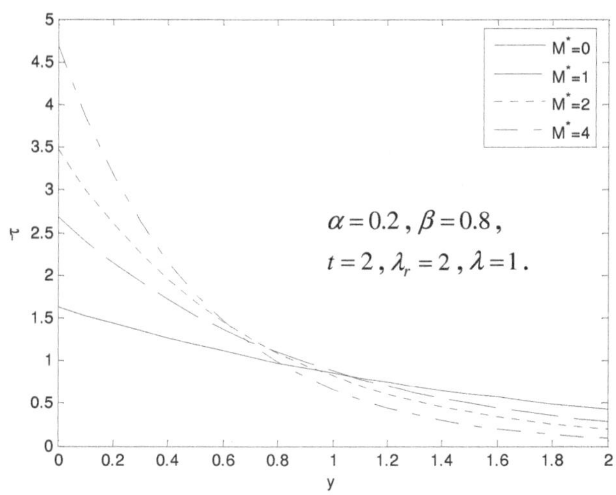

Fig. 4 Stress fields for different values of M^* when keeping α, β, λ_r, λ, t fixed.

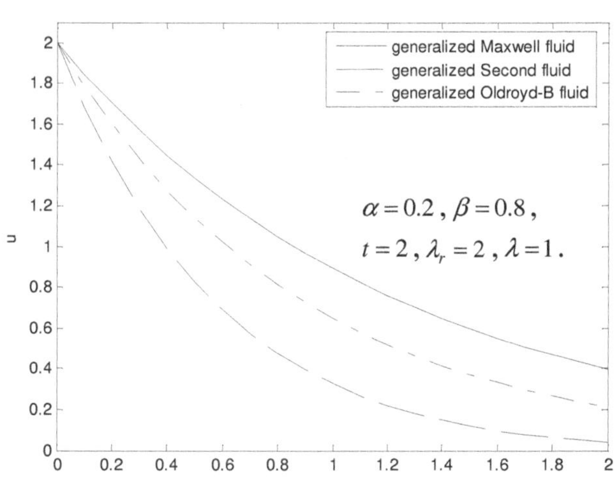

Fig. 5 Velocity fields for three types fluids when keeping α, β, λ_r, λ, t fixed.

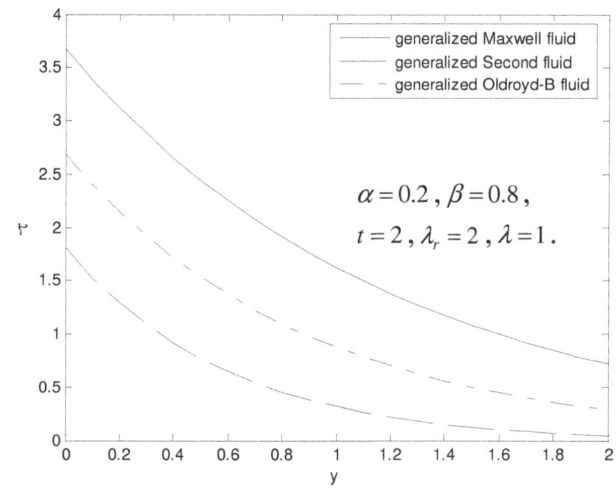

Fig. 6 Stress fields for three types fluids when keeping α, β, λ_r, λ, t fixed.

5. NUMERICAL RESULTS AND DISCUSSION

In this section, we analyze the characteristics of the velocity field and the stress fields by the presented figures. We can see from these figures that the magnetic body force is favorable to the velocity decays, and the more the values of M^* , the more rapidly the velocity decays.

Figures 1 and 2 indicate that the flow velocity as well as the shear stress increase with time or the magnetic body force increase. The strongest shear stress occurs near the plate and the shear stress decreases rapidly with the increase of distance from the plate. If $M^*=0$, the fluid not influenced by magnetic field. Figs.3 and 4 demonstrate the influence of MHD on the velocity and the shear stress, respectively. Fig.3 shows the smaller the values of M^* , the more slowly the velocity decays for the flow. Furthermore, with the increase in magnetic field the velocity decreases. Fig.4 shows the maximum stress occurs near the wall and decreases far away from the wall. Figs. 5-6 illustrate the velocity fields and shear stresses of the generalized Maxwell fluid, second grade fluid and Oldroyd-B fluid. From the figures, we can see that shear stress directly influence velocity field, the Maxwell fluid velocity is the fastest, the next is Oldroyd-B, and second grade fluid velocity is the slowest.

6. CONCLUSIONS

The purpose of this paper is to provide exact solutions for the unsteady flow of a generalized Oldroyd-B fluid due to an infinite accelerating plate. The exact solutions for the velocity and shear stress fields in terms of Fox H-function are obtained by using the Laplace transform. Some previous and classical results, such as the velocity distribution for a viscous Maxwell, Oldroyd-B, and second grade fluid, can be considered as particular cases of our results. It is shown that the fractional constitutive relationship model is more useful than conventional model for describing the properties of viscoelastic fluid.

ACKNOWLEDGEMENTS

The work is supported by the Teaching Reform Project Foundations (No.GDJG201407) of BGD.

The authors wish to thank the Editor and Referees for their helpful comments and suggestions.

REFERENCES

Fetecau C., Prasad Sharat C., Rajagopal K.R., 2007, "A Note on the Flow Induced by a Constantly Accelerating Plate in an Oldroyd-B Fluid," *Applied Mathematical Modeling*, **31**, 647-654. http://dx.doi.org/10.1016/j.apm.2005.11.032

Fetecau Constantin, Fetecau Corina, Kamran M., Vieru D., 2009, "Exact Solutions for the Flow of a Generalized Oldroyd-B Fluid Induced by a Constantly Accelerating Plate Between Two Side Walls Perpendicular to thc Plate," *J. Non-Newtonian Fluid Mech*, **156**, 189-201. http://dx.doi.org/10.1016/j.jnnfm.2008.06.005

Hayat T., Sajid M., 2007, "Homotopy Analysis of MHD Boundary Layer Flow of an Upper-Convected Maxwell Fluid," *International Journal of Engineering Science*, **45**, 393-401. http://dx.doi.org/10.1016/j.ijengsci.2007.04.009

Hayat T., Abbas Z., Sajid M., 2009, "MHD Stagnation-Point Flow of an Upper-Convected Maxwell Fluid Over a Stretching Surface," *Chaos, Solitons and Fractals*, **39**, 840-848. http://dx.doi.org/10.1016/j.chaos.2007.01.067

Hyder Ali Muttaqi Shah S., Khan M., Qi Haitao, 2009, "Exact Solutions for a Viscoelastic Fluid with the Generalized Oldroyd-B Model," *Nonlinear Analysis: Real World Applications*, **10**, 2590-2599. http://dx.doi.org/10.1016/j.nonrwa.2008.03.012

Hyder Ali Muttaqi Shah S., 2009, "Some Accelerated Flows of Generalized Oldroyd-B Fluid Between Two Side Walls Perpendicular to the Plate," *Nonlinear Analysis: Real World Applications,* **10**, 2146-2150. http://dx.doi.org/10.1016/j.nonrwa.2008.03.023

Khan M., HayaT. t, Asghar S., 2006, "Exact Solution for MHD Flow of a Generalized Oldroyd-B Fluid with Modified Darcy's Law," *International Journal of Engineering Science*, **44**, 333-339. http://dx.doi.org/10.1016/j.ijengsci.2005.12.004

Khan M., Hyder Ali S., Qi Haitao, 2009, "Some Accelerated Flows for a Generalized Oldroyd-B Fluid," *Nonlinear Analysis: Real World Applications*, **10**, 980-991. http://dx.doi.org/10.1016/j.nonrwa.2007.11.017

Podlubny I., 1999, *Fractional Differential Equations*. New York: Academic Press.

Qi H.T., Xu M.Y., 2007, "Stokes' First Problem for a Viscoelastic Fluid with the Generalized Oldroyd-B Model," *Acta Mech. Sin.*, **23**, 463-469. http://dx.doi.org/ 10.1007/s10409-007-0093-2

Vieru D., Fetecau Corina, Fetecau C., 2008, "Flow of a Generalized Oldroyd-B Fluid Due to a Constantly Accelerating Plate," *Applied Mathematics and Computation*, **201**, 834-842. http://dx.doi.org/10.1016/j.amc.2007.12.045.

MODELLING AND EXPERIMENTAL VALIDATION OF COMBUSTION IN STRAIGHT INOCULATION COMPRESSION IGNITION ENGINE FUELLED WITH DIESEL AND JATROPHA METHYL ESTER BLEND

Biswajit De*, Rajsekhar Panua

Mechanical Engineering Department, National Institute of Technology, Agartala, Tripura, 799055, India

ABSTRACT

An incorporated arithmetical model has been urbanized and investigated for CFD replication of a solitary cylinder, four stroke, straight inoculation, compressed ignition diesel engine of 3.5 kW for in-cylinder combustion analysis and authenticated under engine simulations at full load functioning conditions with foundation fuel diesel and 10% JME (volume basis) blend with diesel at invariable speed of 1500 rpm. For advancing the exactness of the exertion, a number of sub models, such as species transport model explaining the actual biodiesel energy content and molecular structure as soon as fuel blend is initiated, spray break-up model, wave model and pre-mixed combustion model are also initiated. Consequences of replication and the data from corresponding experimental investigation are evaluated to one another in terms of pressure traces, heat release rates. Excellent concurrences of pressure trace, heat release rate is accomplished involving the experimental and predicted data for all the tested fuels. The outcomes demonstrate that the anticipated mechanism offers reliable predictions of in-cylinder combustion.

Keywords: *CFD; k-ε model; simulation; validation;*

1. INTRODUCTION

In current years, CFD modeling is an effective tool for the analysis of diesel engine combustion and emissions in preference to full experimental study. The main advantages of CFD simulations are that it gives details on spatial and temporal variations of local in-cylinder properties and also saves time, cost and manpower (Andreadis *et al.* and O'Rourke *et al.*). Goldsworthy *et al.* (2006) in his study considered a simplified evaporation and combustion model, fuelled with Heavy fuel oil (HFO), as a mixture of a heavy (residual) and a light (cutter) component, implemented in CFD code Star-CD. During the experiments, it was found that gauged and computed data for ignition delay, burning rate, spray and flame structure were closer to each other. Struckmeier *et al.* (2009) deliberated an additional comprehensive approach of HFO modeling with a KIVA-based code, supplementary embryonic evaporation, ignition and combustion models and originated that a excellent conformity between the simulated data and the experimental data. Kyriakides *et al.* (2009) urbanized a comprehensive model of heavy fuel oil (HFO) with the fuel properties belongings to thermo physical properties, together with viscosity, surface tension, vapour pressure, enthalpy and latent heat of evaporation versus temperature. The model was experienced in an invariable volume combustion chamber through replications of non-evaporating and evaporating sprays. Herrmann *et al.* (2009, 2010) represented the diesel fuel spray experimental results of a large marine Diesel engine by modeling a bulky unvarying volume combustion chamber for presenting the measurements concerning the spray break up and combustion with modification of the swirl concentration, the air temperature and pressure inside the chamber, as well as the injector features (number of holes, orifice diameter and spray orientation). Garaniya *et al.* (2011) studies a comprehensive model of heavy fuel oil (HFO) by making an allowance for the consequences of the high molecular weight hydrocarbons in HFO with the intention of including the tendency to undergo thermal cracking and polymerization. The intention of the present investigation is to authenticate a pre-mixed combustion model in association with the existing dynamic mesh and spray models in conjunction with an experimental data set consisting of diesel and 10% JME (volume basis) blend with diesel at full load functioning circumstances at unvarying speed of 1500 rpm.

2. NUMERICAL PROCEDURE

2.1 CFD code and calculating meshes

In the present study, CFD package, FLUENT 6.3 software is used for modeling the multifaceted combustion happening in compression ignition engine. The injector is modeled at the midpoint of the combustion chamber for symmetrical location and the CFD computations are executed on 0^0 sector mesh. Simulation on in-cylinder flow and combustion processes are contemplated by adding exhaust and intake ports in the computational mesh. Computations commence at Intake Valve open and stop at Exhaust Valve closing. The entire computations are utilized the identical initial and boundary conditions. Computations are done by employing the time step of 0.2 deg of crank angle. Exact number of cells in the mesh is 30225 and 35311 at TDC and BDC, respectively.

2.2 Spray Model

The standard WAVE model (Liu *et al.*, 1993) is employed for the primary and secondary atomization modeling of the consequential droplets. This model includes the enlargement of a primary perturbation on a liquid surface which is related to its wave length and to other

*Corresponding Author. Email: biswajitde62@gmail.com

physical and dynamic parameters of the injected fuel. In this model the growth of an initial perturbation on a liquid surface is linked to its wave length and to other physical and dynamic parameters of the injected fuel and the domain fluid. Nozzle exit diameter is accountable for the injected drop packages with attribute size. Nozzle exit diameter is accountable for the injected drop packages with attribute size. The treatments of the droplets are prepared by the process of heat-up and evaporation including to the assumption of uniform droplet temperature is described in Dukowicz model. Drop parcels are injected with characteristic size equal to the nozzle exit diameter (blob injection). The Dukowicz model was applied for treating the heat-up and evaporation of the droplets, which is described by Dukowicz *et al.* (1979). This model assumes a uniform droplet temperature. In addition, the rate of droplet temperature change is determined by the heat balance, which states that the heat convection from the gas to the droplet either heats up the droplet or supplies heat for vaporization. The spray wall interaction model (Naber *et al.*, 1988) assumes that a droplet, which strikes the wall, is pretentious by rebound or reflection, based on the Weber number. The Shell auto-ignition model (Halstead *et al.*, 1977) is employed for modeling the auto-ignition describing the generic mechanism, 6 generic species for hydrocarbon fuel, oxidizer, total radical pool, branching agent, intermediate species and products.

2.3 The Turbulent Mixing Model

The k-ε approach has been exploited to take account of turbulent effects, while the multifaceted oxidation progression of diesel fuel has been recapitulated by a single step irreversible reaction (Liu *et al.*, 1993 and ICE Physics & Chemistry, 2009). The mean reaction rate has been assessed through the Coherent Flamelet Model (Colin *et al.*, 2004). Intended for a diesel spray, the fuel dewdrops are extremely close to each other and are situated in a region basically made of fuel. Subsequent to the evaporation of the fuel, a sufficient time is wanted for the mixing from the nearly pure fuel region with the surrounding air.

2.4 The Combustion Model

The combustion model is based on the Coherent Flame Model (ECFM-3Z model) differentiates linking all three main regimes relevant in diesel combustion, namely auto-ignition, premixed flame and non-premixed, i.e. diffusion combustion (Colin et al., 2004 and Hélie et al., 2000). The premixed charge of fuel and air is utilized to compute the auto-ignition pre-reactions combining with the ignition delay presided over the local temperature, pressure, fuel/air equivalence ratio and the quantity of remaining gases. Local auto-ignition is afterward the premixed combustion in the fuel/air/residual gas mixture created during the time period between start of injection and auto-ignition onset within the ECFM-3Z modeled according to a flame propagation progression. The third establishment is the one of diffusion combustion where the reaction takes place in a thin zone which detaches fuel and oxidizer. It is implicit that the time needed for the diffusion process is much greater than that of the chemical time in the reaction zone and the rate of reaction throughout the diffusion combustion is computed completely by the intermixing of fuel and oxidizer. This distinct partition of the several ignition/combustion regimes makes the ECFM-3Z model purposely made to conventional as well as alternative diesel combustion modes. For the most part of combustion process of conventional fuel can be presumed as diffusion type but newly initiated substitute conception is that a large amount of fuel is consumed within the premixed combustion.

2.5 Mesh generation

Figure 1(a) illustrates the computational domain of two dimensional combustion chamber geometry of the diesel engine with inlet and exhaust ports and (b) demonstrates the mesh structure of computational domain for model geometry at 605 degree Crank angle.

The mesh organization is detached into two zones; uncomplicated geometry of the indisputable preparation facilitates zones 1 to construct by utilizing the quadrilateral structured mesh. The deformation of the dislodgment volume during compression and expansion strokes and the movements of the inlet and exhaust valve manages zone 2 to acquire the advantage of the triangular structured mesh. The piston and valve movements are arranged specifically by the crank angle based functions. The mesh is composed of approximately 30,225 computational cells when the piston is at the Top-dead centre location, with a resolution of 2 mm.

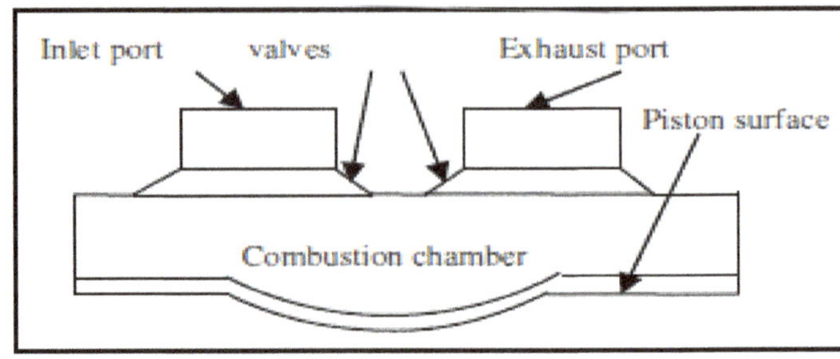

(a)

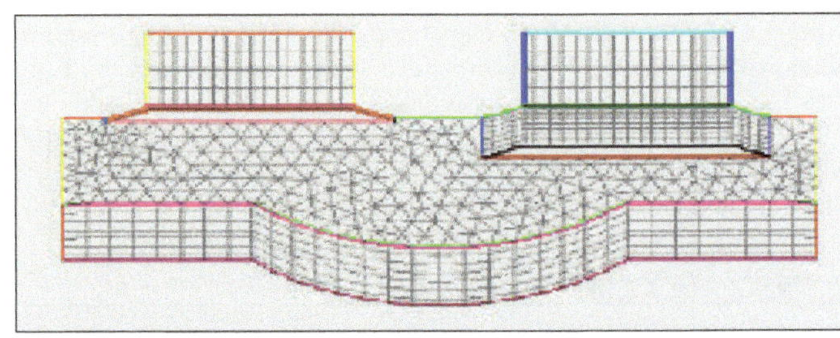

(b)

Fig. Model Geometry of combustion chamber with valves **1(a)** at 0^0 Crank angle and **(b)** Mesh structure of computational domain at 605 degree Crank angle

2.5.1 Gas Exchange Phase

The province for the computation of the gas exchange incorporates the parts of the intake and exhaust channel with something like diameter of 28.2 mm and 28 mm, correspondingly. Initial and boundary circumstances of the computational model are developed from experimental results of engine measurements as well as from two-dimensional engine cycle process simulations using the program code FLUENT 6.3. The channel is represented with quadrilateral cells of h = 4.5 mm size. The size of quadrilateral elements is refined to h = 2 mm, near the valves and to h = 0.11 mm, near the valve seats. On the whole the magnitudes of the computational mesh in the gas exchange phase diverge among 30225 and 33215 thousand elements.

2.5.2 Diesel or JME blend Injection Phase

Making an allowance for the computation of the supersonic diesel or JME blend jet and high velocity gradients at the diesel or JME blend inflow, the resolution of the numerical grid is sophisticated analytically. Diesel or JME blend injection commences at 23^0 BTDC. A sensitivity testing considering the influence of the grid resolution on computational results shows a sufficient discritization of the proposed mesh.

2.5.2 Compression and Expansion Stroke

Seeing that diesel or JME blend inoculation has taken place near the end of compression stroke, so it demands a numerical resolution to a large extent. The magnitude of triangular elements in the combustion chamber, h = 4 mm and the elements are sophisticated to h = 1.2 mm in the province of compression ignition during the combustion phase. The overall number of elements within the domain varies between 33215 and 35311 thousand elements depending on the piston lift.

3 SOLUTION PROCEDURE, GRID INDEPENDENCY TEST AND VALIDATION

3.1 Solution Procedure

3.1.1 Initial and Boundary Conditions

CFD computations considered the gas exchange phase throughout the exhaust and intake stroke, diesel or JME blend injection during the compression stroke and the combustion/ expansion phases. Initial and boundary circumstances of the computational model are developed from experimental results of engine measurements as well as from two-dimensional engine cycle process simulations using the program code FLUENT 6.3. Unsteady pressure and temperature boundary conditions are described at the intake and exhaust port developed from two-dimensional engine cycle simulations. Wall boundary conditions are developed commencing two-dimensional engine cycle simulations utilizing a simple finite volume method with the intention of approximation the mean wall-temperatures of the in-cylinder geometry. Inlet boundary conditions for diesel or JME blend injection are described at the characteristics of the injector. The diesel or JME blend inoculation profile is urbanized by the mass flow inlet boundary condition originated from the injector needle lift measurements and from the values of the averaged mean mass flow. With the supposition of proportionality linking to the needle lift and mass flow, a profile is developed from the start of injection (SOI) and from the end of injection. The measured value of the injected fuel mass is employed to indicate a representative profile which is ultimately used as boundary condition for diesel or JME mass flow rate in the CFD computations. Diesel or JME blend is alleged to be heated up to the level of the coolant temperature within the cylinder head, in view of the fact that the injector is entirely enclosed by the cylinder head geometry and the engine is operated at invariable speed.

3.1.2 Finite Volume Method

Finite volume method is a numerical practice which splits the whole volume of the computational domain into a no. of minute volume enclosing each node point on a mesh for representing and estimating partial differential equations_in the appearance of arithmetic equations.

The common prevailing equations for control volume can be written as:

$$\frac{\partial}{\partial t}(\rho\phi)+\bar{\nabla}.(\rho\bar{v}\phi)=\bar{\nabla}.(\Gamma\bar{\nabla}\phi)+S \quad \ldots (1).$$

Where ϕ is the general variable which can be used to represent the various fluid properties such as temperature or momentum and Γ is the diffusion coefficient, S is the source term.

3.1.3 Numerical Settings

For solving the partial governing equations, the pressure based segregated algorithm SIMPLE is used. The algorithm computes the pressure field utilizing a relationship connecting the velocity and pressure corrections which are based on a number of initial guessed pressure field to put into effect the mass conservation. The preparation is replicated in the fluid field for turbulent and two dimensional flow. SIMPLE is better explanations for convergence than the other algorithm available in Fluent. for momentum and energy are exercised. The girth of time steps is amended for each of the different engine phases.

In the For interpolation of the values at the faces, the body force weighted method for pressure and second order upwind schemes gas exchange phase and in the compression phase the time step is defined as $\Delta\phi = 0.2^0$ CA. The values of relaxation factor are employed in the current study is shown in table 1 and the minimum residuals values for convergence are shown in table 2.

Table.1 Relaxation Factor.

Pressure	Density	Body Force	Momentum	Energy
0.3	1	1	0.7	1

Table 2 Residual limits.

Continuity	X- Momentum	Y- Momentum	Energy
10^{-3}	10^{-3}	10^{-3}	10^{-6}

3.2 Grid Independency Test

The test is conceded by trailing with a variety of grid sizes and surveillance how the solution revolutionizes concerning the changes in grid sizes. Finally, a stage will come when changing the grid spacing will not affect the solution. Solutions are assumed to converge when the following convergence criterion is satisfied at every point in the solution domain. Crevice volume is not considered here and the geometric compression ratio of the test engine is matched by adjusting the clearance volume of the mesh domain The mesh is composed of approximately 30,225 computational cells when the piston is at the Top-dead centre position, with a resolution of 2 mm. This mesh resolution is established to be fabricate the grid independent results, with comparatively high accuracy level and with a lesser computational runtime. Table 3 shows the experimental peak pressure value of the test engine operated with diesel fuel and the various predicted peak pressure values of the simulation when three mesh resolutions are tested for grid independency. The mesh of cell size with 2.5 mm resulted in the highest percentage of error of 2.1% between experimental and predicted values, followed by mesh resolution of 1.5 mm with 1.11% error and mesh resolution of 2 mm with 0.81% error. .

Table 3 Computational conditions of fuel injection for studied cases.

Experimental peak Pressure (bar)	50.236		
Crank angle (degree)	372		
Mesh interval size (mm)	1.5	Mesh interval size (mm)	1.5
Predicted Peak Pressure (bar)	49.68	Predicted Peak Pressure (bar)	49.68
Crank angle (degree)	373	Crank angle (degree)	373
Error in peak Pressure (%)	1.11%	Error in peak Pressure (%)	1.11%

4 RESULTS AND DISCUSSIONS

4.1 Model Validation

The diesel engine utilized for the model validation is a single cylinder, four strokes, water cooled direct injection compression ignition engine. The engine arrangements are prearranged in Table 4 (De *et al.* 2014). The experiments are carried out in an entirely instrumented test engine with CR 17.5 at invariable speed of 1500 rpm, fuelled with diesel and 10% JME (volume basis). The main characteristics of the injection system are listed in Table 5. Table 6 describes the various physical, chemical and thermal properties of Diesel and 10% JME blend. The investigation engine is laden by attaching it with an eddy current sort dynamometer. On line performance assessment of the test engine is achieved by using the Labview based Engine Recital Investigation software package "EnginesoftLV". A piezoelectric pressure transducer (pressure range 0-250 bars) and a charge amplifier is utilized to evaluate the cylinder pressure. Additional analysis is completed by utilizing the pressure data which are transferred and stored in the data acquisition system.

Table 4 Specification of the test engine.

S. No	*Parameters*	*Specification*
1.	General Details	Single cylinder, four stroke compression ignition engine, constant speed, vertical, water cooled, direct injection
2.	Stroke	110 mm
3.	Bore	87.5 mm
4.	Displacement	661 cc
5.	Compression ratio	17.5
6.	Rated output	3.7 KW
7.	Rated speed	1500 rpm

Table 5 Injector fuel system specifications.

S. No	*Injector type*	*Common rail*
1.	Injection pressure Variable	(up to 240 MPa)
2.	Number of nozzle holes	6
3.	Nozzle hole diameter	0.26 mm
4.	Start of injection	23^0 BTDC
5.	Injection duration	24^0 CA
6.	Fuel injected	0.3196 g/cycle

Table 6 Properties of mineral diesel and 10% JME blend.

S. No	*Property*	*Mineral diesel*	*JME10*
1.	Density (Kg/m^3)	835	883.6
2.	Kinematic viscosity at 40^0 C (cSt)	2 - 4	5.74
3.	Flash point (^{0} C)	65	64
4.	Calorific value (MJ/kg)	43.31	39.25
5.	Cetane Number	45-55	NA

Angle measurement is made by utilizing a crank angle encoder (Kistler make) with an accuracy of 1^0. Mass flow meters are employed for quantifying the mass flow rate of JME blends and diesel in litters per minute. At beginning, the experiment engine is operated with standard diesel for looking into the engine performance at full load circumstance and then parallel experiment is carried out with 10% JME which is organized on a volume basis. For all situations, by preserving the injection pressure of 203 bars and 23^0 CA BTDC injection timing at rated speed of 1500 rev/min, the assessments are traced thrice and a mean of these is engaged for comparison. Then the data of experimental tests and the results from parallel simulation are judged against one another in terms of pressure traces, heat release rates.

4.2 Cylinder pressure trace

Figure 2(a) illustrates the simulated and experimental in-cylinder pressure trace with respect to CA for baseline diesel and fig 2(b) for 10%JME blend. In all the injection cases studied, the same amount of fuel is injected in each engine cycle and describes in table 7. The consequence is supported on the supposition of consistent wall temperature 425 K for the cylinder wall, 525 K for the cylinder head and 525 K for the piston top. The predicted pressure data explains a good agreement with experimental data for both diesel and 10% JME blend, even though there is still some dissimilarity which can be seen in Fig. 2.

Table 7 Computational conditions of fuel injection for studied cases.

S. No	*Fuel Injection*	*Parameters*
1.	Total fuel	0.3196 g/cycle
2.	Diesel and JME10 (SOI)	23^0 BTDC
3.	Main duration	24^0 CA

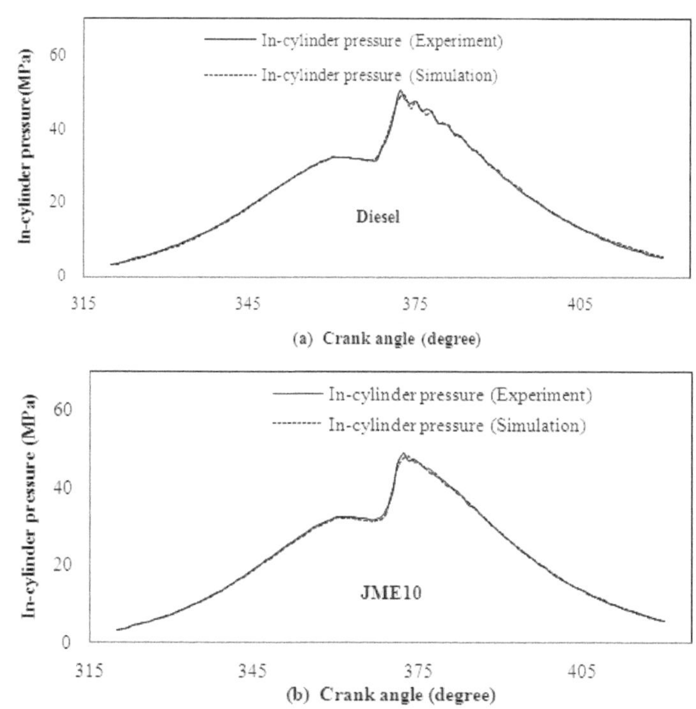

Fig. 2 Comparison of in-cylinder pressure between simulation and experiments: (a) diesel fuel, (b) 10% JME blend (JME10).

These disagreements could be associated with the experimental uncertainties in input parameters to the computations such as the precise injection duration, start of injection timing and gas temperature at IVC. At high engine load, huge amount of fuel is injected (constant rpm engine), hence the heat released in the cylinder and in-cylinder pressure (P_{MEP}) is high. Diesel fuel is a light fuel and is easily atomized, vaporized and mixed with air raising its self-ignition temperature and reduces delay period. Improved combustion of the fuel-air mixture is increased the in-cylinder pressure (P_{MEP}). On the other hand, JME10 blend, a viscous fuel increases the delay period resulting in lower in-cylinder pressure (P_{MEP}) compared to that of diesel fuel. Lower calorific value of JME10 also can be accounted for producing lower peak temperature and pressure of engine cylinder evaluated to that of diesel fuel.

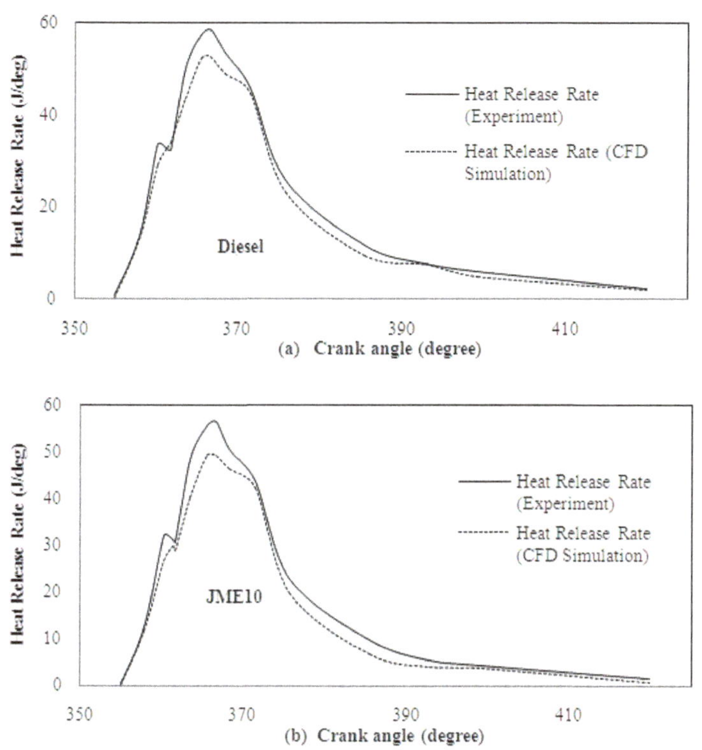

Fig. 3 Comparison of heat release rate between simulation and experiments: (a) diesel fuel, (b) 10% JME blend (JME).

4.2 Heat Release Rate

Figure 3(a) exemplifies the predicted and experimental heat release rate relating to CA for baseline diesel and fig 3(b) for 10%JME blend. The forecasted heat release rate contests thoroughly with the investigational data for both diesel and 10% JME blend, even supposing there are still some variations which can be seen in Fig. 3. These disagreements could be connected to the experimental uncertainties in input parameters to the computations. At elevated engine load, enormous amount of fuel is injected (constant rpm engine), hence the heat released in the cylinder is high. Diesel fuel is a light fuel and the superiority of atomization lessens ignition lag, due to elevated surface volume ratio. Slighter droplet size will have little deepness of infiltration due to less momentum of the droplet and less velocity relative to air from where it has to find oxygen after vaporization. For the reason that this air exploitation factor will be lessened on account of the fuel spray path being shorter. Also with slighter droplets, the aggregate area of inflammation will augment after ignition, resulting in improved heat release rate compared to that of blending fuel. Lower calorific value of JME10 is also may be related to less heat release in the engine cylinder evaluated to that of diesel fuel. Fig. 4(a) represents the contours of static pressure of diesel fuel at 375 degree crank angle and (b) represents the velocity magnitude of 10% JME blend at 600 degree crank angle.

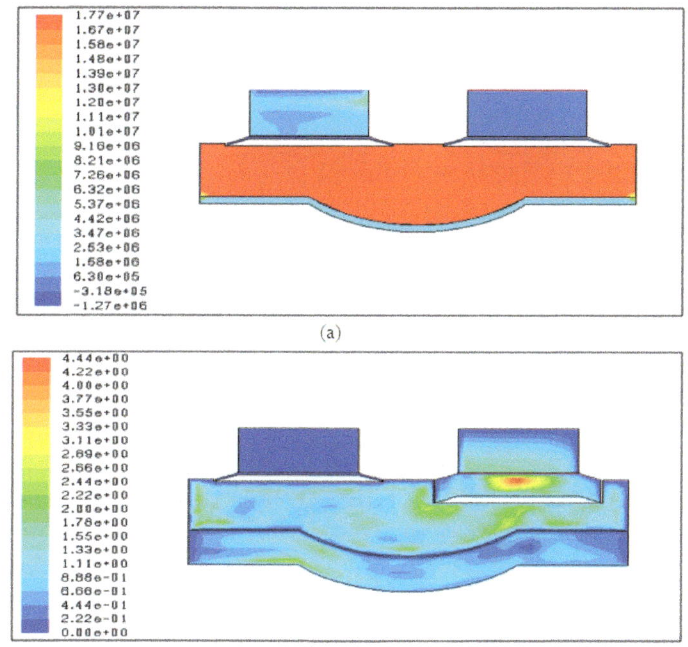

Fig. 4 Contours (a): static pressure of diesel fuel at 375 degree crank angle, (b): velocity magnitude of 10% JME blend at 600 degree crank angle.

5 CONCLUSIONS

1. An integrated numerical model for diesel combustion has been investigated in this study. Good level of agreement is reported between computed and measured data for engine pressure trace and heat release rate for all the test cases, even though there are still some dissimilarities experimental and computed results. These disagreements could be associated with the experimental uncertainties in input parameters to the computations such as the precise injection duration, start of injection timing and gas temperature at IVC.
2. The pressure trace from the simulations is contained by an error fringe of only 10%, where the common trend of premixed and mixing-controlled diesel combustion phases is detained effectively.
3. The consequences of this study is of scrupulous interest to the automotive sector in an endeavour to lessen design-development time and cost. For example, outcomes of novel chamber geometries and functioning approaches could be experimented through this with elevated economical advantage.

NOMENCLATURE

CFD	*Computational Fluid dynamics*
ICE	*Internal combustion engine*
DI	*Direct injection*
CR	*Compression Ratio*
CA	*Crank angle*
JME10	*10% Jatropha Methyl Ester and 90% Diesel oil in blend*
IVC	*Inlet Valve Closing*
P_{MEP}	*Mean Effective Pressure*

REFERENCES

Andreadis, P. A,. Zompanakis, C. and Kaiktsis, L., 2011, "Effects of fuel injection parameters on the performance and emissions in a large-bore marine diesel engine," *International Journal of Engine Research*, 2(1), 14–29. http://dx.doi:10.1243/14680874JER511

O'Rourke, P.J., 1981, "Collective prop Effects on vaporizing liquid Sprays," *PhD Thesis*, University of Princeton.

Goldsworthy, L., 2006, "Computational fluid dynamics modelling of residual fuel oil combustion in the context of marine Diesel engines," *International Journal of Engine Research*, 7, 181–99. http://dx.doi:10.1243/146808705X30620

Struckmeier, D., Tsuru, D., Kawauchi, S., and Tajima, H., 2009, "Multi-component modeling of evaporation, ignition and combustion processes of heavy residual fuel oil," SAE Technical Paper Series 2009-01-2677, *SAE International Powertrains, Fuels and Lubricants Meeting*. San Antonio, Texas.

Kyriakides, N., Chryssakis, C. and Kaiktsis, L., 2009, "Influence of Heavy Fuel properties on spray atomization for marine Diesel engine applications," SAE *Technical Paper Series 2009-01-1858*, SAE International Powertrains, Fuels and Lubricants Meeting, Florence, Italy.

Herrmann, K., Kyrtatos, A., Schulz, R. and Weisser, G., 2009, "Validation and initial application of a novel spray combustion chamber representative of large two-stroke Diesel engine combustion systems," *ICLASS, Vail, Colora.*

Herrmann, K., Rotz, B., Schulz, R., Weisser, G., Boulouchos, K. and Schneider, B., 2010, " Reference data generation of spray characteristics in relation to large 2-stroke marine Diesel engines using a novel spray combustion chamber concept," *ILASS-Europe 2010,* Brno, Czech Republic.

Garaniya, V., Goldsworthy, L., McWilliam, D., 2011, "Thermal cracking and polymerisation modelling of a single heavy fuel oil droplet," *Chemeca 2011,* Sydney, Australia.

Liu, A.B., Reitz, R.D., 1993, "Modeling the Effects of Drop Drag and Break-up on Fuel Sprays," *SAE Paper 930072.*

Dukowicz, J.K., 1979, "Quasi-Steady Droplet Change in the Presence Of Convection," *Informal Report Los Alamos Scientific Laboratory*, LA7997-MS.

Naber, J.D., Reitz, R.D., 1988, "Modeling Engine Spray/Wall Impingement," *SAE Paper 880107.*

Halstead, M., Kirsch, L., Quinn, C., 1977, "The auto ignition of hydrocarbon fueled at high temperatures and pressures-fitting of a mathematical model," *Combust. Flame*, 30, 45–60.

ICE Physics & Chemistry, 2009. "AVL FIRE user Manual," 1.

Colin, O. and Benkenida, A., 2004, "The 3-Zone Extended Coherent Flame Model (ECFM3Z) for Computing premixed/diffusion combustion," *Oil & Gas Science and Technology-Rev, IFP*, 59, 593-609. http://dx.doi.org/10.2516/ogst:2004043

Hélie, J., Trouvé, A., 2000, "A modified coherent flame model to describe turbulent flame propagation in mixtures with variable composition," *Proc. Combust. Inst,* 28, 193–201.

De, B. and Panua, R. S., 2014, "CFD modeling and validation of combustion in direct injection compression ignition engine fuelled with jatropha oil blends with diesel," *Frontiers in Heat and Mass* Transfer, 5(7). http://dx.doi.10. 5098/hmt.5.7

INFLUENCES OF THE PUNCHED DELTA WINGLET VORTEX GENERATORS IN A CIRCULAR TUBE HEAT EXCHANGER ON THERMO-HYDRAULIC PERFORMANCE

Withada Jedsadaratanachai[a] and Amnart Boonloi[b, *]

[a] *Department of Mechanical Engineering, Faculty of Engineering, King Mongkut's Institute of Technology Ladkrabang, Bangkok 10520, Thailand*
[b] *Department of Mechanical Engineering Technology, College of Industrial Technology, King Mongkut's University of Technology North Bangkok, Bangkok 10800, Thailand*

ABSTRACT

Numerical investigations on flow topology, heat transfer behavior and thermal performance evaluation in a circular tube heat exchanger with the punched delta winglet vortex generators (*PDWVG*) inserted in the middle of the test section are presented. The effects of the flow attack angles that converging to the center of the tube; α = 0°, 5°, 10°, 15°, 20°, 25°, and flow directions; winglet tips pointing downstream and upstream, are investigated for the Reynolds numbers; Re = 100 – 2000. The finite volume method and SIMPLE algorithm are used for the current study. The results are presented in terms of flow configuration, heat transfer behavior and thermal performance assessments and also compared with the smooth tube with no *PDWVG*. As numerical results, the use of *PDWVG* inserted in the tube can help to improve heat transfer rate and thermal performance in the heat exchanger by creating the vortex flows through the test section. The rise of the flow attack angle and Reynolds number leads to increase in heat transfer, friction loss and thermal performance. The optimum *TEF* is around 2.33 at Re = 2000, α = 25° for winglet tip pointing downstream.

Keywords: *Compact heat exchanger, Delta winglet, Laminar flow, Thermal enhancement factor.*

1. INTRODUCTION

Many researches on both numerical and experimental focus the improvement on thermal performance and heat transfer rate in the heat exchanger. The improved method is done by use the vortex generators or turbulators insert in the tube or channel heat exchanger. The vortex generators can create the vortex flow or swirling flow in the heat exchanger that helps to enhance the heat transfer rate and thermal performance, but also increase the pressure. The large flow blockage ratio of the vortex generators may increase very large pressure loss in the heating system. The appropriate geometry of the vortex generators will lead to the optimum enhancement on both heat transfer rate and friction loss. The forming method and installation of the vortex generators are another important factors that must investigate. The differences of the forming and installing methods may lead to the variations of the heat and flow behaviors in the heat exchanger.

The numerical and experimental investigations, which focus on heat transfer augmentation with using various types of winglets, had been presented. Khoshvaght-Alianadi *et al.* (2015) presented the experimental investigations on heat transfer and pressure drop in a tube heat exchanger with delta winglets vortex generators with different arrangements. They concluded that the delta winglet vortex generators can help augment heat transfer rate, but also increase in the pressure loss. They also pointed out that the optimum value between the enhancement of the heat transfer rate and friction loss in term of performance evaluation criterion or PEC is around 1.41 at Re = 8715. Gholami *et al.* (2014) investigated heat transfer and pressure loss in a fin-and-tube heat exchanger with wavy rectangular winglet vortex generators by numerical method. They found that the winglet vortex generators can enhance the heat transfer rate with a moderate pressure loss penalty, especially, the wavy-up rectangular winglet. He *et al.* (2013) carried out to investigate the heat transfer, pressure loss and thermal performance improvement in a fin-and-tube heat exchanger with rectangular winglet vortex generators. They studied the effects of the flow attack angle, row-number of the winglet and placement of the winglet on thermo-hydraulic performance. They found that the longitudinal vortices and impinging flow are the main reasons for heat transfer augmentation in the heat exchanger. Caliskan (2014) studied the new type vortex generators; punched triangular vortex generators and punched rectangular vortex generators in a rectangular channel heat exchanger. Caliskan (2014) claimed that the punched triangular vortex generators give the best thermal performance in the heating system. Caliskan (2014) also found that the enhancement of the heat transfer is around 23 – 55% when compared with smooth channel. Wang *et al.* (2015) numerical studied the flow resistance and heat transfer in a channel heat exchanger with delta winglet vortex generators. They reported that the maximum j-factor and f-factor are around 24.96% and 17.61%, respectively. Li *et al.* (2015) presented the heat transfer augmentation in a fin-and-tube heat exchanger with longitudinal vortex generators; rectangular winglet and delta winglet, by numerical simulation. They summarized that the rectangular winglet provides higher heat transfer rate than the delta winglet. They also showed that the flow attack angles of 25° and 45° perform the best thermal performance for rectangular winglet and delta winglet, respectively. Du *et al.* (2014) numerical investigated the influences of the punched longitudinal vortex generators in a fin-and-tube heat exchanger on heat transfer and pressure loss. They stated that the optimum thermal performance in term of the performance evaluation criteria is found at the flow attack angle of 25° for the delta winglet pairs. Lofti *et al.* (2014) presented the three dimensional numerical investigations on

* Corresponding author. Email: amnartb@kmutnb.ac.th

flow configuration and heat transfer characteristic in a fin-and-tube heat exchanger with various types of vortex generators. The rectangular trapezoidal winglet (*RTW*), angle rectangular winglet (*ARW*), curve angle rectangular winglet (*CARW*), wheeler wishbone (*WW*) are types of the vortex generators that presented by Lofti *et al.* (2014). They reported that the *CARW* gives the highest thermal performance for the small attack angle, while the *RTW* performs the best thermal performance at a large attack angle. Du *et al.* (2013) studied the thermal performance assessments in a fin-and-tube heat exchanger with longitudinal vortex generators; punched delta winglet pairs. They showed that the delta winglet vortex generators provide higher heat transfer rate and friction loss around 21 – 60% and 13 – 83%, respectively. They also reported that the average performance evaluation criterion is around 1.31. Gong *et al.* (2015) investigated the punched curve rectangular vortex generators in a fin-and-tube heat exchanger on heat transfer behavior and flow topology. They informed that the vortex generators can help to enhance the local Nussel number on the fin surfaces, which contact to the wake regimes. Ebrahimi *et al.* (2015) informed the effects of the longitudinal vortex generators in a rectangular microchannel on flow structure and heat transfer configuration by numerical investigation. They found that the enhancement of the Nusselt number and friction loss are around 2 – 25% and 4 – 30%, respectively. Deshmukh and Vedula (2014) reported the influences of the projected length ratio, height to tube inner diameter ratio and flow attack angle of curved delta wing vortex generators inserted in a circular tube heat exchanger on thermal performance. They claimed that the Nusselt number for the tube heat exchanger with vortex generators is around 1.3 – 5.0 times higher than the smooth tube with no vortex generators. He *et al.* (2012) studied the heat transfer augmentation in a fin-and tube heat exchanger with punched winglet vortex generators. They found that the heat transfer rate and pressure loss increase with the rise of the flow attack angle. They also summarized that the improvements of the heat transfer rate and friction loss are around 33.8 – 70.6% and 43.4 – 97.2% for the flow attack angle of 30°, respectively. Min *et al.* (2012) numerical simulated on turbulent flow and heat transfer behaviors in a channel heat exchanger with novel longitudinal vortex generators. They presented that the combined longitudinal vortex generators can generate vortices with larger regimes when compared with normal longitudinal vortex generators. Wu *et al.* (2012) experimental studied the punched longitudinal vortex generators; delta winglet pairs, in a fin-and-tube heat exchanger on heat transfer characteristic and flow configuration. They reported that the common-flow-up and common-flow-down of the delta winglet vortex generators give higher heat transfer rate around 16.5% and 28.2%, respectively, when compared with plain fin. Wu and Tao (2012) presented the effects of the longitudinal vortex generators; punched delta winglet, in a rectangular channel heat exchanger on thermal performance. They reported the effect of different punched angles; 15°, 30°, 45° and 60° by numerical and experimental investigations. They informed that the angle of 60° performs higher heat transfer rate and pressure loss than 45°.

As the literature reviews, it can be concluded as follows:

- The use of the winglets in the heat exchangers can help to improve heat transfer rate and thermal performance by generating the vortex flow, longitudinal vortex flow, impinging flow in the test section.
- The presences of the winglet not only increase in heat transfer rate, but also give higher pressure loss, especially in case of the high flow blockage area.
- The rectangular winglet performs higher heat transfer rate and friction factor than the delta winglet.
- The delta winglet can help to reduce the pressure loss in the heating system.
- The numerical investigation can save time and cost to study the thermal performance in the heat exchanger and also help to report the flow and heat transfer structure, which are ways to improve thermal performance.

Therefore, the numerical investigations on heat transfer and flow resistance in a tube heat exchanger with the punched delta winglet vortex generators are presented in the current study. The objective of this work is to enhance the heat transfer rate and thermal performance in the tube heat exchanger with a moderate pressure loss penalty. The comfortable for forming and installing of the vortex generators is another aim in this investigation. The influences of the flow attack angles; 0°, 5°, 10°, 15°, 20° and 25°, and flow directions; winglet tip pointing downstream and upstream, are investigated for the Reynolds numbers, $Re = 100 - 2000$.

2. BOUNDARY CONDITION AND ASSUMPTION

The boundary conditions for the current computational domain are presented as table 1:

Table 1 the boundary condition for the current computational domain.

Zone	*Boundary condition*
Inlet, outlet	Periodic condition
Tube wall	Constant temperature at 310 K
Tested fluid	- Air, constant mass flow rate at 300 K - Constant properties at the average temperature
Perpendicular plate	Adiabatic wall condition
Punched delta winglet	Adiabatic wall condition

The current investigations on both flow configuration and heat transfer behavior are developed under following assumptions:

- Steady three–dimensional fluid flow and heat transfer.
- The flow is laminar and incompressible.
- Constant fluid properties.
- Body forces and viscous dissipation are ignored.
- Negligible radiation heat transfer.

3. MATHEMATICAL FOUNDATION

Based on the above assumptions, the flow in circular tube is governed by the continuity, the Navier–Stokes and the energy equations. In the Cartesian tensor system these equations can be written as follows:

Continuity equation:

$$\frac{\partial}{\partial x_i}(\rho u_i) = 0 \tag{1}$$

Momentum equation:

$$\frac{\partial(\rho u_i u_j)}{\partial x_j} = -\frac{\partial p}{\partial x_i} + \frac{\partial}{\partial x_j}\left[\mu\left(\frac{\partial u_i}{\partial x_j} + \frac{\partial u_j}{\partial x_i}\right)\right] \tag{2}$$

Energy equation:

$$\frac{\partial(\rho u_i T)}{\partial x_i} = \frac{\partial}{\partial x_j}\left(\Gamma\frac{\partial T}{\partial x_j}\right) \tag{3}$$

where, Γ is the thermal diffusivity and is given by

$$\Gamma = \frac{\mu}{\Pr} \tag{4}$$

Apart from the energy equation discretize by the QUICK scheme, the governing equations are discretized by the second order upwind (SOU) scheme, decoupling with the SIMPLE algorithm, and solved by using a finite volume approach (Patankar, 1980). The solutions were considered to be converged when the normalized residual values were less than 10^{-5} for all variables, but less than 10^{-9} only for the energy equation.

Four parameters of interest in the present work are the Reynolds number, friction factor, Nusselt number and thermal enhancement factor. The Reynolds number is defined as:

$$\mathrm{Re} = \rho \bar{u} D / \mu \quad (5)$$

The friction factor, f is computed by pressure drop, Δp, across the length of the periodic tube, L, as

$$f = \frac{(\Delta p / L) D}{\frac{1}{2} \rho \bar{u}^2} \quad (6)$$

The heat transfer is measured by the local Nusselt number which can be written as

$$\mathrm{Nu}_x = \frac{h_x D}{k} \quad (7)$$

The average Nusselt number can be obtained by

$$\mathrm{Nu} = \frac{1}{L} \int \mathrm{Nu}_x \partial x \quad (8)$$

The thermal enhancement factor (*TEF*) is defined as the ratio of the heat transfer coefficient of an augmented surface, h to that of a smooth surface, h_0, at an equal pumping power and given by

$$\mathrm{TEF} = \left.\frac{h}{h_0}\right|_{pp} = \left.\frac{\mathrm{Nu}}{\mathrm{Nu}_0}\right|_{pp} = (\mathrm{Nu}/\mathrm{Nu}_0)/(f/f_0)^{1/3} \quad (9)$$

where Nu_0 and f_0 stand for Nusselt number and friction factor for the smooth tube, respectively.

4. COMPUTATIONAL MODEL

The computational domain of the *PDWVG* in the circular tube heat exchanger is shown in Fig. 1, while the differences of the flow attack angle for the turbulators is presented in Fig. 2. The circular tube diameter, *D*, is set at 0.05 m. The *PDWVG*s are punched out from the perpendicular plates, while the plate width equals to the tube diameter on both vertical and horizontal plates. The length and height of the winglet are set to 0.55*D* and 0.2*D*, respectively. The flow attack angles; 0°, 5°, 10°, 15°, 20°, 25°, flow directions; winglet tip pointing downstream and upstream, are investigated numerically for the Reynolds numbers; *Re* = 100 – 2000.

The numbers of grid are tested by creating the four sets of cells; 80000, 120000, 200000, 300000 for the current computational domain. It is found that the increase of the grid from 120000 to 200000 has no effect on both heat transfer and friction factor. The variations of the Nusselt number and friction loss are around ±0.03% and ±0.1%, respectively. Therefore, the grid cells around 120000 are used for the current computational domain when considering on both time and accuracy result.

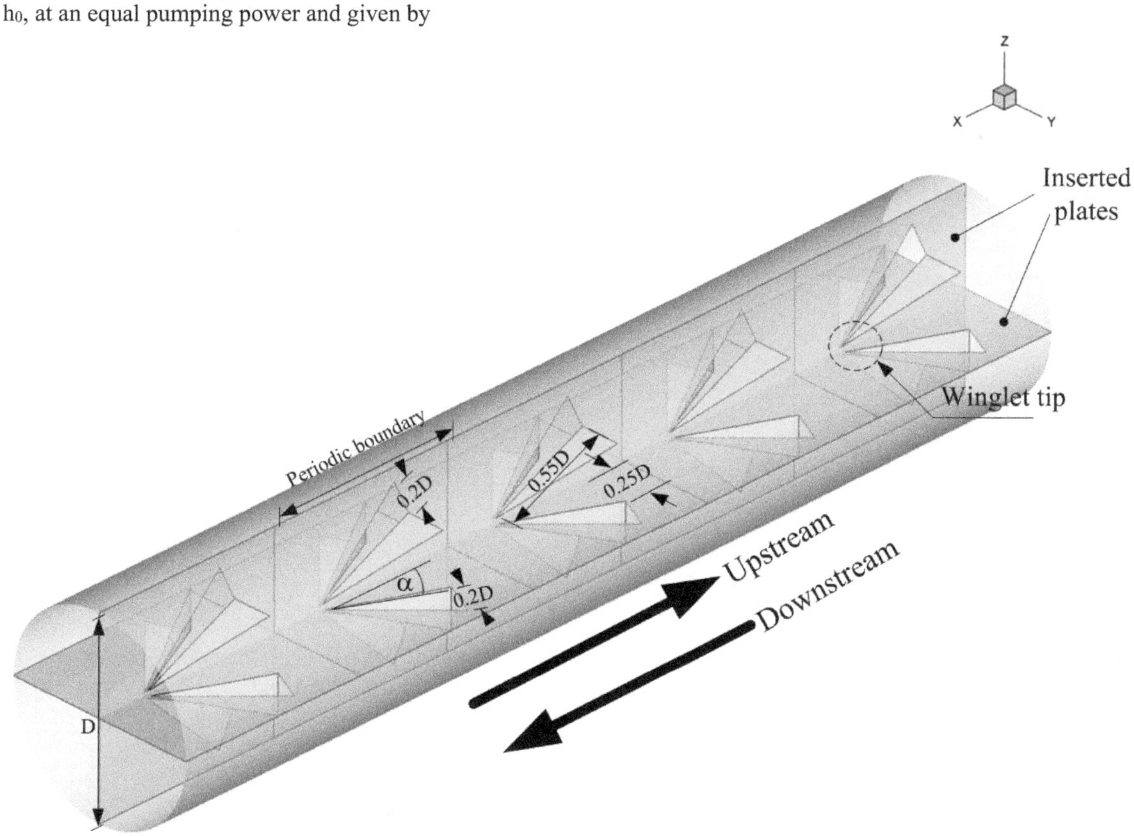

Fig. 1 Computational domain of the tube heat exchanger inserted with the *PDWVG* in the middle of the test tube.

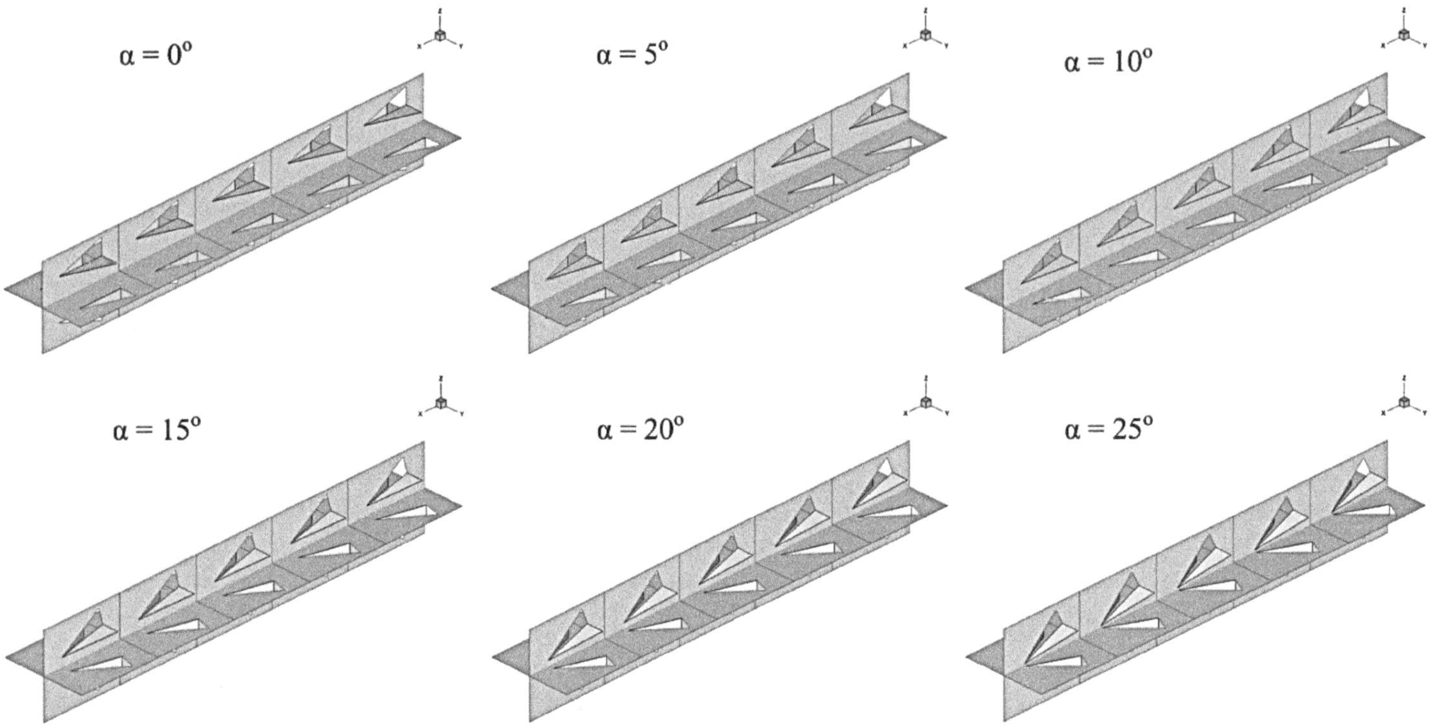

Fig. 2 The punched delta winglet vortex generators with various the flow attack angles.

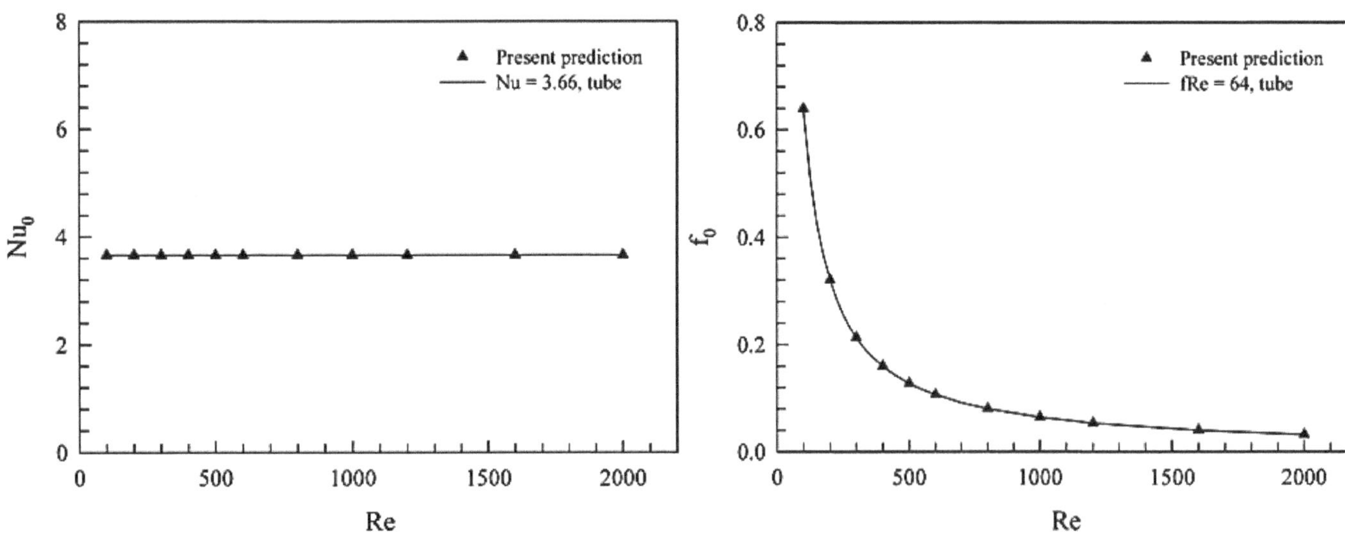

(a) (b)

Fig. 3 Validation of the smooth circular tube for (a) Nusselt number and (b) friction factor.

5. NUMERICAL RESULT AND DISCUSSION

This part can divided into four sections; validation with smooth tube, flow configuration, heat transfer characteristic, thermal assessment.

5.1 Validation with smooth tube

The validation with the smooth circular tube with no winglet is a very important part to make sure that the current computational domain is reliable to predict the flow and heat transfer behaviors in the tube heat exchanger. The verifications are done by compared the present results with the values from the correlations (Incropera, 2006) on both heat transfer and pressure loss in terms of Nusselt number and friction factor, respectively. Figs. 3a and b present the verifications on Nusselt number and friction factor in the tube heat exchanger with no winglet, respectively. As the figures, there are found that the variations on the Nusselt number and friction factor are around ±0.3% and ±0.4%, respectively. Therefore, the current computational domain is reliable to investigate the heat transfer and flow configuration in the tube heat exchanger.

5.2 Flow configuration

The numerical investigation can help to report the flow structure of the fluid in the tube heat exchanger. The understanding of the flow configuration may be advantageous to develop the new design of the compact heat exchanger. The flow configurations for the circular tube heat exchanger with the punched delta winglet vortex generators (*PDWVG*) are plotted in terms of tangential velocity vectors and streamlines in transverse planes.

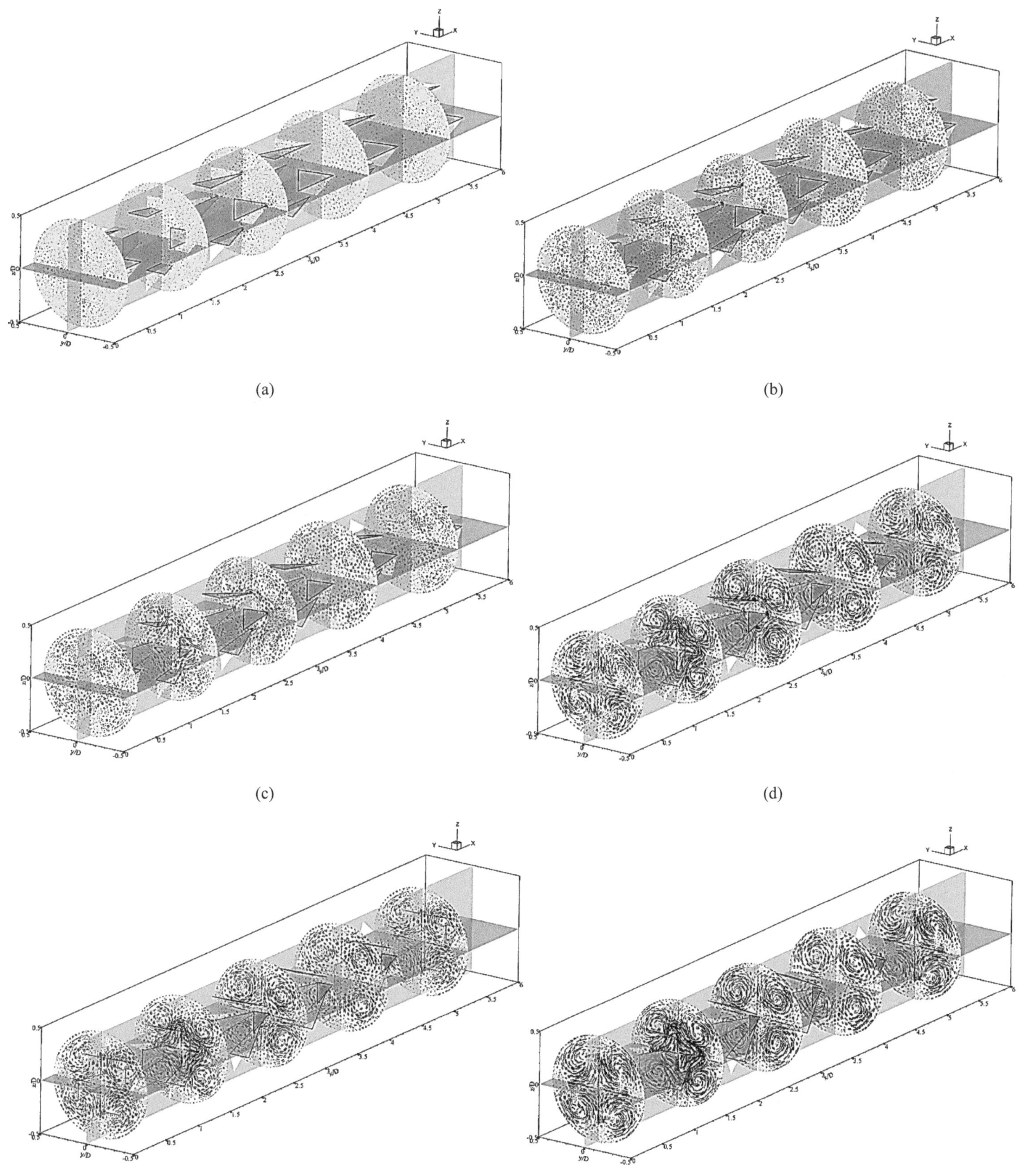

(a) (b) (c) (d) (e) (f)

Fig. 4 The tangential velocity vectors in transverse planes for (a) 0°, (b) 5°, (c) 10°, (d) 15°, (e) 20° and (f) 25° of the *PDWVG* in the tube heat exchanger with winglet tip pointing downstream at $Re = 1000$.

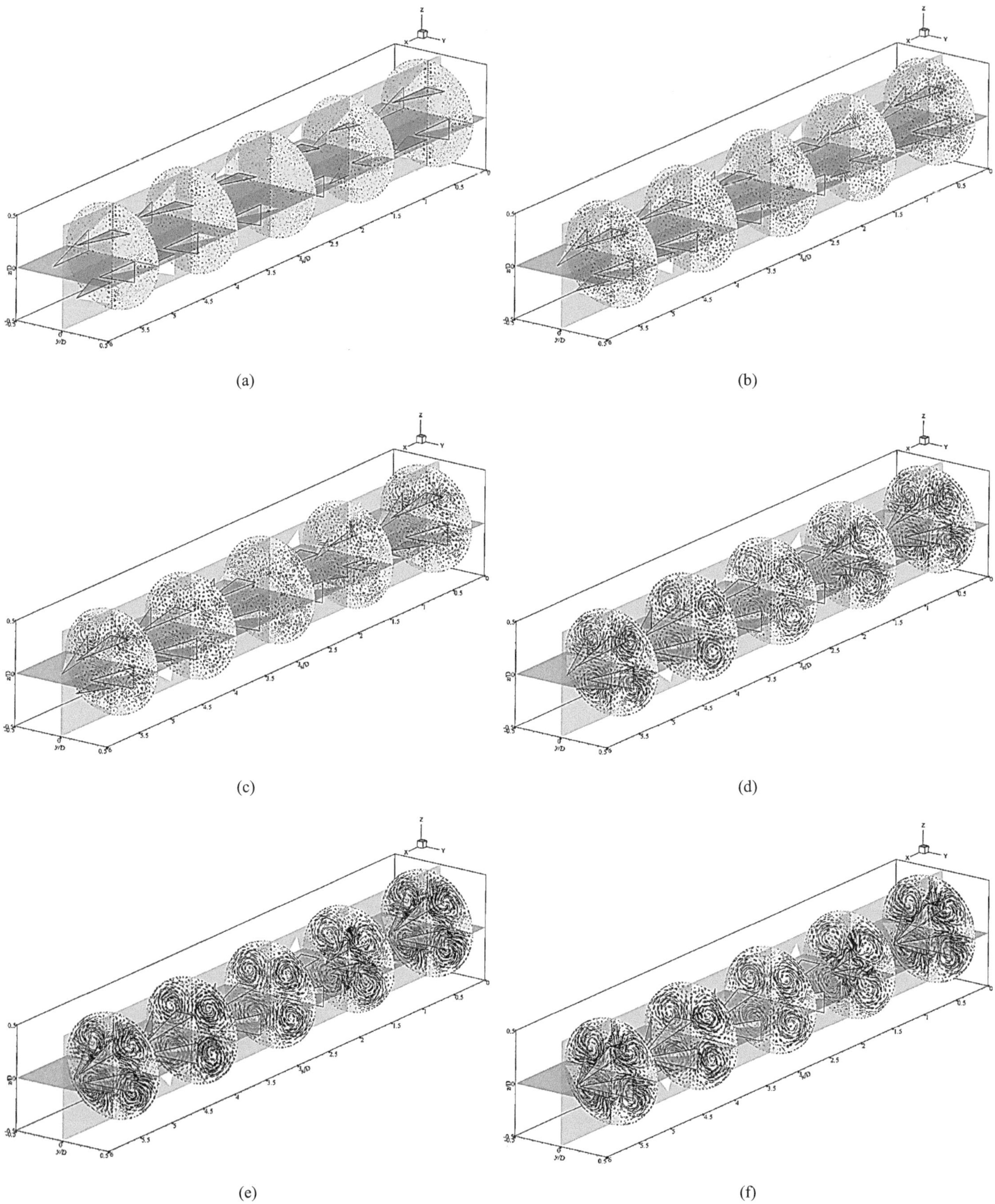

Fig. 5 The tangential velocity vectors in transverse planes for (a) 0°, (b) 5°, (c) 10°, (d) 15°, (e) 20° and (f) 25° of the *PDWVG* in the tube heat exchanger with winglet tip pointing upstream at $Re = 1000$.

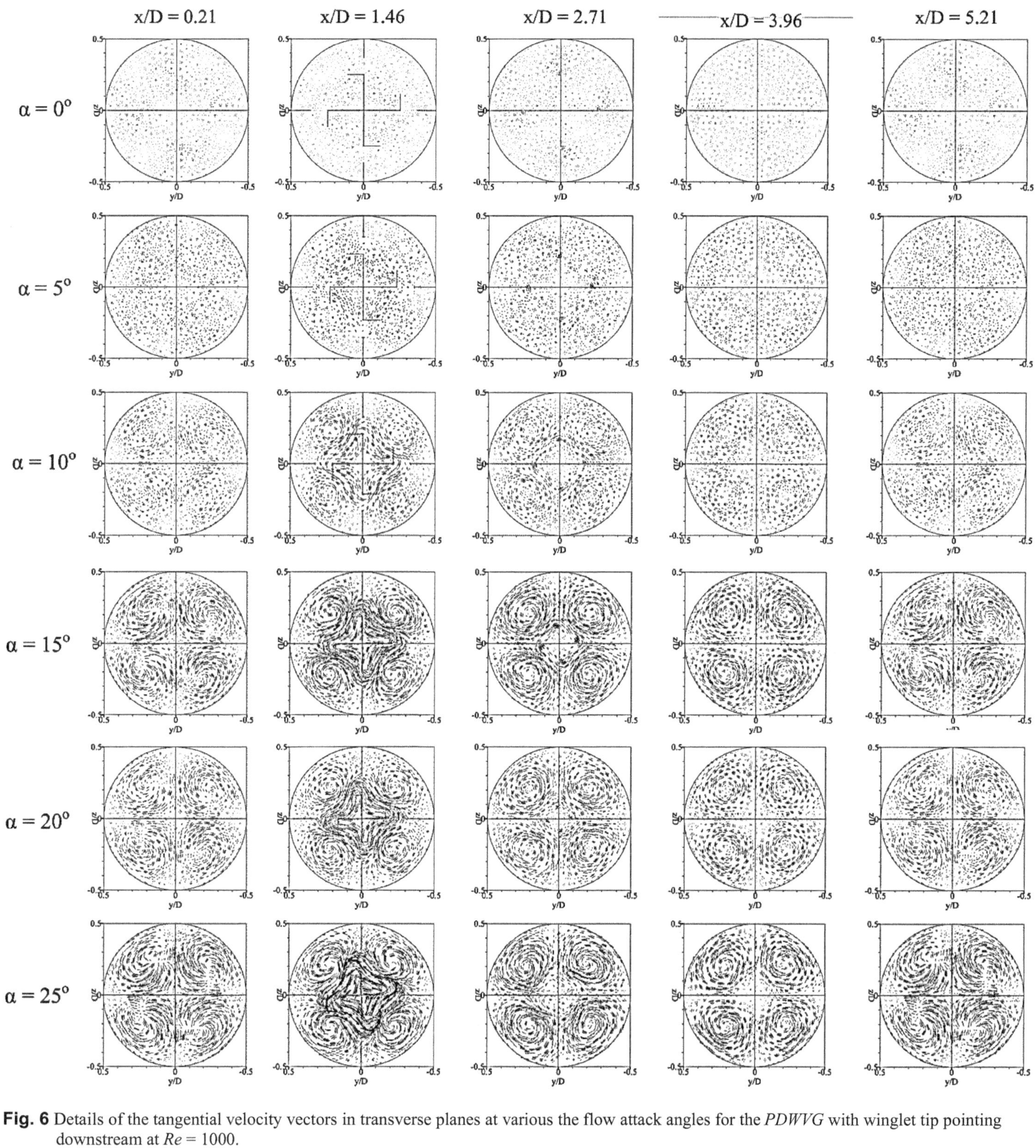

Fig. 6 Details of the tangential velocity vectors in transverse planes at various the flow attack angles for the *PDWVG* with winglet tip pointing downstream at $Re = 1000$.

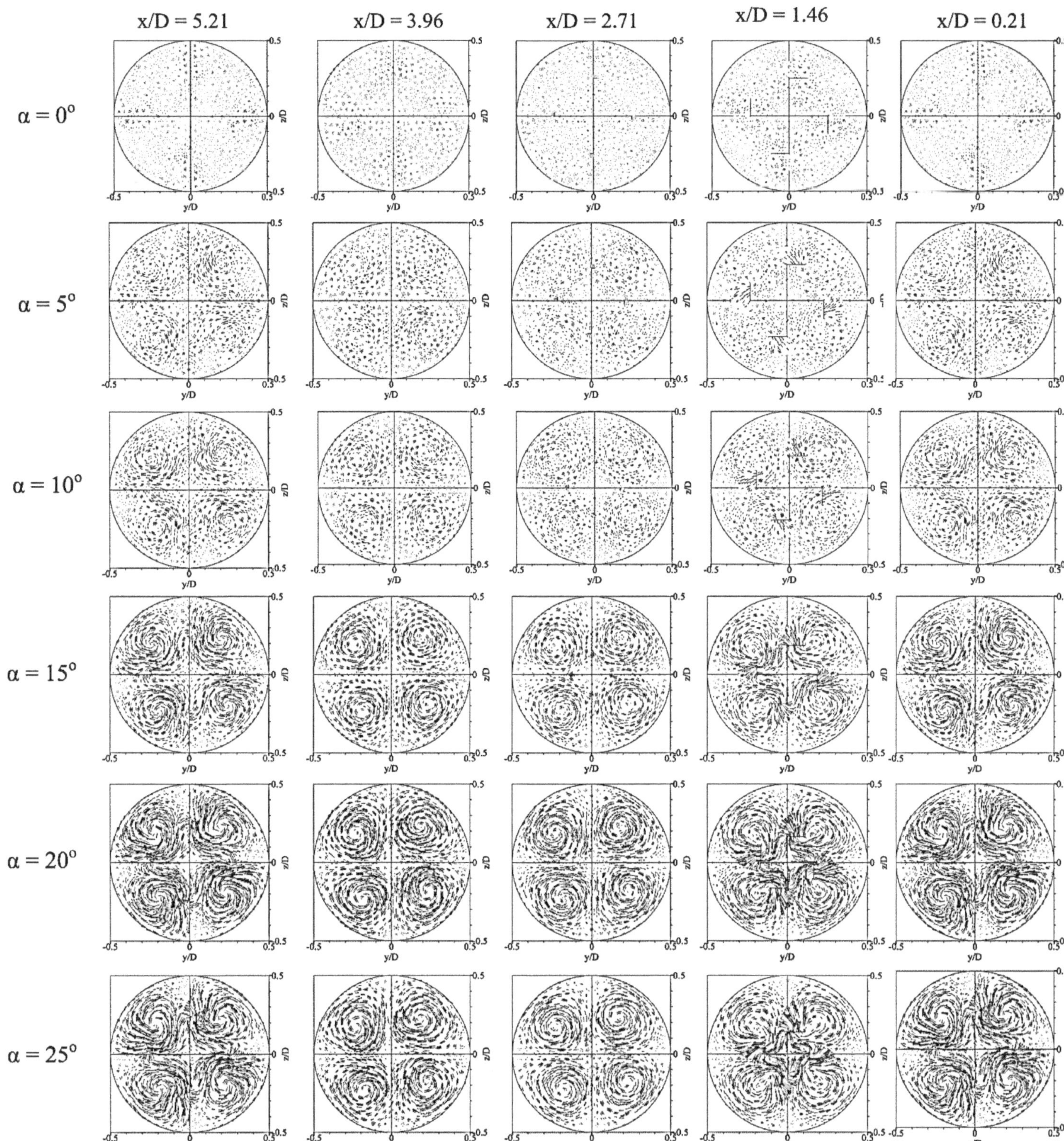

Fig. 7 Details of the tangential velocity vectors in transverse planes at various the flow attack angles for the *PDWVG* with winglet tip pointing upstream at $Re = 1000$.

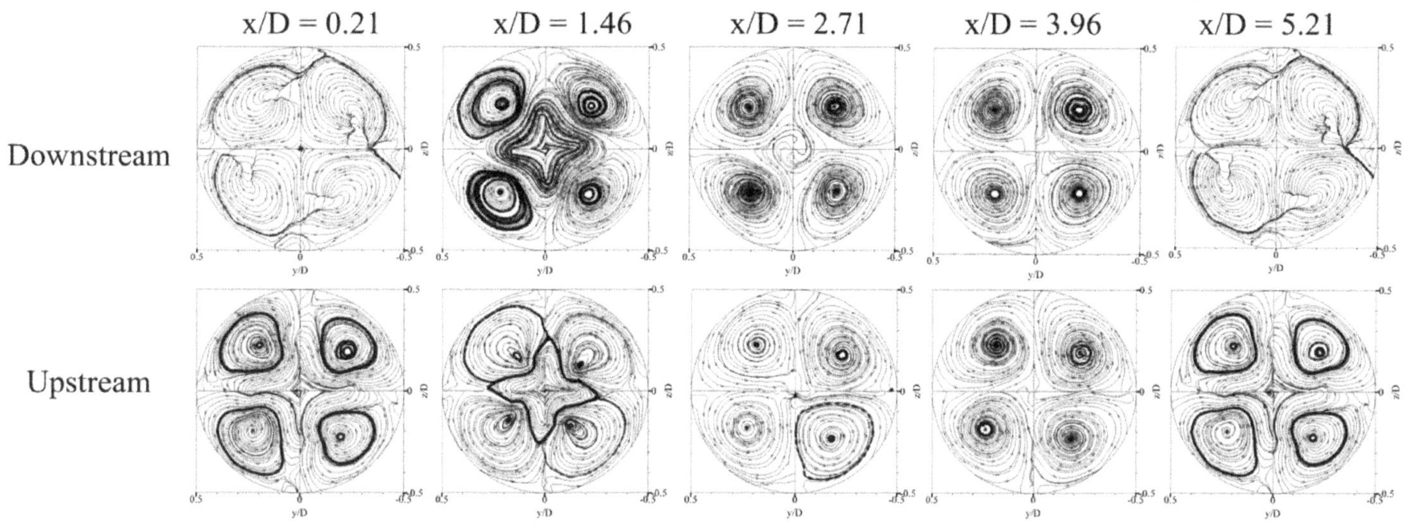

Fig. 8 The streamlines in transverse planes for the *PDWVG* with winglet tips pointing downstream and upstream at $Re = 1000$ and $\alpha = 20^{\circ}$.

Figs. 4*a*, *b*, *c*, *d*, *e* and *f* show the tangential velocity vectors in transverse planes of the tube heat exchanger with *PDWVG* for $\alpha = 0^{\circ}$, 5°, 10°, 15°, 20° and 25°, respectively, at the winglet tip pointing downstream and $Re = 1000$. As the figure, the tangential velocity vectors are found clearly when the flow attack angle is higher than 15°. The flow attack angles of 10° and 15° can create slightly tangential velocity vectors, while the tangential velocity vectors are not found in the flow attack angle of 0°. The tangential velocity vectors are representative of the vortex flow, which are generated by the *PDWVG*. The vortex flows lead to a better mixing of the fluid flow and help to improve the heat transfer rate and thermal performance in the heat exchanger. The similar results are found when using *PDWVG* at the winglet tip pointing upstream as the Figs. 5*a*, *b*, *c*, *d*, *e* and *f* for $\alpha = 0^{\circ}$, 5°, 10°, 15°, 20° and 25°, respectively, $Re = 1000$. The details of the tangential velocity vectors in transverse planes at different locations for the winglet tip pointing downstream and upstream are presented in Figs. 6 and 7, respectively.

Fig. 8 shows the streamlines in transverse planes for the 20° *PDWVG* inserted in the tube heat exchanger with the winglet tips pointing downstream and upstream at various locations and $Re = 1000$. It is found that the use of the *PDWVG* can generate the four main vortex flows on both cases. The winglet tip pointing downstream gives the clockwise vortex flow, while the winglet tip pointing upstream provides the counter clockwise vortex flow.

5.3 Heat transfer characteristic

The temperature distributions in transverse planes and the local Nusselt number distributions on the tube wall are used to present the heat transfer behaviors in the tube heat exchanger when installing with *PDWVG*. Figs. 9*a*, *b*, *c*, *d*, *e* and *f* present the temperature distributions in transverse planes for the *PDWVG* in the tube heat exchanger with $\alpha = 0^{\circ}$, 5°, 10°, 15°, 20° and 25°, respectively, at $Re = 1000$. In general, the use of the *PDWVG* provides a better mixing of the fluid flow between near the tube wall and the core of the test section for all the flow attack angles. Considering at the red layer of the temperature contours near the tube wall, the red layer tends to decrease when increasing the flow attack angles. The flow attack angle of 0° performs the lowest performance of the fluid mixing, while the flow attack angle of 25° provides the best result. The similar trends of the numerical results are found when using the *PDWVG* with the tip pointing upstream as depicted in the Fig.10. The details for each plane of the temperature distributions on both tips pointing downstream and upstream with various locations are reported in the Figs. 11 and 12, respectively, at $Re = 1000$.

Figs. 13*a*, *b*, *c*, *d*, *e* and *f* present the local Nusselt number distributions on the tube wall of *PDWVG* in the tube heat exchanger for $\alpha = 0^{\circ}$, 5°, 10°, 15°, 20° and 25°, respectively, at $Re = 1000$ and the winglet tip pointing downstream. There are clearly seen in the figures that the rise of the flow attack angle leads to increase in the Nusselt number. The flow attack angle of 25° gives the highest heat transfer rate, while the flow attack angle of 0° performs the lowest value. The *PDWVG*s with winglet tip pointing upstream, which plots in the Fig. 14, are performed the related results as the winglet tip pointing downstream for all the flow attack angles. The peaks of heat transfer regimes; red contours of Nusselt number, for both cases are found similarly.

In addition, the flow configuration and heat transfer characteristic in the tube heat exchanger with *PDWVG*s are found to be related. The enhancement of the heat transfer rate is due to the vortex flows that generated from the *PDWVG*s on both the tips pointing downstream and upstream.

5.4 Thermal performance assessment

The thermal performance assessments in the current investigations are plotted as Figs 15 – 17. The heat transfer rate, pressure loss and performance in the tube heat exchanger with *PDWVG*s are presented in terms of the Nusselt number ratio (Nu/Nu_0), friction factor ratio (f/f_0), and thermal enhancement factor (*TEF*) ,respectively.

Figs. 15*a* and *b* present the variations of the Nu/Nu_0 with the Reynolds number at various the flow attack angles for the winglet tips pointing downstream and upstream, respectively. In general, the Nu/Nu_0 tends to increase with the rise of the Reynolds number and flow attack angles on both cases. The Nu/Nu_0 for all the flow attack angles of the winglet tip pointing downstream performs nearly values as the winglet tip pointing upstream. The flow attack angle of 25° gives the highest Nu/Nu_0, while the flow attack angle of 0° performs the lowest heat transfer rate on both downstream and upstream cases. The flow attack angle of 25° gives the maximum higher Nu/Nu_0 than the flow attack angle of 0°, 5°, 10°, 15°, and 20° around 54.36%, 35.34%, 20.29%, 13% and 6.81%, respectively, at $Re = 2000$. In range studies, the use of the *PDWVG* in the tube heat exchanger gives the Nusselt number higher than the smooth tube with no *PDWVG* around 1.6 – 6.31 times.

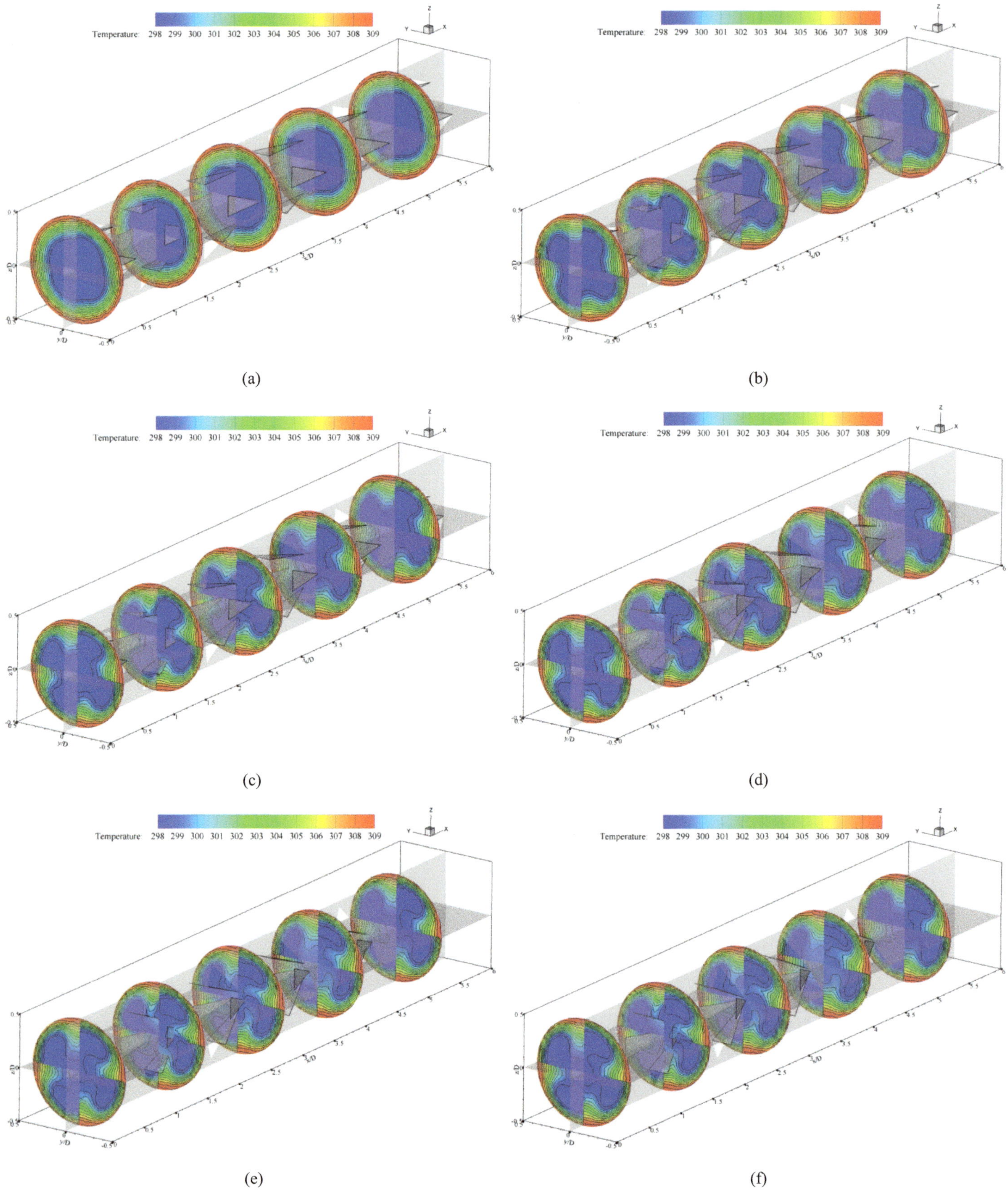

Fig. 9 The temperature distributions in transverse planes for (a) 0°, (b) 5°, (c) 10°, (d) 15°, (e) 20° and (f) 25° of the *PDWVG* in the tube heat exchanger with winglet tip pointing downstream at *Re* = 1000.

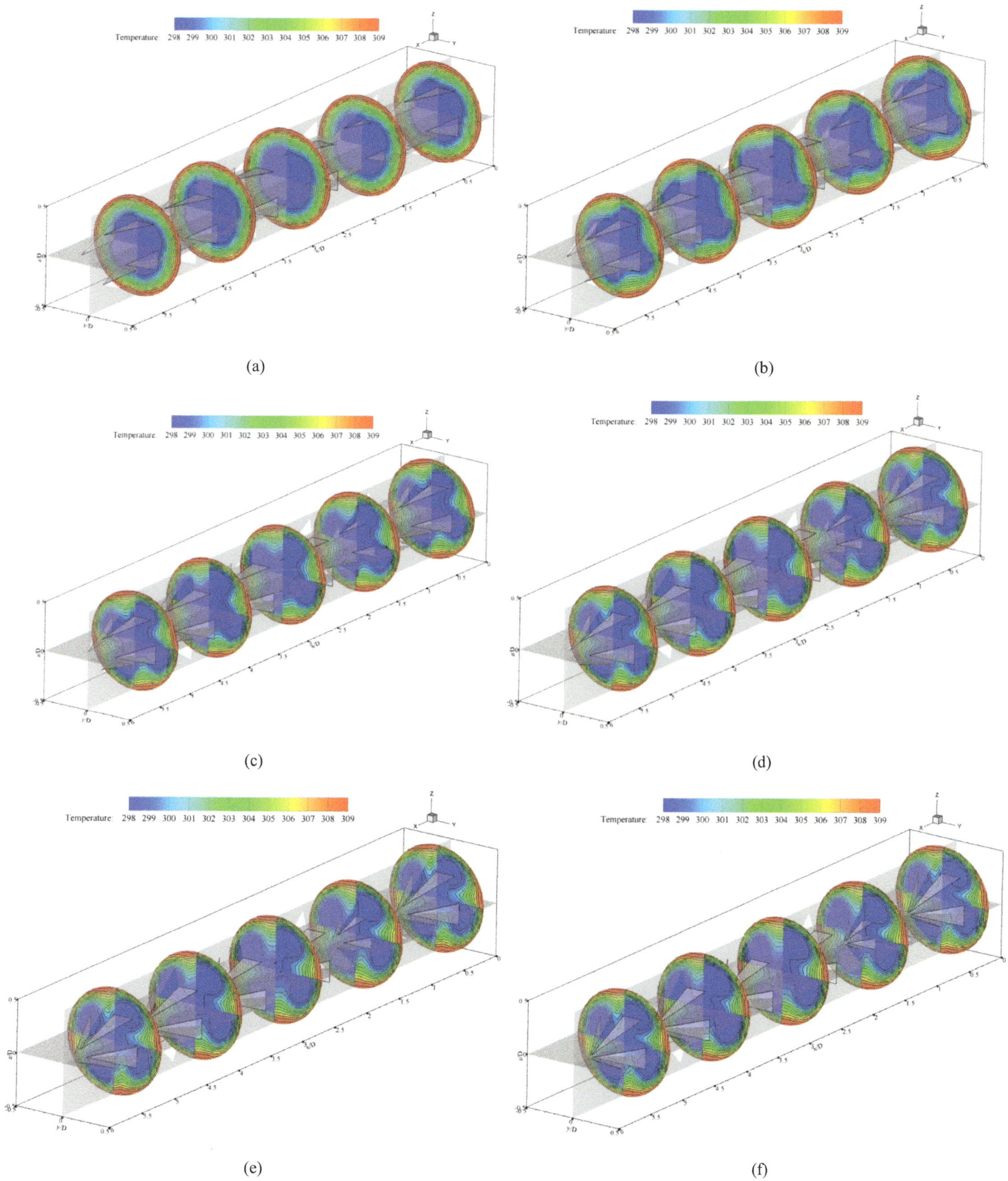

Fig. 10 The temperature distributions in transverse planes for (a) 0°, (b) 5°, (c) 10°, (d) 15°, (e) 20° and (f) 25° of the *PDWVG* in the tube heat exchanger with winglet tip pointing upstream at *Re* = 1000.

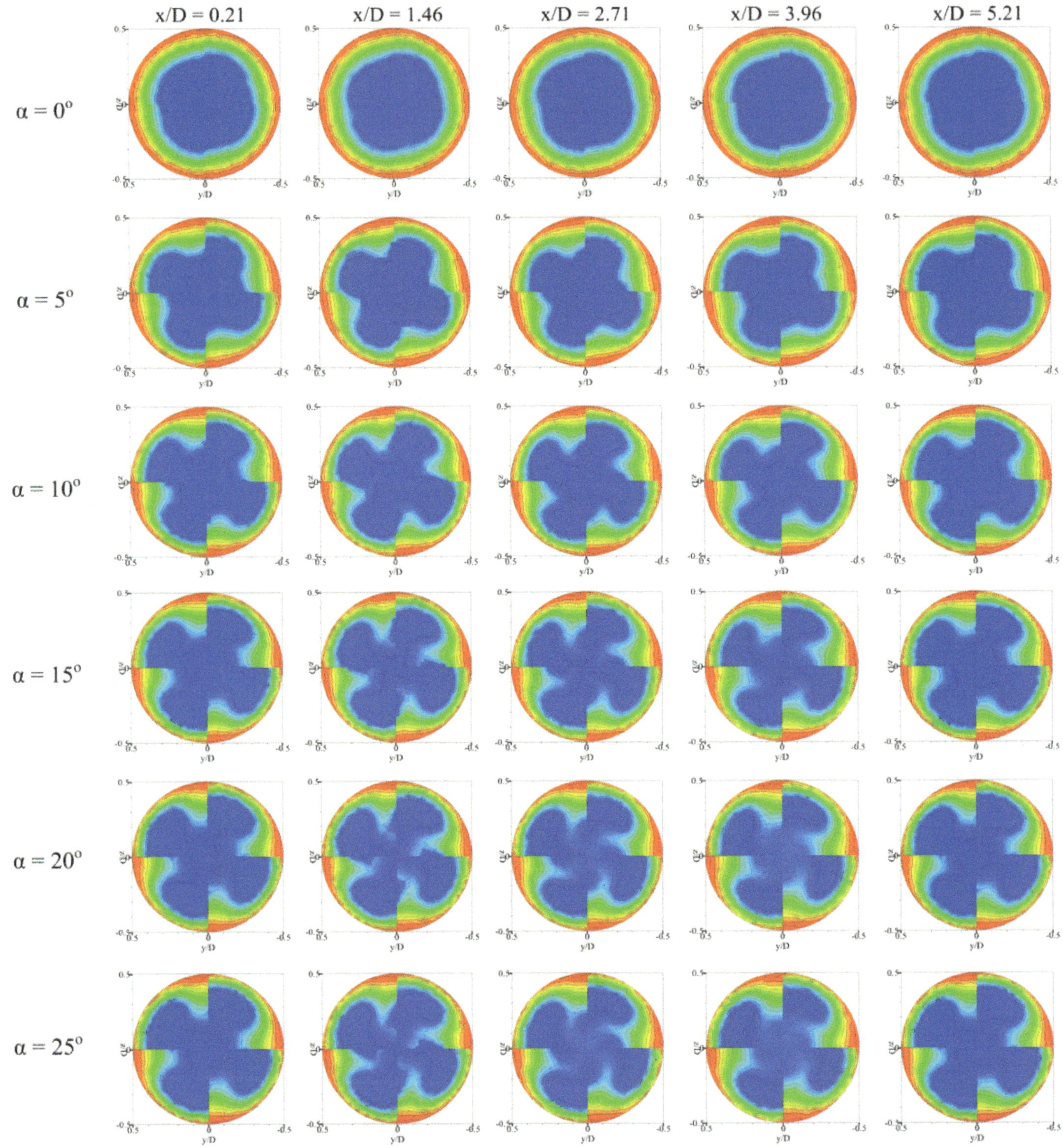

Fig. 11 The details of the temperature distributions in transverse planes at various the flow attack angles for the *PDWVG* with winglet tip pointing downstream at *Re* = 1000.

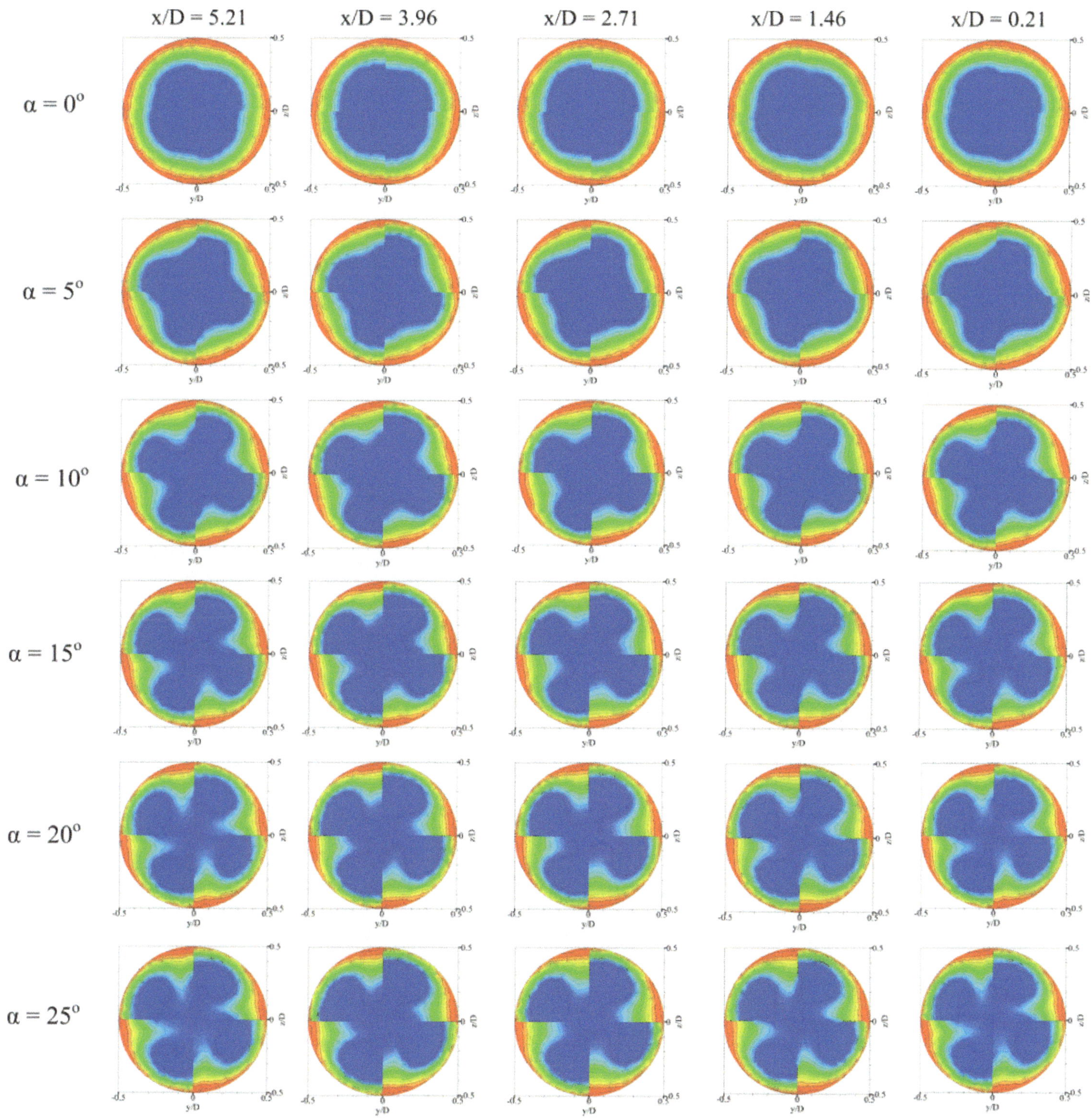

Fig. 12 The details of the temperature distributions in transverse planes at various the flow attack angles for the *PDWVG* with winglet tip pointing upstream at $Re = 1000$.

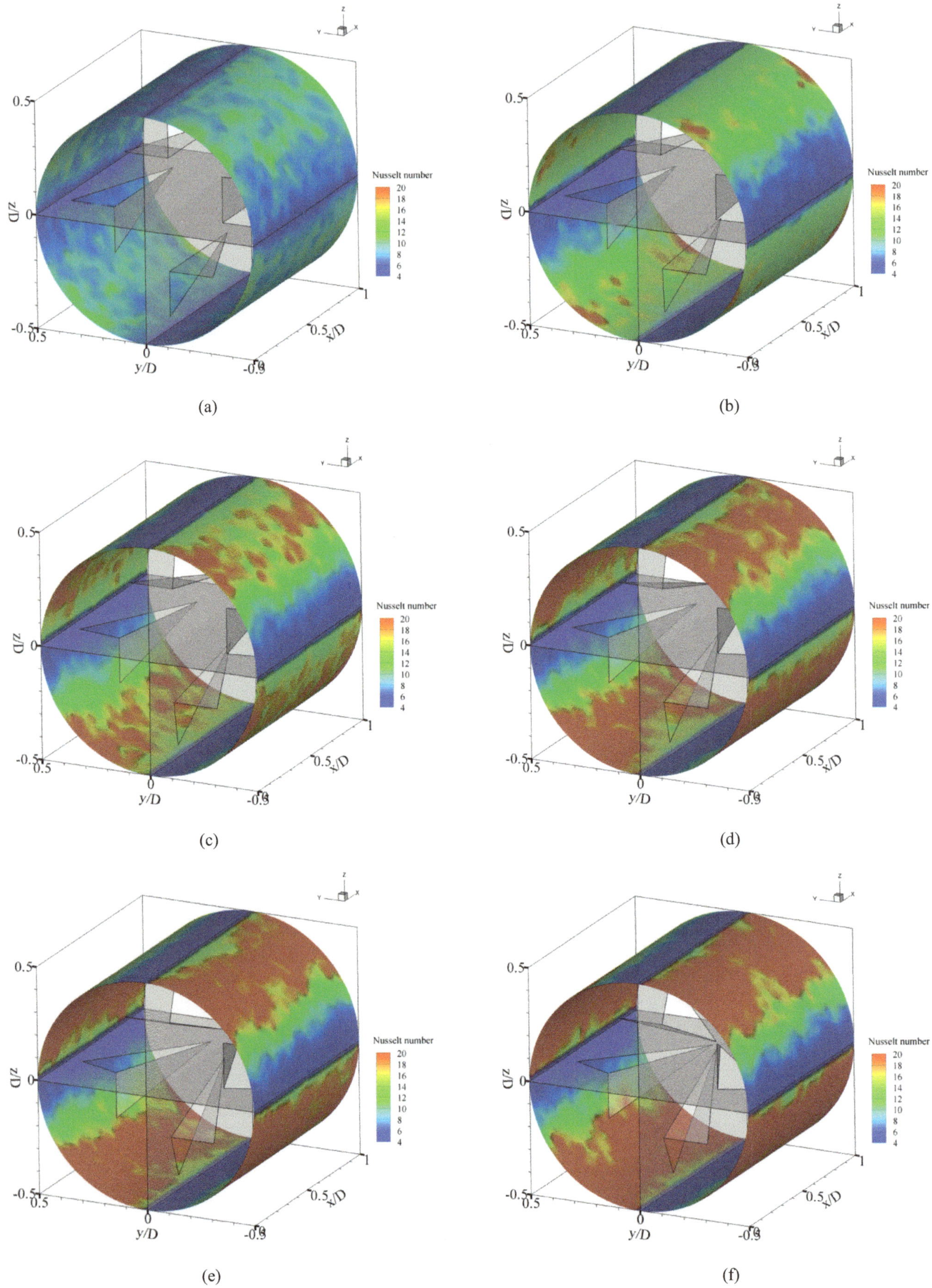

Fig. 13 The local Nusselt number distributions in transverse planes for (a) 0°, (b) 5°, (c) 10°, (d) 15°, (e) 20° and (f) 25° of the *PDWVG* in the tube heat exchanger with winglet tip pointing downstream at $Re = 1000$.

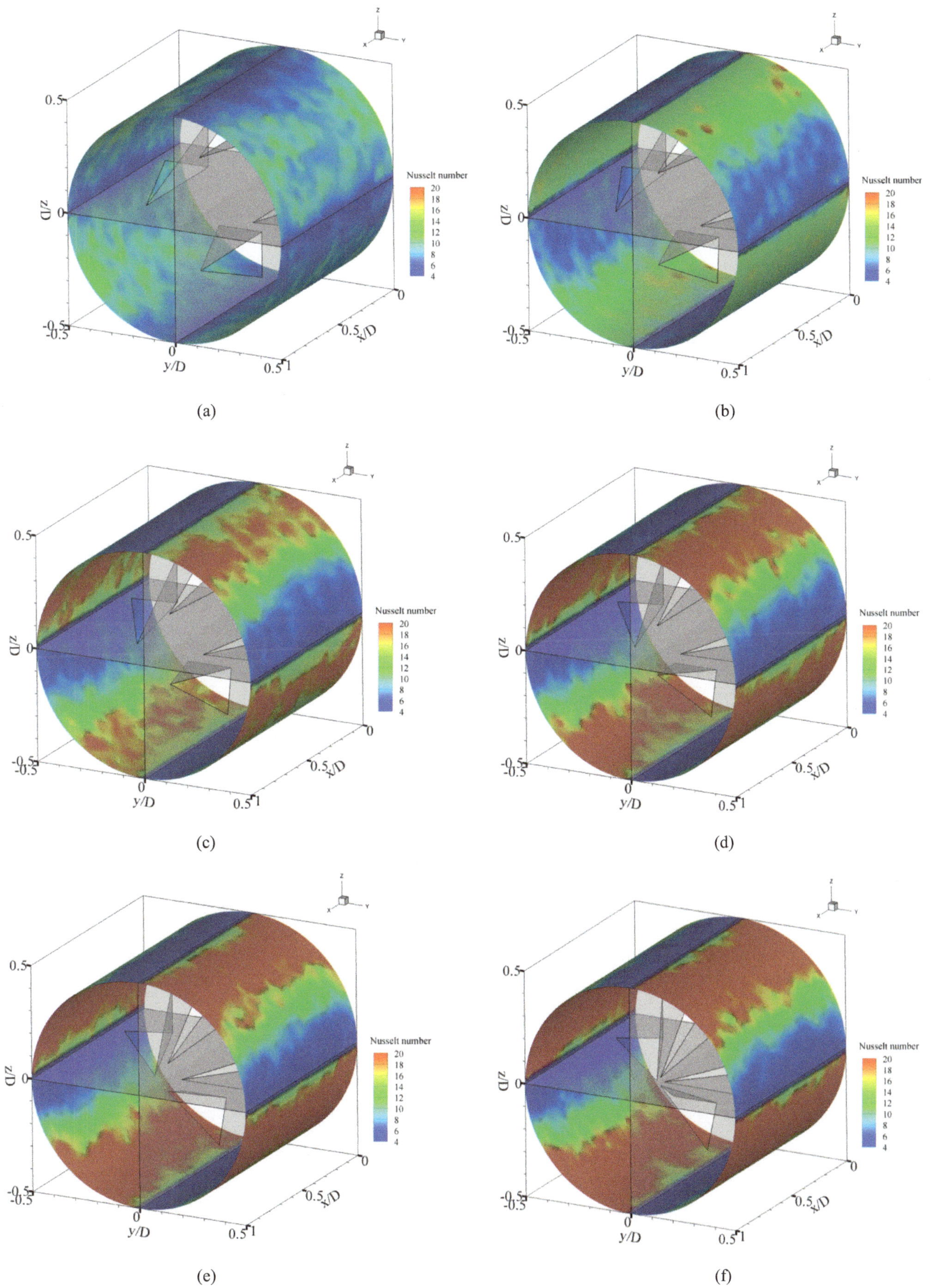

Fig. 14 The local Nusselt number distributions in transverse planes for (a) 0°, (b) 5°, (c) 10°, (d) 15°, (e) 20° and (f) 25° of the *PDWVG* in the tube heat exchanger with winglet tip pointing upstream at *Re* = 1000.

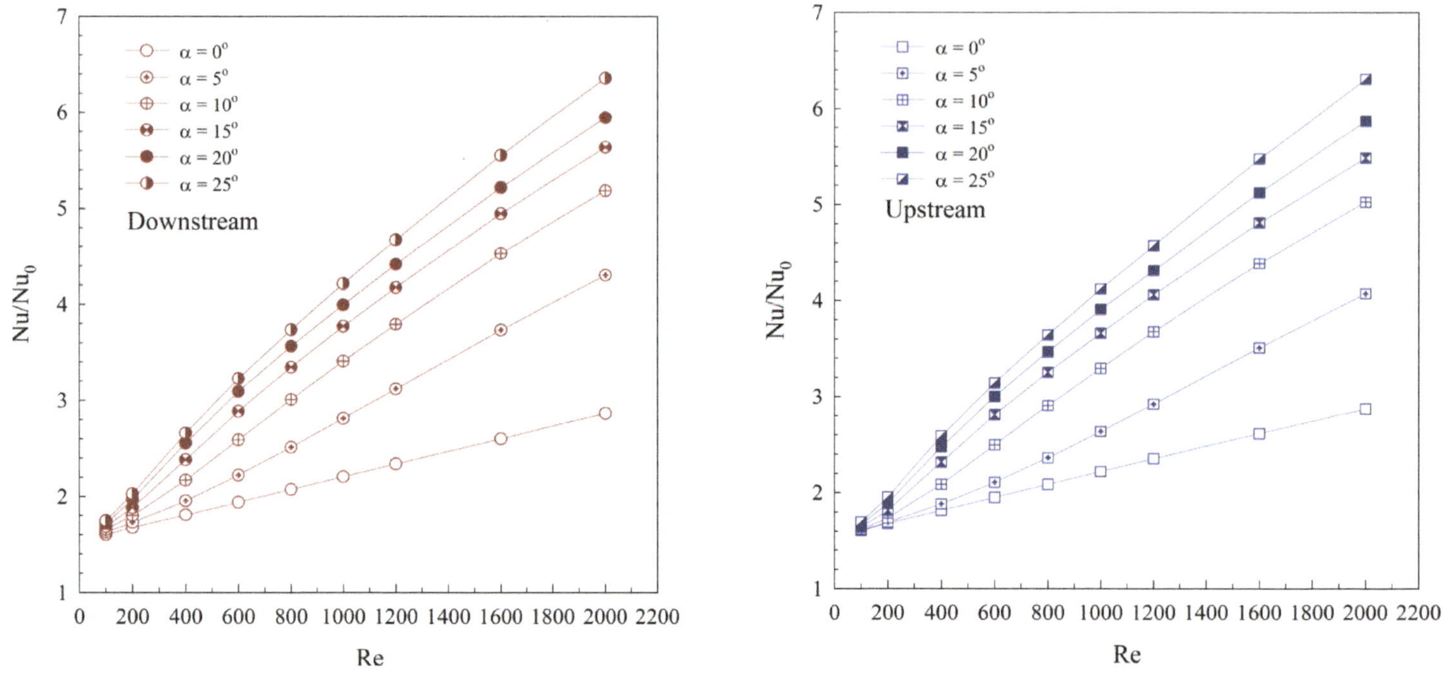

(a) (b)

Fig. 15 The variations of the *Nu*/*Nu*0 with the Reynolds number at various the flow attack angles for (a) downstream and (b) upstream.

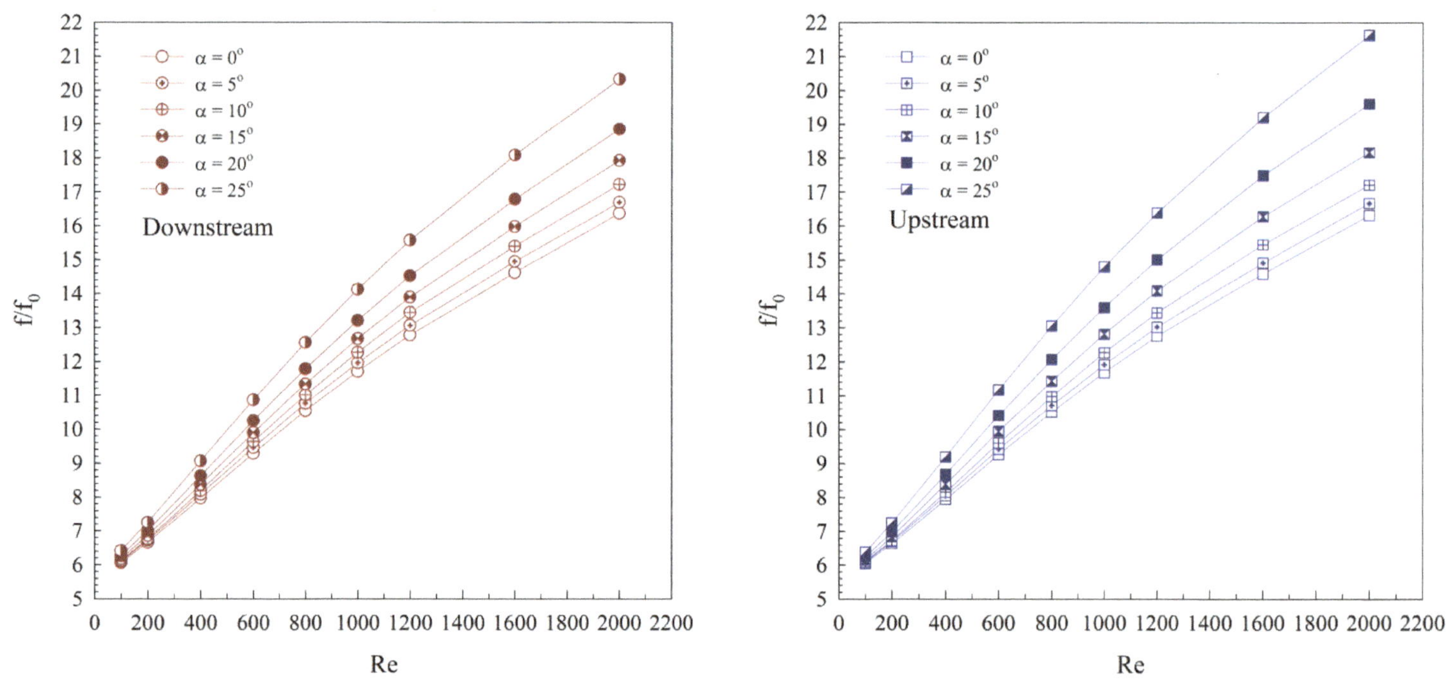

(a) (b)

Fig. 16 The variations of the *f*/*f*0 with the Reynolds number at various the flow attack angles for (a) downstream and (b) upstream.

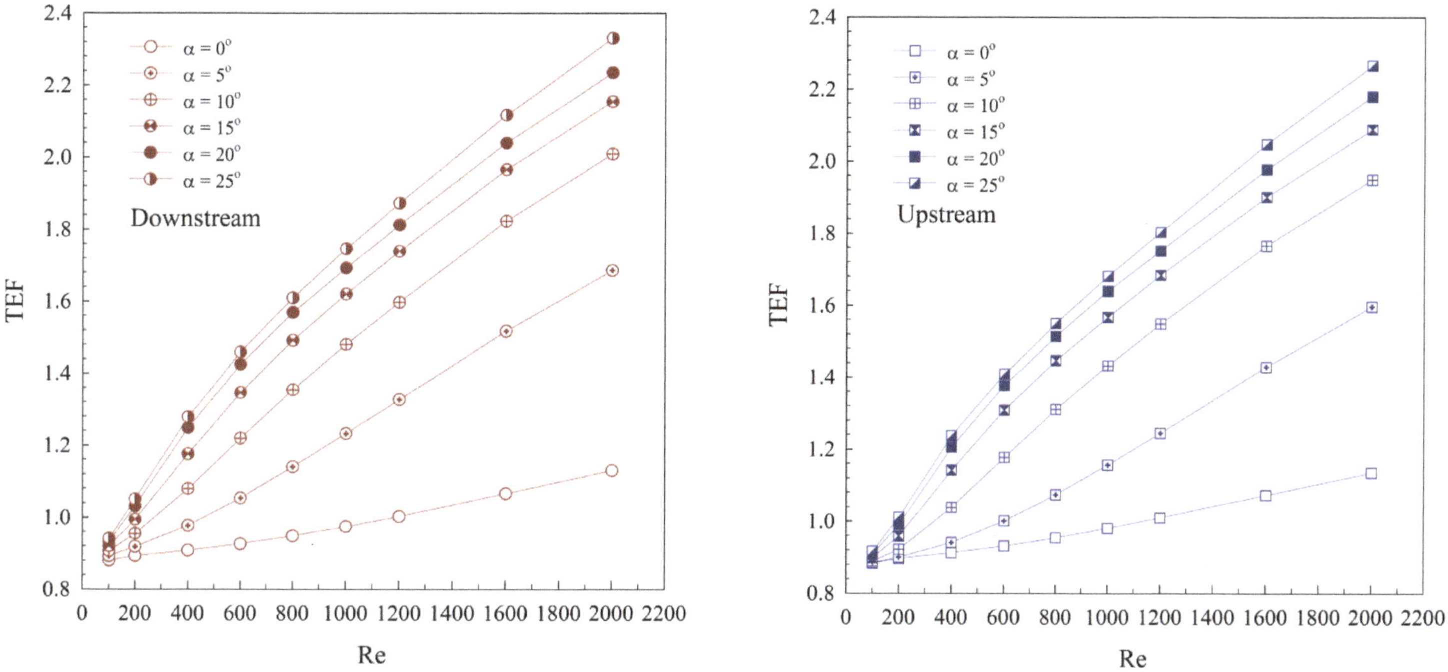

(a) (b)

Fig. 17 The variations of the *TEF* with the Reynolds number at various the flow attack angles for (a) downstream and (b) upstream.

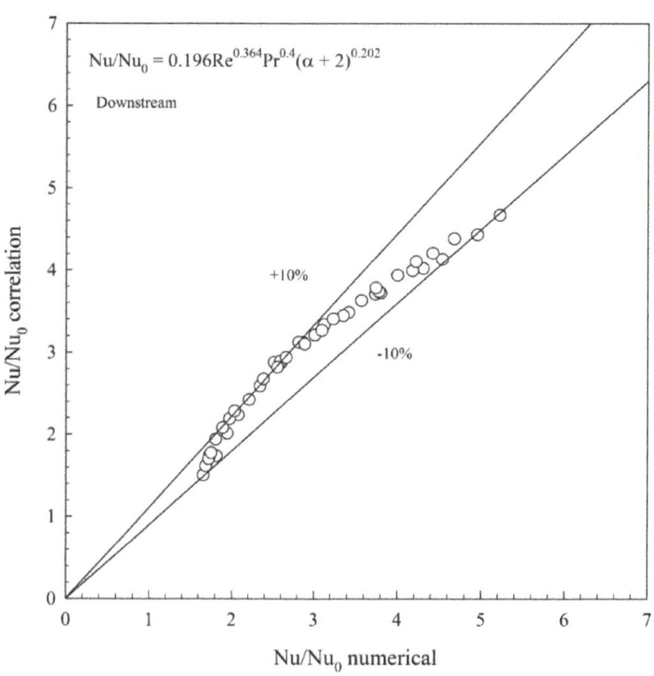

Fig. 18 The correlations of the *Nu*/*Nu*o for downstream of the *PDWVG* in the tube heat exchanger.

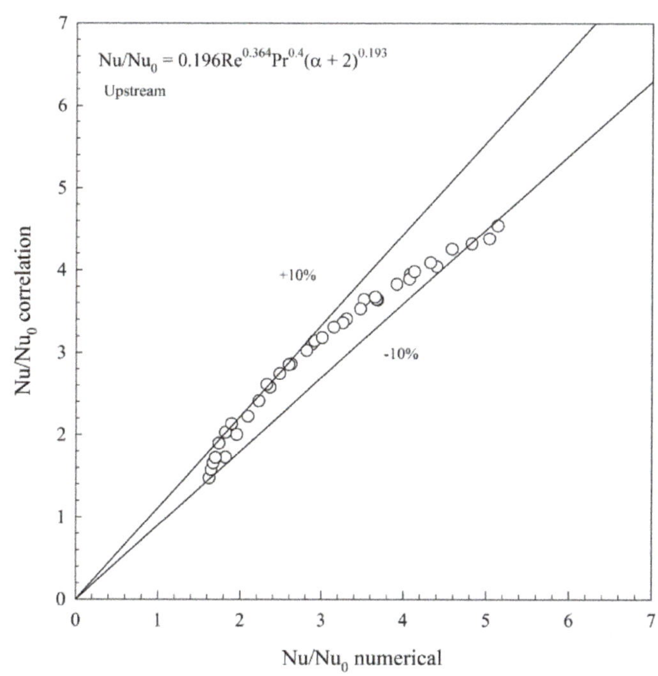

Fig. 19 The correlations of the *Nu*/*Nu*o for upstream of the *PDWVG* in the tube heat exchanger.

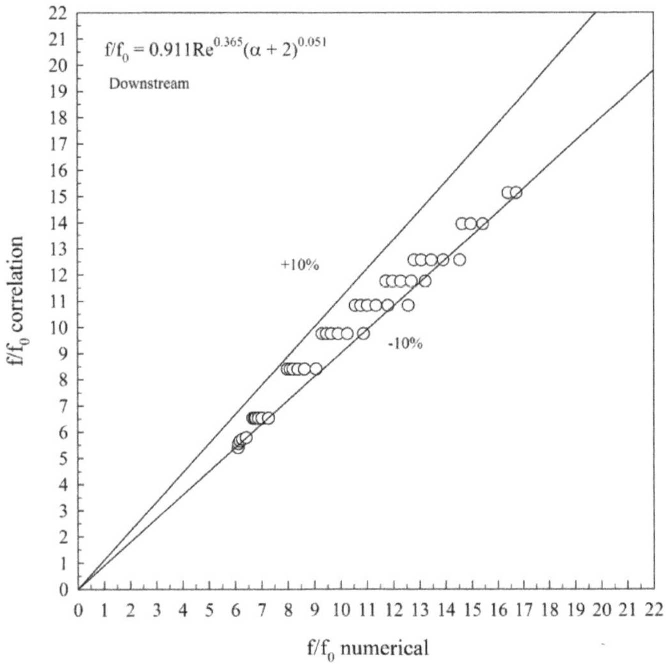

Fig. 20 The correlations of the *f*/*f*0 for downstream of the *PDWVG* in the tube heat exchanger.

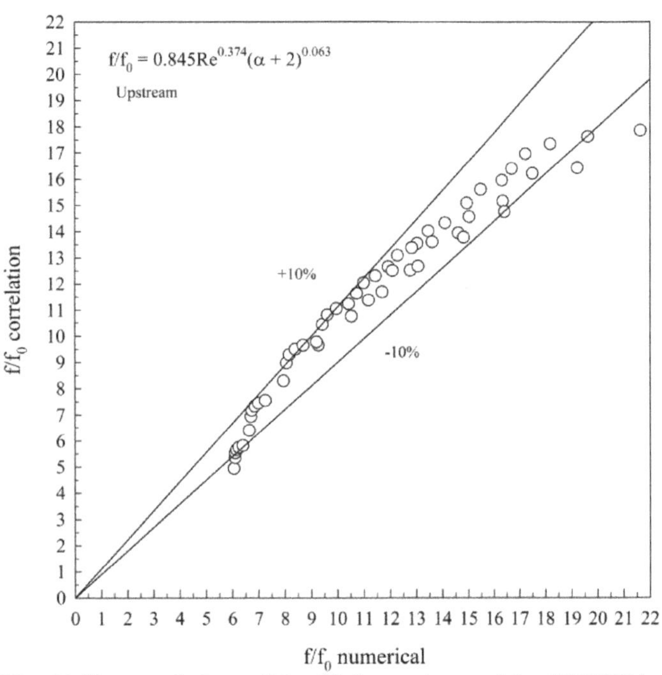

Fig. 21 The correlations of the *f*/*f*0 for upstream of the *PDWVG* in the tube heat exchanger.

The presence of the *PDWVG* in the tube heat exchanger not only increases in heat transfer rate, but also increases the pressure. Figs. 16*a* and *b* present the variations of the *f*/*f*0 with the Reynolds number at various the flow attack angles of the tube heat exchanger with *PDWVG* for the winglet tip pointing downstream and upstream, respectively. Generally, the *f*/*f*0 increase with increasing the Reynolds number and flow attack angle on both cases. The flow attack angle of 25° performs the highest friction loss, while the flow attack angle of 0° provides the reverse result. At α = 0° -10°, the *f*/*f*0 for both cases is found to be nearly values for all Reynolds numbers. When α >10°, the *f*/*f*0 of the winglet tip pointing upstream performs higher than the downstream case. The maximum *f*/*f*0 of the *PDWVG* with the tips pointing downstream and upstream is around 20.33 and 21.64, respectively, at *Re* = 2000 and α = 25°. In addition, the use of *PDWVG* in rang investigates gives the friction factor around 6 – 21.64 times higher than the smooth tube.

The thermal performance in terms of thermal enhancement factor, *TEF*, is reported in Figs. 17*a* and *b*, respectively, for the winglet tip pointing downstream and upstream of the *PDWVG* in the tube heat exchanger. Generally, the use of *PDWVG* gives a higher thermal performance than the smooth circular tube. The *TEF* rises when increasing the Reynolds number and flow attack angle on both cases. The optimum *TEF* is around 2.33, 2.23, 2.15, 2.01,1.69 and 1.13 for the wingtip pointing downstream at the flow attack angles of 25°, 20°, 15°, 10°, 5° and 0°, respectively, and around 2.27, 2.18, 2.09, 1.95, 1.60 and 1.13, respectively, for the upstream case at *Re* = 2000.

6. CONCLUSIONS

Numerical investigations on heat transfer, flow structure and thermal performance in a circular tube heat exchanger with the punched delta winglet vortex generators (*PDWVG*) are presented for *Re* = 100 – 2000. The influences of the flow attack angles, which tip converges to the center of the tube; 0°, 5°, 10°, 15°, 20°, 25°, and flow directions; winglet tips pointing downstream and upstream, are studied. The major findings of the current investigations are as follows:

- The use of the *PDWVG* in the circular tube heat exchanger leads to improve heat transfer rate and thermal performance when compared with the smooth tube with no *PDWVG*. The *PDWVG* can generate the four main vortex flows through the test section.
- The clockwise and counter clockwise vortex flows are found when using *PDWVG* with the tip pointing downstream and upstream, respectively. The vortex flows are clearly found when the flow attack angle of the *PDWVG* higher than 0° for both cases.
- The flow attack angle of 25° performs the highest heat transfer rate, friction factor and thermal enhancement factor, while the flow attack angle of 0° provided the lowest values.
- In range investigates, the heat transfer rate and friction loss are found to be around 1.6 – 6.31 times and 6 – 21.64 times higher than the smooth tube with no *PDWVG*, respectively. The optimum *TEF* is around 2.33 at the flow attack angle of 25°, *Re* = 2000 of the winglet tip pointing downstream.
- The correlations of the *Nu*/*Nu*0 for the *PDWVG* inserted in the tube heat exchanger with the winglet tips pointing downstream and upstream are presented in Figs. 18 and 19, respectively, for *Re* = 100 – 2000 and α = 0° – 25°, while the variations of the *f*/*f*0 are depicted in Figs. 20 and 21 for downstream and upstream, respectively. The correlation equations of the *Nu*/*Nu*0 and *f*/*f*0 of the current investigations are reported as Eqs. (10) – (13).

$$Nu / Nu_0 = 0.196 Re^{0.364} Pr^{0.4} (\alpha + 2)^{0.202}, \text{ Downstream} \quad (10)$$

$$Nu / Nu_0 = 0.196 Re^{0.364} Pr^{0.4} (\alpha + 2)^{0.193}, \text{ Upstream} \quad (11)$$

$$f / f_0 = 0.911 Re^{0.365} (\alpha + 2)^{0.051}, \text{ Downstream} \quad (12)$$

$$f / f_0 = 0.845 Re^{0.374} (\alpha + 2)^{0.063}, \text{ Upstream} \quad (13)$$

ACKNOWLEDGEMENTS

This research was funded by King Mongkut's University of Technology North Bangkok Contract no. KMUTNB-GEN-59-62. The authors would like to thank Assoc. Prof. Dr. Pongjet Promvonge for suggestions.

NOMENCLATURE

D	diameter of a circular tube (m)
f	friction factor
h	convective heat transfer coefficient (W/m^2 K)
k	thermal conductivity (W/m K)
Nu	Nusselt number
p	static pressure (Pa)
Pr	Prandtl number
Re	Reynolds number
T	temperature (K)
u_i	velocity in x_i-direction (m/s)
$\bar{u}$	mean velocity in channel (m/s)

Greek Letter

μ	dynamic viscosity (Pa s)
Γ	thermal diffusivity
α	flow attack angle, (degree)
TEF	thermal enhancement factor ((= $(Nu/Nu_0)(f/f_0)^{-(1/3)}$))
ρ	density (kg/m^3)

Subscripts

in	inlet
0	smooth tube
w	wall
pp	pumping power

REFERENCES

Caliskan, S., 2014, "Experimental Investigation of Heat Transfer in a Channel with New Winglet-type Vortex Generators," *International Journal of Heat and Mass Transfer*, **78**, 604-614. http://dx.doi.org/10.1016/j.ijheatmasstransfer.2014.07.043

Deshmukh, P.W., and Vedula, R.P., 2014, "Heat Transfer and Friction Factor Characteristics of Turbulent Flow Through a Circular Tube Fitted with Vortex Generator Inserts," *International Journal of Heat and Mass Transfer*, **79**, 551-560. http://dx.doi.org/10.1016/j.ijheatmasstransfer.2014.08.042

Du, X., Feng, L., Yang, Y., and Yang, L., 2013, "Experimental Study on Heat Transfer Enhancement of Wavy Finned Flat Tube with Longitudinal Vortex Generators," *Applied Thermal Engineering*, **50**(1), 55-62. http://dx.doi.org/10.1016/j.applthermaleng.2012.05.024

Du, X., Feng, L., Li, L., Yang, L., and Yang, Y., 2014, "Heat Transfer Enhancement of Wavy Finned Flat Tube by Punched Longitudinal Vortex Generators," *International Journal of Heat and Mass Transfer*, **75**, 368-380. http://dx.doi.org/10.1016/j.ijheatmasstransfer.2014.03.081

Ebrahimi, A., Roohi, E., and Kheradmand, S., 2015, "Numerical Study of Liquid Flow and Heat Transfer in Rectangular Microchannel with Longitudinal Vortex Generators," *Applied Thermal Engineering*, **78**(5), 576-583. http://dx.doi.org/10.1016/j.applthermaleng.2014.12.006

Gholami, A.A., Wahid M.A., Mohammed, H.A., 2014, "Heat Transfer Enhancement and Pressure Drop for Fin-and-tube Compact Heat Exchangers with Wavy Rectangular Winglet-type Vortex Generators," *International Communications in Heat and Mass Transfer*, **54**, 132-140. http://dx.doi.org/10.1016/j.icheatmasstransfer.2014.02.016

Gong, B., Wang, L.B., and Lin, Z.M., 2015, "Heat Transfer Characteristics of a Circular Tube Bank Fin Heat Exchanger with Fins Punched Curve Rectangular Vortex Generators in the Wake Regions of the Tubes," *Applied Thermal Engineering*, **75**, 224-238. http://dx.doi.org/10.1016/j.applthermaleng.2014.09.043

He, Y.L., Han, H., Tao, W.Q., and Zhang, Y.W., 2012, "Numerical Study of Heat-transfer Enhancement by Punched Winglet-type Vortex Generator Arrays in Fin-and-tube Heat Exchangers," *International Journal of Heat and Mass Transfer*, **55**(21–22), 5449-5458. http://dx.doi.org/10.1016/j.ijheatmasstransfer.2012.04.059

He, Y.L., Chu, P., Tao, W.Q., Zhang, Y.W., and Xie, T., 2013, "Analysis of Heat Transfer and Pressure Drop for Fin-and-tube Heat Exchangers with Rectangular Winglet-type Vortex Generators," *Applied Thermal Engineering*, **61**(2), 770-783. http://dx.doi.org/10.1016/j.applthermaleng.2012.02.040

Incropera, F., and Dewitt, P.D., 2006, *Introduction to Heat Transfer*, 5rd ed., John Wiley & Sons, Hoboken, NJ.

Khoshvaght-Aliabadi, M., Sartipzadeh, O., and Alizadeh, A., 2015, "An Experimental Study on Vortex-generator Insert with Different Arrangements of Delta-winglets," *Energy*, **82**, 629-639. http://dx.doi.org/10.1016/j.energy.2015.01.072

Li, L., Du, X., Zhang, Y., Yang, L., and Yang, Y., 2015, "Numerical Simulation on Flow and Heat Transfer of Fin-and-tube Heat exchanger with Longitudinal Vortex Generators," *International Journal of Thermal Sciences*, **92**, 85-96. http://dx.doi.org/10.1016/j.ijthermalsci.2015.01.030

Lotfi, B., Zeng, M., Sundén, B., and Wang, Q., 2014, "3D Numerical Investigation of Flow and Heat Transfer Characteristics in Smooth Wavy Fin-and-elliptical Tube Heat Exchangers using New Type Vortex Generators," *Energy*, **73**, 233-257. http://dx.doi.org/10.1016/j.energy.2014.06.016

Min, C., Qi, C., Wang, E., Tian, L., and Qin, Y., 2012, "Numerical Investigation of Turbulent Flow and Heat Transfer in a Channel with Novel Longitudinal Vortex Generators," *International Journal of Heat and Mass Transfer*, **55**(23–24), 7268-7277. http://dx.doi.org/10.1016/j.ijheatmasstransfer.2012.07.055

Patankar, S.V., 1980, *Numerical Heat Transfer and Fluid Flow*, McGraw–Hill, New York.

Wang, Y., He, Y.L., Yang, W.W., and Cheng, Z.D., 2015, "Numerical Analysis of Flow Resistance and Heat Transfer in a Channel with Delta Winglets under Laminar Pulsating Flow," *International Journal of Heat and Mass Transfer*, **82**, 51-65. http://dx.doi.org/10.1016/j.ijheatmasstransfer.2014.11.016

Wu, J.M., and Tao, W.Q., 2012, "Effect of Longitudinal Vortex Generator on Heat Transfer in Rectangular Channels," *Applied Thermal Engineering*, **37**, 67-72. http://dx.doi.org/10.1016/j.applthermaleng.2012.01.002

Wu, J.M., Zhang, H., Yan, C.H., and Wang, Y., 2012, "Experimental Study on the Performance of a Novel Fin-tube Air Heat Exchanger with Punched Longitudinal Vortex Generator," *Energy Conversion and Management*, **57**, 42-48. http://dx.doi.org/10.1016/j.enconman.2011.12.009

27

THERMAL DIFFUSION AND RADIATION EFFECTS ON UNSTEADY FREE CONVECTION FLOW IN THE PRESENCE OF MAGNETIC FIELD FIXED RELATIVE TO THE FLUID OR TO THE PLATE

B. Rushi Kumar*, T. Sravan Kumar, A .G Vijaya Kumar

Fluid Dynamics Division, School of Advanced Sciences, VIT University, Vellore, TN, 632014, India

ABSTRACT

The objective of this study is to investigated thermal diffusion and radiation effects on unsteady free convection chemically reacting fluid flow past an accelerated infinite inclined plate with variable temperature and mass diffusion under the influence of uniform transverse magnetic field when the magnetic lines of force are fixed relative to the fluid or to the plate studied in two cases, (i) when magnetic field fixed relative to the fluid (K=0), (ii) and the magnetic field fixed relative to the plate (K=1) have been considered. A general exact solution of the dimensionless governing partial differential equation is obtained by the usual Laplace transform technique without any restriction. The effect of flow parameters on velocity, temperature, concentration, heat transfer and the rate of mass transfer are studied though graphs and tables.

Keywords: *Soret effect, magnetic field, heat and mass transfer, radiation, free convection, inclined plate and chemical reaction.*

1. INTRODUCTION

The study of magneto hydro dynamics with heat and mass transfer in the presence of radiation and diffusion has attracted the attention of a large number of scholars due to diverse applications. However, when the flow of mass caused by a temperature gradient, the Soret effect is corresponds to species differential equation developing in an initial homogeneous mixer submitted to a thermal gradient and the Dufour corresponds to the heat flux produced by concentration gradient, one cannot neglect the Soret effect. This is also called as thermal-diffusion effect. This type of problem is useful for isotope separation and in mixtures between gasses with very light molecular weight. In free convection flow the study of effect of magnetic field play a major role in liquid metals, electrolytes and ionized gasses etc. In power engineering, the thermal physics of hydro magnetic problems with the mass transfer have enormous applications. Radiative flows are faced with in many industrial and environment processes, heating and cooling chambers, fossil fuel combustion energy processes astrophysical, solar technology, evaporation from large open water reservoirs, In manufacturing industry, for the design of fins, steel rolling , nuclear power plant, gas turbines, various propulsive devices for aircrafts, missiles, satellites, combustion, energy utilization, temperature measurements, remote sensing for Astronomy, in Astrophysics and Geophysics and so on.

Stokes (1851) first studied the flow of an incompressible, viscous fluid past an impulsively started infinite plate in its own plane. The corresponding problem for infinite vertical plate moving impulsively in upward direction was studied by Soundalgekar (1977). MHD effects on impulsively started vertical infinite plate with variable temperature in the presence of transverse magnetic field were studied by Soundalgekar et al. (1981). The effects of transversely applied magnetic field, on the flow of an electrically conducting fluid past an impulsively started infinite isothermal vertical plate were also studied by Soundalgekar et al. (1979). The dimensionless governing equations were solved using Laplace transform technique. Kumari and nath (1999) analyzed the development of a symmetric flow of a viscous electrically conducting fluid in the forward stagnation point region of a two-dimensional body and over a stretching surface was set into impulsive motion from the rest. Muthucumaraswamy and Janakiraman (2006) studied MHD and radiation effects on moving isothermal vertical plate with variable and mass diffusion. The radiative free convection flow of optically thin gray-gas past semi-infinite vertical plate was studied by Soundalgekar and Takhar(1993). Raptis and Perdikis (1999) analyzed radiation and free convection flow past a moving plate. Narahari and Debnath (2013) studied unsteady magneto hydrodynamic free convection flow past an accelerated vertical plate. Tokis (1985) investigated exact solutions of the unsteady magneto hydrodynamic free convection flows. Das et al. (1996) have considered radiation effects on flow past an impulsively started infinite isothermal vertical plate. Kafoussias (1992) studied MHD thermal diffusion effects on free convection and mass transfer flow over an infinite vertical moving plate.

Alam and Sattar (1999) have discussed the thermal-diffusion effect on MHD free convection and mass transfer flow. Jha and Singh (1990) have analyzed the importance of the effect of thermal-diffusion. Alam et al. (2005) studied combined free convection and mass transfer flow past a vertical plate with heat generation and Soret effect through porous medium. Alam et al. (2006) analyzed the Soret effect on unsteady MHD free convection and mass transfer flow past an impulsively started vertical porous plate. Kumar and Varma (2011) studied thermal diffusion and radiation effects on unsteady MHD flow past an impulsively started exponentially accelerated vertical plate with variable temperature and variable mass diffusion. Kumar et al. (2012) studied thermal diffusion and radiation effects on unsteady MHD flow past an exponentially accelerated vertical plate with variable temperature and mass diffusion in the presence of heat source/sink. Rajesh and Varma (2009) studied thermal and radiation effects on MHD flow past an infinite vertical plate with variable temperature and

* Corresponding author. Email: rushikumar@vit.ac.in

mass diffusion. The combined heat and mass transfer problem with chemical reaction are of importance in many processes, and therefore have received a considerable amount of attention in recent years. In processes, such as drying, evaporation at the surface of a water body, energy transfer in a wet cooling tower and the flow in a desert cooler, the heat and mass transfer occurs simultaneously. Chemical reactions can be codified as either homogeneous process or heterogeneous process. A homogeneous reaction is one that occurs uniformly though a given phase. In contrast, a heterogeneous reaction takes place in a restricted region or within the boundary of a phase. A reaction is said to be the first order if the rate of reaction is directly proportional to the concentration itself. In many chemical engineering processes, a chemical reaction between a foreign mass and the fluid flows occur. These processes take place in numerous industrial applications, such as the polymer production, the manufacturing of ceramics or glassware, the food processing etc. Pera and Gebhart (1971) analyzed the stability of laminar plumes by using some numerical solution and experiments. Hossian and Takhar (1996) studied radiation effect on mixed convection along a vertical plate with uniform surface temperature, heat and mass transfer. Raptis and singh (1983) analyzed the free convection flow of a viscous incompressible fluid past an infinite vertical plate embedded in porous medium with constant heat flux in the presence of transverse magnetic field.

However, in the literature we found less attention was paid on unsteady MHD free convection flows past an inclined plate subjected to a variable temperature and uniform mass diffusion with Soret effect when the magnetic lines being fixed relative to the fluid or when the magnetic lines being fixed relative to the plate even though this situation involves in many engineering applications. Chandran et al.(1998) was investigated the effects of magnetic field and buoyancy force on the unsteady free convection laminar flow of an electrically conducting fluid when the flow was generated by uniformly accelerated motion of an infinite vertical plate subjected to constant heat flux. They obtained exact solution with the help of Laplace transform technique and the numerical results were computed with the approximated error functions appeared in the solution.

In this paper, it is proposed to study thermal diffusion and radiation effects on free convection flow past an accelerated infinite inclined plate with variable temperature heat and mass diffusion discussed in two cases. Case (i) when the magnetic field relative to the fluid (*K*=0) and case (ii) the magnetic field fixed relative to the plate (*K*=1) in the presence of transverse applied magnetic field and a first order chemical reaction. The dimensionless governing equations are solved using Laplace transform technique and the solutions are expressed in terms of exponential and complementary error functions.

2. MATHEMATICAL ANALYSIS

In this paper, we study the effects of radiation and thermal-diffusion on unsteady MHD free convection of flow of a viscous incompressible, electrically, conducting, radiating fluid past an impulsively started infinite inclined plate with variable temperature and mass diffusion in the presence of transverse applied magnetic field fixed relative to the fluid or to the plate. The x'-axis is taken along the plate in vertical upward direction and y'-axis is taken normal to it in the direction of applied transverse magnetic field. Initially, it is assumed that the plate and surrounding fluid are at the same temperature and concentration in stationary condition for all the points in entire flow region $y' \geq 0$. At time $t' \geq 0$, the plate is given an impulsive motion with velocity $u = u_0 \exp(a't')$ in its own plane. At the same time, the plate temperature is raised linearly with time t and also the mass is diffused from the plate linearly with time t. A magnetic field of uniform strength B_0 is assumed to be applied normal to the direction of the flow and the induced magnetic field is assumed to be neglected as the magnetic Reynolds number of the flow is taken to be very small.

For free convection flow, it is also assumed that,

- The viscous dissipation is neglected in the energy equation.
- The effects of variation in density (ρ) with temperature and species concentration are considered only on the body force term, in accordance with usual Boussinesq approximation.
- The fluid considered here is gray, absorbing / emitting radiation but a non-scattering medium.
- Since the flow of the fluid is assumed to be in the direction of x' axis, so the physical quantities are functions of the space co-ordinate y' and t' only.

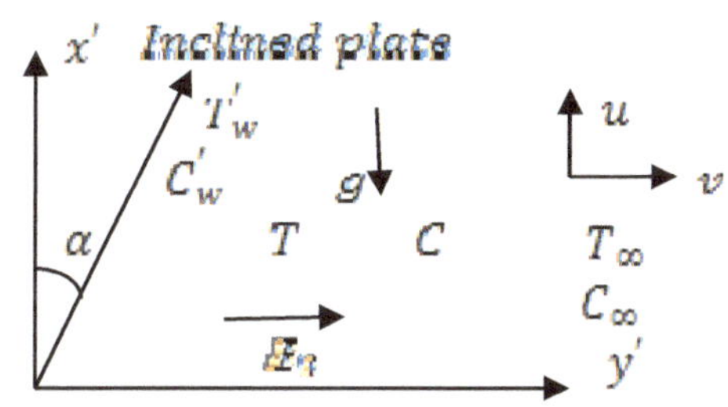

Fig. 1 Physical co-ordinate system

Then by usual Boussinesq's approximation, the flow governed by the following equations:

$$\frac{\partial u'}{\partial t'} = g\beta(T' - T'_\infty)\cos\alpha + g\beta^*(C' - C'_\infty)\cos\alpha + \nu\frac{\partial^2 u'}{\partial y'^2} - \frac{\sigma\beta_0^2 u'}{\rho} \quad (1)$$

$$\rho c_p \frac{\partial T'}{\partial t'} = k\frac{\partial^2 T'}{\partial y'^2} - \frac{\partial q_r}{\partial y'} + Q'(T'_\infty - T') \quad (2)$$

$$\frac{\partial C'}{\partial t'} = D\frac{\partial^2 C'}{\partial y'^2} + D_1\frac{\partial^2 T'}{\partial y'^2} - k'(C' - C'_\infty) \quad (3)$$

If the magnetic field fixed relative to the plate, the momentum equation (1) is replaced by [Camer (1973), Raptis (1983), and Tokies (1985)]

$$\frac{\partial u'}{\partial t'} = g\beta(T' - T'_\infty)\cos\alpha + g\beta^*(C' - C'_\infty)\cos\alpha + \nu\frac{\partial^2 u'}{\partial y'^2} - \frac{\sigma\beta_0^2 u'}{\rho}\left[u' - u_0 f(t')\right] \quad (4)$$

Combining (1) and (4)

$$\frac{\partial u'}{\partial t'} = g\beta(T' - T'_\infty)\cos\alpha + g\beta^*(C' - C'_\infty)\cos\alpha + \nu\frac{\partial^2 u'}{\partial y'^2} - \frac{\sigma\beta_0^2 u'}{\rho}\left[u' - K u_0 f(t')\right] \quad (5)$$

where $K = \begin{cases} 0 & \text{if } B_0 \text{ is fixed relative to the fluid} \\ 1 & \text{if } B_0 \text{ is fixed relative to the plate} \end{cases}$

For an exponentially accelerated plate, $f(t') = \exp(a'_0 t')$,

where a'_0 is dimensional accelerating parameter.

with the following initial and boundary conditions

$t' \leq 0: u' = 0, \quad T' = T'_\infty, \quad C' = C'_\infty, \quad$ for all y'

$t' > 0: u' = u_0 \exp(a't'), T' = T'_\infty + (T'_w - T'_\infty)At', C' = C'_\infty + (C'_w - C'_\infty)At'$

at $y' = 0$

$$u'=0, \quad T' \to T'_\infty, \quad C' \to C'_\infty, \quad \text{as} \quad y' \to \infty \tag{6}$$

where $A=\frac{u_0^2}{v}$ the local radiant for the case of an optically thin gray gas is expressed by $\frac{\partial q_r}{\partial y'}=-4a^*\sigma\left(T_\infty'^4-T'^4\right)$ (7)

It is assumed that the temperature differences within the flow are sufficiently small and that T'^4 may be expressed as a linear function of the temperature. This is obtained by expanding T'^4 in a Taylor series about T'_∞ and neglecting the higher order terms, thus we get

$$T'^4 \cong 4T_\infty'^3 T' - 3T_\infty'^4 \tag{8}$$

From Esq. (5) and (6), Eq. (2) reduces to

$$\rho c_p \frac{\partial T'}{\partial t'} = k\frac{\partial^2 T'}{\partial t'^2} + 16a^*\sigma T_\infty'^3\left(T'_\infty - T'\right) \tag{9}$$

on introducing the following non- dimensional quantities

$$u=\frac{u'}{u_0}, \; t=\frac{t'u_0^2}{v}, \; y=\frac{y'u_0}{v}, \; \theta=\frac{T'-T'_\infty}{T'_w-T'_\infty}, \; k=\frac{u_0^2 k'}{v^2},$$

$$C=\frac{C'-C'_\infty}{C'_w-C'_\infty}, \; Gr=\frac{g\beta v(T'_w-T'_\infty)}{u_0^3}, \; Gm=\frac{g\beta^* v(C'_w-C'_\infty)}{u_0^3}, \; \Pr=\frac{\mu C_\rho}{k}, \; Sc=\frac{v}{D},$$

$$S_o=\frac{D_1(T'_w-T'_\infty)}{v(C'_w-C'_\infty)}, \; M=\frac{\sigma B_0^2 v}{\rho u_0^2}, \; R=\frac{16a^* v^2\sigma T_\infty'^3}{ku_0^2}, \; H=\frac{Q'v^2}{ku_0^2}, \tag{10}$$

we get the following governing equations which are dimensionless.

$$\frac{\partial u}{\partial t} = \frac{\partial^2 u}{\partial y^2} + Gr\theta\cos\alpha + GmC\cos\alpha - M\left(u - Ke^{a_0 t}\right) \tag{11}$$

$$\frac{\partial \theta}{\partial t} = \frac{1}{\Pr}\frac{\partial^2 \theta}{\partial y^2} - \frac{1}{\Pr}(R+H)\theta \tag{12}$$

$$\frac{\partial C}{\partial t} = \frac{1}{Sc}\frac{\partial^2 C}{\partial y^2} + S_o\frac{\partial^2 \theta}{\partial y^2} - kC \tag{13}$$

The initial and boundary conditions in dimensionless form are as follows:

$$t' \le 0: \quad u=0, \quad \theta=0, \quad C=0 \quad \text{for all } y,$$

$$t>0: \quad u=\exp(a_0 t), \quad \theta=t, \quad C=t \quad \text{at} \quad y=0,$$

$$u \to 0, \quad \theta \to 0, \quad C \to 0 \quad as \quad y \to \infty \tag{14}$$

The appeared physical parameters are defined in the nomenclature.

3. METHOD OF SOLUTION

The dimensionless governing equations form (11) to (13), subject to the boundary conditions (14) are solved by usual Laplace transform technique and the solutions are expressed in terms of exponential and complementary error functions

$$\begin{aligned}
C(y,t) = {} & a_7\left[\left(\frac{t}{2}+\frac{y\sqrt{Sc}}{4\sqrt{k}}\right)\exp\left(y\sqrt{kSc}\right)erfc\left(\frac{y\sqrt{Sc}}{2\sqrt{t}}+\sqrt{kt}\right) + \left(\frac{t}{2}-\frac{y\sqrt{Sc}}{4\sqrt{k}}\right)\exp\left(-y\sqrt{kSc}\right)erfc\left(\frac{y\sqrt{Sc}}{2\sqrt{t}}-\sqrt{kt}\right)\right] \\
& + \frac{a_8}{2}\left[\exp\left(y\sqrt{kSc}\right)erfc\left(\frac{y\sqrt{Sc}}{2\sqrt{t}}+\sqrt{kt}\right) + \exp\left(-y\sqrt{kSc}\right)erfc\left(\frac{y\sqrt{Sc}}{2\sqrt{t}}-\sqrt{kt}\right)\right] \\
& - \frac{a_8}{2}\exp(-a_3 t)\left[\exp\left(y\sqrt{(k-a_3)Sc}\right)erfc\left(\frac{y\sqrt{Sc}}{2\sqrt{t}}+\sqrt{(k-a_3)t}\right) + \exp\left(-y\sqrt{(k-a_3)Sc}\right)erfc\left(\frac{y\sqrt{Sc}}{2\sqrt{t}}-\sqrt{(k-a_3)t}\right)\right] \\
& - \frac{a_8}{2}\left[\exp\left(y\sqrt{E}\right)erfc\left(\frac{y\sqrt{\Pr}}{2\sqrt{t}}+\sqrt{\frac{Et}{\Pr}}\right) + \exp\left(-y\sqrt{E}\right)erfc\left(\frac{y\sqrt{\Pr}}{2\sqrt{t}}-\sqrt{\frac{Et}{\Pr}}\right)\right] \\
& + a_5\left[\left(\frac{t}{2}+\frac{y\Pr}{4\sqrt{E}}\right)\exp\left(y\sqrt{E}\right)erfc\left(\frac{y\sqrt{\Pr}}{2\sqrt{t}}+\sqrt{\frac{Et}{\Pr}}\right) + \left(\frac{t}{2}-\frac{y\Pr}{4\sqrt{E}}\right)\exp\left(-y\sqrt{E}\right)erfc\left(\frac{y\sqrt{\Pr}}{2\sqrt{t}}-\sqrt{\frac{Et}{\Pr}}\right)\right] \\
& + \frac{a_8}{2}\exp(-a_3 t)\left[\exp\left(y\sqrt{E-a_3\Pr}\right)erfc\left(\frac{y\sqrt{\Pr}}{2\sqrt{t}}+\sqrt{\frac{Et}{\Pr}-a_3 t}\right) + \exp\left(-y\sqrt{E-a_3\Pr}\right)erfc\left(\frac{y\sqrt{\Pr}}{2\sqrt{t}}-\sqrt{\frac{Et}{\Pr}-a_3 t}\right)\right]
\end{aligned} \tag{15}$$

$$\theta(y,t) = \left[\left(\frac{t}{2}+\frac{y\Pr}{4\sqrt{E}}\right)\exp\left(y\sqrt{E}\right)erfc\left(\frac{y\sqrt{\Pr}}{2\sqrt{t}}+\sqrt{\frac{Et}{\Pr}}\right) + \left(\frac{t}{2}-\frac{y\Pr}{4\sqrt{E}}\right)\exp\left(-y\sqrt{E}\right)erfc\left(\frac{y\sqrt{\Pr}}{2\sqrt{t}}-\sqrt{\frac{Et}{\Pr}}\right)\right] \tag{16}$$

$$\begin{aligned}
u(y,t) = {} & \frac{A_1}{2}\exp(a_0 t)\left[\exp\left(y\sqrt{M+a_0}\right)erfc\left(\frac{y}{2\sqrt{t}}+\sqrt{(M+a_0)t}\right) + \exp\left(-y\sqrt{M+a_0}\right)erfc\left(\frac{y}{2\sqrt{t}}-\sqrt{(M+a_0)t}\right)\right] \\
& + A_2\left[\left(\frac{t}{2}+\frac{y}{4\sqrt{M}}\right)\exp\left(y\sqrt{M}\right)erfc\left(\frac{y}{2\sqrt{t}}+\sqrt{Mt}\right) + \left(\frac{t}{2}-\frac{y}{4\sqrt{M}}\right)\exp\left(-y\sqrt{M}\right)erfc\left(\frac{y}{2\sqrt{t}}-\sqrt{Mt}\right)\right] \\
& + \frac{A_3}{2}\left[\exp\left(y\sqrt{M}\right)erfc\left(\frac{y}{2\sqrt{t}}+\sqrt{Mt}\right) + \exp\left(-y\sqrt{M}\right)erfc\left(\frac{y}{2\sqrt{t}}-\sqrt{Mt}\right)\right] \\
& + \frac{A_4}{2}\exp(-a_3 t)\left[\exp\left(y\sqrt{M-a_3}\right)erfc\left(\frac{y}{2\sqrt{t}}+\sqrt{(M-a_3)t}\right) + \exp\left(-y\sqrt{M-a_3}\right)erfc\left(\frac{y}{2\sqrt{t}}-\sqrt{(M-a_3)t}\right)\right] \\
& + \frac{A_5}{2}\exp(-a_{10} t)\left[\exp\left(y\sqrt{M-a_{10}}\right)erfc\left(\frac{y}{2\sqrt{t}}+\sqrt{(M-a_{10})t}\right) + \exp\left(-y\sqrt{M-a_{10}}\right)erfc\left(\frac{y}{2\sqrt{t}}-\sqrt{(M-a_{10})t}\right)\right] \\
& + \frac{A_6}{2}\exp(-a_{12} t)\left[\exp\left(y\sqrt{M-a_{12}}\right)erfc\left(\frac{y}{2\sqrt{t}}+\sqrt{(M-a_{12})t}\right) + \exp\left(-y\sqrt{M-a_{12}}\right)erfc\left(\frac{y}{2\sqrt{t}}-\sqrt{(M-a_{12})t}\right)\right] \\
& + A_7\exp(-Mt)\,erfc\left(\frac{y}{2\sqrt{t}}\right)
\end{aligned}$$

$$+\frac{A_8}{2}\left[\begin{array}{l}\exp\left(y\sqrt{kSc}\right)erfc\left(\frac{y\sqrt{Sc}}{2\sqrt{t}}+\sqrt{kt}\right)\\+\exp\left(-y\sqrt{kSc}\right)erfc\left(\frac{y\sqrt{Sc}}{2\sqrt{t}}-\sqrt{kt}\right)\end{array}\right]$$

$$+A_9\left[\begin{array}{l}\left(\frac{t}{2}+\frac{y\sqrt{Sc}}{4\sqrt{k}}\right)\exp\left(y\sqrt{kSc}\right)erfc\left(\frac{y\sqrt{Sc}}{2\sqrt{t}}+\sqrt{kt}\right)\\+\left(\frac{t}{2}-\frac{y\sqrt{Sc}}{4\sqrt{k}}\right)\exp\left(-y\sqrt{kSc}\right)erfc\left(\frac{y\sqrt{Sc}}{2\sqrt{t}}-\sqrt{kt}\right)\end{array}\right]$$

$$+\frac{A_{10}}{2}\exp\left(-a_3t\right)\left[\begin{array}{l}\exp\left(y\sqrt{\left(k-a_3\right)Sc}\right)erfc\left(\frac{y\sqrt{Sc}}{2\sqrt{t}}+\sqrt{\left(k-a_3\right)t}\right)\\+\exp\left(-y\sqrt{\left(k-a_3\right)Sc}\right)erfc\left(\frac{y\sqrt{Sc}}{2\sqrt{t}}-\sqrt{\left(k-a_3\right)t}\right)\end{array}\right]$$

$$-\frac{A_6}{2}\exp(-a_{12}t)\left[\begin{array}{l}\exp\left(y\sqrt{(k-a_{12})Sc}\right)erfc\left(\frac{y\sqrt{Sc}}{2\sqrt{t}}+\sqrt{(k-a_{12})t}\right)\\+\exp\left(-y\sqrt{(k-a_{12})Sc}\right)erfc\left(\frac{y\sqrt{Sc}}{2\sqrt{t}}-\sqrt{(k-a_{12})t}\right)\end{array}\right]$$

$$+\frac{A_{11}}{2}\left[\begin{array}{l}\exp\left(y\sqrt{E}\right)erfc\left(\frac{y\sqrt{\Pr}}{2\sqrt{t}}+\sqrt{\frac{Et}{\Pr}}\right)\\+\exp\left(-y\sqrt{E}\right)erfc\left(\frac{y\sqrt{\Pr}}{2\sqrt{t}}-\sqrt{\frac{Et}{\Pr}}\right)\end{array}\right]$$

$$+A_{12}\left[\begin{array}{l}\left(\frac{t}{2}+\frac{y\Pr}{4\sqrt{E}}\right)\exp\left(y\sqrt{E}\right)erfc\left(\frac{y\sqrt{\Pr}}{2\sqrt{t}}+\sqrt{\frac{Et}{\Pr}}\right)\\+\left(\frac{t}{2}-\frac{y\Pr}{4\sqrt{E}}\right)\exp\left(-y\sqrt{E}\right)erfc\left(\frac{y\sqrt{\Pr}}{2\sqrt{t}}-\sqrt{\frac{Et}{\Pr}}\right)\end{array}\right]$$

$$+\frac{A_{13}}{2}\exp\left(-a_3t\right)\left[\begin{array}{l}\exp\left(y\sqrt{E-a_3\Pr}\right)erfc\left(\frac{y\sqrt{\Pr}}{2\sqrt{t}}+\sqrt{\frac{Et}{\Pr}-a_3t}\right)\\+\exp\left(-y\sqrt{E-a_3\Pr}\right)erfc\left(\frac{y\sqrt{\Pr}}{2\sqrt{t}}-\sqrt{\frac{Et}{\Pr}-a_3t}\right)\end{array}\right]$$

$$+\frac{A_{14}}{2}\exp\left(-a_{10}t\right)\left[\begin{array}{l}\exp\left(y\sqrt{E-a_{10}\Pr}\right)erfc\left(\frac{y\sqrt{\Pr}}{2\sqrt{t}}+\sqrt{\frac{Et}{\Pr}-a_{10}t}\right)\\+\exp\left(-y\sqrt{E-a_{10}\Pr}\right)erfc\left(\frac{y\sqrt{\Pr}}{2\sqrt{t}}-\sqrt{\frac{Et}{\Pr}-a_{10}t}\right)\end{array}\right]$$

$$+A_7\exp\left(a_0t\right)\left[\begin{array}{l}\exp\left(y\sqrt{M+a_0}\right)erfc\left(\frac{y}{2\sqrt{t}}+\sqrt{\left(M+a_0\right)t}\right)\\+\exp\left(-y\sqrt{M+a_0}\right)erfc\left(\frac{y}{2\sqrt{t}}-\sqrt{\left(M+a_0\right)t}\right)\end{array}\right] \quad (17)$$

where

E= R+H, $a_1=\frac{-Sc\,S_0}{\Pr(\Pr-Sc)}$, $a_2=\frac{E}{\Pr}$, $a_3=\frac{E-kSc}{\Pr-Sc}$, $a_4=\frac{a_1a_2}{a_3^2}$, $a_5=\frac{a_1a_2}{a_3}$, $a_6=\frac{a_1}{a_3}$, $a_7=(1-a_5)$, $a_8=(a_4-a_6)$,

$a_9=\frac{-Gr\cos\alpha}{\Pr-1}$, $a_{10}=\frac{E-M}{\Pr-1}$, $a_{11}=\frac{Gm\cos\alpha}{Sc-1}$, $a_{12}=\frac{kSc-M}{Sc-1}$,

$a_{13}=\frac{Gm\cos\alpha}{\Pr-1}$, $a_{14}=\frac{a_9}{a_{10}^2}$, $a_{15}=\frac{a_9}{a_{10}}$, $a_{16}=\frac{a_{11}}{a_{12}^2}$

, $a_{17}=\frac{a_{11}}{a_{12}}$, $a_{18}=\frac{a_1a_2a_{11}(a_3+a_{12})}{a_3^2a_{12}^2}$, $a_{19}=\frac{a_1a_2a_{11}}{a_3a_{12}}$,

$a_{20}=\frac{a_1a_2a_{11}}{a_3^2(a_{12}-a_3)}$, $a_{21}=\frac{a_1a_2a_{11}}{a_{12}^2(a_{12}-a_3)}$, $a_{22}=\frac{a_1a_{11}}{a_{12}a_3}$,

$a_{23}=\frac{a_1a_{11}}{a_3(a_{12}-a_3)}$, $a_{24}=\frac{a_1a_{11}}{a_{12}(a_{12}-a_3)}$, $a_{29}=\frac{a_1a_{13}}{a_3a_{10}}$,

$a_{26}=\frac{a_1a_2a_{13}}{a_3a_{10}}$, $a_{25}=\frac{a_1a_2a_{13}(a_3+a_{10})}{a_3^2a_{10}^2}$, $a_{27}=\frac{a_1a_2a_{13}}{a_3^2(a_{10}-a_3)}$,

$a_{28}=\frac{a_1a_2a_{13}}{a_{10}^2(a_{10}-a_3)}$, $a_{30}=\frac{a_1a_{13}}{a_3(a_{10}-a_3)}$, $a_{31}=\frac{a_1a_{13}}{a_{10}(a_{10}-a_3)}$,

$a_{32}=\frac{MK}{M+a_0}$, $A_1=(1-a_{32})$, $A_2=(-a_{15}+a_{17}-a_{19}+a_{26})$

$A_3=(a_{14}-a_{16}+a_{18}-a_{22}-a_{25}+a_{29})$,

$A_4=(-a_{20}+a_{23}+a_{27}-a_{30})$, $A_7=a_{32}$

$A_5=(-a_{14}-a_{28}+a_{31})$, $A_6=(a_{16}+a_{21}-a_{24})$

$A_9=(-a_{17}+a_{19})$, $A_8=(a_{16}-a_{18}+a_{22})$,

$A_{10}=(a_{20}-a_{23})$, $A_{11}=(-a_{14}+a_{25}-a_{29})$, $A_{12}=(a_{15}-a_{26})$,

$A_{13}=(-a_{27}+a_{30})$, $A_{14}=(a_{14}+a_{28}-a_{31})$.

Nusselt number:

From temperature field, the Nusselt number which is given in non-dimensional form as follows:

$$Nu=-\left[\frac{\partial\theta}{\partial y}\right]_{y=0} \quad (18)$$

From eqs. (16) and (18), we get Nusselt number as follows:

$$Nu=\left[t\sqrt{E}\,erf\sqrt{\frac{Et}{\Pr}}+\sqrt{\frac{t\Pr}{\pi}}\exp\left(-\frac{Et}{\Pr}\right)+\frac{\Pr}{2\sqrt{E}}erf\sqrt{\frac{Et}{\Pr}}\right] \quad (19)$$

Sherwood number:

From concentration field, Sherwood Number which is given in non –dimensional form as follows

$$Sh=-\left[\frac{\partial C}{\partial y}\right]_{y=0} \quad (20)$$

From Eqs.(15) and (20) we get Sherwood Number as follows:

$$Sh=a_7\left[\exp(-kt)\sqrt{\frac{tSc}{\pi}}+t\sqrt{kSc}\,efr\left(\sqrt{kt}\right)+\frac{\sqrt{Sc}}{2\sqrt{k}}erf\left(\sqrt{kt}\right)\right]$$

$$+a_8\left[\exp(-kt)\sqrt{\frac{Sc}{\pi t}}-\sqrt{kSc}\,erf\left(\sqrt{kt}\right)\right]$$

$$+a_8\exp(-a_3t)\left[\begin{array}{l}-\exp(kt-a_3t)\sqrt{\frac{Sc}{\pi t}}+\\\sqrt{(k-a_3)Sc}\,erfc(\sqrt{(k-a_3)t})\end{array}\right]$$

$$+a_8\left[\exp\left(-\frac{Et}{\Pr}\right)\sqrt{\frac{\Pr}{\pi t}}+\sqrt{E}\,erf\left(\sqrt{\frac{Et}{\Pr}}\right)\right]$$

$$-a_5\left[\exp\left(-\frac{Et}{\Pr}\right)\sqrt{\frac{\Pr t}{\pi}}+t\sqrt{E}\,erf\left(\frac{Et}{\Pr}\right)+\frac{\Pr}{2\sqrt{E}}erf\left(\sqrt{\frac{Et}{\Pr}}\right)\right] \quad (21)$$

4. RESULT AND DISCUSSION

To get a physical insight into the problem, the numerical evaluation of the analytical results reported in the previous section was performed and as set of results is reported graphically in Figs. 2 to 17. for the case of heating (*Gr*< 0, *Gm*< 0) and cooling (*Gr*> 0, *Gm*> 0) of the plate. The heating and cooling takes place by setting up free convection currents due to the temperature and concentration gradient. These results are obtained to reveal the effect of various physical parameters like radiation parameter (*R*), heat source parameter (*H*) magnetic parameter (*M*), Schmidt parameter (*Sc*), Prandtl number (*Pr*), Soret number (*So*), thermal Grashof number (*Gr*), mass Grashof number (*Gm*) on the velocity, temperature and the concentration profiles.

The velocity profiles for an exponentially accelerated plate are presented from Figs. 2. to 9, when the magnetic field is being fixed to the fluid (K=0) discussed in case (i) and the magnetic field being fixed related though the plate (K=1) discussed in case (ii).The effects of M, So, Gr, Gm, Sc, R, H and t on the velocity field are shown from Fig. 2 to 9, in the presence of radiation and first order chemical reaction respectively, when other parameters being fixed.

Figure 2 reveals that the effect of magnetic field parameter on fluid velocity and it is observed that an increase in magnetic parameter M , the velocity decreases in case of cooling and heating of the plate for Pr =0.71. It is due to the fact that the application of transverse magnetic field will result a Lorentz force similar to drag force, which tends to resist the fluid flow and thus reducing its velocity and Fig. 3, illustrates the effect of magnetic field parameter of the plate velocity and it is noticed that an increase in magnetic parameter M, the velocity increases in both cases of cooling and heating of the plate. It is seen that from Fig. 4, that the velocity increases with an increasing value of thermal diffusion parameter (So) in case of cooling of the plate and a reverse effect is observed in case of heating of the plate, when the magnetic field is being fixed to the fluid (K=0). From Fig. 5, we observe that the velocity suddenly increases with an increasing value of thermal diffusion parameter (So) in case of cooling of the plate and a reverse effect it is noticed in case of heating of the plate, when the magnetic field being fixed related though the plate (K=1).Fig. 6, illustrates the effects Gr and Gm on the velocity field. It is found that the velocity decreases with the increases of Gr and Gm in case of cooling and reverse effect is observed in case of heating, when the magnetic field is being fixed to the fluid (K=0) and Fig. 7, it is seen that the velocity fall down with the increase of Gr and Gm in case of cooling of the plate and reverse effects it is observed in the case of heating of the plate, when the magnetic field being fixed related though the plate (K=1). Fig. 8, reveals the velocity increases with the increases of time t in case of cooling of the plate and the trend is just reversed in the case of heating of the plate, for the magnetic field when it is being fixed to the fluid (K=0) and from Fig. 9, it is observed that the velocity increases with the increases of time t in case of cooling of the plate and reverse effects is noticed in the case of heating of the plate, for the magnetic field being when it is being fixed related though the plate (K=1). In all above figures for both cases of cooling and heating of the plate for case (i) and case (ii) moving for away from the plate, the velocity profiles decreases slowly and tending to zero.

The influence of various flow parameters on the fluid temperature are illustrated in Figs. 10 and 11. From these Figs. 10 and 11, it is observed that as radiation parameter R, heat source parameter H and Prandtl number Pr increases when the temperature of the flow field decreases. Concentration profiles for different values of Sc, So, R, H, and t are presented though Figs. 12, 13 and 14. From these Figs it is seen that the concentration increases with an increases in Sc, or So, or R, or H while it increases with time t. From Fig. 15, it is observed that concentration decreases with an increase in k while it increases with time t. It is seen that from Fig. 16, Nusselt number increases with increasing values of R for both water (Pr=7.0) and air (Pr=0.71). It is also observed that Nusselt number for water is higher than that of air. The reason is that small values of Pr are equivalent to increasing the thermal conductivities and therefore heat is able to diffuse away from the plate more rapidly than higher values of Pr, hence the rate of heat transfer is reduced. Fig. 17, reveals the Sherwood number decreases with the increasing values of So.

The velocity profiles for different values of R and H when K=0 and K=1 in case of cooling and heating of the plate for fixed values of Sc =2.01, Pr=0.71, So=5, k=0.5 and t=0.2, are shown in Table1 and Table 2. It is observed from these tables that when $K=0$ velocity decreases with y for both cooling and heating of the plate and when $K=1$ velocity increases up to certain y value and thereafter velocity decreases for cooling of the plate while velocity decreases gradually in case of heating of the plate.

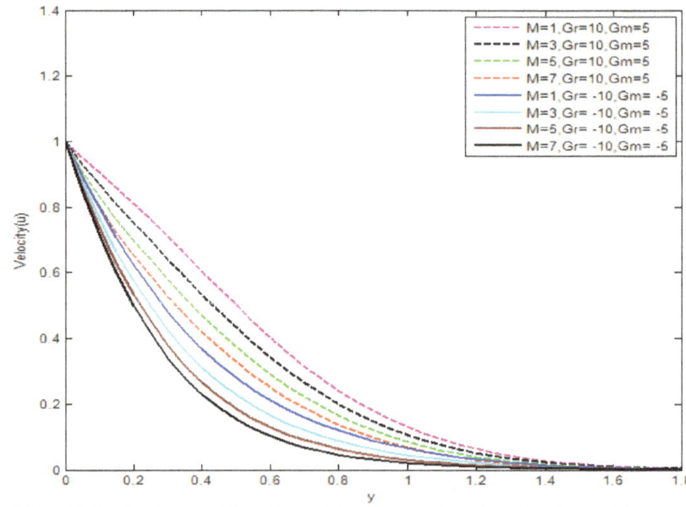

Fig. 2 Velocity profiles for different M when (K=0), Sc =2.01, So =5, Pr =0.71, R =10, a_0 =0 , k =0.5, H =4 and t =0.2

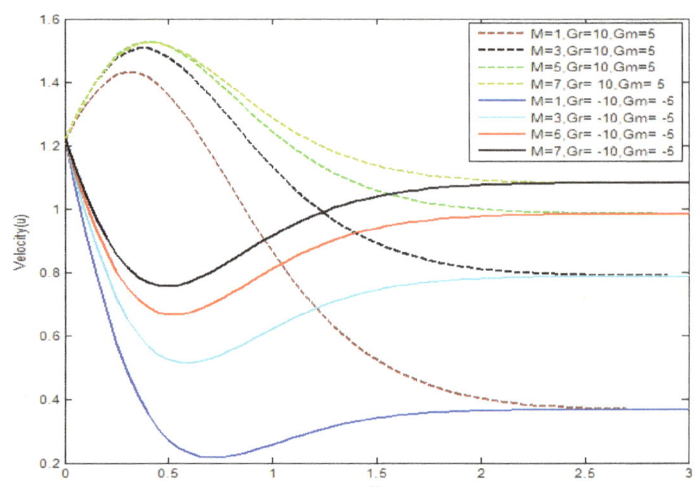

Fig. 3 Velocity profiles for different M when (K=1), Sc =2.01, So =5, Pr =0.71, R =10, a_0=0.5 , k =0.5, H =4 and t =0.4

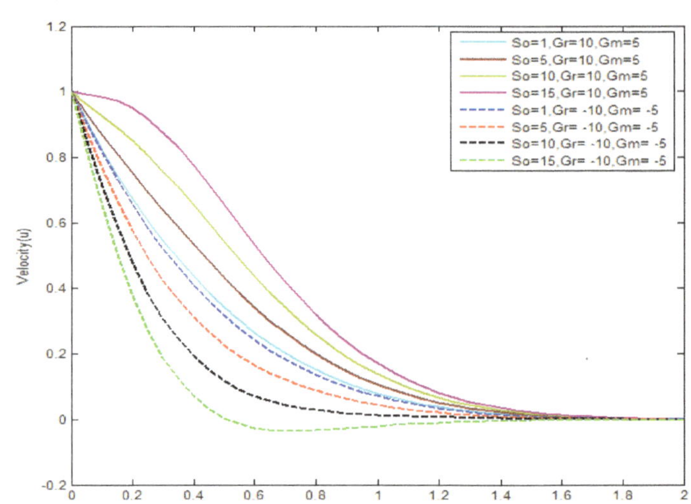

Fig. 4 Velocity profiles for different So when (K=0), Sc =2.01, Pr =0.71, R =10, H=4, M =3, a_0 =0 , k =0.5 and t =0.2

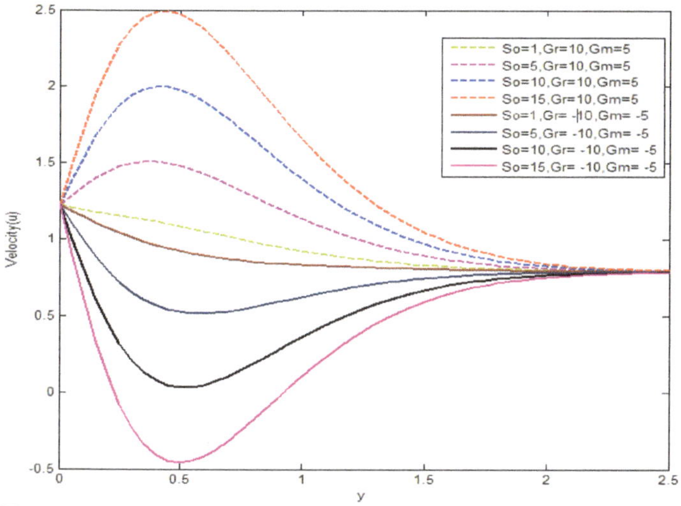

Fig. 5 Velocity profiles for different *So* when (K=1), Sc =2.01, Pr =0.71, R=10, H=4, M=3, a_0=0.5 , k =0.5 and t =0.4.

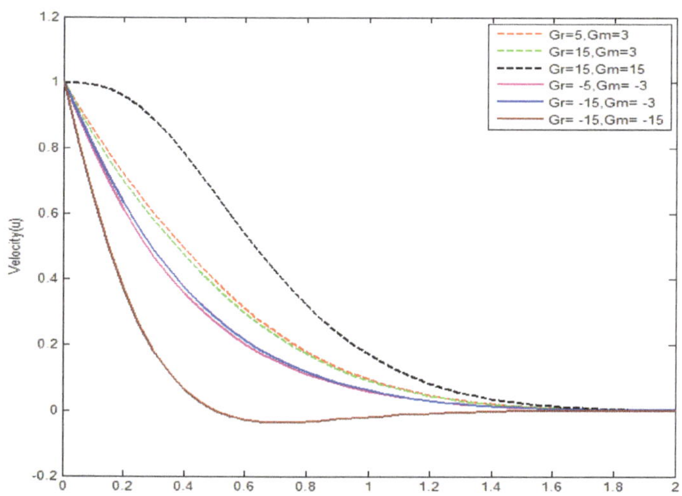

Fig. 6 Velocity profiles for different *Gr* and *Gm* when (K=0),Sc =2.01, So=5, Pr =0.71, R=10, H=4, M =3, a_0=0 , k =0.5 and t =0

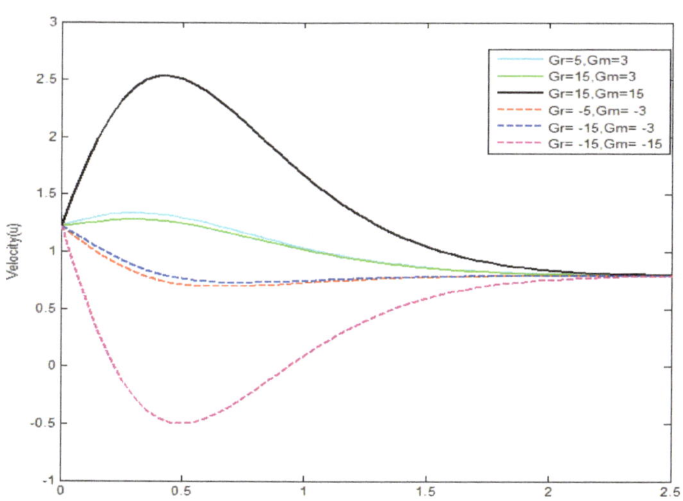

Fig. 7 Velocity profiles for different *Gr* and *Gm* when (K=1), Sc =2.01, So =5 Pr =0.71, R=10, H=4, M=3, a_0=0.5 , k=0.5

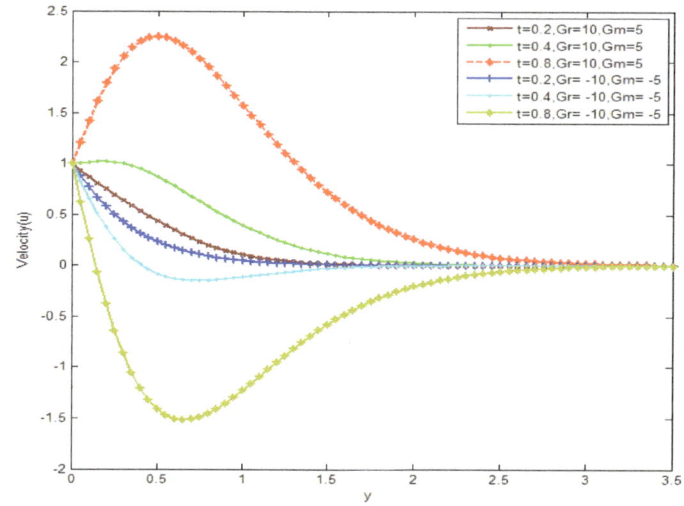

Fig. 8 Velocity profiles for different t when (K=0), Sc =2.01, So =5, Pr =0.71, R=10, H=4, a_0=0, M=3, and k =0.5

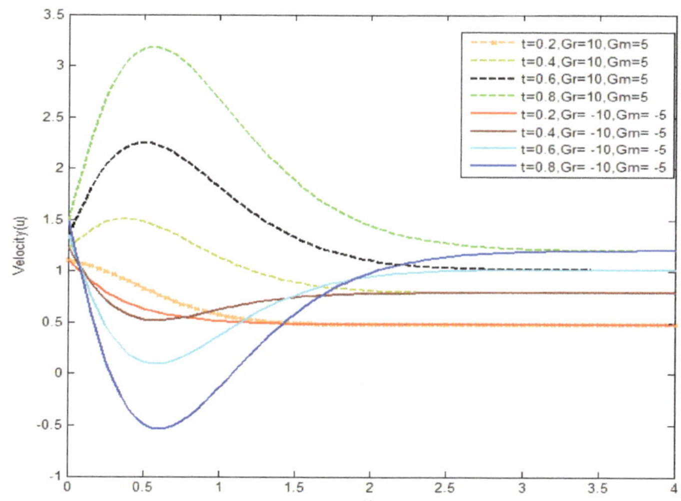

Fig. 9 Velocity profiles for different t when (K=1), Sc =2.01, So =5, Pr =0.71, R =10, H =4, M =3, a_0=0.5 and k =0.5

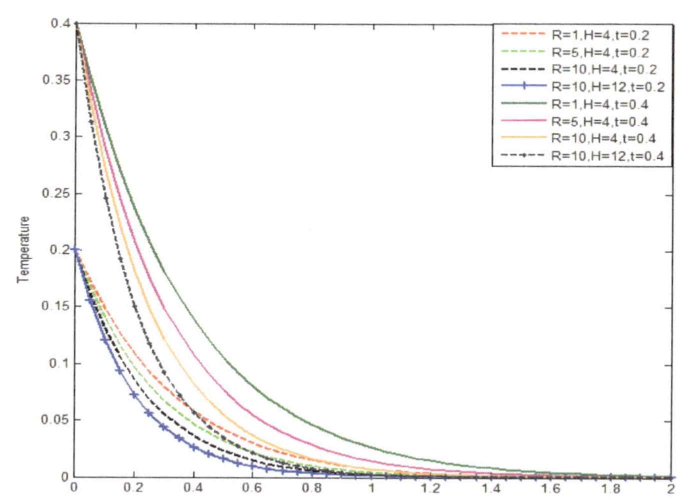

Fig. 10 Temperature profiles for different R and H when Pr =0.7

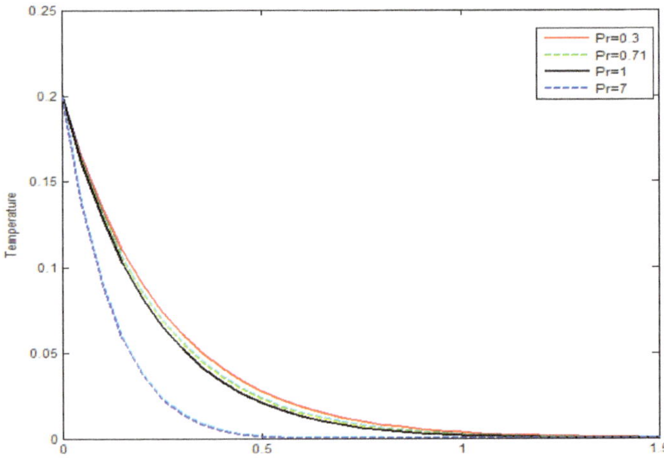

Fig. 11 Temperature profiles for different *Pr* when *R*=10, *H*=4 and *t*=0.2

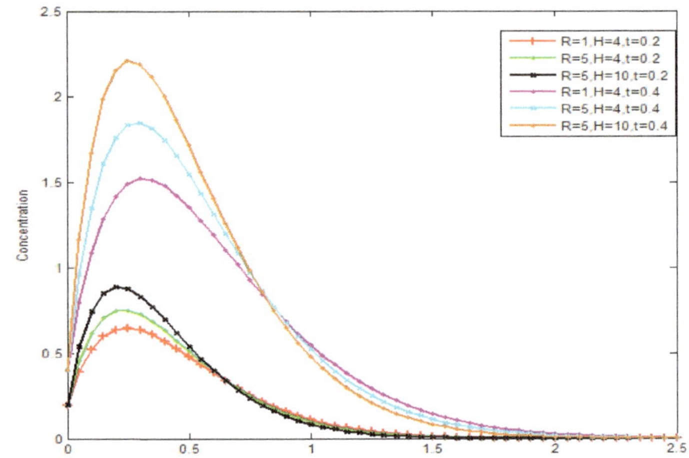

Fig. 14 Concentration profiles for different *R, H* and *t* when *Pr* =0.71, *So* =5, *Sc* =2.01 and *k* =0

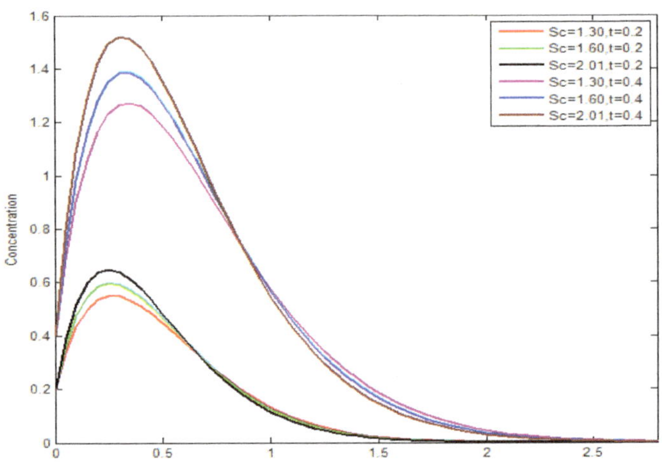

Fig. 12 Concentration profiles for different *Sc* and *t* when *Pr* =0.71, *R*=4, *H*=1, *So*=5, *Sc* =2.01 and *k* =0.5

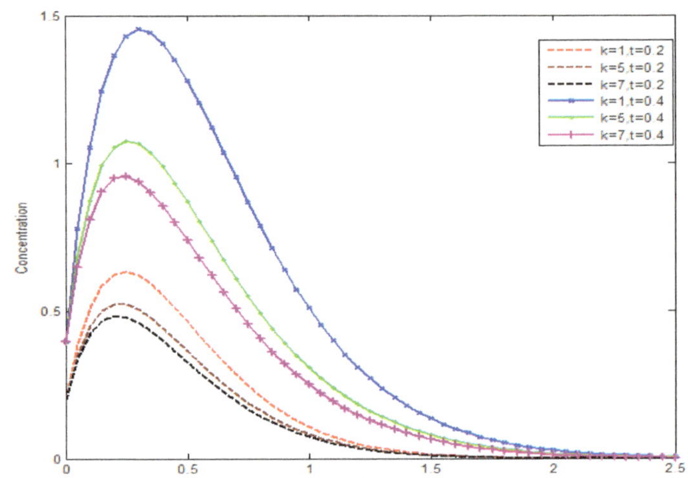

Fig. 15 Concentration profiles for different *k* and *t* when *Pr* =0.71, *So* =5, *Sc* =2.01, *R*=4, *H*=1 and *k* =0.5

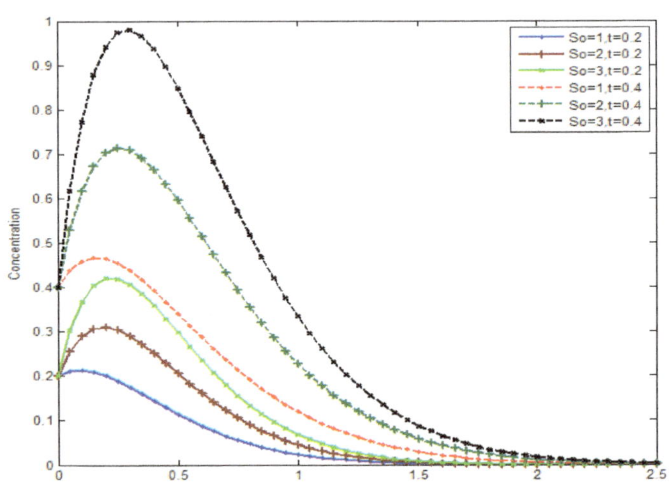

Fig. 13 Concentration profiles for different *So* and *t* when *Pr*=0.71, *R*=4, *H*=1, *So*=5, *Sc*=2.01 and *k*=0.5

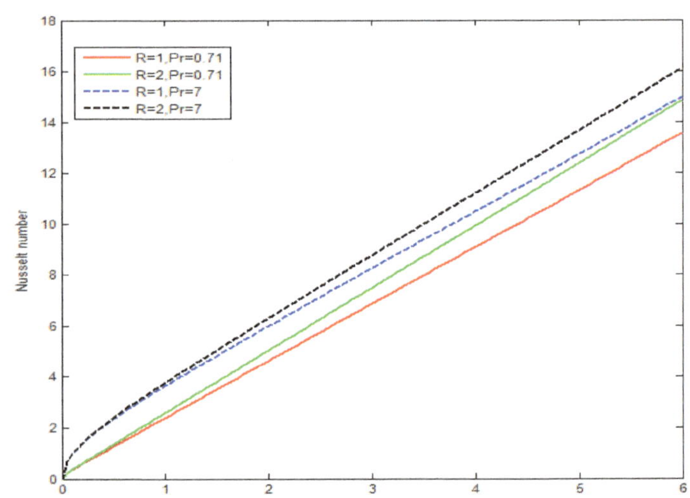

Fig . 16 Nusselt number for different *R* and *Pr*

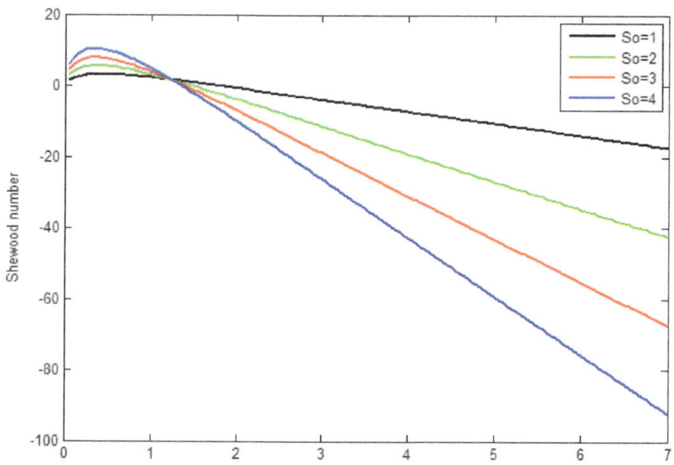

Fig. 17 Sherwood number for different *So*

Table 1 Velocity for different *R* when *Sc*=2.01, *Pr*=0.71, *So* =5, *k*=0.5, and *t*=0.2

		K=0			K=1		
	y	R=1	R=5	R=10	R=1	R=5	R=10
Cooling of the plate	0.0	1.0000	1.0000	1.0000	1.2214	1.2214	1.2214
	0.2	0.7259	0.7391	0.7525	1.3295	1.3892	1.4438
	0.4	0.5081	0.5211	0.5335	1.3682	1.4429	1.5068
	0.6	0.3302	0.3373	0.3435	1.3205	1.3793	1.4257
	0.8	0.1969	0.1991	0.2005	1.2223	1.2567	1.2807
	1.0	0.1075	0.1072	0.1066	1.1113	1.1255	1.1329
Heating of the plate	0.0	1.0000	1.0000	1.0000	1.2214	1.2214	1.2214
	0.2	0.6038	0.5905	0.5771	0.9101	0.8503	0.7957
	0.4	0.3384	0.3254	0.3130	0.7068	0.6322	0.5682
	0.6	0.1805	0.1734	0.1672	0.6223	0.5635	0.5171
	0.8	0.0926	0.0904	0.0890	0.6159	0.5815	0.5575
	1.0	0.0455	0.0458	0.0464	0.6465	0.6322	0.6248

Table 2 Velocity for different *H* when *Sc* =2.01, *Pr*=0.71, *So*=5, *k*=0.5 and *t*=0.2

		K=0			K=1		
	y	H=1	H=5	H=10	H=1	H=5	H=10
Cooling of the plate	0.0	1.0000	1.0000	1.0000	1.2214	1.2214	1.2214
	0.2	0.7449	0.7549	0.7653	1.4132	1.4529	1.4918
	0.4	0.5265	0.5356	0.5446	1.4714	1.5171	1.5595
	0.6	0.3401	0.3445	0.3485	1.4005	1.4327	1.4606
	0.8	0.1998	0.2007	0.2013	1.2681	1.2840	1.2963
	1.0	0.1070	0.1065	0.1058	1.1293	1.1337	1.1357
Heating of the plate	0.0	1.0000	1.0000	1.0000	1.2214	1.2214	1.2214
	0.2	0.5848	0.5748	0.5643	0.8264	0.7867	0.7478
	0.4	0.3200	0.3109	0.3019	0.6036	0.5579	0.5156
	0.6	0.1706	0.1662	0.1623	0.5422	0.5100	0.4822
	0.8	0.0897	0.0888	0.0882	0.5701	0.5541	0.5418
	1.0	0.0460	0.0465	0.0472	0.6284	0.6241	0.6220

5. CONCLUSION

The unsteady free convection chemically reacting fluid flow past an accelerated infinite inclined plate with variable temperature and mass diffusion under the influence of uniform transverse magnetic field when the magnetic lines of force are fixed relative to the fluid or to the plate is studied in two cases. The non-dimensional governing equations are solved with the help of Laplace transform method. The following conclusions have been arrived.

- The fluid velocity decreases as the existence of the magnetic field parameter *M* becomes stronger.
- An increase in *So* or *t* the velocity flow field increases and *Gr* or *Gm* increases the velocity flow field decreases.
- Temperature of the flow field decreases if *R* or *H* or Pr increases.
- Concentration of the flow field increases if *Sc*, or *So*, or *R*, or *H*, or *t* increases and also concentration decreases with maximizing values of *k* .
- *Nu* increments with increasing values of *R* and Pr.
- *Sh* diminishes with enhanced values of *So*

NOMENCLATURE

a^*	Absorption coefficient (cm^{-1})
a	Accelerated parameter ($msec^{-2}$)
H	Heat source parameter (w/m^2K)
B_0	External magnetic field ($A.m^2$)
C'	Species concentration
C'_w	Concentration of the plate
C'_∞	Concentration of the fluid far away from the plate
C	Dimensionless concentration (kg/m^3)
C_p	Specific heat at constant pressure (J/kg.K)
D	Chemical molecular diffusivity (J/kg.K)
D_1	Coefficient of thermal diffusivity
g	Acceleration due to gravity (ms^{-2})
Gr	Thermal Grashof number
Gm	Mass Grashof number
M	Magnetic field parameter (Am^{-1})
Nu	Nusselt number
Pr	Prandtl number
q_r	Radiative heat flux in the y-direction (wm^{-2})
R	Radiative parameter (cm^{-2})
Sc	Schmidt number
So	Soret number
Sh	Sherwood number
T'	Temperature of the fluid near the plate
T'_w	Temperature of the plate
T'_∞	Temperature of the fluid far away from the plate
t'	Time
t	Dimensionless time (s)
u'	Velocity of the fluid in the x' -direction
u_0	Velocity of the plate
u	Dimensionless velocity (ms^{-1})
y'	Co-ordinate axis normal to the plate
y	Dimensionless Co-ordinate axis normal to the plate

Greek symbols

k	Thermal conductivity of the fluid (W/m.K)
α	Thermal diffusivity (Wm^2j^{-1})
β	Volumetric coefficient of thermal expansion
β^*	Volumetric coefficient of expansion with concentration
μ	Coefficient of viscosity (m^2s^{-1})
v	Kinematic viscosity (m^2s^{-1})
ρ	Density of the fluid (kgm^{-3})
σ	Electric conductivity (Sm^{-1})
θ	Dimensionless temperature (K)
erf	Error function
$erfc$	Complementary error function

Subscripts

w	Conditions on the wall
∞	Free stream conditions

REFERENCES

Cramer, K.R., and Pai, S.I.,1973, *Magneto Fluid Dynamics for Engineers and Applied Physicists* (McGraw-Hill Book Company, New York).

Eekert, E.R.G., and Drake, R.M., 1972, *Analysis of Heat and Mass Transfer*, (Mc Graw-Hill Book co, New York).

Gebhart, B., 1971, *Heat Transfer* (2nd edition), (Mc Graw-Hill Book co, New York).

Alam, M.M., and Sattar, M. A., 1999, "Transient MHD Heat and Mass Transfer Flow with Thermal Diffusion in a Rotating System," *J. Energy Heat Mass transfer*. **21** m, 9-21.

Alam, M.S., Rahman, M.M., and Maleque, M.A., 2005, "Local Similarity Solutions for Unsteady MHD Free Convection and Mass Transfer Flow Past an Impulsively Started Vertical Porous Plate with Dufour and Soret effects," *Thammasatint.j.sci.tech*. **10**(3), 1-8.

Alam,M.S., Rahman, M.M., and Samad, M.A., 2006, "Numerical Study of the Combined Free-Forced Convection and Mass Transfer Flow Past a Vertical Porous Plate in a Porous Medium with Heat generation and Thermal Diffusion, Nonlinear". Anal. Model. Control **11**(4), 331-343.

Chandran, P., Sacheti, N.C., and Singh, A.K., 1998, "Unsteady Hydro magnetic Free Convection Flow with Heat Flux and Accelerated Boundary Motion," *J. Phys. Soc. Japan*. **67**, 124–129. http://dx.doi.org/10.1143/JPSJ.67.124

Das, U.N., Deka, R.K., and Soundalgekar, V.M., 1996, "Radiation Effects on Flow Past an Impulsively Started Vertical Infinite Plate," *J.theo. Mech*. **1**, 111-115.

Hossain, M. A., and Takhar, H.S., 1996, "Radiation Effect on Mixed Convection along a Vertical Plate with Uniform Surface Temperature," *Heat and Mass Transfer*. **31**, 243-248. http://dx.doi.org/10.1007/bf02328616 .

Jha,B. K., and Singh, A. K., 1990, "Soret Effects on Free-Convection and Mass Transfer Flow in the Stokes Problem for an Infinite Vertical Plate," *Astrophys. Space Sci*. **173**(2), 251-255. http://dx.doi.org/10.1007/bf00643934

Kafoussias, N. G., 1992, "MHD Thermal Diffusion Effects on Free Convective and Mass Transfer Flow over an Infinite Vertical Moving Plate". *Astrophys. Space Sci*. **192**(1), 11-19. http://dx.doi.org/10.1007/bf00653255

Kumar,A.G.V., and Vijaya Kumar Varma,S., 2011, "Thermal Diffusion and Radiation Effects on Unsteady MHD Flow Past an Impulsively Started Exponentially Accelerated Vertical Plate with Variable Temperature and Variable Mass Diffusion," *Int. J. of Applied Mathematical Analysis and Applications*, *Vol.6*, No.1- 2, pp.169-192.

Kumar, A.G.V., Rajasekarhara Goud, Y., and Vijaya Kumar Varma, S., 2012 ,"Thermal Diffusion and Radiation Effects on Unsteady MHD Flow Past an Exponentially Accelerated Vertical Plate with Variable Temperature and Mass Diffusion in the Presence of Heat Source/Sink". *Elixer Appl. Math*. **49**, 9948-9954

Kumari,M., and Nath ,G., 1999, "Development of Two-Dimensional Boundary Layer with an Applied Magnetic Field due to an Impulsive Motion," *Indian J. pure Appl. Math*. **30**, 695- 708.

Muthucumaraswamy, R., and Janakiraman, B., 2006, "MHD and Radiation Effects on Moving Isothermal Vertical Plate with Variable and Mass Diffusion," *Theo. Appl. Mech*. **33**(1), 17-29. http://dx.doi.org/10.2298/tam0601017m

Narahari, M., and Debnath, L., 2013, "Unsteady Magneto hydrodynamic Free Convection Flow Past an Accelerated Vertical Plate," *Z.Angew.Math. Mech*. **93**, No. 1, 38-49. http://dx.doi.org/10.1002/zamm.201200008

Pera, L., and Gebhart, B., 1971, "On the Stability of Laminar Plums: Some Numerical Solution and Experiments," *Heat Mass Transfer.Int.J.* **14** ,Issu 7, 975 to 984. http://dx.doi.org/10.1016/0017-9310(71)90123-2

Raptis, A., and Singh, A.K., 1983, "MHD Free Convection Flow past an Accelerated Vertical Plate," *Int. Communication Heat and Mass Transfer*. **10**(4), 313–321. http://dx.doi.org/10.1016/0735-1933(83)90016-7

Rajesh, V., and Varma, S.V.K., 2009, "Thermal Diffusion and Radiation Effects on MHD Flow Past an Infinite Vertical Plate with Variable Temperature and Mass Diffusion," *JP Journal of Heat and Mass Transfer*. **3** (1), 1- 72.

Raptis, A., and Perdikis, C., 1999, "Radiation and Free Convection Flow Past a Moving Plate," *Int. J.Appl Mech. Eng*. **4**, 817- 821.

Soundalgekar, V.M., 1977," Free Convection Effects on the Stokes Problem for an Infinite Vertical Plate" *J. Heat Transfer,* **99**(3), 499-501. http://dx.doi.org/10.1115/1.3450729

Soundalgekar, V.M., Gupta,S.K., and Birajdar ,N.S., 1979, " Effects of Mass Transfer and Free Convection Currents on MHD Stokes Problem For a Vertical Plate," *Nuclear Eng. Des*. **53** (3), 339-346. http://dx.doi.org/10.1016/0029-5493(79)90060-8

Soundalgekar, V.M., Patil, M.R., and Jahagirdar, M.D., 1981,"MHD Stokes Problem For a Vertical Plate with Variable Temperature," *Nuclear Eng. Des*. **64** (1), 39-42. http://dx.doi.org/10.1016/0029-5493(81)90030-3

Soundalgekar, V.M. and Takhar, H.S., 1993, "Radiation Effects on Free Convection Flow Past a Semi-infinite Vertical Plate," *Model. Measure. Control,* **B51**, 31-40.

Stokes, G.G., 1851, "On the Effect of the Internal Fraction of Fluids on the Motion of a Pendulum Trans," *Camb. Phil. Soc*. **9**, pp.8-106.

Tokis, J.N., 1985, "A class of Exact Solutions of the Unsteady Magneto hydrodynamic Free Convection Flows," *Astrophysics. Space Sci.* **112**(2), 413–422. http://dx.doi.org/10.1007/bf00653524

28

MHD CONVECTIVE BOUNDARY LAYER FLOW TOWARDS A VERTICAL SURFACE IN A POROUS MEDIUM WITH RADIATION, CHEMICAL REACTION AND INTERNAL HEAT GENERATION

Emmanuel Maurice Arthur*, Timothy Ayando, Yakubu Ibrahim Seini

Faculty of Mathematical Sciences, University for Development Studies, P.O. Box 1350, Tamale, +233, Ghana

ABSTRACT

The combined effects of chemical reaction and viscous dissipation on hydromagnetic mixed convective flow towards a vertical plate embedded in a highly porous medium with radiation and internal heat generation has been examined. The governing boundary layer equations have been transformed to a two-point boundary value problem using a local similarity approach and solved numerically using the Newton Raphson shooting method alongside the Fourth-order Runge - Kutta algorithm. The effects of various embedded parameters on fluid velocity, temperature and concentration have been presented graphically whilst the skin friction coefficient and the rates of heat and mass transfers have been tabulated for varying controlling parameters. The results indicate that the magnetic parameter, the local Grashof numbers and all the controlling parameters identified directly influence the flow around the boundary and hence can be controlled to achieve desired product characteristics.

Keywords: *Viscous Dissipation, Similarity Variable, Permeability, Brinkman Number.*

1. INTRODUCTION

The subject of coupled heat and mass transfer due to hydromagnetic mixed convection flow past a surface embedded in a highly porous medium have gained considerable attention in the recent past due to their diverse applications in engineering systems. MHD applications are common in the cooling of electronic equipment, nuclear reactors; food processing; polymer production; electromagnetic pumps, valves for transportation, batching, stirring metals- which affect the convective heat and mass transfer in crystallization process in metallurgy- and in liquid metal cooling systems for nuclear technology. Furthermore, MHD plays an important role in building of power engineering devices, metallurgical and electrical technologies; MHD relays, communication switches, and units of automatics with a liquid working body.

The analysis of hydromagnetic flows in porous media has been a subject of several recent papers (Kumari et al. (2001); Chamkha (2001); Seddeek (2002); Arthur and Seini (2014a), Christian et al. (2014), Seini and Makinde (2014)). Most studies on hydromagnetic convection in porous media consider the condition of external flows. It is worth mentioning that, stagnation points also occurs at the surface of objects in the flow field, where the fluid attains a zero velocity. Stuart (1959) examined viscous flow near stagnation point when the external flow has uniform vorticity. Tamada (1979) studied two-dimensional stagnation point flow impinging obliquely on a plane wall. Ali et al. (2011) then analyzed the effect of Hall current on MHD mixed convection boundary layer low over a stretched vertical flat plate while Hayat et al. (2004) examined some steady MHD flows of the second order grade fluid. Ishak et al. (2006) studied mixed convection boundary layers in the stagnation-point flow toward a stretching surface and later Ishak et al. (2008) examined mixed convection stagnation point flow of a micropolar fluid towards a stretching sheet. Singh et al. (2010) studied the effects of volumetric heat generation/absorption on mixed convection stagnation point flow on an isothermal vertical plate in a porous media.

Radiative flow of MHD fluid occurs in electrical power generation, astrophysical flows, solar power technology and nuclear reactors. Also radiative heat and mass transfer arises in many geothermal and geophysical processes. Thermal radiation is a characteristic of any flow system at temperatures above the absolute zero and can strongly interact with convection in many situations of engineering interest. The effects of radiation on flow kinematics in the presence or absence of magnetic field under various conditions have been studied by a number of researchers. Chamkha et al. (2002) analyzed the effects of radiative heat transfer on flow and thermal field for horizontal and inclined plates. The effect of thermal radiation on the heat and mass transfer flow of variable viscosity fluid past a vertical porous plate permeated by a transverse magnetic field was reported by Makinde and Ogulu (2008). The hydromagnetic flow in the presence of radiation has been investigated by Bestman and Adiepong (1988), Naroua et al. (1998) and Ouaf (2005) when the induced magnetic field is negligible. Raptis et al. (2003) investigated the effects of radiation in an optically thin gray gas past a vertical infinite plate in the presence of a magnetic field, when the magnetic Reynolds number was negligible and the flow was steady. The effect of thermal radiation in the linearized Rosseland approximation on the heat transfer characteristics of various boundary layer flows was reported by Magyari and Pantokratoras (2003). More recently, Arthur et al. (2014) reported on chemically reacting hydromagnetic flow over a flat surface in the presence of radiation with viscous dissipation and convective boundary conditions.

Viscous dissipation plays an important role in natural convection, geological processes, polymer processing and in strong gravitational field processes on large scales and has been reported extensively in the literature. It plays a practical role in oil products transportation through ducts. Viscous dissipation changes the temperature distribution by playing a role like an energy source, which leads to affected heat transfer

*Corresponding author. Email: jimknights01@yahoo.com

rate and hence needs to be considered in heat transfer problems. Some researchers have examined the effects of viscous dissipation on hydrodynamics with or without magnetic field, notably Pantokratoras (2004), Gangadhar (2012), Ibrahim and Bhashar (2013), Singh (2012), Kazi et al. (2013), Arthur and Seini (2014b), and Imoro et al. (2014).

The objective of this study is to extend the work of Makinde (2011) to include viscous dissipation and chemical reaction. The governing boundary layer equations have been transformed to a two-point boundary value problem using a local similarity approach and solved numerically. The effects of various embedded parameters on fluid velocity, temperature and concentration have been shown graphically. It is hoped that the results obtained will not only provide useful information for application and future research work, but also serve as a complement to the previous study.

2. MATHEMATICAL MODEL

We consider a steady, laminar, stagnation point flow of viscous incompressible hydromagnetic fluid through a porous medium along a vertical isothermal plate in the presence of volumetric rate of heat generation. It is assumed that the fluid property variations due to temperature and chemical species concentration are limited to fluid density. In addition, there is no applied electric field and all of the Hall effects and Joule heating are neglected. Since the magnetic Reynolds number is very small for most fluids used in industrial applications, we assume that the induced magnetic field is negligible. The x-axis is taken along the plate and the y-axis is normal to the plate and the flow is confined in half plane $y>0$ as shown in Fig. 1.

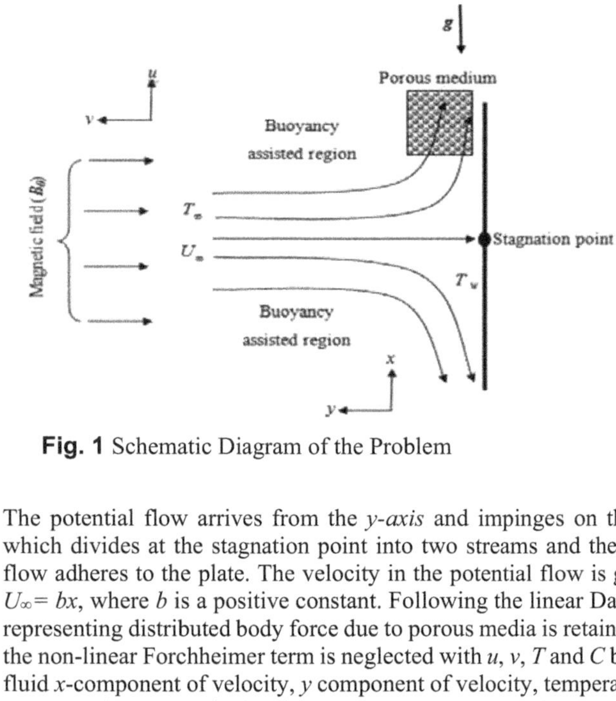

Fig. 1 Schematic Diagram of the Problem

The potential flow arrives from the *y-axis* and impinges on the plate, which divides at the stagnation point into two streams and the viscous flow adheres to the plate. The velocity in the potential flow is given by $U_\infty = bx$, where b is a positive constant. Following the linear Darcy term representing distributed body force due to porous media is retained while the non-linear Forchheimer term is neglected with u, v, T and C being the fluid x-component of velocity, y component of velocity, temperature and concentration respectively, the governing equations of the continuity, momentum, energy, and species concentration are given by:

$$\frac{\partial u}{\partial x}+\frac{\partial v}{\partial y}=0, \tag{1}$$

$$u\frac{\partial u}{\partial x}+v\frac{\partial u}{\partial y}=\upsilon\frac{\partial^2 u}{\partial y^2}+g\beta_T(T-T_\infty)+g\beta_C(C-C_\infty)-\left(\frac{\sigma B_\circ^2}{\rho}+\frac{\upsilon}{\breve{K}}\right)(u-U_\infty)+U_\infty\frac{\partial U_\infty}{\partial x}, \tag{2}$$

$$u\frac{\partial T}{\partial x}+v\frac{\partial T}{\partial y}=\alpha\frac{\partial^2 T}{\partial y^2}+\frac{\upsilon}{c_p}\left(\frac{\partial u}{\partial y}\right)^2-\frac{\alpha}{\kappa}\frac{\partial q_r}{\partial y}+Q(C-C_\infty), \tag{3}$$

$$u\frac{\partial C}{\partial x}+v\frac{\partial C}{\partial y}=D\frac{\partial^2 C}{\partial y^2}-\gamma(C-C_\infty)^n. \tag{4}$$

Subject to the following boundary conditions:

$$u(x,0)=0,\ v(x,0)=0,\ T(x,0)=T_w,\ C(x,0)=C_w,$$
$$u(x,\infty)=U_\infty=bx,\ T(x,\infty)=T_\infty,\ C(x,\infty)=C_\infty. \tag{5}$$

where υ is the kinematic viscosity, T_∞ is the free stream temperature, C_∞ is the free stream concentration, U_∞ is the free stream velocity, α is the thermal diffusivity and D is the mass diffusivity, β_T is the thermal expansion coefficient, β_C is the solutal expansion coefficient, ρ is the fluid density, g is gravitational acceleration, σ is the electrical conductivity, c_p is the specific heat capacity at constant pressure, κ is the thermal conductivity, γ is the rate of chemical reaction, $\breve{K}$ is the permeability of the porous medium, n is the order of chemical reaction, Q is the volumetric rate of heat generation, B_0 is the magnetic field of constant strength, T_w is the temperature of the plate, and C_w is the species concentration at the plate surface.

The velocity components can be expressed in terms of the stream function $\psi(x,y)$ such that

$$u=\frac{\partial\psi}{\partial y},\quad v=-\frac{\partial\psi}{\partial x} \tag{6}$$

We seek for the stream function such that

$$\psi(x,y)=x\sqrt{\upsilon b}\,f(\eta) \tag{7}$$

It may be verified that the continuity equation in Eq.1 is identically satisfied. Using the Rosseland approximation for radiation, Ibrahim and Makinde (2011) simplified the heat flux as:

$$q_r=-\frac{4\sigma^*}{3K'}\frac{\partial T^4}{\partial y} \tag{8}$$

where K' and σ^* are the Stefan-Boltzmann constant and the mean absorption coefficient respectively. We assume that the temperature differences within the flow such as the term T^4 may be expressed as a linear function of temperature. Hence, expanding T^4 in a Taylor series about T_∞ and neglecting higher order terms, we get;

$$T^4\cong 4T_\infty^3T-3T_\infty^4 \tag{9}$$

Also, the following non-dimensional variables can be introduced:

$$\eta=y\sqrt{\frac{b}{\upsilon}},\ \theta(\eta)=\frac{T-T_\infty}{T_w-T_\infty},\ \phi(\eta)=\frac{C-C_\infty}{C_w-T_\infty}. \tag{10}$$

Considering Eq.6, Eq.8 and Eq.10, Eq.2 - Eq.4 can be transformed into the following ordinary nonlinear system of differential equations:

$$f'''+ff''-f'^2+G_T\theta+G_C\phi-(M+K)(f'-1)=-1, \tag{11}$$

$$\left(1+\frac{4}{3}Ra\right)\theta''+\Pr f\theta'+Brf''^2+S\phi=0, \tag{12}$$

$$\phi''+Scf\phi'-Sc\beta\phi''=0. \tag{13}$$

with the transformed boundary conditions as;

$$f'(0)=0,\ f(0)=0,\ \theta(0)=1,\ \phi(0)=1,$$
$$f'(\infty)=1,\ \theta(\infty)=0,\ \phi(\infty)=0. \tag{14}$$

where the prime symbol denotes differentiation with respect to η and

$G_T=\dfrac{g\beta_T(T_w-T_\infty)x^3}{\upsilon^2}$ is the local thermal Grashof number,

$G_C=\dfrac{g\beta_C(C_w-C_\infty)x^3}{\upsilon^2}$ is the local solutal Grashof number,

$Br=\dfrac{\mu U_0^2}{\kappa(T_w-T_\infty)}$ is the Brinkmann number,

$\Pr = \dfrac{\nu}{\alpha}$ is the Prandtl number,

$Sc = \dfrac{\nu}{D}$ is the Schmidt number,

$\beta = \gamma(C_w - C_\infty)^{n-1}/b$ is the chemical reaction parameter,

$K = \dfrac{\upsilon}{b\widetilde{K}}$ is the permeability parameter,

$S = \dfrac{Q(C_w - C_\infty)\upsilon}{b\alpha(T_w - T_\infty)}$ is the heat generation parameter,

$M = \dfrac{\sigma B_0^2}{b\rho}$ is the magnetic field parameter, and

$Ra = \dfrac{4\sigma^* T_\infty^3}{\kappa K'}$ is the thermal radiation parameter.

3. NUMERICAL PROCEDURE

The numerical technique used to solve the coupled ordinary differential Eqs. (11-13) together with the associated transformed boundary conditions in Eq.14 is the standard Newton-Raphson shooting method alongside the fourth-order Runge-Kutta integration algorithm. Let

$$f = x_1,\ f' = x_2,\ f'' = x_3,\ \theta = x_4,\ \theta' = x_5,\ \phi = x_6,\ \phi' = x_7, \tag{15}$$

Equation 15 transforms Eq.11 – Eq.13 to first order differential equations as follows:

$$f' = x_1' = x_2, \qquad f'' = x_2' = x_3,$$

$$f''' = x_3' = -x_1x_3 + x_2^2 - G_T x_4 - Gcx_6 + (M+K)(x_2 - 1) - 1,$$

$$\theta' = x_4' = x_5, \tag{16}$$

$$\theta'' = x_5' = -1/(1 + 4/3Ra)(\Pr x_1x_5 + Brx_3^2 + Sx_6),$$

$$\phi' = x_6' = x_7,$$

$$\phi'' = x_7' = -Scx_1x_7 + Sc\beta x_6{}^n,$$

subject to the boundary conditions

$$x_1(0) = 0,\ x_2(0) = 0,\ x_3(0) = s_1,\ x_4(0) = 1,\ x_5(0) = s_2,\ x_6(0) = 1,$$

$$x_7(0) = s_3,\ x_2(\infty) = 1,\ x_4(\infty) = 0,\ x_6(\infty) = 0. \tag{17}$$

In the shooting method, the unspecified initial conditions; s_1, s_2 and s_3 in Eq.17 were assumed and Eq.16 integrated numerically as an initial valued problem to a given terminal point. The accuracy of the assumed missing initial conditions is checked by comparing the calculated value of the dependent variable at the terminal point with its given value there. If differences exist, improved values of the missing initial conditions are obtained and the process repeated. The computations were done by a written programme which uses a symbolic and computational computer language (MAPLE). A step size of $\Delta\eta = 0.001$ was selected to be satisfactory for a convergence criterion of 10^{-7} in nearly all cases. The maximum value of η_∞ to each group of parameters are determined when the values of unknown boundary conditions at $\eta_\infty = 0$ not change to successful loop with error less than 10^{-7}. From the process of numerical computations, the local skin-friction coefficient, the local Nusselt numbers and the local Sherwood numbers, which are respectively, proportional to $f''(0)$, $-\theta'(0)$ and $-\phi'(0)$ were worked out and their numerical values presented in tables.

Table 1 Comparison of results for different values of *S* for $Pr = G_T = 1$, $G_C = Sc = 0.5$, $M = Ra = Br = n = K = \beta = 0$

	Singh et al. (2010)			Makinde (2011)			Present Study		
S	***-f''(0)***	*-θ'(0)*	*-ϕ'(0)*	***-f''(0)***	*-θ'(0)*	*-ϕ'(0)*	***-f''(0)***	*-θ'(0)*	*-ϕ'(0)*
-1	**1.8444**	1.3908	0.4631	**1.844462**	1.390856	0.463174	**1.844404**	1.390859	0.463124
0	**1.9995**	0.6392	0.4789	**1.999553**	0.639244	0.478964	**1.999554**	0.639244	0.478965
1	**2.1342**	-0.0730	0.4917	**2.134287**	-0.073040	0.491749	**2.134288**	0.073041	0.491750

Table 2 Numerical results of skin friction coefficient, Nusselt number and the Sherwood number

M	*Pr*	*Sc*	G_T	G_C	*Br*	*β*	*N*	*S*	*Ra*	*K*	*f''(0)*	*-θ'(0)*	-ϕ'(0)
1.0	0.71	0.22	0.1	0.1	0.1	0.1	1	1	0.1	1	1.976587	-0.472184	0.354002
3.0	0.71	0.22	0.1	0.1	0.1	0.1	1	1	0.1	1	2.429832	-0.467904	0.359742
0.1	**4.00**	0.22	0.1	0.1	0.1	0.1	1	1	0.1	1	1.712370	0.3037111	0.348490
0.1	**7.10**	0.22	0.1	0.1	0.1	0.1	1	1	0.1	1	1.705917	0.627297	0.348150
0.1	0.71	**0.78**	0.1	0.1	0.1	0.1	1	1	0.1	1	1.718618	-0.208314	0.603392
0.1	0.71	**2.62**	0.1	0.1	0.1	0.1	1	1	0.1	1	1.704956	-0.012482	0.993426
0.1	0.71	0.22	**0.5**	0.1	0.1	0.1	1	1	0.1	1	1.987683	-0.450053	0.360734
0.1	0.71	0.22	**1.0**	0.1	0.1	0.1	1	1	0.1	1	2.284949	-0.426247	0.371958
0.1	0.71	0.22	0.1	**0.5**	0.1	0.1	1	1	0.1	1	1.937509	0.457168	0.357961
0.1	0.71	0.22	0.1	**1.5**	0.1	0.1	1	1	0.1	1	2.424761	-0.420861	0.375171
0.1	0.71	0.22	0.1	0.1	**0.5**	0.1	1	1	0.1	1	1.737424	-0.778013	0.350365
0.1	0.71	0.22	0.1	0.1	**1.5**	0.1	1	1	0.1	1	1.744340	-1.536194	0.350681
0.1	0.71	0.22	0.1	0.1	0.1	**0.5**	1	1	0.1	1	1.730358	-0.407299	0.446052
0.1	0.71	0.22	0.1	0.1	0.1	**1.5**	1	1	0.1	1	1.723057	-0.291312	0.634837
0.1	0.71	0.22	0.1	0.1	0.1	0.1	**3**	1	0.1	1	1.735537	-0.492612	0.337513
0.1	0.71	0.22	0.1	0.1	0.1	0.1	**5**	1	0.1	1	1.735739	-0.495805	0.333211
0.1	0.71	0.22	0.1	0.1	0.1	0.1	1	**0**	0.1	1	1.715682	0.428580	0.348837
0.1	0.71	0.22	0.1	0.1	0.1	0.1	1	**0.5**	0.1	1	1.725223	-0.025583	0.349543
0.1	0.71	0.22	0.1	0.1	0.1	0.1	1	1	**0.5**	1	1.732679	-0.284345	0.350227
0.1	0.71	0.22	0.1	0.1	0.1	0.1	1	1	**1.5**	1	1.730886	-0.100096	0.350273
0.1	0.71	0.22	0.1	0.1	0.1	0.1	1	1	0.1	**3**	2.237210	-0.468874	0.357483
0.1	0.71	0.22	0.1	0.1	0.1	0.1	1	1	0.1	**5**	2.646350	-0.468041	0.362016

4. RESULTS AND DISCUSSION

From the process of numerical computation, the local skin-friction coefficient, the local Nusselt number and the local Sherwood number, which are respectively proportional to $-f''(0)$, $-\theta'(0)$, and $-\phi'(0)$ were computed and their numerical values presented in a tabular form. Table **1** shows the comparison of the works of Singh et al. (2010) and Makinde (2011) with the present study for internal heat generation parameter (S).It is clear from the table that our present study is consistent with the previous published results. The results of varying parameter values on the local skin friction coefficient, the local Nusselt number and the local Sherwood number, are shown in Table 2. It is observed that the local skin friction increases with increasing values of M, G_T, G_C, Br, n, S and K and decreases with increasing values of Pr, Sc, β and Ra. This means that the combined effect of magnetic field, buoyancy forces, viscous dissipation, order of chemical reaction, internal heat generation and porous medium permeability is to increase the local skin friction at the surface of the plate. Similarly, the rate of heat transfer at the plate surface increases with an increase in parameter values of M, G_T, Br, K, Pr, Sc, β and Ra and reduces with increasing values of G_C, n and S. Moreover, it is observed that the rate of mass transfer increases with increasing values of M, G_T, G_C, Br, S, K, Sc, β and Ra and decreases with increasing values of the order of chemical reaction and Prandtl number.

4.1 Effects of Parameter Variation on Velocity Profiles

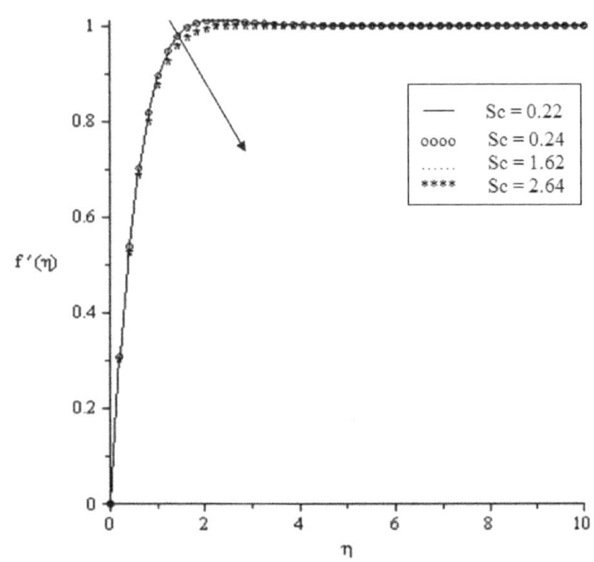

Fig. 2 Velocity profiles for varying values of Schmidt number

The effects of parameter variation on the velocity boundary layers are shown in Fig. 2-9. Normally, the fluid velocity is minimal at the plate surface and increases to the free stream value satisfying the far field boundary conditions. From Fig. 2 and Fig.7, it is observed that increasing values of the Schmidt and Prandtl numbers respectively tend to reduce the velocity profile. This can be attributed to the fact that increasing these two parameters sets momentum diffusion high over heat and mass diffusion at the surface of the plate which makes viscous force dominate the flow and hence, reduces the velocity at the boundary layer. The reaction rate parameter is also observed from Fig. 3 to reduce the velocity profile for obvious reasons. Moreover, it is observed from Fig. 6 that a steady decrease in the longitudinal velocity within the boundary layer accompanies a rise in the magnetic field intensity parameter with all profiles tending asymptotically to the free stream value away from the plate. This means in practice that, the magnetic field tends to decelerate the fluid flow. Meanwhile, in Fig. 8, it is observed that increasing the values of the internal generation parameter increases the velocity at the boundary layer. This can be attributed to the fact that, increase in the internal heat generation increases the fluid temperature which causes the fluid velocity to increase due to buoyancy effect. This fact is also adequate to explain the observed increase in the velocity profile as a result of increasing the thermal and solutal Grashof numbers in Fig. 4 and Fig.9 respectively. We can note here that, increasing buoyancy forces will lead to a better flow kinematics. Moreover increasing the porous medium permeability increases the velocity at the surface of the plate as observed in Fig. 5.

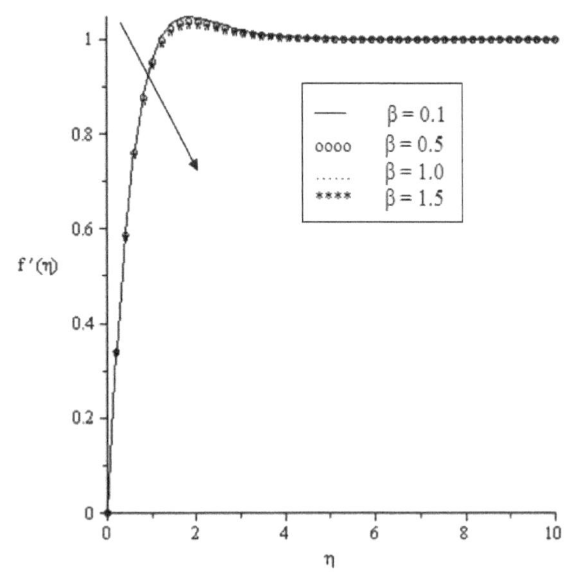

Fig. 3 Velocity profiles for varying values of reaction rate parameter

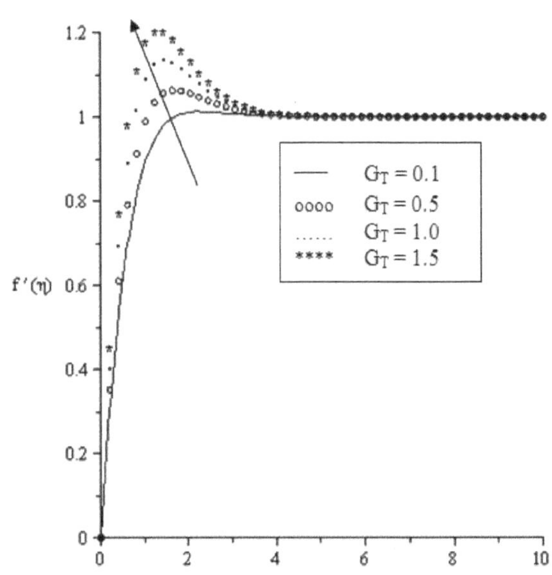

Fig. 4 Velocity profiles for varying values of thermal Grashof number

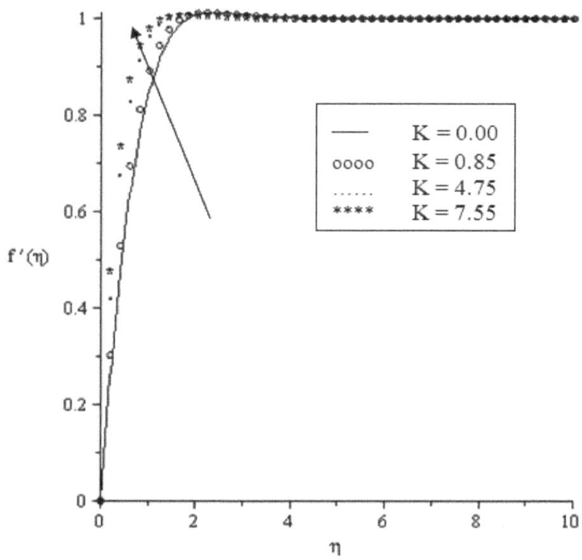

Fig. 5 Velocity profiles for varying values of the permeability parameter

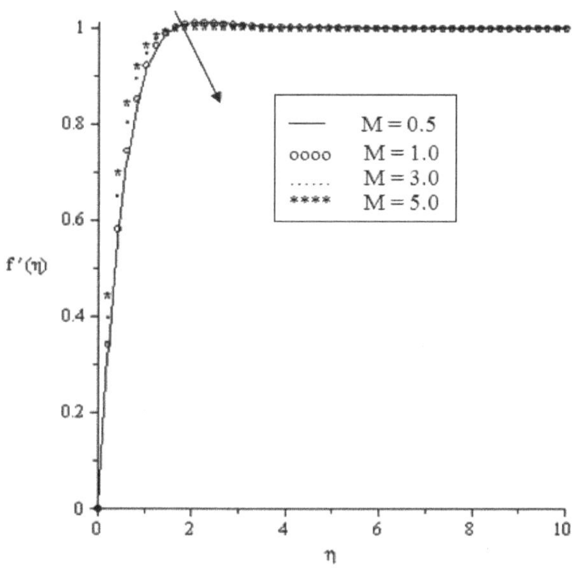

Fig. 6 Velocity profiles for varying values of magnetic field parameter

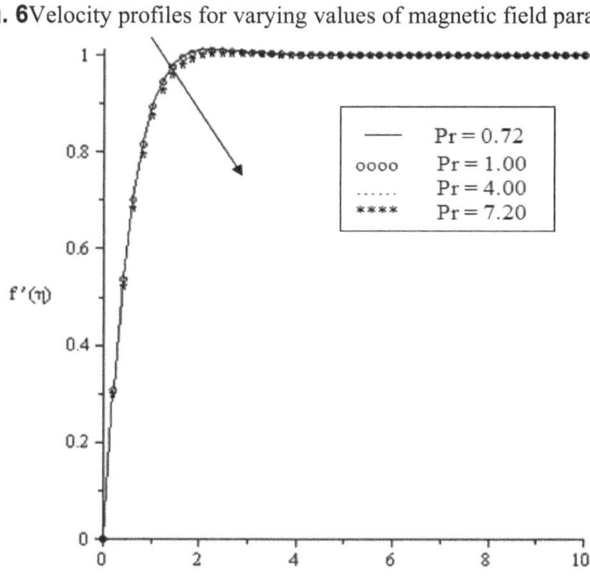

Fig. 7 Velocity profiles for varying values of Prandtl number

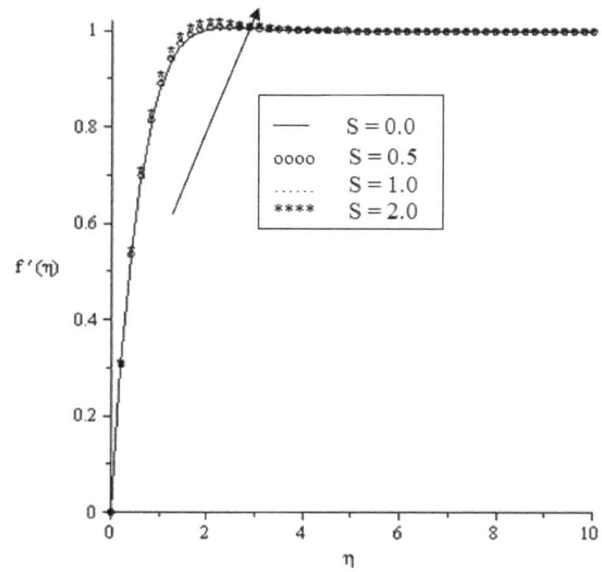

Fig. 8 Velocity profiles for varying values of internal heat generation parameter

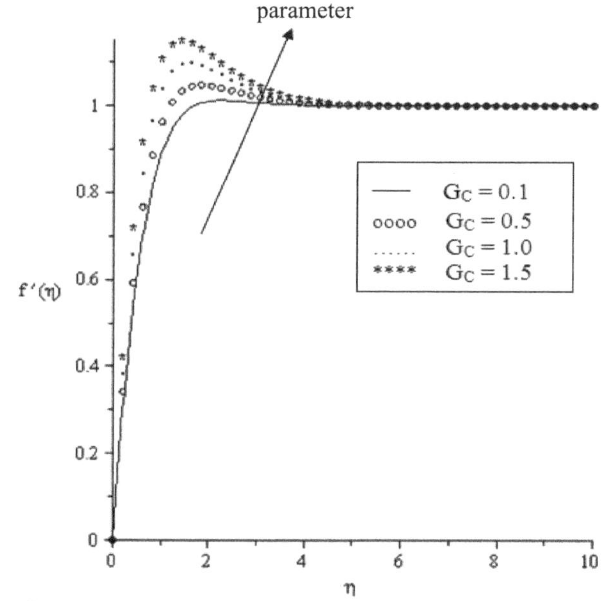

Fig. 9 Velocity profiles for varying values of solute Grashof number

4.2 Effects of Parameter Variation on Temperature Profiles

The effects of parameter variation on temperature profiles are shown in Fig. 10-20. Normally, the fluid temperature is highest at the surface of the plate and shrinks to free stream temperature satisfying the boundary conditions. It is observed in Fig. 10, Fig. 18 and Fig. 19 that increasing viscous dissipation, thermal radiation and internal heat generation respectively increase the temperature profiles. This is due to the fact that as more heat is generated within the fluid, the fluid temperature increases leading to a sharp inclination of the temperature gradient between the plate surface and the fluid. Also, increasing the order of chemical reaction increases the temperature boundary layer thickness at the surface of the plate as observed in Fig. 16. Meanwhile, from Fig. 10, Fig. 14, Fig. 15, Fig. 17 and Fig. 20, it is observed that increasing the values of reaction rate parameter, porous medium permeability, magnetic field intensity, Prandtl number and Schmidt number respectively, reduce the thermal boundary layer thickness. Similarly, from Fig. 12 and Fig. 13, the temperature profiles are reduced as a result of increasing the buoyancy forces (G_C and G_T respectively). This means that increasing the buoyancy forces enhance the cooling process.

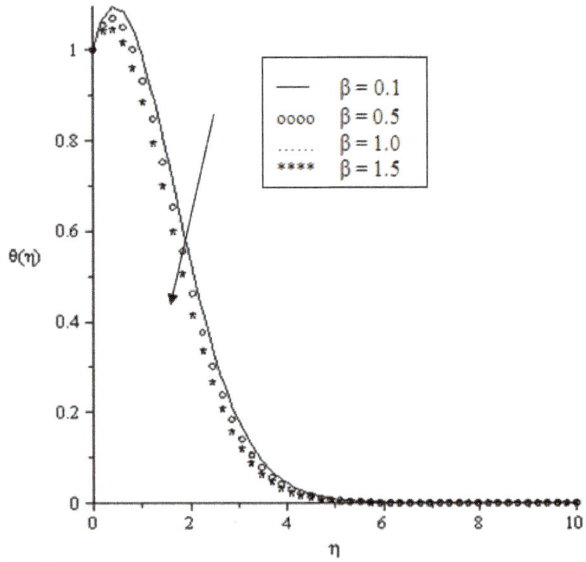

Fig. 10 Temperature profiles for varying values of reaction rate parameter

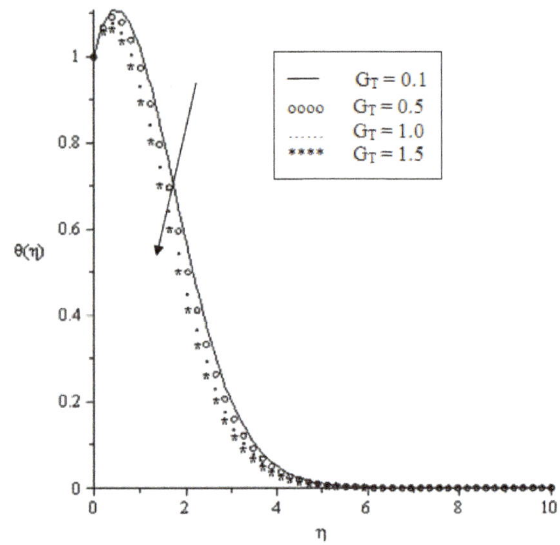

Fig. 13 Temperature profiles for varying values of thermal Grashof number

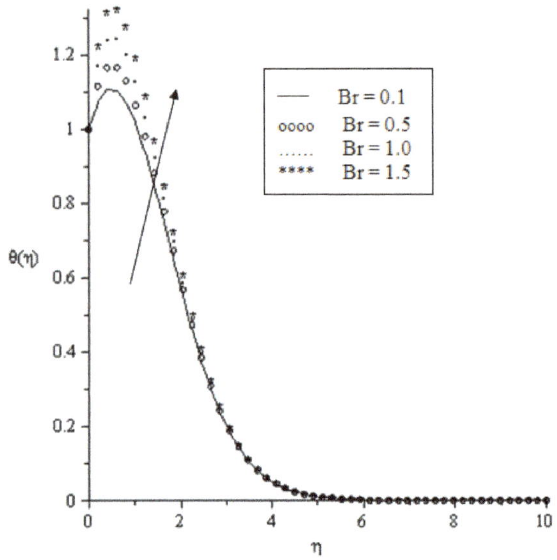

Fig. 11 Temperature profiles for varying values of the Brinkman number

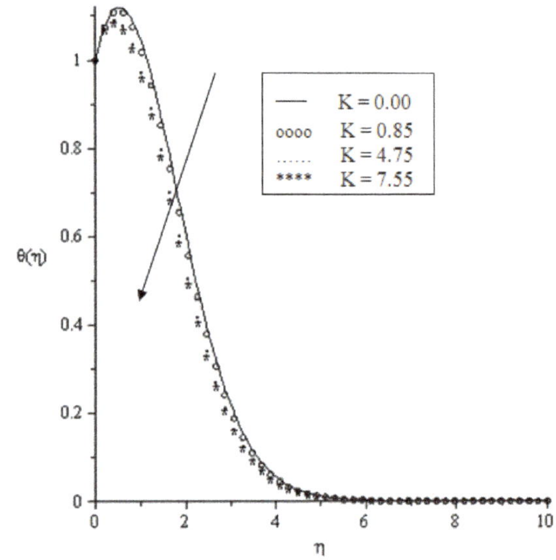

Fig. 14 Temperature profiles for varying values of the permeability parameter

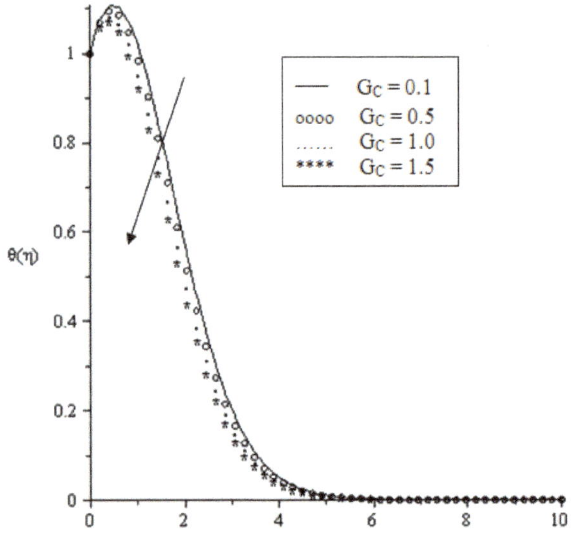

Fig. 12 Temperature profiles for varying values of the solutal Grashof number

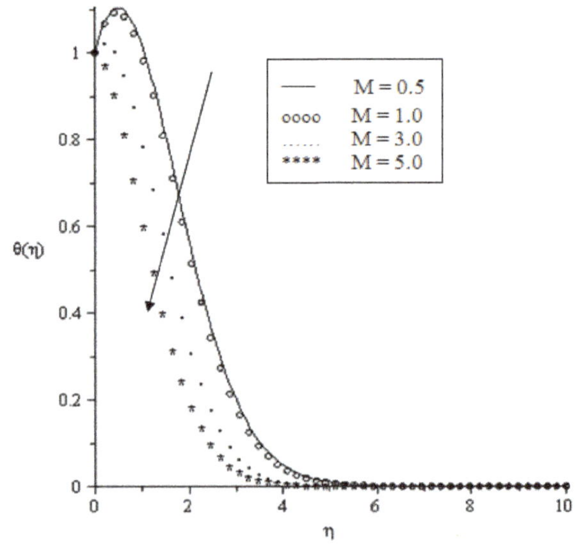

Fig. 15 Temperature profiles for varying values of magnetic field parameter

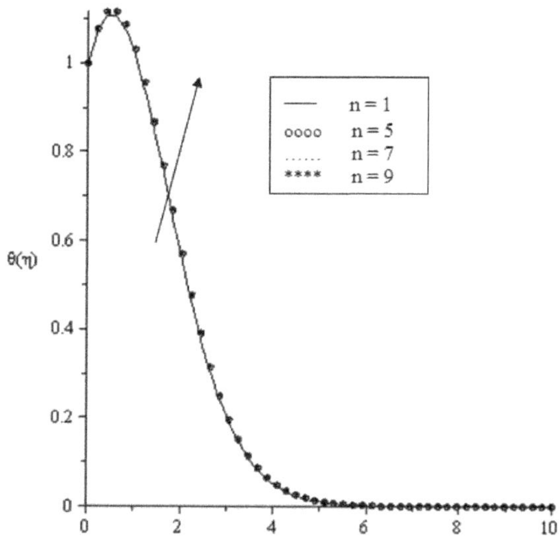

Fig. 16 Temperature profiles for varying values of order of chemical reaction

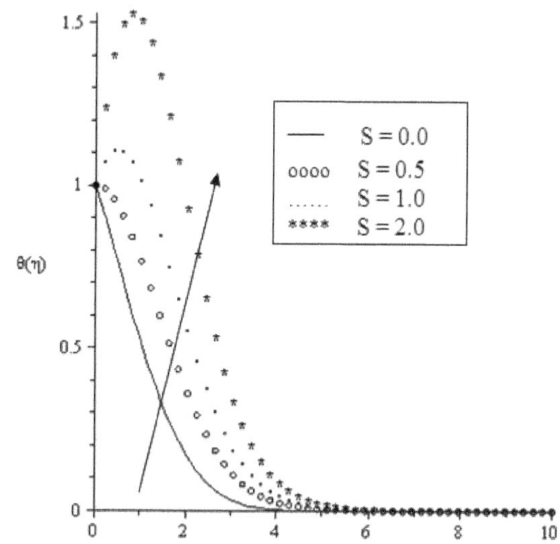

Fig. 19 Temperature profiles for varying values of the internal heat generation parameter

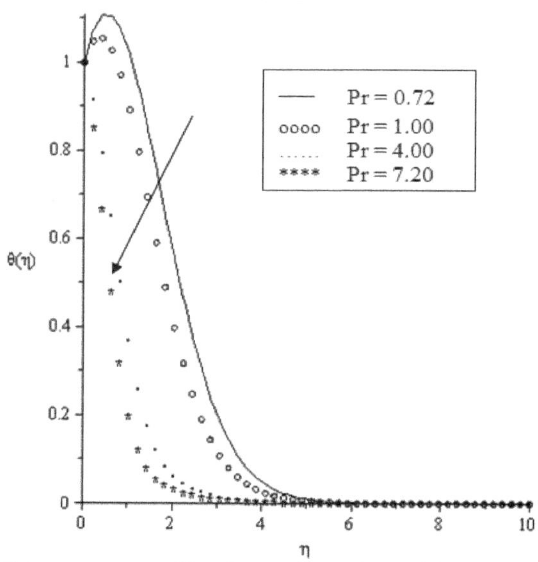

Fig. 17 Temperature profiles for varying values of the Prandtl number

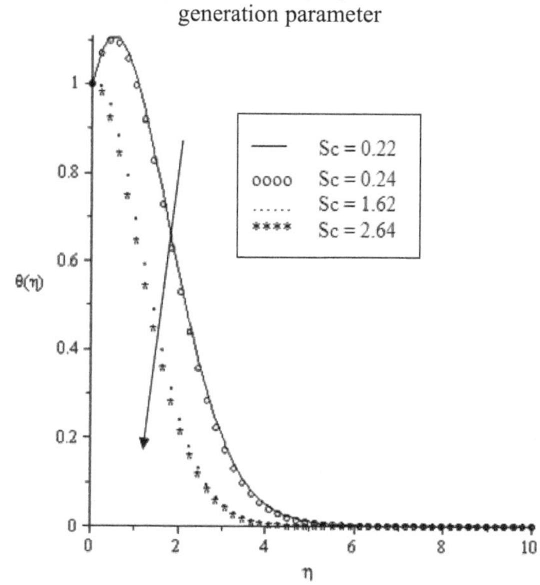

Fig. 20 Temperature profiles for varying values of the Schmidt number

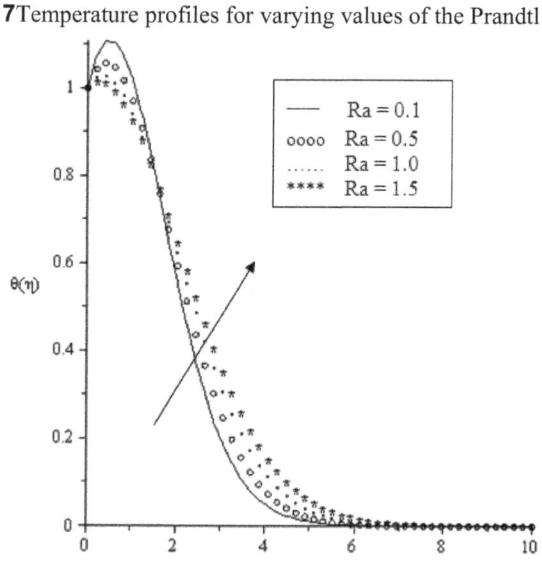

Fig. 18 Temperature profiles for varying values of the radiation parameter

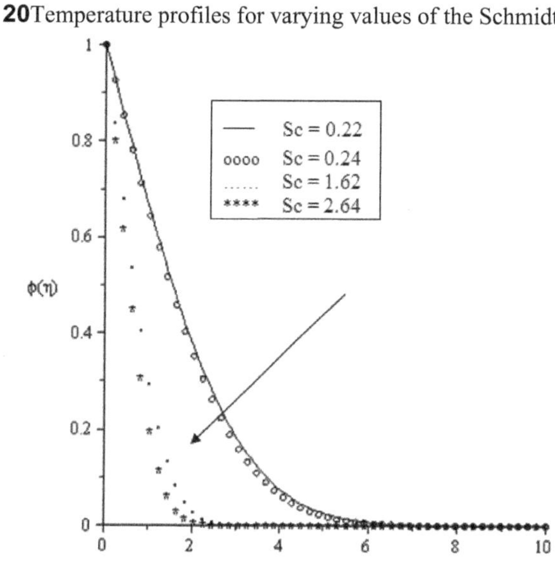

Fig. 21 Concentration profiles for varying values of Schmidt number

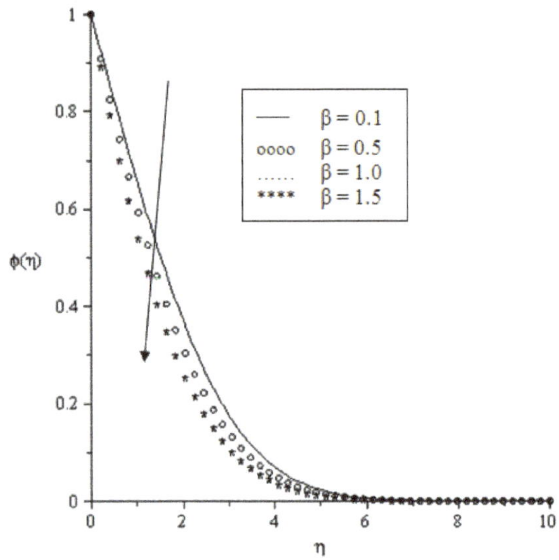

Fig. 22 Concentration profiles for varying values of reaction rate parameter

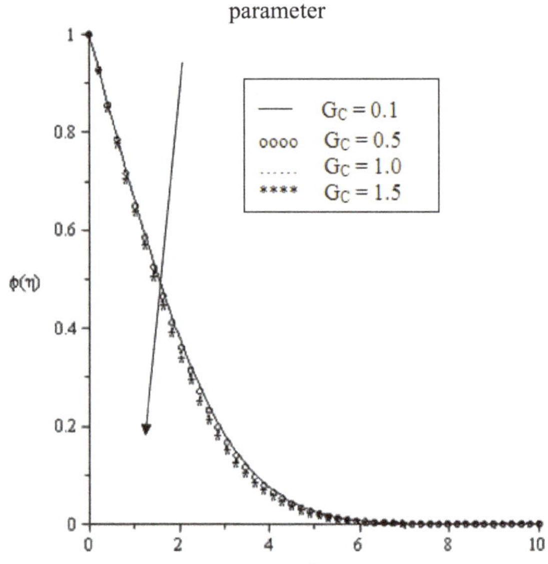

Fig. 23 Concentration profiles for varying values of solutal Grashof number

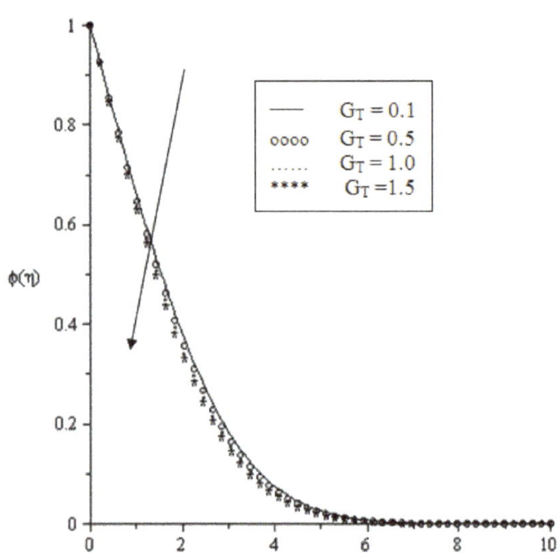

Fig. 24 Concentration profiles for varying values of thermal Grashof number

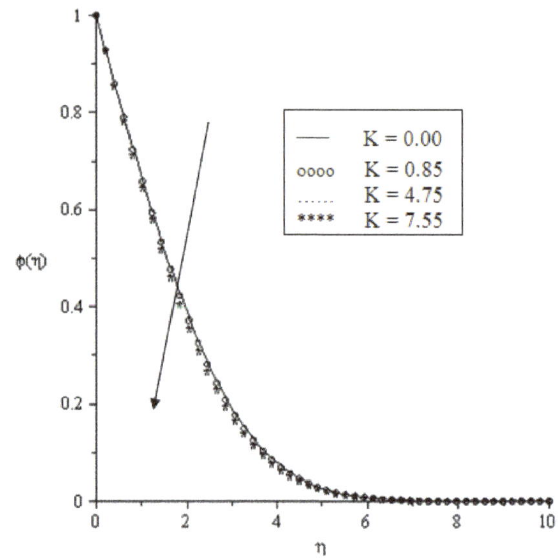

Fig. 25 Concentration profiles for varying values of the permeability parameter

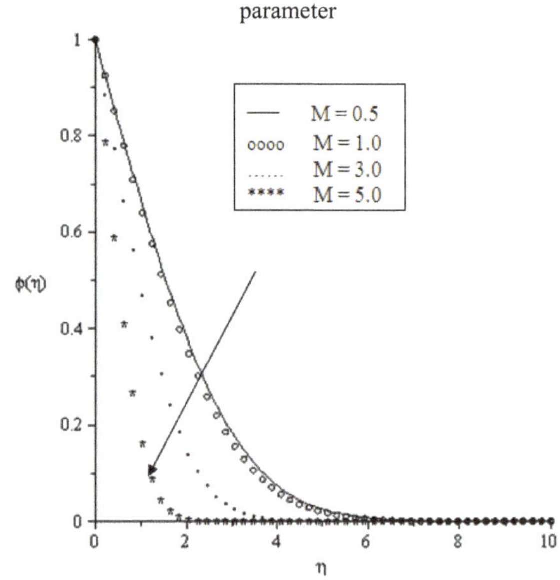

Fig. 26 Concentration profiles for varying values of magnetic field parameter

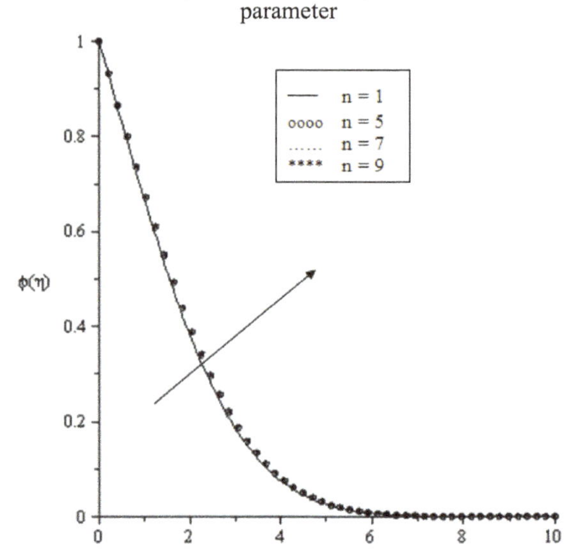

Fig. 27 Concentration profiles for varying values of the order of chemical reaction

4.3 *Effects of Parameter Variation on Concentration Profiles*

Figures 21-27 depict the effects of varying parameters on the concentration boundary layer thickness. Normally, the chemical species concentration is highest at the plate surface and reduces exponentially to the free stream zero value. It is observed that increasing the Schmidt number (Fig. 21), reaction rate (Fig. 22) buoyancy forces (Fig. 23 and Fig. 24), magnetic, porous medium permeability (Fig. 25) and field intensity (Fig. 26) have adverse effect of decaying the concentration boundary layer thickness. Meanwhile, the boundary layer thickness increases when the order of reaction increases as observed in Fig. 27.

5. CONCLUSION

Hydromagnetic fluid flow towards a vertical plate embedded in a saturated porous medium with thermal radiation, viscous dissipation, internal heat generation and n^{th} order chemical reactionhas been studied. Numerical results have been compared to earlier results published in the literature and a perfect agreement was achieved. Among others, our results reveal that:

i. The velocity increases with the increase in M, Sc, G_T, G_C, Br and K. It also decreases with the increase in Pr, β, and S
ii. The temperature reduces with increasing M, Pr, Sc, G_T, G_C, β and K; and also rises with increasing Ra, Br, n and S .
iii. The concentration boundary layer decreases with increasing M, Sc, G_T, G_C, β and K;and increases with increasing n.
iv. The skin friction at the surface increases for the increase in M, G_T, G_C, Br, n, S, and K; and decreases for increasing Pr, Sc, Ra and β.
v. The rate of heat transfer at the surface increases with the increase of M, Pr, Sc, G_T, Br, β, K and Ra; and decreases with increasing G_C, n and S.
vi. The rate of mass transfer at the surface increases with increasing M, Sc, G_T, G_C, Br, β, S, K and Ra; and decreases with increasing Pr and n.

ACKNOWLEDGEMENTS

You may acknowledge the funding source of your work as well as suggestions and comments from, and discussions with colleagues in this section.

NOMENCLATURE

(x, y)	Cartesian coordinates
C_w	Plate surface concentration
C_∞	Free stream concentration
C	Fluid chemical species concentration
D	Diffusion coefficient
f	Dimensionless stream function
Nu	Nusselt number
T_∞	Free stream temperature
Sh	Sherwood number
T	Fluid temperature
Pr	Prandtl number
T_w	Plate surface temperature
(u, v)	Velocity components in Cartesian Coordinates
K′	Mean absorption coefficient
$\check{K}$	Permeability of the porous medium
B_0	Magnetic field of constant strength
q_r	Radiative heat flux
σ^*	Stefan-Boltzmann constant

Greek Symbols

μ	Coefficient of viscosity
ρ	Density of the fluid
ϕ	Dimensionless concentration
θ	Dimensionless temperature
η	Dimensionless variable
υ	Kinematic viscosity
ψ	Stream function
σ	Fluid electrical conductivity
κ	Thermal conductivity coefficient
γ	Reaction rate

REFERENCES

Ali, F., Nazar, R., Arifin, N., and Pop, I., 2011, "Effect of Hall Current on MHD Mixed Convection Boundary Layer Flow over a Stretched Vertical Flat Plate," *Meccanica,* **46**(5), 1103-1112

Arthur, E. M., and Seini, I. Y., 2014b, "Hydromagnetic Stagnation Point Flow over a Porous Stretching Surface in the Presence of Radiation and Viscous Dissipation," *Applied and Computational Mathematics*, **3**(5), 191-196.
http://dx.doi.org/10.11648/j.acm.20140305.11

Arthur, E. M., Seini, I. Y., and Seidu, A., 2014, "On Chemically Reacting Hydromagnetic Flow over a Flat Surface in the Presence of Radiation with Viscous Dissipation and Convective Boundary Conditions," American *Journal of Applied Mathematics*, **2**(5), 179-185. http://dx.doi.org/10.11648/j.ajam.20140205.15

Arthur, E. M., and Seini, Y. I., 2014a, "MHD Thermal Stagnation Point Flow Towards a Stretching Porous Surface," *Mathematical Theory and Modeling*, **4**(5), 163-169.

Bestman, A. R., and Adiepong, S. K., 1988, "Unsteady Hydromagnetic Free-Convection Flow with Radiative Heat Transfer in a Rotating Fluid," *Astrophys Space Sci*, **143**, 73-80.

Chamkha, A. J., and Quadri, M. M. A., 2001, "Heat and Mass Transfer from a Permeable Cylinder in a Porous Medium with Magnetic Field and Heat Generation/Absorption Effects," *Numerical Heat Transfer, Part A, Appl.*, **40**(4), 387-401.
http://dx.doi.org/10.1080/104077801753238167

Chamkha, A. J., Issa, C., and Khanfer, K., 2002, "Natural Convection from an Inclined Plate Embedded in a Variable Porosity Porous Medium Due to Solar Radiation," *Int. J. Therm. Sci.*, **41**, 73-81.
http://dx.doi.org/10.1016/S1290-0729(01)01305-9

Christian, E. J, Seini, Y. I, and Arthur, E. M., 2014, "MHD Boundary Layer Stagnation Point Flow with Radiation and Chemical Reaction towards a Heated Shrinking Porous Surface," *International Journal of Physical Sciences*, **9**(14), 320-328.
http://dx.doi.org/10.5897/IJPS2014.4177

Gangadhar, K., 2012, "Similarity Solution for Natural Convection over a Moving Vertical Plate with Internal Heat Generation and Viscous Dissipation," *Int. J. of Appl. Math and Mech.*, **8** (18), 90-100.

Hayat, T., Hameed, M. I., Asghar, S. S., and Iddiqui, A. M., 2004, "Some steady MHD flows of the second order fluid," Meccanica, **39,** 345-355.
http://dx.doi.org/10.1023/B:MECC.0000029365.39535.28

Ibrahim, M. S., and Bhashar, R. N., 2013, "Similarity Solution of Heat and Mass Transfer for Natural Convection over a Moving Vertical Plate with Internal Heat Generation and a Convective Boundary Condition in the Presence of Thermal Radiation, Viscous Dissipation, and Chemical Reaction," *Hindawi Publishing Corporation ISRN Thermodynamics*, **2013**, Article ID 790604, 10 pages.

Ibrahim, S. Y., and Makinde, O. D., 2011, "Radiation Effect on Chemically Reacting MHD Boundary Layer Flow of Heat and Mass

Transfer through a Porous Vertical Plate," *International Journal of Physical Sciences,* **6**(6), 1508-1516.

Imoro, R., Arthur, E. M., and Seini, Y. I., 2014, "Heat and Mass Transfer over a Vertical Surface with Convective Boundary Conditions in the Presence of Viscous Dissipation and nth Order Chemical Reaction," *International Journal of Computational and Applied Mathematics*, **9**(2), pp. 101-118, ISSN 1819-4966.

Ishak, A., Nazar, R., and Pop, I., 2006, "Mixed Convection Boundary Layers in the Stagnation-Point Flow toward a Stretching Surface," *Meccanica*, **41**:509-518.
http://dx.doi.org/10.1007/s11012-006-0009-4

Ishak, A., Nazar, R., and Pop, I., 2008, "Mixed Convection Stagnation Point Flow of a Micropolar Fluid towards a Stretching Sheet," *Meccanica*, **43**, 4110-418.
http://dx.doi.org/10.1007/s11012-007-9103-5

Kazi, K. H., Alim, M.d. A., and Laek, A. S., 2013, "Effects of Viscous Dissipation on MHD Natural Convection Flow along a Vertical Wavy Surface," *Journal of Theoretical and Applied Physics*, **2013**, 7-31.
http://www.jtaphys.com/content/7/1/31

Kumari, M., Takhar, H. S., and Nath, G., 2001, "Mixed Convection Flow over a Vertical Wedge Embedded in a Highly Porous Medium," *Heat Mass Transf*, **37**, 139- 146.
http://dx.doi.org/10.1007/s002310000154

Magyari, E., and Pantokratoras, A., 2003, "Effect of Radiation in an Optically Thin Gray Gas Flowing Past a Vertical Infinite Plate in the Presence of a Magnetic Field," *heat mass transf*, **39**, 771-773.

Makinde O. D., 2011, "Heat and Mass Transfer by MHD Mixed Convection Stagnation Point Flow toward a Vertical Plate Embedded in a Highly Porous Medium with Radiation and Internal Heat Generation," *Meccanica*, **2012**(47), 1173-1184,
http://dx.doi.org/10.1007/s11012-011-9502-5

Makinde, O. D., and Ogulu, A., 2008, "The Effect of Thermal Radiation on the Heat and Mass Transfer Flow of a Variable Viscosity Fluid Past a Vertical Porous Plate Permeated by a Transverse Magnetic Field," *Chem. Eng Community*, **195**(12), 1575-1584.
http://dx.doi.org/10.1080/00986440802115549

Na, T. Y., 1979, "Computational methods in engineering boundary value problems," Academic Press, New York.

Naroua, H., Ram, P. C., Sambo, A. S., and Takhar, H. S., 1998, "Finite-Element Analysis of Natural Convection Flow in a Rotating Fluid with Radiative Heat Transfer," *J. Magnetohydrodyn plasma Res*, **7**, 257-274.

Neild, D. A., and Bejan, A., 2006, "Convection in porous media," 3rd edn., Springer, New York.

Ouaf, M. E. M., 2005, "Exact Solution of Thermal Radiation on MHD, Flow over a Stretching Porous Sheet," *Appl Math Comput*, **170**(2), 1117-1125.
http://dx.doi.org/10.1016/j.amc.2005.01.010

Pantokratoras, A., 2004, "Effect of Viscous Dissipation in Natural Convection along a Heated Vertical Plate," *Applied Mathematical Modelling*, **29** (2005), 553–564.
http://dx.doi.org/10.1016/j.apm.2004.10.007

Raptis, A., Perdikis, C., and Leontitsis, A., 2003, "Effects of Radiation in an Optically Thin Gray Gas Flowing Past a Vertical Infinite Plate in the Presence of a Magnetic Field," *Heat Mass Transf*, **39**, 771-773.
http://dx.doi.org/10.1007/s00231-002-0317-8

Seddeek, M. A., 2002, "Effects of Magnetic Field and Variable Viscosity on Forced Non-Darcy Flow about a Flat Plate with Variable Wall Temperature in Porous Media in the Presence of Suction and Blowing," *J. Appl Mech Tech Phys*, **43**, 13-17.
http://dx.doi.org/10.1023/A:1013941523513

Singh, G., Sharma, P. R., and Chamkha, A. J., 2010, "Effect of Volumetric Heat Generation/Absorption on Mixed Convection Stagnation Point Flow on An Iso-Thermal Vertical Plate in Porous Media," *Int. Math*, **2**(2), 59-71.

Singh, P. K., 2012, "Effects of Variable Fluid Properties and Viscous Dissipation on Mixed Convection Fluid Flow past a Vertical Plate in Porous Medium," *International Journal of Scientific & Engineering Research*, **3**(7), 2229-5518.

Sparrow, E. M., and Cess, R. D., 1978, "Radiation Heat Transfer, Hemisphere," Washington, Augmented edition.

Stuart, J. T., 1959, "The Viscous Flow near Stagnation Point when the External Flow has Uniform Vorticity," *Aerosp Sci Technol*, **26**, 124-125.
http://dx.doi.org/10.2514/8.7963

Tamada, K., 1979, "Two-Dimensional Stagnation Point Flow Impinging Obliquely on a Plane Wall," *J Phys Soc Jpn*, **46**, 310-311.
http://dx.doi.org/10.1143/JPSJ.46.310

29

TRANSIENT FREE CONVECTION MHD FLOW PAST A VERTICAL PLATE WITH EXPONENTIALLY DECAYING WALL TEMPERATURE AND RADIATION

Rudra Kanta Deka[a], Ashish Paul[b,*], Nityajyoti Kalita[c]

[a] *Department of Mathematics, Gauhati University, Guwahati-781014, India*
[b] *Department of Mathematics, Cotton College State University, Guwahati-781001, India*
[c] *Academic Officer, Board of Secondary Education, Assam, Guwahati-781021, India*

ABSTRACT

A theoretical study of thermal radiation effects on unsteady MHD natural convection flow of an electrically conducting fluid past a vertical plate with variable temperature is considered. It is supposed that the temperature of the plate decays exponentially with time. Exact solutions to the non-dimensionalised coupled linear partial differential equations representing the flow problem are obtained using Laplace transform technique. Effects of different physical parameters involved in the temperature and velocity profiles are investigated, shown graphically and discussed. Skin friction and Nusselt number are also derived and their variations with respect to the parameters are investigated.

Key words: *Vertical plate, free convection, radiation, MHD flow.*

1. INTRODUCTION

Natural convection MHD flow of a viscous incompressible electrically conducting fluid with simultaneous effects of radiation has been drawing considerable attention due to its significant effects on the boundary layer flow control. Numerous studies have been carried out in this field by many researchers under different flow situations. In contrast to the mechanism of heat transfer in fluids, the importance of electromagnetic radiation can never be neglected. Accordingly, in recent years, progress has been considerably made in the study of heat transfer with simultaneous effects of radiation. Radiation and on the optical properties of the emitter, with its internal energy being converted to, is the process of heat propagation by means of electromagnetic waves. Radiative convective flows are encountered in countless industrial and environmental processes, particularly in astrophysical phenomenon and technologies in space dynamics. In space technology applications and at higher operating temperatures, effect of radiations is quite significant. Several investigations have been done which are concerned with radiative MHD flows past flat plate considering different flow situations. Hossain and Takhar (1996) investigated radiation effects on mixed convection along vertical plate with uniform surface temperature. Das et al. (1996) studied radiation effects on flow past impulsively started vertical plate. Later on, Muthucumaraswamy and Kumar (2004) observed the heat and mass transfer effects on moving vertical plate in presence of thermal radiation.

In case of electrically conducting fluid flows, it has been observed that the effect of transversely applied magnetic field plays pronounced role over radiation in the whole flow characteristics. In recent years, more analytical and numerical results on unsteady free convection flow past semi-infinite vertical plate were presented by a good numbers of researchers. Ibrahim et al. (2008) observed the influences of viscous dissipation and radiation on unsteady MHD mixed convection flows of micro-polar fluids, while Mahmoud Mustafa (2007) investigated the flow of fluids with variable viscosity past a moving vertical plate with simultaneous effects of transverse magnetic field and radiation. Ramachandra et al. (2007) studied radiation and mass transfer effects on an unsteady MHD free convective flow past a heated vertical plate in porous medium with viscous dissipation. Studies on flow past plate with variable temperature have been receiving quite remarkable attention by researchers since long back. Temperature variations at the plate may occur with different modes with respect to the source responsible for heating the surface bounding the fluid. Nature of a flow obviously depends on how temperature of the plate varies. Realizing the importance of such flows some researchers had made their investigations on this topic. Das et al. (1999) investigated transient free convection flow past an infinite vertical plate with periodic temperature variation. Pal and Chatterjee (2010) studied heat and mass transfer in MHD non-Darcian flow of a micro-polar fluid over a stretching sheet embedded in a porous medium with non-uniform heat source and thermal radiation.

In the present study we have studied the radiative heat transfer MHD flow past a vertical plate with exponentially decaying plate temperature. Closed form solutions of the flow problem are obtained by using Laplace transform technique. Nusselt number and skin friction are also derived and the effects of different parameters on the flow characteristics are investigated and presented graphically.

* Corresponding author. Email: ashpaul85@gmail.com

2. MATHEMATICAL FORMULATION

Here we have considered a laminar MHD free convection flow of a viscous, incompressible fluid past an infinite vertical plate with thermal radiation. The physical model of the problem is shown in Fig.1. Here the x' - axis is taken along the plate in the vertically upward direction and y' -axis is taken normal to the plate. Initially, at $t' \le 0$, the plate and the fluid is at the same temperature T'_∞. It is assumed that at time $t' > 0$ the plate temperature is raised to $T'_\infty + (T'_w - T'_\infty) + Exp(-ct')$. The viscous dissipation is assumed to be negligible in the energy equation as the motion is due to free convection only. Under these assumptions with usual Boussinesq's approximation the governing boundary layer equations are:

$$\frac{\partial u'}{\partial t'} = g\beta(T' - T'_\infty) + \nu \frac{\partial^2 u'}{\partial y'^2} - \frac{\sigma B_0^2}{\rho} u' \tag{1}$$

$$\rho C_p \frac{\partial T'}{\partial t'} = \kappa \frac{\partial^2 T'}{\partial y'^2} - \frac{\partial q_r}{\partial y'} \tag{2}$$

with initial and boundary conditions as:

$$\left.\begin{array}{lll} t' \le 0: & u' = 0, \quad T' = T'_\infty & \text{for all } y' \\ t' > 0: & u' = 0, \quad T' = T'_\infty + (T'_w - T'_\infty)e^{-ct'} & at\ y' = 0 \\ & \& u' \to 0, \quad T' \to T'_\infty & as\ y' \to \infty \end{array}\right\} \tag{3}$$

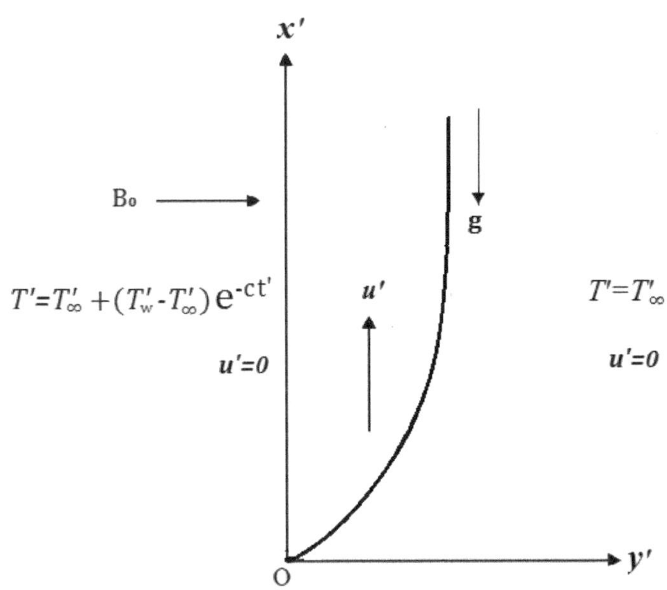

Fig. 1 The physical model and coordinate system.

Here u' is the fluid velocity in x'- direction, T' the temperature, T_w the temperature at the plate, T_∞ the temperature far away from the plate, Bo the magnetic field strength, q_r the radiative flux, C_p the specific heat at constant pressure, t' the time, c is any constant, σ the electrical conductivity, ρ the density, β the coefficient of volume expansion, g the acceleration due to gravity, ν the kinetic viscosity and κ the thermal conductivity.

The term $\frac{\partial q_r}{\partial y'}$ in equation (2) represents the change in the radiative flux with the distance normal to the plate.

For an optically thin gray gas, the local radiant is given by,

$$\frac{\partial q_r}{\partial y'} = -4a^* \sigma^* \left(T'^4_\infty - T'^4\right) \tag{4}$$

If we assume that the temperature difference within the flow is sufficiently small, so that T'^4 may be expressed as a linear function of the temperature. This facilitates to expand T'^4 by Taylor's series about T'_∞. Neglecting the higher order terms in the expansion, we get

$$T'^4 = 4T'^3_\infty T' - 3T'^4_\infty \tag{5}$$

Using (4) and (5) in (2) we get

$$\rho C_p \frac{\partial T'}{\partial t'} = \kappa \frac{\partial^2 T'}{\partial y'^2} - 16a^* \sigma^* T'^3_\infty (T' - T'_\infty) \tag{6}$$

On introducing the following non dimensional quantities,

$$u = \frac{u'}{U_0}, \quad y = \frac{y' U_0}{\nu}, \quad t = \frac{t' U_0^2}{\nu}, \quad \theta = \frac{T' - T'_\infty}{T'_W - T'_\infty}, \quad \Pr = \frac{\mu C_p}{\kappa},$$
$$G = \frac{g\beta\nu(T'_W - T'_\infty)}{U_0^3}, \quad F = \frac{16a^*\sigma^* T'^3_\infty \nu^2}{\kappa U_0^2},$$
$$q = \frac{(c\nu)^{1/2}}{U_0}, \quad M = \frac{\sigma B_0^2 \nu}{\rho U_0^2} \tag{7}$$

equations (1) and (6) respectively become,

$$\frac{\partial u}{\partial t} = \frac{\partial^2 u}{\partial y^2} + G\theta - Mu \tag{8}$$

$$\Pr \frac{\partial \theta}{\partial t} = \frac{\partial^2 \theta}{\partial y^2} - F\theta \tag{9}$$

and the corresponding initial and boundary conditions are,

$$\left.\begin{array}{llll} t' \le 0: & u = 0, & \theta = 0 & \text{for all } y \\ t' > 0: & u = 0, & \theta = e^{-qt} & \text{at } y = 0 \\ & u \to 0, & \theta \to 0 & \text{as } y \to \infty \end{array}\right\} \tag{10}$$

where u, θ and t represents the dimensionless velocity, temperature and time respectively; U_0 is the reference velocity, Pr the Prandtl number G is the Grashof number, F the radiation parameter, M the Hartmann number and q the decay parameter.

3. ANALYTICAL SOLUTION

To solve the unsteady linear governing equations (8) and (9) subject to initial and boundary conditions (10), we apply Laplace transform technique. Accordingly we get,

$$\frac{d^2\bar{u}}{dy^2} - (s+M)\bar{u} + G\bar{\theta} = 0 \tag{11}$$

$$\frac{d^2\bar{\theta}}{dy^2} - (\Pr s + F)\bar{\theta} = 0 \tag{12}$$

Solutions of the coupled equations (11) and (12) subject to the Laplace transformations of boundary conditions (10) give the velocity and temperature profiles as [using Hetnarski (1964) and Hetnarski (1975)]

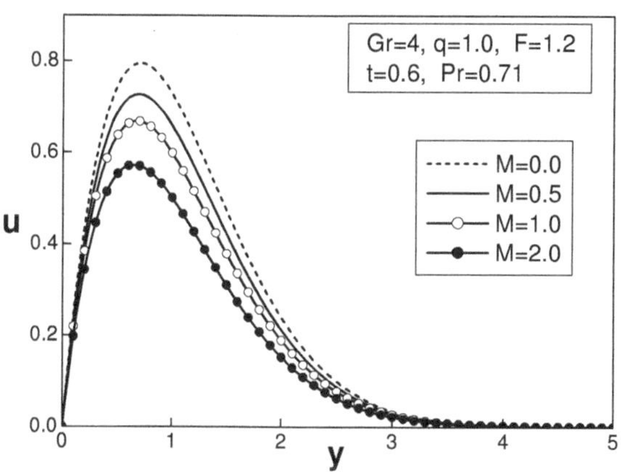

Fig. 2 Magnetic field effect on velocity profile against *y*.

$$u = \frac{Gr}{2(F-M)}\left[e^{-y\sqrt{M}} erfc\left(\frac{y}{2\sqrt{t}} - \sqrt{Mt}\right) + e^{y\sqrt{M}} erfc\left(\frac{y}{2\sqrt{t}} + \sqrt{Mt}\right) - e^{-y\sqrt{F}} erfc\left(\frac{y\sqrt{\Pr}}{2\sqrt{t}} - \sqrt{\frac{F}{\Pr}t}\right) - e^{y\sqrt{F}} erfc\left(\frac{y\sqrt{\Pr}}{2\sqrt{t}} + \sqrt{\frac{F}{\Pr}t}\right)\right]$$

$$+ \frac{Gre^{-qt}}{2(F-M-q(\Pr-1))}\left[e^{-y\sqrt{M-q}} erfc\left(\frac{y}{2\sqrt{t}} - \sqrt{(M-q)t}\right) + e^{y\sqrt{M-q}} erfc\left(\frac{y}{2\sqrt{t}} + \sqrt{(M-q)t}\right) - e^{-y\sqrt{F-q\Pr}} erfc\left(\frac{y\sqrt{\Pr}}{2\sqrt{t}} - \sqrt{\left(\frac{F}{\Pr}-q\right)t}\right) - e^{y\sqrt{F-q\Pr}} erfc\left(\frac{y\sqrt{\Pr}}{2\sqrt{t}} + \sqrt{\left(\frac{F}{\Pr}-q\right)t}\right)\right]$$

$$-\frac{Gr}{2}\left\{\frac{1}{F-M} + \frac{1}{F-M-q(\Pr-1)}\right\}e^{-t\frac{F-M}{\Pr-1}}\left[e^{-y\sqrt{\frac{M\Pr-F}{\Pr-1}}} erfc\left(\frac{y}{2\sqrt{t}} - \sqrt{\frac{M\Pr-F}{\Pr-1}t}\right) + e^{y\sqrt{\frac{M\Pr-F}{\Pr-1}}} erfc\left(\frac{y}{2\sqrt{t}} + \sqrt{\frac{M\Pr-F}{\Pr-1}t}\right) - e^{-y\sqrt{\frac{M\Pr-F}{\Pr-1}}} erfc\left(\frac{y\sqrt{\Pr}}{2\sqrt{t}} - \sqrt{\frac{M\Pr-F}{(\Pr-1)\Pr}t}\right) - e^{y\sqrt{\frac{M\Pr-F}{\Pr-1}}} erfc\left(\frac{y\sqrt{\Pr}}{2\sqrt{t}} + \sqrt{\frac{M\Pr-F}{(\Pr-1)\Pr}t}\right)\right] \tag{13}$$

and

$$\theta = \frac{e^{-qt}}{2}\left\{e^{-y\sqrt{F-q\Pr}} erfc\left(\frac{y\sqrt{\Pr}}{2\sqrt{t}} - \sqrt{(F-q\Pr)t}\right) + e^{y\sqrt{(F-q\Pr)\Pr}} erfc\left(\frac{y\sqrt{\Pr}}{2\sqrt{t}} + \sqrt{(F-q\Pr)t}\right)\right\}$$

$$+\frac{1}{2}\left\{e^{-y\sqrt{F}} erfc\left(\frac{y\sqrt{\Pr}}{2\sqrt{t}} - \sqrt{\frac{F}{\Pr}t}\right) + e^{y\sqrt{F}} erfc\left(\frac{y\sqrt{\Pr}}{2\sqrt{t}} + \sqrt{\frac{F}{\Pr}t}\right)\right\} \tag{14}$$

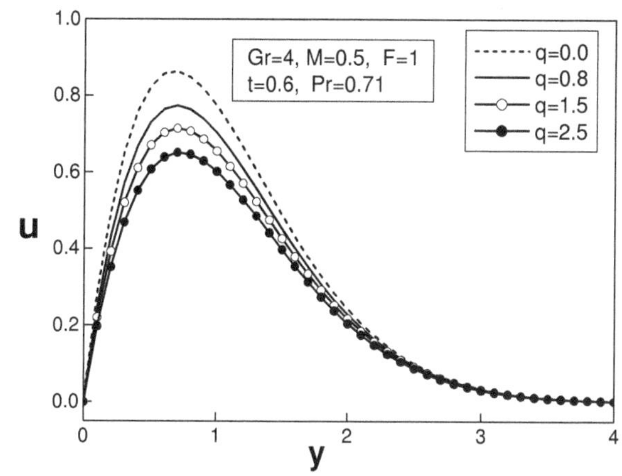

Fig. 3 Effect of *q* on velocity profile against *y*.

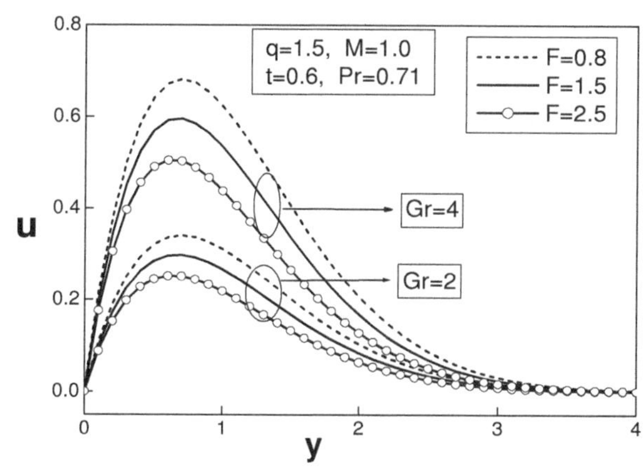

Fig. 4 Effects of *F* and *Gr* on velocity against *y*.

3.1 Skin Friction

The non-dimensional skin-friction $\tau = -\left.\frac{du}{dy}\right|_{y=0}$ obtained from expression of velocity profile as,

$$\tau = \frac{Gr}{2(F-M)}\left[\sqrt{M}\left\{erfc\left(-\sqrt{Mt}\right) - erfc\left(\sqrt{Mt}\right)\right\} - \sqrt{F}\left\{erfc\left(-\sqrt{\frac{Ft}{\Pr}}\right) - erfc\left(\sqrt{\frac{Ft}{\Pr}}\right)\right\}\right] + \frac{Gre^{-qt}}{2(F-M-q(\Pr-1))}$$

$$\left[\sqrt{M-q}\left\{erfc\left(-\sqrt{(M-q)t}\right) - erfc\left(\sqrt{(M-q)t}\right)\right\} - \sqrt{F-q\Pr}\left\{erfc\left(-\sqrt{\frac{(F-q\Pr)t}{\Pr}}\right) - erfc\left(\sqrt{\frac{(F-q\Pr)t}{\Pr}}\right)\right\}\right]$$

$$-\frac{Gr}{2}\left\{\frac{1}{F-M}+\frac{1}{F-M-q(\Pr-1)}\right\}e^{-t\frac{F-M}{\Pr-1}}\sqrt{\frac{M\Pr-F}{\Pr-1}}$$

$$\left\{erfc\left(-\sqrt{\frac{(M\Pr-F)t}{\Pr-1}}\right)-erfc\left(\sqrt{\frac{(M\Pr-F)t}{\Pr-1}}\right)\right.$$

$$\left.-erfc\left(-\sqrt{\frac{(M\Pr-F)t}{(\Pr-1)\Pr}}\right)+erfc\left(\sqrt{\frac{(M\Pr-F)t}{(\Pr-1)\Pr}}\right)\right\} \qquad (15)$$

3.2 Nusselt number

The non-dimensional Nusselt number $Nu=-\left.\frac{d\theta}{dy}\right|_{y=0}$ obtained from the expression of temperature profile (14) as,

$$Nu=\frac{2\sqrt{\Pr}}{\sqrt{\pi t}}e^{-\frac{Ft}{\Pr}}+\frac{1}{2}e^{-qt}\sqrt{F-q\Pr}\left\{erfc\left(-\sqrt{\frac{(F-q\Pr)t}{\Pr}}\right)\right.$$

$$\left.-erfc\left(\sqrt{\frac{(F-q\Pr)t}{\Pr}}\right)\right\}+\frac{\sqrt{F}}{2}\left\{erfc\left(-\sqrt{\frac{Ft}{\Pr}}\right)-erfc\left(\sqrt{\frac{Ft}{\Pr}}\right)\right\} \qquad (16)$$

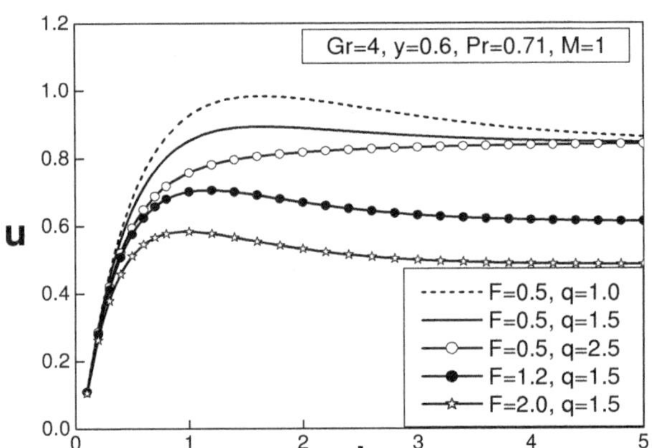

Fig. 5 Effects of F and q on velocity profile against t.

4. RESULTS AND DISCUSSIONS

In this section we try to grasp the problem physically. To make the physical interpretation of the flow characteristics, figures are plotted for temperature, velocity, Nusselt number and skin friction, showing the effects of different parameters involved in the problem. Fig. 2 shows the effect of transverse magnetic field on velocity against y. It is seen that magnitude of velocity goes down as the strength of the transverse magnetic field is increased. Lorentz's force, which grows up as the magnitude of M increases is responsible for this. In fact, Lorentz's force acts as a resisting force on the motion. Velocity profiles against y for different values of q are displayed in Fig. 3. In the figure it is seen that magnitude of velocity gets declined as q increases. A combined effect of Grashof number (Gr) and radiation parameter (F) on the flow can be observed from Fig. 4. Grashof number is responsible for the enhancement of buoyancy force. Buoyancy force increases as magnitude of Gr increases and as such velocity goes up. The same figure shows that magnitude of velocity gets reduced as F increases.

Velocity profile against time under the influences of F and q is displayed in Fig. 5. Observation shows that magnitude of velocity decreases under the effect of both the parameters. In case of q, it is interesting to observe that although magnitude of velocity initially suffers a fall for larger value of q, as time increases, it is seen that q cannot sustain its effect on the velocity for longer period, as we see in the figure that all the three curves (for q =1, 1.5, 2.5) nearly get merged at t = 5.

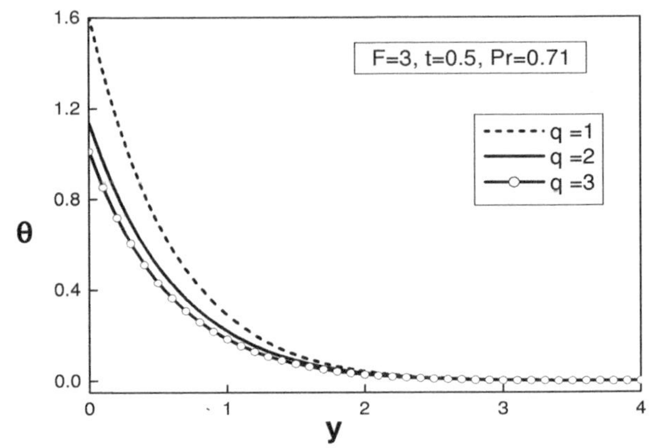

Fig. 6 Effect of q on temperature profile against y

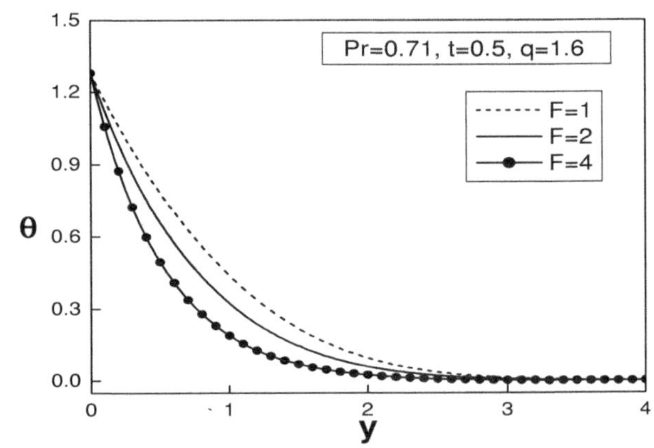

Fig. 7 Effect of radiation on temperature against y.

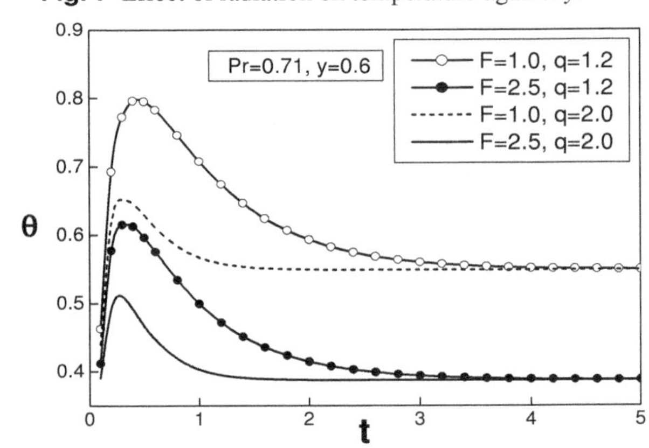

Fig. 8 Effects of F and q on temperature against t.

Temperature profiles versus y for different values of q are demonstrated in Fig. 6 and it is seen that q affects adversely on temperature. Similar effect is also observed in Fig. 7, where effect of radiation (F) on temperature is displayed. Time dependent temperature profile effected by q, shown in Fig. 8 indicates that at the vicinity of the plate, effect of q is quite significant, but as time increases the temperature profiles get closer and closer and finally between t = 3 and t = 4 each graph attains the ambient temperature. Whereas a different variation is seen in the same figure, when effect of radiation on temperature is considered. It is witnessed that effect of radiation remains quite significant till the fluid mass is far away from the plate. Thus variation of radiation creates a long-term effect on temperature.

Skin friction on the plate varying with F and q can be observed in Fig. 9. Skin friction at the plate increases with q, of course it becomes steady after some time. Secondly skin friction at the plate significantly grows as F increases. Further Fig. 10 demonstrates that skin friction at the plate is very low in absence of the transverse magnetic field (M=0), but it develops in a drastic manner as soon as the transverse magnetic field is imposed and goes on increasing with increasing values of M. Considering the Nusselt number profile for different values of q and F, in Fig. 11, observe that Nusselt number (rate of heat transfer) at the plate is enhanced due to the increase of both q and F.

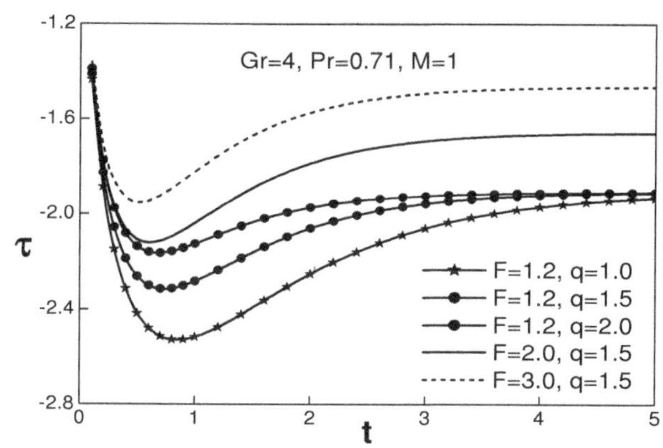

Fig. 9 Skin friction profiles for different values of F and q

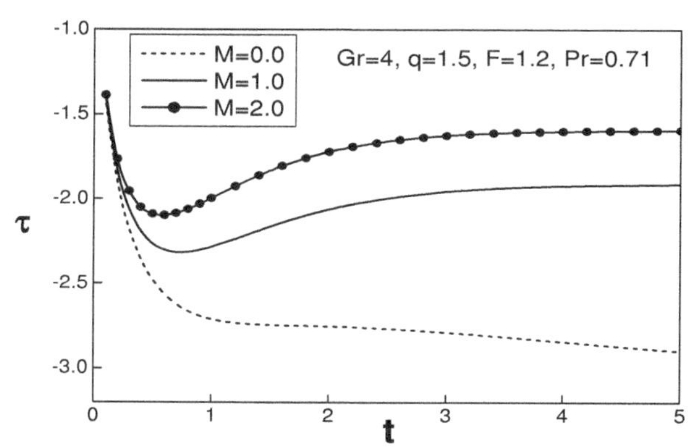

Fig. 10 Effect of magnetic field on skin friction.

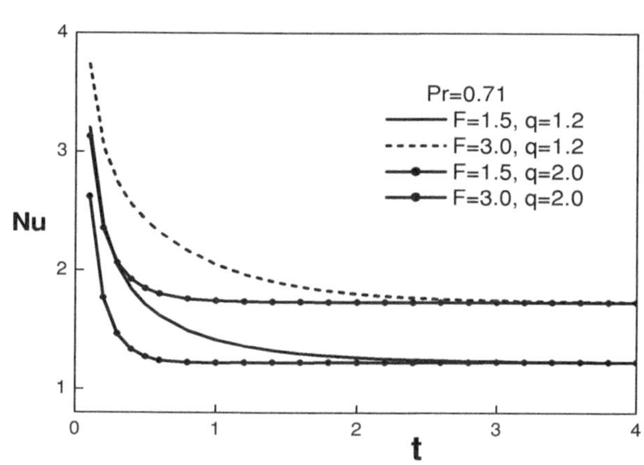

Fig. 11 Variation of Nusselt number profiles with respect to F and q

5. Conclusions

On the basis of our present study and graphical analysis we summarize the conclusions as follows:

i. Fluid velocity is adversely affected by transversely applied magnetic field as well as radiation.
ii. Temperature of the fluid decreases with the increase in radiation parameter.
iii. Magnitude of skin friction and Nusselt number are enhanced due to the increase in the radiation parameter.

REFERENCES

Das, U. N., Deka, R. K., and Soundalgekar, V. M., 1996, "Radiation Effects on Flow Past an Impulsively Started Vertical Plate," *Journal of Theoretical Mechanics, **1**(2), 111-115.*

Das, U. N., Deka, R. K., and Soundalgekar, V. M., 1999, "Transient Free Convection Flow Past an Infinite Vertical Plate with Periodic Temperature Variation," *Journal of Heat Transfer, **121(4)**, 1091-1094.* http://dx.doi.org/10.1115/1.2826063

Hetnarski, R. B., 1964, "On Inverting the Laplace Transforms Connected with the Error Function," *Zastosowania Matematyki, VII, 398-403.*

Hetnarski, R. B., 1975, "An Algorithm for Generating Some Inverse Laplace Transforms of Exponential Form," *ZAMP, **26(2)**, 248-253.* http://dx.doi.org/10.1007/BF01591514

Hossain, M. A., and Takhar, H. S., 1996, "Radiation Effects on Mixed Convection along a Vertical Plate with Uniform Surface Temperature," *Heat and Mass Transfer, **31(4)**, 243-248.* http://dx.doi.org/10.1007/BF02328616.

Ibrahim, F. S., Elaiw, A. M., and Bakr, A. A., 2008, "Influence of Viscous Dissipation and Radiation on Unsteady MHD Mixed Convection Flow of Micro-Polar Fluids," *Applied Mathematics & Information Sciences, **2**(2), 143-162.*

Muthucumaraswamy, R., and Kumar, S, 2004, "Heat and Mass Transfer Effects on Moving Vertical Plate in the Presence of Thermal Radiation," *Theoret. Appl. Mech., **31**(1), 35-46.* http://dx.doi.org/10.2298/TAM0401035M

Mahmoud Muatafa, A., 2007, "Variable Viscosity Effect of Hydro-Magnetic Boundary Layer Flow along a Continuously Moving Vertical Plate in Presence of Radiation," *Applied Mathematical Sciences, **1** (17), 799-814.*

Pal, D., and Chatterjee, S., *2010,* "Heat and Mass Transfer in MHD Non-Darcian Flow of a Micro-Polar Fluid over a Stretching Sheet Embedded in a Porous Media with Non-Uniform Heat Source and Thermal Radiation," *Communications in Nonlinear Science and Numerical Simulation, **15**(7), 1843-1857. doi:10.1016/j.cnsns.2009.07.024*

Ramachandra, P. V., and Bhaskar Reddy, N., 2007, "Radiation and Mass Transfer Effect on an Unsteady MHD Free Convective Flow Past a Heated Vertical Plate in Porous Medium with Viscous Dissipation," *Theoretical. Applied Mechanics **34**(2), 135-160.* http://dx.doi.org/10.2298/TAM0702135P

Permissions

All chapters in this book were first published in FHMT, by Global Digital Central; hereby published with permission under the Creative Commons Attribution License or equivalent. Every chapter published in this book has been scrutinized by our experts. Their significance has been extensively debated. The topics covered herein carry significant findings which will fuel the growth of the discipline. They may even be implemented as practical applications or may be referred to as a beginning point for another development.

The contributors of this book come from diverse backgrounds, making this book a truly international effort. This book will bring forth new frontiers with its revolutionizing research information and detailed analysis of the nascent developments around the world.

We would like to thank all the contributing authors for lending their expertise to make the book truly unique. They have played a crucial role in the development of this book. Without their invaluable contributions this book wouldn't have been possible. They have made vital efforts to compile up to date information on the varied aspects of this subject to make this book a valuable addition to the collection of many professionals and students.

This book was conceptualized with the vision of imparting up-to-date information and advanced data in this field. To ensure the same, a matchless editorial board was set up. Every individual on the board went through rigorous rounds of assessment to prove their worth. After which they invested a large part of their time researching and compiling the most relevant data for our readers.

The editorial board has been involved in producing this book since its inception. They have spent rigorous hours researching and exploring the diverse topics which have resulted in the successful publishing of this book. They have passed on their knowledge of decades through this book. To expedite this challenging task, the publisher supported the team at every step. A small team of assistant editors was also appointed to further simplify the editing procedure and attain best results for the readers.

Apart from the editorial board, the designing team has also invested a significant amount of their time in understanding the subject and creating the most relevant covers. They scrutinized every image to scout for the most suitable representation of the subject and create an appropriate cover for the book.

The publishing team has been an ardent support to the editorial, designing and production team. Their endless efforts to recruit the best for this project, has resulted in the accomplishment of this book. They are a veteran in the field of academics and their pool of knowledge is as vast as their experience in printing. Their expertise and guidance has proved useful at every step. Their uncompromising quality standards have made this book an exceptional effort. Their encouragement from time to time has been an inspiration for everyone.

The publisher and the editorial board hope that this book will prove to be a valuable piece of knowledge for researchers, students, practitioners and scholars across the globe.

List of Contributors

Zhen Tian and Wenbo Jin
Petroleum and Natural Gas Engineering College, Southwest Petroleum University, Chengdu, Sichuan, 610500, China

Lei Wang
The Second Oil Production Plant of Qinghai Oilfield, Haixi/Qinghai, 817500, China

Zhi Jin
Research Institute of Shaanxi Yanchang Petroleum(Group) Co, LTD, Xi'an/Shaanxi, 710075, China

Benjamin Lepers, Guido Link and John Jelonneka
Karlsruhe Institute of Technology, Institute for Pulsed Power and Microwave Technology, WEggenstein Leopoldshafen, 76344, Germany

Aditya Putranto
Monash University, Dept. of Chemical Engineering, Clayton Campus, Melbourne, Victoria 3800, Australia
Parahyangan Catholic University, Dept. of Chemical Engineering, Jolan Ciumbuleuit 94, Bandung, Indonesia

Martin Umminger
Karlsruhe Institute of Technology, Institute of Concrete Structures and Building Materials, Karlsruhe, 76131, Germany

Zan Wu and Bengt Sundén
Department of Energy Sciences, Lund University, Lund, SE-22100, Sweden

Zhaozan Feng
Department of Energy Engineering, Zhejiang University, Hangzhou, 310027, China

Lars Wadsö
Division of Building Materials, Lund University, Lund, SE-22100, Sweden

Huang, Vitaly E. Gruzdev, Yuwen Zhang and J. K. Chen
Department of Mechanical and Aerospace Engineering, University of Missouri, Columbia MO 65211, USA

J. Srinivas and J. V. Ramana Murthy
Department of Mathematics, NIT Warangal 506004, Telangana State, India

Manoj Kr. Triveni, Dipak Sen and RajSekhar Panua
National Institute of Technology, Agartala, Tripura, 799055, India

Edoardo Gino Macchi, Giovanni Braschi and Mario Gallati
Università di Pavia, Dipartimento di Ingegneria Civile ed Architettura, via Ferrata 3, 27100 Pavia (PV), Italy

Amnart Boonloi
Department of Mechanical Engineering Technology, College of Industrial Technology, King Mongkut's University of Technology North Bangkok, Bangkok 10800, Thailand

Withada Jedasadaratanachai
Department of Mechanical Engineering, Faculty of Engineering, King Mongkut's Institute of Technology Ladkrabang, Bangkok 10520, Thailand

A. K. M. Sadrul Islama
Department of Mechanical Engineering, Islamic University of Technology (IUT), Board Bazar, Gazipur-1704, Bangladesh

Arafat A. Bhuiyan
Department of Mechanical Engineering, Islamic University of Technology (IUT), Board Bazar, Gazipur-1704, Bangladesh
Faculty of Science Engineering and Technology (FSET), Swinburne University of Technology, Victoria-3122, Australia

M. Ruhul Amin
Department of Mechanical and Industrial Engineering, Montana State University, Bozeman, Montana, MT 59717, USA

Jamal Naser
Faculty of Science Engineering and Technology (FSET), Swinburne University of Technology, Victoria-3122, Australia

Mitiku Daba, P. Devaraj and S. V. Subhashini
Department of Mathematics, College of Engineering Guindy Campus, Anna University, Chennai, Tamil Nadu, Chennai-600 025, India

Anand Shukla
Faculty of Engineering & Technology, Mahatma Gandhi Chitrakoot Gramodaya, Vishwavidyalaya, Satna-485331,India

Alok Chaube
Department of Mechanical Engineering, Jabalpur Engineering College, Jabalpur-482011, India

Shailesh Gupta and Arvind Sirsath
Department of Mechanical Engineering, Hitkarini College of Engineering & Technology, Jabalpur-482005, India

Makoto Shibahara
Graduate School of Maritime Sciences, Kobe University, Kobe, 658-0022, Japan

J. N. Chung and Kun Yuan
Department of Mechanical and Aerospace Engineering, University of Florida Gainesville, FL. 32611-6300, USA

R. R. Kairi
Department of Mathematics, Islampur College, Islampur, West Bengal, India - 733 202

Ch. RamReddy
Department of Mathematics, National Institute of Technology Warangal, India - 506 004

Srinath Ekkad
Dept. of Mechanical Engineering, Virginia Tech, Blacksburg, VA 24061, USA

Je-Chin Han
Dept. of Mechanical Engineering, Texas A&M University, College Station, TX 77843, USA

Amnart Boonloi
Department of Mechanical Engineering Technology, College of Industrial Technology, King Mongkut's University of Technology North Bangkok, Bangkok, 10800, Thailand

Junjie Chen, Xuhui Gao and Deguang Xu
School of Mechanical and Power Engineering, Henan Polytechnic University, Jiaozuo, Henan, 454000, China

D. Srinivasacharya and O. Surender
a Department of Mathematics, National Institute of Technology, Warangal - 506004, India

A. J. Chamkha
Manufacturing Engineering Department, The Public Authority for Applied Education and Training, Shuweikh 70654, Kuwait

A. M. Rashad
Department of Mathematics, Aswan University, Faculty of Science, 81528, Egypt
Department of Mathematics, Salman bin Abdul Aziz University, College of Science and Humanity Studies, AL-Kharj, Saudi Arabia

Jian Zhuang, Changqing Huang and Hui Jiang
College of Mechanical and Electrical Engineering, Beijing University of Chemical Technology, Beijing, 100029, China
Polymer Material Processing Equipment Engineering Research Center of The Ministry of Education, Beijing , 100029, China

Daming Wu and Ying Liu
College of Mechanical and Electrical Engineering, Beijing University of Chemical Technology, Beijing, 100029, China
Polymer Material Processing Equipment Engineering Research Center of The Ministry of Education, Beijing , 100029, China
State Key Laboratory of Organic-Inorganic Composites, Beijing University of Chemical Technology, Beijing , 100029, China

Zhuohuan Hu, Lulu Wang and Mo Yang
School of Energy and Power Engineering, University of Shanghai for Science and Technology, Shanghai, 200093, China

Hui Wang
National Computer Network and Information Safety Administration Center Shanghai Network Access Point, Shanghai, 201315, China

M. Umamaheswar and M. C. Raju
Department of Humanities and Sciences, Annamacharya Institute of Technology and Sciences Rajampet (Autonomous), Rajampet, Andhra Pradesh, 516126 India

S. V. K. Varma
Department of Mathematics, S. V. University, Tirupati, 517502, Andhra Pradesh, India

M. Y. Abdollahzadeh Jamalabadi
Chabahar Maritime University, Chabahar 99717-56499, Iran

Benjamin Lepers, Thomas Seitz and Guido Link
Karlsruhe Institute of Technology, Institute for Pulsed Power and Microwave Technology (IHM), Eggenstein Leopoldshafen, 76344, Germany

John Jelonnek
Karlsruhe Institute of Technology, Institute for Pulsed Power and Microwave Technology (IHM), Eggenstein Leopoldshafen, 76344, Germany
Karlsruhe Institute of Technology, Institute for Highfrequency Techniques and Electronic (IHE), Karlsruhe, 71131, Germany

Mark Zink
TU Bergakademie Freiberg, Institute of Energy Process Engineering and Chemical Engineering (IEC), Fuchsmuehlenweg 9, 09599 Freiberg, Germany

Yaqing Liu
Gengdan Institute of Beijing University of Technology, Beijing 101301, China

Jinyu Ma
NCPC Hebei huamin pharmaceutical Co., Ltd, Shijiazhuang 052165,China

Biswajit De and Rajsekhar Panua
Mechanical Engineering Department, National Institute of Technology, Agartala, Tripura, 799055, India

Withada Jedsadaratanachaia
Department of Mechanical Engineering, Faculty of Engineering, King Mongkut's Institute of Technology Ladkrabang, Bangkok 10520, Thailand

Amnart Boonloi
Department of Mechanical Engineering Technology, College of Industrial Technology, King Mongkut's University of Technology North Bangkok, Bangkok 10800, Thailand

B. Rushi Kumar, T. Sravan Kumar and A .G Vijaya Kumar
Fluid Dynamics Division, School of Advanced Sciences, VIT University, Vellore, TN, 632014, India

Emmanuel Maurice Arthur, Timothy Ayando and Yakubu Ibrahim Seini
Faculty of Mathematical Sciences, University for Development Studies, P.O. Box 1350, Tamale, +233, Ghana

Rudra Kanta Deka
Department of Mathematics, Gauhati University, Guwahati-781014, India

Ashish Paul
Department of Mathematics, Cotton College State University, Guwahati-781001, India

Nityajyoti Kalita
Academic Officer, Board of Secondary Education, Assam, Guwahati-781021, India